MASS
SPECTROMETRY

MASS SPECTROMETRY

Applications in Science and Engineering

Frederick A. White

Professor Emeritus of Nuclear Engineering and Engineering Science
Rensselaer Polytechnic Institute
Troy, New York

Adjunct Professor of Physics
State University of New York at Albany
Albany, New York

George M. Wood

Senior Research Scientist
National Aeronautics and Space Administration
Langley Research Center
Hampton, Virginia

A Wiley-Interscience Publication

JOHN WILEY & SONS

New York / Chichester / Brisbane / Toronto / Singapore

Library of Congress Cataloging in Publication Data:

White, Frederick Andrew, 1918–
  Mass spectrometry.

  "A Wiley-Interscience publication."
  Includes index.
  1. Mass spectrometry. I. Wood, George M. (George
Marshall), 1933–   . II. Title.

QC454.M3W48 1986   539'.6'028   86-4126
ISBN 0-471-09236-3

Printed in the United States of America

10 9 8 7 6 5 4 3 2 1

Dedicated to

PROFESSOR A. O. NIER

Pioneer in mass spectrometry
and its many applications in the
physical and life sciences

PREFACE

Analytical instrumentation is more than an ancillary tool of science and engineering; it is an integral part of science itself. Visible portions of the electromagnetic spectrum are captured by the telescope and the microscope, and the infrared spectrometer permits the determination of energy levels in molecules and semiconductors. The x-ray spectrometer reveals the internal symmetry of crystals, and the radiofrequency portion of the spectrum can probe both outer space and the inner structure of the atomic nucleus. The laser is also of great analytical significance, as it makes possible the generation of coherent radiation of high intensity and narrow bandwidth.

However, few analytical instruments have a broader range of application than mass spectrometers that can identify all the atoms and molecules that comprise gases, liquids, and solids. Even the so-called fourth and fifth states of matter, plasmas and clusters, can be characterized by mass spectral techniques. In university laboratories throughout the world, mass spectrometers are being applied to studies in chemistry, physics, biology, pharmacology, agriculture, geology, metallurgy, environmental science, and to many advanced engineering investigations. The technological impact of mass spectrometry in industrial research and development is equally profound. The prediction made over a century ago by the eminent philosopher, Auguste Comte, has even been refuted—"there are some things of which the human race must remain forever in ignorance, for example, the chemical constitution of the heavenly bodies"—because mass spectrometers have revealed the composition of the lunar surface and the planetary atmospheres. Thus, mass spectrometry now stands at the crossroads of virtually every scientific discipline.

This book has two principal objectives. The first is to provide an interdisciplinary overview of modern mass spectrometry in its many and diverse forms. The second is to have it serve as a catalyst for generating additional and innovative mass spectrometric measurements. It is further hoped that this monograph will complement the many excellent publications that focus on organic chemistry.

A majority of applications and examples cited in this work refer to the recent international literature, but a limited amount of material has been borrowed from a former book (F. A. White, *Mass Spectrometry in Science and Technology*, Wiley, New York, 1968). Furthermore, in order to provide a broader perspective of the role of mass spectrometry in science and engineering, some material has been included that is not circumscribed by mass spectrometry, per se.

The organization of the book affords options for its use in seminars, formal courses, or independent study. The discussion of any single topic is brief, but the references at the end of each chapter should be useful. Part 1 begins with an introduction that points out the relevance of mass spectrometry in industrial research and development, traces its historical evolution, and reviews some early salient measurements. Six chapters follow relating to instrumentation: the production, analysis, and detection of ions, data processing, and chromatograph-spectrometer systems. Part 2 focuses on applications in engineering and the physical sciences: geochemistry, atmospheric and space science, structural and electronic materials, electrophysics, and the utilization of mass spectrometry as a diagnostic tool for energy-generating and conversion systems. A chapter on "on-line" applications is also included, because mass spectrometric instrumentation is becoming increasingly important in high technology manufacturing and process control.

Part 3 presents mass spectrometry as the common denominator for selected topics in the life sciences. Areas in which it is an analytical imperative include environmental monitoring, agriculture and food science, clinical medicine, pharmacology, toxicology, and forensic investigations. A final chapter envisions some new challenges for both mass spectrometry and ion beam technology. This new frontier includes the characterization of very large molecules and of charged particles in the MeV range. Energetic and highly focused ion beams also offer the potential for synthesizing materials and for producing a new generation of large-scale integrated circuits by means of ion beam lithography.

FREDERICK A. WHITE
GEORGE M. WOOD

ACKNOWLEDGMENTS

It is a privilege to dedicate this book to Professor A. O. Nier, whose professional concerns have impacted so many areas of science and whose personal influence in the mass spectrometric community has been so profound. Direct and indirect inputs to this monograph have come from many sources. Some of our colleagues have stimulated our own mass spectrometric endeavors for several decades. Others have given generously of their time to this specific assignment, hosted us at their laboratories, furnished published and unpublished data, edited selected portions of the manuscript, and provided a perspective relative to many diverse topics. We are also grateful to the thousands of scientists whose research is referenced in this book, because collectively they are its co-authors.

Organizations and staff members that have extended an unusual degree of encouragement and assistance include the following: A T & T Bell Laboratories: D. J. Malm, W. D. Reents, Jr., M. J. Vasile; Analog Technology: C. M. Judson, J. L. Lawrence; Argonne National Laboratory: C. E. Johnson, S. J. Riley; California Institute of Technology, Jet Propulsion Laboratory: S. P. Sinha; Cambridge Mass Spectrometry: B. W. Griffiths; Cameca: H. N. Migeon; Chemicon: B. T. Upchurch; Consolidated Electrodynamic Consultants: D. B. Hughes; Cornell University: F. W. McLafferty; E. I. DuPont: T. W. Whitehead, Jr.; Eastman Kodak: R. Gohlke, S. T. Lee, J. D. Walling; Extrel: W. L. Fite; Finnigan/Mat: R. E. Finnigan; University of Florida: R. A. Yost; Ford Motor: D. Schuetzle; Galileo Electro-Optics: C. H. Tosswill; General Electric: W. D. Davis, W. Katz, W. V. Ligon, Jr., G. L. Merrill, J. L. Mewherter, J. D. Stein; General Foods: M. G. Kolor; General Ionex: K. H. Purser; General Motors: T. M. Sloane; University of Georgia: A. A. Giardini, C. E. Melton; IBM: B. Crowder; Iowa State University: H. J. Svec; Jeol, Ltd.: F. Kunihiro; Johns Hopkins University: C. C. Fenselau; Hughes Research Laboratories: R. G. Wilson; Hoffman-LaRoche: H. T. Gordon; University of Illinois: K. L. Reinhart, Jr.; Lawrence Livermore National Laboratory: C. M. Wong; Leybold-Heraeus: F. Thum; University of Southern Mississippi: G. H. Rayborn; National Aeronautics and Space Administration: A. Copeland, R. J. Duckett, G. H. Gouger, J. S. Heyman, H. B. Neimann, S. K. Seward, J. J. Singh,

A. C. Smith, E. L. Wildner, P. R. Yeager; Naval Research Laboratory: J. E. Campana, J. J. De Corpo, M. M. Ross, J. R. Wyatt; University of New Orleans: G. E. Ioup, J. W. Ioup; State University of New York at Albany: W. M. Gibson, W. A. Lanford; Nicolet Instruments: S. Ghaderi; Northern Analytical Laboratory: R. J. Guidoni, F. D. Leipziger; Old Dominion University: K. G. Brown; Osaka University: K. Gamo, H. Matsuda, R. Shimizu; Perkin-Elmer: V. A. Adams, M. E. Koslin, M. R. Ruecker, C. F. Sawin; Purdue University: R. G. Cooks; RCA Laboratories: R. E. Honig, C. W. Magee; Rensselaer Polytechnic Institute: G. Krycuk, B. K. Malaviya, D. Steiner; University of Rochester: D. Elmore, H. E. Gove; Rochester General Hospital: E. D. Fugo, M.D., F.A.C.S.; Rochester Institute of Technology: R. D. Sperduto; Sandia National Laboratories: S. T. Picraux; Sciex: D. J. Douglas, A. M. Lovett, B. Shushan, B. A. Thomson; Spire: P. Sioshansi; Sterling Winthrop Research Institute: S. Clemans; University of Texas at Dallas: J. Hoffman; Texas A & M University: R. D. Macfarlane; U.S. Food and Drug Administration: T. Cairns; University of Utah: J. Michl, J. A. McCloskey, H. L. C. Meuzelaar; Virginia-Tidewater Forensic Laboratory: R. J. Campbell; Virginia Institute of Marine Science: R. H. Bieri; Weizmann Institute of Science: J. Yinon; Westinghouse Electric: W. M. Hickam; and Xerox: M. M. Shahin.

We are also indebted to numerous technical societies and publishers that have granted permission to use material from their journals; to Beatrice Shube, editor at John Wiley and Sons, for initiating and guiding this project; to Evelyn Cook and Pat Merriman for manuscript typings; and to Ronald LeVan for a substantial portion of the graphics. Finally, we are grateful to our wives, Janet White and Nancy Wood, for invaluable assistance in connection with this as well as previous publications.

FREDERICK A. WHITE
GEORGE M. WOOD

CONTENTS

MASS
SPECTROMETRY

PART ONE

INSTRUMENTATION

CHAPTER ONE

INTRODUCTION: ATOMS, MOLECULES, IONS, AND ISOTOPES

MASS SPECTROMETRY IN INDUSTRIAL RESEARCH

The petroleum industry was the first to adopt mass spectrometry as a practical tool for quantitative analysis. In 1943, a nine-component hydrocarbon mixture was assayed in less than 1 h of instrument time, compared to an alternative procedure requiring 240 h. This was the same year that the Consolidated Engineering Corporation in Pasadena sold its first commercial mass spectrometer to the Atlantic Refining Company. Today, mass spectrometers are providing a substantial return on investment in corporate R&D centers throughout the world. Virtually every high-technology industry (e.g., aircraft, communications, food, petrochemical, pharmaceutical, photographic, semiconductor, etc.) employs mass spectrometry in some facet of research, product development, or process control. Furthermore, there are few if any analytical instruments that surpass the mass spectrometer in the diversity of its applications in both basic and applied science (e.g., biochemistry, environmental science, genetic engineering, geochronology, materials, plasmas, and space research). This seemingly unlimited potential of modern mass spectrometry can be made plausible on the basis of three general considerations.

First, by means of precise mass measurements, it is possible to explore the microscopic domain of atoms, ions, and molecules. The binding energies of the nuclear atom can be determined unambiguously, and molecular structure can be better understood. It is even possible to assign a position to a specific atom in a molecule. Collectively, these highly detailed data permit us to predict the macroscopic observables of fission and fusion, the velocity of sound in gases, the elasticity of polymers, the color of pigments, the flavors of synthetic beverages, and the efficacy of drugs.

Second, via isotopic ratio measurements, the mass spectrometer is unmatched for determining the chemistry of gases, liquids, and solids. The approximately 300 naturally occurring isotopes of the 92 chemical elements represent a vast data base, and the almost invariant isotopic ratios of these elements in nature permit us to identify materials that have been extracted from the earth's crust and compounded into countless useful forms. However, variations in isotopic abundances are potentially an even greater information source. In the same sense that an electromagnetic carrier-frequency transfers information only when it is modulated, so information retrieval from mass spectra depends upon the countless variations from the "normal" isotopic ratios that characterize elements in nature. Thus, isotopic ratio measurements can be used analytically to detect trace metals at 10^{-15} g levels in a transistor, assay trace elements in biological systems, or reveal the geochemistry of our solar system.

Third, the mass spectrometer has recently been developed into a dynamic interactive system. By utilizing tandem magnets, quadrupoles, and ion beams having a wide range of kinetic energies, detailed spatial and time-dependent information can be obtained on the interaction of ions with matter. Mass-resolved ion beams have been used to produce, as well as analyze, impurity-doped profiles in semiconductors. Other studies have focused on sputtering, secondary electron yields, "channeling" in single crystals, charge exchange, diffusion, plasma diagnostics, lifetimes of excited atomic states, and the detection of other transient phenomena. Laser desorption, fast-atom bombardment, mass-analyzed ion kinetics, and collision-induced dissociation are also among the many dynamic schemes that have evolved to elucidate molecular structure and decomposition mechanisms and identify chemical intermediates.

EARLY CONCEPTS AND EXPERIMENTS

Although many applications of mass spectrometry have emerged only recently, the concept of the isotopic structure of chemical elements was envisioned a century ago. Sir William Crookes, in an address to the Chemical Section of the British Association at Birmingham in 1886 stated:

> I conceive, therefore, that when we say the atomic weight of, for instance, calcium is 40, we really express the fact that, while the majority of calcium atoms have an actual atomic weight of 40, there are not a few which are represented by 39 or 41, a less number by 38 or 42, and so on. Is it not possible, or even feasible, that these heavier and lighter atoms may have been in some cases subsequently sorted out by a process resembling chemical fractionation? This sorting out may have taken place in part while atomic matter was condensing from the primal state of intense ignition, but also it may have been partly affected in geological ages by successive solutions and reprecipitations of the various earths. This may seem an audacious speculation, but I do not think it beyond the power of chemistry to test its feasibility.

Shortly thereafter, an observation was made by Eugene Goldstein, a German physicist, who was investigating electrical discharges at low pressure. In a rudimentary precursor to a cathode ray tube, he noticed streamers of light that appeared to be emanating from holes in a perforated cathode disc. He correctly surmised that these luminous streamers must be associated with rays or ions (from the Greek ιὸν, for travelers) that moved in a direction opposite to the usual cathode rays. And in 1898, using a similar apparatus, Wilhelm Wien showed that these ions could be deflected in a strong magnetic field. During this same period, Sir J. J. Thomson was at work in the Cavendish Laboratory at Cambridge, England, examining in detail the trajectories of both electrons and positive ions. Using photographic plates to supplement his fluorescent detector, he clearly saw that the recorded ion streams were discrete, sharply defined parabolas of particles and not mere blurs. This work constituted an important experimental proof that the individual atoms of the same element had approximately the same mass. It also marked the advent of a completely new field of spectroscopy, *the spectroscopy of mass*. Accurately forecasting the future, Thomson said, "I feel sure that there are many problems in chemistry which could be solved with far greater ease by this than by any other method. The method is surprisingly sensitive, more so even than that of spectrum analysis [optical spectroscopy], requires an infinitesimal amount of material and does not require this to be specially purified. The technique is not difficult if appliances for producing high vacuum are available."

RADIOACTIVITY AND THE ISOTOPES OF LEAD

Concurrent with these early experiments on the nature of "positive rays" were studies of radioactivity and atomic structure, for example, the discovery of radioactivity of uranium by Henri Becquerel (1896), the identification of the alpha particle as a helium nucleus by Lord Rutherford (1903), and the mass-energy equivalence proposed by Albert Einstein (1905). However, many chemists were also spending long hours examining the heavy elements with new tools. In 1906, B. Boltwood at Yale discovered a new element that he termed ionium, which he identified as having the properties of thorium. Subsequent work conducted in Germany suggested that ionium and thorium were chemically inseparable. In 1907, mesothorium was discovered by Otto Hahn, and this element was subsequently found to be chemically indistinguishable from radium.

These and other studies seemed to point up similar chemical identities for the radioactive species, and certain regularities in radioactive decay chains were noticed by other workers (Aston, 1922). Collectively these caused Frederick Soddy (an early coworker and golf partner of Rutherford) to issue a memorable report on radioactivity in 1910. He concluded:

These regularities may prove to be the beginning of some embracing generalization, which will throw light, not only on radioactive processes, but on elements in general and the Periodic Law. The recognition that *elements of different atomic weights may possess identical properties seems destined to have its most important application in the region of inactive elements, where the absence of a second radioactive nature makes it impossible for chemical identity to be individually detected.* Chemical homogeneity is no longer a guarantee that any supposed element is not a mixture of several of different atomic weights, or that any atomic weight is not merely a mean number. The constancy of atomic weight, whatever the source of the material, is not a complete proof of homogeneity, for, as in the radio elements, genetic relationships might have resulted in an initial constancy of proportion between the several individuals, which no subsequent natural or artificial chemical process would be able to disturb. If this is the case, the absence of simple numerical relationships between the atomic weights becomes a matter of course rather than one of surprise.

This brilliant exposition by Professor Soddy provided an answer to the many anomalies in atomic weights; he coined a word of Greek derivation, *isotope* (*iso*, equal; *tope*, place), that found its way quickly into the scientific literature:

> *The same algebraic sum of the positive and negative charges in the nucleus when the arithmetical sum is different gives what I call isotopes or isotopic elements because they occupy the same place in the periodic table.* They are chemically identical, and save only as regards the relatively few physical properties which depend upon mass directly, physically identical also.

By about 1915, the collective results of many investigators seemed to indicate that the final decay product of all heavy elements was lead. It was also known that the emission of discrete alpha and beta particles would correspond to an integral change in atomic number (nuclear charge) and atomic weight. Thus, thorium subjected to 6 alpha decays should result in a net change of 24 units (i.e., from 232 to 208). But lead from nonradioactive sources was known to be approximately 207.2—far from an integral value. A plausible explanation was advanced by Professor Soddy, who suggested that the atomic weight of any lead sample might be related to the source—and thus its relative concentration of uranium or thorium. He reported a value of greater than 207.7 for thorite mined in Ceylon. In 1916, T. W. Richards of Harvard analyzed lead from uranium rich ore from Norway that yielded a value as low as 206.1. The results were conclusive. Several decay series of these heavy elements were responsible for the multiplicity of daughter products. The final stable nuclei that remained were those of lead, but the atomic weights were indeed quantized; they were ^{206}Pb, ^{207}Pb, and ^{208}Pb.

IONS AND ISOTOPES OF NEON

Continued research on gaseous discharges and refinements of the parabolic ray analyzer led to another discovery. By the summer of 1912, Thomson could resolve and clearly distinguish ion trajectories corresponding to atomic mass differences of less than 10%. Several gases were subjected to analysis; in November, a small purified sample of neon gas was examined—this at a time when "there was probably less than one gramme in existence."

In a subsequent address to the Royal Institution in January 1913, Thomson said:

> I now turn to the photograph of the light constituents; here we find the lines of helium, of neon (very strong), of argon, and in addition there is a line corresponding to an atomic weight 22, which cannot be identified with the line due to any known gas. I thought at first that this line, since its atomic weight is one-half that of CO_2, must be due to a carbonic acid molecule with a double charge of electricity, and on some of the plates a faint line at 44 could be detected. On passing the gas slowly through tubes immersed in liquid air the line at 44 completely disappeared, while the brightness of the one at 22 was not affected.

Although Thomson conceded that mass 22 might be attributed to the molecule, NeH_2, there was evidence of the isotopic structure of matter, completely independent of any correlation with nuclides appearing as daughter products of radioactive decay. In 1919, F. W. Aston, a colleague of Thomson at the Cavendish Laboratory, provided additional evidence for the isotopic nature of neon. He measured an abundance ratio of 10:1 for the ion beam intensities at masses 20 and 22, a ratio that was in good agreement with the known molecular weight of neon, 20.18.

ASTON'S MASS SPECTROGRAPH

The continuing research of Aston was monumental with respect to: (1) the instrumentation that he developed, (2) his systematic survey of many elements to determine their isotopic structure, and (3) the establishment of an atomic mass scale. His new mass spectrograph (Aston, 1919) was conceived from an optical analogue; he proposed that an electrical and magnetic field should be arranged so as to act successively upon a positive ray, or ion, in a fashion similar to the refraction of a light beam by two prisms. Thus, consider the unresolved ion beam in Figure 1.1 passing through collimating slits S_1 and S_2 so that it enters the electric field established by the parallel plate electrodes P_1 and P_2. The positive ions will undergo a clockwise deflection dependent upon their kinetic energies (and charge), and a portion of the beam will pass through the aperture at S_3. The ions then pass through

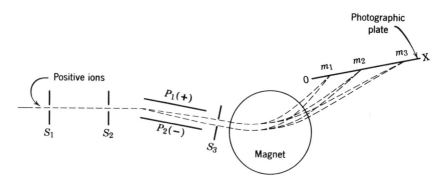

Figure 1.1 Aston's mass spectrograph.

a magnetic field region in a counterclockwise fashion; they are brought to a focus according to their mass, on a line OX, whose specific angle with respect to the incident beam depends on several parameters. Aston showed that for small angles, this arrangement of opposing electrical and magnetic deflections could bring a "chromatic" mass spectrum to a focus. Use of a photographic plate then permitted a convenient calibration of a mass scale, with the relative isotopic abundance being crudely measured by the intensity of line blackening.

With this instrument, Aston proceeded to survey much of the Periodic Table. Clearly, isotopic species might be expected for light elements such as chlorine (atomic weight, 35.5) and magnesium (atomic weight, 24.3), because their weight was quite different from an integer value. Indeed, the chlorine isotopes 35 and 37 were soon identified. In 1922 Aston received the Nobel prize in chemistry for his work, and by 1924 he had characterized the isotopic structure of krypton, xenon, mercury, silicon, potassium, rubidium, etc.—a total of about 50 elements. About this same time, Professor A. J. Dempster in the United States was reporting his independent observations on lithium, magnesium, and zinc.

The impact of these early studies can hardly be overemphasized. It led to the ultimate discovery that *most, rather than a few, of the elements were isotopic*, which justified a "whole number rule" that had always been so appealing to both philosopher and physicist (Aston, 1942). It also generated intensive efforts toward improved instrumentation that allowed hypotheses of nuclear structure to be validated or rejected.

THE MASS SCALE

By 1927, Aston had completed the construction of his second mass spectrograph; it had a resolution approaching 600. In this same year, he reported his landmark "packing fraction" curve that displayed the general manner in which nuclear stability varies with mass number. Other physicists were also building instruments that would accurately measure atomic masses as well as determine isotopic ratios. At the University of Chicago, Professor Dempster (1918) designed the first 180° mass spectrograph, an instrument that was to serve as the prototype for many subsequent analyzers. Although Aston's first instrument possessed both velocity and mass focusing properties, it was restricted to narrow collimated beams. The 180° Dempster magnet possessed direction focusing properties (i.e., it focused ion beams having a large angular divergence). Furthermore, it utilized a quadrant electrometer as an ion current monitor; thus, it was the first mass spectrometer to employ an electrical means of detection.

At the Bartol Foundation of the Franklin Institute, Bainbridge (1933a) also built a 180° instrument, but it used a velocity filter for ions entering the magnet; it also used a photographic plate for accurately determining relative atomic masses. With this apparatus, precise mass determinations were made of the atoms involved in the nuclear reaction $^1H + {}^7Li \rightarrow 2\,^4He$ (Bainbridge, 1933b), providing the first experimental proof of the Einstein mass-energy relationship.

Important new refinements were included in double focusing instruments designed by Mattauch and Herzog (1934), Dempster (1935), and Bainbridge and Jordan (1936). All three of these instruments employed photographic plate detectors for making precise mass measurements; the atomic masses being correlated with the position of lines on the plates. The double focusing spectrometer of Nier and Roberts (1951) at the University of Minnesota permitted even more precise measurements by determining small differences of "doublets" (two ion species having the same nominal mass-to-charge ratio) and by using electronic circuitry. These measurements and those of other groups in Canada, England, Germany, and the United States led to discussions between chemists and physicists with respect to a unified scale of nuclidic masses. Chemists had used a mass scale based on the average atomic mass of the isotopic mixture of "natural" oxygen, while the "physical scale" was referenced to $^{16}O = 16$. A proposal to use ^{12}C as a new reference nuclide was made independently by Nier in the United States and by A. Olander, a chemist, in Sweden (Kohman et al., 1958). This proposal was subsequently adopted by the International Union of Pure and Applied Physicists in 1960 and by the International Union of Pure and Applied Chemists in 1961.

This new mass scale based upon ^{12}C resolved several problems. First, it corrected Aston's oxygen scale, which was established prior to the discovery of the ^{17}O and ^{18}O isotopes in 1929. Second, it produced a scale in which the isotopes of all the elements have close to integer values. Third, the scale is independent of the small natural geographic variations in isotopic composition that are known to occur. Finally, it provided the foundation whereby physicists could calculate nuclear reactions from the Einstein mass-energy postulate and chemists could confidently identify compounds and determine molecular structure.

THE DISCOVERY OF DEUTERIUM

Conclusive evidence for a second isotope of hydrogen was obtained by Professor Urey at Columbia University on Thanksgiving day in 1931, a discovery for which he was awarded the Nobel prize in chemistry in 1934. As early as 1913, however, careful measurements at New York University (Lamb and Lee, 1913) indicated that pure water was not characterized by an invariant density. These measurements were sensitive to 2×10^{-7} g/cm^3, and density fluctuations were observed as great as 8×10^{-7} g/cm^3 among various samples. The possibility that these density fluctuations might be related to something as fundamental as isotopic structure was, of course, not envisioned. Urey's first identification of deuterium was made on an optical spectrograph by noting faint photoplate lines corresponding to a unit mass shift in the Balmer series formula. Subsequent confirmation of the heavy isotope and its relative abundance ($\sim$1 part in 6500) was forthcoming from early mass spectrometric observations. Heavy hydrogen enrichments, undertaken by distillation and electrolysis, were also monitored mass spectrometrically.

The impact of the ^{2}H discovery was profound. In 1932, the neutron was discovered, and by 1934 over 200 papers had been published relating to deuterium and its chemical compounds. Mass-dependent properties of physical and chemical systems were explored, and the potential of deuterium as an isotopic tracer was recognized immediately in research relating to metabolism and complex biological processes. Deuterium also assumed an important role as a bombarding particle in cyclotrons and various accelerators for producing nuclear reactions.

ISOTOPIC RATIO MEASUREMENTS: 1930–1940

Some of the same spectrometers that were built to establish a mass scale were also employed to find new isotopes and make isotopic ratio measurements. For such measurements, however, an electrical detection system was preferred over photoplate recording; an isotopic ratio could be determined more precisely, and a higher sensitivity was achieved ultimately for small ion currents and "minor" isotopes. Furthermore, data could be acquired in real time, and the spectrometer could be programmed over the entire mass scale or any selected portion of it. Thus, the decade 1930–1940 witnessed not only additional isotopic discoveries, but it was marked by salient new isotopic ratio measurements and their interpretation.

Following the discovery of deuterium in 1931, Professor Nier (1935) showed that potassium possessed a third isotope, ^{40}K, having an abundance of only 1/8600 of ^{39}K. The known radioactivity of potassium was later attributed to this minor isotope. The 180° instrument of Nier (1937) was also significant, as it utilized a bakeable analyzer tube so that very low pressures could be obtained. The electromagnet was stabilized to 1 part in 10,000 and ion beam currents were monitored using electrometers or electrometer tube amplifiers. New isotopes discovered by Professor Nier at Harvard (1936–1938) included ^{36}S, ^{46}Ca, ^{48}Ca, and ^{184}Cs. However, of even greater significance during this period was Nier's introduction of isotopic ratio measurements to the field of geochronology (Nier, 1938). He also showed that ^{207}Pb and ^{206}Pb are the end-products of decay of ^{235}U and ^{238}U, respectively, and that the radioactive decay rates of the two uranium isotopes are markedly different (Nier, 1939). Additional accurate isotope abundance measurements were made by Bleakney, Sampson, Blewitt, and Hipple at Princeton.

Other milestones in isotopic ratio measurements by Nier and collaborators at the University of Minnesota included ^{13}C/^{12}C abundances (Nier and Gulbransen, 1939). Plants were exposed to ^{13}C-enriched CO_2 atmospheres to study photosynthesis, and other related experiments were designed to study the assimilation of CO_2 by bacteria (Wood et al., 1941). The discovery of nuclear fission in 1939 raised many fundamental questions regarding atomic structure and the nuclear parameters of isotopes; in 1940, Nier et al., identified ^{235}U as the nuclide responsible for the slow neutron-induced fission of uranium.

WORLD WAR II TO 1960

With the advent of World War II, mass spectrometry entered a new era. Analogous to the development of microwave radar, it emerged as an increasingly powerful tool in surveillance and diagnostics. Specifically, extensive product monitoring of enriched uranium (UF_6) was required as soon as the United States decided that an all-out effort should be made to produce ^{235}U for atomic bombs. For this task, Professor Nier's 60°, 6 in. radius magnetic sector instrument (1940) was selected as the prototype from which dozens of spectrometers were patterned and supplied to the Oak Ridge plant by the General Electric Company. Furthermore, the high priority assigned to the synthetic rubber program and demands for high octane aviation fuel were additional stimuli for both the further development and use of mass spectrometers within the petrochemical industry.

The post-war decade 1950–1960 was characterized by an explosive growth in high-resolution and isotope mass spectrometry, the study of fragmentation patterns and ionization, and theoretical molecular modeling. It was also a decade in which important new spectrometers and multistage systems made their appearance. Analyzers of high resolution were developed in Britain, Canada, Germany, and the United States; in Japan, Ogata and Matsuda (1955) designed a double focusing spectrometer having a resolution approaching 500,000. Important commercial instruments were also made available by Metropolitan Vickers Electrical Company Ltd. (now Kratos Analytical Instruments) and by Consolidated Electrodynamics Corporation. The cyclotron principle was applied to mass spectrometry (Hipple et al., 1949; Sommer et al., 1949, 1951) and termed an omegatron; this device was the forerunner of the modern ion cyclotron resonance spectrometer. The quadrupole analyzer of Paul and Steinwedel (1953) represented another important and practical option to mass filtering, and the time-of-flight spectrometer of Wiley and McLaren (1955), using a pulsed ion source, provided a real-time capability of extended mass range. In 1957, Holmes and Morrell described a combination gas chromatograph-mass spectrometer that was to impact many areas of chemical analysis, with the spectrometer functioning as a sophisticated detector for the chromatograph. At the Bell Laboratories, Hannay (1954) devised a spark source mass spectrograph; and at the RCA Laboratories, one of the first secondary ion mass spectrometers was constructed for characterizing the surface composition of electronic materials (Honig, 1958).

Several tandem mass spectrometers were also built that were precursors of commercial and specialized instruments. The first tandem magnet system was designed by Inghram and Hayden (1954). In this instrument, ions were mass analyzed in the first magnetic sector, given an acceleration in a region between the two magnets, and reanalyzed in a second magnetic sector. Subsequently, White and Collins (1954) constructed a tandem analyzer comprising two 90° magnets, stabilized by nuclear magnetic resonance circuitry, with a field-free region between the magnets. With this system, an isotopic abundance sensitivity of $10^8/1$ was achieved, and a new stable isotope in nature (^{180}Ta) was discovered. An even larger multistage instrument was then built (White et al., 1958) comprising two 90°, 50.8 cm radius-of-curvature magnets and a 90° electrostatic lens of the same radius. This apparatus, in which a large magnetic sector was followed by a "reverse geometry" double focusing system, made possible "on-line" determinations of molecular dissociation energies, even though the primary ion beam that interacted with a gaseous "target" was of high kinetic energy (Rourke et al., 1959). This general type of tandem instrumentation, in its many subsequent forms, also made it feasible to select primary ions and to independently observe products of ion-molecule interactions.

By 1960, the growing availability of commercial instruments stimulated university and government research in Canada, England, France, Germany, Japan, Russia, Sweden, and the United States. Mass spectrometry had also become firmly established as an indispensable adjunct to the chemical and petroleum industries and to interdisciplinary R&D programs at corporate laboratories such as Bell Laboratories, Eastman Kodak, General Electric, RCA, and Westinghouse.

CONTEMPORARY DEVELOPMENTS

The last 3 decades have witnessed dramatic developments in instrumentation and the scope of mass spectrometric applications. A new generation of ion sources has emerged, utilizing high-powered lasers, high electrical fields, energetic ions, neutral molecules, charge exchange, and fission fragments. Analyzers have included multistage magnets and quadrupoles, laser/time-of-flight systems, and the cyclotron resonance-Fourier transform spectrometer. Improved ion detection schemes have been collinear with the advent of new electron multipliers, microchannel plates, p-n junctions, fiber optics, and solid-state charge-coupled arrays. Research on collision-induced reactions has become so extensive that it now dominates much of the mass spectral literature in organic chemistry, and mass fragmentography is now a major subfield of mass spectrometry.

The study of ion clusters is providing new insights relative to the solid-liquid-gas interface, and mass spectrometry is essential for measuring the high-charge states of atoms produced by multiphoton ionization. Mass-resolved ion implantation has also become a powerful technique for creating new materials, as well as being a quantitative adjunct to their analyses. In addition, two ion beam techniques appear destined to become an integral part of mass spectrometry. The first, Rutherford backscattering (RBS), provides a means for assaying the surface impurities of solids and the impurity profiles of near-surface atom layers. In some instances the assay is unambiguous, and detailed spatial information can be obtained that is unmatched by any other technique. A second salient development is the accelerator-mass spectrometer method for charge stripping and for identifying chemical species or isotopes at subparts per billion concentrations. This new extension of mass spectrometry permits the detection of many radioactive nuclides at a sensitivity greater than with radiation detectors; a discrimination factor of $\sim 10^{15}$ has been achieved for atoms having the same mass number (e.g., ^{14}C and ^{14}N). Some of these new frontiers in mass spectrometry are discussed further in the concluding chapter of this book.

REFERENCES

Aston, F. W., *Phil. Mag., 38*, 709 (1919).

Aston, F. W., *Isotopes*, Arnold, London (1922), p. 7.

Aston, F. W., *Proc. Roy. Soc.*, A, *115*, 487 (1927).

Aston, F. W., *Mass Spectra and Isotopes*, Arnold, London (1942).

Bainbridge, K. T., *J. Franklin Inst., 215*, 509 (1933a).

Bainbridge, K. T., *Phys. Rev., 44*, 123 (1933b).

Bainbridge, K. T., and E. B. Jordon, *Phys. Rev.*, 50, 282 (1936).

Boltwood, B. B., *Am. J. Sci., 22*, 537 (1906).

Crookes, W., *Nature, 34*, 423 (1886).

Dempster, A. J., *Phys. Rev., 11*, 316 (1918).

Dempster, A. J., *Proc. Am. Phil. Soc., 75*, 755 (1935).

Hannay, N. B., *Rev. Sci. Instr., 25*, 644 (1954).

Hipple, J. A., H. Sommer, and H. A. Thomas, *Phys. Rev., 76*, 1877 (1949).

Holmes, J. C., and F. A. Morrell, *Appl. Spectry., 11*, 86 (1957).

Honig, R. E., *J. Appl. Phys., 29*, 549 (1958).

Inghram, M. G., and R. J. Hayden, *A Handbook of Mass Spectrometry,* National Academy of Sciences, National Research Council, Nuclear Science Series, Report No. 14, Washington, D. C. (1954).

Kohman, T. P., J. H. E. Mattauch, and A. H. Wapstra, *Science, 127*, 1431 (1958).

Lamb, A. B., and R. E. Lee, *J. Am. Chem. Soc., 35*, Part 2, 1666 (1913).

Mattauch, J., and R. Herzog, *Z. Physik, 89*, 786 (1934).

Nier, A. O., *Phys. Rev., 48*, 283 (1935).

Nier, A. O., *Phys. Rev., 50*, 1041 (1936).

Nier, A. O., *Phys. Rev., 52*, 933 (1937).

Nier, A. O., *J. Am. Chem. Soc., 60*, 1571 (1938).

Nier, A. O., *Phys. Rev., 55*, 153 (1939).

Nier, A. O., and E. A. Gulbransen, *J. Am. Chem. Soc., 61*, 697 (1939).

Nier, A. O., *Rev. Sci. Instr., 11*, 212 (1940).

Nier, A. O., J. R. Dunning, E. T. Booth, and A. V. Grosse, *Phys. Rev., 57*, 546 (1940).

Nier, A. O., and T. R. Roberts, *Phys. Rev., 81*, 507 (1951).

Ogata, K., and H. Matsuda, *Z. Naturforsch., 10a*, 843 (1955).

Paul, W., and H. Steinwedel, *Z. Naturforsch, 8a*, 448 (1953).

Richards, T. W., and C. Wadsworth, *J. Am. Chem. Soc., 38*, 2613 (1916).

Rourke, F. M., J. C. Sheffield, W. D. Davis, and F. A. White, *J. Chem. Phys., 31*, 193 (1959).

Sommer, H., H. A. Thomas, and J. R. Hipple, *Phys. Rev., 76*, 1877 (1949).

Sommer, H., H. A. Thomas, and J. R. Hipple, *Phys. Rev., 82*, 697 (1951).

Thomson, J. J., in *Isotopes*, by F. W. Aston, Arnold, London (1922), p. 34.

White, F. A. and T. L. Collins, *Appl. Spectry., 8*, 169 (1954).

White, F. A., J. C. Sheffield, and F. M. Rourke, *Appl. Spectry., 12*, No. 2, 46 (1958).

Wiley, W. C., and I. H. McLaren, *Rev. Sci. Instr., 26*, 1150 (1955).

Wood, H. G., C. H. Werkman, and A. O. Nier, *J. Biol. Chem., 139*, 365 (1941).

CHAPTER TWO

ION SOURCES

Ionization is only one of several excited states that can occur when energetic electrons, atoms, or photons interact with gases and condensed matter. A molecule can be ionized and dissociated in a single collision, metastable ions can be formed, and under certain conditions a bombarding electron can be captured within the electronic structure of a target molecule to produce a negative ion. Energetic ions and neutral atoms can also be used not only to characterize solids and involatile organic compounds, but to reveal fragmentation patterns and the existence of atomic clusters. In addition, the advent of sharply focused pulsed laser beams has provided a new means for analyzing particulates with time-of-flight spectroscopy. Thus, the selection of an appropriate ion source will be conditioned not only by the kind of analysis, but by the type of mass spectrometer. The choice of a source may also be limited further by the sample size.

For magnetic analyzers, some general criteria can also be enumerated, based on the equation relating to the image "line width" at the detector focal plane. Specifically, for a symmetric magnetic lens, the relationship between the object slit width S_0, the half-angle, α, of rays emerging from the defining slit, the potential through which ions have been accelerated V, the spread in ion energy ΔV, the line width W (at the image plane), and the radius of curvature of the magnetic sector R is

$$W = S_0 + R \left(\alpha^2 + \frac{\Delta V}{V} \right)$$

If the half-angle of divergence α, is expressed in radians, a substitution of values for S_0, R, and $\Delta V/V$ will yield an approximate value for the image line width. This equation and other considerations suggest the generally desirable properties of ion sources and the ion beams they produce:

1. The supply of ions should be sufficiently intense to be compatible with the analyzer geometry and detector sensitivity.

2. The beam should have an energy spread that is small compared with the total accelerating voltage.

3. The ion source slit width should be small compared with the radius of curvature of the analyzer.

4. The half-angle of divergence should be commensurate with the image line width—as expressed in the equation above (unless special magnetic pole shaping of the analyzer is employed that can minimize the $\alpha^2 R$ contribution).

5. If the sample to be analyzed is very small, it is important that a high percentage of sample atoms become ionized.

6. Ideally, the source should be selective against unwanted ions that might appear in the same portion of the mass spectrum.

7. The source should not have a "memory" that would yield ions from the contamination of prior analyses.

8. Ion emission should be reasonably stable with time.

9. The source should minimize the chemical procedures required prior to mass spectral analysis.

10. For time-of-flight systems, minimizing the ion energy spread is important, but other parameters of both time-of-flight and quadrupole instruments relate to ion source apertures, transmission, and dynamic electric fields.

Other general considerations relate to beam spreading caused by coulomb repulsion (for intense beams), gas scattering, molecular dissociation, charge exchange, and mass discrimination. In most instances, of course, the prerequisite characteristics of the source depend upon the experimental objectives. A mass spectrometric study, for example, might be primarily concerned with the magnitude of an ion beam energy spread. Furthermore, there are situations in which a multiplicity of ions of the same mass, but in several charge states, provides a more positive elemental identification than can be obtained from a single atomic group.

The most widely used ion sources for all types of spectrometers are reviewed below. They include (1) electron, ion, and neutral atom bombardment; (2) surface, chemical, and photoionization; (3) arc-spark and field emission; and (4) several special types.

ELECTRON BOMBARDMENT

The production of positive ions by electron impact is a widely employed technique, as it can be utilized for the analysis of nearly all gases, volatile compounds, and metallic vapors. The ion beam current can be accurately controlled because the ionizing electron beam is generally space-charge limited. The kinetic energy of the electron beam can also be varied precisely. Thus, ionized species of complex molecules can be produced both with and without fragmentation so as to reveal details relating to molecular structure. For the case where there is no fragmentation, the ions generated by an electron impact source can have an energy spread as low as 0.1 eV or less. Because of the large mass difference between electrons and target molecules, the latter will acquire only a small energy increment on impact; also, the ionizing electron beam usually will be highly collimated and orthogonal to the ion beam trajectory. With such constraints, the electron-bombardment source is applicable to single-stage spectrometers in which there is no energy focusing.

A typical source will always include (1) an electron-producing filament and electron trap, (2) high-voltage electrodes for accelerating the positive ions, and (3) collimating slits, focusing and beam-centering electrodes to achieve high-beam definition and transmission. The schematic diagram of Figure 2.1 shows these essential elements.

It might be presumed that the energy of the ionizing electrons need only exceed the first ionization potential of the sample gas to produce ions. For achieving a maximum ionization efficiency, however, the electron beam energy must always substantially exceed this value. Figure 2.2 indicates the relative yield of ions as a function of electron energy (Brown, 1959).

It will be noted from Figure 2.2 that although there are large differences in the positive ion yield of various gases, most ionization maxima occur in a region of 50–100 eV, and a practical operating voltage for the electron beam is about 70 V. The number of positive ions produced by the electron beam is directly proportional to the density of gas molecules at pressures less than 10^{-4} torr (1 torr = 1 mm Hg = 133.32 Pa). At higher pressures, the production of ions is not linearly related to gas density, owing to space-charge effects and recombination phenomena. The bombarding electrons

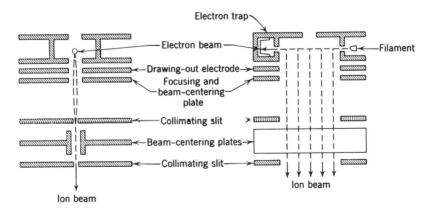

Figure 2.1 Schematic diagram of an electron-bombardment ion source.

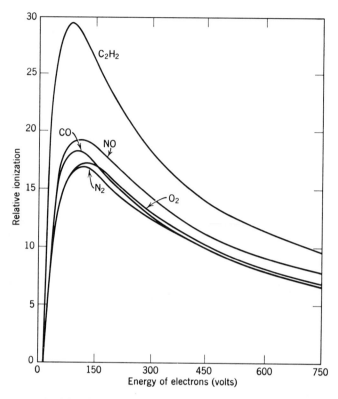

Figure 2.2 Probability of ionization as a function of electron energy (Brown, 1959).

will also produce atoms and molecules in many excited states. Atomic species will lose energy by radiation or by energy exchange in collisons. Molecular ions, however, can also undergo deexcitation by dissociation, giving rise to complex spectra that include the fragmented ions of lighter mass as well as ions of the parent molecule.

In a typical source, the electron beam is orthogonal to the trajectory of the ions, and the region from which ions are drawn into the accelerating region is small. Tungsten and rhenium filaments supply a copious electron current (of an order of magnitude ~ 1 mA), and this current is stabilized by appropriate circuitry in order to obtain a constant ion beam intensity. Electron beam stability may approach 1 part in 10^4, although this condition rarely results in a comparable stability with respect to the ion beam. The magnitude of the ion beam current will depend on several parameters—the specific gas, pressure, size of collimating slits, etc., in addition to the electron-bombarding current. Currents of 10^{-7} to 10^{-14} A are monitored in typical analyzers. The spread in energy of ions produced by typical electron-bombardment sources may be only a few electron volts, with suitable collimating slits and ion optics. For high-resolution work, however, a double focusing system—or some type of energy discrimination—is essential.

The sensitivity of this type of source varies over many orders of magnitude. Milligram amounts of a sample may be needed, depending on the requirements of precision and other factors. The minimum gas sample is restricted primarily by the background spectrum of residual gases, the pumping system, and the detector sensitivity. Reynolds (1956), using a small analyzing tube and employing a static vacuum system (no pumping) was able to detect xenon and argon with only $\sim 10^7$ atoms of sample. Such measurements, however, are quite beyond the capabilities of conventional analytical instruments. In fact, background gases often seriously limit the detection of trace gases because of the "memory" that an ion source retains for occluded gases on walls and surfaces. Despite such difficulties, detection limits of 1 part in 10^7 for trace gases have often been observed.

Precision values ± 0.1 to $\pm 0.01\%$ in isotopic abundance ratios can be achieved, but few investigators claim that absolute accuracies of conventional analyses approach the latter value. Utilization of standards, statistical methods, and knowledge of instrumental biases are prerequisites for achieving best values, even with an electron-bombardment-type source. Negative as well as positive ions can be generated by electron-impact phenomena for compound species (e.g., $XY + e \rightarrow X^+ + Y^- + e$), and electron-capture processes present an alternative mode of formation.

CHEMICAL IONIZATION

The technique of chemical ionization (CI) for mass spectrometry was first introduced by Munson and Field (1966). At present, it is one of the most widely used ionization methods for structural elucidation, and a definitive monograph relating to this topic has been prepared by Harrison (1983). Chemical ionization is accomplished by an ion-molecule charge transfer occurring between a preionized reactant gas and the sample gas. The reaction takes place at a relatively high pressure and with the transfer of a small amount of energy. Consequently, little internal energy is imparted to the ionized molecule, so that fragmentation is substantially less than by electron impact. Hydrocarbons, such as methane, propane, and isobutane, etc., have been used as reactant gases in which primary ions are formed by electron-impact. Secondary ions are then formed as indicated in Table 2.1. These secondary ions then react with the neutral molecules of the sample gas, which is incorporated in the reactant gas at a low concentration ($< 1\%$). Tertiary ions are then formed that are extracted from the ion source and

TABLE 2.1 **Principal Secondary Ions in Reactant Gases**

Reactant Gas	Reactant Ions Produced
Methane	CH_5^+ (44%); $C_2H_5^+$ (30%)
Propane	$C_2H_5^+$ (15%); $C_3H_7^+$ (60%)
Isobutane	$C_3H_7^+$ (12%); $C_4H_9^+$ (64%)

Source: Johnstone, 1972.

mass analyzed. Specifically, if methane is ionized at about 1 torr and used as a reactant gas, the following sequence takes place:

$$CH_4 + e \rightarrow CH_4^{+\cdot} + 2e$$

$$CH_4^{+\cdot} \rightarrow CH_3^+ + {}^{\cdot}H$$

$$CH_4^{+\cdot} + CH_4 \rightarrow CH_5^+ + {}^{\cdot}CH_3$$

$$CH_3^+ + CH_4 \rightarrow C_2H_5^+ + H_2$$

Both CH_5^+ and $C_2H_5^+$ will easily transfer an H^+ to a good proton acceptor, so that these secondary ions may protonate sample gas molecules:

$$CH_5^+ + M \rightarrow MH^+ + CH_4$$

$$C_2H_5^+ + M \rightarrow MH^+ + C_2H_4$$

If the sample gas is not a good proton acceptor, other ion-molecule reactions may occur, for example:

$$CH_5^+ + C_nH_{2n+2} \rightarrow [C_nH_{2n+1}]^+ + CH_4 + H_2$$

$$C_2H_5^+ + C_nH_{2n+2} \rightarrow [C_nH_{2n+1}]^+ + C_2H_6$$

Chemical ionization is also useful for producing negative ions or $(M - 1)^-$ species; and in favorable cases, the sensitivity of the negative ion technique exceeds that of the positive ion technique. For example, using 1 torr of methane moderator/reagent gas in a CI source, negative ion mass spectra of organic nitriles have been obtained with abundant $(M - 1)^-$ ions and minimal fragmentation (Bush et al., 1981). Rare gases have also been used as reactant gases. In such cases, ionization is effected by charge exchange or charge transfer, that is, a reactant ion accepts an electron to form an odd electron species. The resulting mass spectrum will be similar to an electron-impact spectrum, and some fragment ions will also appear.

Several specific ion sources have been developed to extend the chemical ionization technique. An atomizing sample introduction device has been reported by Hirata et al. (1979) that yields reproducible CI mass spectra of various involatile compounds. In addition, a direct chemical ionization (DCI) method uses a direct insertion probe within a chemical ionization reagent gas plasma. Reinhold and Carr (1982) have also used a nichrome wire insertion probe that was coated with a polyimide material and have employed a two-step procedure using an electric current to induce rapid polymerization. By avoiding metal-sample interactions, compounds of greater polarity and higher molecular weight have been analyzed.

SURFACE IONIZATION

The surface or thermal ionization method for producing positive ions was first applied to mass spectrometry by Dempster (1918). This source can be used for the assay of many metals, and the spurious spectrum of background gases is virtually eliminated because there is no ionizing electron beam. Samples in the range of 10^{-6} g to 10^{-12} g are amenable to analysis, and with suitable design and operation, the spread in ion energy from such a source can be limited to about 0.2 eV. Basically, when atoms or molecules are evaporated from a heated metal surface, the emission of neutral vapor may be accompanied by positive ions; that is, a fraction of the atoms or molecules will leave the surface in an electron-deficient state. This phenomenon can occur if the metal surface has a higher affinity for retaining an orbital electron from an escaping atom than does the atom itself. The quantitative relationship for predicting the ratio of ionized-to-neutral species was first suggested by Langmuir and Kingdon (1925) as:

$$\frac{n^+}{n^0} \propto \exp\left[\frac{e(W - IP)}{kT}\right] \propto \exp\left[\frac{11,606\,(W - IP)}{T}\right],$$

where n^+ = number of positive ions, n^0 = number of neutral atoms, e = electronic charge (1.6×10^{-19} C), IP = ionization potential (in volts), W = surface work function (in volts), k = Boltzmann constant (1.38×10^{-23} J/K), and T = surface temperature (Kelvin).

Thus, the efficiency of ion production for a hot-filament- or surface-ionization-type source is dependent on (1) the work function of the metal filament, (2) the filament temperature, and (3) the first ionization potential of the element that is being evaporated. It will be noted that:

$$\ln \frac{n^+}{n^0} \propto T, \qquad (\text{for } IP > W),$$

$$\ln \frac{n^+}{n^0} \propto \frac{1}{T}, \qquad (\text{for } IP < W).$$

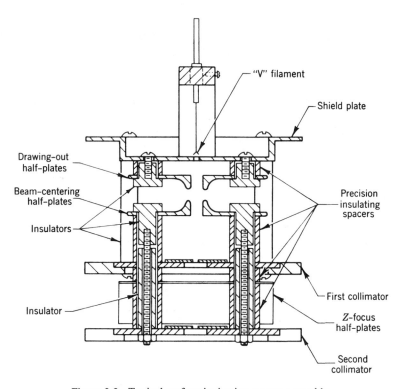

Figure 2.3 Typical surface-ionization source assembly.

The above relationships indicate that a higher fraction of atoms evaporates as ions at high temperatures if the ionization potential is greater than the filament work function. Indeed, the surface-ionization source is ideally suited to the production of all elements of low ionization potential. Figure 2.3 shows a typical surface-ionization source assembly, complete with filament, and with accelerating and focusing electrodes.

The selection of a filament material is usually limited to metals with high work functions, desirable mechanical properties, and which can be operated over a wide temperature range. Table 2.2 indicates some of the metals that have found considerable use as surface-ionization filaments.

The hot-filament source is also useful for the production of negative ions. In this case, the relative yield of negative ions to neutrals is:

$$\frac{n^-}{n^0} \propto \exp \frac{e(EA - W)}{kT} \propto \exp \frac{11,606 \ (EA - W)}{T}$$

In the above expression, *EA* is the electron affinity in volts, and negative ions can be produced for elements having an electron affinity greater than about 1 V.

TABLE 2.2   Metals Used for Surface-Ionization
Filaments

Metal	Work Function (V)	Melting Point (°C)
Nickel	4.50–5.24	1453
Niobium	3.96–4.01	2468
Platinum	5.08–6.27	1769
Rhenium	4.74–5.10	3180
Tantalum	4.03–4.19	2996
Tungsten	4.25–5.01	3410

Salient advantages of the surface-ionization source include (1) low ion energy spread, (2) simplicity in sample preparation, (3) selectivity of overlapping isobars and impurities, and (4) the fact that all ions are singly charged. Thus, for many isotopic analyses, a single magnetic sector analyzer is adequate—especially with the availability of sensitive ion detectors. In practical cases, the greatest energy spread may arise from the voltage drop across the filament—caused by resistance heating (i.e., a few volts). If the surface-ionization filament is heated thermally by some indirect method, this voltage spread can be virtually eliminated. However, there will always be some finite spread in the energies of ions leaving a thermal source, as small variations in the work function over the hot surface always exist—especially when deposition of the sample may give rise to various oxidation states.

In many instances, the sample preparation will be simple. For very small samples, however, complete preheating and outgassing of filament impurities is mandatory. As sample size decreases, the need for reagents of very high purity also becomes crucial, especially in isotopic dilution analyses; but for general assay and sample amounts of the order of 10^{-6} to 10^{-3} g, the preparation of a dilute solution is relatively easy. A weak acid or distilled water solution of the sample (usually a salt) is prepared, micropipetted onto the filament, and evaporated to dryness. In a majority of instances, the elements or simple metal oxide will be observed; but with some materials that have high ionization potentials, it is desirable to prepare special compounds. For example, sodium borate heated on tungsten will yield molecular ions of $(Na_2BO_2)^+$, and a quantitative mass spectrum can be obtained for boron at mass positions 88 and 89 instead of at the atomic mass positions of 10 and 11.

The geometrical form of a surface-ionization filament has many variations. In its simplest configuration, it is merely a ribbon filament that is about 10 mm long, 1 mm wide, and 0.025 mm in thickness; a "V" or canoe-shaped single filament has also been used extensively. This ribbon filament essentially replaces the ionizing electron beam of the electron-bombardment source, and the temperature of the filament is controlled by varying the current passing through the filament. The actual operating temperature of such a filament will be a compromise between the efficiency of ionization and the evaporation rate or "lifetime" of the sample being analyzed. This limitation is effectively bypassed by the use of a multiple filament source, whereby one filament is used to evaporate neutral vapor that impinges upon a second higher temperature-ionizing filament. This multifilament scheme originally introduced by Inghram and Chupka (1953), with independent control of vaporization and ionization, has permitted many elements and compounds to be analyzed; many ingenious geometries have been reported by various investigators. As a result, elements of reasonably high ionization potential can be ionized, molecular or atomic species can be selectively enhanced, and it is often possible to discriminate against impurity and spurious ions, or overlapping isobars, in isotopic ratio measurements.

Pulsed thermal sources of neutral atoms, in conjunction with laser ionization, represents an extension of the surface-ionization source (Fassett et al., 1984). Basically, neutral atoms are bombarded by an intense photon flux that is tuned to electronic transitions of an element, thereby providing an enhanced measure of selectivity from isobaric interferences. The surface treatment of filaments to enhance ion production has also been investigated extensively. For example, the surface ionization of both polar and nonpolar organic molecules has been reported by coating a tungsten wire with a LiI layer and applying a high electric field (Borchers et al., 1977). Carburized rhenium has also been used extensively to improve the thermal ionization of small samples of uranium and plutonium, even though the details of metal ion formation are not completely understood (Pallmer et al., 1980). The work function variation of rhenium has also been studied as a function of various surface layer and thermal regimes (Zandberg et al., 1981).

In a few instances, the surface-ionization source can furnish ion currents that are substantially greater than those obtained by electron-bombardment, even for organic molecules. For example, Fujii (1984) has reported an $(M - H)^+$ ion current of 9×10^{-6} A, at a pressure of $\sim 1 \times 10^{-7}$ torr of trimethylamine. In this experiment, O_2 gas was supplied separately to the heated Re filament to increase its work function. Fujii (1984) also suggests that some ionization anomalies can be explained by the fact that the dissociation of molecules impinging upon the filament, and their subsequent ionization, may occur faster than the surface-ionization process of the intact molecules.

A pulsed surface-ionization source, utilizing a platinum filament and a rapid electric field reversal (powered by an MOS transistor driver) has been used with time-of-flight spectrometry (Möller and Holmlid, 1984). This technique may be useful for the assay of alkali metals and for applications where only low-mass resolution is required.

SURFACE IONIZATION-DIFFUSION

The surface ionization-diffusion source is an important development relative to trace element analysis in both physical and biological systems (Bourgeois and White, 1981; Rec et al., 1974). In contrast to measurements of organic species, some trace metal analyses require the complete suppression of molecular and multiply charged ions, so that only atomic, single-charged species reach the detector focal plane. The surface ionization-diffusion technique uses an "encapsulation" of the sample by a thin film of high work function material. Essentially, after a sub-microgram sample has been loaded onto the surface of a rhenium ribbon, the surface is coated with a thin film (~5000 Å) of high work function material by sputtering or electroplating. This encapsulation provides several specific advantages over the usual thermal ionization technique:

1. The encapsulation prevents the prompt loss of neutral sample atoms. In some instances with a conventional filament, much of the sample is immediately lost upon raising a filament to an ionizing temperature.

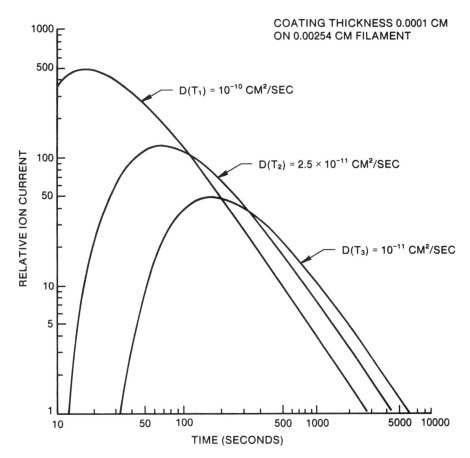

Figure 2.4 Theoretical plot of ion current vs. time for sample atoms being emitted from a surface ionization-diffusion source (Rec, 1975).

2. Sample atoms must diffuse through the thin film, and the diffusion process totally dissociates molecular species so that only atomic ions (or molecules formed at the surface) appear in the mass spectrum.

3. The filament can be operated at a higher temperature so that when sample atoms reach the surface, their probability of ionization is enhanced.

4. There is greater control of ion emission with the diffusion process, that is, the analysis can be programmed to accommodate constraints due to sample size and ion detector response.

Figure 2.4 is a theoretical plot of ion current vs. time for sample atoms diffusing through a thin encapsulation film for three different diffusion coefficients (Rec, 1975).

For thicker coatings and a smaller diffusion coefficient (low-temperature operation), longer times will be required to reach a maximum ion emission rate from the surface. The decay rate will also be dependent upon coating thickness and diffusion coefficient, so that both parameters are important as they affect the quality of the recorded data. For precise isotopic analysis and if sample size is adequate, a stable, very slowly decreasing ion current is desirable. With this technique, sensitivities of better than 10^{-15} g have been obtained, and ionization efficiencies greater than 10% are possible for some elements. Other investigators have successfully used electroplating, instead of RF sputtering, to obtain accurate isotopic ratios for uranium samples (Rokop et al., 1982), and the general methodology is applicable to many metallic elements. ^{240}Pu:^{239}Pu ratios have recently been measured for nanogram samples to 0.15% precision and accuracy at the 95% confidence level; detection limits of 2×10^5 atoms of plutonium have been obtained by Perrin et al. (1984).

THE VACUUM SPARK

The high-frequency spark source continues to be an important method of ion production for the bulk analysis of metals, powders, semiconductors, and insulators despite the advent of many other types. Its advantages include simplicity, universality, minimum selectivity, the generation of multiple-charged ions, and high ion yield. In certain applications, some of these advantages are also disadvantages, and analyses with this source are generally characterized by poor reproducibility, low precision, and a large ion energy spread, ΔV. However, a recent study of the ion energy spread in a vacuum spark discharge suggests that it initially may be only ~100 eV (at 50% of maximum intensity), and that much of the large energy spread that is experimentally observed occurs during the formation of the ion beam. Essentially, after a plasma is formed by the spark discharge, the plasma expands

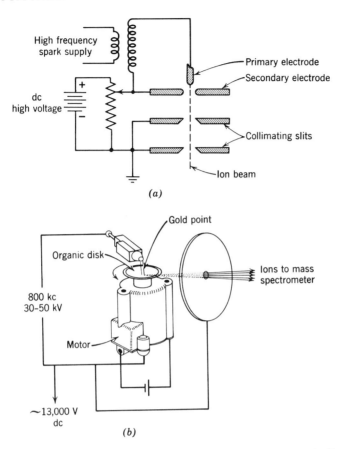

High frequency
spark supply

Primary electrode

Secondary electrode

dc
high voltage

Collimating slits

Ion beam

(a)

Gold point

Organic disk

Ions to mass
spectrometer

800 kc
30–50 kV

Motor

~13,000 V
dc

(b)

Figure 2.5 Schematic of (a) conventional vacuum spark source and (b) spinning electrode device (Hickam, 1966).

and the ions acquire energy by several different mechanisms (Ramendik et al., 1981). In any event, the ion energy spread is large compared to other ion sources, so that energy-focusing spectrometers are required. Nevertheless, a resolving power approaching 10,000 has been attained in energy- and direction-focusing instruments, and a minimum spot size of about 10 μm has been obtained when this source has been used as a "spark milliprobe" (Ramendik, 1980).

The vacuum spark source produces ions by the high-voltage breakdown across two electrodes, one of which consists of or includes the sample material. The terms arc and spark are often loosely used interchangeably (Honig et al., 1966), but a spark is generally characterized by a short-term, high-current flow—followed by a prompt recovery of original electrode potentials and current quenching. The basic spark source is shown schematically in Figure 2.5. As the potential across the two electrodes is increased, prebreakdown currents occur that are solely a function of voltage and pressure. At a sufficiently high voltage, a spark discharge will take place that results in both a region of negative resistance and an interelectrode voltage reduction of several orders of magnitude. If the external circuit possesses sufficient stored energy and a low impedance, a high-current arc may exist for many microseconds. At high-field gradients ($\sim 10^5$ V/cm), current increases are also attributed to field emission from "whiskers" or projections (< 1 μm) that exist on any cathode surface. As the voltage is further raised, some whiskers may vaporize and contribute cathode atoms to the interelectrode region. Electron-bombardment of the anode will generate neutral atoms that will be immediately ionized by the cathode electron stream. Ions from the anode region may also cause cathode sputtering—thus aiding the process of current multiplication. During the spark, singly and multiply charged ions will be produced by the copious vaporization of the electrodes and their subsequent ionization by electrons in a discharge.

In a typical spark source, the spark duration will be of the order of 20 μs at a maximum voltage of 30–40 kV, and the repetition rate can be varied. Ion detection by means of a photographic plate is traditional; with a Mattauch-Herzog geometry, masses from 7 to 252 can be recorded simultaneously (Murugaiyan, 1982).

PHOTOIONIZATION AND RESONANCE IONIZATION

Short wavelength electromagnetic radiation has long been used to generate ions in mass spectrometer ion sources, but until recently, limitations of intensity and photon energy have restricted this source from general applicability. The Planck-Einstein relationship of photon wavelength λ, and energy E, is given by

$$\lambda = \frac{hc}{E} = \frac{12{,}399}{E(\text{eV})} \qquad \text{(in angstroms)}$$

where h is Planck's constant and c is the velocity of light. Many ionization processes require ~ 10 eV in energy, corresponding to a photon wavelength of 1240 Å. This is near the cutoff wavelength of optical windows that are needed to isolate the photon source and the ion slit system. However, monochromers have been used to determine ionization potentials in conjunction with photoionization sources, and such sources produce simple mass spectra that are easy to interpret (Steiner et al., 1961). Photoionization has also provided the basis for photoelectron-photoion coincidence spectroscopy, whereby the excited states of a molecular ion can be related to both a mass spectrum and the measurement of the kinetic energies of ejected electrons.

Essentially, when a gas molecule is ionized by an ultraviolet photon of frequency v, an electron is ejected with a kinetic energy:

$$KE = hv - I$$

where I is one of the molecular ionization potentials. A positive ion results, containing an internal excitation energy:

$$E^* = I - I_0$$

where I_0 is the lowest molecular ionization potential (Eland, 1972).

However, with high-powered lasers, the ionization of atoms and molecules is no longer limited to radiations whose photon energy exceeds that of the ionization potential. If a sufficiently large flux of photons is available, ionization can take place by a multiphoton process, whereby the combined energy of several photons raises the atom or molecule to an ionized state. Furthermore, if a bound excited state exists at an energy level that is resonant with the absorption of an integral number of photons, the ionization probability is significantly enhanced. This phenomenon is termed *resonance ionization spectroscopy* (RIS); it is a unique method for detecting atoms of a selected quantum state in the presence of a large background of other atoms. With sufficient flux density, the RIS process, in principle, can be saturated so that target atoms intercepted by the laser beam will be ionized with unit probability; the RIS technique can be applied to most elements (Beekman et al., 1980).

This process has been used to demonstrate the laser separation of uranium isotopes, and isotopic ratio measurements have been made for other elements using time-of-flight mass analysis. The primary advantage of the RIS technique is its selectivity, whereby specific atoms in gases and vaporized solids can be detected in the presence of high concentrations of low ionization impurities. The selective ionization of mixture components has also been demonstrated for organic isomers $C_{10}H_8$, naphthalene ($IP = 8.12$ eV), and azulene ($IP = 7.42$ eV). Naphthalene is transparent to radiation at 532 nm (2.33 eV), but azulene is readily ionized by the absorption of four photons at this wavelength (Lubman et al., 1980). The relative multiphoton ionization rates for Xe, NO, NO_2, N_2O, O_2, N_2, CO, CO_2, H_2O, SO_2, CF_3Cl, and C_2F_5 have also been measured as a function of laser power at 193 and 248 nm (Hodges et al., 1981). With the advent of very high-power lasers, the RIS method has recently been applied to isotopic ratio measurements. Miller et al. (1984) have even compared mass spectral data to radioactive counting of lutetium isotopes ([173]Lu and [174]Lu), and they reported an agreement of approximately 1%. For the [173]Lu isotope, the measurement was performed with only $\sim 10^8$ atoms (27 fg).

INDUCTIVELY COUPLED PLASMAS

Inductively coupled plasmas (ICP) have been extensively used in emission spectroscopy, where photons from the excited atoms are detected in an optical system and specific wavelengths characterize the elements. Houk et al. (1980, 1981) showed that an ICP could also function as a mass spectrometer ion source. Gray and Date (1983) further demonstrated that metals having an ionization potential of less than 10 eV were fully ionized in such a source, where a temperature exceeding 8000 K was observed in the plasma core. Ionized gas was extracted at a temperature of ~7500 K. Also, in this particular source, the solution samples were introduced as a finely dispersed mist by a pneumatic nebulizer to an argon ICP operating at 1200 W.

A further development of the ICP and its application to quadrupole mass spectrometry has been reported by Douglas et al. (1983). These investigators have used mechanical and cryogenic pumps at very high pumping speeds so that large apertures (0.75 mm) could be used between the atmospheric pressure plasma region, the ion accelerating section, and the analyzing region. As a result, ion transmission is enhanced and the orifices do not plug, even with sample solutions of comparatively high salt content (5000 µg/ml).

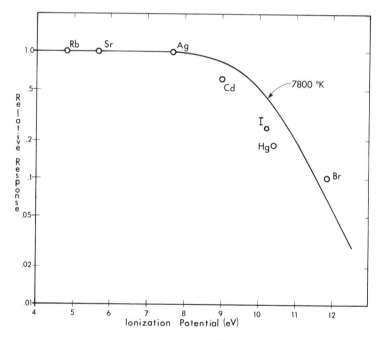

Figure 2.6 Response vs. ionization potential of an ICP-MS system for several elements. The data are commensurate with a plasma temperature of 7800 K. (Douglas et al., 1983; courtesy of D. J. Douglas and Sciex, Inc.)

TABLE 2.3   ICP-MS Data for NBS SRM 1643a
(10 Runs)[a]

Element	Ion	NBS	ICP-MS $\bar{x}$	RSD (%)
Beryllium	$^{9}Be^{+}$	19	21	20
Vanadium	$^{51}V^{+}$	54	52	6
Chromium	$^{52}Cr^{+}$	17	18	12
Manganese	$^{55}Mn^{+}$	32	34	5
Cobalt	$^{59}Co^{+}$	19	21	7
Zinc	$^{66}Zn^{+}$	69	57	11
Arsenic	$^{75}As^{+}$	77	76	5
Strontium	$^{88}Sr^{+}$	243	297	7
Molybdenum	$^{98}Mo^{+}$	97	134	9
Silver	$^{107}Ag^{+}$	2.8	3.5	16
Cadmium	$^{114}Cd^{+}$	10	13	22
Barium	$^{138}Ba^{+}$	47	74	17
Lead	$^{208}Pb^{+}$	27	31	8

Source: Gray and Date, 1984.
[a] Data in $ng \cdot mL^{-1}$.

Figure 2.6 shows the response of their ICP-MS system to elements of different ionization potentials. A specific advantage of this source is the rapid throughput of samples—at intervals of 2 min or less. The sensitivity of this source is also quite high, and the simplicity of the mass spectra is attractive. When used with a quadrupole analyzer and a channel electron multiplier, detection limits have been reported in the range of 0.5–5 ng/ml for 90% of the elements. Important uses for this ion source include trace element analysis in biological matrices and the semiquantitative assay of water samples. Table 2.3 shows the high elemental sensitivity in water-calibrated samples reported by Gray and Date (1984).

Although the ICP is primarily useful for determining elemental concentrations, Smith et al. (1984) have recently applied this source for determining Pb-isotopic ratios in geochemical studies.

LASER MICROPROBE

The first use of laser beams as a mass spectrometer ion source was reported by Honig and Woolston (1963), who demonstrated that sufficiently high surface temperatures were generated to produce vaporization and ionization from metals, semiconductors, and sintered insulators. This work was followed by the work of Fenner and Daly (1966), who used a ruby laser that produced a power level of 10 mJ in 30 ns pulses in the time-of-flight system

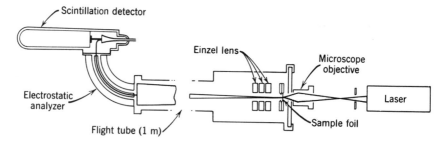

Figure 2.7 Laser ion source applied to a time-of-flight analyzer (Fenner and Daly, 1966).

shown in Figure 2.7. The laser beam was focused through a microscope objective onto a foil area of only 2×10^{-5} cm^2. Each laser pulse produced about 2×10^{13} atoms from a foil of 3×10^{17} atoms/cm^2, and an energy spread of up to 500 eV was observed. Metal foils of Li, Be, B, C, Al, Fe, Cu, Ag, Pb, and Au yielded comparable numbers of ions. This foil method provides a higher signal-to-noise signal than the evaporation of bulk samples, and it serves as the basis for contemporary instrumentation. In addition, the geometry is favorable in that photons can be incident upon one side of the specimen and ions can be extracted from the opposite face. However, the large ion energy spread requires an energy selector (in the time-of-flight system) or a double focusing arrangement when magnetic analysis is employed.

The laser microprobe is now an important ion source for the analysis of metals, polymers, inorganics, and biological specimens, and a review of applications and geometrical configurations, etc., has been given by Conzemius and Cappellen (1980). Also, models of the vaporization/ionization phenomenon in solids have been analyzed by Hercules et al. (1982), who

TABLE 2.4 Average Depth of a Layer Ablated from a Substance by Laser Radiation[a]

Substance	Depth (μm)	Substance	Depth (μm)
Graphite	3.22	Copper	0.21
Magnesium	1.08	Zinc	0.53
Aluminum	0.70	Germanium	1.92
Silicon	1.49	Zirconium	0.37
Vanadium	0.29	Niobium	0.16
Chromium	0.25	Molybdenum	0.17
Iron	0.42	Silver	0.51
Nickel	0.45	Cadmium	0.54

Source: Kovalev et al., 1978.

[a] Radiation energy = 1 J; $S = 1 \times 10^{-2}$ cm^2; pulse duration = 40 ns.

have suggested a five-step process: "(1) direct ionization of the solid by the laser beam, (2) desorption and ionization of the solid in the region immediately adjacent to the laser pulse, (3) surface ionization, (4) gas-phase (ion-molecule) reactions, and (5) emission of neutral particles vs. ion formation." At low laser power densities ($< 10^8$ W/cm^2), the ionization efficiency is about 10^{-5}; but at power densities in the range of 10^9 to 10^{10} W/cm^2, ionization efficiencies have been reported in the range of 0.01–0.1%.

A limitation of the laser probe for analytical work on bulk solids is the large variation in the ionization probability of vaporized samples, plus the difference in the rate of layer-by-layer ablation of material for various substances. Table 2.4 shows different depths removed by a 1 J laser for a focused beam spot, $S = 10^{-2}$ cm^2, on various substances. Despite such a limitation, the laser microprobe is a powerful analytical tool. Not only has a spatial resolution of less than 1 μm been obtained with this source, but Simons (1983) has reported isotopic ratio measurements on nanogram samples, with relative accuracies of ~3%.

In the laser desorption source described by Hardin et al. (1984), samples were sprayed "on-line" onto a moving stainless steel ribbon with a thermospray vaporizer. The samples were then transported through a differentially pumped vacuum lock and ionized in the mass spectrometer with 45 ns, 10^8 W/cm^2 laser pulses. With the ribbon advancing at 0.25 cm/s, small samples of thermally labile biomolecules (less than one monolayer) were removed after a single pass under the laser. In general, the laser source is operated at low-power densities for molecular analysis and at high-power densities for the elemental characterization of solid materials. Conzemius et al. (1984) have discussed the general problem of optimization of the laser ion source with respect to focus, reproduction of power densities during analysis, etc., and have provided data on relative ion production rates for various metals and alloys.

ION BOMBARDMENT

The production of secondary ions by ion bombardment or sputtering is one of the most important methods for the analysis of solids. When a solid surface is bombarded by energetic (keV) ions, there will be an emission of secondary particles (positive and negative ions, neutrals, and electrons) and photons. The mass analysis of the secondary ions is the basis of secondary ion mass spectrometry (SIMS). The generation of ions by ions can also be attained in gases, but the real potential for ion impact sources relates to solids. Several advantages can be stated explicitly:

1. The secondary ion energy spread is reasonably low. Some secondaries will be ejected with considerable energy, but the energy spread of most secondaries will be less than 100 eV.

2. The sample may be analyzed at room temperature.
3. Only surface atoms are ionized (~ 20 Å depth, depending upon the primary ion mass and energy).
4. Ions can be produced for all elements.
5. High sample sensitivity is achieved for surface atoms of many elements by selecting the proper bombarding ion.
6. The sputtered mass spectrum yields species of atoms, compounds, and clusters.
7. The ion bombardment source can be operated in either the "static" or "dynamic mode." In the former, the total "dose" of primary ions impinging on a solid surface is limited to $\sim 10^{13}/cm^2$ (Magee, 1983). In the latter, current densities are $> 10^{-6}$ A/cm^2 and such ion beams provide depth profiling by "milling" away the sample.
8. The ion impact source may be used in the steady-state or pulsed mode, so that it is applicable to magnetic analyzers, quadrupole mass filters, and time-of-flight spectrometers.
9. Advances in ion optics have made this source the basis of the high-resolution ion microprobe (Chapter 3).
10. Very small samples and sample concentrations are amenable to analysis.

The ion impact source for SIMS also has its limitations, and some of these have been included in a review paper on surface analysis by Turner and Colton (1982):

1. The elemental sensitivity for SIMS varies by up to four orders of magnitude, and the sensitivity is also affected by the sample matrix and the presence of oxygen. This makes SIMS difficult to quantify.
2. At high sputtering rates (> 50 Å/min), the analysis may be affected by the ion implantation of the primary ions into the sample and by differential sputtering; nonconducting samples may acquire a charge build-up.
3. If a sample is to be analyzed at room temperature, the vacuum should be high enough so that the flux of the bombarding beam is large compared to the time rate of arrival of residual gas atoms.
4. In order to detect true surface monolayers, the ion impact source must be operated with a broad beam; and in the "static" mode, with low ion beam currents and a corresponding loss in signal intensity.

A comprehensive account of the development of SIMS and its many inorganic and organic applications has been prepared by Honig (1985), who was the first to devise an ion sputter source for the analysis of semiconductors (Honig, 1958, 1959). The basic elements of the ion impact source

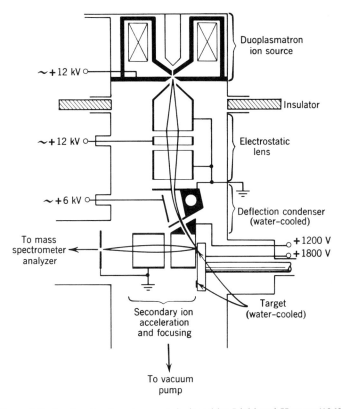

Figure 2.8   Ion bombardment source designed by Liebl and Herzog (1963).

are shown in Figure 2.8. A primary ion beam is provided by a duoplasmatron. These ions are then accelerated by a drawing-out potential, focused by an electrostatic lens, and deflected by a condenser lens. Sputtered secondary ions are then accelerated into the mass analyzing system (Chapter 3). Present-day ion sources in commercial instruments utilize Ar^+, O^-, O_2^+ and Cs^+ ion beams. The cesium source introduces no heavy gas load, as does a gas discharge source, and the heavier ions have a shorter range than lighter ions of the same energy. Specifically, the Cs^+ ions have a projected range in solid materials of about 60% of that of O_2^+ of the same energy (Bentz and Liebl, 1982). In addition, the negative ion yields of certain elements are enhanced by Cs^+, so that a reduction in sample consumption can be realized for the same mass-resolved ion counting rate.

The ion bombardment source, using keV particles, also causes the emission of polyatomic or cluster ions, whose mass, energy, angle of emission, and intensity can help to characterize the secondary emission process (Barlak et al., 1983). Furthermore, this source is useful for tandem mass spectrometry, where an objective is to determine characteristic fragmentation patterns of the ions generated. For example, Todd et al. (1984) have designed a source for a triple-sector instrument that produces an intense secondary ion beam, and that permits the routine tandem mass spectrometry of many involatile analytes.

FAST ATOM BOMBARDMENT

The fast-atom bombardment (FAB) source is a development evolving from SIMS, and it has greatly extended the scope of analyses that have been performed by means of SIMS, ^{252}Cf plasma desorption, field desorption, and other techniques. In France in the 1970s, Devienne and coworkers studied the sputtering of both inorganic and organic materials by neutral beams (Devienne and Roustan, 1982). Barber and associates, at the University of Manchester, developed several new improvements leading to a practical FAB source that could be applied to existing mass spectrometers (Barber et al., 1981). In this latter work, fast argon atoms (1.2×10^5 m/s = 3 keV energy) were produced by resonant charge neutralization of an Ar^+ beam passing through Ar gas. Residual ions in the neutralized beam were removed by electrostatic deflection, and the resultant beam bombarded a sample in the source chamber of a double focusing spectrometer. This neutral beam technique circumvented problems of charge build-up on insulating materials that were being analyzed by SIMS, and it permitted the primary ion/neutral and secondary ion source potentials to be adjusted independently. It also led to important applications of FAB in the assay of organic molecules (Magee, 1983).

A distinguishing feature between FAB and other ion production methods is the use of glycerol containing the sample molecules instead of assaying thin solid films or coatings. With glycerol, the liquid surface appears to be continuously refreshed with sample molecules, and ions continue to be produced until evaporation of the glycerol is complete. Recent work by Ligon (1983), however, who experimented with fully deuterated glycerol, indicates that surface renewal by diffusion alone may be an insufficient explanation, and that several other phenomena may contribute to the longevity of the sample.

In one commercial FAB source, the glycerol is mounted on a copper probe tip whose area is in the range of $0.04-0.2$ cm^2 (Williams et al., 1981). This probe is inserted into the source through a conventional vacuum lock, and the operating pressure is about 10^{-5} torr. The sample is then bombarded by a 4–6 keV argon beam (~40 μA) that has been neutralized by charge exchange through argon gas. Compounds studied with this particular source have included organic salts, polar antibiotics, nucleoside phosphates, and underivatized peptides. Molecular weight determinations have also been routinely made in the range of 300–2000 amu with less than 1 nmol of material, and sequence data have been deduced from samples of 2–50 nmol.

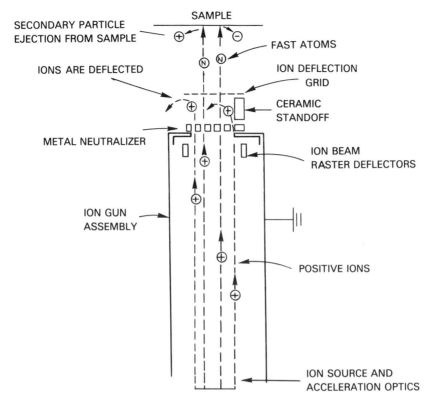

Figure 2.9 Schematic diagram of one type of fast-atom bombardment source. Ion/electron recombination occurs in the vicinity of the ion beam-neutralizing metal. (Ross et al., 1983; courtesy of M. M. Ross.)

Another commercial ion gun that has been modified to generate a diffuse fast-atom beam employs a different ion beam neutralizer (Ross et al., 1983). As indicated in Figure 2.9, a neutralizing metal, such as aluminum, can be used effectively.

Rudat (1982) has reported on a minature fast-atom source that can be attached to a standard mass spectrometer ion source; they observed FAB spectra at neutral beam energies from 2–6 keV. The atom beam impacts the sample probe tip (a solid 3 mm steel rod) at an angle of 70° from the surface normal. The author has suggested that this probe is applicable to either magnetic or quadrupole instruments, and that intensities are adequate to record computerized spectra at scan speeds of 3–5 s/decade. FAB sources can now be procured from a number of instrument manufacturers.

The FAB source has now attained widespread use because of its several important advantages (Barber et al., 1982):

1.   Ions can be generated from samples at room temperature; volatili-
     zation is not required.

2. The mass spectrum can include both positive and negative ions, and the spectrum comprises both molecular weight and fragmentation information. Molecular weight data are usually obtained from $(MH)^+$ ions and $(M - H)^-$ ions in positive and negative ion spectra, respectively.

3. The lifetime of the sample is long ($\sim$20 min), and beam currents are relatively stable.

4. Sample sensitivity is high.

5. The source is applicable to a wide range of analyses, but it is especially effective for biomolecules of high molecular weight.

6. The instrumentation is relatively simple, and source voltages of < 10 kV are sufficient to produce energetic ion-neutral beams.

Fast-atom bombardment high-resolution spectrometry is being further developed by many research groups, and the extensive applications of FAB in organic analysis have been reviewed by Fenselau (1983). In industrial analyses, fast-atom bombardment has been especially effective for numerous types of polar polymeric chemicals, including rubber and plastic additives, and organometallic compounds (Lattimer, 1983–1984). Secondary ion yields from the bombardment of samples in a glycerol matrix have been observed as high as 10^{-10} A, and a mass resolution of 1 part in 10,000 has been obtained with this ion source for molecules in the 1000–2000 amu range (Fenselau et al., 1983). Furthermore, Ligon and Dorn (1984) have demonstrated that surfactants can be used to enhance the response of an analyte relative to a matrix such as glycerol, thereby providing an improved signal-to-background ratio.

CALIFORNIUM-252 PLASMA DESORPTION

It is interesting to note that mass spectrometry, previously used to determine the masses and charge states of fission products of the transuranic elements, is now utilizing these fission fragments as a unique ion source. This source, first developed by Macfarlane and Torgerson (1976a), is an important one because it has permitted the assay of biologically important molecules that are nonvolatile and thermally unstable. Contemporary applications now range from investigating antitumor drugs (Jungclas et al., 1982) to complex mixtures of molecules with widely varying molecular weights, such as coal (Lytle et al., 1982).

Fission fragments are energetic particles whose mass can range from approximately 80–160 amu; approximately 200 MeV of kinetic energy is released from a fissioning nucleus. In the case of ^{252}Cf, a typical fission fragment pair is $^{142}\text{Ba}^{+18}$ and $^{106}\text{Tc}^{+22}$, having kinetic energies of approximately

79 MeV and 104 MeV, respectively. In contrast to some other transuranic nuclides, ^{252}Cf also has a short half-life; a very high percentage of its decay takes place by spontaneous fission (~3%). These fission fragments also have a very short range in solids, and this property has been exploited in (1) fission track detectors, where the fission track can be made visible by etching, and (2) shallow silicon detectors, where a 100 MeV particle generates ~2.8 × 10^7 electron-hole pairs. For the production of ions in mass spectrometry, these fission fragments dissipate their high kinetic energy within a sample that is mounted on a very thin substrate. Thus, within ~10^{-12} s, a single fission particle will produce a highly localized "hot spot," resulting in volatilization and some ionization.

To obtain a mass spectrum by ^{252}Cf desorption, the sample is dissolved in an appropriate solvent and mounted on a thin (~1 × 10^{-3} mm) nickel foil. This sample foil is then precisely aligned in close proximity to the ^{252}Cf source of fission fragments (Macfarlane and Torgerson, 1976b; Macfarlane, 1983). Fission fragments penetrate the sample, and ions emerge from the opposite face and are accelerated into a time-of-flight spectrometer. Both positive and negative ions are generated, and masses have been measured to within 0.5 × 10^{-3} amu. This precision is made possible because one of the fragments in binary fission (traveling in the opposite direction to the particle that penetrates the sample) is detected and serves as a zero time marker. By this means, fragments can be correlated within 10^{-9} s, and spectral data is obtained in the pulsed mode with a high signal-to-noise ratio. This type of source is now being used for the identification of molecules in the mass range approaching 10,000, as large molecules are not always dissociated. Furthermore, although this is a rather specialized source, the ^{252}Cf isotope is now being produced at the rate of 500 mg/yr by the Transplutonium Processing Plant at the Oak Ridge National Laboratory (Keller et al., 1984), so that adequate amounts of material are available for research.

A reasonably intense source used with the Rockefeller University fission fragment ionization spectrometer produces a count rate of 5000 ions/s; this has permitted molecular weight distributions of certain polyethers to be determined up to 2000 amu (Chait et al., 1984).

FIELD IONIZATION AND DESORPTION

The field ionization source was first applied to mass spectrometry by Inghram and Gomer (1954), and a comprehensive treatment of field ionization sources was subsequently prepared by Beckey, a pioneer in the development of field desorption (Beckey, 1971). Field ionization (FI) utilizes high local electrical fields (10^7 to 5 × 10^8 V/cm) to generate ions from the tips of thin filaments, the edges of metal blades, or grown microstructures. Ionization is assumed to occur at a distance of several angstroms in front of the tip; the minimum distance to generate FI is given as:

$$d_{\min} \cong \frac{IP - W}{\epsilon}$$

where ϵ = field strength, IP = ionization potential, and W is the work function. For an organic molecule that has an ionization potential of 10 V, a surface whose work function is 5 V, and a field of 0.5 V/Å, $d_{\min}$ is approximately 10 Å. A theoretical explanation has been proposed for the hydrogen atom based on quantum mechanical considerations. At the gas-solid interface of the ionizing electrode, it is assumed that a finite probability exists for tunneling of the electron from the hydrogen atom to the metal. This electronic transfer will occur only if an electron in a ground state is raised (by the high external field) to an energy level corresponding to the Fermi level of the metal. This mode of ion formation results in an ion energy spread of only 1 or 2 eV.

The gases that are to be analyzed are introduced at a pressure of approximately 10^{-4} torr or less in order to prevent electrical breakdown. The small ion energy spread of the field ion source provides better mass resolution than is possible with a spark source, and ion currents as high as 10^{-9} A have been obtained with field strengths of 10^8 V/cm. An especially advantageous feature of this source relates to its selectivity for organic molecules. In hydrocarbon spectra, parent ions are observed to be more abundant by several orders of magnitude than are fragment ions, so that a less complex spectrum results. This same discrimination allows the detection of free radicals at lower concentrations than can be observed with some other techniques. Because FI involves small amounts of energy transfer, weakly bound molecules can be analyzed without appreciable dissociation and ion-molecule reactions can be observed.

In field desorption (FD), the sample is coated onto the emitting surface and ionized directly from the solid state. With FD $[M]^+$ and/or $[M + H]^+$ ions will be formed depending upon the nature of the compound. This type of ion source has special applicability to the analysis of biologically important compounds, large molecules, and samples that are thermally involatile (Beckey, 1969).

The development of organic semiconducting microneedles by field polymerization has been an important development, because the field strength at the surface of smooth wires with radii of even 10^{-4} cm is insufficient to ionize many organic compounds. This process of needle growth for FI and FD sources is called "activation," and the process is dependent upon the chemical nature of the field-ionized substance, the electrical field strength, the pressure, the temperature, and emitter radius (Migahed and Beckey, 1971). The use of benzonitrile as an activator generates a very large increase in the surface area of the emitter, and the compound can be applied directly

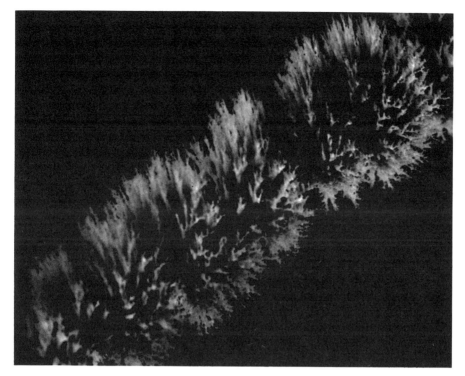

Figure 2.10   Scanning electron micrograph of an "activated" field desorption emitter. Dendritic carbon whiskers are grown on a 10 μm diameter tungsten wire. (Courtesy of W. V. Ligon.)

to this surface. Figure 2.10 is a photomicrograph of such an "activated" emitter. Silicon whiskers have also been grown on tungsten wire emitters for the analysis of inorganic salts. In one investigation, the ions were extracted from the emitters by a potential of 6 kV and at temperatures close to the melting point of various compounds. This technique has provided an ion source that has been used for assaying the isotopic composition of Cu, Sn, Ag, Te, Cd, and Sb in meteorites and terrestrial geological samples (Katakuse et al., 1979). Metals such as Al, Cs, Cr, Co, Cu, and Rb have also been identified in physiological fluids in the nanogram-to-picogram range by FD, with the Rb^+ ion detection limit being measured at $< 10^{-15}$ g (Schulten et al., 1981).

Pulsed field desorption (PFD) is a variation of the FI and FD methods. Negative voltage pulses of approximately 100 ns duration are applied to the cathode of an FI source after periods of adsorption in the gas phase in the nanosecond range (Röllgen and Beckey, 1973). This technique permits the external electrical field where ionization takes place to be kept relatively low, and as a result, PFD mass spectra have a reduced fragmentation pattern. The pulsed field ion source also yields information on surface reactions and the formation of cluster ions. Cluster ions have been observed during laser pulsing with photon energy densities in the range of 0.1–1 J/pulse/cm² and a 50 Hz repetition rate (Jentsch et al., 1981). This photon-induced FI at high laser irradiance greatly enhances evaporation and desorption, and multiply charged species have also been observed in time-of-flight mass measurements. Recent developments in laser-assisted FD have further extended its use for molecular weight determinations, structural elucidation, and purity analysis of inorganic and organometallic compounds. In an application to the latter group, the isotopes of iron were measured in a hemoglobin (molecular weight, about 65,000) when the sample was directly deposited on the emitter, and the isotopic abundance of magnesium was measured in chlorophyll (Schulten et al., 1982).

PYROLYSIS

Pyrolysis mass spectrometry presently encompasses several types of ion sources that include (1) the direct probe and oven pyrolysis, (2) laser pyrolysis, and (3) Curie-point pyrolysis—in combination with ionization techniques of electron-impact, chemical ionization, FD, and FI. Oven pyrolysis is usually characterized by slow heating rates (< 1°C/s), low pyrolysis temperatures (< 400°C), and a relatively long residence time of the pyrolysis products in the hot reaction zone (Haverkamp and Kistemaker, 1982). By thermal activation, molecules tend to fragment via specific mechanisms, and this thermal fragmentation pattern can be observed. Oven pyrolysis has the advantage in that slow-scanning magnetic sector instruments can be used in analysis. The more recently developed laser pyrolysis, however, provides some desirable options. First, it permits a very rapid heating of the sample while avoiding heating of a substrate. Second, the high spatial resolution of laser pyrolysis permits a detailed examination of many different types of samples, and it has led to the development of the laser microprobe. However, there remain practical problems relating to controlling the energy absorbed per unit sample volume.

The Curie-point pyrolysis technique involves a simple procedure. A thin ferromagnetic filament is placed inside an induction coil that is connected to a high-frequency power supply. The filament is then inductively heated to its Curie-point temperature, where the filament loses its ferromagneticity and becomes paramagnetic. This change results in a sharp decrease in the energy absorption by the filament from the induction coil. If filament dimensions and the induction coil geometry and other parameters are suitably matched, the filament will quickly reach an equilibrium temperature that is close to the Curie-point temperature; residual energy absorption by eddy currents will be balanced by loss of heat through radiation and conduction (Bühler and Simon, 1970). Curie-point temperatures are 358°C for Ni, 770°C for Fe, and 1128°C for Co. Temperature-time profiles of Ni, Fe, and Co

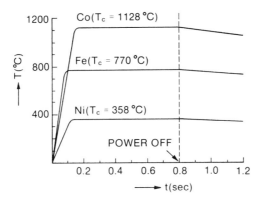

Figure 2.11 Time-temperature profiles and Curie-point temperatures for Ni, Fe, and Co filaments (diameter = 0.5 mm) when using a 1.5 kW, 1.1 MHz power supply. (Meuzelaar et al., 1982; courtesy of H. L. C. Meuzelaar.)

filaments, when powered by a 1.5 kW, 1.1 MHz power supply, are shown in Figure 2.11. Thus, a very accurate reproduction of a temperature-time profile is possible for sample analysis, although many other factors involved in sample preparation can also influence the ionization and pyrolysis pattern.

The definitive monograph by Meuzelaar et al. (1982) is an excellent document, encompassing the broad field of pyrolysis mass spectrometry and its many applications. A recent review (Meuzelaar, 1984) also focuses on the pyrolysis mass spectrometry of complex organic materials.

MISCELLANEOUS SOURCES

^{63}Ni Beta Emission

A ^{63}Ni beta ray emitter can serve as an ionization source at atmospheric pressure. Essentially, a sample molecule can capture an electron via the reaction:

$$M + e \rightleftarrows M^-$$

to form a negative ion. Electron affinities (EA) of most organic compounds are reported to be 1 eV, and the EA of halide ions and some nitrocompounds is in the range of 2–4 eV (Kim and Karasek, 1982).

Microwave Discharge

A microwave discharge source has been designed for use in a chemical ionization system (Siegel, 1980). Microwave power is coupled to a source volume of about 1 cm^3 via a capacitive electrode. Microwave power dissipation of ~ 1 W to the active volume is reported to function as a direct replacement for the conventional electron beam source in a tandem EI/CI system. The simplicity and reliability of this type of ionization source are cited as factors that may lead to its usefulness for a wide range of carrier gases and to its general use in GC/MS systems.

Electrohydrodynamic

Electrohydrodynamic (EHD) liquid metal ion sources of high intensity and small spot size have been developed for use in both scanning ion microprobes and ion beam lithography (Chapter 20). Typical of these specialized sources is a liquid Ga ion source that produces a 0.1–5 μm Ga^+ ion beam at energies ranging from 2–20 keV, and with 20 pA to 8 nA beam currents (Ishitani et al., 1982). A needle-type EHD-Ga ion source is employed; a small amount of Ga is retained at the end of a hairpin filament to which a tungsten emitter (120 μm in diameter) is spot-welded. This emitter element is combined with a control electrode, extractor, and three-electrode lens to produce submicron ion beams for secondary ion mass spectrometry. Gallium sources are attractive for their high brightness, and measurements have been made of their energy distributions. Full width at half maximum (FWHM) values range from 5–30 eV for total currents of 1–30 μA (Swanson et al., 1980).

Gold has also been employed as a liquid-metal ion source, and lifetimes up to 200 h at 10 μA have been reported (Komuro and Kawakatsu, 1981). Aluminum liquid metal sources have also been investigated, with ion source currents being in the range of 10–20 μA (Torii and Yamada, 1982). Both gold and aluminum are important in the microcircuit fabrication process, but the ion sources with lower mass elements appear to produce ion beams with a smaller energy spread for the same source current.

Collision-Induced Dissociation

The use of collision-induced dissociation (CID) to produce a secondary mass spectrum is one of the most important recent ionization methods that has evolved into a major subfield of mass spectrometry (Cooks, 1978). The sensitivity and selectivity of the technique is so great that it can be properly classified as an additional *primary* ion source. Positive and negative ions can be generated from the interaction of beams of energetic ions with gaseous atoms, neutral molecules, and solid surfaces, and by dissociation in passing through thin (~ 100 Å) metallic and polymeric films. The charge stripping (of many electrons) for MeV ions can be considered a special case of *collisional activation*, whereby combined accelerator-mass spectrometer systems can identify nuclides at subparts per billion concentrations (Chapter 4). For more conventional analysis, a fairly simple collisional dissociation scheme has been reported (Glish and Todd, 1982) that uses a gas flow-through hypodermic needle located ~5 mm behind the mass-resolving slit. This scheme approximates a very thin collision cell, it provides a good fragmentation efficiency, and it can be placed in the flight path of existing instruments.

THE ISOTOPIC DILUTION METHOD

The isotopic dilution method is an important analytical technique that is used with many ion sources for the assay of gases, liquids, and solids. It permits the analysis of very small samples, and an attractive feature is that the sample often does not need to be chemically pure. However, contamination either from reagents or from the atmosphere is an ultimate limitation, which may determine its sensitivity for a particular element. The method applies to all chemical elements that have two or more naturally occurring isotopes. For monoisotopic elements, the method may still be considered if a long-lived radioisotope of the element is made available. A definitive monograph dealing with the many variations of the isotopic dilution method (using either stable or radioisotopes) has been prepared by Tölgyessy et al. (1972).

The principle of the method is to mix or blend the unknown quantity of an element (sample) with a spike (or tracer) that has an isotopic composition substantially different from that of the sample. The isotopic composition of the sample must be known or measured, and both the isotopic composition and the amount of the spike must be known before blending. The element should also be present in the same chemical form in the sample and in the spike. For an element of two isotopes, the following equation yields the unknown weight W_x, of the sample:

$$W_x = W_t \frac{(R_t - R_m)(R_s M_1 + M_2)}{(R_m - R_s)(R_t M_1 + M_2)},$$

where W_t is the known weight of the spike (tracer), R_s is the isotopic ratio of isotope 1 to isotope 2 of the sample element, R_t is the isotopic ratio of isotope 1 to isotope 2 of the spike, R_m is the isotopic ratio of isotope 1 to isotope 2 of the mixture, and M_1 and M_2 are the atomic weights of isotope 1 and 2, respectively. In this equation, R_s is usually the natural occurring ratio and is known; R_t is also known or measurable, and R_m is the ratio of the mixture determined by mass spectrometry. With a known spike weight W_t, the unknown weight of the sample element can be calculated.

Boron is an element having two isotopes, and the application of the isotopic dilution method for its assay is illustrated in Figure 2.12. This element is important for producing p-type regions in silicon-based electronic devices. Boron is also an important element (the [10]B isotope) for controlling the neu-

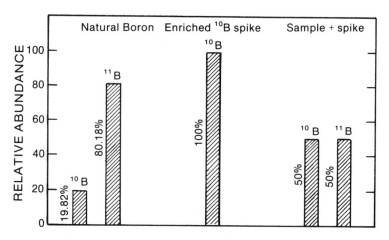

Figure 2.12   Application of the isotopic dilution method to boron, using ^{10}B as the tracer.

tron flux and power level of nuclear power generating plants. Furthermore, an estrogen containing ^{10}B atoms has recently been synthesized for the radiation treatment of cancerous tissues. In this example, boron, whose isotopic abundance is ^{10}B (19.82%), ^{11}B (80.18%), can be spiked with pure ^{10}B, which is the minor isotope. Table 2.5 is presented simply to indicate the approximate mass spectrometrically observed ^{10}B/^{11}B ratios of the mixture R_m that would correspond to the ratios of the weight of the sample W_x to the spike weight W_t. Such a tabulation suggests that if a 1×10^{-6} g spike of ^{10}B leads to a mass spectrometrically observed ^{10}B/^{11}B ratio of 1000 for the blend or mixture, the unknown weight of the boron is approximately 1.3 $\times 10^{-9}$ g.

It is also assumed that (1) the mass spectrometer is capable of measuring such ratios at high sensitivity, and (2) that reagent or other impurities will not lead to errors in the analysis. Such assumptions are not always justified. For the example cited, it is often desirable to measure ^{10}B/^{11}B abundances

TABLE 2.5   Isotopic Ratios for Boron Corresponding to Sample-Spike Weights, with a Pure ^{10}B Tracer (1 ppm ^{11}B)

R_m (^{10}B/^{11}B)	Sample-Spike wt (W_x/W_t)
1	1.78
10	1.38×10^{-1}
50	2.71×10^{-2}
100	1.35×10^{-2}
1000	1.34×10^{-3}

not at mass numbers 10 and 11, but at a higher mass number using compound ions (e.g., $Na_2BO_2^+$ at mass numbers 88 and 89) in order to minimize discrimination in the ion source.

For samples comprising elements of more than two isotopes, the isotope dilution equation may still be used if a normalization is made to account for the other isotopes; generalized analytical expressions can also be employed (Webster, 1959). The method has the great advantages of eliminating the necessity for quantitative recovery and, in favorable instances, the need for any prior chemistry. The mass spectrometer itself will often discriminate against other chemical species. Even if impurity mass peaks overlap the mass spectrum of the sample, it is sometimes possible to reduce spurious mass peaks to very low values. For example, large differences between the sample and impurities, with respect to volatility or the probability of ion formation (in the case of thermal ionization sources), permit a differentiation of several orders of magnitude.

Finally, it should be emphasized that the isotopic dilution method requires that during the time of actual analysis, the spike and sample element must be in the same chemical form. When the spike and sample are first mixed, they may not be in the same chemical state; an appropriate chemical treatment must then be carried out.

REFERENCES

Barber, M., R. S. Bordoli, R. D. Sedgwick, and L. W. Tetler, *Organic Mass Spectrom., 16,* No. 6, 256 (1981).

Barber, M., R. S. Bordoli, G. J. Elliott, R. D. Sedgwick, and A. N. Tyler, *Anal. Chem., 54,* No. 4, 645A (1982).

Barlak, T. M., J. E. Campana, J. R. Wyatt, B. I. Dunlap, and R. J. Colton, *J. Phys. Chem., 87,* 3441 (1983).

Beckey, H. D., *Int. J. Mass Spectrom. Ion Phys., 2,* 500 (1969).

Beckey, H. D., *Field Ionization Mass Spectrometry,* Pergamon, Oxford (1971).

Beekman, D. W., T. A. Callcot, S. D. Kramer, E. T. Arakawa, G. S. Hurst, and E. Nussbaum, *Int. J. Mass Spectrom. Ion Phys., 34,* 89 (1980).

Bentz, B. L., and H. Liebl, in *Secondary Ion Mass Spectrometry SIMS III,* eds. A. Benninghoven, J. Giber, J. László, M. Riedel, and H. W. Werner, Springer-Verlag, Berlin (1982).

Borchers, F., U. Giessmann, and F. W. Röllgen, *Organic Mass Spectrom., 12,* No. 9, 539 (1977).

Bourgeois, M. A., and F. A. White, *Ann. Conf. on Mass Spectrom. and Allied Topics,* Minneapolis (May 1981), p. 253.

Brown, S. C., *Basic Data of Plasma Physics,* Wiley, New York (1959), p. 116

Bühler, Ch., and W. Simon, *J. Chromatogr. Sci., 8,* 323 (1970).

Bush, K. L., C. E. Parker, D. J. Harvan, M. M. Bursey, and J. R. Hass, *Appl. Spectros., 35,* No. 1, 85 (1981).

Chait, R. T., J. Shpungin, and F. H. Field, *Int. J. Mass Spectrom. and Ion Processes, 58,* 121 (1984).

Conzemius, R. J., and J. M. Capellen, *Int. J. Mass Spectrom. Ion Phys., 34,* 197 (1980).

Conzemius, R. J., S. Zhao, R. S. Houk, and H. J. Svec, *Int. J. Mass Spectrom. and Ion Processes, 61,* 277 (1984).

Cooks, R. G., ed., *Collision Spectroscopy,* Plenum Press, New York (1978).

Dempster, A. J., *Phys. Rev., 11,* 316 (1918).

Devienne, F. M., and J. C. Roustan, *Organic Mass Spectrom., 17,* No. 4, 173 (1982).

Douglas, D. J., G. Rosenblatt, E. S. K. Quan, and J. B. French, *Proc. 6th Int. Symposium on Plasma Chemistry,* Montreal, ed. M. I. Boulos, and R. J. Munz, *2,* 444 (1983).

Eland, J. H. D., *Int. J. Mass Spectrom. Ion Phys., 8,* 143 (1972).

Fassett, J. D., L. J. Moore, R. W. Shideler, and J. C. Travis, *Anal. Chem., 56,* 203 (1984).

Fenner, N. C., and N. R. Daly, *Rev. Sci. Instr., 37,* 1068 (1966).

Fenselau, C., "Fast Atom Bombardment (Review)," in *Ion Formation From Organic Solids,* ed. A. Benninghoven, Springer-Verlag, New York (1983), pp. 90–100.

Fenselau, C., R. J. Cotter, D. Heller, and J. Yergey, *J. Chromatogr., 271,* 3 (1983).

Fujii, T., *Int. J. Mass Spectrom. and Ion Processes, 57,* 63 (1984).

Glish, G. L., and P. J. Todd, *Anal. Chem., 54,* 842 (1982).

Gray, A. L., and A. R. Date, *Int. J. Mass Spectrom. Ion Phys., 46,* 7 (1983).

Gray, A. L., and A. R. Date, BGS (NERC)/EEC Project Report 1984, Contract MSM 104 UK(4).

Hardin, E. D., T. P. Fan, C. R. Blakley, and M. L. Vestal, *Anal. Chem., 56,* 2 (1984).

Harrison, A. G., *Chemical Ionization Mass Spectrometry,* CRC Press, Boca Raton (1983).

Haverkamp, J., and P. G. Kistemaker, *Int. J. Mass Spectrom. Ion Phys., 45,* 275 (1982).

Hercules, D. M., R. J. Day, K. Balasanmugam, T. A. Dang, and C. P. Li, *Anal. Chem., 54,* No. 2, 280A (1982).

Hickam, W. M., in *Mass Spectrometric Analysis of Solids,* ed. A. J. Ahearn, Elsevier, Amsterdam, 1966, p. 147.

Hirata, Y., T. Tareuchi, S. Tsuge, and Y. Yoshida, *Organic Mass Spectrom., 14,* No. 3, 128 (1979).

Hodges, R. V., L. C. Lee, and J. T. Moseley, *Int. J. Mass Spectrom. Ion Phys., 39,* 133 (1981).

Honig, R. E., *J. Appl. Phys., 29,* 549 (1958).

Honig, R. E., in *Advances in Mass Spectrometry,* ed. J. D. Waldron, Pergamon, New York (1959), p. 162.

Honig, R. E., and J. R. Woolston, *Appl. Phys. Lett., 2,* 138 (1963).

Honig, R. E., J. W. Guthrie, W. M. Hickam, and G. G. Sweeney, in *Mass Spectrometric Analysis of Solids,* ed. A. J. Ahearn, Elsevier, Amsterdam (1966), chapter 2.

Honig, R. E., "The Development of Secondary Ion Mass Spectrometry (SIMS): A Retrospective," *Int. J. Mass Spectrum. Ion Phys., 66,* 31 (1985).

Houk, R. S., V. A. Fassel, G. D. Flesch, H. J. Svec, A. L. Gray, and C. E. Taylor, *Anal. Chem., 52,* 2283 (1980).

Houk, R. S., H. J. Svec, and V. A. Fassel, *Appl. Spectros., 35,* No. 4, 380 (1981).

Inghram, M. G., and W. A. Chupka, *Rev. Sci. Instr., 24,* 518 (1953).

Inghram, M. G., and R. Gomer, *J. Chem. Phys., 22,* 1279 (1954).

Ishitani, T., H. Tamura, and H. Todokoro, *J. Vac. Sci. Technol., 20,* No. 1, 80 (1982).

Jentsch, Th., W. Drachsel, and J. H. Block, *Int. J. Mass Spectrom. Ion Phys., 38,* 215 (1981).

Johnstone, R. A. W., *Mass Spectrometry for Organic Chemists,* Cambridge University Press, London (1972), p. 164.

Jungclas, H., H. Danigel, L. Schmidt, and J. Bellbrügge, *Organic Mass Spectrom.*, *17*, No. 10, 499 (1982).

Katakuse, I., T. Matsuo, W. Wollnik, and H. Matsuda, *Int. J. Mass Spectrom. Ion Phys.*, *32*, 87 (1979).

Keller, O. L., Jr., D. C. Hoffman, R. A. Penneman, and G. R. Choppin, *Physics Today, March,* 35 (1984).

Kim, S. H., and F. W. Karasek, *J. Chromatogr.*, *234*, 13 (1982).

Komuro, M., and H. Kawakatsu, *J. Appl. Phys.*, *52*, No. 4, 2642 (1981).

Kovalev, I. D., G. A. Maksimov, A. I. Suchkov, and N. V. Larin, *Int. J. Mass Spectrom. Ion Phys.*, *27*, 101 (1978).

Langmuir, I., and K. H. Kingdon, *Proc. Roy. Soc., London*, *107*, 61 (1925).

Lattimer, R. P., *Int. J. Mass Spectrom. and Ion Processes*, *55*, 221 (1983–1984).

Liebl, H. J. and R. F. K. Herzog, *J. Appl. Phys.*, *34*, 2893 (1963).

Ligon, W. V., Jr., *Int. J. Mass Spectrom. Ion Phys.*, *52*, 183 (1983).

Ligon, W. V., Jr., and S. B. Dorn, *Int. J. Mass Spectrom. and Ion Processes*, *61*, 113 (1984).

Lubman, D. M., R. Naaman, and R. N. Zare, *J. Chem. Phys.*, *72*, 3034 (1980).

Lytle, J. M., G. L. Tingey, and R. D. Macfarlane, *Anal. Chem.*, *54*, 1881 (1982).

Macfarlane, R. D., and D. F. Torgerson, *Int. J. Mass Spectrom. Ion Phys.*, *21*, 81 (1976a).

Macfarlane, R. D., and D. F. Torgerson, *Science, 191*, 920 (1976b).

Macfarlane, R. D., *Anal. Chem.*, *55*, No. 12, 1247A (1983).

Magee, C. W., *Int. J. Mass Spectrom. Ion Phys.*, *49*, 211 (1983).

Meuzelaar, H. L. C., J. Haverkamp, and F. D. Hileman, *Pyrolysis Mass Spectrometry of Recent and Fossil Biomaterials*, Elsevier, Amsterdam (1982), p. 34.

Meuzelaar, H. L. C., W. Windig, A. M. Harper, S. M. Huff, W. H. McClennen, and J. M. Richards, *Science, 226*, 268 (1984).

Migahed, M. D., and H. D. Beckey, *Int. J. Mass Spectrom. Ion Phys.*, *7*, 1 (1971).

Miller, C. M., N. S. Nogar, and S. W. Downey, "Analytical Developments in Resonance Ionization Mass Spectrometry," presented at the *32nd Ann. Conf. on Mass Spectrom. and Allied Topics*, San Antonio (May 27 to June 1, 1984), p. 368.

Möller, K., and L. Holmlid, *Int. J. Mass Spectrom. and Ion Processes*, *61*, 323 (1984).

Munson, M. S. B., and F. H. Field, *J. Am. Chem. Soc.*, *88*, 2621 (1966).

Murugaiyan, P., *Pure and Appl. Chem.*, *54*, No. 4, 835 (1982).

Pallmer, P. G., R. L. Gordon, and M. J. Dresser, *J. Appl. Phys.*, *51*, No. 7, 3776 (1980).

Perrin, R. E., G. W. Knobeloch, V. M. Armijo, and D. W. Efurd, Los Alamos National Laboratory Report No. LA-10013-MS, (June 1984).

Ramendik, G. I., in *Advances in Mass Spectrometry*, Vol. 8A, ed. A. Quayle, Heyden and Son Ltd., London (1980), p. 408.

Ramendik, G. I., V. I. Derzhiev, Y. A. Surkov, V. F. Ivanova, A. V. Grechishnikov, A. A. Sysoev, V. A. Oleinikov, and V. A. Alekmandrov, *Int. J. Mass Spectrom. Ion Phys.*, *37*, 331 (1981).

Rec, J. R., W. G. Myers, and F. A. White, *Anal. Chem.*, *54*, No. 9, 1234 (1974).

Rec, J. R., Ph.D. Thesis, Rennselaer Polytechnic Institute (1975).

Reinhold, V. N., and S. A. Carr, *Anal. Chem.*, *54*, 499 (1982).

Reynolds, J. H., *Rev. Sci. Instr.*, *27*, 928 (1956).

Rokop, D. J., R. E. Perrin, G. W. Knobeloch, V. M. Armijo, and W. R. Shields, *Anal. Chem.*, *54*, 957 (1982).

Röllgen, F. W., and H. L. Beckey, *Int. J. Mass Spectrom. Ion Phys.*, *12*, 465 (1973).

Ross, M. M., J. R. Wyatt, R. J. Colton, and J. E. Campana, *Int. J. Mass Spectrom. and Ion Processes, 54,* 237 (1983).

Rudat, M. A., *Anal. Chem., 54,* 1917 (1982).

Schulten, H. R., B. Bohl, U. Bahr, R. Müller, and R. Palavinskas, *Int. J. Mass Spectrom. Ion Phys., 38,* 281 (1981).

Schulten, H. R., P. B. Monkhouse, and R. Müller, *Anal. Chem., 54,* 654 (1982).

Siegel, M. W., in *Advances in Mass Spectrometry*, Vol. 8B, ed. A. Quayle, Heyden and Son Ltd., London (1980), p. 1655.

Simons, D. S., *Int. J. Mass Spectrom. and Ion Processes, 55,* 15 (1983–1984).

Smith, R. G., E. J. Brooker, D. J. Douglas, E. S. K. Quan, and G. Rosenblatt, *J. Geochem. Exploration, 21,* 385 (1984).

Steiner, B., C. F. Giese, and M. G. Inghram, *J. Chem. Phys., 34,* 189 (1961).

Swanson, L. W., G. A. Schwind, and A. E. Bell, *J. Appl. Phys., 51,* No. 7, 3453 (1980).

Todd, P. J., G. L. Glish, and W. H. Christie, *Int. J. Mass Spectrom. and Ion Processes, 61,* 215 (1984).

Tölgyessy, J., T. Braun, and M. Kyrš, *Isotope Dilution Analysis*, Pergamon, New York (1972).

Torii, Y., and H. Yamada, *Japanese J. Appl. Phys., 21,* No. 3, L132 (1982).

Turner, N. H., and R. J. Colton, *Anal. Chem., 54,* 293R (1982).

Webster, R. K., "Isotope Dilution Analysis" in *Advances in Mass Spectrometry*, Vol. I, ed. J. D. Waldron, Macmillan, New York (1959), p. 103.

Williams, D. H., C. Bradley, G. Bojesen, S. Santikarn, and L. C. E. Taylor, *J. Am. Chem. Soc., 103,* 5700 (1981).

Zandberg, E. Y., E. G. Nazarov, and Y. K. Rasulev, *Sov. Phys. Tech. Phys., 26,* No. 6, 706 (1981).

CHAPTER THREE

TYPES OF SPECTROMETERS

Major types of mass analyzers now include magnetic, quadrupole, time-of-flight, and Fourier transform ion cyclotron resonance. Various combinations of magnetic sectors, electrostatic lenses, and time-of-flight schemes are also incorporated in double focusing, ion microprobe/microscope, and laser microprobe instrumentation. This chapter presents only an introduction to these analyzers, which are now essential tools for characterizing materials, chemical reactions, electronic devices, etc. Multistage magnetic and quadrupole systems are discussed in Chapter 4, together with some specialized apparatus for charged particle analysis.

SINGLE MAGNETIC ANALYZERS

The 180° Sector

The semicircular, homogeneous field design was first reported by Dempster (1918); and it is still widely employed either as a single magnetic analyzer or in conjunction with electrostatic lenses. Typical ion trajectories for the 180° magnet arrangement are shown in Figure 3.1, assuming a highly collimated ion beam.

Let singly charged ions be produced and accelerated through a potential V, thus acquiring a kinetic energy eV. If this accelerating potential is large compared with the initial energy distribution of the ions, we can assume that all ions enter the magnetic field with a discrete velocity, which is given by:

$$eV = \tfrac{1}{2}mv^2$$

where m is the mass of the ion, e is the electronic charge, and v is the terminal velocity of the ion after acceleration. For positive ions having a charge state n ($n = 1, 2, 3 \ldots$), where n denotes the number of electrons stripped from a neutral atom, the more general expression results:

$$neV = \tfrac{1}{2}mv^2.$$

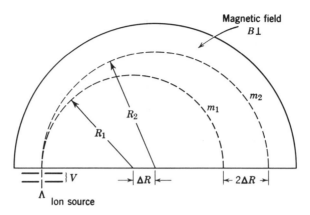

Figure 3.1   Ion trajectories in a homogeneous magnetic field 180° sector.

If the magnetic field B is perpendicular to the velocity vector of the ions, the ions will be deflected into a circular orbit, resulting from the balancing of centrifugal and centripetal forces:

$$Bnev = \frac{mv^2}{R},$$

where R is the radius of curvature. Eliminating v from the above two equations yields the general expression relating the radius of curvature of an ion in a homogeneous magnetic field to the several other parameters:

$$R = \frac{1}{B} \left(\frac{2mV}{ne} \right)^{1/2}$$

If the magnetic field strength B is expressed in gauss, m is given in atomic mass units, V is the accelerating potential in volts, and n is the multiplicity of electronic charge, then the radius of curvature, in centimeters, is approximated by the relation:

$$R = \frac{144}{B} \left(\frac{mV}{n} \right)^{1/2}$$

An alternative expression relating to these same parameters, but with B expressed in tesla (1 tesla $= 10^4$ G), is:

$$\frac{m}{n} = 5 \times 10^3 \frac{R^2 B^2}{V}$$

These equations indicate that for ions accelerated through the same potential, those of larger mass number will have a larger radius of curvature. It will also be noted that an ambiguity remains with respect to the mass number and charge state. A doubly charged magnesium ion, $^{24}Mg^{2+}$, will have approximately the same radius of curvature as will a carbon atom, $^{12}C^+$, that is only singly ionized. Therefore, unless the magnetic analyzer can resolve the small momentum difference corresponding to these two species, only a single peak will be observed at the $m = 12$ spectral position. One equation can also be differentiated to yield an expression of the general form:

$$\frac{2\Delta R}{R} = \frac{\Delta m}{m} + \frac{\Delta V}{V} - \frac{2\Delta B}{B}.$$

For a homogeneous magnetic field and ions having a negligible energy spread, the last two terms will vanish and the expression reduces to:

$$\frac{\Delta m}{m} R = 2\Delta R$$

The quantity $(\Delta m/m)R$ gives a measure of the mass dispersion or the separation of resolved ions along the focal plane. Thus, in Figure 3.1, if R_1 and R_2 are the radii of curvature of masses m_1 and m_2, respectively, the mass dispersion of these two isotopes will be

$$\left(\frac{m_1 - m_2}{m_1}\right)R_1 = 2(R_1 - R_2).$$

For isotopes that differ by only one mass number ($\Delta m = 1$), the mass dispersion D, along the focal plane will have a magnitude:

$$D = c\frac{R}{m}$$

where the constant c will depend upon the magnetic sector angle and upon ion entrance and exit angles. (In several ion optical schemes, the ion entrance angle is not normal to the magnet pole face.)

Thus, the distance along the 180° focal plane between adjacent isotopes will increase directly with the radius of curvature, and it will be inversely proportional to the mass number. Hence, analyzers with small radii of curvature and having a reasonable mass separation for light gases may be inadequate for the analysis of high mass species.

Although the 180° analyzer possesses good directional focusing proper-

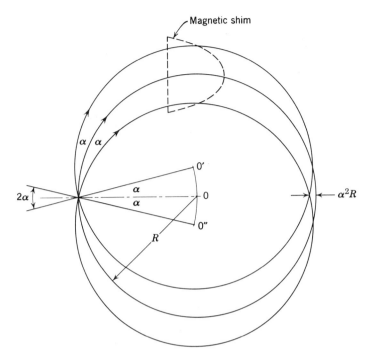

Figure 3.2 Angular aberration in the 180° sector.

ties, there are definite limitations on the sharpness of the line image, even
if the ions are monoenergetic and the magnetic field is perfectly homogeneous. Consider Figure 3.2, in which ions leaving a source have a half-angle
of divergence α and rays are drawn with equal radii from centers of circular
orbits from points 0, 0', and 0". When α is large, the central ray will not
coincide with the peripheral rays; and at the 180° boundary, the image width
will be of magnitude $\alpha^2 R$. (At the 360° boundary, i.e., initial source position,
this aberration will vanish, but so will all mass dispersion.)

In most mass spectrometers, α is restricted to small angles so that the
image in the focal plane remains reasonably sharp. Even with large angles
of divergence, special methods may be used to reduce the $\alpha^2 R$ line image.
From Figure 3.2, it would appear that better focusing might result if one
could decrease slightly the effective radius of curvature of the central ion
trajectory relative to the paraxial rays. Such a perturbation can, in fact, be
accomplished by the introduction of a thin magnetic shim (Balestrini and
White, 1960), appropriately contoured and indicated schematically by the
dotted outline. Other schemes involve providing special pole-piece contours
at the magnet boundaries. For high-resolution work, however, combinations
of electrostatic and magnetic analyzers provide a better general solution.

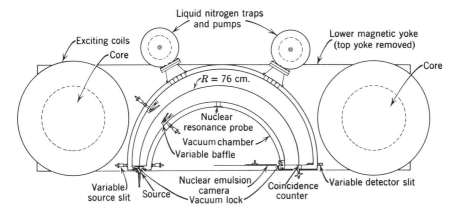

Figure 3.3 Large radius of curvature 180° mass spectrometer and alpha particle analyzer (White et al., 1958).

The resolution or resolving power RP of a single magnetic sector can be determined approximately by the expression:

$$RP = \frac{R}{S_1 + S_2 + R\left(\dfrac{\Delta V}{V}\right) + R\alpha^2}$$

where R is the radius of curvature, S_1 and S_2 are the respective widths of the entrance and exit slits, V is the accelerating potential, ΔV is the spread in ion energy, and α is the half-angle of divergence. Several other experimental factors will affect the actual resolution, including beam broadening due to elastic scattering and resonance charge exchange with residual molecules within the analyzer tube. This effect is minimal at pressures below 10^{-7} torr, but resolving power has been observed to decrease exponentially in a pressure range of 10^{-5} to 10^{-4} torr (Ghoshal et al., 1975).

There are several desirable features to the 180° geometry. The ion trajectories lie completely within the analyzing field; thus there are no field boundary effects and the theoretical resolution of an instrument can usually be approximated in practice. Ions of all masses are focused along the 180° boundary so that the simultaneous readout of mass spectra can be conveniently recorded by photographic plates. The vacuum chamber may be separate from the magnet, or the top and bottom pole pieces can function as an integral part of the analyzing vacuum housing. If the magnet is of large dimensions, the 180° analyzer may also serve as a wide-angle alpha particle analyzer. A schematic diagram of a 76 cm, 180° radius-of-curvature spectrometer, used as both a high-resolution alpha particle spectrometer and a mass analyzer, is shown in Figure 3.3. In this instance, the upper and lower

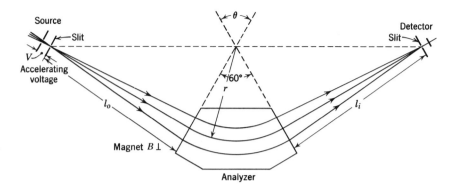

Figure 3.4 The 60° magnetic analyzer.

magnetic yokes served as the top and bottom surfaces of the vacuum analyzing chamber (White et al., 1958). Also, by adjusting the accelerating potential of the ion source in this instrument so that an ion beam of known mass and a 5.3 MeV alpha particle (^{4}He^{2+}) traversed an *identical* trajectory, it was possible to establish an accurate energy scale without a magnetic field measurement (see Chapter 13).

The 60° Sector

The 60° magnetic sector was introduced in the late 1930s by Nier (1940). Many of the instruments built for the Oak Ridge National Laboratory during World War II were patterned after Nier's design; for several decades, this configuration was the standard in hundreds of commercial instruments. It continues to be an important analyzer; and for equal radii of curvature, such magnets require considerably less material than the 180° type. Figure 3.4 indicates the ion trajectories for this wedge-shaped design, which provides direction focusing for divergent ions, with the source, detector slit, and the apex of the magnet on a common line. Both symmetric and asymmetric geometries have been investigated, but the symmetric (equal object-image distances from the magnetic boundary) type was the one widely utilized in early spectrometers. A mathematical treatment for the general case of an ion entering and leaving a magnetic sector at right angles has been developed by Herzog (1934).

If l_{om} is the object distance (source slit to magnetic boundary), l_{im} is the image distance, r is the effective radius of curvature of the ion trajectory, and θ is the angle of the magnetic sector, then the directional focus equation has the form:

$$(l_{om} - g_m)(l_{im} - g_m) = f_m^2,$$

where $f_m = r/\sin \theta$ and $g_m = f_m \cos \theta$. Here f_m is the focal length of the system, as in the optical case. In the symmetric situation, $l_{om} = l_{im}$ so that

$$l_{om} = l_{im} = f_m + g_m$$

or

$$l_{om} = l_{im} = r\,(csc\ \theta + \cot \theta).$$

For an angle θ of 60°, $l_{om} = l_{im} = r\sqrt{3}$. The ion source and detector slits are thus placed at a distance of $1.732r$ from the magnetic boundaries. This sector angle has proved to be a convenient one for instruments, the radii of which have ranged from 5–40 cm. There is adequate space to accommodate an ion source housing and appropriate differential pumping. Also, the magnetic shielding of electron multiplier detectors is simplified. For large radii of curvature instruments, the fringing field at the detector may be negligible, so that magnetic shielding may be unnecessary.

The 90° Sector

The symmetric 90° magnetic sector represents another useful configuration that is available in many commercial forms. Analyzers having small radii of curvature (~5 cm) have found extensive use as partial pressure gas analyzers, leak detectors, breath analyzers, and in space-related applications, etc. Davis and Vanderslice (1960) were among the earlier investigators who used 90° 5 cm radius magnets, a Nier-type ion source, a 10-stage electron multiplier, and a bakeable analyzing tube to measure high pressures at fast scanning rates (1.5 μs/amu) and low pressures in the range of 10^{-12} torr. Subsequently, Davis (1962) used these small radius-of-curvature magnet analyzers to measure the partial pressure of gases down to 10^{-16} torr. Such a pressure corresponds to a density approximating only 1 molecule/cm^3, a pressure presumably comparable to that of outer space. Of course, such a pressure can only be obtained in the laboratory for short periods with cryogenic pumping.

The 90° magnetic sector is also used in many large radius-of-curvature instruments, either as a single, directional focusing analyzer or in conjunction with double focusing spectrometers. Note that $l_{om} = l_{im} = r\,(csc\ 90° + \cot 90°) = r$ for this analyzer. The path length of an ion is thus intermediate between the 180° and 60° instruments having a comparable mass resolution. Thus, the probability that an ion will undergo scattering by residual gas molecules in the vacuum-analyzing chamber is less than in the 60° case—but greater than for the 180° deflection—for the same radius of curvature. Figure 3.5 shows a typical configuration for a 90° instrument, and it indicates a practical correction for the fringing field that must be made with most

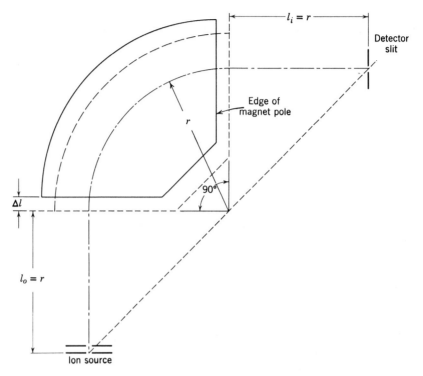

Figure 3.5 A 90° magnetic sector displaced from its theoretical position to correct for the fringing field.

sector magnets for proper ion focusing. The general ion optics equations presume a sharply defined magnetic boundary from which object and image distances can be computed. Of course, no such well-defined boundary exists. In practice, it is convenient to assume that the magnetic fringing field extends about one gap width beyond the physical edge of the magnetic sector (Duckworth, 1960). This assumption appears to be quite valid for cases where the gap is small compared to the magnet radius. Thus, a magnet should be displaced a distance Δl from the position indicated by the dotted line, which represents the idealized situation. In some analytical instruments, however, a final "best focus" position is often achieved experimentally, with the magnet being displaced in small increments with respect to the analyzing tube. There is no reason why sector magnets of any angle cannot be constructed, but the 180°, 60°, and 90° ones have become somewhat standard. It is also unlikely that very small angular sectors ($< 10°$) would yield comparable performance, as the inhomogeneous fringing field would begin to have dimensions comparable to that of the primary homogeneous field region. In modern instruments, optimum ion entrance angles and trajectories are determined by fringing field mapping and computer calculations.

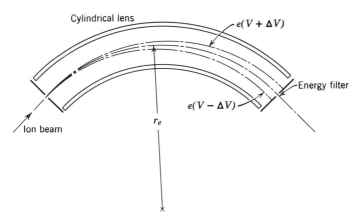

Figure 3.6 Electrostatic lens providing energy filtering of an ion beam.

ELECTROSTATIC ANALYZERS

In contrast to magnetic analyzers where the mass-to-charge ratio m/z of a particle can be determined, electrostatic lenses or "prisms" provide a means of energy filtering, or focusing, or for determining the charge state of an ion beam. Thus, an electrostatic lens can furnish a monoenergetic ion group to a subsequent magnet analyzer to yield a highly resolved mass spectrum. Another option is to use electrostatic sectors in conjunction with double focusing systems, as discussed in the following section. The most widely used method for achieving energy or velocity focusing is to employ a radial electrostatic field, such as can be established with the cylindrical lens shown in Figure 3.6. If a voltage difference is established across the inner and outer plates, ions will be deflected in traversing the lens; and a simple relationship exists between the radius of curvature r_e, the potential V through which the ions have been accelerated, and the electrostatic field strength ϵ existing between the two cylindrical electrodes. If r_e is in centimeters, V is in volts, and ϵ is in volts per centimeter, then

$$r_e = \frac{2V}{\epsilon}$$

This is a mass-independent relationship that holds for electrons, protons, ions, or any charged particle. It is important in mass spectrometry because it permits the selection of ions having a limited energy spread. Thus, in Figure 3.6, if narrow collimating slits are used at the ends of a cylindrical condenser, ions having a kinetic energy $e(V + \Delta V)$ and $e(V - \Delta V)$ will be filtered out of the beam, thus allowing the transmission of ions having a limited energy spread. Spherical electrostatic analyzers are also now available in which

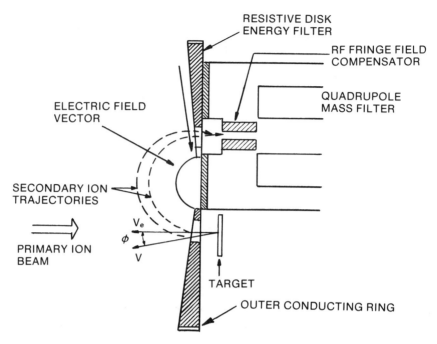

Figure 3.7 High-transmission energy analyzer for secondary ion mass spectrometry. (Siegel and Vasile, 1981; courtesy of M. J. Vasile.)

ions within a predetermined energy range are transmitted between the inner and outer surfaces and focused upon an exit aperture. Thus, by programming the lens voltage, one can obtain an ion energy profile. Or if the input of ions is pulsed, the time-resolved output provides a means of mass identification. A detailed analysis of end-plate corrections, charged particle entry angle, etc. for the 127° analyzer has been made by Ramos and Rayborn (1984).

By using time-of-flight techniques, "an all electrostatic mass spectrometer capable of very high mass resolution, broadband mass analysis and high sensitivity" has been proposed (Kilius et al., 1981). This system (for use with a 3 MeV tandem accelerator) makes use of a flight detector in which ion flight times are essentially isochronous, with respect to small variations, incident angle, and initial displacement at a "start" detector. While this reported development is somewhat specialized, it is indicative of the increasing use of pulsed ion techniques, which in conjunction with electrostatic lenses permit these normally mass-independent analyzers to be appropriated for mass assignments.

Another electrostatic lens of special interest is the unique type reported by Siegel and Vasile (1981), where a "conically tapered resistive disk with a single conducting hemisphere is used to effect the same energy selection as a pair of concentric hemispheres." Figure 3.7 is a schematic diagram of this very wide angle device, which is capable of collecting essentially every secondary produced ion within its energy bandwidth. In this scheme a radial electric current is produced in a shaped resistive disk to establish the electric field, so that the angular acceptance aperture of this energy filter is nearly 2π steradians. When used with a quadrupole, this device is an excellent analyzer both for studying secondary ion energy distributions and in SIMS applications where very high transmissions are required.

DOUBLE FOCUSING SPECTROMETERS

The single magnetic analyzer has no provision for focusing ions of different energy; but by using a suitable combination of electrostatic and magnetic fields, one may obtain directional focusing of a given mass-to-charge ratio ion beam, even if it is heterogeneous in energy. The term *double focusing* is thus ascribed to mass spectrometers in which both the angular aberrations and energy or velocity aberrations effectively cancel. The general equations that provide this double focusing have been developed by Mattauch and Herzog (1934) and Mattauch (1936). First, the relationship for the electrostatic lens is similar to the magnetic case. Thus, letting l_{oe} and l_{ie} denote the object and image distances, r_e the mean radius of curvature of the electrostatic analyzer, ϕ the sector angle, and f_e the focal length of the lens:

$$g_e = \frac{r_e \cot(\sqrt{2}\phi)}{2} = f_e \cos(\sqrt{2}\phi),$$

and

$$f_e = \frac{r_e}{\sqrt{2}\,\sin{(\sqrt{2}\phi)}},$$

For the special case where

$$l_{oe} = l_{ie},$$

and

$$l_{oe} - g_e = f_e,$$

$$l_{oe} = l_{ie} = g_e + f_e = f_e \cos(\sqrt{2}\phi) + f_e$$

$$= \frac{r_e}{\sqrt{2}} \frac{1 + \cos{(\sqrt{2}\phi)}}{\sin{(\sqrt{2}\phi)}} = \frac{r_e}{\sqrt{2}} [\csc{(\sqrt{2}\phi)} + \cot(\sqrt{2}\phi)].$$

Two of the most widely used electrostatic sector angles in mass spectro-

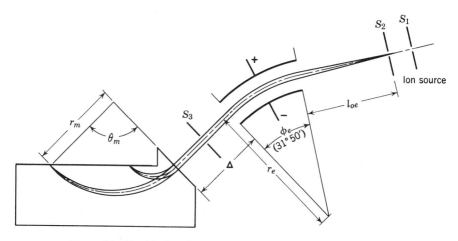

Figure 3.8   Double focusing system of Mattauch and Herzog (1934).

metry have been $\phi = 31° \, 50'$ and $\phi = 90°$. The symmetrical object or image distance from the electrostatic boundary is $0.35 \, r_e$ for $\phi = 90°$.

The particular combination of electrostatic and magnetic field parameters that provide double focusing is given by satisfying the equations cited above for g_e and f_e, and the expression:

$$r_e(1 - \cos\sqrt{2}\phi) + \sqrt{2}l_{ie}\sin\sqrt{2}\phi$$

$$= \pm\{r_m(1 - \cos\theta) + l_{om}[\sin\theta + \tan\delta\,(1 - \cos\theta)]\}.$$

The angle δ is the angle between the velocity vector of the ion and the normal-to-magnetic field boundary. The positive sign is employed when the ion is deflected in the same sense in both electrostatic and magnetic fields; the negative sign is used when the ion undergoes successive deflections in opposing directions.

Two important examples of double focusing spectrometers are the consecutive electrostatic-magnetic field configuration of Mattauch and Herzog (1934) and the general type developed by Nier and his collaborators (Nier and Roberts, 1951). Inasmuch as both of these instruments are of historical importance, a brief description of these analyzers is outlined.

The Mattauch-Herzog design has the remarkable property of possessing double focusing for all masses. Figure 3.8 is a schematic diagram of the system that comprises a $\pi/4 \, \sqrt{2}$ (31° 50′) electrostatic sector followed by a 90° magnetic analyzer. The ion source is located at the principal focus of the electrostatic lens so that ions emerge from the energy analyzer in a parallel beam. This is consistent with the equation which indicates that if $l_{oe} = r_e/\sqrt{2}$, $l_{ie} = \infty$. The ion beam thus enters the magnetic sector as a bundle of parallel rays, which is equivalent to having $l_{om} = \infty$. Hence, ions of all masses will be focused along a line that is coincident with the second magnetic field boundary. Because there is no intermediate focal point, the distance Δ between the electrostatic and magnetic field boundaries is indeterminate and may be chosen at will. The range of m/z values that can be simultaneously focused on a photographic plate will depend on the range values of r_m allowed by the magnet configuration. In some contemporary commercial instruments, a mass range extending from boron to uranium can be monitored by a photographic plate with a single setting of the magnetic field.

A collimating slit or a variable aperture between the electrostatic and magnetic analyzer permits acceptance into the magnet of an ion beam having either a small or large energy distribution. The combined mass and energy focusing of this system is excellent, providing a reasonably sharp mass spectrum, even if the spread in ion energies is an appreciable fraction ($\sim 5\%$) of the total accelerating potential . The resolution and mass dispersion depend, respectively, on the selection of values of r_e and r_m. Many versions of this basic spectrometer have been built and successfully used for mass measurements, isotopic abundance measurements, and general analytical work. Mass spectrometric doublets can be measured with precision, and the instrument can be used with virtually any ion source.

The high-resolution spectrometers built by Nier and coworkers at the University of Minnesota represented improved versions of the somewhat similar ion optics system devised by Dempster (1935) and by Bainbridge and Jordan (1936). The first instrument (Nier and Roberts, 1951) had the parameters $r_e = 18.87$ cm and $r_m = 15.24$ cm. An enlarged version of this tandem arrangement built by Nier's group (Quisenberry et al., 1956) had parameters $r_e = 50.31$ cm and $r_m = 40.64$ cm. A schematic diagram of this instrument is shown in Figure 3.9. It will be noted that the electrostatic lens is symmetric with respect to object and image slits S_1 and S_3. This arrangement provided not only first-order energy focusing, but both first- and second-order directional focusing, thereby allowing wide limits on the half-angle of divergence α that can be used without angular aberration. These two instruments have yielded mass measurements of the highest precision both for an exceedingly large number of elements and isotopes and in the calibration of the present mass scale.

Important features of the above instruments included (1) highly stabilized accelerating potentials and magnetic fields, (2) a method of "peak-matching" ions of nearly equal masses by bringing them sequentially into focus along identical ion trajectories, and (3) a complete electrical detection system replacing the photographic plate. As shown in Figure 3.9, the magnetic sector contains an auxiliary mass spectrometer whose output, from a pair of ion beam collectors, is connected to a differential amplifier and linked to the voltage supplies of the ion source and electrostatic analyzer. If the accelerating potential or the magnetic field changes for any reason, the differential

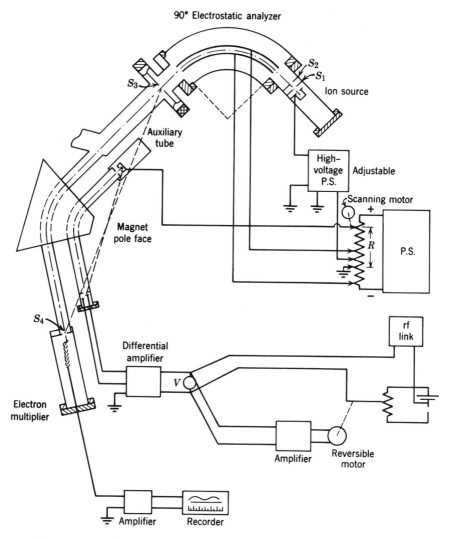

Figure 3.9 Double focusing spectrometer designed by Nier and Roberts (1951).

amplifier senses this change in the auxiliary tube and compensation is provided for in a feedback loop. The sequential focusing of two ions of a mass doublet, having small mass differences, has proved to be quite successful. A voltage scan corresponding to a small fraction of the mass scale is related directly to a small change in resistance ΔR of a voltage divider. By measuring the precise resistance change $\Delta R/R$, it can be shown that ions differing by a small fractional mass, ΔM, are related by $\Delta R/R = k\,(\Delta M/M)$.

All contemporary high-resolution mass spectrometers now utilize some

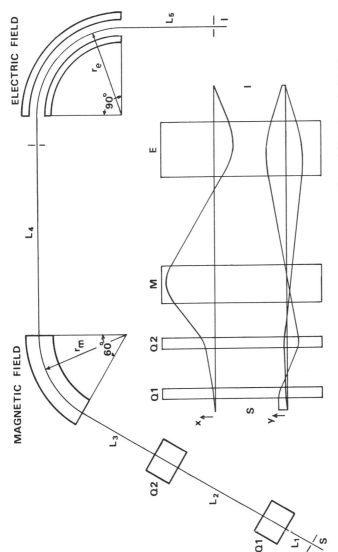

Figure 3.10 Double focusing spectrometer with quadrupole lenses. Horizontal (x) and vertical (y) ion trajectories are indicated qualitatively by the insert. (Matsuda, 1981a; courtesy of H. Matsuda.)

65

type of double focusing ion optics. For example, a design by Matsuda (1981a) includes a 60° sector magnet, a 90° electrostatic analyzer, and a pair of electrical quadrupole lenses; a schematic is shown in Figure 3.10. This instrument has a relatively small magnet ($r_m = 25$ cm), but a resolving power of 5000 at 5000 amu is reported. The instrument described by Devienne et al. (1983) is another double focusing instrument designed for the analysis of ions of high molecular weight (8000 amu) or cluster ions. This spectrometer has a 90° electrostatic analyzer of 37.1 cm and a magnet of 30 cm that can produce a maximum field of approximately 13,000 G. Following the double focusing electrostatic/magnetic system, dissociated ions can be further analyzed via a 20 cm electrostatic sector. Cluster ions containing up to 37 atoms of uranium have been resolved with this apparatus. Many commercial double focusing instruments have now also been adapted to operate in the negative as well as positive ionization mode, so that mass spectra from various compounds can be compared (Yinon and Boettger, 1972). Other Mattauch-Herzog-type instruments have been modified for the study of gas phase kinetics, and virtually all ion microprobe spectrometers utilize some type of double focusing ion optics for the in-depth profile analysis of solids.

A general review of special double focusing ion optics schemes for obtaining high transmission and high resolving power (up to 100,000) has been prepared by Matsuda (1981b; 1983). Included is a description of the unique Osaka University large-dispersion spectrometer, having a maximum resolving power of one million, so that doublets such as ^{40}Ca-^{40}Ar, having a mass difference of only 208.2×10^{-6} amu, are clearly resolved.

QUADRUPOLE MASS FILTERS

Strictly speaking, the quadrupole is a mass filter rather than an analyzer because it transmits ions having only a small range of m/z values, and there is no mass dispersion or focusing as in magnetic analyzers. Thus, it is analogous to a narrow-band pass electrical filter that transmits signals within a finite frequency bandwidth, and a trade-off is made between transmission and resolution. For the quadrupole, the transmission T and resolution R are easily adjusted electronically, and the product of these two parameters, TR, tends to remain a constant. The initial development of the quadrupole was in relation to its potential as an isotope separator; in this role, it has been recently used for the isotopic enrichment of ^{26}Mg (Finlan and Sunderland, 1982). However, the quadrupole mass filter as conceived by Paul and Steinwedel (1953) was quickly adopted in mass spectrometry and applied to a broad range of measurements. The basic concept of this device is to provide a potential field distribution, periodic in time and symmetric with respect to the axis of an ion beam, which will transmit a selected mass group and cause ions of improper mass to be deflected away from the axis. Figure 3.11a indicates the idealized hyperbolic electrode configuration, and Figure

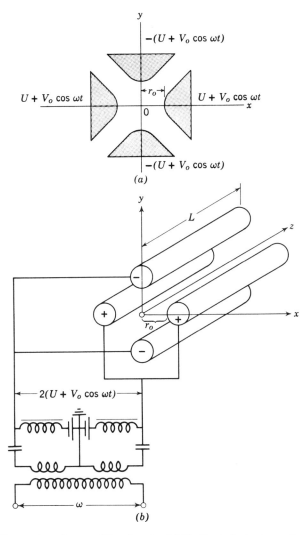

Figure 3.11 The quadrupole mass filter showing (a) idealized electrode configuration and (b) cylindrical rod assembly.

3.11b shows an assembly of four cylindrical rods and associated RF circuitry. This mass selection scheme employs a combination of DC potentials plus a radiofrequency potential, so that the transit time of the ions is long compared to the RF period. By proper programming of potentials and frequency, an ion of desired mass can be made to pass through the system, while unwanted masses will undergo an oscillating trajectory of increasing amplitude perpendicular to the major axis so that they will ultimately be collected on one of the electrodes.

If the voltage applied to the quadrupole consists of a fixed voltage U and the radiofrequency component $V_0 \cos \omega t$, the potential Φ may be expressed in terms of the rectilinear coordinates and the constant r_0, which is a measure of the electrode spacing (Blauth, 1966). The potential will have the form

$$\Phi = \frac{(U + V_0 \cos \omega t)(x^2 - y^2)}{r_0^2},$$

so that the motion of an ion having a charge-to-mass ratio ne/m and moving in the z direction can be represented by the equations:

$$\ddot{x} + \frac{2ne}{mr_0^2} (U + V_0 \cos \omega t)x = 0,$$

$$\ddot{y} - \frac{2ne}{mr_0^2} (U + V_0 \cos \omega t)y = 0,$$

$$\ddot{z} = 0;$$

by letting

$$\omega t = 2\zeta; \qquad \frac{8neU}{mr_0^2\omega^2} = a; \qquad \frac{4neV}{mr_0^2\omega^2} = q$$

the orthogonal set of equations in x and y become:

$$\frac{d^2x}{d\zeta^2} + (a + 2q \cos 2\zeta)x = 0,$$

$$\frac{d^2y}{d\zeta^2} - (a + 2q \cos 2\zeta)y = 0.$$

Taken together, these relations represent an oscillating system in which the restoring force is periodic in time. The general solution of these equations takes an exponential form, in which there exist finite limits on the range of values of a and q if the amplitude of an ion oscillation is to remain bounded. Hence, it is customary to construct a stability diagram that reveals the range of a/q values that are consistent with a real solution.

In principle, for a given potential field configuration, it is only the mass and ionic change that determines whether an ion trajectory is stable or unstable. The initial position and velocity in the z direction does not enter into the theoretical considerations for a stable ion trajectory. In practice, however, there are very real limits on the angular divergence of an entering ion beam and the x, y, and z components of its initial velocity. The performance

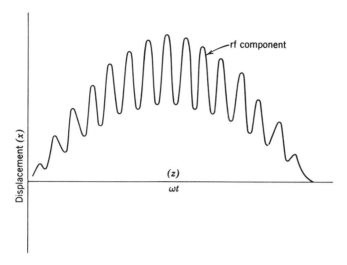

Figure 3.12   Trajectory of an ion in the x-z plane of a quadrupole mass filter.

of a quadrupole is also dependent upon the fringing fields at the entrance
and exit apertures (Dawson, 1971).

If the particles injected into the quadrupole are positive ions, the potential
lines established by the DC field alone will have some focusing effect toward
the z axis, and a slightly divergent ion constrained to the x-z plane will
undergo simple harmonic oscillations. The addition of the RF field will then
superpose a high-frequency oscillation. The DC voltages in the y-electrodes
tend to cause some defocusing, but the superposition of an RF voltage (which
is usually high compared to the DC potentials) has the effect of a focusing
action toward the z axis. This results from the fact that the mean trajectory
of an ion with time (from the z axis) is determined by the difference in the
opposing accelerations induced by the DC and RF potentials. A qualitative
representation of an ion in the x-z plane is sketched in Figure 3.12.

In general, the greater the length of a quadrupole, the better the resolution.
Also, in practical designs, cylindrical rods are chosen instead of hyperbolic-
contoured surfaces because of the ease of manufacture. The cylindrical rod
assembly provides approximately the same field distributions, provided the
radius of the rods, r, relative to the value of r_0 is related by $r = 1.147\, r_0$
(Denison, 1971).

At present, the quadrupole is extensively employed in both research laboratory and on-line processing applications, and it is increasingly being used
as a component in multistage analytical systems (Chap. 4). It is doubtful that
quadrupoles will replace double focusing magnetic instruments for very high-
resolution work, but their fast scanning capability, linear mass scale, extended mass range, programmability, compactness, and cost have made
quadrupoles the dominant commercial type. Several manufacturers offer the

quadrupole with ion source options, pulsed ion counting detectors, a mass range up to 300 amu or greater, microcomputer-controlled programming, and sensitivities in the parts per billion range for some analyses. Using pulsed ion sources, both positive and negative ion spectra can also be recorded. Applications include process gas stream and air pollution monitoring, combustion monitoring, medical diagnostics, isotopic ratio measurements, etc.

The quadrupole ion storage trap (Quistor) is a somewhat specialized three-dimensional quadrupole in which ions are retained within potential wells by a combination of RF and DC fields; they are subsequently extracted after a preset storage time (typically 1 ms). It has been employed in investigations of ion-molecule reactions, and a general review of this type has been prepared by Todd (1981).

TIME-OF-FLIGHT

The use of the time-of-flight (TOF) principle for mass separation was first suggested by Stephens (1946), and a prototype of the modern TOF spectrometer was subsequently developed by Wiley and McLaren (1955). The essential components of this type of analyzer include (1) an electron gun for the production of ions, (2) a grid system for accelerating ions to uniform velocities in a pulsed mode, (3) an ion drift region, and (4) a sensitive fast-response ion detector. Suitable electronic circuitry is also required for translating the time-dependent arrival of ions of different velocities into a time base that is related to mass number. Thus, ions that have been accelerated through a potential V acquire a velocity v of magnitude

$$v = \left(\frac{2neV}{m}\right)^{1/2}$$

where ne is the charge and m is the ion mass. If the drift region length is L, the ion flight time, t, will be

$$t = \frac{L}{v} = L\left(\frac{m}{2neV}\right)^{1/2}$$

The difference in transit time Δt for two ions of masses m_1 and m_2 will be:

$$\Delta t = \frac{L(\sqrt{m_1} - \sqrt{m_2})}{\sqrt{2neV}}$$

Also, the mass resolution can be approximated by $m/\Delta m = t/2\Delta t$. From these expressions, it is clear that time resolution will increase with increased

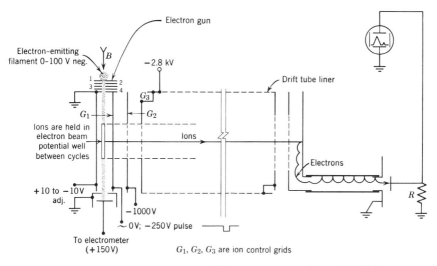

Figure 3.13 Schematic of time-of-flight spectrometer (Damoth, 1964).

drift tube length and will decrease with an increased accelerating potential. Figure 3.13 is a schematic diagram of one version of TOF instrument.

A prime consideration with the TOF instrument is the matter of ion injection. What is desired is the restriction of an "ion bunch" to the drift region that is neither spatially dispersed nor inhomogeneous in energy. If ions leave the source region at the improper time or with a spread in energy, the resolution will substantially decrease and a background or "noise" spectrum will result.

Several methods for providing fairly discrete "ion bunches" have been demonstrated: (1) the pulsed mode (2) the "continuous" ionization mode, and (3) the ion reflection technique. In the pulsed mode, the ionizing electron beam is pulsed, biased off, and the ions formed are initially accelerated by the draw-out grid. A second grid provides the additional accelerating potential. This two-grid system has been shown to minimize both initial spatial dispersion and inhomogeneities in ion energy. By immediately applying a drawing-out pulse to the source region after ion formation, the two-field grid system allows slow ions to catch up with those of higher velocity—thus substantially "bunching" an ion group in time.

The "continuum" mode of operation is continuous only in a single sense. The electron beam is left on for a very large fraction of a cycle (e.g., 50–100 μs). During this interval, the electrons are continually generating ions, and it is assumed that a "potential well" is formed between the backing plate and the first drawing-out grid G_1. This well serves as a trap for lower energy ions, but more energetic ions escape and are attracted to the backing plate by a slight bias. Under proper operating conditions, the subsequent pulsing of the G_1 grid results in a spectrum of improved resolution. An

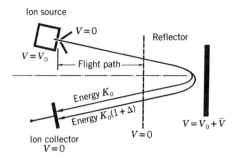

Figure 3.14 Reflection method for bunching ions of differing initial energy in time-of-flight spectrometry (Mamyrin et al., 1973).

additional advantage of the "continuous" mode is the increase in sensitivity (10 to 100), owing to the longer duty cycle that results from having ions immediately available. This factor is an important one because in obtaining very fast mass spectral scans, one is often limited with respect to the total number of recorded ions.

The ion reflection technique is an alternative method for "bunching" ions of different kinetic energy, which is shown schematically in Figure 3.14. Mamyrin et al. (1973) showed that a kind of double focusing could be achieved, in addition to increasing the resolution, via a doubling of the flight path. Many variations of this method have been reported recently in connection with various ion sources and detectors. For example, Negra and LeBeyec (1984) report a resolution of 3000 for light organic molecules in a TOF instrument equipped with an electrostatic mirror that provides time focusing.

The TOF instrument occupies a unique place in mass spectrometry, as it provides a simple, rapid measurement of the abundance of various isotopes or elements comprising a sample. In practice, 10,000–100,000 spectra can be scanned per second. With the aid of suitable electronic circuitry, it is thus possible to monitor reaction rates and to investigate reaction profiles of only 100 μs duration. Increased drift tube lengths greater than 1 m have also contributed to improved mass resolution, and the very high scanning rates have made possible improved ion statistics, that is, a scan can be made covering 0–900 amu in ~1 s. In order to increase the dynamic range of the instrument, a special technique has also been developed. For example, to prevent multiplier saturation when very large ion peaks are present in the presence of smaller adjacent peaks, appropriate "gating" pulses can be applied to the multiplier. Thus, it is possible to suppress mass 40 without interfering with the recording of masses 39 or 41. This type of refinement has extended the practical range of sensitivity in identifying gas chromatograph effluents by several orders of magnitude.

One recent TOF design utilizes a pulsed primary ion beam as a source for generating secondary ions (Chait and Field, 1984). Essentially, a primary ion beam, incident upon a target, is pulsed by electrostatic deflection plates. Bursts of secondary (target) ions are then accelerated into a 2 m flight tube. This method of pulsing provides ion bursts of the order of 5 ns, and the system can be operated at a 3 kHz repetition rate.

Thus, despite the fact that TOF systems cannot provide the mass resolution of some other analyzers, they are crucial for certain measurements where the ion formation time is very short. "Pulsed" ion sources and analyzers that utilize the TOF principle include: ion-electron coincidence spectrometry, the laser microprobe, ^{252}Cf spectral analysis, photoionization and photodecomposition, shock wave pyrolysis, energy distribution in plasmas, the analysis of clusters and the velocity distribution of particles, and collision phenomena and ion reaction rates. In one recent study, the dissociation pathways of molecules were unambiguously determined by utilizing infrared multiple photon dissociation in combination with a TOF mass filter (Chou et al., 1983).

Several other theoretical and experimental studies have also contributed to improved resolution for TOF instrumentation. For example, a method of "impulse field focusing" has been reported that employs time-dependent draw-out fields for ion bunching (Browder et al., 1981); and a "new look at time-of-flight spectrometers" has been presented in a review paper by Baril (1981). Many specialized TOF systems have been used with other apparatus or major facilities. These facilities or apparatus include (1) an induction accelerator (Kunibe et al., 1982) for charged particle diagnostics, (2) a heavy ion accelerator (Lennard et al., 1982) for ion energy loss measurements in thin films, and (3) a double-pass cylindrical mirror analyzer (Traum and Woodruff, 1980) for measuring positively charged ions emitted from a surface by electron-stimulated desorption. In addition, TOF measurements have been made in conjunction with magnetic, electrostatic, and quadrupole analyzers, and with ion microscope studies of surface layers (Ahmad and Tsong, 1984).

The advent of improved timing circuits and automated data acquisition has also led to a wider use of TOF systems, with multichannel counters having resolving times as short as 6 ns (Reinmüller, 1983).

FOURIER TRANSFORM ION CYCLOTRON RESONANCE

Fourier transform mass spectrometry (FTMS) is a technique of mass analysis whereby the cyclotron frequencies of ions of different species are detected. In this sense, it is quite distinct from the more traditional mass spectrometers, in which the trajectories of various ions are resolved. However, there are some salient advantages to this type of instrumentation that is now available in commercial systems:

1. Ion generation and mass analysis occur within the same region.
2. All ions are detected simultaneously.

3. Negative or positive ion spectra can be obtained with equal ease and rapid switching between the two.

4. Ion formation and detection are separated in time.

5. Complex ion optics and high voltages are not needed.

6. Ion-molecule reactions can be studied at pressures as low as 10^{-9} torr—far below pressures usually required for chemical ionization.

7. A resolution exceeding 1,000,000 is possible.

8. Samples may be run without an internal mass standard, once the instrument is calibrated.

9. Sensitivity increases with resolution, in contrast to conventional instruments.

Although the FTMS is a fairly recent addition to analytical mass spectrometry, many instruments have been built on the cyclotron principle that was first conceived by Lawrence and Livingston (1931) to accelerate ions to high energies. Lawrence demonstrated that charged particles would follow a spiral path if an appropriate radio frequency voltage was applied to semicircular electrodes, and if the ions were also constrained by a magnetic field. For an ion of charge q, the angular frequency ω_c of the RF field was shown to have the relation:

$$\omega_c = \frac{qB}{m}$$

and this principle was subsequently applied to mass spectrometry in a device termed the omegatron (Sommer et al., 1951). In this device, it was demonstrated that if the frequency of the RF field $E = E_0 \sin \omega_0 t$ was equal to the cyclotron frequency ω_c of the ion, an ion would trace a spiral trajectory until it was intercepted by a collector. The omegatron was used mainly as a residual gas analyzer.

A major advance in the practical use of ion cyclotron resonance (ICR) for mass spectrometry came with the introduction by McIver of a pulsed technique, together with a specially designed analyzer cell (McIver, 1970, 1971). A further development by Comisarow and Marshall (1974) was a Fourier transform (FT) method, so that an entire mass spectrum could be obtained in the time previously required to observe only a single peak.

A schematic diagram of the cubic trapped-ion cell, as used in FTMS analysis, is shown in Figure 3.15. To obtain an electron-impact (EI) mass spectrum, the sample is introduced into the cell and ionized by a short electron beam pulse. The ions thus formed undergo a circular (orbital) motion known as cyclotron motion. This motion is due to the thermal energy of the ions and the applied magnetic field and is orthogonal to the magnetic field (B).

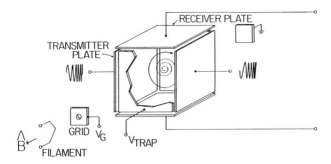

Figure 3.15 Schematic of the cubic ion cyclotron resonance cell. Ions are formed by an electron beam pulse, and the coherent motion of ions undergoing resonance is detected by image currents induced in the receiver plates. (Nicolet Analytical Instruments.)

Ion motion parallel to the magnetic flux lines is restricted by electrostatic potentials applied to the trapping plates. The polarity of the ions that are trapped is determined by the polarity of the DC electric potential applied to the trapping plates. To detect these trapped ions, the cyclotron motion of the desired mass ions in the cell are coherently excited (Comisarow, 1980) to larger circular orbits. For an ion having a charge-to-mass ratio q/m, the resonance frequency f (in Hz) is related to the magnetic field strength **B** expressed in tesla (1 tesla $= 10^4$ G) by the expression:

$$f = 1.5357 \times 10^7 \frac{q\mathbf{B}}{m}$$

The excitation of ions is accomplished by applying a swept frequency electric field to the "transmitter" plates of the cell, forcing all ions of a given mass into larger orbits "coherently" or in-phase as the swept frequency passes through the resonant frequency of the ions. This sweep of the excitation frequency is very fast (typically 1 ms). The cyclotron motion of the ions persists after the excitation, and their coherent motion induces an analog signal at the resonance frequency in the "receiver" plates.

This analog signal is amplified, converted to a digital signal, and added into the memory of the computer. A quench pulse at the end of each scan clears the cell from the previously detected ions. This sequence of events can be repeated as fast as 100 times per second, and it is shown in Figure 3.16. The pulsing and recording of induced signals can be repeated as often as may be required to obtain an adequate signal-to-noise ratio. The resultant summed digital data is then subjected to a Fourier transformation to produce a frequency domain spectrum. The amplitude of each frequency component will correspond to the number of ions of the related frequency. A frequency-to-mass conversion then yields the corresponding mass spectrum.

The mass resolution R of this type of spectrometer can be expressed in

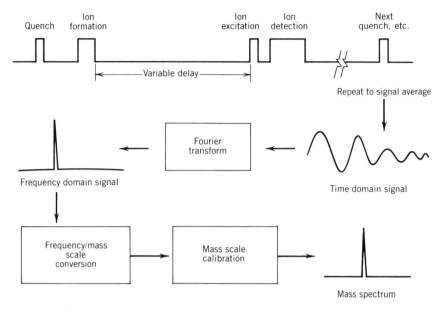

Figure 3.16 FTMS sequence of ion formation, ion detection, and conversion of a time domain signal to a mass spectrum.

terms of the time constant τ of the decay of the ion signal. This decay occurs primarily due to collisions of ions with neutral molecules, resulting in the loss of phase coherence (Ledford et al., 1980):

$$R \cong \tfrac{1}{2} \omega \tau = \pi f_o t$$

Thus, mass resolution is simply π times the number of orbits traversed by the ions within the time constant for ion signal decay. It can also be shown that in terms of more conventional parameters, mass resolution varies directly with the strength of the magnetic field and is inversely proportion to ion mass:

$$\frac{m}{\Delta m} = \frac{\omega}{\Delta\omega} = \frac{q\mathbf{B}}{m\Delta\omega}$$

To obtain a chemical ionization (CI) mass spectrum or to study the ion-molecule reaction of a given chemical system in gas phase, it is possible to insert a variable delay time between the ionization and ion excitation events. During this delay time, which could vary between a few milliseconds and several seconds, sample or reagent gas (CI) ions can react with sample neutral molecules to produce product ions. The extent of the reaction can be easily controlled by adjusting the pressure and/or delay (reaction) time.

Resolution (FWHH): 1,000,000

on doublet at nominal mass 152

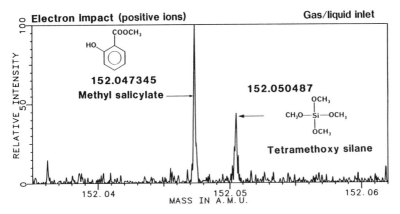

Figure 3.17 Fourier transform spectrum of a doublet at mass 152. A resolution of 1,000,000 is achieved at a magnetic field strength of 3 tesla. (Nicolet Analytical Instruments)

When only sample ions react (no separate reagent gas is used), the phenomenon is termed self-chemical ionization (self-CI).

One of the unique features of FTMS is the capability for selective ion excitation or ion ejection. Ions of a given mass or mass range can be selectively and efficiently excited so that their cyclotron orbit size becomes larger than the size of the cell. Those ions will be neutralized upon impact with the cell walls and, therefore, are ejected from the cell. Application of the selected ion ejection is used in experiments such as MS/MS, GC/MS, and dynamic range enhancement.

Ion detection in FTMS is based upon electronic amplification of the ion image current—not ion impact upon the collector of an electron multiplier—so FTMS does not benefit from its high sensitivity and dynamic range. However, a combination of very high ion transmission efficiency, rapid signal averaging, and capability for selective ion ejection in FTMS makes it possible to achieve sensitivity and dynamic range comparable to many conventional mass spectrometers.

Advances in two technological areas have had an important impact on translating the FTMS concept to laboratory instrumentation. One is the development of superconducting magnets. High fields yield substantial improvements; resolution of $m/\Delta m = 4 \times 10^6$ at $m/q = 131$ at 3 tesla (Nicolet Analytical Instruments) or 1×10^7 at $m/q = 18$ and trapping times of more than 12 h at 4.7 tesla have been reported (Allemann et al., 1980). Figure 3.17 shows a resolution of 1×10^6 at mass 152 at a magnetic field strength of 3 tesla. The second area is computer technology. In order to obtain high resolution over an extended mass range, digital data must be processed at very high speeds. Ledford et al. (1980) point out that for a spectral analysis at a magnetic field strength of 1.2 tesla, and for ions of $m/q = 15$ to $m/q = 1000$, the frequencies will range from about 1.2 MHz to 18kHz, respectively. Furthermore, according to the Nyquist theorem, the analog/digital converter must digitize at a rate at least twice that of the highest frequency component. Thus, for a 1.2 MHz ion transient signal of 100 ms duration, 240 kilobytes of fast memory are required to register the transient signal. Advances in semiconductor technology have made such memories available at a reasonable price. Hence, FTMS spectrometry has grown to its present form largely because of advances in data acquisition systems.

FTMS can be expected to emerge rapidly as a major subfield of mass spectrometry. It has already demonstrated its capability as an analytical mass spectrometer. In the area of mass measurements, for example, Reents (1984) has determined the mass of m/z 466, $C_{10}F_{16}N_3^+$ to be 465.98370 within 4 ppm. Such accuracies are obtainable without an internal mass calibration standard; mass measurement in the sub-ppm range is possible when the instrument is operated in the high-resolution mode. In contrast with conventional mass spectrometers, ions can also be trapped for very long times at very high pressures. This enables chemists to investigate ion-molecule reactions in an ideal gas phase environment. The programmability of the system in relation to ion formation, ion storage, ion excitation, and ion detection permits many diverse experiments to be performed (Gross and Rempel, 1984); among them are MS/MS and collision-induced dissociation studies.

Taking full advantage of the capabilities of the FTMS technique, however, has been limited to applications where sample pressure is low and under control. The one-region cell design described above is inherently vulnerable to high sample or background pressure conditions, leading to collisional damping of the ion signal and its inherent degradation of mass resolution and sensitivity. The advent of the differentially pumped dual cell (Nicolet, 1984) provides a solution to this problem and offers some additional options. In a dual configuration, it is possible to maintain a very low operating pressure in the analyzer cell, even if a substantially higher pressure is present in the sample introduction/ionization (source) region. A simplified schematic diagram of this dual cell system is shown in Figure 3.18. This design contains a conductance limit plate between the source and the analyzer cells, and it allows for a separation in space between ion formation and ion detection. Due to the higher sample pressure in the source, ions are formed mainly in that region and are then transferred through the conductance limit hole into the analyzer region. This ion transfer is due to the thermal energy of the ions and their free oscillation parallel to the magnetic flux lines. The extent of this ion exchange between the two cell compartments can be manipulated electronically by controlling the potential bias of the conductance limit plate. The differentially pumped dual cell also allows for interfacing of the mass

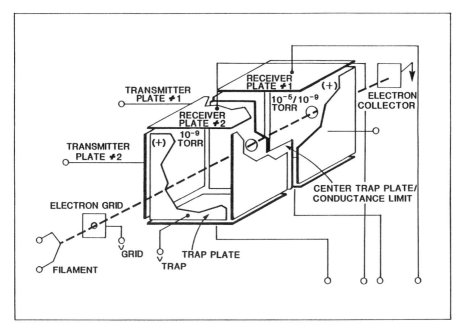

Figure 3.18 Schematic diagram of a differentially pumped FTMS dual cell. (Nicolet Analytical Instruments)

spectrometer to gas chromatographs (GC/FTMS), and for greater versatility in the MS/MS mode of operation.

Another recent development is the tandem quadrupole-Fourier transform spectrometer reported by McIver et al. (1984). In this configuration, ions pass through a first quadrupole, are focused by the second quadrupole, are decelerated, and enter the FTMS analyzer cell. A mass resolution exceeding 100,000 has been reported, and the system can be used with the second quadrupole functioning as a collision cell to produce daughter ion spectra. Finally, the pulsed mode of operation of FTMS makes it compatible with any pulsed ionization method, such as laser desorption and pulsed Cs^+-SIMS. Using these ionization methods, solid samples and surfaces may be analyzed, with the inherently high mass resolution associated with FTMS.

ION MICROPROBE/MICROSCOPE ANALYZERS

The modern ion microprobe analyzer is a highly sophisticated instrument for the secondary ion mass spectrometry (SIMS) of surface layers. Historically, the study of the interaction of particles with solids dates back to the early observations of Gove (1852), in which the cathode in a discharge source was eroded by gas ions in the tube. Thus, when a sample or target is bombarded by a beam of primary ions having energies in the kiloelectron volt range, secondary particles are sputtered from the surface due to primary ion impact; and these secondary particles can include neutral atoms (ground state or excited state), ions (positive or negative), molecular species in various charge states, and atomic clusters. This secondary mass spectrum is matrix-dependent, and the elemental sensitivity can vary over several orders of magnitude. However, all elements from hydrogen to uranium can be detected by SIMS, and the minimum detectable concentration for many metallic elements is typically between 1 ppm to 1 ppb. This minimum detectable concentration will depend upon the primary ion current density, the ratio of ions to neutrals of the sputtered element, plus many parameters that relate to the sample and the analyzing instrument. Many types of spectrometers have been used in SIMS (Liebl, 1981), but only the dedicated ion microprobe or ion microscope analyzer can provide multielement detection at high sensitivity, with lateral spatial resolution approaching 1 μm and a depth resolution of $\sim$ 10 Å. This is the depth of the surface layer from which most sputtered particles emerge.

A distinction is usually made in SIMS depending on the primary ion beam intensity or total number of incident primary ions upon the sample. The term *static* secondary ion mass spectrometry (SSIMS) is applied to very low primary ion currents where there is very limited sputtering or erosion of the sample. This mode of analysis is used when a "nondestructive" surface assay is desired, or for investigating chemisorbed and physisorbed layers. For example, adsorbed methanol molecules have been analyzed on Cu substrates with a 0.8 keV primary argon ion current density as low as 10^{-9} A/cm^2 and 10^{13} ions/cm^2 total dose (Marien et al., 1983). With such low primary ion currents, the lifetime of monolayers can be extended to hours, perturbations produced by the sputtering process are minimized, and the secondary ion spectrum closely reflects the state of the sample *prior* to analysis. Also, many molecular surface structures have been explored by this static mode. The more conventional use of the ion microprobe for materials analysis is to use high-brightness ion sources of energy ranging from 1–20 keV that can be focused to micron-sized beams, which can then be employed for faster scanning and depth profiling. This *dynamic* SIMS is characterized by a considerable amount of sample erosion, some spatial relocation of atoms due to collision cascades, and the breaking of molecular bonds (Rudenauer, 1982). Differential sputtering rates of elements in the sample will also introduce uncertainties in the analysis. In either static or dynamic modes, the secondary ions are accelerated into a double focusing system that provides an elemental and spatially resolved image of the sample surface layers. Elements with low ionization potentials or high electron affinities (H^+, Li^+, Na^+, K^+, O^-, Cl^-, Br^-, F^-, etc.) can be detected at 10^{-14} g levels, and cluster ions have been observed having masses approaching 20,000. A critical review of the several parameters relating depth resolution, sensitivity, and dynamic range has been prepared by Magee et al. (1982).

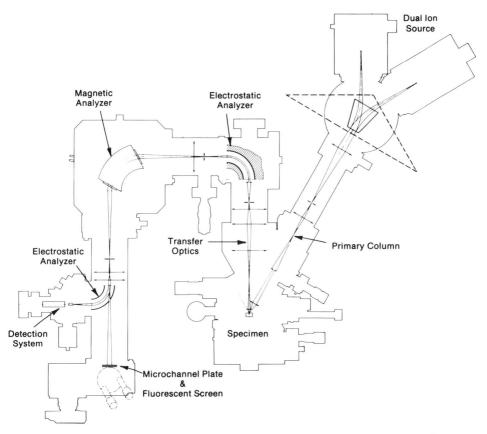

Figure 3.19 Schematic diagram of a high-resolution ion microprobe analyzer. The dual primary ion source permits the analysis of both electropositive and electronegative elements within the same sample area. (Cameca Instruments, Inc.)

A dual primary ion source plus a primary mass filter in the modern microanalyzer arises from (1) the need to analyze both electropositive or electronegative elements within the same sample area, and (2) the need for primary ion beam purity (LeGoux and Migeon, 1982). Also, the ability to switch primary ion sources for achieving the best ionization conditions saves time in sample assays, and the primary mass filter improves the detection limit for elements that may be present in both the ion beam and sample.

A schematic diagram of a widely used commercial instrument (Cameca IMS-3f) is shown in Figure 3.19. The instrument contains a dual ion source: (1) cesium, and (2) a duoplasmatron (oxygen/argon). A three-lens primary ion column is able to produce probe diameters as small as 1 μm, and current densities of 150 mA/cm^2 (for a 3μm probe diameter). Secondary ions are accelerated to ~4500 eV and enter the transfer optics that are matched to the double focusing spectrometer, which produces fully stigmatic and achromatic images. This instrument is most commonly used in the microscope mode where the sample is illuminated by a large primary ion beam. The sample area from which secondary ions are imaged can be varied from 400 × 400 μm to an area approaching a few square micrometers. The ion detection system includes a Faraday cup, an electron multiplier and a fluorescent screen. Lateral stepscanning is accomplished by synchronizing a displacement of the sampled area (in μm steps) with appropriate control voltages in the spectrometer via a computer. Depth profiles for any element can also be obtained at very high resolution.

An additional unique feature of the modern ion microprobe is that the highly localized primary ion beam can also produce an ion-induced electromagnetic spectrum that is qualitatively useful. Some of this radiation is emitted within the solid from luminescence, but an important contribution is made by the sputtered particles that are free from the influence of the surface. Consequently, the characteristic line spectra of atoms or the band spectra of molecules can be observed. Indeed, using a glass specimen as a target, Guenther et al. (1982) found that both the SIMS and the ion-induced radiation (IIR) spectrum provided complementary data of elements and oxides during ion depth profiling. Also, at least one commercial ion microprobe analyzer has been modified and equipped with an auxilary laser source for exciting surface species by either pulsed or CW radiation (Furman and Evans, 1982). The laser beam is focused by a lens with a focal length of approximately 1 m to a spot size of ~1 mm. The beam enters the apparatus through a quartz window to impinge on the sample at 45°. By means of this arrangement, ion spectra are directly compared that are generated by Cs^+ ion bombardment and laser irradiation. Furthermore, thermal cycling of the sample permits time-dependent measurements to be made in situ relating to diffusion and desorption.

At present, there appears to be no limit to the types of measurements that are being made with ion microprobes, which include: (1) basic studies of secondary ion emission; (2) the chemistry of grain boundaries; (3) the distribution of impurities and phases in ferrous metals; (4) profiles in semiconductor materials and devices; (5) oxidation, corrosion, and diffusion phenomena; (6) airborne microparticle analysis; (7) elemental and isotopic measurements of meteorites; (8) detection of defects in cladding and plating; (9) trace elements in teeth enamel, bones, blood cells, and biological tissue; (10) forensic studies involving trace elements; (11) the assay of antireflection optical coatings; (12) spatially resolved (micrometer) ion implantations; and (13) the optical spectroscopy of ion-induced radiations. Some of these topics are discussed in later chapters.

LASER MICROPROBE INSTRUMENTATION

The laser microprobe provides an important alternative to the ion microprobe for the qualitative analysis of metals, polymers, trace contaminants, and localized distributions of the elements in various matrices. In the field of biology alone, laser mass spectrometry has been applied to the assay of hair, blood serum, nerve fiber, skeletal tissue, etc.; and the field of forensic science is becoming increasingly dependent upon this mode of microanalysis (Conzemius and Capellen, 1980). Laser-induced ionization requires a minimum of sample preparation, irradiated areas less than 1 μm^2 can be sampled by focusing the laser beam through a microscope objective, and the amount of energy deposited in the sample area can be controlled easily by varying the laser intensity. Wavelengths range from 10.6 μm for the CO_2 laser to 265 nm for the frequency-quadrupoled Nd:YAG laser, and irradiation time can vary between continuous-wave lasers to Q-switched pulse lasers ($\tau <$ 10 ns). Because of the wide limits of sampled areas (100 to 1 μm diameters) and laser power density (10^6 to 10^2 W/cm^2), two modes of laser-induced ionization have been identified: (1) a *plasma* mode, and (2) a *laser-induced desorption* (LD) mode (Heinen, 1981). The *plasma* mode generates atomic and small molecular fragment ions at high-power densities, while the LD mode is used to characterize surface species at low excitation levels and with a minimum degree of dissociation.

The laser microprobe can be used in conjunction with any type of mass analyzer or filter. However, the TOF technique has two salient advantages: (1) a complete mass spectrum can be recorded from a single laser pulse, and (2) the high transmittance of the TOF geometry provides high sensitivity, despite the small microvolume that is irradiated. In addition, the TOF system is useful in photodissociation and other studies where the absorption of a photon of known energy can be clearly related to both time and energy variables (Dunbar and Armentrout, 1977), and for observing laser-induced positive and negative molecular ions emitted from thin foils (Fürstenau et al., 1979).

As mentioned in Chapter 2, the evolution of the modern laser microprobe began with Honig and Woolston (1963), who used a pulsed ruby laser with a 150 μm diameter beam, a pulse length of 50 μs, and a double focusing mass spectrometer. This work was followed by a Q-switched ruby laser and TOF system with a path length of 1 m (Fenner and Daly, 1966). The pulse length was in the nanosecond range and the beam was focused to 20 μm; this system also included an electrostatic energy selector and a scintillation ion detector. Subsequently, Hillenkamp et al. (1975) used microscope optics to reduce the beam diameter to $\sim$ 1 μm, and a detection limit as low as 10^{-19} g was observed for some elemental samples that were suspended in thin (0.1–1.0 μm) epoxy resin films. A further improvement by Wechsung et al. (1978) substantially increased mass resolution (to $m/\Delta m = 800$ at $m/z = 200$) by introducing a "time focusing" ion reflector in the TOF system.

A detailed description of the modern *transmission* bilaser-TOF system has been reported by Denoyer et al. (1982). The sample particles are supported on a thin organic film ($\sim$0.1 μm) on a moveable x-y stage that can be viewed through a thin quartz window. The pilot laser is a low power (2

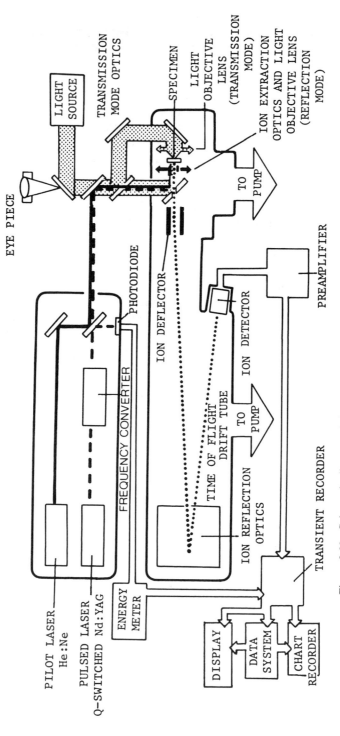

Figure 3.20 Schematic diagram of the laser microprobe (LAMMA 500) time-of-flight spectrometer. (Leybold-Heraus Co.)

84

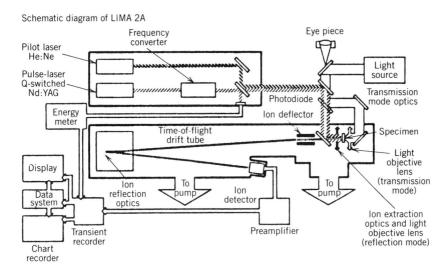

Schematic diagram of LIMA 2A

Figure 3.21 Schematic diagram of laser microprobe for sample analysis in either the transmission or reflection mode. (Cambridge Mass Spectrometry, Ltd.)

mW), visible (λ = 613 nm), He-Ne search laser that is focused to illuminate the sample. The second is a Q-switched (15 ns) Nd:YAG laser quadrupled in frequency to provide a high-power density (10^{10} to 10^{11} W/cm^2) for a 0.5 μm diameter spot. Figure 3.20 is a general schematic. The power laser vaporizes a sample particle to produce a microplasma containing elementary, molecular, and fragment ions. For the highest laser power densities, the ion energy spread is in the 0–60 eV range, but this spread is effectively reduced by ion filtering and reflecting ion optics. Transmissions of 50% are reported for a 1.8 m TOF instrument (Denoyer et al., 1982).

Some of the most interesting applications of the laser ion microprobe have included biological studies and the identification of specific products within organisms. However, the absolute detection limit for some trace metals is also impressive. The detection limit for potassium is 1 × 10^{-20} g (~150 atoms). Other metallic elements having detection sensitivities in the 10^{-18} to 10^{-20} g range include Li, Na, Mg, Al, Ca, Cu, Rb, Cs, Sr, Ag, Ba, Pb, and V (Hercules et al., 1982). An evaluation of the dynamic range, abundance sensitivity, accuracy for isotopic ratio measurements, etc. for the LAMMA instrumentation depicted in Figure 3.20 has been reported by Simons (1983).

In the instrumentation cited above, sample particles are usually mounted on thin films and the sample specimens must be less than a few microns in thickness. Recently, however, new instrumentation has been developed for thick specimens in which the laser beam, ion focusing lens, and high-resolution observation objective can be positioned on the same side of the sample (Feigl et al., 1983). A schematic of this type of laser microprobe is shown in Figure 3.21. The Nd:YAG laser is focused through a specially designed objective onto the bulk surface at an angle of 30° to the surface normal, and the ions are extracted perpendicular to the sample surface. The ions are then imaged by an ion lens through a TOF instrument. At detection threshold, the amount of analyzed bulk sample is of the order of a few picograms. It is also reported that by using multiple laser pulses on the same spot, a layer-by-layer analysis can be made. For the depth resolution of inorganics, it is unlikely that laser instrumentation will approach that of the ion microprobe. However, the flexibility, short sample changing time (< 10 min), and scope of applications has made the laser microprobe an important new tool in the field of mass spectrometry.

At present, laser microprobe instrumentation is now available, operating in either the transmission or reflection mode, so that either small particles or a microvolume of a bulk sample can be analyzed. In the reflection mode, a lateral resolution of 1–3 μm can be attained; a depth-profiling resolution less than 1–10 μm is also possible with repeated laser pulsing on the same sample area.

REFERENCES

Ahmad, M., and T. T. Tsong, *Appl. Phys. Lett.*, *44*, No. 1, 40 (1984).

Allemann, M., Hp. Kellerhals, and K. P. Wanczek, *Chem. Phys. Lett.*, *75*, No. 2, 328 (1980).

Bainbridge, K. T., and E. B. Jordan, *Phys. Rev.*, *50*, 282 (1936).

Balestrini, S. J., and F. A. White, *Rev. Sci. Instr.*, *31*, 633 (1960).

Baril, M., *Nucl. Instr. Methods, 187*, 153, (1981).

Blauth, E. W., *Dynamic Mass Spectrometers*, Elsevier, Amsterdam (1966), p. 28.

Browder, J. A., R. L. Miller, W. A. Thomas, and G. Sanzone, *Int. J. Mass Spectrom. Ion Phys.*, *37*, 99 (1981).

Chait, B. T., and F. H. Field, "A Highly Sensitive Pulsed Ion Bombardment TOF Mass Spectrometer" presented at the *32nd Ann. Conf. on Mass Spectrom. and Allied Topics*, San Antonio (May 27 to June 1, 1984), p. 237.

Chou, J. S., D. Sumida, and C. Wittig, *Chem. Phys. Lett.*, *100*, No. 3, 209 (1983).

Comisarow, M. B., in *Advances in Mass Spectrometry*, Vol. 8B, ed. A. Quayle, Heyden and Son Ltd., London (1980), p. 1698.

Comisarow, M. B., and A. C. Marshall, *Chem. Phys. Lett.*, *25*, 282 (1974).

Conzemius, R. J., and J. M. Capellen, *Int. J. Mass Spectrom. Ion Phys.*, *34*, 197 (1980).

Damoth, D. C., in *Advances in Analytical Chemistry and Instrumentation*, Vol. 4, ed. C. N. Reilley, Wiley, New York (1964), p. 371.

Davis, W. D., in *Trans. of the 10th National Symposium of Vacuum Technology*, Pergamon, New York (1962), p. 363.

Davis, W. D., and T. A. Vanderslice, *Trans. of the 8th National Symposium of Vacuum Technology* (1960), p. 417.

Dawson, P. H., *Int. J. Mass Spectrom. Ion Phys.*, *6*, 33 (1971).

Dempster, A. J., *Phys. Rev.*, *11*, 316 (1918).

Dempster, A. J., *Proc. Am. Phil. Soc.*, *75*, 755 (1935).

Denison, D. R., *J. Vac. Sci. Technol.*, *8*, 266 (1971).

Denoyer, E., R. Van Grieken, F. Adams, and D. F. S. Natusch, *Anal. Chem.*, *54*, No. 1, 26A (1982).

Devienne, M., M. Repoux, and J. C. Roustan, *Int. J. Mass Spectrom. Ion Phys.*, *46*, 143 (1983).

Duckworth, H. E., *Mass Spectroscopy*, Cambridge University Press, London (1966), p. 26.

Dunbar, R. C., and P. Armentrout, *Int. J. Mass Spectrom. Ion Phys.*, *24*, 465 (1977).

Feigl, P., B. Schueler, and F. Hillenkamp, *Int. J. Mass Spectrom. Ion Phys.*, *47*, 15 (1983).

Fenner, N. C., and N. R. Daly, *Rev. Sci. Instrum.*, *37*, 1068 (1966).

Finlan, M. F., and R. F. Sunderland, *Nucl. Instr. and Methods*, *195*, 447 (1982).

Furman, B. K., and C. A. Evans, Jr., in *Secondary Ion Mass Spectrometry SIMS III*, eds. A. Benninghoven, J. Giber, J. László, M. Riedel, and H. W. Werner, Springer-Verlag, Berlin (1982), p. 88.

Fürstenau, N., F. Hillenkamp, and R. Nitsche, *Int. J. Mass Spectrom. Ion Phys.*, *31*, 85 (1979).

Ghoshal, S. N., M. Das, and N. N. Mitra, *Int. J. Mass Spectrom. Ion Phys.*, *17*, 67 (1975).

Gove, W. R., *Trans. Faraday Soc.*, *142*, 87 (1852).

Gross, M. L., and D. L. Rempel, *Science*, *226*, 261 (1984).

Guenther, K. H., E. Hauser, G. Hobi, P. G. Wierer, and E. Brandstaetter, in *Secondary Ion Mass Spectrometry SIMS III*, eds. A. Benninghoven, J. Giber, J. László, M. Riedel, and H. W. Werner, Springer-Verlag, Berlin (1982), p. 97.

Heinen, H. J., *Int. J. Mass Spectrom. Ion Phys.*, *38*, 309 (1981).

Hercules, D. M., R. J. Day, K. Balasanmugam, T. A. Dang, and C. P. Li, *Anal. Chem.*, *54*, No. 2, 280A (1982).

Herzog, R., *Z. Physik*, *89*, 447 (1934).

Hillenkamp, F., R. Kauffmann, R., Nitsche, and E. Unsöld, *Appl. Phys.*, *8*, 341 (1975).

Honig, R. E., and J. R. Woolston, *Appl. Phys. Lett.*, *2*, 138 (1963).

Kilius, L. R., E. I. Hallin, K. H. Chang, and A. E. Litherland, *Nucl. Instr. and Methods*, *191*, 27 (1981).

Kunibe, T., Y. Kubota, and A. Miyahara, *Japanese J. Appl. Phys.*, *21*, No. 12, 1788 (1982).

Lawrence, E. O., and M. S. Livingston, *Phys. Rev.*, *37*, 1707 (1931).

Ledford, E. B., Jr., S. Ghaderi, C. L. Wilkins, and M. L. Gross, in *Advances in Mass Spectrometry*, Vol. 8B, ed. A. Quayle, Heyden and Son Ltd., London (1980), p. 1707.

LeGoux, J. J., and H. N. Migeon, in *Secondary Ion Mass Spectrometry SIMS III*, eds. A. Benninghoven, J. Giber, J. László, M. Riedel, and H. W. Werner, Springer-Verlag, Berlin (1982), p. 52.

Lennard, W. N., H. R. Andrews, M. Freeman, I. V. Mitchell, D. Phillips, D. A. S. Walker, and D. Ward, *Nucl. Instr. and Methods*, *203*, 565 (1982).

Liebl, H., *Nucl. Instr. and Methods*, *191*, 183 (1981).

Magee, C. W., R. E. Honig, and C. A. Evans, Jr., in *Secondary Ion Mass Spectrometry SIMS III*, eds. A. Benninghoven, J. Giber, J. László, M. Reidel, and H. W. Werner, Springer-Verlag, Berlin (1982), p. 172.

Mamyrin, B. A., V. J. Karatajev, D. V. Shmikk, and V. A. Zagulin, *Sov. Phys.,-JETP*, *(Engl. Transl.)*, *37*, 45 (1973).

Marien, J., E. DePauw, and G. Peizer, *J. Phys. Chem.*, *87*, 4344 (1983).

Matsuda, H., *Japanese J. Appl. Phys.*, *20*, No. 11, L825 (1981a).

Matsuda, H., *Nucl. Instr. Methods*, *187*, 127 (1981b).

Matsuda, H., *Mass Spectrom. Rev.*, *2*, 299 (1983).

Mattauch, J., and R. Herzog, *Z. Physik*, *89*, 786 (1934).

Mattauch, J., *Phys. Rev.*, *50*, 617 (1936).

McIver, R. T., Jr., *Rev. Sci. Instrum., 41,* 555 (1970).

McIver, R. T., Jr., Ph.D. Thesis, Stanford University (1971).

McIver, R. T., Jr., R. L. Hunter, and W. D. Bowers, "Recent Developments with the Tandem Quadrupole-Fourier Transform Mass Spectrometer," presented at the *32nd Ann. Conf. on Mass Spectrom. and Allied Topics,* San Antonio (May 27 to June 1, 1984), p. 593.

Negra, D. S., and Y. LeBeyec, *Int. J. Mass Spectrom. and Ion Processes, 61,* 21 (1984).

Nier, A. O., *Rev. Sci. Instr., 11,* 212 (1940).

Nier, A. O., and T. R. Roberts, *Phys. Rev., 81,* 507 (1951).

Paul, W., and H. Steinwedel, *Z. Naturforch., 8A,* 448 (1953).

Quisenberry, K. S., T. T. Scolman, and A. O. Nier, *Phys. Rev., 102,* 1071 (1956).

Ramos, E., and G. H. Rayborn, *Bull. Am. Phys. Soc., 29,* No. 9, 1503 (1984).

Reents, W. D., Jr., "High Accuracy Measurement with a Fourier Transform Mass Spectrometer," presented at the *32nd Ann. Conf. on Mass Spectrom. and Allied Topics,* San Antonio (May 27 to June 1, 1984), p. 530.

Reinmüller, H., *J. Phys. E.; Sci. Instrum., 16,* 1228 (1983).

Rudenauer, F. G., *Int. J. Mass Spectrom. Ion Phys., 45,* 355 (1982).

Siegel, M. W., and M. J. Vasile, *Rev. Sci. Instr., 52,* No. 11, 1603 (1981).

Simons, D. S., *Int. J. Mass Spectrom. and Ion Processes, 55,* No. 1, 15 (1983).

Sommer, H., H. A. Thomas, and J. R. Hipple, *Phys. Rev., 82,* 697 (1951).

Stephens, W. E., *Phys. Rev., 69,* 691, (1946).

Todd, J. F. J., in *Dynamic Mass Spectrometry,* eds. D. Price, and J. F. J. Todd, Heyden and Son, Ltd., London (1981), chapter 4.

Traum, M. M., and D. P. Woodruff, *J. Vac. Sci. Technol., 17,* No. 5, 1202 (1980).

Wechsung, R., F. Hillenkamp, K. Kaufmann, R. Nitsche, and H. Vogt, *Micros. Acta, Suppl. 2,* 611 (1978).

White, F. A., F. M. Rourke, J. C. Sheffield, R. Schuman, and J. R. Huizenga, *Phys. Rev., 109,* 437 (1958).

Wiley, W. C., and I. H. McLaren, *Rev. Sci. Instr., 26,* 1150 (1955).

Yinon, J., and H. G. Boettger, *Int. J. Mass Spectrom. Ion Phys., 10,* 161 (1972).

CHAPTER FOUR

TANDEM SYSTEMS AND SPECIAL TYPES

Tandem mass spectrometry now occupies a dominant position in research that relates to molecular structure elucidation and the analysis of complex mixtures. The recent publication edited by McLafferty, titled *Tandem Mass Spectrometry* (Wiley, New York, 1983), contains contributions by 52 authors representing 32 research groups that are active in this field. Hence, the reader is referred to this monograph for an authoritative overview of MS/MS instrumentation and its many applications in organic chemistry. A growing number of hybrid mass spectrometers are also making their appearance, comprising magnetic sector, electric sector, and quadrupole combinations. This chapter includes a description of only a few multistage systems, plus several specialized instruments that have been developed for charged particle analysis.

TANDEM MAGNETIC ANALYZERS

While multistage magnet systems are now employed in a wide variety of investigations, the first generation of tandem magnetic spectrometers was designed for high-sensitivity isotopic abundance measurements. The first such tandem magnet system was reported by Inghram and Hayden (1954). In this instrument, ions were mass analyzed in the first magnet, given an additional acceleration in a region between the two magnets, and reanalyzed in a magnetic sector that deflected the ion beam in a direction opposite to that of the first analyzer. The general orientation of the magnets led to this scheme being identified as an "S"-type tandem analyzer. Subsequently, White and Collins (1954) reported on the construction of a two-stage magnetic analyzer comprising two 90° sectors with 30.5 cm radius of curvature arranged in a "C"-type configuration so that deflection of the ion beam proceeded in the same sense. No additional acceleration was given to the ions in the region between the two magnets, this being a "drift" tube only.

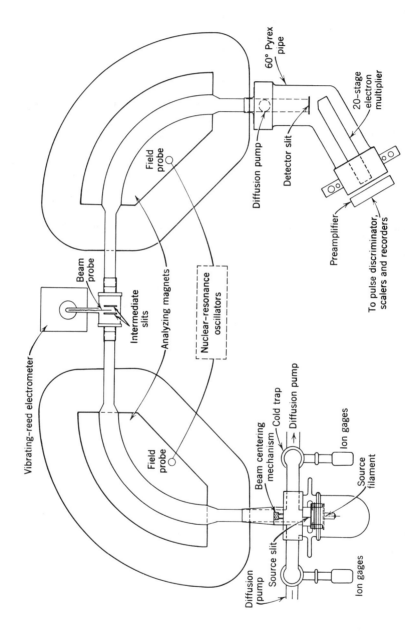

Figure 4.1 Two-stage magnetic analyzer with "C"-shaped geometry (White and Collins, 1954).

In this latter instrument (Fig. 4.1), nuclear magnetic resonance (NMR) was also introduced to provide both short- and long-term stabilization of the tandem magnet system (Sheffield and White, 1958). A high-gain electron multiplier was also used on this system for measuring mass-resolved ions in the pulse-counting mode; an abundance sensitivity of approximately 10^8 was realized with this instrument. In addition, this early multistage instrument was used to discover the stable isotope ^{180}Ta, and for establishing new "upper limits" on the possible existence of other isotopes of low abundance for many elements (White et al., 1956).

Tandem magnetic systems have now been widely adopted in both research and commercial analytical instruments for isotopic measurements, where it is desirable to measure large ratios, that is, $> 10^6 : 1$. In conventional single-magnet spectrometers, isotopic abundance ratios for adjacent mass numbers (in the mass range of approximately 100) can rarely be measured for ratios greater than $10^5 : 1$. Even if resolution is high, peak-to-valley ratios are limited by phenomena such as small-angle scattering from residual gas molecules in the analyzer (Ioanoviciu, 1973), reflections from metal surfaces, imperfections in ion optics, and charge exchange. However, if an ion beam is constrained to pass through two identical momentum analyzers, allowing but a single mass number to enter into the second magnet, considerable discrimination can be achieved with respect to spurious ions. The second magnet will reanalyze the spectrum with a mass discrimination approximating the first, providing the collimating slit that allows the ion beam to enter the second magnet is small compared to the mass dispersion. In other words, if adjacent isotopes can be filtered through a single-momentum analyzer, so that $10^4 : 1$ ratios can be observed, two magnets—operating as independent filters—might be expected to provide a $10^8 : 1$ discrimination from adjacent isotopes. The actual performance of single-stage and tandem double focusing instruments for "C" and "S" configurations has been reviewed by Gall et al. (1980), together with an analytical expression for making "peak tail" calculations.

High abundance sensitivities have been realized for tandem magnets of small and large radii of curvature, provided that the source of ions has a restricted ion energy spread. Hence, the surface-ionization source has been used extensively for the isotopic analysis of many metals in such dual-magnet systems. Other dual-stage magnets have been employed for special measurements, such as the instrument designed by Svec and Flesch (1968) for the simultaneous collection of positive and negative ions from a single ion source, and for measuring ionization efficiencies.

Most tandem magnet systems that do not include an electrostatic focusing element have used 90° sectors. However, one unique design includes two 180° trajectories within a *single* inhomogeneous pole piece (Zashkvara, 1981). The inhomogeneous ion optics allows the ion source and final detector to be located remotely from the magnet boundary, and an electrostatic "mirror" is used to generate the second 180° trajectory.

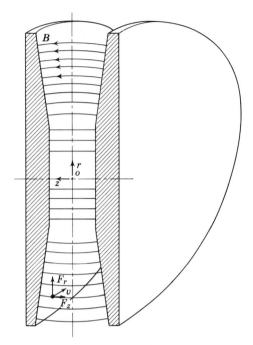

Figure 4.2 Forces on a charged particle moving in an inhomogeneous magnetic field.

INHOMOGENEOUS FIELD MAGNETS

The 255° Sector

Inhomogeneous magnetic fields can provide both radial and axial forces on a charged particle, as indicated in Figure 4.2. Siegbahn and Svartholm (1946) developed specific expressions for the focal angle, dispersion, and resolving power of a magnetic field given as:

$$B(r) = B_o(r_o/r)^n$$

where $B(r)$ is the midplane magnetic field as a function of radius, r_o is the radius of the central orbit, B_o is the midplane magnetic field strength at r_o and n is the inhomogeneous field index. Their analysis indicated that both radial (ϕ_r) and azimuthal (ϕ_z) focusing occurs when:

$$\phi_r = \phi_z = \pi \sqrt{2} \cong 255°$$

and that at this focal angle, and for $n = 0.5$, spherical aberration is minimized.

The focusing properties of the 255° system are shown in Figure 4.3, where both the radial and the azimuthal or axial trajectories converge to the same

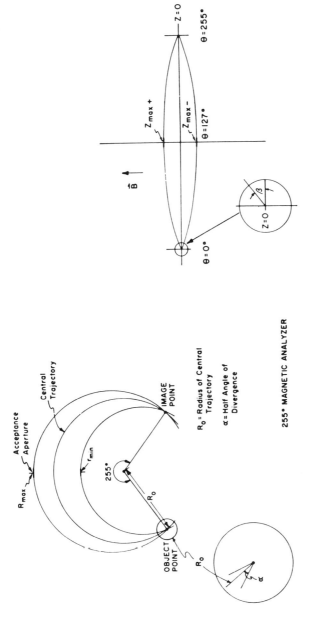

Figure 4.3 Radial and axial focusing properties of the 255° inhomogeneous magnetic sector (Spaniol et al., 1974).

image point. As a result, the transmission of a 255° analyzer can approach 100% for reasonable ion source limits relating to angular divergence and ion energy spread. The mass dispersion for an inhomogeneous field analyzer is also greater than that of one having a uniform field. For a magnet having a basic radius of curvature r_0, with the object and image symmetrically located, the dispersion D is given by:

$$D = \frac{r_0 \Delta m}{m_0(1 - n)}$$

where n is the index of inhomogeneity of the magnetic field. Thus, for a 0.5 index the dispersion is doubled. A very large 255° analyzer having this index has been reported by Spaniol et al. (1974). This magnet for charged particle analysis is designed to operate with an energy resolution of 1000, with a transmission factor of 0.3% of 4π steradians over a $\pm 10\%$ energy range. Alternatively, as an isotope analyzer, it has a transmission approaching 100% for several types of ion sources. The mean radius of curvature of this 75 ton electromagnet is 76 cm, the pole-piece gap width at this radius is 10 cm, and the mass dispersion at mass 100 is ~ 1.5 cm. The extent and intensity of the radial field is also sufficient to permit the magnetic analysis of alpha particles from transuranic nuclides in the energy range of 4.5–5.5 MeV. Hence, for the analysis of transuranic nuclides, such an analyzer can be used with a solid-state detector at the focal point to measure particle energy, while the magnetic field measurement simultaneously determines the momentum of the particle. This suggested dual or "coincidence" determination of an energetic particle could, in principle, virtually eliminate all spurious background counts.

Recently, inhomogeneous analyzers have been used in connection with experiments on collision cross-sections, scattering, and charge transfer phenomena because of the superior ion optics of these analyzers for low-intensity beams (Chakrabarty et al., 1982). The possibility of using very small ($r_0 \leq 5$ cm) 255° analyzers has also been considered for space-related and other applications, where the excellent transmission, lighter weight, and enhanced dispersion of this inhomogeneous sector are important parameters.

Multistage Systems

Few multistage inhomogeneous magnet systems have been built despite their several advantages. However, a three-stage 180° system is shown in the schematic diagram of Figure 4.4. This magnetic analyzer was designed to function as either a high-purity isotope separator or mass spectrometer so that the physical dimensions are large (Whitehead and White, 1972). The nominal radius of curvature is 61 cm and the field index of inhomogeneity is 0.5, so that the mass dispersion of a single magnet is equivalent to that of a 122 cm uniform field magnet. The distance of the focal plane d from

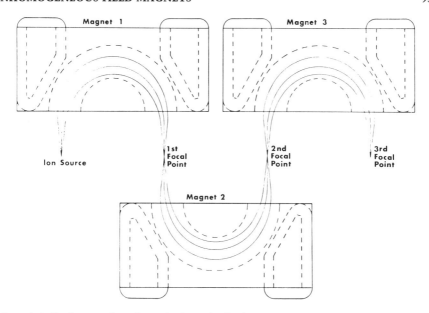

Figure 4.4 Ion beam trajectories and schematic of a three-stage, 180° inhomogeneous magnetic analyzer (Whitehead and White, 1972).

the magnetic boundary also differs from that of a uniform field instrument, and it can be calculated from the expression:

$$d/r_0 = \tan[\pi/2 - (1 - n)^{1/2} \phi/2]/(1 - n)^{1/2}$$

where r_0 is the mean radius, n is the field index, and ϕ is the sector angle (Dagenhart and Whitehead, 1970). A plot of this relationship is shown in Figure 4.5. Thus, for the 180° sector, the focal plane is remote from the magnet boundary, so that adequate space is available for sources, detectors, auxiliary electrostatic lenses or filters, defining slits, and reaction cells. With the layout shown in Figure 4.4, there is also ample space for laser or high-voltage sources, ion detectors at intermediate focal points, or special apparatus after the third focal point. The magnet-to-focal plane dimension is ~50 cm, the pole-piece gap height is 5.1 cm at the basic radius, and the ion optics were designed for an angular acceptance of ±6°. The three 17 ton electromagnets have a maximum induction of 6000 G, and focal lengths can be matched through a limited adjustment of the magnets on roller-bearing supports.

Overall, this general type of multistage inhomogeneous arrangement is attractive because of the variety of investigations that can be pursued. First, the very high transmission is useful for all types of isotopic separations, collision-induced reactions, and scattering experiments. Each stage is capable of providing an isotopic enrichment of 10^3, so that a 10^9:1 isotopic

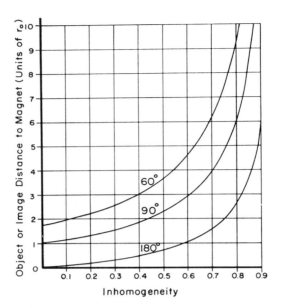

Figure 4.5 Focal plane distance from the magnetic boundary for symmetric object-image positions in units of the basic radius, for a variable field index *n,* and for sector angles of 60°, 90°, and 180°.

purity should, in principle, be obtained. Second, a three-stage system lends itself to differential pumping schemes, where the first, second, and third sections could be pumped to 10^{-6}, 10^{-8}, and 10^{-10} torr, respectively—and thus make feasible surface studies that require high vacua. Finally, future solid state research will certainly include ion implantations that can be quantified in situ, in which the isotopic dilution method is applied via an *ion beam isotopic spike.* For example, if a sample is to be analyzed at the 10^{-18} g level for copper, ^{63}Cu ions can be pulse counted at the first focal plane and ^{65}Cu atoms can be quantitatively implanted in the sample at the third focal point. Furthermore, future research will certainly include the fabrication of very thin (~10Å) films by mass-resolved ion beams as well as by conventional epitaxial growth.

REVERSE GEOMETRY AND MULTITRAJECTORY INSTRUMENTS

One of the first multistage spectrometers to be used in a "reverse geometry" was the three-stage instrument constructed by White et al. (1958a), comprising two 90°, 50.8 cm radius-of-curvature magnets followed by a 90° electrostatic lens. A schematic diagram of this apparatus is shown in Figure 4.6. The second magnet and electrostatic lens functioned as an integral double focusing system after a mass separation by the first magnet. Both magnets

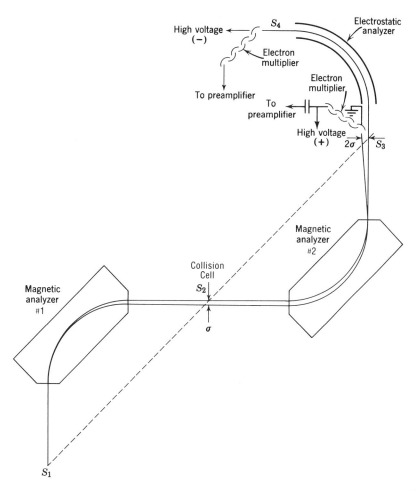

Figure 4.6 Schematic diagram of "reverse geometry" three-stage spectrometer (White et al., 1958a).

were designed for a field strength of 10,000 G, adjustable slits were at all focal points, and ion beam detection was in the pulse counting mode. The ion source voltage was programmable from 5000–50,000 V and the magnetic fields were stabilized by NMR. Although the instrument was designed primarily for isotopic measurements of the heavy elements, this multistage spectrometer was quickly demonstrated to be useful for:

1. Isotopic measurements requiring ultra-high abundance sensitivities, that is, 10^{10}:1.
2. The generation of pure atomic beams.

3. Energy loss and charge exchange measurements.
4. Collision-induced dissociation (CID) phenomena and the determination of molecular dissociation energies.

The very-high abundance sensitivity was achieved because small-angle elastic gas scattering and inelastic encounters of the ion beam were largely eliminated. Even in the high-mass range, it was difficult to detect any contribution to a minor peak from an adjacent major peak, as graphically illustrated in the isotopic spectrum of irradiated uranium (Fig. 13.12). In this instrument, "upper limits" less than 1 ppb were also established for ^{233}U, ^{236}U, and ^{237}U in uranium samples from samples of Belgian Congo pitchblende and Colorado carnetite (Davis and Mewherter 1962). The concept of producing a pure atomic beam was also demonstrated with this instrument by passing molecular beams through very thin (100–250Å) foils (White et al., 1958b). Basically, an energetic molecular specie can be totally dissociated in passing through thin films so that the mass spectrum is totally atomic—with no interfering molecular ions. Both energy loss (dE/dx) and charge exchange measurement were made with this "reverse geometry" arrangement, and equilibrium charge exchange measurements were made for C, O, Ne, He, Ar, N, Li, and Na in various charge states for both nickel and vinyl thin films.

Finally, molecular dissociation phenomena were observed in this apparatus for high-energy molecules interacting with helium in a collision cell, and experiments were performed to show that a kinetic energy spectrum could lead directly to a calculation of molecular dissociation energies (Rourke et al., 1959). Specifically, dissociation energy measurements with this early apparatus were made for CO^+, H_2^+, NO^+, N_2^+, and $C_3H_7^+$. In the general case, this type of analysis also permits some interpretation to be made relative to the dissociation mode.

At present, kinetic energy spectrometry is now a major research area, and the monograph titled *Collision Spectroscopy,* edited by Cooks (1978), provides a comprehensive overview of theory, instrumentation, and measurement techniques. "Reverse geometry" magnetic instruments are now available from various instrument manufacturers, some of which have been specifically designed for ion kinetic energy spectrometry.

A limited number of specialized analyzers have been designed that contain more than a single ion trajectory. The spectrometer of Svec and Flesch (1968) for measuring simultaneously both positive and negative ions is an example. Another example is the "cascade" analyzer, which accommodates three distinct ion trajectories within a single magnetic sector (White and Stein, 1970). As indicated in Figure 4.7, the magnet pole piece is shaped to contain (1) a Mattauch-Herzog trajectory, (2) an 80° trajectory, and (3) a tandem or "cascade" 180° ion path. The Mattauch-Herzog configuration was used to simultaneously detect multiple ion beams in a pulse-counting mode with reverse-biased p-n junction detectors (Chap. 5). The 80° sector has a suf-

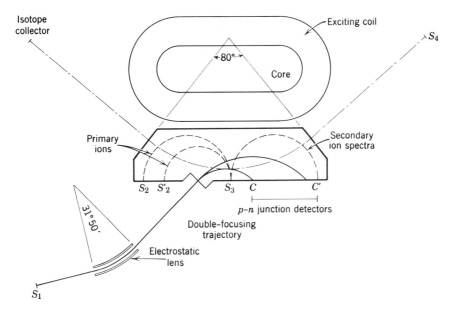

Figure 4.7 An analyzing magnet that accommodates: (1) a Mattauch-Herzog trajectory, (2) an 80° trajectory, and (3) a tandem or "cascade" ion path (White and Stein, 1970).

ficiently large radius (30 cm) so that 5 MeV alpha particles as well as ions can be analyzed. Finally, the "cascade" arrangements makes possible experiments whereby primary ions can be (1) reflected from an electrostatic mirror, (2) sputtered from a target, or (3) ion implanted and reemitted from a surface-ionization filament (White et al., 1963).

TANDEM DOUBLE FOCUSING AND HYBRID SYSTEMS

As in the case of tandem sector magnets, the first tandem double focusing systems of large radii of curvature were constructed for isotopic measurements and physics research. The instrument reported by White and Forman (1967) comprised two 50.8 cm radius-of-curvature 90° magnets and two 90° electrostatic cylindrical lenses arranged in a "C"-shaped configuration. In this instrument, a very high-abundance sensitivity was achieved, and a peak-to-valley ratio of $10^{11}:1$ was obtained in the light mass range. A larger but similar double focusing system (122 cm radius) was subsequently reported by Rasekhi and White (1972), which was used for the analysis of energetic particles, charge exchange, and stopping power (dE/dx) in thin films. This latter instrument had a source-to-detector dimension of approximately 25 m, and a schematic is shown in Figure 4.8.

For recent MS/MS studies, the tandem double focusing system at Cornell

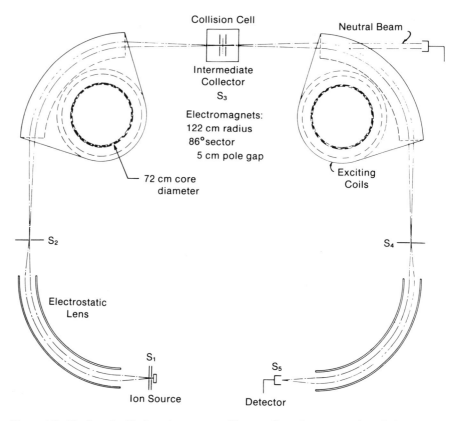

Figure 4.8 Tandem double focusing system of large radius of curvature, in a C-shaped configuration (Rasekhi and White, 1972).

University provides many options (McLafferty, 1983). It provides high resolution in both parent ion selection and in daughter ion dispersion, being designed to analyze 2.5 keV primary ions of 40,000 amu with unit resolution and collision-activated products with 10,000 resolution. This tandem system, shown schematically in Figure 4.9, can also be programmed in various "linked-scan" modes. The MS-I double focusing system is comprised of a 50 cm radius, 70° electrostatic lens followed by a 40 cm radius, 70° magnetic sector. The MS-II double focusing system comprises a 31.8 cm, 90° electrostatic lens followed by a 36.9 cm, 60° inhomogeneous field magnet. Furthermore, precursor ions of high molecular weight can be mass selected at low kinetic energies by MS-I and accelerated to much higher energies for collision. The interface region between these two systems includes a collision chamber, defining slits, etc. The laminated core and low-resistance coils of the MS-II magnet also permits fast mass scanning (0.2 s/decade). Ions can also be detected at several points of the tandem double focusing trajectory by channeltron electron multipliers.

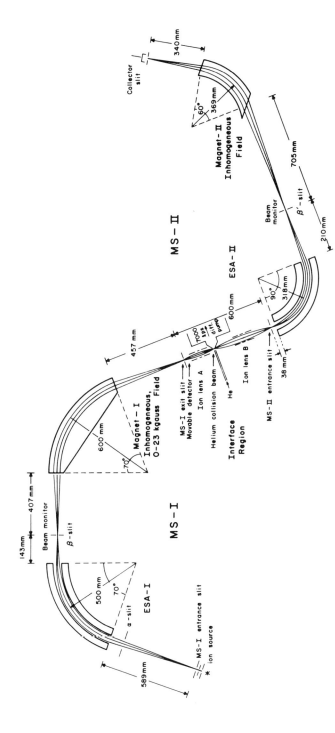

Figure 4.9 Schematic of Cornell tandem double focusing mass spectrometer. (McLafferty, 1983; courtesy of F. W. McLafferty.)

101

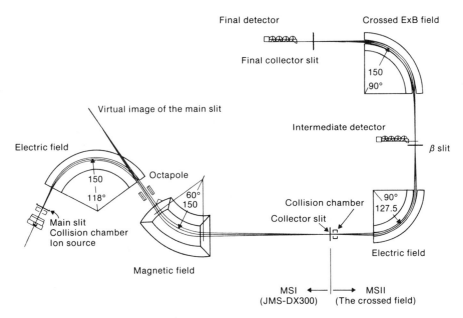

Figure 4.10 Ion optical geometry of the MS/MS system with a crossed ExB field analyzer. (Kunihiro et al., 1984; courtesy of F. Kunihiro and Joel, Ltd.)

The tandem double focusing system reported by Kunihiro et al. (1984) is another advanced system for obtaining precursor and daughter ion spectra at high resolution. As indicated in Figure 4.10, the ion optical components include: (1) MS-I, a 118° electrostatic lens and a 60° magnet, and (2) MS-II, a 90° electrostatic lens plus a crossed ExB field analyzer of 90°. Other tandem double focusing systems are now available from several manufacturers.

The hybrid mass spectrometer designed by the Purdue University-Finnigan MAT staffs (Schoen et al., 1985) has some very unique features. It consists of a high-resolution double focusing (B-magnetic sector/E-electrical sector) system that is interfaced to a double quadrupole (QQ) by a deceleration/zoom lens system. Figure 4.11 is a diagram of this "BEQQ" configuration. Figure 4.12 is the corresponding potential diagram for ion kinetic energies for both the magnet/electrical sector (kiloelectron volt range) and quadrupoles (< 100 eV). The deceleration/zoom lens has a dual function (Cooks et al, 1982, 1983). The first section of the lens transfers the ion beam from the exit slit of the high-energy/high-resolution double focusing system to the DC quadrupole and reduces its energy to approximately 200 eV. This quadrupole also shapes the ion beam. A second zoom lens decelerates the 200 eV ion beam to the desired collision energy.

Furthermore, the four reaction regions of this BEQQ configuration permit: (1) low- and high-energy MS/MS experiments, (2) exact mass measurements, and (3) the detection of sequential ion-molecule reactions.

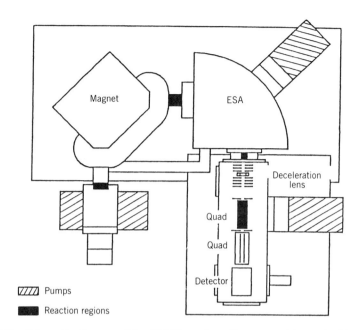

Figure 4.11 Schematic diagram of the BEQQ configuration, comprising a double focusing spectrometer interfaced to a double quadrupole. (Schoen et al., 1985; courtesy of R. G. Cooks.)

Hence, this system has the capability for recording parent, daughter, and neutral loss spectra associated with reactions in the collision quadrupole, as well as those in the first and second field-free regions.

Clearly, this BEQQ spectrometer can be operated in numerous scan modes. For example, the (BE) section can provide highly resolved parent ions for quadrupole MS/MS experiments, or the quadrupole can be set to transmit a specific daughter ion, and a B^2/E parent scan can be programmed.

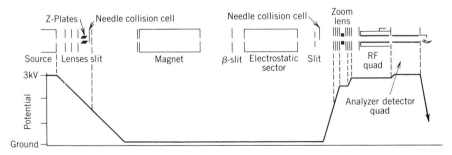

Figure 4.12 Potential diagram for the double focusing section, deceleration lens, and dual quadrupoles of the BEQQ spectrometer. (Schoen et al., 1985; courtesy of R. G. Cooks.)

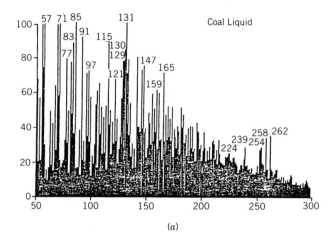

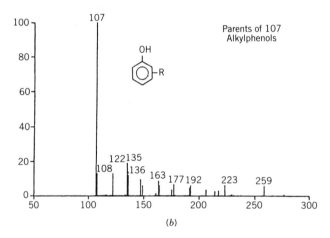

Figure 4.13 (*a*) Electron impact spectrum of a coal-liquid sample, and (*b*) the parent spectrum for ions for which 107$^+$ is a characteristic fragment. (Ciupek et al., 1985; courtesy of R. G. Cooks.)

The performance of this hybrid spectrometer has been described by Ciupek et al. (1985) for transmission measurements, daughter spectra and energy-resolved spectra, ion-molecule reactions, high-resolution MS/MS, high-energy collisions, angle-resolved spectrometry, consecutive collision experiments, and applications in chemical analysis. Figure 4.13 illustrates the use of this spectrometer to characterize a coal-liquid sample. A portion of the complex electron impact spectrum is shown in Figure 4.13*a* and a parent spectrum is shown in Figure 4.13*b* for ions in which 107$^+$ is a characteristic fragment. A very high signal-to-noise ratio is achieved for this latter alkylphenol series. Overall, this type of hybrid mass spectrometer combines the many advantageous features of magnetic sector instruments with the flexibility and speed of quadrupoles.

Several other hybrid instruments for producing fragment-ion spectra have been fabricated from components of conventional spectrometers. The electric sector/quadrupole combination reported by Harris et al. (1984) is an example. With computer control, scanning procedures can be selected that provide fragment-ion spectra of a preselected parent, with simultaneous information on the translational energies released. Finally, the hybrid instrument proposed by Matsuda merits special mention (Matsuda, 1985). This new design consists of a configuration of QQHQC (quadrupole, quadrupole, homogeneous magnetic sector, quadrupole, cylindrical electric sector). This is a very high-resolution and high-transmission spectrometer, with a narrow magnet gap, a very small image magnification, and small second- and third-order aberrations. The high transmission through the narrow magnet gap is achieved by the introduction of electric quadrupole lenses, and these lenses also produce a very small image magnification that leads to high resolution. This new ion optical scheme permits the use of relatively small magnetic sector angles and a wide source slit. A resolving power of 10,000 has been calculated for a slit width of 0.2 mm and a magnetic radius of curvature of 50 cm.

TANDEM QUADRUPOLES

In 1978 and 1979, a multistage quadrupole system was reported (Yost and Enke, 1978, 1979; Yost et al., 1979). Today tandem quadrupoles, having the general configuration shown in Figure 4.14, are among the most widely used in sequential mass spectrometry; commercial versions have been developed by Extranuclear Laboratories, Inc., Finnigan/Mat, and Sciex, Inc. These systems provide a powerful means for analyzing mixtures, with or without chromatographic separation, and for obtaining molecular structure information without resort to high-resolution magnetic systems. The collision energies in quadrupole MS/MS instruments are only on the order of tens of volts rather than kiloelectron volts in magnetic tandems, but collision dissociation cross-sections are often large (Dawson et al., 1982).

Other salient features of these modern triple quadrupoles include: (1) ion source options, (2) programmability, (3) varying resolution and mass range, (4) ease of interfacing to data acquisition and storage systems, and (5) compactness. These instruments can operate in several modes and perform at least four categories of experiments (Slayback and Story, 1981):

1. In daughter experiments, the first quadrupole, $Q1$, filters a single mass or series of masses that undergo collision in $Q2$. Daughter ions are then analyzed by a full scan or by selected ion monitoring in $Q3$.

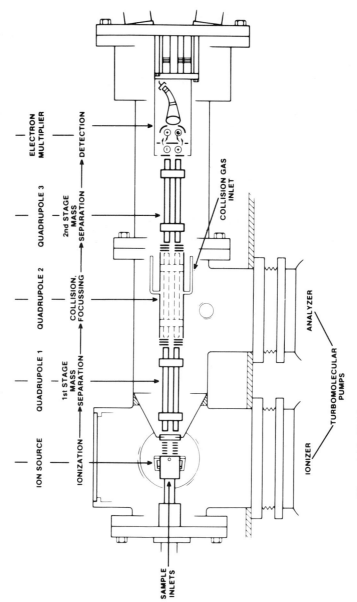

Figure 4.14 Triple-stage quadrupole GC/MS/MS system (Finnigan/Mat Corp.).

2. For parent experiments, all masses of interest are scanned in $Q1$, and collision-activated species are formed in $Q2$. These ions are then mass filtered through $Q3$ in order to identify all parent ions that decompose to yield daughter ions detected in $Q3$.

3. A third type of experiment involves programming $Q1$ and $Q3$ to detect the loss of a neutral molecule (of constant mass). Since neutral losses often represent specified fragments from parent ions (e.g., $-CHO$, $-OH$, $-SO_2$ etc.), neutral loss scanning may quickly reveal a series of similar compounds.

4. A fourth category relates to using the tandem quadrupole as a standard GC/MS system. In this case, the collision cell $Q2$ is unpressurized, and $Q1$ and $Q2$ provide only sequential mass filtering.

Other hybrid tandem systems have been developed for research and analysis, combining quadrupole, magnetic, and electrostatic components. A review of their use in organic chemical investigations has been prepared by Beynon et al. (1982). Basically, the quadrupole selects ions on the basis of the ion mass-to-charge ratio (m/z), magnetic sectors analyze on a momentum-to-charge ratio (mv/z), and electrostatic sectors provide filtering on the basis of a kinetic energy-to-charge ratio (mv^2/z). Thus, combinations of these analyzers can be used in the vast spectrum of studies that encompass isotopic abundance measurements, molecular structure, charge states, and ion kinetic energy related phenomena.

STATIC GAS ANALYZERS

In the usual mass spectrometric analysis of gases, a large fraction of the gas sample makes no contribution to the measurement, but is merely pumped away by the vacuum system. In the "static" mode of operation first introduced by Reynolds (1956), the gas sample is introduced into the analyzing tube, and the system is then isolated from all dynamic pumping. Under appropriate conditions, the gas sample may last for hours, a much larger fraction of the gas molecules can be ionized, and samples as small as 10^7 atoms of noble gases can be analyzed. In connection with $^{12}C/^{13}C$ ratios, static mode measurements have also been made of CH_4 and CD_4 ions (Fallick et al., 1980). Prior to sample introduction, of course, the spectrometer must be pumped down to approximately 10^{-10} torr and adsorbed gases must be removed by an extended bake-out of the analyzer tube at an elevated temperature. An all-metal vacuum system is required, with bakeable valves, flanges, and electrical feed-throughs.

The work reported by Merrill (1974) is of special interest because it essentially combines an "ion implantation" technique with the "static" analysis of Xe and Kr. Basically, the experiment involved determining the

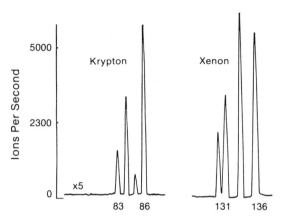

Figure 4.15 "Static" ^{235}U fission product spectrum of Xe and Kr, after 7 h without dynamic pumping of a mass spectrometer (Merrill, 1974).

fractional number of ^{235}U atoms in a small sample (~1 μg) that were converted to fission product gases upon neutron irradiation. The sample was first placed in aluminum foil in a manner so that upon neutron irradiation, the Xe and Kr fission products were completely captured or "ion implanted" in the Al foil. Subsequent to the irradiation, the Al foil was placed into a "static" 6 in. radius-of-curvature spectrometer equipped with an electron multiplier. After pump-down and bake-out, the Al foil was totally evaporated by a heated filament, thereby releasing all fission product gases. Figure 4-15 shows the resulting Xe and Kr spectra, virtually free of background peaks and after 7 h without dynamic pumping. This technique thus appears to provide an important option for the study of noble gases relating to dating, cosmology, and meteoritic analysis, or even for ion implantation experiments where a "partial pressure" of a countable number of implanted atoms can be released by heating from a host substrate.

ATMOSPHERIC PRESSURE ANALYZERS

Both magnetic and quadrupole spectrometers can be used as "atmospheric pressure" analyzers, provided they are equipped with an appropriate ionization source and an adequate pumping system. The magnetic instrument of Davis (1977), utilizing a capillary inlet and a heated filament for analyzing atmospheric particulates by surface ioniziation, is in this category (see Chapter 15). Other magnetic sector instruments have also been used in pioneering studies of combustion, ions formed in flames (Knewstubb and Sugden, 1958), the atomic and molecular composition of the upper atmosphere (Nier et al., 1964), etc. The advent of quadrupoles, however, greatly stimulated the de-

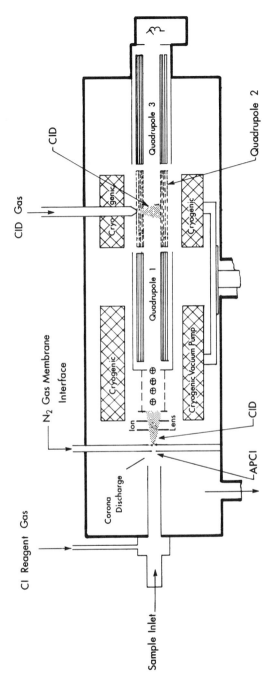

Figure 4.16 Schematic diagram of a triple quadrupole atmospheric pressure ionization mass spectrometer, with corona discharge source and a high-capacity cryogenic pumping system. (Sciex, Inc.)

velopment of commercial atmospheric pressure ionization mass spectrometers (API/MS) for chromatography, trace atmospheric studies, breath analysis, and other applications.

In most contemporary instruments, the API source is either a thin foil ^{63}Ni beta ray emitter or a corona discharge at the tip of a needle. The ^{63}Ni foil, sometimes shaped in a cylindrical configuration, emits electrons with a broad distribution in energy. The "most probable" energy will be of the order of 20 keV, so that there is limited control of the ionization region in the source. With the corona discharge, there is greater spatial definition of the ionization volume, and there exists the option of operating the discharge in a pulsed mode. In the direct air-sampling chemical ionization (CI) source (Sciex, Model 1400), the discharge occurs between a needle and an open-mesh screen electrode; it is also both current limited between 0.5 and 3 mA and computer controlled. The sample is also introduced directly through the on-axis discharge zone. In the general case, samples may be introduced directly from the air, via a carrier gas stream or from a liquid chromatographic column. In particular, API/MS has made possible the real-time detection and quantitation of many atmospheric pollutants.

The interfacing of the API source to the spectrometer is accomplished by several techniques (Proctor and Todd, 1983). One scheme uses a single pinhole orifice of ~30 μm diameter. Such a small aperture limits the vacuum pumping load, and as few as ~1 in 10^4 ions may actually enter the spectrometer. An alternative sampling method utilizes a fairly large sampling orifice (~0.1 mm), a two-stage closed-loop cryogenic pumping system of high capacity (40,000 L/s), plus a nitrogen "gaseous membrane," collimators, and ion optics—so as to substantially increase the number of ions that are available for analysis (Lane, 1982). A schematic diagram of a triple quadrupole system with a corona discharge ion source and a cryogenic pumping system is shown in Figure. 4.16.

ION SCATTERING SPECTROMETRY

Ion-scattering spectrometry (ISS) must be included within the general context of mass spectrometry, and surface-profiling capabilities of both ISS and secondary ion mass spectrometry (SIMS) have been incorporated within a limited number of instruments (Vasile and Malm, 1976). Whereas SIMS spectra are generated by the sputtered atoms that leave a surface, ISS spectra are produced by the measurement of the energies of scattered noble gas atoms that are used to probe a surface layer. At a fixed angle of incidence and for a scattering angle of 90°, the mass of an incident ion M_o, of energy E_o, is related to the energy of the scattered noble gas atom E_1 and the mass M_s of a surface atom by the simple expression:

$$\frac{E_1}{E_o} = \frac{M_s - M_o}{M_s + M_o}$$

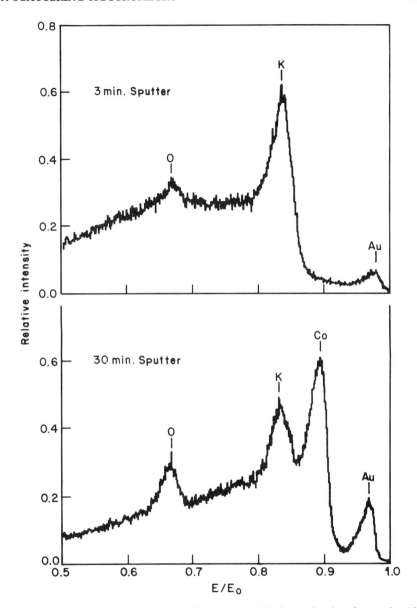

Figure 4.17 Ion scattering spectrum of a Co-hardened gold-electroplated surface, using 1500 eV ^{4}He ions as a probe. (Malm, 1982; courtesy of D. L. Malm.)

Thus, because M_o and E_o are known, a measurement of E_1 can identify the mass of the surface atom M_s that produced the ion scattering. Then by plotting $N(E)$ vs. E_1/E_o, it is possible to identify a multiplicity of surface species that contribute to the ISS spectrum. Figure 4.17 shows such a spec-

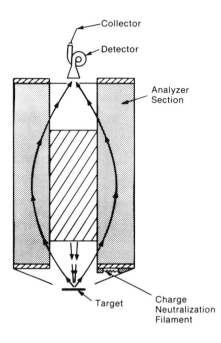

Figure 4.18 Schematic diagram of ion scattering spectrometer of high transmission for detecting surface impurities. (Malm, 1982; courtesy of D. L. Malm.)

trum for a surface using a 1500 eV ^{4}He ion beam as a probe. In this instance, a Co-hardened gold-electroplated surface was sputtered, and the low-energy ion scattering revealed relative impurity concentrations as a function of depth (Malm, 1982).

There are limitations and advantages to ISS. Surface measurements must be made under ultra-high vacuum conditions, and the elemental sensitivity is in the range of 1–0.01% of a single monolayer. In general, an incident ion beam of 1 mm or greater in diameter is used to obtain an adequate signal, and a typical spectrum can be obtained within a few minutes. The depth resolution is one monolayer, and it is independent of the surface matrix or the element. However, the sensitivity is determined by the scattering cross-section, which is a slowly varying function of the atomic number of the target element. The common probe gases include ^{4}He, ^{20}Ne, and ^{40}Ar, with the heavier ions being used to resolve the elements of high atomic number and mass. In principle, ISS can be used to detect surface isotopes (Goff and Smith, 1970), but much greater mass resolution is realized by SIMS. As with SIMS, depth-profiling rates can vary over wide limits—about one monolayer per hour (virtually nondestructive)—up to ~3000 monolayers per hour. Signal-to-background ratios for ISS as high as 10^3:1 have been attained under ideal conditions, but 10^2:1 or 10:1 ratios are more representative in the spectra obtained in general surface analyses. Factors affecting the selectivity and sensitivity of this type of surface analysis have been reviewed by Swartzfager (1984).

An ion analyzer used by Malm (1982) for obtaining data on surface impurities, etc. is shown in Figure 4.18. In this arrangement, the scattering angle is 138°, and the backscattered ions are focused onto a channeltron electron multiplier detector.

PHOTOELECTRON-PHOTOION COINCIDENCE SPECTROMETERS

Photoelectron-photoion coincidence measurements provide detailed information of the reaction:

$$M + h\nu \rightarrow M + e^-$$

by detecting mass-selected ions in delayed coincidence with energy-analyzed photoelectrons. Specifically, if the photon energy $h\nu$ is known and the kinetic energy of the emitted electron is KE_e, the excitation energy of the ionized molecule $E^*_{m^+}$ will be:

$$E^*_{m^+} = h\nu - KE_e - I + E_{th}$$

where I is the molecular ionization potential and E_{th} is the original thermal energy of the molecule. E_{th} is assumed to be zero at low temperatures, and the experimentally determined quantity $h\nu - KE_e$ is then a measure of the energy transferred in ionization. Furthermore, the coincidence signal identifies the specific photoionization events from which both the ion and photoelectron have been detected (Eland, 1980).

The spectrometric apparatus for such coincidence measurements consists of components borrowed directly from conventional mass spectrometric instrumentation. Figure 4.19 is a schematic diagram of the instrumentation used by Parr et al. (1980) and Butler et al. (1984), but other investigators have used quadrupoles or magnetic sectors for ion analysis. An ionizing light source generates the electron-ion pairs, and the photoelectrons and photoions are accelerated by a weak electrostatic field in the source region. The electrons are energy analyzed in the 127° ($\pi/\sqrt{2}$) cylindrical electrostatic analyzer, and the ions are mass analyzed by time-of-flight (TOF). The ion drift tube is about 22 cm in length, a trigger pulse is applied to draw-out and accelerate the ion, and an electronic delay is introduced in order to provide a coincident signal between the photoelectron and ion from which it came.

There are basically two modes of data acquisition for this type of spectrometry. The first uses ionizing light of a fixed wavelength (usually the 58.4 nm line from a helium discharge lamp), and the molecular internal ion energies are determined by varying the electrostatic field strength—and hence by defining the kinetic energies of the electrons. In the second mode of analysis, the electrostatic lens transmits only *threshold* electrons (zero energy electrons, which are then accelerated to only a few electron volts), and

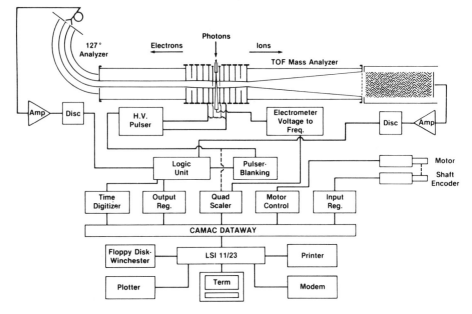

Figure 4.19 Schematic diagram and associated circuitry of a coincidence electron and time-of-flight ion analyzer (Butler et al., 1984).

the energy resolution can be set for about 0.02 eV. In this case, the photon source must be of variable wavelength.

Inghram et al. (1980) have utilized photoelectron-photoion spectroscopy to study (1) threshold photoelectron spectra, (2) fragmentation breakdown curves, (3) time-dependent breakdown curves, and (4) kinetic energy release upon fragmentation. In recent photoelectron studies of reactive transient species by Morris et al. (1984), a multichannel electron detector system was used. This system permitted both the detection of a range of photoelectron energies simultaneously and a substantial reduction in data acquisition time compared to measurements made with a single detector.

RESONANCE IONIZATION MASS SPECTROMETRY

As noted in Chapter 2, resonance ionization is a powerful technique for producing ions selectively and at a high sensitivity. By tuning a laser to frequencies corresponding to discrete, resonant electronic transitions of an element, the ionization probability can be increased manyfold. An especially attractive scheme is to combine the laser source with a pulsed mode for sample vaporization, such as sputtering. If the duty cycles of the laser and sputter source are synchronized, a very high sensitivity can be achieved. Fassett et al. (1983) have discussed several possibilities for resonance ion-

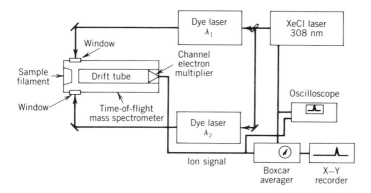

Figure 4.20 Schematic diagram of resonance ionization spectrometer (Downey et al., 1984).

ization mass spectrometry, and they indicated that a tunable ultraviolet laser plus a modified thermal source can be used for the assay of 47 elements. Surface analysis by laser multiphoton techniques has at least one advantage over conventional SIMS, inasmuch as the desorption step is separated spatially and temporally from the ionization (Zare, 1984). In some assays of gas samples and gas flows, individual atoms of many elements can be counted one by one.

The spectrometer reported by Downey et al. (1984), whose schematic is shown in Figure 4.20, utilizes (1) thermal volatilization of sample atoms from a filament, (2) multiphoton ionization, and (3) TOF mass analysis. In this work, designed to detect trace amounts of technetium (a rare element with no stable isotopes), Tc atoms are evaporated from a heated rhenium filament (1650°C). The output from the XeCl excimer laser is used to pump simultaneously two tunable dye lasers whose output pulses ($\leq$ 10 ns, and having an energy of 1–4 mJ at 20 Hz) then provide a multistep photoionization of the evaporated sample atoms. Typical flight times for ^{99}Tc ions in the drift tube are 28 μs; under typical operating conditions, 3–10 photogenerated ions per pulse are detected by the channel electron multiplier. This particular application of resonance spectrometry is of practical significance, because trace amounts of technetium that may exist in the earth's crust are expected to be accompanied by molybdenum and ruthenium. These two elements could generate isobaric interferences with conventional mass spectral analysis.

ACCELERATOR-SPECTROMETER SYSTEMS

The idea of using a high-voltage (MeV) accelerator of charged particles as an ultrasensitive mass spectrometer was conceived independently by Purser (1976) and Muller (1977). The concept of charge stripping for increasing ion

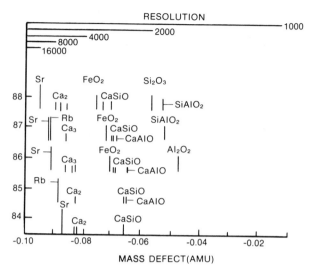

Figure 4.21 Resolution required to resolve molecular species on the basis of mass defect, with a mass range of 84–88. (Purser et al., 1979; courtesy of K. H. Purser.)

energies in a DC accelerator was envisioned by Alvarez at an even earlier date (Alvarez and Cornog, 1939; Alvarez, 1982). With conventional spectrometry, molecules must be resolved at high resolution, and because source slit apertures must be limited to obtain this resolution, ion collection efficiency is reduced. At high ion energies (~2 MeV), molecular ions can be removed completely from a mass spectrum by stripping to a high charge state ($q > 3$), which has the effect of dissociating the molecule. Basically, while atoms can lose many of their electrons and remain stable, molecular species with three binding electrons missing have a very short lifetime and suffer a "Coulomb explosion" in approximately 10^{-9} s. Furthermore, at high ion energy, effects such as ion scattering, charge exchange, charge ambiguity, and isobaric interference can be either eliminated or substantially reduced. Ion collection efficiency can also be enhanced in going from keV to MeV energies, as the effective solid angle of an ion beam varies as $1/E$. At very high kinetic energies, ions can also be in a multidimensional spectrum—using several simultaneous detection techniques such as range discrimination, energy loss (dE/dx), TOF, and coincidence methods. As a practical consequence, SIMS can be extended further, if DC accelerators are appropriately coupled to charged-particle analyzing systems (Purser et al., 1979). Figure 4.21 serves to illustrate the resolution required to resolve molecular impurities around mass 88 on the basis of conventional instrumentation. With an accelerator-spectrometer system, these molecular species can be removed from the spectrum, and only an atomic spectrum remains.

Figure 4.22 is a schematic diagram of the University of Rochester ac-

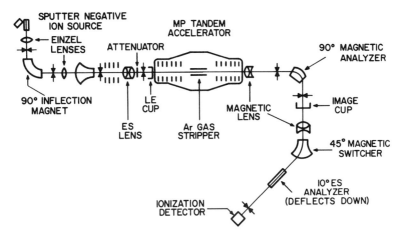

Figure 4.22 Schematic diagram of the University of Rochester tandem Van de Graff accelerator with magnetic and electrostatic analyzers that permit it to function as an ultrasensitive mass spectrometer. (Gove et al., 1980; Elmore, 1985; courtesy of D. Elmore.)

celerator that has been used extensively as an ultrasensitive mass spectrometer for the detection of long-lived radioisotopes (Gove et al., 1980; Elmore, 1985). A 20 keV cesium beam bombards the sample to be analyzed; and the negative ions that are produced are extracted and momentum analyzed in a 90° inflection magnet with a mass resolution of $M/\Delta M = 200$. The ions are then further accelerated and injected into the tandem Van de Graff accelerator. A differentially pumped gas cell within the accelerator serves to strip molecules and ions of several electrons, which then causes the molecules to dissociate. These multiply charged positive ions are then accelerated in the second half of the tandem facility, and the 90° analyzing magnet is programmed to select a specific mass-energy product ME/Q^2. An additional charged-particle filtering is provided by the 45° deflecting magnet (that eliminates scattered ions), and the final electrostatic deflection ($E/\Delta E \sim 200$) serves to limit E/Q ratios so that the ions reaching the final detector have unique velocities and M/Q values.

Such accelerator-spectrometer systems have recently become especially powerful tools for detecting trace radioisotopes, and important applications are cited in Chapters 8 and 20.

REFERENCES

Alvarez, L. W., *Physics Today, January*, 25 (1982).

Alvarez, L. W., and R. Cornog, *Phys. Rev.*, 56, 379 (1939).

Beynon, J. H., F. M. Harris, B. N. Green, and R. H. Bateman, *Org. Mass Spectrom.*, *17*, No. 2, 55 (1982).

Butler, J. J., D. M. P. Holland, A. C. Parr, and R. Stockbauer, *Int. J. Mass Spectrom. and Ion Processes, 58*, 1 (1984).

Chakrabarty, T., S. Kundu, and S. B. Karmohapatro, *Nucl. Instr. Methods, 199*, 631 (1982).

Ciupek, J. D., J. W. Amy, R. G. Cooks, and A. E. Schoen, *Int. J. Mass Spectrom. Ion Phys., 65*, 141 (1985).

Cooks, R. G., ed. *Collision Spectroscopy*, Plenum, New York, (1978).

Cooks, R. G., M. S. Story, G. Jung, and P. Dobberstein, Finnigan-MAT, patent pending, Germany, 1982, United States, 1983.

Dagenhart, W. K., and T. W. Whitehead, *Nucl. Instr. Methods, 85*, 215 (1970).

Davis, W. D., and J. L. Mewherter, *Cosmochimica Acta, 26*, 681 (1962).

Davis, W. D., *Envir. Sci. Technol., 11*, 593 (1977).

Dawson, P. H., J. B. French, J. A. Buckley, D. J. Douglas, and D. Simmons, *Org. Mass Spectrom., 17*, No. 5, 205 (1982).

Downey, S. W., N. S. Nogar, and C. M. Miller, *Int. J. Mass Spectrom. and Ion Processes, 61*, 337 (1984).

Eland, J. H. D., in *Advances in Mass Spectrometry*, Vol. 8A, ed. A. Quayle, Heyden and Son Ltd., London (1980), p. 17.

Elmore, D., private communication (1985).

Fallick, A. E., L. R. Gardiner, A. J. T. Jull, and C. T. Pillinger, in *Advances in Mass Spectrometry*, Vol. 8A, ed. A. Quayle, Heyden and Son Ltd., London (1980), p. 309.

Fassett, J. D., J. C. Travis, L. J. Moore, and F. E. Lytle, *Anal. Chem., 55*, 765 (1983).

Gall, R. N., N. S. Pliss, and A. P. Shcherbakov, in *Advances in Mass Spectrometry*, Vol. 8A, ed. A. Quayle, Heyden and Sons Ltd., London (1980), p. 1893.

Goff, R. F., and D. P. Smith, *J. Vac. Sci. Technol., 7*, 72 (1970).

Gove, H. E., D. Elmore, R. Ferraro, R. P. Beukens, K. H. Chang, L. R. Kilius, H. W. Lee, A. E. Litherland, and K. H. Purser, *Nuc. Instr. and Methods, 168*, 425 (1980).

Harris, F. M., G. A. Keenan, P. D. Bolton, S. B. Davies, S. Singh, and J. H. Beynon, *Int. J. Mass Spectrom. and Ion Processes, 58*, 273 (1984).

Inghram, M. G., and R. J. Hayden, *A Handbook in Mass Spectrometry*, National Academy of Science, National Research Council, Nuclear Science Series, Report No. 14, Washington, D.C. (1954).

Inghram, M. G., G. R. Hanson, and R. Stockbauer, *Int. J. Mass Spectrom. Ion Phys., 33*, 253 (1980).

Ioanoviciu, D., *Int. J. Mass Spectrom. Ion Phys., 12*, 115 (1973).

Knewstubb, P. F., and T. M. Sugden, *Nature, 181*, 474 (1958).

Kunihiro, F., M. Naito, Y. Naito, Y. Kommel, and Y. Itagaki, *32nd Ann. Conf. on Mass Spectrom. and Allied Topics*, San Antonio (May 27 to June 1, 1984), p. 523.

Lane, D. A., *Envir. Sci. Technol., 16*, 38A (1982).

Malm, D. L., *24th Rocky Mt. Conf.*, Denver (Aug. 1982).

Matsuda, H., *Int. J. Mass Spectrom. and Ion Processes, 66*, 209 (1985).

McLafferty, F. W., ed. *Tandem Mass Spectrometry*, Wiley, New York (1983), p. 271.

Merrill, G. L., Jr., *Int. J. Mass Spectrom. Ion Phys., 13*, 281 (1974).

Morris, A., N. Jonathan, J. M. Dyke, P. D. Francis, N. Keddar, and J. D. Mills, *Rev. Sci. Instrum., 55*, No. 2, 172 (1984).

Muller, R. A., *Science, 196*, 489 (1977).

Nier, A. O., J. H. Hoffman, C. Y. Johnson, and J. C. Holmes, *J. Geophys. Res., 69*, 986 (1964); *69*, 4629 (1964).

Parr, A. C., A. J. Jason, R. Stockbauer, and K. McCulloh, in *Advances in Mass Spectrometry*, Vol. 8A, ed. A. Quayle, Heyden and Son Ltd., London London (1980), p. 62.

Proctor, C. J., and J. F. J. Todd, *Org. Mass Spectrom., 18*, No. 12, 509 (1983).

Purser, K. H., U. S. Patent 4,037,100 (filed March 1, 1976; issued July 19, 1977).

Purser, K. H., A. E. Litherland, and J. C. Rucklidge, *Surface and Interface Analysis, 1*, No. 1, 12 (1979).

Rasekhi, H., and F. A. White, *Int. J. Mass Spectrom. Ion Phys., 8*, 277 (1972).

Reynolds, J. H., *Rev. Sci. Instrum., 27*, 928 (1956).

Rourke, F. M., J. C. Sheffield, W. D. Davis, and F. A. White, *J. Chem. Phys., 31*, No. 1, 193 (1959).

Schoen, A. E., J. W. Amy, J. D. Ciupek, R. G. Cooks, P. Dobberstein, and G. Jung, *Int. J. Mass Spectrom. Ion Phys., 65*, 125 (1985).

Sheffield, J. C., and F. A. White, *Appl. Spectrom., 12*, No. 1, 12 (1958).

Siegbahn, K., and N. Svartholm, *Nature, 156*, 872 (1946).

Slayback, J. R. B., and M. S. Story, *Ind. Res. and Development*, February, 129 (1981).

Spaniol, G. C., T. W. Whitehead, and F. A. White, *Nucl. Instr. Methods, 117*, 363 (1974).

Svec, H. J., and G. D. Flesch, *Int. J. Mass Spectrom. Ion Phys., 1*, 41 (1968).

Swartzfager, D. G., *Anal. Chem., 56*, 55 (1984).

Vasile, M. J., and D. L. Malm, *Int. J. Mass Spectrom. Ion Phys., 21*, 145 (1976).

White, F. A., and T. L. Collins, *Appl. Spectry., 8*, 169 (1954).

White, F. A., T. L. Collins, and F. M. Rourke, *Phys. Rev., 101*, 1786 (1956).

White, F. A., F. M. Rourke, and J. C. Sheffield, *Appl. Spectry., 12*, No. 2, 46 (1958a).

White, F. A., F. M. Rourke, and J. C. Sheffield, *Rev. Sci. Instr., 29*, No. 2, 182 (1958b).

White, F. A., J. C. Sheffield, and F. M. Rourke, *Appl. Spectry., 17*, No. 2, 39 (1963).

White, F. A., and L. Forman, *Rev. Sci. Instr., 38*, No. 3, 355 (1967).

White, F. A., and J. D. Stein, *Int. J. Mass Spectrom. Ion Phys., 5*, 205 (1970).

Whitehead, T. W., and F. A. White, *Nucl. Instr. Methods, 103*, 437 (1972).

Yost, R. A., and C. G. Enke, *J. Am. Chem. Soc., 100*, 2274 (1978).

Yost, R. A., C. G. Enke, E. McGilvery, D. Smith, and J. D. Morrison, *Int. J. Mass Spectrom. Ion Phys., 30*, 127 (1979).

Yost, R. A., and C. G. Enke, *Anal. Chem., 51*, 1251A (1979).

Zare, R. N., *Science, 226*, 298 (1984).

Zashkvara, V. V., *Sov. Phys. Tech. Phys., 25*, No. 7, 852 (1981).

CHAPTER FIVE

DETECTION
OF ION BEAMS

With the advent of improved primary sensors and solid-state circuitry, it is now possible to monitor several mass-resolved ion beams simultaneously. A high-resolution "electronic photographic plate" has not yet made its appearance, but p-n junction arrays, microchannel plates, and charge-coupled devices make it possible to obtain real-time measurements for several isotopes and/or elements. This chapter presents a brief review of detectors for monitoring ion beams in either the DC or pulse-counting mode. Collectively, these detectors permit mass spectral data to be acquired at high sensitivities and over a wide dynamic range.

SINGLE AND MULTIPLE FARADAY COLLECTORS

The basic technique for detecting ion currents in the range 10^{-6} to 10^{-13} A is to allow the beam to stop in a hollow conducting electrode. This type of beam monitor is often termed a "Faraday cage" after the eminent nineteenth-century physicist Michael Faraday, who employed a metal bucket for collecting electric charges. In a Faraday cage, a mass-resolved ion beam, defined by a slit, passes into the interior of a small cylinder that is appropriately insulated. The slit edges should be sharp to prevent ion reflection or scattering, and the dimensions of the cylinder are usually chosen so that secondary electrons, which are generated by stopping of the ion beam, are effectively trapped within this electrode. A suppressor electrode with a negative bias of about 20 V is often used to prevent secondary electrons from drifting out of the cylinder. It is important that the collecting electrode be shielded from light to prevent photoelectron emission or ionization from spurious ions that may drift to the cylinder, and also from varying magnetic or electric fields that can induce a sizeable voltage in any electrode of small capacitance. For optimum results, these electrodes should be connected directly to a suitable preamplifier. The development of current-to-frequency

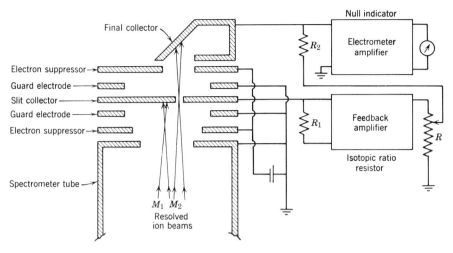

Figure 5.1 Schematic of dual ion beam collector (Nier et al., 1947).

converters has also made possible more accurate DC current measurements, with minimum drift, improved stability, and reduced response time.

There are many instances in which greatly improved isotopic ratio measurements may be achieved by the simultaneous collection of two or more resolved ion beams. A dual collector may be used to (1) circumvent errors due to variations in ion beam production; (2) detect ratio changes arising from the preferential depletion of isobaric impurities; (3) conserve samples; (4) decrease the time required for analysis; and (5) improve precision generally by the cancellation of certain time-dependent errors that are associated with the analysis.

There are probably as many variations of multiple-beam recording as there are laboratories that employ the method. The basic scheme, however, as devised by Nier et al. (1947), is shown in Figure 5.1. The multiple-slit arrangement permits one ion beam to proceed through a narrow defining slit to one collecting electrode; a second beam is intercepted on a plate. Both electrodes are close to parallel plate suppressors, the function of which is to turn back secondary electrons produced by the incidence of the high-energy ions. A negative bias of 20 V is sufficient to repel most secondaries. This potential is also sufficiently low so that the primary ions are not deflected—even if slight electrostatic asymmetries occur. If the two electrodes are connected to two stable amplifiers with variable input resistors (R_1 and R_2), a voltage balance can be achieved by suitable resistor values; and a null reading will indicate an ion beam ratio that is proportional to R_2/R_1, but which will also depend upon amplifier gain and feedback loop voltages. For accurate analysis, isotopic standards must be employed to achieve normalization.

It will be noted in Figure 5.1 that if the polarity of the secondary electron suppressor is reversed and made positive, some amplification can be achieved. (This method is usually used for a single collector only.) The amplification ratio is equal to $(I_p + I_s)/I_p$, where I_p is the positive ion current and I_s is the secondary electron current leaving the electrode, thus effectively producing an additional positive current. A multiplication of about 3 may be achieved in this fashion, but this artifice is only resorted to for marginal situations. In very precise work, account must also be taken of possible errors arising from the ejection of secondary positive ions, although this effect is usually neglected because it is small ($< 0.1\%$ of the secondary electron current).

While dual collectors comprise the majority of the multiple ion beam detector schemes, systems having many elements have been designed and constructed. A 180° mass spectrometer at the Lawrence Radiation Laboratory used 94 equally spaced cups placed along the magnetic focal plane that served as collecting electrodes (Ruby et al., 1965). The objective was to record a complete mass spectrum from only a few pulses of a high-current ion source. The collector cups were fastened mechanically to the vacuum analyzing chamber, but were electrically insulated from it. Each cup was connected to a 1000 pF silver-mica capacitor having an exceptionally low leakage current. The recording of collected charge was accomplished by a trolley pickup electrometer programmed to sense the potentials across the 94 capacitors. Such a multichannel scheme does not lend itself to great precision, but spectra that included ions of H^+, D^+, C^+, O^+, Ti^{2+}, and Ti^+ were monitored with this integral-current type of method.

More recently, a multiple Faraday cup (five collector) system, which has been used for determining argon isotope ratios in a static mass spectrometer, points out advantages of simultaneously measuring several ion beams in special situations (Stacey et al., 1981). Ions at masses 36–40 were collected on discrete collectors, each with its own electrometer amplifier and analog-to-digital measuring channel. Isotope ratios were observed to stabilize in approximately 10 s after sample admission, and errors normally attributed to variations in ion beam intensity were minimized. Changes in measured ratios were monitored as being linear, typically within $\pm 0.02\%$, and the time required for a complete analysis was reduced to less than 10 min. Commercial assemblies of Faraday cups for multiple-isotope monitoring are now available that can be adjusted to the exact mass separation of several isotopes of an element, using micrometers that are external to the vacuum housing of the detector.

ION-SENSITIVE EMULSIONS

The characterization of ion-sensitive photoplates for mass spectrometry has been reviewed in the classic paper by Honig et al. (1967) and a companion paper by Woolston et al. (1967). Most emulsions consist of a suspension of silver bromide grains in a gelatin spread on glass plates or other supporting substrates. The gelatin is somewhat permeable and allows the developing solution to find its way to the bromide crystals and to react with them. Trace amounts of sulfur in the gelatin also play an active role in the sensitizing process by making available sulfide sensitivity specks. The threshold energy necessary to render a grain developable differs for photons and ions. The band gap of silver bromide is 2.6 eV, so that the absorption of one photon having a wavelength less than about 4600 Å is sufficient to raise an electron from the valence band to the conduction band and to create an electron-hole pair. However, absorption of some 10–20 photons appears to be the threshold for producing an aggregate of silver atoms that will make a crystal grain developable.

There are less data available for ions with respect to the minimum absorption energy needed to produce one conduction electron in a silver halide grain. However, it has been suggested by several investigators that an order-of-magnitude increase in energy is required of ions over photons. The energy density to produce a detectable image is also high. Owens (1966) has reported that the minimum detectable line image (~0.1 mm^2 in area) requires about 10^4 ions of 15 keV energy; this figure is for ions of about 100 amu on Ilford Q_2 emulsions. In a practical case, it is difficult to estimate the number of ions of a given energy that are required to produce a developable grain. Grains near the emulsion surface can conceivably be made developable from the impact of a single energetic ion. Grains at greater depths in the emulsion may require successive ion collisions, or they may be made developable by a weak luminescence of the gelatin.

There exists some choice as to the type of emulsion best suited for ion detection. A high sensitivity requires large silver halide crystals, a relatively low gelatin content, and a thin emulsion layer on the backing plate. Large grains, however, result in spectral lines having a granular appearance, so that for high-resolution spectrometry, a finer grain emulsion may be a preferred choice. The gelatin content is an important parameter, as it will determine what fraction (on the average) of the kinetic energy of the incident ion is given up to the gelatin, or to the bromide crystal in a primary collision. In the usual type of emulsion, many ions will be stopped in the gelatin alone because of the very limited range of heavy ions. Ilford Q-type emulsions have a high concentration of silver bromide crystals at the emulsion surface, so that there is an increased probability for producing a latent image from a direct ion impact.

More recently, ion-sensitive plates have been made available commercially that are produced by the direct vapor deposition of a thin silver bromide film on a glass substrate (Masters, 1969). An electron-microscopic examination of these films shows the surface to be a continuous layer of silver bromide crystals with contiguous boundaries and free of voids. These evaporated films are very thin (0.2 μm), but this dimension exceeds the range

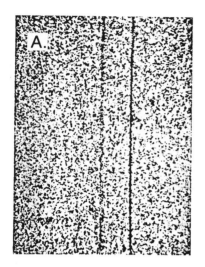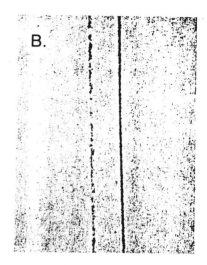

Figure 5.2 Microdensitometer comparison of (*a*) conventional emulsion and (*b*) vapor-deposited AgBr film. (Masters, 1969.)

of most ions so that maximum sensitivity is realized. The absence of the gelatin binder improves the dimensional stability of the ion line image, because swelling is avoided and the thin emulsion reduces fog and enhances resolution. Figure 5.2 compares line sharpness and signal-to-noise of a conventional emulsion plate and the direct vapor deposition of an AgBr film for identical ion exposures at $m/z = 311$ and 312.

Other important considerations relevant to emulsions relate to their latitude or range of measurable ion densities, mass dependence, and calibration. In the range of ion intensities that can be recorded, photographic emulsions excel. If the minimum detectable number of ions is 10^4 and a strong spectral line is caused by 10^{13} ions, this represents a sensitivity range of 1 billion for a given element. There will also be a mass dependence on sensitivity, which should be taken into account; in general, emulsion sensitivity decreases for increasing mass number. Burlefinger and Ewald (1963) have reported a sensitivity of Ilford Q_1 emulsions to be proportional to $(2/M)^{\frac{1}{2}}$. Also, monatomic ions and ions comprised of a small number of atoms produce a denser image than do ions of the same mass and energy, but of a more polyatomic nature (Vourdos et al., 1972). In any practical case, however, rather extensive calibrations are imperative for quantitative work, and standards of metals or semiconductors of known composition must be available for comparison. The readout of data stored on ion-sensitive emulsions is analogous to that for optical and x-ray spectrometry. Commercial dual-beam microdensitometers are available that provide analog-to-digital conversion, with suitable options for molecular weight determinations and isotopic measurements. The spatial resolution will depend upon the density of the emulsion, as well as on parameters associated with the spectrometer, but resolutions in the micron range can be obtained. A microdensitometer and the digital image processing of photoplates used in spark source mass spectrometry have been described by Naudin et al. (1983).

ELECTRON MULTIPLIERS

Electron multipliers of several designs have greatly extended the scope of measurements that are presently being made in mass spectrometry. Their high-current gain, nanosecond response time and use in multiple detector configurations have not only made possible the detection of many transient phenomena, but these devices have also conditioned the design of a new generation of mass/charge/energy analyzers that are characterized by single-particle counting and that can be applied to

1. Lifetime measurement of nuclear, atomic, and molecular excited states.
2. Ion-ion, ion-electron, and photon-ion studies.
3. Coincidence ion-electron spectroscopy.
4. Laser-induced chemical reactions.
5. Ion kinetic energy spectrometry.
6. Time-resolved photodissociation kinetics.
7. Laser and ion microprobe assays of solids.
8. Velocity distributions of particles from supersonic nozzles.

In general, the need for very fast-response ion detectors has also been prompted by the advent of new modes of ionization. With conventional electron-impact (EI) sources, the residence time of ions in an ion source is about 1 μs, which is long compared to the vibrational periods of chemical bonds (10^{-13} to 10^{-14} s). Thus, the EI source tends to give an integrated view of the many reactions that transpire in a time span of 10^{-14} to 10^{-6} s (Nibbering, 1982), and fast reaction events are obscured. With collisional activation, lasers, and ^{252}Cf fission fragment-induced ionization, much shorter time-resolved studies are possible.

Electrostatic Focusing

Electrostatic-focusing multipliers are widely used in spectrometry as DC amplifiers and for pulse counting. Figure 5.3 shows two configurations that are available commercially with an optional number of electron-multiplying stages or dynodes. The overall gain G is

$$G = pq^n$$

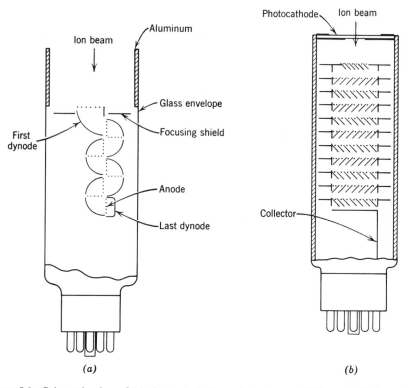

Figure 5.3 Schematic view of (*a*) electrostatic-focused electron multiplier and (*b*) "venetian blind" configuration.

where p is the number of secondary electrons produced on the first dynode by ion impact, q is the average secondary electron yield per dynode, and n is the number of multiplying stages. The parameter p is dependent on the dynode material, its surface state and degree of oxidation, the ion kinetic energy, and the ion specie. The secondary electron yield per stage will depend on the dynode material and the interstage voltage. Almost any good commercial multiplier can be used for integral-current measurements, but the gain will usually decrease markedly after exposure to the atmosphere.

For very low-current measurements, there are many advantages to using multipliers in a pulsed mode; and for the isotopic measurement of small samples, ion counting is now routine (Ihle and Neubert, 1971). In general, electrostatic-focusing multipliers can be successfully operated in a pulsed counting mode for quantitative mass spectral work provided that

1. The first dynode has a suitable ion-secondary electron yield.
2. A sufficient number of electron-multiplying stages are used with reasonable gain per stage.

3. The multiplier is shielded from magnetic fields.
4. The arrival rate of ions is restricted to values commensurate with the resolving times of both multiplier and associated circuitry.

In the usual multiplier, the total transit time for an electron will be between 10^{-7} and 10^{-8} s, and the pulse rise time will be in the 10^{-9} s range.

Some immediate advantages arise from counting these ion-induced pulses. First, much of the "dark current" of a multiplier may be eliminated. A discriminator will usually be used to reject all but the large pulses arising from the first or second dynode. The discriminator will thus bias out all pulses arising from thermionically emitted electrons of the latter stages of the multiplier. An improvement of at least 10 in signal-to-noise ratio is usually achieved. Second, the multiplier is not so sensitive to voltage fluctuations of the multiplier voltage source as when operated in a DC fashion. Low-frequency noise is filtered out, and recording of pulsed currents is generally much more convenient and amenable to statistical analysis. BeCu or AgMg are dynode materials that provide adequate gain, and their work function (~ 4 eV) is sufficiently high so their thermionic emission is low at room temperatures. In some mass spectrometers, the use of an electrostatic lens between the final collector slit and the electron multiplier provides an improved signal-to-noise ratio by filtering out scattered ions and neutrals (Brenton et al., 1979). As discussed in a subsequent section, scintillation/photomultipliers are also excellent ion counters, but a very high voltage (~ 40 kV) is required for pulse counting.

In some instances, a single-electron multiplier can be operated in two modes of operation. For example, the scintillation ion counter of Daly et al. (1973) has been used in one mode to record an entire mass spectrum; but by means of a fast-switching system, it can also be used to respond only to ions whose energies are within a preselected range. This second mode of operation permits a unique means for observing metastable decompositions.

Channel Electron Multipliers

The continuous dynode or channel electron multiplier is widely used in quadrupole and magnetic sector spectrometers. One dynode configuration is shown in the schematic of Figure 5.4. This type has several advantages over discrete dynode structures in some applications: (1) dark currents are limited to ~ 0.5/s; (2) the resistive surface coating is stable and can be exposed to air without appreciable degradation; and (3) the physical size can be made much smaller than most discrete, multistage multipliers. Power consumption is also low, and multipliers can be made that have a narrow gain distribution of output pulses. The dynode structure is made of glass that is heavily lead-doped, and the secondary emissive surface is activated and stabilized by a suitable heat treatment. The resistive surfaces exhibit resistances of 10^8 to

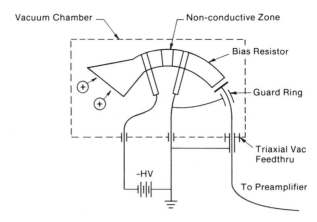

Figure 5.4 Continuous dynode (Channeltron®) electron multiplier designed for use in mass spectrometry. (Galileo Electro-Optics Corp.)

$10^9\ \Omega$, and electric contacts (usually chromium) are deposited on both ends of the channel for making connections to a voltage source.

Channel multipliers designed to operate in the analog mode have typical bias currents of 40 μA at 2000 V, so that large input currents can be drawn. In the pulse-counting mode, the bias currents will be much smaller; a gain of 10^8 can be realized with an operating voltage of 2000 V and a bias current of only ~3 μA. A helical-type Channeltron® multiplier with electric connections for pulse counting is shown in Figure 5.5. When used in conjunction with a coaxial anode and cable that minimize pulse reflections, this type of detector has a rise time of about 0.5 ns. Gain, count rate, and general operating characteristics of these devices have been reported by Kurz (1979). The absolute efficiences for particle counting of H^+, He^+, Ne^+, Ar^+, Kr^+, and Xe^+ ions for energies in the range of 4–15 keV have been reported by Fricke et al. (1980). The gain and pulse height distributions are affected by

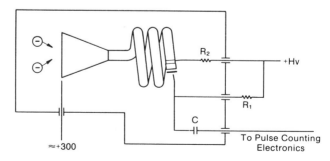

Figure 5.5 Helical-type Channeltron® designed for use in the pulse-counting mode. (Galileo Electro-Optics Corp.)

magnetic fields, but these detectors have been used in magnetic fields up to 300 G (Hahn et al., 1972).

Magnetic

Magnetic or crossed-field multipliers were first developed commercially as the basic detector for Bendix time-of-flight (TOF) spectrometers (Wiley and McLaren, 1955). This device has a thin metallic oxide film coated on glass-backing plates of approximately 10 cm in length. When a high voltage is applied across the terminals of this high-resistance strip, the high-voltage gradient causes electrons to multiply in cascade along the length of the electrode. Another crossed-field magnetic multiplier of high gain for its physical size used dynodes made of bulk semiconducting silicon (White et al., 1961a). Magnetic multipliers are now available commercially with gains of 10^3 or greater to feed into analog electronics, and with gains of 10^7 and greater for use in a pulse-counting mode. The gain is easily variable by changing the voltage or magnetic field. Typical operating characteristics for a voltage of 2000 V are: (1) a gain of 3×10^6; (2) a bias current of 10 μA; and (3) a maximum average output current of 0.5 μA. The dynode strip input cathode material is tungsten with an input area of ~0.5 cm^2.

SCINTILLATION/PHOTOMULTIPLIER DETECTOR

The ion-counting scheme first devised by Daly (1960) continues to be highly useful for low current detection. After emerging from the detector slit of a mass spectrometer, ions are accelerated and impinge upon a secondary emitting electrode that is maintained at a high negative potential (~40 kV).These secondary electrons are accelerated by the same electrostatic field and strike a phosphor that is optically coupled to a photomultiplier external to the vacuum system (Fig. 5.6). This scheme has several advantages over scintillation methods that were attempted by earlier investigators to detect heavy ions directly. First, the phosphor is impacted by electrons that produce less damage than the beam of primary ions. Second, the conversion efficiency to light quanta in an organic phosphor is many times larger for an electron than for a heavy positive ion. Finally, a high secondary electron emission is obtained because of the high ion energy. As a result, it has been possible to achieve a very high signal-to-noise ratio. Daly (1960) has reported that when a thallium ion releases five secondary electrons, which are subsequently accelerated through 40 kV and deliver a total kinetic energy of 200 keV to the phosphor, about 90 photoelectrons are released. It is thus possible to operate the phototube at low gain, and effective noise levels as low as 4 $\times$ 10^{-20} A have been achieved. A good phosphor has a short decay (3 ns) (Young et al., 1967), so there is no sacrifice in the ion-counting rates that may be observed. Another important advantage of this method is that the

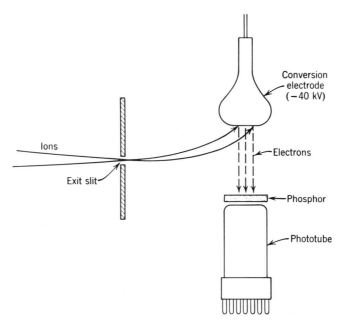

Figure 5.6 Scintillation/photomultiplier ion-counting scheme (Daly, 1960).

activated surfaces of the photomultipliers are never exposed to the atmosphere, thus greatly extending the useful tube life.

This scintillation/photomultiplier detector has also been coupled with a second low-gain photomultiplier (Stoffels et al., 1978). In its normal mode of operation, it is used for pulse-counting ion currents of 10^{-20} to 10^{-13} A, as in the original Daly arrangement. However, the dynamic range of this system is extended by having the scintillator light-coupled to a second photomultiplier whose output is fed to an electrometer. This second photomultiplier is operated with a gain of $\sim 10^3$, and the input impedance to the electrometer is reduced to $10^7 \, \Omega$. As a result, the response time is reasonably short, and the dynamic range of the system is increased several orders of magnitude without sacrificing sensitivity. Commerical versions of the Daly detector are also now available with a Faraday cup collector, so that ion currents can be monitored in both the DC and pulse-counting modes.

ION-TO-ELECTRON CONVERTERS

A surface that functions to convert primary ions to secondary electrons for further amplification via secondary electron production should be characterized by (1) high secondary electron yield; (2) low thermionic emission; and (3) long-term stability. For these and other reasons, the converter surface should possess a reasonably high work function. Thus, some very high-yield compounds such as porous KCl are not attractive. However, thin films of Al_2O_3 have provided a stable yield of seven to nine secondary electrons/ incident ion, as measured for monoatomic alkali ions with 20 keV impact energy; and the polyatomic ion $Na_2BO_2^+$ yields about 16 electrons/incident ion at this same impact energy (Dietz, 1965). BeCu also has a very stable, high-work function converter surface. Its electron yields for 4.5 keV ions of 35 elements (including atomic, monoxide, dioxide, and trioxide negative ions) have been studied by Rudat and Morrison (1979) using a CAMECA ion microanalyzer. They report a yield that generally increases as the number of atoms in the ion increases, and the mass dependence on yield for each ion type follows a $m^{-\frac{1}{2}}$ function, suggesting that the secondary yield is primarily due to kinetic emission. The isotope effect on ion-electron emission has also been reported by Fehn (1974), and a further discussion of secondary electron yields is contained in Chapter 12. The usefulness of a statistical model in diagnostic studies of secondary yields has also been reported by Dietz (1967).

The physical geometry or configuration of the ion-to-electron converter may be even more important than the secondary electron yield. For the case of a single-channel detector (i.e., mass scanning with a single-electron multiplier), the optimum configuration will depend upon the type of spectrometer, the effect of the fringing magnetic field, the spacing required for the high-voltage electrode, etc. The simplest case, of course, is to have the first dynode of an electron multiplier function as the conversion stage. If the detector is operating in a fringing magnetic field, it may be desirable to design the converter so the secondary electron trajectories are collinear with the magnetic field. In weak magnetic fields, the trajectory of the primary ions will be more affected by the high local electric field of the converter, but the secondary electron "image" can undergo a significant displacement or defocusing. Also, in special situations, it may be desirable to make the mass of the ion-to-electron emitting surface quite small, even though the converter must have large enough dimensions to prevent high-voltage breakdown. If this is done, the converter surface temperature can then be easily lowered (via a thermoelectric junction and suitable high-voltage isolation) to reduce thermionic emission.

The ion-to-electron converter for simultaneously registering an extended mass spectrum presents a greater challenge. If isotopic abundance measurements only are required, the converter geometry of Figure 5.7 can be used, which shows only a cross-sectional view (Stein and Peterson, 1979). The actual dynode, however, can be a continuous one extending to more than 1 m in length, when used in conjunction with reversed bias p-n junction detectors—and an inhomogeneous magnetic sector that allows the focal plane to be remote from the physical magnetic sector boundary (see Chapter 4). For reasonable mass resolution, in which only a few centimeters of a focal plane are to be monitored, it is desirable to have the ions impact the

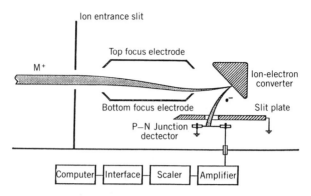

Figure 5.7 Cross-sectional view of an ion-to-electron converter with focusing electrodes (Stein and Peterson, 1979).

converter normally, as with a photoplate detector. The secondary electrons will emerge from the detector in the same direction as the incident ions. The ion-to-electron converter may then be: (1) a very thin foil; (2) a semitransparent "grid" or multiwire array; or (3) a microchannel plate—or equivalent converter-amplifier. Thin foils of ~100 Å can be fabricated that effectively serve as a converter, with electrons emerging from the rear surface—but this option is only feasible for light ions (H, Li, or C), even for ions of 10–15 keV energies due to the exceedingly short ion range. Experimental multiwire grids have also been fabricated, but it is difficult to build such grids that both intercept a major portion of the ion beam and transmit secondary electrons while maintaining a high degree of spatial resolution. The microchannel plate appears to be the best contemporary option, as it both provides amplification and maintains a tight collimation of the secondaries during the amplification process. Such plates also have a very large capillary surface area where ions can be transformed into electrons. For example, in a commercial microchannel plate of a few centimeters in diameter, with a channel diameter of 10 μm and 10^7 channels, the channel surface area will be ~3 × 10^3 cm². Only a small fraction of this large area serves to convert ions to electrons, but several interesting detector options are available, especially if the device is followed by a second amplification stage—as in a "chevron" configuration. Furthermore, very thin films can be placed over the front surface of a microchannel plate; these films can then generate secondary electrons in those applications where the primary ion energy is sufficient for the ion to penetrate the thin films.

In conjunction with ion-to-electron conversion, many schemes have been used to accelerate the electrons onto either a phosphor or nuclear emulsions. If a phosphor is used, a vidicon camera recording permits real-time data to be acquired; this type of system has been successfully applied to spark source spectrometry (Donohue et al., 1980) and other applications. The conversion of ions to electrons via a microchannel plate plus post-acceleration onto a photographic plate also enhances sensitivity, as approximately 10^4 ions are required to produce a line image of measurable intensity. However, the loss of spatial resolution is often not an acceptable trade-off.

COUNTING RATE LOSSES

Pulse counting of ions is now routine, with ion currents being recorded over a wide dynamic range. However, even for fast detector circuitry, counting losses may occur at average counting rates of about 1 MHz. Furthermore, ion currents do not consist of evenly spaced pulses; some ions will arrive at the detector separated by times that are shorter than the resolving time of the circuitry. Thus, in isotopic ratio measurements, it is important to evaluate possible counting rate losses and to place limits on this source of error. The true counting rate can be related to the observed counting rate by the expression:

$$n = \frac{n_0}{1 - n_0\tau}$$

where n and n_0 are the true and observed counting rates, respectively, and τ is the resolving time of the circuitry.

Now a fairly accurate measure of the resolution time τ can be made without recourse to sophisticated measurements. It is possible simply to select an element having two isotopes that differ in abundance by a factor of about 100. At high counting rates of the major isotope, it is possible to observe a departure from the known isotopic ratio. This ratio change results from counting-rate losses of the major isotope. If R and R_0 are the known and observed isotopic ratios for the major and minor isotopes, and n_0 is the observed counting rate of the major isotope, it can be shown that τ may be calculated from

$$\tau = \frac{R_0 - R}{n_0(1 - R)}$$

Thus it becomes a fairly simple matter to compute counting-rate losses, provided that these losses do not exceed some reasonable value (about 5%).

Dead-time corrections for secondary ion mass spectrometry (SIMS) measurements with raster scanning require a more complex analysis. Traxlmayr et al. (1984) have given correction parameters for various primary-beam intensity distributions for a circular geometry and aperture, and these corrections have been verified on a CAMECA IMS-3F instrument. The effective dead times were determined by measuring four titanium isotopic ratios relative to ^{48}Ti.

"ZERO" BACKGROUND ION DETECTION

For isotopic ratio measurements of very small samples, it is essential to detect ion beam currents in the pulse-counting mode and to acquire data with a detector that has virtually a "zero" background. It is possible to operate electron multipliers at low temperatures and thereby to decrease noise from thermionic emission; and several other schemes have been proposed to maximize signal-to-noise ratio. However, a simple ion-to-electron converter scheme can be used for the case where only a single mass-analyzed ion beam is to be monitored. Figure 5.8 is a schematic arrangement that can be used with commercially available electron multipliers, channeltrons, or microchannel plates (MCPs). Only simple electron optics are required, and the method is attractive because it can be used for detecting ions whose kinetic energies are comparable to those used in mass spectrometry. Referring to Figure 5.8, let the ions enter normally through a small slit in the

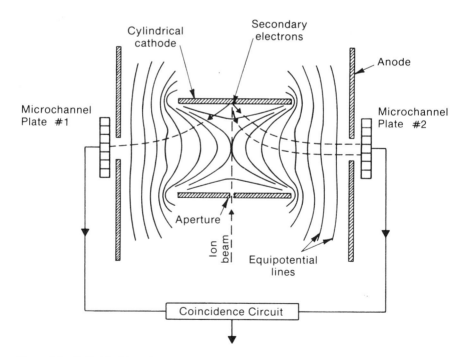

Figure 5.8 Cylindrical ion-electron converter for coincidence detection of secondary electrons produced by a single ion (White et al., 1967).

wall of the cylinder and impinge upon the inner surface. If this surface has been activated, several secondary electrons will be emitted for each incident ion, and these secondaries will have a mean kinetic energy of only a few electron volts. If accelerating electrodes (biased a few hundred volts positive) are placed at ends of the cylinder, symmetric-fringing electrostatic fields will reach into the interior of the cylinder, effectively dividing the cylinder. Hence, the secondary electrons will be focused into multipliers 1 or 2, depending upon their initial directions of motion. Coincident output pulses will be obtained if both multipliers are triggered by electrons generated from a single ion impact.

The inherent advantage of this method is its simplicity, with a virtual guarantee of zero noise pulses. With reasonably good secondary electron focusing, acceptable counting efficiencies can be expected; and there may be some situations in which the investigator is willing to trade off counting efficiency for a low background. For example, in an x-ray or high-temperature environment, secondary electrons may be generated in each individual detector, resulting in spurious pulses; but in this fast coincidence scheme, such events would have little probability of being recorded. Hence, with this or a similar coincidence scheme, it is suggested that future spectrometer ion-counting rates may have significance not only in terms of 1 ion/s (i.e., 1.6×10^{-19} A), but also in terms of a few ions per hour ($\sim 10^{-22}$ A).

P-N JUNCTIONS

Reverse bias p-n junctions are important solid-state detectors that provide excellent energy resolution for protons, alpha and beta particles, fission fragments, and ions of high kinetic energy. Their use for detecting mass spectrometric ion beam currents was first reported by White et al. (1961b). This sensor is now being used in research spectrometers as a single-ion counter or in arrays to record simultaneously mass-resolved ion beams of many isotopes. Basically, the reverse bias junction acts as a solid-state ionization chamber, with one electron-hole pair being formed per 3.6 eV of energy expended in a silicon crystal. A major difficulty in adapting junctions for ion detection is the very restricted range of charged atoms in matter. Only very light ions of high energy (> 30 keV) have sufficient range to penetrate the silicon junction "dead layer," and expend a large fraction of their energy in the "depletion region," thereby generating a large amount of ionization. Post-acceleration of mass-resolved ions is possible, but very high voltages would be required for heavy ions. An alternative is to cause the ions to strike a converter plate that will generate secondary electrons and accelerate these secondaries into the junction. As in the case of the scintillation counter, there is the further advantage of charge amplification in the secondary process itself.

A silicon p-n junction, operating under conditions of reverse bias, is an

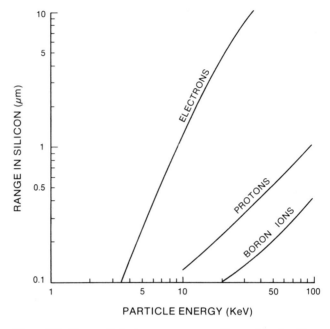

Figure 5.9 Range of electrons, protons, and boron ions in silicon.

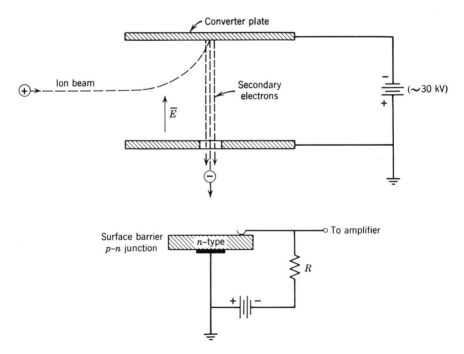

Figure 5.10 Use of a p-n junction as an ion detector.

exceedingly shallow ionization chamber. The depth of this silicon ionization chamber, or "depletion zone," can be derived from Poisson's equation—the current continuity equation—and various parameters of the material (Dearnaley and Northrup, 1966). The results, however, can be expressed simply as

$$x = 3.2 \times 10^{-5}(\rho V)^{\frac{1}{2}} \text{ cm} \qquad \text{(p-type)}$$
$$x = 5.3 \times 10^{-5}(\rho V)^{\frac{1}{2}} \text{ cm} \qquad \text{(n-type)}$$

where x is the depletion depth, V is the reverse bias (in volts) applied to the p-n junction, and ρ is the resistivity of the silicon expressed in $\Omega - \text{cm}$. For high-resistivity silicon (3000 $\Omega - \text{cm}$) and a reverse bias of only 50 V, the effective depth of the depletion zone would be approximately 200 μm. The range of electrons and ions in silicon has been found to follow the approximate relationship shown in Figure 5.9. It is thus concluded that a junction may easily be formed under conditions of reverse bias—having a depth comparable to the range of secondary electrons that are incident normally upon it for kinetic energies up to 50 keV.

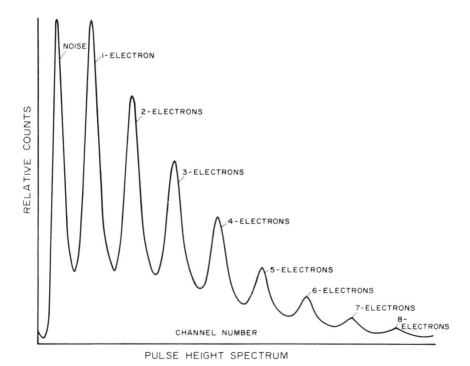

Figure 5.11 Secondary electron yield from ions as detected in a p-n junction (Krycuk and Walling, 1973).

The schematic diagram in Figure 5.10 shows the use of a reverse bias p-n junction for detecting positive ions in the pulse-counting mode. The ions are focused onto a high-voltage electrode and secondary ions are accelerated onto the junction. The secondary electrons generate electron-hole pairs in the depletion region of the junction to produce the spectrum shown in Figure 5.11, which was obtained with a multichannel analyzer. The multiple peaks correspond to the ionization produced by the incidence of 1, 2, 3 . . . n electrons, corresponding to ~30, 60, 90 . . . keV energies. It will be noted that this detection scheme is well above noise levels for even single electrons with the use of fast electronic circuitry (Krycuk and Walling, 1973).

This basic arrangement has been used with several geometrical modifications for various types of mass spectrometers (Stein and White, 1972). The p-n junctions are not affected by magnetic fields, and their small size (< 1 mm in width) allows them to be used in arrays for pulse-counting multiple ion beams. These arrays also permit *time-dependent* spectra of transient phenomena to be obtained, as the response time of these solid-state sensors is very short.

MICROCHANNEL PLATES

The microchannel plate (MCP) is an important primary sensor for detecting ultraviolet radiation, soft x-rays, electrons, and ions. Essentially, an MCP is an array of 10^4 to 10^7 capillary-type electron multipliers oriented parallel to each other and having channel diameters in the range of 10–100 μm. Figure 5.12 is a schematic view of the disc-shaped device whose overall dimensions can be tailored to specific applications; but a typical disc diameter and thickness will be 3 cm and 1 mm, respectively. This disc is fabricated from lead

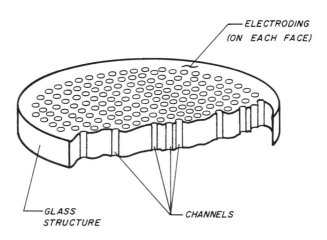

Figure 5.12 Cutaway view of a microchannel plate. (Galileo Electro-Optics Corp.)

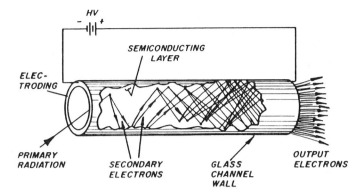

Figure 5.13 Schematic view of a single channel of a microchannel plate.

glass capillaries and is heat-treated to (1) render the channel walls semi-conducting; and (2) optimize the secondary emission characteristics of a channel. Each capillary or channel can then function as a continuous dynode structure, with the semiconducting layer of every channel providing a voltage gradient when potentials are applied across the 1 mm thick disc. Electric contacts are provided by nichrome or inconel metallic films on the opposite surfaces of the MCP, and the total resistance between these electrodes is on the order of 10^9 Ω. Figure 5.13 is a schematic of a single channel of an MCP.

The salient features of these devices have been reviewed by Wiza (1979), together with a description of the several configurations that have been fabricated for their use as ion detectors in mass spectrometry. Important features are:

1. Ultra-high time resolution (< 100 ps).
2. Spatial resolution limited only by channel dimensions and spacings. Center-to-center spacings as small as 10 μm are now available, and 5 μm spacings appear feasible (Tosswill, 1984).
3. Gains of 10^3 to 10^4 are typical for straight channels, with applied potentials of 1000 V.
4. Curved channels provide even higher gain (10^6), and curved channels tend to suppress ion feedback.
5. Detection efficiency for positive ions in the pulse-counting mode is very high (> 60%) for many ion species with kinetic energies in the range of 5–10 keV.
6. The devices are relatively insensitive to low magnetic fields.

The manufacturing technology to produce these high-gain multiplier arrays is quite remarkable. Single channels are drawn from glass fibers having

Figure 5.14 Scanning electron microscope photograph of a microchannel plate (*a*) before etching and (*b*) after etching and processing (Galileo Electro-Optics Corp.).

two components: (1) a core glass soluble in a chemical etchant; and (2) a lead glass cladding that is not soluble in the core glass etchant. Large numbers of these two-component fibers are packed together, fused, and drawn to reduce the fiber diameters to the desired dimensions. Up to 10^7 fibers are fused into a matrix, which is then sliced normal to the microchannel axis to produce the disc or plate that is shown schematically in Figure 5.12. These discs are then subjected to etching, activation, and final processing steps. The uniformity of the capillaries or microchannels produced by this technology is shown in the scanning electron microscope photograph in Figure 5.14, where the multifiber array is seen before and after etching and processing.

These devices, of course, have their limitations despite their high gain, fast response, simple geometry, low applied voltage, and relative insensitivity to low magnetic fields. The gain of MCPs has been observed to increase with an increase in the charge state of incident ions regardless of the ionic

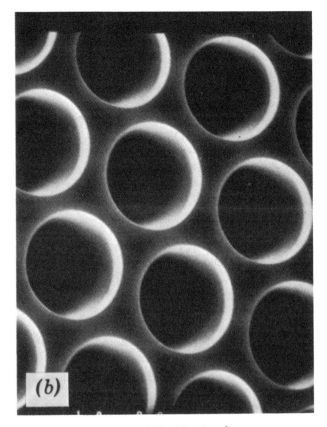

Figure 5.14   (*Continued*)

species (Takagi et al., 1983), but this is hardly a limitation. More important is a consideration of recovery time. The "recovery time" of a single microchannel is quite long ($\sim 10^{-2} - 10^{-3}$ s), because each channel capacitance must be recharged through a very high channel resistance. However, for the detection of very low ion beam currents, the MCP can function as a fast counter because a very large number of channels are available ($\sim 10^6$), and the probability of ions impacting the same channel within 10^{-3} s may be very low. Clearly, this time response constraint will be dependent upon ion beam densities and specific applications. However, for the increasing number of investigations that require the detection of single ions with high spatial resolution, the availability of 10^6 electron multipliers contained within a 3 cm wafer provides some new detection and imaging options. When even higher gains are required than can be realized in a single MCP, two MCPs have been assembled in the back-to-back configuration shown in Figure 5.15. This arrangement not only provides a very high secondary electron gain (up to 10^7), but ion feedback is almost totally suppressed. The plates are oriented

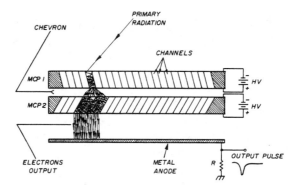

Figure 5.15   Chevron microchannel plate assembly (Galileo Electro-Optics Corp.)

so that the 8° channel angles of the two plates are in opposition, and the two MCPs are separated by about 150 μm. This chevron arrangement has been utilized extensively when a high electron output has been desired to activate a phosphor.

A recent innovation in MCP development is that of a *volume-conducting* MCP reported by Tosswill (1984). In this new design, the whole volume of glass is of uniform conductivity, but the total current is of the same magnitude as existing surface-conducting MCPs. Salient advantages include the following:

1. The direction of current flow, and therefore also the direction of the electric vector, can be controlled independently of the direction of the channel axes. Accordingly, one can establish the channel electric field at an angle to the channel axis. As a result, any ions formed by electron collision with residual gas molecules can be effectively trapped, and the MCP can deliver large uniform pulses without the onset of spurious after-pulses associated with ion-generated breakdown.

2. Spurious electron emission is reduced from local electric fields associated with surface defects.

3. The interconnection in a volume-conducting MCP of the whole array of microchannels shortens the recovery time of an individual channel following an electron avalanche. The net effect is to increase the dynamic range of these detectors when used in the pulse-counting mode.

POSITION-SENSITIVE DETECTORS

Many attempts have been made to devise the electronic equivalent of the photographic plate for the detection of both high- and low-energy-charged

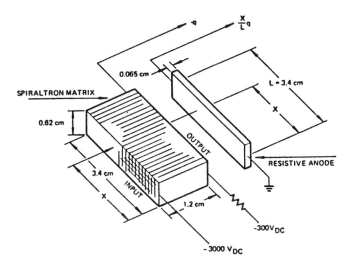

Figure 5.16 Position-sensitive detector, using 7100 channel electron multipliers and a linear-resistive anode (Carrico et al., 1973).

particles. At MeV energies, solid-state detection schemes can provide real-time, spatially resolved data of single particles. Also, fission track detectors afford another option for registering alpha particles, fission fragments, and energetic ions. These detectors make use of the fact that such particles produce narrow trails of damage when they impact certain minerals, glasses, or polymers; the localized damaged sites can then be rendered visible in an optical microscope by using a suitable etchant that will preferentially attack the damaged region (Fleischer et al., 1965).

For low-energy-charged particles, position-sensitive detection is more difficult, but considerable progress has been made in recent years. For dispersion-type mass spectrometers, advantages of real-time, position-sensitive detection include (1) a reduction in the complexity of mass-scanning systems; (2) a decrease in the data acquisition time; (3) an increase in sample sensitivity; (4) a potential increase in the accuracy of certain isotopic ratio measurements; and (5) an enhanced ability to detect transient phenomena. Among the first position-sensitive detectors to be devised for the detection of single ions in a Mattauch-Herzog type spectrometry was the matrix array of 7100 channel electron multipliers having inside diameters of 0.024 cm and center-to-center spacings of 0.036 cm (Carrico et al., 1973). The secondary electron output of this array was then incident upon a linear-resistive anode, as shown in Figure 5.16. This arrangement was first tested with a collimated ultraviolet light beam, and displacements of a few tenths of a millimeter could be resolved. For ion detection, the amplified charge is focused on the resistive anode at a distance X from the grounded anode. As indicated in the schematic in Figure 5.16, the amount of charge collected at the ungrounded end of the resistor anode is directly related to X; when terminated by a sufficiently low impedance charge amplifier, the anode pulse is proportional to Xq/L. This anode pulse is then fed to a multichannel pulse height analyzer to give the position at which the secondary electrons from a channeltron impact the resistive anode.

In contemporary position-sensitive detectors, the various components have been changed or modified, but the common elements are analogous to those in Figure 5.16, namely

1.   A charge amplification (preamplifier) matrix that generates many secondary electrons while maintaining a high degree of spatial resolution.
2.   An anode, phosphor, or other collector/converter that registers the secondary electron avalanche emitted from the preamplifier.
3.   A readout register that provides both time and coordinate data.

Position-sensitive (resistive anode) plates are now available from commercial sources (Aberth, 1981; Surface Science Laboratories, Inc.) that can provide both x and y coordinate information in both digital and analog form. The electronic charge from an MCP is incident upon the resistive plate having four leads; and the amplitude of the voltage signals, with an appropriate multichannel analyzer, provides a dynamic two-dimensional image of the incident ion beam.

PHOTODIODE ARRAYS AND CHARGE-COUPLED DEVICES

The charge-coupled device (CCD) was invented by Boyle and Smith (1970) at the Bell Laboratories in connection with a search for an alternative to magnetic bubble memories. Thus, the CCD evolved as part of a unique signal-processing scheme, using silicon technology, that is presently used for analog memory, digital memory, time delay and integration, multiplexing, and other functions. Many CCDs are fabricated from a monolithic silicon chip containing a high-number density of photodiodes, so that a CCD can also serve as an optical imaging device of high sensitivity, speed, and resolution. It is this dual capability—of signal processing and imaging—that has made the CCD such an important primary sensor, with applications in astronomy, infrared technology, optical spectroscopy, etc. Charge-coupled devices have also been used to directly image energetic electrons, 20–100 keV (Bross, 1982), and soft x-rays; but their dominant imaging applications have been in the infrared to ultraviolet photon range.

The photodiodes respond to radiation with the generation of electron-hole pairs in the depletion region of these devices; and photodiode arrays can now be fabricated with as many as 10^6 *pixels*, or picture elements, per square centimeter. "Packing densities" as high as $10^7/cm^2$ are being projected by

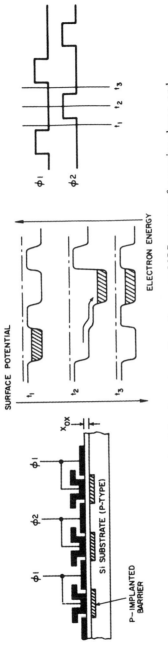

Figure 5.17 Schematic of the electrode configuration on an MOS structure for moving charge, and the formation of potential wells. (Kosonocky and Sauer, 1975; courtesy of W. F. Kosonocky.)

1990 in connection with CCDs and real-time imaging applications. The basic CCD concept of detecting the charge that is generated in these photodiodes is to have them also function as a string or matrix of closely spaced metal-oxide semiconductor (MOS) charge storage capacitors. By applying appropriate voltages to electrodes at the Si-SiO_2 interface, potential wells can be created; and by moving the potential minima, charge can be transferred to an adjacent capacitor. Thus, by applying suitable step potentials to "gate" electrodes on the surface of a high-density matrix of conductor-insulator-semiconductor capacitors, it is possible to store or move a charge that has been introduced into the CCD device—either electrically or optically (Kosonocky and Sauer, 1975). Figure 5.17 is a schematic diagram of a few elements of a CCD structure, and of the potential wells that are generated and moved by a two-phase clock voltage. The efficiency of charge transfer from one MOS capacitor to an adjacent capacitor is very high; and in many modern devices, the charge readout to external circuitry from a single pixel can be greater than 99%. A detailed discussion of CCDs and parameters relating to their performance has been provided in the monograph edited by Barbe (1980).

ION IMAGING WITH MCP/PHOSPHOR/CCD COMBINATIONS

A number of recent investigations have utilized the MCP with a CCD for multichannel ion detection and imaging. Figure 5.18 is a general schematic

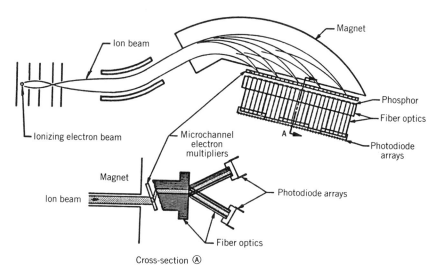

Figure 5.18 Schematic of electro-optical imaging system using a microchannel plate, phosphor screen, fiber optics, and a charge-coupled photodiode array (Giffin et al., 1979).

for detecting mass-resolved ion beams over an extended mass range (Giffin et al., 1979) with an electro-optical system that includes (1) an MCP; (2) an accelerating section for increasing the kinetic energy of the secondary electrons; (3) a phosphor electron-photon converter; (4) a fiber optics bundle for interfacing the phosphor to the CCD; and (5) the CCD photodiode array. This arrangement permits a video recording of the ion-electron image outside of the vacuum system; an option exists for changing the magnification of the electron-photon image that is transferred to the CCD. Such systems are now available from commercial sources. In the electro-optical system used by Louter et al. (1983) for the simultaneous detection of fragment mass spectra, an MCP of 75 × 15 mm was used as the ion-electron amplifier, and the CCD was comprised of a linear array of 1024 photodiodes—each diode having dimensions of 25 × 430 μm. This detection system was used with a tandem mass spectrometer with variable dispersion. The ratio of the highest and lowest masses in a simultaneously detected spectrum was variable from 1.06:1 to 12:1, with typical resolution values being 60 and > 1000, respectively. In another MCP/phosphor/CCD system for isotopic abundance measurements (Donohue et al., 1980), an extended mass range of 10–20 mass units was monitored using two adjacent MCPs, with each having a 5 × 45 mm active area, 25 μm diameter channels, channel bias angles of 15°, and a channel length-to-diameter ratio of 40.

In commercially available CCDs, the photodiode array is usually packaged and sealed with a quartz window of ~20 mils thickness. Furthermore, a silicon dioxide layer of ~3 μm thick covers the active region of the device so that even if the protective quartz window is removed (for x-ray and electron-sensing applications), 20 keV electrons will be stopped in the 3 μm SiO_2 layer. This layer, however, can be reduced to ~1 μm (the approximate "range" of a 10 keV electron) for special particle-sensing applications.

Charge-coupled photodiode arrays have also been employed directly as multiwavelength detectors in chromatography. A recording of the complete ultraviolet spectrum for each component in the effluent from a reverse-phased column has been reported, having low detector noise levels and a fairly linear response (Amita et al., 1982). Charge-coupled devices have also been employed for x-ray detection in three basic x-ray imaging topologies: (1) x-ray-visible light photon conversion; (2) x-ray photon-electron conversion; and (3) direct x-ray detection (Allison, 1982). In some instances, it is desirable (or necessary) to employ a "wavelength shifter" to optimize the generation of electron-hole pairs in a CCD. The spectral response of a typical silicon CCD extends from about 3500–11,000 Å (0.35–1.1 μm), with a maximum response at about 7000–8000 Å. Hence, compounds such as coronene ($C_{24}H_{12}$), which is excited by ultraviolet radiation of < 3800 Å but fluoresces near 5000 Å, have been used to provide an enhanced quantum efficiency for short wavelength radiation; a useful response with this hydrocarbon converter extends to approximately 500 Å (Blouke et al., 1980).

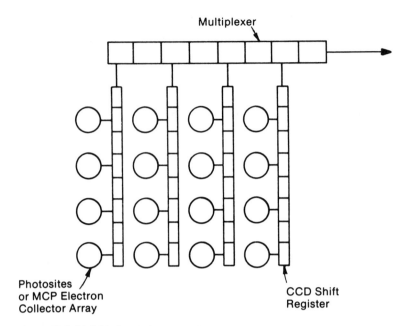

Figure 5.19 Hybrid CCD focal plane array in which separate photosites or electron-sensing elements are used for signal charge collection prior to charge transfer to the shift register (adapted from Baker, 1980).

MCP/CCD-HYBRID ION DETECTOR

While the MCP/CCD device combination with a scintillator is an important development, this system requires a high-voltage acceleration of electrons to produce photons that can then be sensed by the CCD. Ideally, one should not have to convert the charge output from an MCP to an optical signal and then convert the optical signal back to an electronic charge. A mass spectrometric detector, operating in the pulse-counting mode, should be required only to provide a real-time/spatial registration of the burst of electrons that are emitted from the rear surface of an MCP for each incident primary ion. A conversion to photons is presently employed because the CCDs are commercially designed for optical imaging.

Specifically, it should be possible to use CCDs in their hybrid configuration, where the sensing elements (*charge* rather than *photon* sensing elements) are distinct from the shift register circuitry. Hybrid configurations have already been fabricated for focal plane arrays, wherein separate photosites are used for signal charge collection and integration prior to charge transfer to the shift register (Baker, 1980) as indicated in Figure 5.19. This configuration has permitted the discrete sensing elements to be optimized for use as infrared or photovoltaic detectors, while retaining the basic advantages of the CCD for signal processing. For example, a hybrid pyro-

Hybrid MCP/CCD

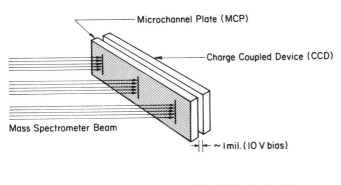

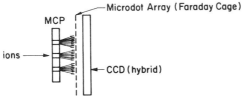

Figure 5.20 Hybrid MCP/CCD configuration for direct sensing of the output charge of a microchannel plate.

electric-silicon structure has also been theoretically evaluated for thermal imaging, where a pyroelectric detector array is bonded to input pads of a CCD (Watton et al., 1982). In an analogous fashion, it has been suggested that an MCP/CCD-hybrid could serve as a mass spectrometer detector where the photosites or *pixels* (picture-sensing elements) would be replaced by *exels* (electron-sensing elements), which would essentially consist of a microdot array of Faraday charge collectors (White, 1982). A schematic of the MCP/CCD-hybrid for mass spectrometric detection is shown in Figure 5.20. The microdot array can be directly coupled to the CCD register, and it can be engineered to be the front surface layer of a multilayered circuit. Eventually, it may even be possible to directly fabricate the CCD (or some other multiplexing circuitry) directly on the rear face of the MCP. In the interim, the MCP can be closely spaced to the CCD with a very small bias voltage (~10 V) to suppress secondary electrons. Several advantages of this hybrid circuitry approach can be stated explicitly:

1. No high voltage is required.
2. Limitations of phosphor optical lifetimes and image spreading (blooming) are eliminated, that is, higher sensitivity phosphors have a longer persistence and there are problems of spectral mismatch between phosphors and CCD.

3. MCP and CCD densities are closely matched, that is, $\sim 10^6/cm^2$.

4. Good signal-to-noise ratio, that is, a "dark current" of ~ 500 electrons per diode is small compared to the 10^5 electrons from an MCP for a single ion.

5. It is not essential that perfect registration be achieved between the MCP microchannels and the microdot array, as any mismatch can be accommodated via signal processing (Kronenberg, 1982).

6. Close spacing of the MCP to the CCD should minimize spatial electron dispersion, and electron optics techniques can be used to retain secondary electron image definition.

7. An extended focal plane detector can be fabricated by interconnecting multiple detectors in a modular fashion.

8. Alternate multiplexing circuitry to the CCD can also be devised, using the basic microdot array.

Finally, with CCD/MOS capacitor densities projected to be $10^7/cm^2$ by 1990 and microchannel center-to-center spacings already in the 5–10 μm range, it is realistic to envision such *charge-* vs. *photon*-sensitive devices being useful in many future mass spectrometric applications. Also, alternative ion-electron/CCD configurations are possible that, with signal-processing techniques, should ultimately approach the resolution limits of many mass spectrometers.

DETECTION OF NEUTRAL BEAMS

At high kinetic energies (~ 10 keV), neutral atoms and molecules can be detected with electron multipliers, inasmuch as kinetic secondary electron emission is virtually the same for an ion or a neutral particle. However, for helium ions and neutrals in the 2–5 keV range, recent measurements with a Bendix detector show a lower detection efficiency for neutrals ($\sim 60\%$) compared to ions (Dev and Boers, 1983). Furthermore, at very low energies, no simple mechanism exists for converting low-intensity neutral beams to charged particles, which can then be amplified and measured. Nevertheless, there exists a need for measuring the ratio of ions to neutrals in charge exchange experiments, plasmas, and the important phenomenon of sputtering. Also, in space-related systems, satellites collide with neutral molecules in the upper atmosphere at energies of a few electron volts. In this instance, it is also desirable to measure the flux of neutrals, and the ion-to-neutral ratio, as a function of other parameters.

It appears that at least one approach for monitoring neutral beam flux, particle velocity, and mean particle mass is to utilize a microphonic sensor to register the impact of discrete pulses of particles on the membrane surface. Conceptually, the feasibility of such a device is directly related to the phenomenon of human hearing (White, 1975). Studies have shown that the human ear is capable of detecting sound (a modulation of a neutral molecular ensemble) on the order of 10^{-16} W/cm^2. Indeed, at the threshold of hearing, the displacement of the tympanic membrane has been extrapolated to be less than 10^{-9} cm, that is, less than the radius of a hydrogen molecule. Obviously, the sensing of such a small displacement would be quite difficult. However, there are in existence commercial transducers of high sensitivity that can detect "chopped" neutral beams of sufficient flux density; these can be further developed using very thin membranes and specialized electronic circuitry. Essentially, we are interested in the response of a momentum (mv) sensor, which undergoes a displacement in contrast to a charge-sensing element.

Consider the equivalent circuit for a condenser microphone in which C is the microphone capacitance, C_s is the stray capacitance, V_p is the polarization voltage, R_p is the polarization resistance, and R is the preamplifier input resistance. If R_p is very large, the output voltage signal ΔV_0, is

$$\Delta V_0 = k \, Q_0 \Delta x \, \frac{C}{C + C_s[1 + \Delta x/x_0]}$$

where k is a constant of calibration, Q_0 is an initial charge on C, x_0 is the equilibrium separation for C, and Δx is the condenser microphone membrane displacement. To determine whether such a sensor could be used to detect a flux of neutral particles, a 0.5 in. diameter electret condenser microphone was placed at the focal point of a tandem mass spectrometer of large radius and a ^{23}Na$^+$ beam of ions was chopped at 1 kHz to simulate neutrals having energies of 12, 14, and 16 keV (White and Szatkowski, 1978). Even though the condenser microphone had a membrane thickness of 0.5 mil ($\sim$1.27 $\times$ 10^{-3} cm), it was possible to detect the momentum transfer of only 10^4 atoms per burst. Furthermore, on the basis of classical mechanics, it can be shown that a membrane displacement is inversely related to the third power of its thickness t (Timoshenko, 1960). Thus, if a microphone were designed with a 300 Å membrane (3 $\times$ 10^{-6} cm thickness vs. 1.27 $\times$ 10^{-3} cm), a nine-order of magnitude increase in sensitivity (180 dB) would result, that is,

$$\Delta V_0 \propto \Delta x \propto \frac{1}{t^3}$$

Films of this thickness (or even 100 Å) can indeed be made so that it does not seem unrealistic to suggest that such a sensor will be developed that can monitor either (1) very low fluxes of energetic neutrals; or (2) larger fluxes of epithermal molecules. Furthermore, an mv sensor should be able to monitor massive energetic particles or clusters. Also, for single-particle analysis, if the velocity is known, the mass can (in principle) be determined or vice versa.

Finally, it might be noted that the vacuum system of a mass spectrometer is an ideal environment for a microphonic sensor. While the human ear is exposed to thermal noise of Brownian motion, these random fluctuations of molecular bombardment on each side of the membrane are substantially reduced at 10^{-10} atm. In addition, a single thin membrane could be used as a common element in two capacitor-modulated circuits, so that a coincidence scheme could be employed to further suppress spurious signals.

REFERENCES

Aberth, W., *Int. J. Mass Spectrom. Ion Phys., 37*, 379 (1981).

Allison, N. M., *Nucl. Instr. Methods, 201*, 53 (1982).

Amita, T., M. Ichise, and T. Kojima, *J. Chromatogr., 234*, 89 (1982).

Baker, W. D., in *Charge-Coupled Devices*, ed. D. F. Barbe, Springer-Verlag, New York (1980), p. 27.

Barbe, D. F., ed., *Charge-Coupled Devices*, Springer-Verlag, New York (1980).

Blouke, M. M., M. W. Cowens, J. E. Hall, J. A. Westphal, and A. B. Christensen, *Appl. Optics, 19*, No. 19, 3318 (1980).

Boyle, W. S., and G. E. Smith, *Bell Syst. Tech. J., 49*, 587 (1970).

Brenton, A. G., J. H. Beynon, and R. P. Morgan, *Int. J. Mass Spectrom. Ion Phys., 31*, 51 (1979).

Bross, A., *Nucl. Instr. Methods, 201*, 391 (1982).

Burlefinger, E., and H. Ewald, *Z. Naturforch, 18A*, 1116 (1963).

Carrico, J. P., M. C. Johnson, and T. A. Somer, *Int. J. Mass Spectrom. Ion Phys., 11*, 409 (1973).

Daly, N. R., *Rev. Sci. Instr., 31*, 264 (1960).

Daly, N. R., A. McCormick, R. E. Powell, and R. Hayes, *Int. J. Mass Spectrom. Ion Phys., 11*, 255 (1973).

Dearnaley, G., and D. C. Northrup, *Semiconductors for Nuclear Radiations*, Wiley, New York (1966), p. 128.

Dev, B., and A. L. Boers, *J. Phys. E. Sci. Instr., 16*, 1015 (1983).

Dietz, L. A., *Rev. Sci. Instr., 36*, No. 12, 1763 (1965).

Dietz, L. A., *Rev. Sci. Instr., 38*, No. 9, 1332 (1967).

Donohue, D. L., J. A. Carter, and G. Mamantov, *Int. J. Mass Spectrom. Ion Phys., 33*, 45 (1980).

Fehn, U., *Int. J. Mass Spectrom. Ion Phys., 15*, 391 (1974).

Fleischer, R. L., P. B. Price, and R. M. Walker, *Science, 149*, 383 (1965).

Fricke, J., A. Müller, and E. Salzborn, *Nucl. Instr. Methods, 175*, 379 (1980).

Giffin, C., R. Britten, H. Boettger, J. Conley, and D. Norris, in *Ann. Conf. on Mass Spectrom. and Allied Topics*, Seattle (1979).

Hahn, Y. B., R. E. Hebner, Jr., D. R. Kastelein, and K. J. Nygoard, *Rev. Sci. Instr., 43*, No. 4, 695 (1972).

Honig, R. E., J. R. Woolston, and D. A. Kramer, *Rev. Sci. Instr., 38*, No. 12, 1703 (1967).

Ihle, H. R., and A. Neubert, *Int. J. Mass Spectrom. Ion Phys., 7*, 189 (1971).

Kosonocky, W. F., and D. J. Sauer, *Electron Des., 23*, 58 (1975).

Kronenberg, S., private communication (1982).

Krycuk, G., and J. D. Walling, *IEEE Trans. Nucl. Science, NS-20*, No. 1, 228 (1973).

Kurz, E. A., *Am. Laboratory, March* (1979).

Louter, G. J., A. N. Buijserd, and A. J. H. Boerboom, *Int. J. Mass Spectrom. Ion Phys., 46*, 131 (1983).

Masters, J. I., *Nature, 223*, 611 (1969).

Naudin, G., J. P. Billou, and M. Leclerc, *Int. J. Mass Spectrom. Ion Phys., 46*, 107 (1983).

Nibbering, N. M. M., *Int. J. Mass Spectrom. Ion Phys., 45*, 343 (1982).

Nier, A. O., E. P. Ney, and M. G. Inghram, *Rev. Sci. Instr., 18*, 294 (1947).

Owens, E. B., in *Mass Spectrometric Analysis of Solids*, ed. A. J. Ahearn, Elsevier, Amsterdam (1966), Chapter 3.

Ruby, L., J. G. Kramasz, Jr., W. G. Pone, and T. Vuletich, *Nucl. Instr. Methods, 37*, 293 (1965).

Rudat, M. A., and G. H. Morrison, *Int. J. Mass Spectrom. Ion Phys., 29*, 1 (1979).

Stacey, J. S., W. D. Sherrill, G. B. Dalrymple, M. A. Lanphere, and N. V. Carpenter, *Int. J. Mass Spectrom. Ion Phys., 39*, 167 (1981).

Stein, J. D., and F. A. White, *J. Appl. Phys., 43*, No. 6, 2617 (1972).

Stein, J. D., and C. L. Peterson, in *Ann. Conf. on Mass Spectrom. and Allied Topics*, Seattle (1979), p. 719.

Stoffels, J. J., C. R. Lagergren, and P. J. Hof, *Int. J. Mass Spectrom. Ion Phys., 28*, 159 (1978).

Surface Science Laboratories, Inc., 1206 Charleston Road, Mountain View, CA, 94043.

Takagi, S., T. Iwai, Y. Kaneko, M. Kimura, N. Kobayashi, A. Matsumoto, S. Ohtani, K. Okuno, H. Tawara, and S. Tsurubuchi, *Nucl. Instr. Methods, 215*, 207 (1983).

Timoshenko, S., *Strength of Materials*, Part II, Van Nostrand, New York (1960), p. 104.

Tosswill, C. H., Galileo Electro-Optics Corp., private communication (1984).

Traxlmayr, U., K. Riedling, and E. Sinner, *Int. J. Mass Spectrom. Ion Processes, 61*, 261 (1984).

Vourdos, P., D. M. Desiderio, J. G. Leferink, T. J. Odionne, and J. A. McCloskey, *Int. J. Mass Spectrom. Ion Phys., 10*, 133 (1972).

Watton, R., P. Manning, and D. Burgess, *Infrared Phys., 22*, 259 (1982).

White, F. A., *Our Acoustic Environment*, Wiley, New York (1975), Chapter 5.

White, F. A., "The Real-Time Detection of Mass Resolved Ion Beams," in *Progress Report (July 1982) for NASA-Langley Research Center*, Hampton, VA.

White, F. A., J. C. Sheffield, and W. D. Davis, *Nucleonics, 19*, 58 (1961a).

White, F. A., J. C. Sheffield, and J. W. Mayer, *Electronics, 34*, 74 (1961b).

White, F. A., J. D. Walling, and A. J. Schwabenbauer, in *15th Ann. Conf. on Mass Spectrom.*, Denver (May 1967), p. 379.

White, F. A., and G. P. Szatkowski, *IEEE Int. Conf. on Plasma Science*, Monterey, CA (May 15–17, 1978).

Wiley, W. C., and I. H. McLaren, *Rev. Sci. Instr., 26*, 1150 (1955).

Wiza, J. L., *Nucl. Instr. Methods, 162*, 587 (1979)

Woolston, J. R., R. E. Honig, and E. M. Botnick, *Rev. Sci. Instr.*, No. 12, 1708 (1967).

Young, W. A. P., R. C. Ridley, and N. R. Daly, *Nucl. Instr. Methods, 51*, 257 (1967).

COMPUTER-AIDED DATA PROCESSING

In the 40's and 50's—at Exxon's Bayway refinery in New Jersey—some 20 scientists and specialists, working two and three shifts each day, analyzed thousands of gas and product samples each year. The group included eight people who did *nothing* but make calculations and one soul who did *nothing* but measure peak heights with a ruler! (David, 1984)

As in most other fields, the ubiquitous computer has increasingly become an integrated part of the mass spectrometer. This is evidenced by the almost routine inclusion of microcomputers or minicomputers in even moderately priced commercial instrumentation, and today, there are many mass spectrometrists who have never used (or even seen) an instrument without a data system attached. A major contribution of the integrated computer has been to extend the range of applicability of the mass spectrometer beyond the research facility into industrial process control and routine quantitative analysis. Indeed, without this development, it is likely that the mass spectrometer might still be only a tool of the trained research physicist or chemist. In virtually all areas of research in which mass spectrometry is the analytical method of choice, the computer will increasingly be used. In only the most routine of these applications, however, will the total operation be accomplished in the foreseeable future without direct interaction of the original, albeit slower, computer—the knowledgeable mass spectrometrist!

Current trends indicate that for limited or repetitive routine tasks, the use of small, dedicated microcomputers will continue to increase, while the larger minicomputer will also be used for artificial intelligence-based control or for off-line data processing. The modern computerized control and data acquisition system (CDAS) commonly provides some degree of interactive control based on operator input, records and/or displays the data output and the operating parameters used to conduct the analyses, and provides on-line or off-line reduction of the data consistent with experiment or measurement requirements. The design and implementation of such a system requires an in-depth understanding of the operating characteristics and limitations of both the mass spectrometer and the computer, of the effects of noise and the methods used to extract information from the signal, and of the application in which it is to be used (e.g., McMurray, 1971). Perone et al. (1975) discussed the design parameters that should be considered in developing an automated CDAS, and they outlined a systematic procedure based on three major steps: user-level specification of objectives and constraints; preparation of a hardware-independent functional design; and implementation of the design with specific hardware/software components. These principles were applied to the automation of a gas chromatograph/mass spectrometer that was to be used in a research and development facility for nonroutine qualitative and semiquantitative analyses of complex organic mixtures having concentrations ranging from parts per million to percent levels. A functional outline of their CDAS is shown in Figure 6.1. The complete functional

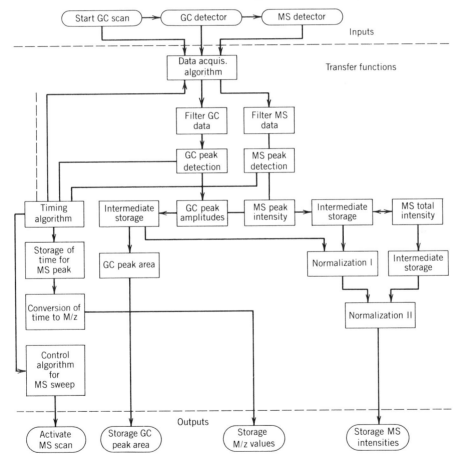

Figure 6.1 Functional design of a CDAS for a GC/MS system. (Perone et al., 1975; copyright ASTM.)

TABLE 6.1 Functional Design Path for Acquiring Mass Spectrometer Signals

Function[a]	Description	Comments
1. MS data acquisition	Sampling and transfer of MS detector analog output to CPU	Linear mass scan of 3 s over 400 mass units; time per peak = 7.5 ms; for 50 data points per peak, the required data sampling rate = 6.7 kHz; conversion rate should be fast relative to sampling rate
2. Filtering	Algorithms for digital filters (finite memory filter?)	S/N enhancement of smoothing filters depends on the polynomial fit equation and the data window size; analog prefiltering with 500 Hz BW is to be considered; Bessel filter recommended for minimum peak shifts; amplitude distortion of up to 10% can be tolerated if uniform for all peaks
3. Peak detection	Procedure to detect peak from sampled and smoothed data	Sampling rate, 6.7 kHz; peak detection must be correlated to clocktime; clocktime at peak must be saved
4. Peak intensity	Algorithm for determining the peak intensity of each MS peak	Peak profile does not need to be saved
5. Intermediate storage	Algorithm for transfer of peak intensity and associated clocktime to temporary storage	Maximum of 400 intensity and 400 clocktime values must be stored for each spectrum, one pair at a time; transfer rate is 267 data points/s

156

6. Normalization I	Procedure for correcting MS peak intensities for change in GC peak amplitude; includes correction for time delay	During the 3 s MS scan, the GC peak amplitude and thus the instantaneous concentration in MS change; this requires normalization to constant GC amplitude; the time delay between concentration maxima in GC and MS must be determined experimentally and input to computer by operator
7. Total MS intensity	Algorithm for summing all intensity values	Original individual values must be saved for the next step
8. Mass scale	Algorithm to convert clocktimes to m/z values	For linear scan: $M = c_1 + c_2 l$; constants in the function must be evaluated from calibration run; functional relationship must be stored in memory
9. Normalization II and output	Procedure to express peak intensities as percent of largest peak in spectrum; transfer of reduced peak intensities and associated m/z values to storage	Requires search for largest peak in spectrum; maximum number of GC peaks is 20, maximum number of MS peaks is 400 plus 400 m/z values; total storage requirement per experimental run is about 16,000 locations; before reinitialization for next run, transfer of data to bulk storage may be required

Source: Perone et al. (1975).

[a] Functions 1 through 5 must be executed in real time in this design (continuous scan). Functions 6 through 9 are also executed in real time after each spectrum scan is completed and before the next MS scan is initiated at the peak of the next GC peak. A minimum of 3 s is available to perform the functions. System flexibility in specifications requires that MS range and time are variables set by the operator.

design, however, includes considerable additional information for each pathway, such as maximum data generation and transfer rates, time response characteristics, and references to procedures or algorithms. Some details of the functional design path for the real-time acquisition of the mass spectrometer signal are listed in Table 6.1. A similar table is developed for each identifiable function and is incorporated into the overall documentation.

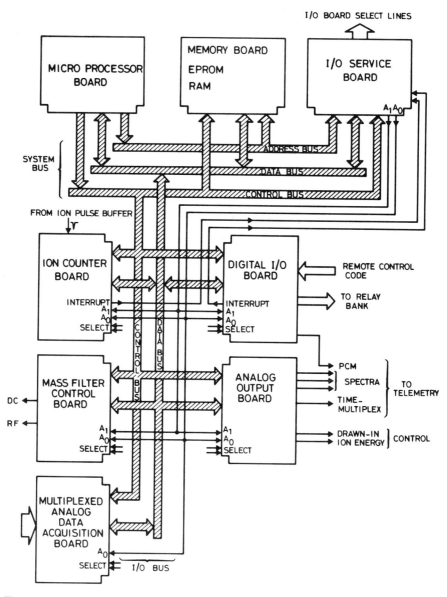

Figure 6.2 Schematic of a microprocessor-based data acquisition and control system for a balloon-borne quadrupole mass spectrometer. (Nevejans et al., 1982; courtesy E. Arijs.)

The integrated CDAS has found particular application in industrial control processes (Chapter 14), and also in space applications (Chapter 9) where remote control via a telemetry link is essential. Nevejans et al. (1982), for example, have described the CDAS shown schematically in Figure 6.2, which is integrated with a balloon-borne quadrupole mass spectrometer used to measure ions in the stratosphere. Their CDAS uses an eight-bit microprocessor coupled with a 24-bit, high-speed ion counter to provide ground-directed data management and instrument control during the course of the flight. The system combines a number of functions, including mass scan control, signal acquisition, spectrum accumulation, multiple analog data storage and transmission, multiplexed analog-to-digital conversion, pulse code modulation encoding for serial digital data transmission, and interfacing of telemetered control signals. Instructions for subsequent mass scans can be accumulated and processed without interfering with the scan in progress, and mass spectra also are stored on-board for later playback if necessary to prevent loss of data due to telemetry problems.

A number of systems using more than one microprocessor dedicated to

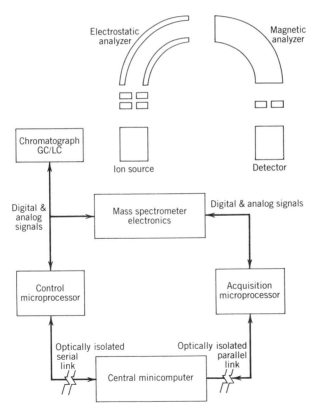

Figure 6.3 CDAS with distributed microprocessors for a chromatograph/high-resolution double focusing mass spectrometer. (Stradling et al., 1983; copyright 1983. Reprinted by permission of John Wiley & Sons, Ltd.)

specific tasks have recently been developed and reported in the literature. To illustrate, Stradling et al. (1983) described a CDAS for controlling a high-resolution, double focusing mass spectrometer coupled with a gas chromatograph (Fig. 6.3). One of the microprocessors is dedicated to controlling the electronic and electromechanical functions, while a second controls the scan and data acquisition and processes the data prior to storage. Both microprocessors are under control of a central minicomputer, wherein final data processing is accomplished. The system is interactive and operates under priority interrupts, so that certain tasks (e.g., control of GC oven temperatures) are suspended when necessary.

Wong et al. (1983, 1984a, 1984b, 1984c) have also developed a three-computer system (Fig. 6.4) for control of, and data acquisition from, a triple-quadrupole mass spectrometer (TQMS) used for mixture analyses and structure elucidation in fossil fuel research. Rule-based heuristic techniques of artificial intelligence (AI) are used to optimize tuning over the entire mass range and to provide real-time modification of the operating parameters.

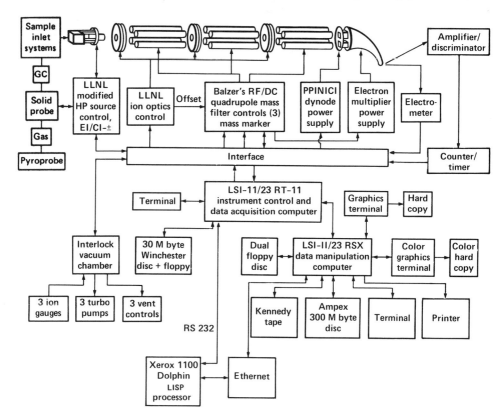

Figure 6.4 CDAS with distributed microprocessors for a TQMS using rule-based heuristic techniques of artificial intelligence. (Wong et al., 1984a; courtesy C. M. Wong, by permission UC/LLNL, copyright © 1984 IEEE.)

Instrument control and data acquisition are accomplished with the first computer, time-shared multitasked data processing, manipulation, and storage are done with the second, and the AI system resides within a knowledge representation language in the third. These AI methods differ from the traditional algorithmic techniques of totally defined, stepwise problem solving in that logical inference heuristics or "reasoning" are used to achieve a result. In this case, the tuning procedure that is encoded includes heuristics to describe a self-adaptive feedback control process for real-time optimization throughout the data collection process. Because there are approximately 30 adjustable parameters for the TQMS, optimal tuning is time consuming, yielding at best an "average" spectrum having decreased resolution and sensitivity at each end of the mass range. With AI procedures, once the initial tuning file is established, which requires approximately 11 minutes per mass range, subsequent on-line changes to optimize the operating conditions over the entire mass range are completed in 7.7 milliseconds. The large quantities of multidimensional data generated by the TQMS are efficiently recorded in real time for subsequent analysis with computerized database management techniques developed for the system by Crawford et al. (1984). In addition to rapid storage and retrieval of the data, the database stores all initial operating parameters of the instrument, selects the independent variables pertinent to the particular experiment, and generates selected X-Y data pairs.

In addition to the TQMS, development of other advanced instrumental methods has also progressed along with the collinear improvement and availability of the computer-based CDAS. The advent of Fourier transform mass spectrometry, for example, can be directly attributed to a large degree to the availability of these high-speed data acquisition and processing systems. As indicated in Chapter 3, the concept of ion cyclotron resonance is, in itself, not new, but it required fast computers to translate this physical phenomenon into a powerful new form of analysis.

DATA ACQUISITION

The signal produced by the mass spectrometer is a peak in an incrementally continuous spectrum whose position indicates the mass-to-charge ratio (m/z) of an ion and whose intensity is proportional to the rate of arrival of ions at the ion current detector. The width, shape, and size of the peak in response to a given input is a function of the instrument type (e.g., quadrupole, magnetic sector, etc.) and configuration, and of the adjustment of the various voltages and other controlling parameters.

Whatever the means by which they are formed, the positively charged ions will be one of several kinds: (1) *molecular (or atomic) ions* formed by the loss of a single electron ($m/z = m$); (2) *multiply charged ions* formed by the loss of two or more electrons ($m/z = m/2, m/3, \ldots$); (3) *fragment ions* formed by the loss of an atom or group of atoms from a molecular or larger fragment ion having sufficient internal energy to further decompose; or (4) *metastable ions* that result when ions decompose to a measured nonintegral mass in the field-free region between the ion source and detector in a magnetic deflection mass spectrometer. Under some conditions, ions formed by a rearrangement mechanism may also be observed (McLafferty, 1980), and negatively charged ions are sometimes also measured. Because the ionizing energy is statistically imparted to the neutral molecules and gas phase collisions are rare at normal ion source pressures, the rate and extent of fragmentation are determined by the internal molecular energy and are essentially reproducible in any instrument operating under a given set of conditions. The resulting series of peaks form a *cracking (fragmentation) pattern* or *mass spectrum* that is indicative of the molecular ion and its structure. Isomers can also be identified when the cracking pattern for isomeric forms of the same molecule are sufficiently different.

The resolved ion beams may be measured simultaneously using multidetection methods described in Chapter 5, or they may be sequentially imposed on a single detector by variation of one or more of the operating parameters of the mass spectrometer—a procedure known as *mass scanning*. The scan may be generated in a continuously increasing or decreasing manner, or it may consist of a series of steps to selected m/z by stepwise variation of voltages to achieve *peak switching*. The peaks are most often cataloged in a table of relative intensities after normalizing to the largest (base) peak in the spectrum, or in a bar-graph representation. The normalization removes the effect of source pressure, sample throughput, and amplifier gain; and it facilitates the comparison of an unknown sample that has been normalized in the same manner. For a mixture, the *mass spectrum* becomes a summation at each mass of the intensities representative of each of the components, as determined by both the instrument sensitivity and the concentration.

A typical high mass spectrum of air taken on a magnetic deflection mass spectrometer is shown in Chapter 15 (see Fig. 15.3), with the mass peaks identified. The mass values shown are the integer or nominal mass. In point of fact on the $^{12}C = 12.000$ atomic scale, the elements have fractional mass (Table 6.2); therefore, all ions other than ^{12}C appear at nonintegral values. In most cases, use of the nominal masses is sufficient to identify and calculate the abundance of the species. This is particularly true in the analysis of low molecular weight compounds or in the computer-assisted identification of molecules with the library search methods discussed later in this chapter. Molecular weight measured with low-resolution data will not, in itself, umambiguously define the molecular formula, since many compounds may appear at the same nominal mass. However, examination of the fragment pattern will often provide insight into the structure, elemental composition, and class of compound (e.g., aliphatic, aromatic, etc.) to which the substance belongs. Hill (1972) and McLafferty (1980), among others, have provided

TABLE 6.2   Exact Mass and Natural
Abundances of Some Elements Useful in
Mass Spectrometry

Element	Mass	% Abundance
^{1}H	1.0078	99.985
^{2}H	2.0141	0.015
^{12}C	12.0000	98.889
^{13}C	13.0034	1.11
^{14}N	14.0031	99.634
^{15}N	15.0001	0.366
^{16}O	15.9949	99.763
^{17}O	16.9991	0.037
^{18}O	17.9992	0.200
^{19}F	18.9984	100.0
^{20}Ne	19.9924	90.51
^{21}Ne	20.9940	0.27
^{22}Ne	21.9914	9.22
^{32}S	31.9721	95.02
^{33}S	32.9715	0.75
^{34}S	33.9679	4.21
^{36}S	35.9671	0.02
^{35}Cl	34.9689	75.77
^{37}Cl	36.9659	24.23
^{36}Ar	35.9676	0.34
^{38}Ar	37.9627	0.06
^{40}Ar	39.9624	99.60
^{79}Br	78.9184	50.69
^{81}Br	80.9164	49.31
^{78}Kr	77.9204	0.36
^{80}Kr	79.9164	2.28
^{82}Kr	81.9135	11.58
^{83}Kr	82.9141	11.52
^{84}Kr	83.9115	56.96
^{86}Kr	85.9106	17.30

detailed procedures for interpreting spectra based primarily on examination
of the nominal mass pattern. One important aspect of this fragmentation
analysis is the ability to differentiate between isomeric analogs, since the
chemical properties are often quite different. If, on the other hand, the exact
mass can be determined with precision and if the molecular ion does not
totally fragment before reaching the detector, then the molecular weight can
be determined and the empirical formula developed (e.g., Rose and Johnstone, 1982).

The resolving power—or resolution—is a measure of the smallest increment in mass that can be identified in the mass spectrometer output. By

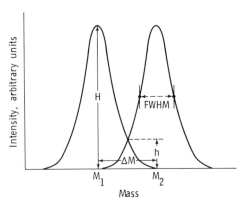

Figure 6.5 Two overlapping peaks of equal height from which resolution can be defined.

convention, this is defined in terms of $m/\Delta m$ for two adjacent mass peaks of equal intensity having a valley between them that is 10% of the peak height (Fig. 6.5); that is, $h/H = 0.1$, at which point the height contribution of one to another becomes negligibly small. If, for example, the two peaks resulted from nitrogen ($^{14}N_2 = 28.00614$) and carbon monoxide ($^{12}C^{16}O = 27.99491$), the resolving power of the instrument is approximately 2500; conversely, this is the resolving power required to separate the two species. It should be noted that the required resolution is directly proportional to the mass, so that resolution of the higher masses becomes increasingly difficult.

A measure of the resolution may also be obtained from a single peak if the shape is approximately Gaussian (Litzow and Spalding, 1973). The mass difference Δm, calculated in the two-peak case above, is equal to the width of a single Gaussian peak at 5% of its height above baseline, and equal to 2.08 times the full width at half maximum height (FWHM). Therefore, it is possible to estimate the resolution at any point in the spectrum without introducing a mixture containing equal amounts of two species, with the accuracy of the calculation being dependent upon how well the peak approximates a Gaussian.

The resolving power for a particular instrument is determined by factors influencing ion beam geometry and by the frequency response of the detection circuitry. There are, however, finite limitations to the reduction of Δm by physical or electronic means. Consider, for example, the single focusing magnetic deflection mass spectrometer, in which the ion beam geometry is primarily determined by the exit and entrance slits. Even if the slits were infinitesimally narrow, the mass peaks would still have a finite width due to ions entering at an angle ($\pm \alpha$) to the perpendicular and to variations in electrostatic (ΔV) or magnetic (ΔB) fields; hence, the resolution in all cases has a limit determined by instrumental broadening of the measured signal. Therefore, any further improvement of instrument resolution requires the application of computer-assisted computational methods, such as those described in the following section on data enhancement.

Below about 50 eV, the number of ions produced by electron bombardment of a particular atom or molecule is strongly dependent upon the energy of the ionizing electrons. This fact may be used to eliminate or reduce the contribution of some constituents to a particular mass peak, thus enhancing the effective resolution. The procedure has been used for many years with high-resolution mass spectrometry in the analysis of monodeuterated or unsaturated hydrocarbons, and of higher boiling point aromatics and olefins in petroleum (Honig, 1950; Field and Hastings, 1956; Lumpkin, 1958); and it has been shown to be useful with low- or moderate-resolution mass spectrometers as well (Wood et al., 1978). In the classic work of Field and Franklin (1970), which also contains the extensive table of ionization potentials and ionic heats of formation compiled by Franklin et al. (1969), the effects of ion source conditions and other parameters on the measured ionization efficiency curves are discussed in detail. In general, however, it may not be necessary to completely eliminate the interfering fragment, nor is it necessary to measure the exact appearance potentials for the ions. For this procedure, it is usually sufficient to measure the ionizing voltage at which a known singlet peak disappears (e.g., the actual appearance potential for the doubly charged $Ar^{2+} = 43.3$ eV) to establish a correction to the tabulated values of ionization potentials for other relevant ions. These corrected values may then be used as an aide to identify an ion or to eliminate or reduce its contribution to a mass peak in the spectrum. The procedure is illustrated in Figure 6.6, which shows relative ionization efficiency curves normalized to the intensity at 70 eV for the ions indicated. Resolution of the $^{40}Ar^{2+}$ (19.9812) from $^{20}Ne^{+}$ (19.9924) requires a resolving power of 1785; however, if the ionizing electron energy is reduced to 42 eV, the Ar^{2+} ion disappears and about 45% of the Ne^{+} signal remains. Analysis of methane in air may also be accomplished if the voltage is reduced to about 20 eV, at which point the interfering contribution at mass 16 from atomic oxygen or water has been significantly reduced with respect to the CH_4. The method may also be useful in simplifying the spectrum and in determining details concerning the ionization pathway.

The sensitivity of the mass spectrometer depends upon several parameters, including (1) the operating conditions for the ion source; (2) the ionization cross-section for the particular species; (3) the gain and response of the detector; and (4) the overall transmission efficiency. In general, higher resolution reduces the available sensitivity, and it becomes necessary to balance one against the other. When adequate gain cannot be retained while resolving the species of interest, electronic and computational methods to increase the signal-to-noise ratio may be useful.

Except when exceedingly high data rates requiring real-time analog recording of large quantities of data are encountered, high-speed digital circuitry permits on-line digitization of the ion current and other signals for computer-assisted acquisition and reduction of the data. A block diagram of a typical single computer-based CDAS is shown in Figure 6.7. The mul-

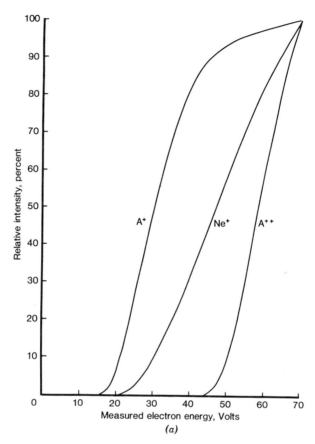

Figure 6.6   Ionization efficiency curves for several species, illustrating (a) the analysis of $^{20}Ne^+$ in the presence of $^{40}Ar^{2+}$ and (b) of methane in the presence of oxygen (Wood et al., 1978).

tiplexer enables a number of inputs such as total and partial ion currents, GC detector output, peak defining parameters, and other operational information to be sequentially sampled by the analog-to-digital converter (ADC). The ADC converts the analog signal to a series of encoded digital signals, each proportional to the magnitude of the input signal during the digitization cycle. Since the ADC has a finite cycle time during which the magnitude of the ion current may be rapidly changing, increased accuracy is obtained by sampling over short intervals and by preceding the ADC with a sample-and-hold amplifier, which maintains the signal at a constant level at the time of digitization (Chapman, 1978). After digitization, the data signals may be compared to a threshold value by the discriminator, and only those exceeding the threshold are passed to the computer.

This signal processing removes a portion of the baseline noise, prevents the processing of useless information, and frees the computer to reduce data

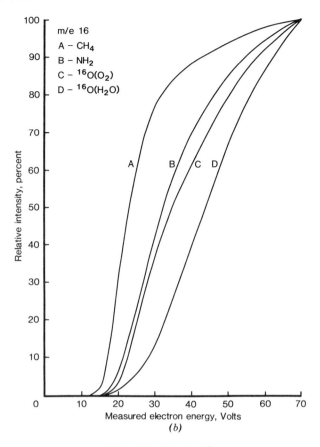

Figure 6.6 *(Continued)*

or to perform control duties in the interval between the appearance of peaks. This function is often performed by the computer when peak switching is used in lieu of mass scanning. In this case, a baseline value may also be subtracted from the signal prior to comparison to the threshold value. The analog signal is thus transformed into a series of digital values with magnitudes approximating the shape of the input signal curve; the fidelity of the approximation is determined by the sampling rate, that is, the number of points taken as determined by the width of the mass peak and ADC cycle time. The larger the number of points taken across the peak, the more accurate the representation of the true peak shape (Fig. 6.8). Perone and Jones (1973) have suggested that a minimum sampling frequency of 10 times the bandwidth of the waveform is necessary to faithfully reproduce the peak shape, which for mass spectra leads to the accepted number of 10–20 points/amu to ensure accurate mass measurement (Chapman, 1978). Considering the exponential scan mode for the magnetic mass spectrometer noted in

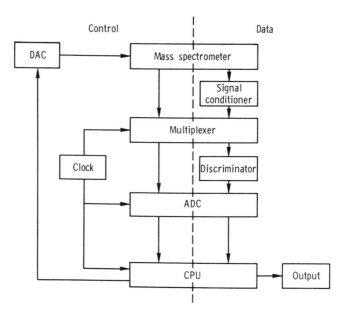

Figure 6.7 Block diagram of a typical single-processor-based CDAS for data acquisition and control of a mass spectrometer.

Chapter 7, the mass peak width in terms of time (t_p) is given by $t_p = 0.43t_{10}/R$, where t_{10} is the scan rate in seconds over a decade in mass and R is the resolution of the instrument. The sampling rate in hertz necessary to yield N points/amu is then given by $S = N/t_p$. For a scan rate of 1 s/decade at a resolution of 1000, $t_p = 0.43$ ms; hence, 20 points/amu requires a sampling rate of 46 kHz, well within the capability of current ADC technology. However, as the resolution is increased to higher values (e.g., $R > 10^4$), the scan rate may be limited by the digitizing rate of the ADC.

A data measurement problem exists for the Fourier transform mass spec-

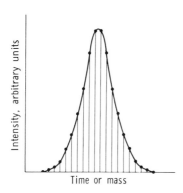

Figure 6.8 Points taken during digital sampling at equal time intervals across a mass peak.

trometer (FTMS), which while providing high resolution lacks precision in peak shape, height, and exact peak centroid location due to the limited number of points that can be taken across the peak. The primary measured parameter is frequency, so that stability and accuracy can be achieved; but this requires that narrow-band detection methods be used, which in turn prevent the simultaneous detection of all ion beams. Since the data are discrete, however, numerical analysis methods for fitting to theoretical peak shapes (Giancaspro and Comisarow, 1983) and for reducing the spacing between data points by "zero-filling" (Comisarow and Melka, 1979) can be applied to improve the accuracy of peak identification. Francl et al. (1983) have developed a "zoom-transform" procedure implemented by applying a parabolic fitting routine and limited zero-filling, along with forward and reverse Fourier transformations, to data stored in the computer. With this offline technique, they reported mass measurement accuracies of 0.8 ppm along with significantly reduced computation time. The procedure can be applied anywhere in the mass scale, an important factor since in FTMS the number of points taken per amu varies inversely as the square of the mass.

In general, ion current intensity data, along with the mass defining signals, are stored in the computer in one of several ways. If adequate storage capacity exists, it is preferable that all of the sampled points taken during a continuous scan be retained, since this provides information concerning peak shape as well as intensity. The onset and termination of a peak are detected by threshold comparison, generally after digital smoothing of the signal, or by comparison to the previous values (Hites and Biemann, 1967; Holmes et al., 1971). The mass/charge value of the peak is identified at the point of maximum ion current intensity or by calculating the peak centroid $C = \Sigma v_i f_i / \Sigma v_i$, where v_i is the output and f_i is a function in terms of which m/z is defined (e.g., time, voltage, or magnetic field strength). C is then compared to a mass calibration scale that is also in terms of f. The mass scale calibration table where $m/z = f(t)$, etc. is derived from the measurement of the mass peaks of reference compounds, most commonly perfluorokerosene or heptacosafluorotributylamine. These may be introduced along with the sample being analyzed and the unknown peaks compared with those of the standard, or the calibration data may be stored in the computer for later comparison.

Peak-matching, a dynamic version of the first method, determines the exact mass ratio of measured mass/charge between the sample and reference peaks by precise variation of the acceleration voltage V in the magnetic sector mass spectrometer. Since the variation of m/z at the focal point is linear with voltage, the unknown mass is precisely determined by the product of the known mass and the ratio of the voltages. In a high-resolution mass spectrometer the peak-matching method can determine the unknown mass with a high degree of precision. Tondeur et al. (1984), for example, have recently described a computer-assisted peak-matching system in which mass measurement errors of 0.3–5.9 ppm and standard deviations of 2 ppm were obtained on picogram samples of tetrachlorodibenzodioxin. The masses used in *peak-matching* are usually closely aligned to increase measurement precision of the acceleration voltage change. However, Ligon (1980a) has described a variation of the traditional method for the identification of high molecular weight species ($m/z > 1000$) where the absolute masses of two ions were unknown, but Δm could be accurately determined. In this case, the lower mass was determined by $m_1 = (\Delta m)/(R - 1)$, where R is the ratio of the two masses, and the accuracy is increased as Δm becomes large.

Because the linear mass scan of the quadrupole mass spectrometer (QMS) can be precisely controlled by the computer through digital-to-analog conversion of voltages, a simple calibration can often be used and the remainder of the mass scale can be predicted from the theoretical scan law. As an example of this approach, a recently developed, computer controlled QMS having a mass range of 1–200 incorporated a six-point interpolation algorithm for determining the mass scale (Judson et al., 1976, 1978; Wood and Yeager, 1980). When commanded to perform a mass calibration, the computer searches a calibration gas (e.g., octafluorocyclobutane) for specified masses, stores the voltages corresponding to the top of the mass peak in memory, and calculates the corrected mass scale by interpolation along the line passing through the points. Once the mass scale is determined, movement of a cursor on a computer readout terminal (CRT) identifies mass peaks generated by the scan; and keyboard entry of a nominal mass sets exact operating parameters for single mass monitoring or for sequentially stepping to the tops of up to 40 selected mass peaks.

If the intensities taken at each point sampled across the peak are not all stored, then the peak centroid must be calculated in real time; or, peak-identifying information (e.g., RF and DC voltages, magnetic field strength and acceleration voltage, or time from the start of the scan) must be input to the computer along with the ion intensity data. This can be accomplished by causing the mass spectrometer to step to a peak location, or by continuously monitoring the ion current during the scan to determine when the top of a peak has been reached, at which time the mass and intensity data are transmitted to the computer. There are various algorithms for detecting the top of a peak, one of which is simply to compare each newly digitized value to several previous ones. If the new value is greater than the others, it is added to the list and the oldest value is dropped. If it is lower than the previous values for a selected number of cycles, the location of the top of the peak is calculated and the data stored. A more accurate method described by Carrick (1969) takes the analog first derivative of the ion current signals to locate the top of the peak. At the peak maximum, dI/dt is independent of peak height and is precisely equal to zero. When this condition is detected, the digitizing circuitry is triggered to record both intensity and mass data.

The centroid of the peak and hence its mass may also be computationally determined by positional interpolation from known reference peaks. Snelling et al. (1984), for example, have calculated the real mass of an unknown (m_x) in terms of known nearby peaks at higher and lower mass values (m_h, m_l)

and an interpolated value (m_i) from $m_x = (m_h m_l)/(m_h + m_l - m_i)$, where assuming a linear mass/time relationship, m_i is a function of simple interpolation from the centroid time intervals from scan initiation t_x, t_h, and t_l. The interpolated mass is

$$m_i = m_l + \frac{(t_x - t_l)(m_h - m_l)}{(t_h - t_l)}$$

and therefore the unknown mass is determined by

$$m_x = \frac{m_h m_l (t_h - t_l)}{m_h (t_h - t_x)} + m_l (t_x - t_l).$$

Ion abundances are measured from the height or area of a peak after determining instrument sensitivity for a particular ion. The difference in intensity between the maximum and minimum detectable peaks in any given spectrum may often be as much as 10^6. While this may be within the dynamic range of many conventional linear electrometer DC amplifiers, it requires a large number of conversion bits ($10^6 \approx 2^{20}$) and is beyond the range of most ADCs. Obtaining a wide dynamic range thus requires range switching or that specialized measurement or computational techniques be employed. Auto-ranging amplifiers can be used, but these require some means of identifying the selected range for the computer, and they present the additional possibility that data will be distorted or lost during the switching operation. Range switching can be avoided through the use of logarithmic amplifiers; however, this reduces measurement precision and complicates data interpretation. In many instances, the signal is passed through a multiplexer to parallel amplifiers or attenuators. The signal from the most intense output may be selected for digitization and processing or the parallel outputs representing several gains (e.g., $x1$, $x10$, $x100$, $x1000$) recorded simultaneously. The latter has the advantage of retaining all of the output for subsequent reduction, but it significantly increases the amount of data storage capacity required.

An effective method for extending the dynamic range is voltage-to-frequency or current-to-frequency conversion followed by frequency-to-digital conversion. A block diagram of an efficient current-to-frequency converter having a dynamic range of 10^6 Hz over each of three ranges corresponding to 10^{-12}, 10^{-13}, and 10^{-14} A/Hz is shown in Figure 6.9. Input currents of either polarity are linearly converted into voltage pulses at a rate determined by the magnitude of the current. The voltage pulses are then converted into parallel-binary or eight-digit binary-coded decimal (BCD) signals for computer input, or they are output as voltage pulses suitable for sampled data systems in standard pulse-counting and digital display instrumentation.

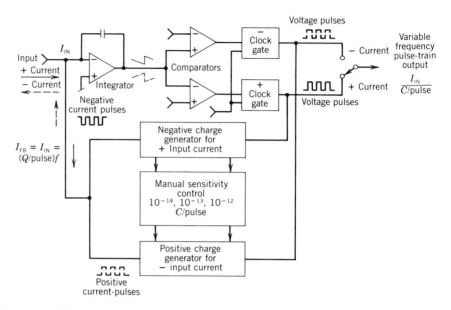

Figure 6.9 Wide range bi-polar current-to-frequency converter for high-speed data acquisition (Analog Technology Corp.; courtesy J. L. Lawrence.)

DATA ENHANCEMENT

The signal generated by the mass spectrometer consists of a DC component quantitatively indicative of the rate of arrival of ions at the detector, along with any DC off-set, and an AC component resulting from fundamental and environmental sources of noise. *Fundamental noise* arises from the quantization of energy and the thermal motion of atoms at temperatures above absolute zero, while *environmental noise* results from externally imposed inputs; for example, (1) pressure, temperature, or line current fluctuations; (2) charge build-up on insulating surfaces; (3) mechanical vibrations; and (4) 60 Hz coupling. With care, the latter can be made arbitrarily small; however, *fundamental noise* has a magnitude below which it cannot be reduced, and so it imposes a lower limit on ion current detection. The impressed noise is further categorized (Fig. 6.10) as (1) *white noise*, which is a random mixture of frequencies, amplitudes, and phase angles having a relatively constant power density that generally predominates at the higher frequencies; (2) f^{-1} *noise*, which includes *environmental noise* and which increases approximately as the reciprocal of the frequency at low frequencies; and (3) *interference noise*, which occurs at discrete frequencies, and which is typified by AC coupling and by transient inputs generated by switch operations or other impulse-generating actions. Each of these types of noise is present in the mass spectrometer signal, and each must be addressed in its own way.

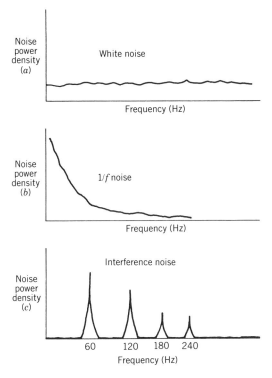

Figure 6.10 Classes of noise. (Malmstadt et al., 1981; copyright Benjamin–Cummings.)

In order to make full use of the powerful analytical capability of the mass spectrometer, it is almost always necessary to provide some degree of signal conditioning; and, except when large ion currents are being measured, some method to increase the signal-to-noise ratio (SNR = peak intensity/rms noise) may be required. There are many generalized electronic and mathematical approaches to signal conditioning and enhancement, with the method selected being determined by the type and amount of noise to be removed. A few of these methods most applicable to mass spectrometry are described below; a more complete discussion of the theory and applications of signal processing is found in the literature (e.g., see Coor, 1968; Goodyear, 1971; Oliver and Cage, 1971; O'Haver, 1972; Vassos and Ewing, 1972; Beauchamp, 1973; Ott, 1976; Beauchamp and Yuen, 1979; Liteanu and Rica, 1980; Arbel, 1980; and Malmstadt et al., 1981).

Some improvement of the SNR can be obtained by straightforward low-pass filtering (LPF) of the analog output prior to digitizing, in which simple RC (or LC) integrating network circuitry similar to that shown in Figure 6.11 limits the noise at frequencies higher than that of the signal. The filter output voltage V_f, which lags the input V_0 by the phase angle $(2\pi - \theta)$, is approximately equal to V_0 at low frequencies; it begins to decrease at 20

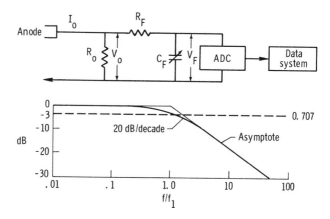

Figure 6.11   Low-pass filtering for improving the signal-to-noise ratio.

dB/decade at the cutoff frequency $f_c = (2\pi RC)^{-1}$, at which point $V_f = 0.707$ V_0. Sharper rolloff characteristics at cutoff are obtained with higher order (multiple) filters, and loading effects can be minimized with active filters incorporating operational amplifiers. The utility of LPF when the mass spectrometer is operating in a scanning mode is limited due to the RC time constant (τ). This results in a distortion of the peak even when τ is much less than the time (t) required to scan across the peak; and in a loss in peak magnitude when $\tau \geq t$. A certain amount of LPF will occur due to stray capacitance coupled with the large (10^7 to 10^{10} Ω) input resistors across which the output voltage is developed. Since the peak distortions are passed unchanged to the data system and are not restored in the computation, reduction of this inadvertent filtering becomes very important.

Further improvements in the SNR are limited by baseline drift as well as by other sources of f^{-1} noise, which cause the frequency spectrum of the signal and the noise to overlap. As noted in Figure 6.10, most of the noise occurs at moderate frequencies and below; and the higher frequency region is relatively noise-free. It is thus advantageous to transpose the DC signal into an AC signal having a frequency in the low-noise region, and then to amplify and measure this AC signal. The transposition is accomplished by *beam modulation*, in which the ion beam is electronically pulsed or mechanically chopped prior to interacting with the detector. This results in a nearly square waveform proportional to the difference between the zero (no current) baseline and the magnitude of the ion current. This waveform can be used to amplitude-modulate a carrier wave, or it can be AC-amplified if the pulse frequency is sufficiently high. The AC signal is then demodulated to DC for recording or digitized for input to the data system. It should be noted that noise components occurring prior to modulation will be transposed along with the signal, and only those occurring after will be removed. Modulating between the detector and amplifier, for example, would be less effective because f^{-1} noise associated with the detector would be passed, and only that associated with the DC amplifier would be removed. The improvement could still be significant however due to the large noise and drift associated with dc amplification.

If the modulation must take place following the detector, *chopper amplification*—a procedure in which pulse amplitude modulation of the signal is followed by internally synchronized demodulation and LPF—can be used to improve the SNR (e.g., Vassos and Ewing, 1972). Additive noise generated in the AC amplifier is rejected since it does not further modulate the signal, but only causes deviations about the baseline that are removed in the detection and filtering stage.

Amplification of an AC signal can be achieved with straightforward narrow-band-pass amplification; however, this approach is limited by the frequency stability of the AC amplifier. The stability problem is alleviated by *phase-lock amplification* (PLA), in which the preamplified modulated carrier is selectively amplified and synchronously demodulated by multiplication with a square wave having an equivalent frequency and phase to the modulated input signal (Coor, 1968; O'Haver, 1972; Malmstadt et al., 1981). The DC output is then low-pass filtered, producing an amplified form of the original signal without the additive noise component (Fig. 6.12).

None of these procedures will effectively reject white noise occurring within the modulation frequency bandwidth or multiplicative noise, such as that arising from drift in the AC amplifier gain or the statistical fluctuations in electron multiplier response. In these cases, computational methods are also required to extract the weak signal from the noise. One of the most effective is *time-based signal averaging* (TBSA) (e.g., Beauchamp, 1973;

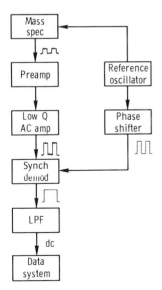

Figure 6.12 Phase-lock amplification to remove the effects of low-frequency noise.

Kraus, 1975; Arbel, 1980). The procedure is sometimes called ensemble or multichannel averaging, and it may be useful with or without pulse modulation of the signal.

In TBSA, the noisy signal is repeatedly sampled and digitized at fixed time intervals and the time-correlated magnitudes stored in separate memory locations within the computer. Successive repetitions are algebraically summed at each location, so that the coherent data signal increases linearly with the number of sampling cycles, while non-time-correlated noise is summed on a statistical basis and therefore averaged to a smaller value due to its incoherence. If the noise distribution is Gaussian with a mean value of zero and the variance of the noise (σ_n^2) is constant over the total sampling time, then the rms value of the noise will be the square root of the variance and the SNR will increase approximately as the square root of the number of samples taken. For random nonwhite noise, however, the $n^{\frac{1}{2}}$ improvement in SNR does not hold precisely since the noise values may or may not be correlated. Ernst (1965) considered these deviations in detail and showed that if the frequency dependence of the noise is given by $f^{-\lambda}$ ($0 \leq \lambda \leq 1$), then the proportionality to $n^{\frac{1}{2}}$ holds exactly, which is the most general case. With each repetition, the signal will increase with respect to the noise; or, if the signals are normalized to the number of samples taken, the noise will disappear into the baseline. A noisy spectrum of atmospheric krypton in natural abundance (1.1 ppm) taken with the small computer-controlled QMS mentioned earlier is shown in Figure 6.13a. Krypton has several isotopes, the most abundant of which are at mass 82 (11.56%), 83 (11.55%), 84

(a)

(b)

82   83   84   85   86   m/z

Figure 6.13 Time-based signal averaging of the mass spectrum of naturally occurring krypton at approximately 1.1 ppm (Wood and Yeager, 1980).

(56.90%), and 86 (17.37%). The TBSA spectrum (Fig. 6.13b) clearly shows the krypton isotopes in the correct ratio, and it establishes a detection limit for the instrument of about 40 ppb by volume when sampling against atmospheric pressure. The peak appearing at m/z 85 is not due to krypton, but to the $CClF_2$ fragment from a trace amount of freon-114 ($C_2Cl_2F_4$) simultaneously introduced into the sample. While the independent variable used in signal averaging is most often time, equally useful are magnetic field intensity, acceleration voltage, or quadrupole RF/DC voltages. Any of the mass spectrometer parameters relating to generating a repetitive mass scan may therefore be used, provided it can be synchronously coupled to the start of the time-averaging computer.

Signal averaging is particularly useful in acquiring weak or transient mass spectra associated with rapid scanning across eluting GC peaks. It has also been successfully applied with field desorption studies (Ligon, 1980b), fast-atom bombardment, and chemical ionization. By and large, signal averaging systems are integrated into a particular mass spectrometer, but Snelling et al. (1984) have described a multichannel signal analyzer using a minicomputer and interface that are independent of the type of mass spectrometer or the ionization method. The system provides for centroid and peak area calculations as well as for Savitzky-Golay (Savitzky and Golay, 1964) smoothing of the data.

Identification of the overlapping or low-intensity peaks often requires enhanced resolution as well as sensitivity. A resolution-enhancing procedure used for many years is *derivative mass spectrometry* (e.g., Beynon et al., 1958; Liteanu and Rica, 1980), in which the first or second derivative of the ion current signal with respect to mass—or to any independent variable indicative of the mass—is taken. The derivative can be obtained electronically; but to minimize losses in the SNR, it is more common to digitize the signal and obtain the derivative by computer-assisted numerical methods. As is apparent from Figure 6.14a, the derivative plot increases the precision with which the top of a resolved peak (and hence the exact mass) can be determined, since (as pointed out earlier) it becomes necessary only to determine the magnitude of the independent variable for which $dy/dx = 0$. In addition, hidden peaks can sometimes be resolved. Figure 6.14b shows the increased resolution obtained by taking the first and second derivatives of a signal representing the summation of two overlapping Gaussian peaks. Quantitative analysis based on the derivative is also valid, and the form of the transfer (calibration) functions remain unchanged.

Differentiation of the signal is also useful when continuously varying measurement conditions occur. An example of such an application is the measurement of the rate of gaseous desorption from solids by single mass monitoring with static mass spectrometry (Chapter 4), during which the total system pressure increases as a function of the temperature of the adsorbent (Kornelsen, 1970; van Gorkum and Kornelsen, 1981; Kornelsen and van Gorkum, 1981). This kind of data appears as a more or less continuously

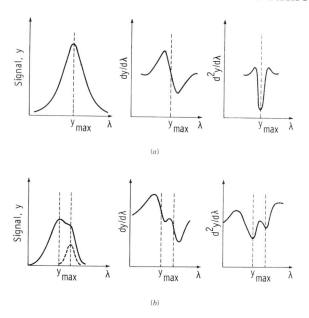

Figure 6.14   Derivatives of Gaussian signals. (Liteanu and Rica, 1980; copyright Ellis Horwood, Ltd.)

increasing signal representative of the species being monitored, with little indication of variation in the desorption rate. In this case, differentiation of the signal (sometimes coupled with PLA) will be reflected as relatively large changes in slope when a deviation from a steady-state increase occurs, even though it is not readily observable in the much larger total signal. Figure 6.15 is an example of data derived in this manner. The time derivative of the signal at $m/z = 4$, resulting from the desorption of helium from crystalline

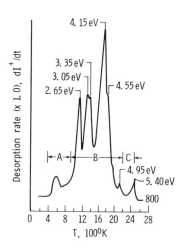

Figure 6.15   Desorption rate of helium ion implanted in a (100) tungsten surface measured by differentiation of a continuously increasing signal from a static mass spectrometer. (Adapted from Kornelsen, 1970; with permission of *Canadian Journal of Physics.*)

tungsten, shows a series of peaks when plotted against the temperature of the crystal. These peaks, which are not measurable in the undifferentiated signal, are indicative of desorption from the surface (A) and from different sites within the bulk (B) having the activation energies for desorption as shown. Therefore, differentiation of the signal provides additional information about the process that would not be realized by simply plotting the partial pressure.

Deconvolution

The most effective computational method to reduce instrumental broadening imposed on the signal is deconvolution; however, it is mathematically difficult to apply, and some noise is also restored along with the data. The results are such, however, that properly applied and performed deconvolution of the signal will yield an effective resolution much greater than that obtainable by careful tuning of the instrument and enhancing of the signal alone—sometimes reaching two to three times the theoretical limit of the resolving power.

If the instrument response does not change appreciably during the measurement, it is called *shift invariant*, and the relationship between the measured and actual signal is defined by the convolution integral (e.g., Bracewell, 1978; Wood et al., 1981; McGillem and Cooper, 1984)

$$h(x) = \int_{-\infty}^{\infty} f(y)g(x - y)dy = f * g$$

where h is the measured signal from the instrument, f is the true signal that would describe the physical process, g is the instrument response function representing all of the broadening effects, and $f * g$ is a notational convenience for the integral. If the Fourier transforms are obtained, the convolution integral $h(x) = f * g$ becomes a simple multiplication, $H(S) = F(S)G(S)$. The transform or frequency domain variable (S) is, in fact, the reciprocal of the function domain-independent variable (x), which may be any m/z-defining parameter sampled at equal intervals. Deconvolution in the transform domain is given by $F(S) = H(S)/G(S)$; and in the function domain, by obtaining the inverse Fourier transform $f(x)$ of $F(S)$.

Two limitations of this operation must be considered. As is evident from Figure 6.16, at the cutoff frequency (S_c) beyond which the instrument will not respond, $G(S) - 0$; and it is useful to define a principal solution F_p such that $F_p(S) = 0$ past that point (Bracewell and Roberts, 1954; Frieden, 1979; Ioup and Ioup, 1983). Even in the absence of noise this abrupt truncation of F_p produces Gibbs oscillations or sidelobes in f_p, which in some cases may be mistaken for mass peaks. Second, noise cannot be ignored, so that in actuality, for additive noise $F_p(S) = [H(S) + N(S)]/G(S)$ where $N(S)$ is the transform of the noise. Since noise predominates as the cutoff frequency

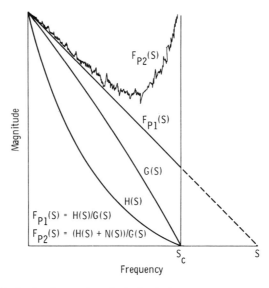

Figure 6.16 Effects of noise on the principal solution in the Fourier transform domain.

is approached, the high-frequency part of the transform domain spectrum is often deleted prior to obtaining the inverse transform, which is a process known as *ideal low-pass inverse filtering*. This reduces the effects of noise, but does not eliminate the Gibbs oscillations.

A number of computer-oriented computational methods to solve the convolution integral have evolved (e.g., Dromey and Morrison, 1970; Frieden, 1979; Carley and Joyner, 1979; Jannson, 1984; McGillem and Cooper, 1984 and Blass and Halsey, 1981). Two of these methods that have been recently applied to mass spectrometric data are (1) Fourier transform deconvolution with transform domain function continuation, and (2) function domain iterative deconvolution, both of which address the effects of noise and retain high frequency information (Ioup et al., 1980, 1983–1984; Wood et al., 1981; Howard, 1984; Rayborn et al., 1984). The higher frequencies are often important in enhancing resolution, and also may contain usable mass spectral information.

The function continuation method was devised to reduce the Gibbs oscillations and other effects of noise while extending the useful range of the principal solution to smaller values of $G(S)$ in the noise-dominated region (Howard, 1984). It was initially applied to gas chromatography (Howard and Rayborn, 1980), and to time-of-flight (TOF) mass spectrometry (Wood et al., 1981). Using the shape of an isolated, fully resolved peak as a guide, a simple instrument function, usually a symmetric or skewed Gaussian, is found. An initial solution f_1 is found by straightforward low-pass inverse filtering of the data, and the approximate size and location of the peaks are determined. The instrument function is skewed slightly, and the procedure is repeated until an optimum-filtered spectrum is obtained. This function is then used to construct a noise-free artificial representation a of f_1 with Gaussian peaks of equal height, width, and location as the optimized filtered solution. The Fourier transforms $F_1(S)$ and $A(S)$ are then obtained and their coefficients are numerically compared. A new transform domain function $F_2(S)$ is formed by replacing the coefficients of $F_1(S)$ in the high frequency region (where noise predominates) with the noise-free coefficients of $A(S)$ as shown in Figure 6.17. The low-frequency terms in $F_2(S)$, which determine the principal features of the inverse-filtered solution $f_2(x)$, are of course unchanged.

Iterative deconvolution, on the other hand, allows the general application of physically motivated constraints while iteratively solving the convolution sum

$$h_i = \sum_{j=-\infty}^{j=\infty} f_j g_{i-j} \equiv f * g$$

An iterative method for noise removal developed by Morrison (1963) to prepare data for deconvolution first smooths the data and then gradually restores the non-noisy data and the compatible noise (Ioup, 1968). The initial smoothing iteration produces data $h_1 = h * g$. The successive iterations are restorative, with the nth iteration given by

$$h_n = h_{n-1} + (h - h_{n-1}) * g \qquad n > 1$$

After the smoothed data have been restored the selected number of iterations, van Cittert's (1931) deconvolution or unfolding method is applied. The first approximation is assumed to be h (e.g., $f_0 = h$), while the nth unfolding is given by

$$f_n = f_{n-1} + (h - f_{n-1} * g) \qquad n > 0$$

A major difficulty of both of the iterative procedures is nonconvergence for many response functions, and some special methods have been developed to assure convergence (Kawata and Ichioaka, 1980; Ioup, 1981; Whitehorn, 1981; Lacoste, 1982). Iterative deconvolution is somewhat empirical, and the usual procedure is to compute several iterations and compare the output from each. In general, after the initial smoothing, fewer restorations and unfoldings should be used if the data are noisy since the noise will also be amplified.

Figure 6.18 shows original and deconvolved peaks at m/z 16 in a mixture of methane ($^{12}CH_4^+ = 16.0313$) and a small amount of oxygen ($^{16}O^+ = 15.9949$) taken on a 180° magnetic deflection mass spectrometer, which was

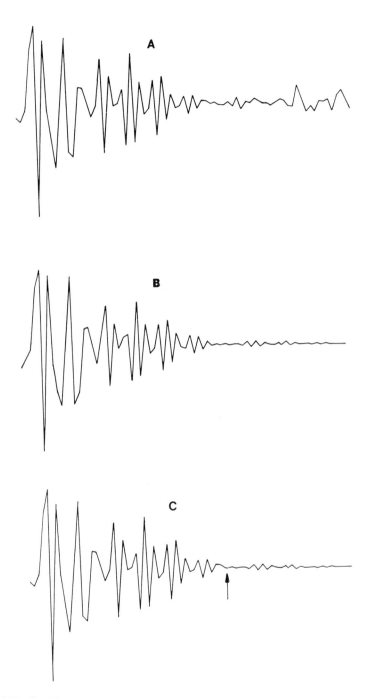

Figure 6.17 Combining the transform of the noisy function (*a*) with that of the artificial function (*b*) where indicated by the arrow to obtain the hybrid low-noise function (*c*). (Wood et al., 1981; copyright © 1981 IEEE.)

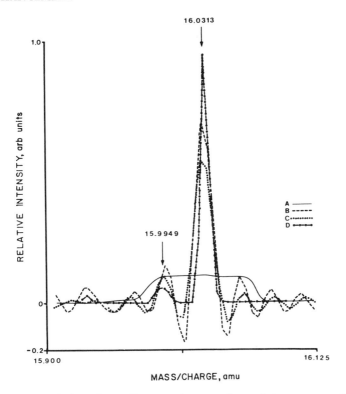

Figure 6.18 Deconvolution of a small oxygen fragmentation peak in an excess of methane, (*A*) with the mass spectrometer adjusted to completely overlap the two peaks. (*B*) Results of low-pass filtering alone. (*C*) Function-continuation solution using a Gaussian artificial function, and (*D*) iterative deconvolution with 30 smoothings and 150 unfoldings (Ioup et al., 1983–1984; copyright Elsevier Science Publishers B.V.)

purposely adjusted to have the peaks completely overlap (curve *A*). A Gaussian was used as the artificial function *a*, and the response function *g* was determined from the isolated Ar^{2+} peak at *m/z* 19.9812. Curve *B* is the result of ideal low-pass inverse filtering alone, where the methane peak is seen to be clearly resolved. Although the first positive sidelobe to the left of the methane is larger than the one on the right, the magnitude of the Gibbs oscillations makes positive identification of the oxygen peak difficult. Considerable improvement is seen in the function continuation result in curve *C*; however, enough sidelobe effect remains to cause some difficulty in interpretation and precludes quantification. Further reduction in the magnitude of the oscillations would have been realized by making small changes in the shape of the instrument function, for example, from a symmetric Gaussian to one that more accurately fits the shape of the isolated peak, and repeating the calculation. Iterative deconvolution using 30 smoothings and 150 unfoldings is shown in curve *D*. Gibbs oscillations have been suppressed, and the clearly resolved oxygen peak can be quantified with some confidence.

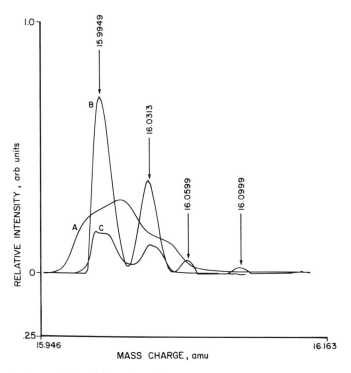

Figure 6.19 Deconvolution of the methane ion peak from a predominant oxygen fragmentation peak. Curve *A* is the unresolved spectra, *B* is the iterative deconvolution with 30 smoothings and 50 unfoldings, and *C* is the data taken with the spectra resolved by the instrument (Ioup et al., 1983–1984).

A second case in which oxygen is the predominant species is shown in Figure 6.19, where the oscillations have not been suppressed by iterative deconvolution. Curve *A* is again the unresolved data, and curve *C* is the spectrum with the instrument adjusted, so that the peaks are resolved. The major peaks in curve *B* can be unambiguously identified; however, the small peaks at mass numbers 16.0599 and 16.0999 cannot be readily explained from a mass spectrometric standpoint. The latter is almost certainly an uncancelled Gibbs oscillation. The peak at 16.0599 may be a similar oscillation, or it may be an artifact resulting from a marked dissimilarity in the peak shapes of the methane (which closely matched that of the doubly charged argon) and the atomic oxygen (which did not because of a long tail on the high-mass side). The response function as defined by the argon peak would not necessarily be correct for the oxygen in this case, since the concept of shift invariance assumes that identical broadening applies to both peaks. This, in itself, does not invalidate the deconvolution of the oxygen peak. There is, however, an additional broadening effect that corresponds to a change in input and is not removed by the convolution. Therefore, some care must be used in defining g, and a close examination of any small peaks occurring in the deconvolved result is required.

An analogous procedure that improves the SNR (but not necessarily the resolution) is function (time) domain or frequency (transform) domain digital filtering (e.g., Horlick, 1972; Beauchamp, 1973; Stearns, 1975; Beauchamp and Yuen, 1979; Bromba and Ziegler, 1981, 1983; Bush, 1983). In this case, the response function in a convolution equation is the impulse response fuction of a low-pass, high-pass, or band-pass filter, that is, the response of the filter to an infinitesimally narrow input. Assuming stationary white noise, the noise reduction is given by the inverse square root of the sum of the squares of the coefficients of the filter function (Hamming, 1977). The filtered signal in the time domain is then obtained by iterative deconvolution. The computational methods are most often based on the procedures of Savitzky and Golay (1964), in which they showed that the convolution of polynomial functions was equivalent to moving-average smoothing filters, with added information being obtained from the signal by least squares minimization of random error. As before, the filter function in the function domain is related to that in the frequency domain by the Fourier transform. The filtered solution can be obtained from the inverse transformation, since the filtering is performed as a multiplication in the transform domain. Low-pass transfer functions for the filter again exhibit Gibbs oscillations due to abrupt truncation of the Fourier transform; these are sometimes smoothed themselves by multiplication with an appropriate window to remove effects caused by sharp discontinuities at the end-points.

Digital signal-processing devices generally allow the use of symmetric filter functions, are less susceptible to drift, and can provide time constants that are not available with their analog counterparts. Incorrect selection of gain or time constant, however, results in amplitude quantization errors or distorted output. Thompson and Dessey (1984) have addressed these considerations and have described experimental methods for determining output signal fidelity and for reducing aliasing of the noise. Aliasing often results from undersampling of the data, and is characterized as higher frequency components reflected into and, appearing as, overlapping low frequency components (Bracewell, 1978; McGillem and Cooper, 1984).

DATA REDUCTION

The interpretation and reduction of data acquired by mass spectrometry falls into three very broad categories: (1) identification and quantization of atomic and molecular species; (2) measurement of isotopic ratios; and (3) the detection of time- or temperature-dependent phenomena relating to the interaction of ions with matter. Recently, the characterization of high atomic charge states has also become important in connection with laser multiphoton ionization and ions of mega electron volt energies.

Qualitative analysis by mass spectrometry is somewhat straightforward if pure compounds of low molecular weight are involved. The identification of the components in a mixture or the analysis of the higher molecular weight species often encountered in organic chemistry is generally more difficult. The *fragmentation* or *cracking pattern* of a pure compound is a series of peaks, each normalized to the most intense (or "base") peak, at masses that are indicative of the molecular weight and structure of the compound.

Comparison of a similarly normalized spectrum of an unknown to a previously compiled library of the spectra of pure compounds should then identify the unknown. This is true in principle; in fact, it is how the analyses are often carried out. It should be apparent, however, that as the number of atoms increase, the number of pure spectra to be compared becomes quite large and the uniqueness of a spectrum is diminished.

Identifying the components in a mixture is more difficult in that each must be sequentially identified, generally beginning with the highest molecular weight; then the contribution to the total spectrum at each mass must be subtracted. The process is then repeated with the remaining peaks until all of the species have been identified. The successful identification of the species in the mixture generally requires that at least one resolved singlet for each provide sufficient leverage to calculate the fractional composition of the remainder from the calibration cracking pattern, and that "pattern recognition" and intuitive reasonings of the mass spectrometrist be applied.

To facilitate the identification procedure, collections or "libraries" of tables of normalized mass spectra of pure compounds have been compiled with which unknown spectra can be compared. Use of these tables in qualitative analysis can be very effective. Quantitation, on the other hand, requires that special calibration procedures be followed to determine the instrument sensitivity for each species, and that the analyses be performed systematically.

A number of the collections and the procedures and algorithms for searching them have been reviewed (e.g., Chapman, 1978; Henneberg, 1980; Martinson, 1981; Heller, 1984; Karasek and Clement, 1985), and an extensive literature has developed over the last decade. There are presently three major computer-based collections that are readily available or accessible: (1) the Wiley Registry of Mass Spectral Data, which contains approximately 80,000 different spectra of 68,000 compounds (McLafferty and Stauffer, 1984); (2) the MSDC Database of 25,000 spectra (the Aldermaston Library) compiled by the Mass Spectrometry Data Center of Great Britain (Jager et al., 1979); and (3) the NIH/EPA/NBS/MSDC Mass Spectral Search System (MSSS) compiled through a joint effort of the United States and British governments, which holds 60,000 spectra—each identified by the Chemical Abstracts Service (CAS) registry number for the compound (Heller, 1984).

The MSSS database has been incorporated into the NIH/EPA Chemical Information System (CIS), which consists of a network of chemical databases (e.g., nuclear magnetic resonance, magnetic resonance, x-ray diffraction, toxicity, etc.) linked by a central structure and nomenclature search system (SANSS). This allows the user to search each of the databases independently or interactively to identify a compound (Milne and Heller, 1980; Heller and Milne, 1980a, b).

A number of operating parameters for the mass spectrometer may have an effect on the observed cracking pattern, particularly when higher molecular weight spectra are obtained by GC/MS. Speck et al. (1978) have developed a procedure for calculating a quality index (QI) for the spectra based on a number of factors. This index, currently the product of nine individual quality factors, has been applied to MSSS—where it was found that if the spectra received a low QI, it was most often due to peaks appearing at masses above the molecular ion, to truncation of the lower masses, or to illogical loss of fragments. Databases of chemical ionization spectra, mostly obtained using methane as the reagent gas, are also being compiled by the MSDC and CIS; and MSDC has begun a compilation of spectra obtained by fast-ion-bombardment (FAB). A QI algorithm for chemical ionization is also being developed by modification of the electron-impact algorithm to account for differences in the two ionization processes.

In addition to the libraries of complete spectra, the MSDC publishes an "Eight Peak Index," also available in printed form, of the eight most intense peaks in approximately 31,000 spectra. Cataloged in order of increasing molecular weight, elemental composition, and most abundant ions, the Index is extremely useful in preliminary or prefiltering searches, or for manual identification of a compound.

The major collections are all accessible through external communication links for information retrieval or for interactive searching. Currently, this method is preferred for generalized searching and is the one most often used. For the most part, the versatility and speed of the main frame computer and the size and centralized compiling of the major collections make them both efficient national repositories and the most effective means of data analysis. Clearly, however, the availability of the megabyte hard disc-based memories in small computers will make practical the compilation of segments of the major databases to form smaller, user-specific libraries; papers relating to this approach are now appearing in the literature. To cite only one example, Koyama et al. (1982) have described a database containing only hydrocarbons (C_nH_m) and single-oxygen compounds ($C_nH_mO_1$) extracted from MSSS and stored on standard-sized floppy discs for use with a small personal computer. By limiting the stored masses to the 10 to 13 most intense peaks having relative intensities above 10% of the base peak, approximately 2500 species can be stored on a single disc when peak height information is not included. Grotch (1973) showed that mass spectra are highly overdetermined; and that even if intensity is ignored, good identification could still be obtained if only the mass positions above a given threshold were known. Using this, Koyama et al. (1982) were able to correctly identify species and to correlate base peaks to molecular structure using only 5–10 major peaks. The rapid development in data storage capabilities will soon reduce the need for truncation of the reference spectra, thus retaining interpretively significant information and permitting the use of the more advanced search methods.

Database Search Methods

A number of computerized methods for interpreting mass spectra have been reported (e.g., Chapman, 1978; Zupan, 1978; Dromey, 1979; Rasmussen et al., 1979; Martinsen, 1981). Many of these are specialized programs used in a research environment or in routine analyses in the laboratories in which they were developed. Others were developed for use with particular types of instruments such as GC/MS (Dromey et al., 1976) or MS/MS (Cross and Enke, 1984).

The most highly developed methods, which are also generally available, are those related to the major databases. Of these, the most interactive is the PEAK search method resident in MSSS, in which selected mass and peak intensity data from the unknown are compared one at a time to all spectra in the library, and the number of compounds containing a peak within the specified range of relative intensities are reported (Heller, 1984). As subsequent peaks and intensities are entered, the number of compounds containing all of them decreases. When a small enough number remain, the CAS registry number, QI, molecular weight, molecular formulas, and name of each candidate species are listed. The database can also be queried for a listing of all molecules or molecular fragments having a specified mass. For example, a list of compounds containing a fixed number of chlorine or bromine atoms, which also have a base peak at a specified m/z, can be generated.

The PEAK search method permits maximum use of the intuitive skills of the mass spectrometrist. It also requires the most time, and is not the optimum method for processing large quantities of data such as those generated in the normal operation of the GC/MS. Several less interactive methods have evolved that compare a selected number of peaks in unknown and reference spectra and compute a similarity index indicating the probability that the unknown is correctly identified. The degree of confidence in the match is as much or more a function of which peaks or peak clusters are matched than how many. The mass spectra of both the reference and the unknown are generally abbreviated by selecting the n most intense peaks within either a set of m masses or within mass unit intervals Δm. Both of these abbreviation methods reduce storage requirements and computation time; however, the latter generally deletes less structural information. Biemann and coworkers selected the two most intense peaks in a Δm of 14 amu, corresponding to the CH_2 homologous series, and started the intervals at $m/z = 6$ (i.e., 6–19, 20–33, etc.) to avoid splitting common peak clusters (Hertz et al., 1971). They found that interpretively significant peaks in the upper end of the spectrum were retained, including the molecular peak if it existed.

Furthermore, this method is less susceptible to distorted spectra, that might result from an improperly tuned QMS or from obtaining data on the side of a GC peak, since the two largest peaks in each interval should be the same irrespective of their relationships to peaks in adjacent intervals.

The search methods are classified as retrieval systems that attempt to identify the molecule by point-to-point comparison of the unknown and reference spectra, and as interpretive systems that attempt to identify features such as molecular weight, structures, substructures, and elements present in the molecule (McLafferty, 1978). The Biemann (KB) peak and intensity search system (Hertz et al., 1971) is resident on MSSS (Heller, 1972), and it is one of the most successful and widely used of the retrieval methods. The reference spectrum and the unknown are both abbreviated by the Δm = 14 amu method described above. A similarity index is calculated based on three factors: (1) the ratios of peaks common to both spectra; (2) the total intensity of all peaks in the unknown; and (3) the fraction of the total intensity due to peaks that do not have a corresponding peak in the other spectrum. Before matching, prescreening eliminates widely dissimilar spectra in order to reduce search times.

The KB system is a forward search method, that is, all peaks in the abbreviated spectrum of the unknown are compared to those in the reference. The method is straightforward and is generally successful when the unknown compound is in the library. It will also retrieve homologous compounds or isomers when the unknown itself is not actually listed, which is an important consideration since it effectively expands the database.

The KB system is less successful when the unknown is a mixture, as is encountered with incompletely resolved GC fractions. In this case, reverse searching methods are used (Abramson, 1975) in which the peaks in the abbreviated reference spectrum are compared to that of the unknown. Unmatched peaks in the unknown are not considered; hence, they do not affect calculation of the similarity index. The most commonly used reverse search method is the probability-based matching (PBM) system developed by McLafferty et al. (1974), with many subsequent modifications and improvements (e.g., Pesyna et al., 1976; McLafferty, 1978; Mun et al., 1981a, 1981b; McLafferty et al., 1983; Buttrill et al., 1983). With PBM, an abbreviated reference spectrum of up to 15 peaks is assembled for each compound, with peak selection being based on the uniqueness of the peak in the spectrum as well as its abundance. The uniqueness values (U and A) for each of the selected masses are based on the probability of occurrence at abundances above a selected value in spectra of randomly chosen compounds. As first proposed by Grotch (1969), the probability of occurrence of relative abundances follows a log-normal distribution. The probability of finding a peak of a particular mass at any abundance level (AL) is defined as $2^{-(U+A)}$ where $A = 0$ for an AL of 1%. At higher abundances, A increases because of the decreasing probability of finding a unique peak.

The peaks in each reference spectrum are ordered according to the U +

A values, and those having equal $U + A$ are subordered by decreasing m/z. If the reference peaks are not found in the unknown spectrum, the compound is rejected and the next reference spectrum is entered. When a compound is identified, a confidence factor K, which is also weighted by U and A values, is calculated and compared to a threshold K_{th}. The identified compound is then accepted if $K \geq K_{th}$. After additional checking, the next reference spectrum is entered and the process is repeated until all compounds in the mixture have been identified.

As with the KB forward search, PBM makes use of the nominal (low resolution) mass of the ion. Spencer and Stauffer (1984) have reported on an enhanced PBM method (EPBM) using a library of nominal mass tables to analyze high-resolution mass spectra of an unknown. In this procedure, the exact mass of an assigned reference peak is calculated and compared with the measured accurate mass of the unknown ion. If the calculated mass differs from the measured mass by more than experimental error, then the reference peak is flagged as not being present in the unknown. It is therefore possible to discriminate between ions having the same nominal mass, which cannot be done with straightforward PBM. Using this method, they reported a 40% decrease in the number of possible compounds listed for 3,5-dimethylpyrazole ($C_5H_8N_2$) in an analysis of FTMS data.

Yasuhara et al. (1983) described a PBM search (NEIS/MSLS) in which a five-step prefiltering procedure based on interpretive characteristics of mass spectra is used. The algorithm also calculates a measure of confidence based on both McLafferty U and A values and a proportion contributed from other spectra using a method similar to that of Pesyna et al. (1975, 1976). In an analysis of 54 spectra of two-component mixtures, correct identification of both components was reported in 70% of the cases. The system was found to be very selective for alcohols and fatty acids, but was less so for aliphatic hydrocarbons.

Damen et al. (1978) and Henneberg (1980) described a search system (SISCOM) for identifying components in a mixture even when the unknown compound is not in the library. The spectrum is abbreviated by requiring that the isotope-corrected intensity of the selected peak exceed both a relative threshold (i.e., 2% of the base peak) and the arithmetic mean of that of the next higher and next lower homologous neighbors (± 14 amu) if the peak is to be considered characteristic. This procedure succeeds in selecting certain ions, including small ones, in the neighborhood of more intense but less characteristic fragments. For example, abundant ions of low molecular weight in simple alkane fragmentation patterns, which provide little analytical information, are usually selected by the maximum m in n or Δm methods, but are not selected by SISCOM. The calculated similarity index is based on (1) a modified correlation coefficient determined from the correspondence of the pattern of the peaks common to both the unknown and the reference; (2) the number of the remaining unmatched peaks common to both the unknown and reference; and (3) the sum of the intensities of unmatched characteristic peaks normalized to the total intensity of all characteristic peaks in the unknown and reference, respectively. A high value for the correlation coefficient and a low number of unmatched peaks indicate a good match, while a high number of unmatched peaks indicates that the compound is in a mixture.

Pattern recognition methods for the classification of compounds based on mass spectral features have been under development for some time (Isenhour and Jurs, 1971; Kowalski, 1975; Martinsen, 1981; Varmuza, 1984). These methods retrieve similar compounds from the library in the absence of an exact match through recognition of structural features. Pattern recognition methods are complex, and they are based on (1) representing a set of reference spectra as points in multidimensional space (2) developing a classification algorithm with a training procedure based on known structures through which the computer learns to classify the patterns, and (3) applying the algorithm to classify structures from an unknown compound. For example, a planar surface in multidimensional space can be found that separates compounds with a particular feature (e.g., oxygen, sulfur, etc.) from all others. The axes in spectral space can be defined as increments in m/z with a point representing an ion located by intensity. In structural space, the axes correspond to molecular structural features with the representative point located by the frequency of occurrence in the molecule. Pattern recognition then maps the molecule from spectral to structural space (forward mapping) or generates mass spectra from candidate structures, which are then matched to the measured unknown spectrum (reverse mapping). An important pattern recognition method is cluster analysis, which is based on the fact that compounds of similar structure will be represented by points clustered in the same region in multidimensional space. An effective method for cluster analysis is the K-nearest neighbor (KNN), $K = 1, 2, 3, \ldots . n$, in which the distance between unknown and reference points is calculated, the K-nearest spectra is retrieved, and the predominant class is assigned to the unknown. The KNN method has been found to have the highest predictive reliability (Justice and Isenhour, 1974; McGill and Kowalski, 1978; Chien, 1985). Several other methods used under certain circumstances, such as the learning machine method for linearly separable data, digital learning networks, and factor analysis, have been described in detail by Chapman (1978) and reviewed by Martinsen (1981) and others.

The self-training interpretive and retrieval system (STIRS) developed by McLafferty and coworkers (Kwok et al., 1973; McLafferty, 1978; McLafferty and Venkataraghaven, 1978) is the most analytically efficient of the interpretive pattern recognition methods. The STIRS algorithm currently predicts the presence of any of 600 preselected substructures in the unknown spectrum (Stauffer et al., 1983); however, it cannot identify a substructure that is not present in some compound in the reference file. The absence of a predicted substructure, therefore, does not absolutely determine its absence in the unknown.

The structural features of a compound are represented by characteristic fragmentation patterns of relative abundances or mass differences. Fifteen data classes indicative of different compound types or fragment structure are defined in STIRS. These are based on (1) homologous, aromatic, or amine series; (2) characteristic ions in selected portions of the mass range; and (3) neutral losses. STIRS trains itself to interpret the unknown spectrum by matching peaks for each of the classes against the corresponding peaks in all of the reference spectra, and it indicates similarity by calculating individual and overall match factors based on measured intensities and specified weighting factors. The 15 compounds having the highest overall match factors are then searched for the presence of each of the preselected substructures; those found in a substantial number of the least-matching reference spectra have a corresponding high probability of being in the unknown. STIRS also contains subroutines for identification of the molecular ion to determine elemental composition using isotopic ratios and to crosscheck for an identical spectrum in the reference file. Compared to the KNN method, in one study, STIRS was found to show a factor of 4 reduction in false-positives with a corresponding increase from 64% to 84% in correct identifications (Isenhour et al., 1974). Furthermore, KNN predicts substructures for which it has been pretrained, while STIRS will predict any structural feature that is present in the database.

The best example of a heuristic interpretive or artificial intelligence method that attempts to find a structure by matching a hypothetical mass spectrum with that of an unknown are the DENDRAL (*Dendr*itic *Al*gorithm) programs developed at Stanford (Buchanan et al., 1971; Smith et al., 1972, 1973; Lederberg, 1972; Carhart et al., 1975; Smith and Carhart, 1978). As opposed to the library search methods, DENDRAL incorporates rules by which chemically plausible structures can be generated, more closely emulating the inductive reasoning and procedures of the mass spectrometrist in performing the analysis. The success of the system is largely due to the extensive amount of heuristic information it contains and its interactive approach. DENDRAL is actually a collection of subprograms that recognize the molecular ion, infer structure from the unknown spectrum, generate all possible structures within limitations imposed from consideration of the spectra, and test the generated structures against the unknown. The unknown mass spectra and the empirical formula of the molecular ion are entered. Based on fragmentation theory, a list of functional groups is then generated by a subprogram that contains a list of structural fragments, each having a characteristic set of spectral values and relationships associated with it. Any functional groups considered to be present are placed in a GOODLIST, and those considered to be absent because of missing peaks are placed in a BADLIST. All possible candidate isomeric structures that both contain the elements in GOODLIST and lack the elements in BADLIST are then constructed by a structure-generating program. Additional constraints inferred from the unknown spectrum are applied to force the selection of plausibly stable and nonredundant structures (Lederberg, 1972). Each of the candidate structures is examined using fundamental ionization processes and rules of mass spectrometry, and each structure is discarded if the predicted spectrum is incompatible with the unknown. The remaining predicted structures are then ranked in descending order of compatibility.

QUANTITATIVE ANALYSIS

If a suitable calibration can be obtained, the mass spectrometer provides a highly sensitive means for quantitative analysis. Aside from instrumental and ionization factors, the intensity of the ion beam is determined by the number density of the species in the ion source. The instrument sensitivity for any pure species is generally expressed in terms of the intensity of the ion current for the base peak normalized to the total pressure in the ion source, or to the pressure measured at a point past which the pumping speed and pressure drop are constant. Sensitivity is therefore given by $S = I^+/P$, where I^+ has units of intensity and P has units of pressure. If, on the other hand, the source pressure is constrained to remain constant, such as in the case of an inlet designed to continuously sample against atmospheric pressure, the sensitivity may be expressed in terms of intensity per unit of concentration. Quantitative analysis of a pure compound is therefore straightforward once instrument sensitivity has been determined, but it is somewhat more complicated for a mixture.

The ion current i^+ at any mass due to any given component is $i^+ = ap$, where a is the relative sensitivity per unit pressure (or concentration) for that particular mass and p is the partial pressure if the component is part of a mixture. It should be noted that a applies to fragment as well as molecular ions, and therefore defines the fragmentation or cracking pattern for any compound. For any given m/z, the total intensity i_n^+ is the sum of the contributions from each component in the mixture, $i_n^+ = a_1 p_1 + a_2 p_2 + \ldots$ and so on. Each component in a mixture, therefore, contributes to the measured mass spectrum according to its fragmentation pattern and concentration. For a simple mixture consisting of a few components, the spectrum often has isolated peaks from which computational leverage can be obtained. In this case, it is relatively straightforward to calculate the contribution at each of the remaining masses from the known fragmentation pattern associated with a component, and to eliminate the components from the mixture one at a time. In a simple air mixture, for example, the concentration of CO_2 would be determined based on the m/z 44 peak, and the contributions at m/z 28, 16, and 12 would be subtracted from the remaining spectrum. The procedure would be repeated until all of the components were identified and their concentrations calculated. For more complex mixtures, the total set of data may be represented as a set of linear simultaneous equations

$$i_1^+ = a_{11}p_1 + a_{12}p_2 + a_{13}p_3 + \cdots + a_{1n}p_n$$

$$i_2^+ = a_{21}p_1 + a_{22}p_2 + a_{23}p_3 + \cdots + a_{2n}p_n$$

$$\vdots$$

$$i_m^+ = a_{m1}p_1 + a_{m2}p_2 + a_{m3}p_3 + \cdots + a_{mn}p_n$$

which must be solved to perform the analysis. If a characteristic mass or a series of masses for each species can be selected, the set can be reduced to a subset of n equations with n unknowns, which may be expressed as a square matrix, $I = AP$. It should be noted that some care must be taken in the selection of the masses in the matrix to avoid errors. The partial pressure (or concentration) is then obtained from the product of the intensity matrix (I) and the inverse matrix of the fragmentation coefficients (A), $P = IA^{-1}$. The elements of the inverse matrix can be entered from tables or can be determined by direct calibration prior to inverting. The inverted matrix method forms the basis for quantitative analysis of selected species in several computer-controlled mass spectrometer systems used in process control or as automated trace gas analyzers to measure contaminants in air. In these applications, it is usually sufficient to measure only a limited number of species, which simplifies the construction of the fragmentation matrix as long as characteristic masses having minimal interferences from other species can be found.

For many applications, measurement of relative rather than absolute concentrations is sufficient to define a process or event. It is then necessary only to measure changes in the ion current of a characteristic ion while holding total pressure and other ion source operating conditions constant. Alternatively, the measured ion current may be normalized to that of a tracer present in known concentration in the sample, in which case changing ion source conditions are accounted for. The tracer is most often an inert gas to avoid depletion through reaction with other components in the sample; for example, argon may be used if the primary gas mixture is normal atmospheric air, and more than one tracer may be used if necessary. A recent experiment exemplifying this approach was the measurement of mixing of small amounts of nitrogen injected into a stream of superheated air flowing at approximately Mach 7 in a high-enthalpy wind tunnel (Wood et al., 1983). Argon in natural abundance (0.93%) was used as a measure of the nitrogen concentration (78.1%) in the air, and 1% neon was mixed with the nitrogen additive as a tracer. The mixing ratio was then determined from the ratio of the Ne^+ (m/z 20) to Ar^+ (m/z 40) intensities taken with the ionizing electron energy at 35 eV to eliminate interference from the doubly charged argon at m/z 20.

A particularly sensitive method for measuring small quantities of material is isotope dilution (Chapter 2), where an isotopically enriched elemental tracer is added to the sample and the weight of the corresponding element is determined by measuring isotopic ratios. Once the sample and tracer have been mixed, chemical extraction methods can be carried out without loss of analytical integrity, since the isotopic effects are small and can be accounted for.

The natural isotopic abundances of stable isotopes of some of the elements useful in mass spectrometry are listed in Table 6.2. Unless enriched, the mass spectrum of an atomic ion will simply reflect the natural abundances shown, that is, the ^{22}Ne peak will be 9.2% of the peak at ^{20}Ne, and so on. For molecular ions, however, the spectrum must be calculated from the products of the binomial expansions of probability equations $(a + b + c + \ldots)^n$ for each atom in the ion, where a, b, c, etc. are the fractional abundances of each isotope and n is the number of each type of atom in the ion—a task handled much more efficiently by the computer when more than two or three isotopes or atoms are involved (e.g., O'Malley, 1982). The methane (CH_4) molecular ion, for example, is represented by $(a + b)(c + d)^4$, where a through d corresponds to the ^{12}C, ^{13}C, ^1H, and ^2H fractional abundances, respectively. The m/z 17 peak in the methane spectrum due to the ^{13}C isotope will be 1.1% of the $^{12}CH_4$ at m/z 16, corresponding to the $^{13}C/^{12}C$ ratio. On the other hand, the contribution to the same peak from $^{12}C^1H_3{}^2H$ corresponding to the $4c^3d$ term in the expansion of $(c + d)^4$ will be only about 0.06%.

Isotope labeling is applicable to all elements having two or more isotopes. The natural isotopic contribution at some masses is extremely small, so that adding small amounts of a labeled tracer can often provide a means of unambiguously determining reaction products and, in some cases, calculating the chemical kinetics in a carefully designed experiment. The data shown in Figure 6.20 are typical of that obtained from this type of experiment, and they show dissociation and recombination of oxygen as a function of the temperature on a heated SiO_2 surface. The test gas, in this case, was a 1% mixture of $^{18}O_2$ and $^{16}O_2$ in a neon carrier, where the reaction was indicated by the appearance of the $^{16}O^{18}O$ peak at m/z 34.

The isotopic dilution method has been mentioned above in terms of an internal standard, and elsewhere throughout this volume. It is, for example, an important technique for geological and archaeological age determinations, and it is a powerful method for chemical analysis It is also an effective method for determining molecular structure and in tracing reaction mechanisms (Grostic and Rinehart, 1971). Carter et al. (1978) have reported the use of isotopic dilution methods in computerized spark source mass spectrometry for the quantitative analysis of trace levels of cadmium and molybdenum spiked with ^{106}Cd and ^{97}Mo, respectively. They also extended the technique to dry samples in the analysis of powdered coal and fly ash for metallic elements.

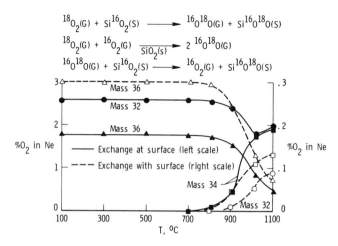

Figure 6.20 Measurement of dissociation and recombination of oxygen on a heated quartz surface by stable isotope mass spectrometry (Batten et al., 1985).

For a hydrocarbon, the contribution to the $m + 1$ isotopic peak is approximately equal to the number of carbon atoms, which is useful in verifying the molecular ion formula from the isotopic ratio. Atoms whose major isotope appears at $m + 2$, for example, chlorine (^{35}Cl, ^{37}Cl) and bromine (^{79}Br, ^{81}Br), have characteristic intensity ratios at $m + 2$, $m + 4$, etc. that are strongly indicative of their presence. The abundance ratios of the molecular ion in the groups associated with each of these isotopes is also indicative of the number of these atoms in the molecule.

REFERENCES

Abramson, F. P., *Anal. Chem.*, *47*, 45 (1975).

Arbel, A. F., *Analog Signal Processing and Instrumentation*, Cambridge University Press, New York (1980).

Batten, C. E., B. T. Upchurch, G. M. Wood, R. T. Swann, R. F. Hoyt, and G. J. Allen, *189th National Meeting*, American Chemical Society, Miami (April 1985).

Beauchamp, K. G., *Signal Processing*, Wiley, New York (1973).

Beauchamp, K. G., and C. Yuen, *Digital Methods for Signal Analysis*, George Allen and Unwin, London (1979).

Beynon, J. H., S. Clough, and A. E. Williams, *J. Sci. Instr.*, *35*, 164 (1958).

Blass, W. E., and G. W. Halsey, *Deconvolution of Absorption Spectra*, Academic Press, New York (1981).

Bracewell, R. N., *The Fourier Transform and Its Application*, 2nd ed., McGraw-Hill, New York (1978).

Bracewell, R. N., and J. A. Roberts, *Austr. J. Phys.*, *7*, 615 (1954).

Bromba, M. U. A., and H. Ziegler, *Anal. Chem.*, *53*, 1583 (1981).

Bromba, M. U. A., and H. Ziegler, *Anal. Chem.*, *55*, 648 (1983).

Buchanan, B. H., A. M. Duffield, and A. V. Robertson, in *Mass Spectrometry: Techniques and Applications*, ed. G. W. A. Milne, Wiley-Interscience, New York (1971), p. 121.

Bush, I. E., *Anal. Chem.*, *55*, 2353 (1983).

Buttrill, W. H., D. P. Martinsen, B. H. Song, and F. L. Tobin, presented at the *31st Ann. Conf. on Mass Spectrom. and Allied Topics*, Boston (1983), p. 552.

Carhart, R. E., D. H. Smith, H. Brown, and C. Djerassi, *J. Am. Chem. Soc.*, *97*, 5755 (1975).

Carley, A. F., and R. W. Joyner, *J. Electron Spec. and Related Phenomena*, *16*, 1 (1979).

Carrick, A., *Int. J. Mass Spectrom. Ion Phys.*, *2*, 333 (1969).

Carter, J. A., J. C. Franklin, and D. L. Donohue, in *High Performance Mass Spectrometry: Chemical Application*, ed. M. L. Gross, ACS Symp. Series 70, American Chemical Society, Washington, D.C. (1978), p. 299.

Chapman, J. R., *Computers in Mass Spectrometry*, Academic Press, London (1978).

Chien, M., *Anal. Chem.*, *57*, 348 (1985).

Comisarow, M. B., and J. D. Melka, *Anal. Chem.*, *51*, 2198 (1979).

Coor, T., *J. Chem. Ed.*, *45*, A533, A583 (1968).

Crawford, R. W., H. R. Brand, C. M. Wong, H. R. Gregg, P. A. Hoffman, and C. G. Enke, *Anal. Chem.*, *56*, 1121 (1984).

Cross, K. P., and C. G. Enke, presented at the *32nd Ann. Conf. on Mass Spectrom. and Allied Topics*, San Antonio (1984), p. 646.

Damen, H., D. Henneberg, and B. Weimann, *Anal. Chem., Acta*, *103*, 289 (1978).

David, E. E., keynote address at *32nd Ann. Conf. on Mass Spectrom. and Allied Topics*, San Antonio (1984), p. 858.

Dromey, R. G., *Anal. Chem.*, *51*, 229 (1979).

Dromey, R. G., and J. D. Morrison, *Int. J. Mass Spectrom. Ion Phys.*, *4*, 475 (1970).

Dromey, R. G., M. J. Stefik, T. C. Rindfleish, and A. M. Duffield, *Anal. Chem.*, *48*, 1368 (1976).

Ernst, R. R., *Rev. Sci. Instr.*, *36*, 1689 (1965).

Field, F. H., and S. H. Hastings, *Anal. Chem.*, *28*, 1248 (1956).

Field, F. H., and J. L. Franklin, *Electron Impact Phenomena and Properties of Gaseous Ions*, Academic Press, New York (1970).

Francl, T. J., R. L. Hunter, and R. T. McIver, Jr., *Anal. Chem.*, *55*, 2094 (1983).

Franklin, J. S., J. G. Dillard, H. M. Rosenstock, J. T. Herron, K. Draxl, and F. H. Field, *Ionization Potentials, Appearance Potentials, and Heats of Formation of Gaseous Positive Ions*, NSRDS-NBS Publ. 26, Washington, D.C. (1969).

Frieden, B. R., in Picture Processing and Digital Filtering: *Topics in Applied Physics*, Vol. 6, 2nd ed., ed. T. S. Huang, Springer-Verlag, Berlin (1979), p. 177.

Giancaspro, C., and M. B. Comisarow, *Appl. Spec.*, *37*, 153 (1983).

Goodyear, C. C., *Signals and Information*, Wiley-Interscience, New York (1971).

Grostic, M. F., and K. L. Rinehart, in *Mass Spectrometry: Techniques and Applications*, ed. G. W. A. Milne, Wiley-Interscience, New York (1971).

Grotch, S. L., in *17th Ann. Conf. on Mass Spectrom. and Allied Topics*, Dallas (1969), p. 459.

Grotch, S. L., *Anal. Chem.*, *45*, 2 (1973).

Hamming, R. W., *Digital Filters*, Prentice-Hall, Englewood Cliffs, NJ (1977).

Heller, S. R., *Anal. Chem.*, *44*, 1951 (1972).

Heller, S. R., *Kemia-Kemi*, *1*, 15 (1984).

Heller, S. R., and G. W. A. Milne, *Am. Laboratory, March*, 33 (1980a).

Heller, S. R., and G. W. A. Milne, *Anal. Chem. Acta*, *122*, 117 (1980b).

Henneberg, D., in *Advances in Mass Spectrometry*, Vol. 8B, ed. A. Quayle, Heyden and Son Ltd., London (1980), p. 1511.

Hertz, H. S., R. A. Hites, and K. Biemann, *Anal. Chem.*, *43*, 681 (1971).

Hill, H. C., *Introduction to Mass Spectrometry*, 2nd ed., revised by A. G. Loudon, Heyden and Son Ltd., London (1972).

Hites, R. A., and K. Biemann, *Anal. Chem.*, *39*, 965 (1967).

Holmes, W. F., W. H. Holland, and J. A. Parker, *Anal. Chem.*, *43*, 1806 (1971).

Honig, R. E., *Anal. Chem.*, *22*, 1474 (1950).

Horlick, G., *Anal. Chem.*, *44*, 943 (1972).

Howard, S. J., and G. H. Rayborn, *Deconvolution of Gas Chromatographic Data*, NASA Contractor Report CR-3229 (1980), Washington, D.C.

Howard, S. J., in *Deconvolution with Applications to Spectroscopy*, ed. P. A. Jansson, Academic Press, Orlando (1984), p. 262.

Ioup, G. E., Ph.D. Dissertation, University of Florida (1968).

Ioup, G. E., *Bull. Am. Phys. Soc.*, *26*, 1213 (1981).

Ioup, G. E., and J. W. Ioup, *Geophysics*, *48*, 1287 (1983).

Ioup, J. W., G. M. Wood, B. T. Upchurch, G. E. Ioup, and G. H. Rayborn, in *28th Ann. Conf. on Mass Spectrom. and Allied Topics*, New York (1980), p. 682.

Ioup, J. W., G. E. Ioup, G. H. Rayborn, G. M. Wood, and B. T. Upchurch, *Int. J. Mass Spectrom. and Ion Processes*, *55*, 93 (1983–1984).

Isenhour, T. L., and P. C. Jurs, *Anal. Chem.*, *43*, 20A (1971).

Isenhour, T. L., B. R. Kowalski, and P. C. Jurs, *Crit. Rev. Anal. Chem.*, *4*, 1 (1974).

Jager, H. D. M., D. C. Maxwell, and A. McCormick, in *Practical Mass Spectrometry*, ed. B. S. Middleditch, Plenum Press, New York (1979), p. 179.

Jansson, P. A., ed., *Deconvolution with Applications to Spectroscopy*, Academic Press, Orlando (1984).

Judson, C. M., G. M. Wood, and P. R. Yeager, in *24th Ann. Conf. on Mass Spectrom. and Allied Topics*, San Diego (1976), p. 537.

Judson, C. M., J. L. Lawrence, C. S. Josias, and R. A. Suchter, in *28th Ann. Conf. on Mass Spectrom. and Allied Topics*, St. Louis (1978), p. 466.

Justice, J. B., and T. L. Isenhour, *Anal. Chem.*, *46*, 223 (1974).

Karasek, F. W. and R. E. Clements, in *Mass Spectrometry in Environmental Sciences*, eds. F. W. Karasek, O. Hutzinger, and S. Safe, Plenum Press, New York (1985), p. 123.

Kawata, S., and Y. Ichioka, *J. Opt. Sci. Am.*, *70*, 762, 768 (1980).

Kornelsen, E. V., *Can. J. Phys.*, *48*, 2812 (1970).

Kornelsen, E. V., and A. A. van Gorkum, *Vacuum*, *31*, 99 (1981).

Kowalski, B. R., *Anal. Chem.*, *47*, 1152A (1975).

Koyama, Y., T. Tazawa, and K. Maeda, in *30th Ann. Conf. on Mass Spectrom. and Allied Topics*, Honolulu (1982), p. 835.

Kraus, J. B., *Industr. Res.*, June, 62 (1975).

Kwok, K., R. Venkataraghaven, and F. W. McLafferty, *J. Am. Chem. Soc.*, *95*, 4185 (1973).

LaCoste, L. J. B., *Geophysics*, *47*, 1724 (1982).

Lederberg, J., in *Biochemical Applications in Mass Spectrometry*, ed. G. R. Waller, Wiley-Interscience, New York (1972), p. 193.

Ligon, W. V., *Int. J. Mass Spectrom. Ion Phys.*, *34*, 193 (1980a).

Ligon, W. V., in *28th Ann. Conf. on Mass Spectrom. and Allied Topics*, New York (1980b), p. 490.

Liteanu, C., and I. Rica, *Statistical Theory and Methodology of Trace Analysis*, Ellis Horwood Ltd., Chichester (1980) (Wiley, distributor).

Litzow, M. R., and T. R. Spalding, *Mass Spectrometry of Inorganic and Organometallic Compounds*, Elsevier, Amsterdam (1973), p. 24.

Lumpkin, J. E., *Anal. Chem.*, *30*, 321 (1958).

Malmstadt, H. V., C. G. Enke, and S. R. Crouch, *Electronics and Instrumentation for Scientists*, Benjamin/Cummings, Menlo Park, CA (1981).

Martinson, D. P., *Appl. Spectros.*, *35*, 255 (1981).

McGill, J. R., and B. R. Kowalski, *J. Chem. Inf. Comp. Sci.*, *18*, 52 (1978).

McGillem, C. D., and G. R. Cooper, *Continuous and Discrete Signal and System Analysis*, Holt Rinehart & Winston, New York (1984).

McLafferty, F. W., *Pure and Appl. Chem.*, *50*, 197 (1978).

McLafferty, F. W., R. H. Hertel, and R. D. Villwock, *Org. Mass Spectrom.*, *9*, 690 (1974).

McLafferty, F. W., *Interpretation of Mass Spectra*, 3rd ed., University Science Books, Mill Valley, CA (1980).

McLafferty, F. W., and R. Venkataraghaven, in *High Performance Mass Spectrometry*, ed. M. L. Gross, ACS Symp. Series 70, American Chemical Society, Washington, D.C. (1978), p. 310.

McLafferty, F. W., S. Cheng, K. M. Dully, C. Guo, I. K. Mun, D. W. Petersen, S. O. Russo, D. A. Salvucci, J. W. Serum, W. Staedeli, and D. B. Stauffer, *Int. J. Mass Spectrom. Ion Phys.*, *47*, 317 (1983).

McLafferty, F. W., and D. B. Stauffer, *Int. J. Mass Spectrom. and Ion Processes*, *58*, 139 (1984).

McMurray, W. J., in *Mass Spectrometry: Techniques and Applications*, ed. G. W. A. Milne, Wiley-Interscience, New York (1971), p. 43.

Milne, G. W. A., and S. R. Heller, *J. Chem. Infor. Comp. Sci.*, *20*, 204 (1980).

Milne, G. W. A., W. L. Budde, S. R. Heller, D. P. Martinson, and R. G. Oldham, *Org. Mass Spectrom.*, *17*, 547 (1982).

Morrison, J. D., *J. Chem. Phys.*, *39*, 200 (1963).

Mun, I. K., D. R. Bartholomew, D. B. Stauffer, and F. W. McLafferty, *Anal. Chem.*, *53*, 1938 (1981a).

Mun, I. K., D. B. Stauffer, R. G. Dromey, S. O. Russo, and F. W. McLafferty, in *29th Ann. Conf. Mass Spectrom. and Allied Topics*, Minneapolis (1984b), p. 437.

Nevejans, D., P. Frederick, and E. Arijs, *Bulletin de La Classe des Sciences*, Académi Royale de Belgique, *68*, No. 5, 314 (1982).

O'Haver, T. C., *J. Chem. Ed.*, *49*, A131, A211 (1972).

Oliver, B. M., and J. M. Cage, *Electronic Measurements and Instrumentation*, McGraw-Hill, New York (1971).

O'Malley, R. M., *J. Chem. Ed.*, *59*, 1073 (1982).

Ott, J. W., *Noise Reduction Techniques in Electronic Systems*, Wiley-Interscience, New York (1976).

Perone, S. B., and D. O. Jones, *Digital Computers in Scientific Instruments*, McGraw-Hill, New York (1973).

Perone, S. B., K. Ernst, H. R. Brand, and J. W. Frazer, in *Computerized Laboratory Systems*, ASTM STP 578 (1975), p. 25.

Pesyna, G. M., F. W. McLafferty, R. Venkataraghaven, and H. E. Dayringer, *Anal. Chem.*, *47*, 1161 (1975).

Pesyna, G. M., R. Venkataraghaven, H. E. Dayringer, and F. W. McLafferty, *Anal. Chem.*, *48*, 1362 (1976).

Rasmussen, G. T., T. L. Isenhour, and J. C. Marshall, *J. Chem. Inf. Comput. Sci.*, *19*, 98 (1979).

Rayborn, G. H., S. J. Howard, G. M. Wood, and B. T. Upchurch, invited paper at *23rd Eastern Analytical Symposium*, New York (1984).

Rose, M. E., and R. A. W. Johnstone, *Mass Spectrometry for Chemists and Biochemists*, Cambridge University Press, Cambridge (1982).

Savitzky, A., and M. J. E. Golay, *Anal. Chem.*, *36*, 1627 (1964).

Smith, D. H., B. G. Buchanan, R. S. Englemore, A. M. Duffield, A. Yeo, E. A. Feigenbaum, J. Lederberg, and C. Djerassi, *J. Am. Chem. Soc.*, *94*, 5962 (1972).

Smith, D. H., B. G. Buchanan, W. C. White, E. A. Feigenbaum, J. Lederberg, and C. Djerassi, *Tetrahedron*, *29*, 3117 (1973).

Smith, D. H., and R. E. Carhart, in *High Performance Mass Spectrometry: Chemical Applications*, ed. M. L. Gross, ACS Symp. Series 70, American Chemical Society, Washington, D.C. (1978), p. 325.

Snelling, C. R., Jr., J. C. Cook, R. M. Milberg, M. E. Hemling, and K. L. Rinehart, Jr., *Anal. Chem.*, *56*, 1474 (1984).

Speck, D. D., R. Venkataraghaven, and F. W. McLafferty, *Org. Mass Spectrom.*, *13*, 209 (1978).

Spencer, D. B., and D. B. Stauffer, in *32nd Ann. Conf. on Mass Spectrom. and Allied Topics*, San Antonio (1984), p. 658.

Stauffer, D. B., M. A. Sharaf, R. G. Dromey, C. J. Guo, and F. W. McLafferty, in *31st Ann. Conf. on Mass Spectrom. and Allied Topics*, Boston (1983), p. 558.

Stearns, S. D., *Digital Signal Analysis*, Hayden, Rochelle Park, NJ (1975).

Stradling, R. S., P. A. Ryan, and J. D. Wood, *Comput. Enh. Spectros.*, *1*, 25 (1983).

Thompson, M. R., and R. E. Dessy, *Anal. Chem.*, *56*, 583 (1984).

Tondeur, Y., J. R. Hass, D. J. Harvan, and P. W. Albro, *Anal. Chem.*, *56*, 373 (1984).

van Cittert, P. H., *Z. Physik*, *69*, 298 (1931).

van Gorkum, A. A., and E. V. Kornelsen, *Vacuum*, *31*, 89 (1981).

Varmuza, K., *Anal. Chem. Symp. Ser.-Mod. Trends in Anal. Chem. Pt. B*, *18*, 99 (1984).

Vassos, B. H., and G. W. Ewing, *Analog and Digital Electronics for Scientists*, Wiley-Interscience, New York (1972), p. 105.

Whitehorn, M. A., M.S. Thesis, University of New Orleans (1981).

Wong, C. M., R. W. Crawford, V. C. Barton, H. R. Brand, K. W. Neufeld, and J. E. Bowman, *Rev. Sci. Instr.*, *54*, 996 (1983).

Wong, C. M., R. W. Crawford, J. C. Kunz, and T. P. Kehler, *IEEE Trans. Nucl. Sci.*, *NS-31*, 804 (1984a).

Wong, C. M., S. M. Lanning, R. W. Crawford, and H. R. Brand, in *32nd Ann. Conf. on Mass Spectrom. and Allied Topics*, San Antonio (1984b), p. 642.

Wong, C. M., and R. W. Crawford, *Int. J. Mass Spectrom. and Ion Processes*, *60*, 107 (1984c).

Wood, G. M., B. T. Upchurch, and D. B. Hughes, in *28th Ann. Conf. on Mass Spectrom. and Allied Topics*, St. Louis (1978), p. 469.

Wood, G. M., and P. R. Yeager, in *Environmental Impact of Coal Utilization*, eds. J. J. Singh and A. Deepak, Academic Press, New York (1980), p. 187.

Wood, G. M., G. H. Rayborn, J. W. Ioup, G. E. Ioup, B. T. Upchurch, and S. J. Howard, in

9th Int. Cong. on Instrum. in Aerospace Simulation Facilities, Dayton, OH, *ICIASF '81 Record* (1981), p. 25.

Wood, G. M., B. W. Lewis, B. T. Upchurch, R. J. Nowak, D. G. Eide, and P. A. Paulin, in *10th Int. Cong. on Instrum. in Aerospace Simulation Facilities*, St. Louis, France, *ICIASF '83 Record* (1983), p. 259.

Yasuhara, A., J. Shindo, H. Ito, and T. Mizoguchi, *Comput. Enh. Spectros.*, *1*, 117 (1983).

Zupan, J., *Anal. Chimica Acta*, *103*, 273 (1978).

CHAPTER SEVEN

CHROMATOGRAPHY/MASS SPECTROMETRY

One of the attributes of the mass spectrometer is the ability to identify and measure minute quantities of material; one of its limitations is the difficulty in resolving a multicomponent mixture when ions of nominally the same mass-to-charge ratio occur in the composite spectrum. The signal representing an ion is a superposition of signals from each component in the mixture having a molecular or fragment ion at that mass; and for most instruments, the resolution simply is not high enough to identify the species through identification of the exact mass alone. The difficulty is increased when one attempts either a qualitative analysis of a mixture of high molecular weight species or the identification of isomers of a species in the same compound. In these instances, it is not only desirable, but it may be essential that the sample be pretreated in order to perform the analysis. Hence, this chapter is intended to provide a very brief discussion of the principles of chromatography, and to furnish references to some widely used chromatograph-spectrometer interfaces.

GAS-LIQUID CHROMATOGRAPHY/MASS SPECTROMETRY

Of all the methods used to simplify the mass spectrum of a mixture, the most powerful and the most widely applied is gas-liquid chromatographic preseparation followed by mass analysis. Since its introduction in 1952, gas-liquid chromatography (GLC) has played a crucial role in the chemical analysis of mixtures. In gas-liquid chromatography, the chromatographic process consists of vapor pressure equilibration between a mobile gas phase and a stationary liquid phase. The gas passes through a column, which is usually a small diameter tube installed in an oven, as shown schematically in Figure 7.1. The column contains the stationary liquid phase, which may be coated on the tube wall or on an inert support such as Teflon or diato-

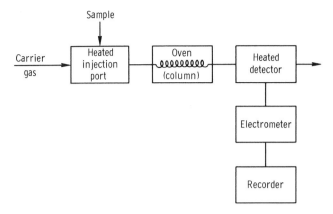

Figure 7.1   Components of the gas chromatograph.

maceous earth. The liquid phase is dictated by the class of compounds to be analyzed, and the chemical rule of "like-dissolves-like" provides a convenient basis for selection (Simpson, 1972).

The sample is introduced into a heated port where it is rapidly vaporized if it is a liquid, and it is swept into the column by a carrier gas. It then proceeds through the column, where the components are absorbed into the stationary phase. As the flow continues, the concentration of the component remaining in the carrier gas falls below the equilibrium vapor pressure dictated by the column temperature and the liquid phase, and some amount then desorbs back into the gas stream. This absorption and desorption equilibrium continues throughout the column, with the sample components moving forward only when in the gas phase, and with those most readily absorbed in the liquid phase being retained in the column longer. Since each component, therefore, moves through the column at a different rate, they emerge from the column and enter the detector separated in time.

The efficiency of the column is expressed quantitatively by the number of absorption/desorption zones, called theoretical plates N, where

$$N = 5.54(t_r/w)^2$$

This equation relates the peak width at half-peak height w to the length of time t_r that the component has been in the column (Willard et al., 1981), a large number of theoretical plates corresponding to higher efficiency. Another measure of performance is the resolution

$$R = 2\Delta t_r/(w_1 + w_2)$$

which compares the retention times of two peaks to the peak width at the bases, where each peak width represents 4σ of the normal error curve (Fig.

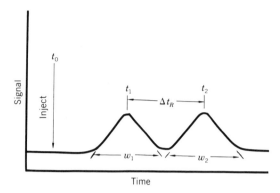

Figure 7.2 Resolution of chromatographic peaks.

7.2). For a pair of peaks of equal area, the degree of resolution required to achieve essentially complete separation is $R = 1.5$.

It should be noted that the gas chromatograph is an excellent quantitative instrument. It is less so qualitatively because (1) complete separations cannot always be achieved; (2) retention times can be altered by the column operating parameters; and (3) measurement precision may be insufficient to identify species with nominally equal retention times (McFadden, 1973). Therefore, chromatography and mass spectrometry are truly complementary. If resolution of the mixture constituents can be accomplished chromatographically, qualitative identification is obtained from the mass spectrum and the quantitative measurement from the area under the corresponding peak in the chromatogram. Even if complete chromatographic separation cannot be obtained, reduction and analysis of the mass spectra may still be considerably simplified.

Measuring the Chromatographic Signal

A number of schemes based on physical or chemical properties have been used to measure the sample eluting from the column. However, most are specific for one or more classes of compounds, and only three of the detection methods have become standard: (1) thermal conductivity (TCD); (2) flame ionization (FID); and (3) electron-capture (ECD). The oldest and most universal method, the TCD, depends upon differences in thermal conductivity of the carrier gas and the sample to produce a signal proportional to concentration. With the TCD, parallel streams of carrier gas, one void of sample, pass over opposing sides of a Wheatstone bridge detector circuit, so that a change in concentration of the solute produces a corresponding change in resistance of the sensing side. The TCD is a nondestructive universal detector of medium sensitivity with a detection limit of about 10^{-8} g/s (Perry, 1981), or 1 ng/μl in a 10 μl sample for most substances (Okamura and Sawyer, 1978). Because thermal conductivity is species-dependent, the TCD requires calibration if the analysis is to be quantitatively accurate.

A more sensitive but less universal method is flame ionization (FID), in which organic molecules mixed with hydrogen are combusted in air to form ions and free electrons, the resulting current being measured with an electrometer. The method is insensitive to rare gases, to N_2, O_2, H_2O, and CO_2, and to many inorganic and highly oxygenated compounds. The sensitivity decreases with increasing substitution of halogens, amines, or hydroxyl groups; but within limits, the response per oxidizable carbon is nearly constant, so that separate calibrations for pure hydrocarbons are generally not necessary. The response is proportional to the number of ions reaching the detector; hence, the FID measures mass flow. The signal is also inversely proportional to flow rate, with rates of 10–50 ml/min used to maximize signal-to-noise. The sensitivity is several orders of magnitude higher than the TCD, with a detection limit of approximately 10^{-12} g/s and a dynamic range of 10^7 (Perry, 1981). The FID will measure all hydrocarbon constituents in the gas stream, and it is therefore essential that high-purity (99.999% or greater) hydrogen, helium, and air be used to reduce background.

Electron-capture (ECD) is selective for compounds having high electron affinities. It is therefore particularly useful for detecting halogenated compounds, conjugated carbonyls, nitriles, nitrates, and metal organics, while at the same time being insensitive to hydrocarbons (Szepsey, 1970). A beta emitter (e.g., ^{63}Ni, ^{3}H) produces free electrons, which migrate to a charged electrode. When an electronegative compound enters the detector, electrons are captured, thus reducing the measured steady-state current. The detector is highly sensitive, with a detection limit of about 10^{-14} g/s for compounds with several halogen atoms, and it has a dynamic range of 10^3 to 10^5.

GC MEASUREMENTS WITH THE MASS SPECTROMETER

The mass spectrometer is both sensitive and spectrally specific; consequently, it is a highly efficient and universal GC detector. Data can be recorded as spectra of relevant masses in conjunction with chromatograms obtained with a standard GC detector, or as mass chromatograms obtained with the MS detector alone. The mass chromatograms are generated by monitoring total or selected ion currents, or by computer reconstruction of the sums of all of the resolved ion currents. It is almost always useful to use more than one recording method, since all have limitations that may cause loss of information. Monitoring the sample with the GC detector as it elutes from the column, for example, relates the spectral scan to a position on an eluate peak, and is considered to be required in all but the most routine computerized GC/MS analyses (McFadden, 1973).

Measurements with Linked GC and MS Detectors

If a portion of the effluent from the column is diverted to the mass spectrometer, a mass spectrum of the chromatographic peak may be obtained. It should be noted that the GC peak constitutes an essentially Gaussian pressure pulse in the mass spectrometer ion source. Reproducible mass spectra depend upon a constant source pressure, and it is essential that the effects of changing pressure be minimized. In general, this requires rapid cyclic scanning or that the scan be initiated as near as possible to the apex of the GC peak; and that it should be completed before the pressure significantly decreases. Because of the necessity for rapid scanning, the quadrupole and, to a lesser extent, the time-of-flight (TOF) spectrometers have often been used when high resolution is not an important requirement. If resolution is important, then magnetic deflection instruments, generally double focusing, are preferred because of their analytical superiority. Rapidly scanning over a wide mass range by changing the magnetic field is preferred, since it minimizes the mass-dependent variation in electron multiplier response that accompanies voltage scanning.

Repetitive scanning with the magnetic sector is sometimes difficult because of the inertia of the electromagnet, thereby limiting the rate at which spectra can be generated in this manner. However, newly designed magnets fabricated from contemporary materials have faster recovery times, thus increasing the utility of the method. The quadrupole and TOF, on the other hand, depend entirely upon electric fields, so that rapid repetitive scanning is easily achieved. For the quadrupole, this requires only that the RF and DC voltages be changed while maintaining a constant RF:DC ratio. The TOF is not scanned in the usual sense, with mass spectra being generated by controlled gating of the ion source and electron multiplier.

The methods by which the instruments are scanned result in different mass/time relationships or "scan functions." The function that produces equal peak widths over the mass range is designated the "natural function" and is preferred for GC/MS applications (McFadden, 1973). The resolution $R = m/\Delta m$ of the magnetic sector is essentially constant throughout the mass range, and the exponential natural scan function results in a corresponding decrease in separation between adjacent peaks with increasing mass. The mass scale for the quadrupole is linear so that mass separation at high or low masses is the same, while that for the TOF decreases proportionally to $(\Delta m)^{\frac{1}{2}}$. In the latter two cases, the time spent on a peak is dependent upon the starting mass, so that maintaining an equal peak width requires a change in resolution during the scan. Scan rates are usually designated in terms of seconds per mass decade, t_{10}. Because the exponential scan function is the only one independent of starting mass, it is also the only one having an equal t_{10} regardless of the mass decade scanned. These relationships and an illustrative t_{10} are listed in Table 7.1, where m_0 is the

TABLE 7.1   Scan Functions and Scan Rate Relationships

MS Type	Scan Mode	Scan Function	Scan Rate (t_{10})	$t_{10}(40-400)$ (s)
Magnetic	Exponential	$m = m_0 e^{kt}$	$2.3/k$	4
Quadrupole	Linear	$m = m_0 + kt$	$9\, m_0/k$	8
TOF	Hyperbolic	$m = m_0 + kt^2$	$3\,(m_0/k)^{1/2}$	5.7

Source: Compiled from McFadden (1973).

starting mass, m is the mass at time t, k is the scan constant, and t_{10} (40–400) is the time required to scan from 40–400 amu when t_{10} (20–200) is 4 s.

Scanning at the top of the chromatographic peak may be accomplished by several methods. When a separate GC detector is used, peak maxima may be noted and the scan initiated after allowing for the delay due to transit through the interconnecting tubing between the stream splitter and the mass spectrometer. Spectra obtained in this manner by scanning both unresolved and resolved chromatographic peaks of a 1.7:1 mixture of n-hexane ($CH_3[CH_2]_4CH_3$) and 2-methylpentane ($CH_3CHCH_3[CH_2]_2CH_3$) are illustrated in Figure 7.3. Because both species contribute to the same mass peaks, the relative intensities of the m/z 43, 57, 71, and 86 peaks in Figure 7.3*a* provide an initial clue that more than one isomer may be present. When the resolved chromatogram is scanned at the top of each of the peaks, then the differences in the cracking patterns for the major peaks in Figures 7.3*b* and 7.3*c* can be used to identify both of the C_6H_{14} hydrocarbons.

If a fast-scanning spectrometer is used, several complete spectra can be obtained during the passage of the eluate through the ion source. If the GC peak results from a single species, then the successive mass spectra will simply exhibit the cracking pattern for that species distorted by the pressure change as the eluate passes through the ion source. This effect is illustrated in Figure 7.4, where the spectra are those obtained by scanning repeatedly across a GC peak of myrcene ($C_{10}H_{16}$). It should be noted that the scan taken at the top of the peak produces a spectrum approximating that obtained isobarically; therefore, it is the most useful for qualitative analysis. On the other hand, if the GC peak contains two or more unresolved components, then the mass peaks obtained from successive scans will have intensities proportional to total source pressure and to the quantity of each species present in the source at the time of the scan.

Incomplete separation of peaks in GC is not uncommon. Data reduction of the pressure-distorted spectra can be simplified if very rapid scanning with a quadrupole or TOF mass spectrometer is used to obtain 10–20 scans/ GC peak. This will usually indicate the presence of multiple compounds and will often provide positive identification of the species. At a minimum, effectiveness of the chromatographic separation can be observed and a subsequent operational strategy devised (Junk, 1972).

Rapid scanning places demands on the response of the detector that may

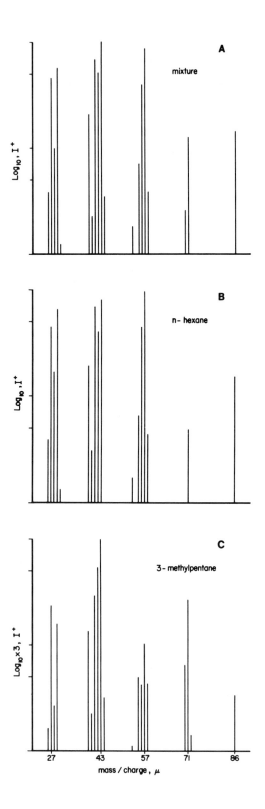

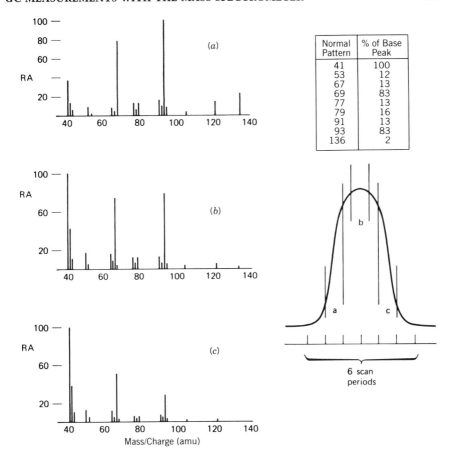

Normal Pattern	% of Base Peak
41	100
53	12
67	13
69	83
77	13
79	16
91	13
93	83
136	2

Figure 7.4 Distortion of the mass spectrum of myrcene ($C_{10}H_{16}$) during scanning (a) up, (b) across, and (c) down the chromatographic peak. (McFadden, 1973; copyright John Wiley & Sons.)

diminish resolution, sensitivity, and signal-to-noise ratio (SNR). In general, detector response frequencies of 1.5–2 kHz will suffice if low resolution and a scan rate of 3–5 s/decade is satisfactory for the analysis. At the same scan rate, high resolution will require a correspondingly higher frequency response. When the scan is exponential, the relationship between the bandwidth (B), resolution (R), scan speed (t_{10}), and the time spent between $\pm 95\%$ of peak maximum (t_p) is approximated by B (Hz) $= 7.3/t_{10}$ and t_p (ms) $= 0.430 t_{10}/R$ (Millard, 1978). Table 7.2 summarizes these parameters for four values of R at three values of t_{10}. The large amount of data generated by

←

Figure 7.3 Mass spectra of (A) unresolved and (B, C) resolved GC peaks of two C_6H_{14} hydrocarbons.

TABLE 7.2 Relationships Between Bandwidth, Resolution, Scan Speed, and Time Spent on a Peak

Resolution (R)	Time on Peak (t_p) (ms)			Bandwidth (B) (Hz)		
	$t_{10} = 1$	$t_{10} = 3$	$t_{10} = 5$	$t_{10} = 1$	$t_{10} = 3$	$t_{10} = 5$
1000	0.43	1.29	2.15	7300	2433	1460
1500	0.29	0.86	1.43	10,950	3650	2190
3000	0.14	0.43	0.72	21,900	7300	4380
5000	0.09	0.26	0.43	36,500	12,166	7300

Source: Adapted from Millard (1978).

the rapid multiple scanning method may also require computer assistance (Chapter 6) in order to effectively acquire and analyze the data.

Total Ion Current Monitoring

The intensity of the mass peak depends upon the number density of each species in the ion source. A well-designed ion source is linear over its entire range, so that the total number of ions produced is also indicative of the total pressure. An alternative method to remove the effects of pressure on the mass spectrum, while at the same time retaining a slower scan rate across the GC peak, is to simultaneously monitor the total ion current and to normalize the mass-resolved signal to it. Ion current variations due to changes in pressure alone will be removed from the data, and only those due to relative concentrations will be recorded (Brunee et al., 1962).

Methods used to monitor the total ion current depend upon the instrument type. Because the magnetic sector generally monitors only one mass at a time, a collector that continuously intercepts part of the beam or periodically has the beam deflected to it is placed between the ion source and the entrance to the magnetic field. This can be an effective measurement technique, but it has a somewhat low sensitivity because the measurement is made without the electron multiplier. High-speed digital recording and computer reconstruction of the total ion current permits use of the electron multiplier; the simultaneous ion beam detection methods discussed in Chapter 5 may also increase sensitivity.

A similar collector can be placed between the ion source and the entrance to the rods for the quadrupole mass spectrometer (QMS). Alternatively, the resolution can be momentarily reduced so that most of the masses pass through the quadrupole field and are detected, thus providing an indication of source pressure. This method requires only a simple switching of the DC power supply, but it generally does not measure all species in the source. Alternatively, if the quadrupole is rapidly scanned, the output can be monitored and summed with an integrating digital voltmeter or similar device in addition to obtaining spectral data. The TOF mass spectrometer monitors selected ion peaks with a series of electronic gates at the exit from the electron multiplier, and it is a simple matter to periodically direct all of the electrons to a single gate as a measure of total ion current. The TOF has an exceedingly high scan rate, making it ideally suited for use as a GC detector (Gohlke, 1959).

Recording a chromatographic trace by monitoring total ion current as a function of time produces a chromatogram that is comparable in sensitivity to the FID while retaining all of the universality of the TCD. Total ion current monitoring requires that all of the ions formed in the source are representative of the sample. However, a significant part of the signal may actually result from species that are not pertinent to the analysis, such as those from background or from a carrier gas such as helium. These should be minimized to obtain the highest sensitivity for the sample and to reduce the dynamic range required of the recording system. Helium ions may be eliminated by simply reducing the ionizing electron energy to about 20 eV, which is below the ionization potential of 24.6 eV for helium, but is sufficiently high to obtain an adequate signal from most species of interest (Chapter 6). The ionizing energy is usually returned to 70 eV prior to obtaining the mass spectrum if small quantities of samples are to be analyzed (Fenselau, 1974). The QMS excludes all species outside of the region of stability, so that it is only necessary to adjust the potentials to prevent the transmission of helium ions to the detector. In the TOF mass spectrometer, predynode gating deflects the carrier gas and other selected ions from the entrance to the electron multiplier, reducing background and increasing sensitivity (Merritt, 1970).

If the spectrometer is connected to a storage device large enough to contain all of the data, then ion chromatograms can later be reconstructed. Ions not pertinent to the analysis can be excluded from the displayed or recorded spectra by instructions to the computer. The method has the advantage that data can be manipulated to assist in the analysis, but it has the disadvantage that the data are not accessible or displayed in real time.

Selective Ion Monitoring

Scanning the mass spectrometer over a wide mass range in conjunction with each chromatographic peak is most useful for the identification of unknown species occurring in a mixture. For many analyses, however, it is only necessary to establish the presence of a particular compound or class of compounds. In such cases, a complete mass spectrum is not required, the GC retention times and mass peaks indicative of the compound are known, and it is sufficient that the presence or absence of selected masses be confirmed (e.g., Ten Noever de Brauw, 1979). Since the amount of data handling is significantly reduced, this method is often used in analyzing blood or urine to detect the presence of medications or illegal drugs, or in following the progress of a chemical reaction. The selected ion may represent either a

TABLE 7.3   Ions Indicative of Functional Groups

Mass	Class	Structure
30	Aliphatic amines	CH_2NH_2
47	Sulfides	CH_2SH
73	Trimethylsilates	$Si(CH_3)_3$
91	Benzyls	$C_6H_5CH_2$

molecular or fragment ion, as long as it represents the species. Examples of some peaks that are indicative of specific functional groups are listed in Table 7.3.

Single-ion monitoring is the most sensitive of the MS detection methods because the signal is measured with maximum amplification and long integration times, and because nonpertinent ions are not measured if their mass differs from that being monitored. When the relationship between peak height and concentration is linear, sample sizes of less than 50 pg can be reliably and readily detected (Fenselau, 1974). Stable isotope-labeled analogs or homologous internal standards eluting separately from the sample provide a means of accurate analysis even when the linearity relationship does not exist. Internal standards invariably improve the quality and sensitivity of the assay, with reliable measurements of picogram quantities having been reported (Fenselau, 1978; Watson, 1985).

It is often expedient to make use of a chemical blank (Millard, 1978) in which the mass spectrum of a sample known to be free of the compound being sought is examined to determine which ions should be monitored. For example, an extract of a urine specimen known to be free of a particular drug might be used to examine an identical extract of a specimen in which it is suspected to be present.

If the chemical blank cannot be obtained, quantitatively determining the interferences becomes more difficult, often requiring extended calculations. A reliable method is to mass resolve the representative and interfering peaks if the resolving power of the instrument can be adjusted to do so. Reducing interferences in this manner will result in lower sensitivity, which must be balanced against the maximum resolution that can be used. Kimble (1978) has discussed the principles of and use of GC/high-resolution mass spectrometry, and has addressed the effects of peak shape and scan speed on the analysis.

Selective Monitoring of Multiple Ions

Selective ion monitoring of a single mass is limited by the availability of isolated peaks. Extending the method to monitoring two or more ions increases the probability that a specie in the mixture is correctly identified, and Millard (1978) has shown that taking relative sensitivities into account, only very few peaks are needed even at low resolution. Indeed, 1.4×10^{10} compounds can theoretically be identified by monitoring only four ions and scanning to mass 350! The nonrandom distribution of ions in a spectrum prevents even approaching this theoretical limit; but, in general, only two or three carefully selected ions may be needed to identify a single compound, and perhaps twice that may be sufficient to resolve a mixture. The Eight Peak Index of Mass Spectra (Mass Spectrometry Data Centre, 1974), for

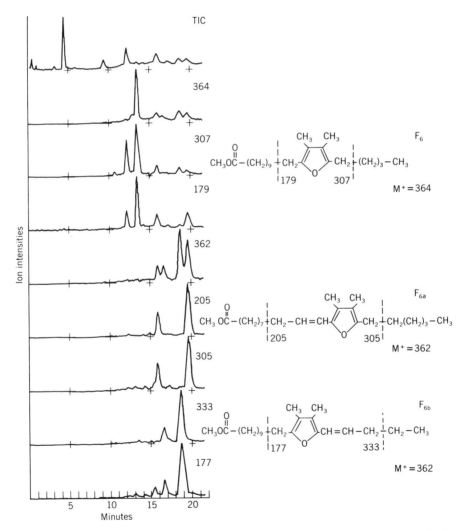

Figure 7.5 Identification by selected ion monitoring of furan fatty acids extracted from freshwater fish (Krick et al., 1978).

example, lists approximately 31,000 compounds that may be identified by the eight most intense peaks in each spectrum (Chapter 6).

Mass chromatograms of multiple ions can be obtained by simultaneous recording, peak switching, or computer reconstruction from repetitively scanned GC peaks. To illustrate, the total and single ion chromatograms from selective ion monitoring of eight peaks for the identification of three furan fatty acids extracted from freshwater fish are shown in Figure 7.5. At 20 eV ionizing electron energy, about 85% of the total ionization results from both the molecular ion and ions due to cleavage allylic to the double bonds. The three major ions indicate the molecular weight and position of the furan ring, so that any specific furan fatty acid can be identified. Selective ion monitoring is less sensitive for multiple ions than single-ion monitoring because less time is spent on each peak, but detection limits are still well within the picogram range.

THE GC/MS INTERFACE

Small quantities of volatile effluents can be trapped in various cryogenically cooled devices or by adsorption on a suitable medium, and can be independently introduced later into the mass spectrometer (Safe and Hutzinger, 1973). From a sampling standpoint, this method is highly efficient but time-consuming, and it is generally more expedient that the two instruments be interfaced directly. A twofold incompatibility exists, however, that must be resolved by the GC/MS interface: (1) chromatographic resolution must be maintained, while at the same time the pressure must be reduced to 10^{-4} torr or less to minimize intermolecular collisions in the ion source and mass analyzer; and (2) if small samples are to be analyzed, enrichment with respect to the carrier gas is often necessary to take advantage of the dynamic range of the mass spectrometer detector.

An ideal interface would remove all carrier gas from the stream while passing an unaltered sample into the mass spectrometer. The yield, or efficiency, that designates the amount of sample entering the spectrometer from the interface is expressed by

$$Y = \frac{Q_m}{Q_c} \times 100\%$$

and the enrichment factor, which relates the increase in sample concentration, is expressed by

$$N = \frac{Q_m V_c}{Q_c V_m} = \frac{Y V_c}{100 \ V_m}$$

where Q_m and V_m are the amounts of sample and carrrier flow entering the mass spectrometer and Q_c and V_c are the amounts of sample and carrier flow leaving the chromatograph (Grayson and Wolf, 1967).

A number of enrichment devices that successfully integrate the two instruments have been reviewed (Rhyage and Wikstrom, 1971; Simpson, 1972; Junk, 1972; McFadden, 1973, 1979; Gudzinowicz et al., 1977; Colby, 1978; Greenway and Simpson, 1980; Watson, 1985). While each device has advantages for certain classes of compounds, those most generally used with packed columns are the diffusion jet, the effusive tube, and the semipermeable membrane. Methods for directly coupling capillary columns have recently evolved along with concurrent improvements in high-speed turbomolecular vacuum pumps and in capillary column manufacturing technology (Grob, 1979; Chapman, 1982). Virtually all contemporary GC/MS systems make use of this highly efficient direct interface for the analysis of hydrocarbons, particularly for organic compounds of high molecular weight.

Jet Separators

The most commonly used method for interfacing the packed column with the mass spectrometer is the diffusion jet separator. The operating principles of the jet separator are illustrated in Figure 7.6. A mixture of the organic sample and carrier gas enters the device in viscous flow and exits through a small orifice as an expanding jet. The greater diffusivity of the low atomic weight carrier (e.g., helium) causes it to expand into the chamber to be pumped away, while the heavier sample molecules tend to continue on, passing through a small skimmer orifice located a short distance from the jet. The residual mixture directed to the mass spectrometer has a higher sample-to-carrier ratio that is dependent upon both flow rate and molecular weight, but is relatively independent of temperature. Rhyage (1964, 1967) developed the first operational jet separator—a highly efficient two-stage stainless steel device capable of interfacing a wide range of compounds at carrier flow rates as high as 60 ml/min with an enrichment factor of about

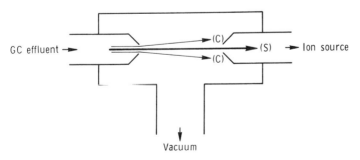

Figure 7.6 The diffusion jet separator showing diffusive separation of the carrier gas (C) from the sample (S).

100. Using modern, high-speed, vacuum-pumping technology, a single-stage separator is generally adequate to sufficiently remove the carrier. The interaction of susceptible compounds with the stainless steel can be reduced by silanization of a glass surface; and a silanized glass, single-stage separator has been used to analyze polar high-boiling point compounds at carrier gas flows of about 10 ml/min (Bonelli et al., 1971). Fabrication and alignment of the glass jet and orifice are more difficult, and the enrichment factor is limited to about 25.

Effusion Separators

Almost simultaneously with Rhyage's development of the jet separator, Watson and Biemann (1964) described an all glass device with a porous tube through which carrier gas atoms would effuse (Fig. 7.7). The pressure of the mixture of sample and helium carrier entering the device is reduced by a constriction sufficiently small enough to ensure that the mean-free-path of the gas is at least an order of magnitude larger than the pores of the tube. Under these conditions, the flow of the gas through the pores is molecular, that is, it is directly proportional to the difference in pressure and is inversely proportional to the square root of the molecular or atomic weight. Therefore, the low atomic weight carrier is preferentially removed, and the ratio of the concentration of the sample to that of the carrier increases as the mixture flows through the separator to the mass spectrometer. The gas exits from the device through a restriction that is dimensioned to maintain both viscous flow and an acceptable pressure in the ion source. The glass device has an enrichment factor of about 5–50. Other versions of the effusive separator use porous stainless steel tubing (Krueger and McCloskey, 1969), silver membranes (Cree, 1967; Blumer, 1968; Copet and Evans, 1970; Grayson and Levey, 1971; Grayson and Bellina, 1973), and a variable conductance (Brunnee et al., 1969).

Membrane Separators

In contrast to removing the carrier from the gas stream, the membrane separator preferentially removes the sample (Llewellyn and Littlejohn, 1966). The method makes use of the fact that organic compounds will readily enter into solution in a dimethyl silicone membrane while the carrier gas will not. Diffusion will occur if there is a pressure difference established across the membrane, with the permeability (P) equal to the product of the diffusion rate (D) and the solubility coefficient (S) of the compound in the membrane, that is, $P = DS$ (Watson, 1985). Since the permeability for organic compounds is generally much greater than it is for helium or other inorganic carrier gases, the enrichment factor and yield are both high. McFadden

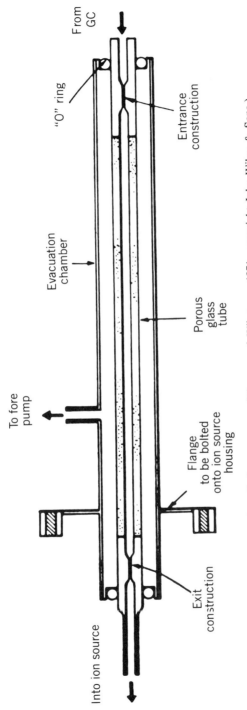

Figure 7.7 The effusion separator. (Rhyage and Wikstrom, 1971; copyright John Wiley & Sons.)

TABLE 7.4   Permeabilities (P_n) of Organic Compounds in Dimethyl Silicone Normalized to That of Helium

Compound	P_n	Compound	P_n	Compound	P_n
He	1	H_2S	78	CH_4	3
N_2	25	N_2O	12	C_2H_2	73
O_2	50	CS_2	250	C_2H_4	4
Ar	50	CCl_4	195	C_2H_6	7
CO_2	270	CH_2O	31	C_4H_{10}	25
H_2O	100	$COCl_2$	41	C_5H_{12}	56
Xe	6	CCl_3F	43	C_6H_{14}	26
C_5H_5N	53	CH_3OH	39	C_8H_{18}	24
$C_2Cl_2F_2$	7	C_3H_6O	16	$C_{10}H_{22}$	12

(1979) has shown that the yield can be expressed as a function of permeability, membrane dimensions, and flow rate by

$$Y = 1 - \exp-[(AP)/(lV_c)]$$

where A is the surface area, P is the permeability, l is the membrane thickness, and V_c is the volumetric flow from the GC. Relative permeabilities of some representative compounds normalized to that of helium are listed in Table 7.4.

In a typical membrane separator (Fig. 7.8), the gas mixture at atmospheric pressure passes over a 2.5–4 μm thick membrane. The membrane is supported on the vacuum (ion source) side by a frit, and the enriched mixture enters directly into the ion source. The membrane temperature should be ideally maintained 50–75°C below the boiling point of the permeating compound, presenting a problem in optimization since solubility is inversely

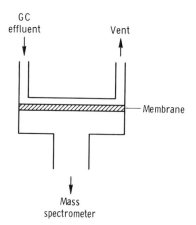

Figure 7.8   The membrane separator.

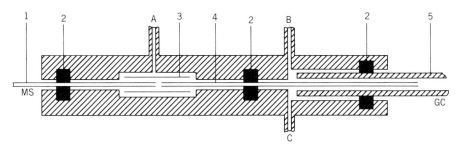

Figure 7.9 Schematic of an open-split coupling interface: A, B: He inlet; C: He outlet; 1: transfer line; 2: sealings; 3: guiding capillary; 4: coupling capillary; 5: GC column (Kuster and Wetzel, 1983; copyright Elsevier Science Publishers B.V.)

proportional and diffusion rate is directly proportional to temperature. Helium permeation is essentially independent of flow rate, but it becomes excessive as membrane temperature approaches 200°C. In this case, it becomes necessary to use a two-stage device with intermediate pumping. Most of the two-stage devices use two supported silicone membranes, while Grayson and Wolf (1970) have described a device that uses a supported dimethylsilicone membrane and a silver frit.

Open-Split Coupling

In open-split coupling, the column effluent is vented to the atmosphere and a portion is diverted through a suitable restriction to the mass spectrometer. The sampling rate is determined by the leak dimensions and by the ion source pumping speed (Stern and Abraham, 1978; Henneberg et al., 1978; Arrendale et al., 1984). If the column flow is less than the sampling rate, additional carrier is added to the gas stream to prevent air from being drawn into the ion source. If it exceeds the sampling rate, the excess is vented to keep the inlet pressure constant, in which case the sampling efficiency falls below 100%. The GC operating conditions are not affected. However, no enrichment occurs, which for some samples may be a disadvantage. A number of open-split devices have been devised, and a representative version capable of either quasidirect or open-split coupling is shown in Figure 7.9. For open-split operation, port C is opened to the atmosphere, and optimized GC operating conditions are retained.

Direct Coupling of Capillary Columns

Whenever sample size or column flow rates permit, the gas chromatograph may be directly coupled to the mass spectrometer. The first interfacing of the two instruments, in fact, used a long capillary to reduce the pressure and to introduce the sample directly into the ion source (Holmes and Morrell,

1957; Gohlke, 1959). Broadening of the chromatographic peak due to reduced column pressure is minimal for a wall-coated, open-tubular capillary column, while the speed of analysis is increased. The shortened analysis time for vacuum outlet operation is treated in detail by Leclercq et al. (1982). They considered both thin film- and thick film-coated capillaries, with the latter having increased sample capacity and resolution of low-boiling compounds; little loss in column efficiency was found with the reduced analysis time. Hatch and Parrish (1978) examined the effects of vacuum on short (12 and 23 m), directly coupled capillaries that were below atmospheric pressure over their entire length. They found that column efficiency decreased 30% or more at reduced pressure with normal optimum flow conditions, but it increased dramatically at very high flow rates.

Direct coupling with fused silica capillary columns minimizes adsorption losses and decomposition of thermally labile compounds. These columns are most often coupled to mass spectrometers having high-speed differential pumping and chemical (McFadden, 1979) or atmospheric pressure (Mitchum and Korfmacher, 1983) ionization sources.

GC/GC/MS Coupling

Two fundamental problems related to introduction of the sample occur in the analysis of unknown compounds. First, it is often necessary to empirically determine the optimum sample size to obtain the maximum usable signal without saturating the detector amplifiers or diminishing chromatographic resolution through column overload. This is accomplished through one or more chromatographic runs with appropriately scaled samples prior to introduction into the GC/MS. For costly or very small samples, such empirical determinations may not be possible. Second, because of the difference in output representing major and minor constituents in the sample, the detection limit is determined by the maximum signal generated by the most abundant constituent and the dynamic range of the instrument. While this factor can be addressed electronically with auto-ranging or multiple amplifiers having different full-scale ranges (Chapter 6), interpretation of the data becomes somewhat laborious and the problem of column overload remains.

Both problems can be addressed simultaneously through selective attenuation of the major constituents by proportional control of the sample entering the GC/MS column. One efficient method of accomplishing this is sequential chromatography, in which appropriate quantities of sample eluting from the first column, as measured by the primary GC detector, are diverted to the column interfaced with the mass spectrometer. Chromatographic efficiency is increased by the multiple separation, since the demands on the second column are lower and the need for careful, repeatable control of sample size is reduced.

Sequential chromatography often uses relatively sophisticated valving to control sample transfer to the second chromatograph in an on-off manner (Schomburg et al., 1975). Ligon and May (1980) have described a simple, fully proportional, automated stream splitter for qualitative analysis, in which center-weighted fractions of each peak eluting from the primary chromatograph flow into a cold trap located in the oven of the second chromatograph. The amount of eluate diverted is determined by the primary chromatograph detector output, with a doubling in output halving the flow to the cold trap. The flow is equal to the total elution below a given threshold, and it is completely terminated as the detector nears saturation.

After completion of a primary separation, the cold trap contains generally normalized quantities of each component. Carrier flow is then directed through the cold trap, and the trap is heated by temperature programming of the secondary chromatograph oven. Quantitative analysis of the sample can also be accomplished through the use of internal standards or isotopic dilution methods.

LIQUID CHROMATOGRAPHY/MASS SPECTROMETRY

High-performance liquid chromatography (HPLC), like its GLC counterpart, consists of the separation of components in a column under controlled conditions. The carrier gas is replaced by a solvent moving through the column under high pressure, and partitioning of the sample occurs in the liquid phase. The sample components need not be volatile as in GLC; and because HPLC does not require elevated temperatures, it is a viable method for resolving thermally labile substances. Analogous to temperature programming in GLC, HPLC solvent programming (gradient elution) is a stepwise variation of the composition of the mobile solvent phase. The liquid phase can be a mixture of different ionic solutions or miscible solvents with different polarities, as long as neither is a solvent for a stationary phase liquid if one is used (Ewing, 1973). A detailed introduction to HPLC is found in Willard et al. (1981).

In liquid-liquid HPLC, separation of the organic solute occurs between a liquid mobile phase and a stationary liquid phase adsorbed on or chemically bonded with a solid support. The phases must be immiscible and, hence, have significantly differing polarities. In "normal phase" HPLC, the stationary phase is a polar (usually hydrophilic) liquid compound such as an alcohol held on a porous silica or alumina support, and the mobile phase is a nonpolar (usually hydrophobic) liquid such as hexane or carbon tetrachloride. In "reverse phase" HPLC, which is most often used, the polarities of the phases are interchanged, with the nonpolar hydrophobic solvent bonded to the support. Compounds are selectively retained on the stationary liquid phase as a function of decreasing polarity, that is, nonpolar hydrocarbons move through the column more slowly than do the polar alcohols.

If the sample is completely soluble in moderately polar organic solvents, separation can often be achieved without a stationary liquid. In this case, the solvent forms the mobile phase, and a porous solid such as silica gel or alumina forms the stationary phase. Separation is governed by successive adsorptions on the solid surface followed by release of the solute back into the solvent. Preferential adsorption results from the electrostatic interaction of polar functional groups of the solute with reactive hydroxyl groups on the silica or alumina surface. Both nonpolar and polar compounds can be resolved, with the order of adsorption increasing with polarity. Because selectivity is more dependent on functional groups than on molecular weight, complex mixtures can be sorted into classes having similar chemical characteristics.

High molecular weight molecules are best resolved by exclusion chromatography HPLC, often referred to as gel filtration or gel permeation chromatography, in which the solid stationary phase is a cross-linked polymer gel or silica with a controlled range of pore dimensions. Separation is based on molecular size or shape instead of on partitioning between phases, since the larger molecules enter fewer of the pores or interstices and thus exit from the column first. Typical pore sizes ranging from 40–2500 Å are capable of separating molecules over the molecular weight range of 10^2 to 10^8 amu.

HPLC Detectors

As with GLC, detection of an eluate from an HPLC column is based on the measurement of a physical property. The most common are ultraviolet or infrared absorption photometry, which measure the absorbance of the solute; or differential refractometry, which measures the difference in refractive index of the mobile phase when the solute is present. The first is sensitive to nanogram levels, but is restricted to those compounds that absorb at the operating wavelength of the photometer, while the latter is several orders of magnitude less sensitive but detects all species. The ultraviolet wavelengths of 254 or 280 nm are most commonly used, since almost all species have at least some absorption there—the exception being hydrocarbons, most of which do not absorb in the ultraviolet. These compounds, along with those having only saturated alkyl, alcohol, or ether groups, are transparent in the 200–1000 nm region; hence, they are useful as solvents. The differential refractometer measures changes in reflected light at the gas-liquid interface (Fresnel's law) or in the refraction of a monochromatic beam of light (Snell's law). In either case, the detection limit is about 1 μg.

THE HPLC/MS INTERFACE

The requirements for interfacing the HPLC to the mass spectrometer and various methods used to meet these requirements are reviewed in detail by McFadden (1979, 1982), Arpino and Guiochon (1979), Snyder and Kirkland

(1979), Games (1980), and Games et al. (1983). At present, only a moving sample transport belt for electron-impact or chemical ionization and several variations of direct liquid injection with chemical ionization have been successful enough to be commercially available. On-line coupling of the HPLC is limited by two substantial yet fundamental incompatibilities: (1) the effective high-flow rates in the HPLC column; and (2) the introduction of nonvolatile or thermally labile compounds into the mass spectrometer ion source. For these reasons, sample-enriching interfaces suitable for the GLC are not generally applicable to the HPLC. Depending upon the solvent, the typical 1 ml/min of liquid flow in the column expands to 10^2 to 10^3 ml·min^{-1} when vaporized. Maximum tolerable flow rates for chemical ionization are typically of the order of 20 ml·min^{-1} and are an order of magnitude less for electron-impact. While higher input can be handled effectively by increasing ion source pumping speed, there is a limit dictated by conductances; further improvement must be achieved with reduced column flow rates through microbore capillary tubing. The solvent can serve as the reagent gas for chemical ionization, but it should be removed for electron-impact ionization, with enrichment factors as high as 10^4 being desirable.

Manual Collection and Injection

The analysis of many thermally labile or moderately volatile compounds can be accomplished through manual collection of isolated fractions followed by injection into the mass spectrometer (Huber et al., 1973; Heresch et al., 1980). The collected fraction is generally evaporated to 25–50 µl and is transferred to the MS injection probe with a microsyringe. Typically, 5–10 ng samples are used to characterize the specimen; however, quantities as small as 1 ng can be detected and identified under favorable circumstances. The off-line collection method has the advantage that both the HPLC and the MS can be operated under optimum conditions, and the most favorable ionization method can be used for the analysis. The disadvantages are that an independent HPLC detector and relatively broad peaks are required to facilitate sample collection, and that the procedure becomes lengthy and tedious if a complex mixture is to be analyzed.

Moving Belt Interfaces

Scott et al. (1974) introduced a moving wire interface in which the column effluent was deposited on a thin (0.13 mm) stainless steel wire separated from the ion source by two vacuum interlock chambers. The solvent was evaporated in these chambers and the solute was thermally vaporized in the ion source region with a sample transfer efficiency of about 1%. McFadden et al. (1976) obtained efficiencies as high as 40% by replacing the wire with a narrow (3.2 mm) continuous belt of stainless steel or polyamide and using heaters to aid in solvent evaporation. A typical moving belt interface is

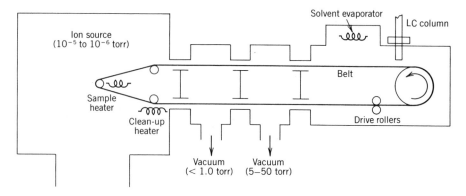

Figure 7.10 Schematic of a moving belt interface for LC/MS.

shown schematically in Figure 7.10. The effluent is deposited directly on the belt, where it forms a thin (0.2 mm) film, and then passes through the preliminary solvent evaporation zone and into the interlocks, where most of the remaining solvent is removed. The eluate is flash-evaporated either adjacent to (McFadden et al., 1976) or directly in the ion source (Millington et al., 1980; Games et al., 1983; Dobberstein et al., 1983; Hardin et al., 1984). Because of the very rapid desorption, decomposition of thermally labile compounds is minimized. Any residual materials are then removed with a clean-up heater. The amount of solvent entering the ion source is on the order of 10^{-8} g·sec^{-1}, so that either chemical or electron-impact ionization is possible; and, if properly insulated, magnetic deflection as well as QMS may be used. Flows of 2 ml·min^{-1} can be tolerated for nonpolar solvents; however, aqueous solutions are limited to lower flow rates to prevent water droplets from entering the ion source, and highly polar solvents require the use of a zero-dead-volume splitter. Transfer efficiencies as high as 80% compared to the direct introduction probe have been reported (Millington et al., 1980), and packed microbore glass-lined columns have been used to reduce problems associated with highly aqueous mobile phases (Lant et al., 1983).

Atmospheric Pressure Ionization

Atmospheric pressure ionization (API) is an emerging method for interfacing GC/MS as well as for HPLC applications, and it has recently been reviewed by Mitchum and Korfmacher (1983). In API, a carrier gas and the total effluent enter the ion source, where positive and negative ions are formed at atmospheric pressure by ion-molecule reactions in a plasma generated by a ^{63}Ni beta emitter or by a corona discharge (Fig. 4.16) (Horning et al., 1973; Carroll et al., 1975; Lane et al., 1980; Tsuchiya and Taira, 1980). A portion of the ions thus formed are extracted through a small (20–50 μm) orifice and are focused by a lens system into a QMS.

Molecular, protonated, and cluster ions are all formed in the plasma. Either positive or negative ions derived from the solute are useful in detecting and identifying sample components with normal or reverse phase HPLC (Arpino and Guiochon, 1979). In the apparatus of Tsuchiya and Taira (1980), the sample ions are formed by interaction with a carrier gas that has been ionized in a corona discharge. The mass spectra observed are derived mainly from the solute; and because its proton affinity is higher, protonation of the solute from the solvent occurs. Cluster ions are dissociated by increasing the draw-out potential on the orifice from 20 V to about 100 V, resulting in simple fragment ions related to the structure of the molecule in the sample. The method therefore provides information concerning structure as well as molecular weight.

Direct Liquid Introduction

The introduction of a small (1%) quantity of the column effluent into a chemical ionization mass spectrometer was first described by Baldwin and McLafferty (1973). Since then, several different methods have evolved to directly introduce a sample from the HPLC into the ion source. Splitting the effluent and formation of a liquid jet are accomplished by a capillary splitter or small-diameter orifice, and by vacuum nebulization in which the effluent is nebulized from the capillary tip or diaphragm by a flow of gas into a differentially pumped vacuum chamber (Arpino et al., 1974a, 1974b, 1981; Melera, 1980; Chapman et al., 1983; Voyksner and Bursey, 1984). Dedieu et al. (1982) have detailed methods for calculating the average droplet diameter and jet speed through small apertures, and also the flow velocity resulting from evaporation of the effluent in vacuum as applied to HPLC/ MS coupling.

Water cooling is generally, but not always, used to prevent vaporization of the sample as it passes near the heated ion source; and a small back pressure is maintained to generate a stable jet of liquid droplets through the orifice into an ionization chamber from where the vapor enters the chemical ionization (CI) source. The direct liquid introduction (DLI) method has been extended to microbore HPLC, which makes use of the total column effluent (e.g., Henion and Maylin, 1980; Henion and Wacks, 1981; Games et al., 1982; Levsen and Schafer, 1983; Apffel et al., 1983) and a typical device is shown in Figure 7.11. The DLI has the disadvantage that nonvolatile solutes cannot be analyzed. On the other hand, the advantages are: (1) simplified CI spectra are realized; (2) both negative and positive CI are possible; (3) the use of strongly polar or highly aqueous solvents are permitted; and (4) reverse phase chromatography can be used.

Thermospray

In order to analyze relatively nonvolatile but thermally labile molecules, Vestal and coworkers developed an innovative inlet system, in which the

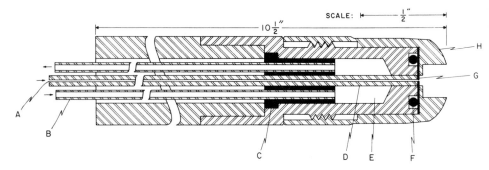

Figure 7.11 Direct liquid introduction interface. (Henion and Wachs, 1981; Reprinted with permission from *Anal. Chem.*, *53*, 1963 (1981). Copyright 1981 by American Chemical Society.)

column effluent is rapidly vaporized at the end of a capillary embedded in a copper cylinder heated to about 1000°C with a laser or an oxy-hydrogen flame (Blakley et al., 1980a, 1980b), or by direct or indirect electric heating (Vestal, 1983, 1984; Blakley and Vestel, 1983). The exiting jet of vapor and aerosol expands adiabatically, and a portion passes through a skimmer into the ion source, where it is ionized by conventional electron-impact or CI methods. The thermally labile molecules do not decompose, because the liquid is heated from ambient to the vaporization point in several milliseconds and is expelled in less than 1 ms after the vapor is formed. In the course of this development, they discovered that under certain conditions, ions were produced in the molecular beam even when the other ionization methods were not used, and that the mass spectra are more similar to those resulting from field desorption than from normal CI (Blakley et al., 1980c; Alexander and Kebarle, 1986). Thermospray ionization produces molecular ions for large nonvolatile molecules that are ionized in solution, and approximately equal numbers of positive and negative ions are observed.

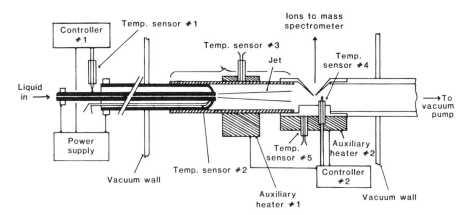

Figure 7.12 Schematic of the thermospray apparatus. (Vestal, 1984; copyright 1984 by the AAAS.)

A schematic of an electrically heated thermospray ionization device is shown in Figure 7.12. In this arrangement, a hot, dry, very intense vapor jet containing a mist of droplets emerges from the capillary at supersonic velocity with a flow rate of about 0.01 ml·min^{-1} when the copper cylinder is heated to 250°C. The jet traverses the ion source and enters a high-speed (300 ml·min^{-1}) pump-out line, with a portion of the ion beam extracted through the conical aperture and focused into the QMS. Nonvolatile molecules are preferentially retained in the mist droplets, which are positively or negatively charged. Molecular ions clustered with solvent molecules evaporate from the superheated droplets aided by high local fields associated with the charge of about 5×10^{-12} C on the droplet. The cluster ions rapidly equilibrate with the vapor in the ion source, resulting in the degree of solvation commensurate with local temperature and pressure. If exothermic, then ion-molecule reactions will also occur. Ionization efficiency is sensitive to both temperature and flow; however, relatively intense (10^{-8} A) total ion currents at flow rates of 1–1.5 ml·min^{-1} can be generated.

Thermospray HPLC/MS does not require splitting of the HPLC effluent or thermal desorption of the eluate into the ion source. The ionization process, however, does require that a volatile buffer such as ammonium acetate

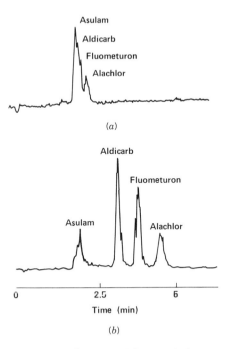

Figure 7.13 Total ion current traces for (*a*) on-column and (*b*) post-column addition of ammonium acetate buffer. (Adapted from Voyksner et al., 1984; courtesy R. D. Voyksner, Reprinted with permission from *Anal. Chem.*, 56, 1507 (1984). Copyright 1984 by American Chemical Society.)

TABLE 7.5   Full-Scan Detection Limits by Using the Base Peak, Retention Time, and Peak Width at Half-Height for the Pesticides Analyzed

Compound	Retention Time (s)	Peak Width at $\frac{1}{2}$-Ht (s)	Ion (m/z)	Detection Limit (ng)
Asulam	126	6	248	20–25
Aldicarb	295	10	208	3–5
Propoxur	303	14	227	4–8
Carbofuran	320	14	239	1–4
Propachlor	425	16	212	10–15
Fluometuron	445	14	233	15–20
BPMC	457	13	225	3–5
Carbaryl	468	12	219	3–5
Diuron	544	17	233	50–80
Alachlor	600	18	270	10–20
Linuron	655	17	266	4–8
Desmedipham	700	23	318	70–100
Phenmedipham	710	23	185	5–8
Benzoylprop ethyl	824	26	366	60–90

Source: Voyksner et al. (1984).

be present in the HPLC effluent. Adding the buffer before the column will, in some cases, reduce chromatographic resolution; and in gradient elution analysis, variations in the nonsoluble buffer concentration also result in changes in sensitivity. Voyksner et al. (1984) have described a method for post-column addition of the buffer to the effluent in order to retain chromatographic resolution and to enhance the thermospray efficiency. Total ion current traces of an analysis of a mixture of four thermally labile carbamate and urea pesticides obtained under isocratic, HPLC conditions are shown in Figure 7.13. Considerable reduction in chromatographic resolution was observed when 0.1 M ammonium acetate was added prior to the column (Fig. 7.13a). Post-column addition of the buffer through a coaxial tee, on the other hand, retained chromatographic resolution and significantly increased the measured ion intensities (Fig. 7.13b). A gradient elution analysis, with post-column addition of buffer, of the 14 pesticides listed in Table 7.5 showed characteristic spectra consisting primarily of $(M + H)^+$ and $(M + NH_4)^+$ ions with no excessive baseline drift. Detection limits listed in the table are dependent upon peak width and solvent composition, but are below those obtained with DLI or moving belt methods, or with ultraviolet detection.

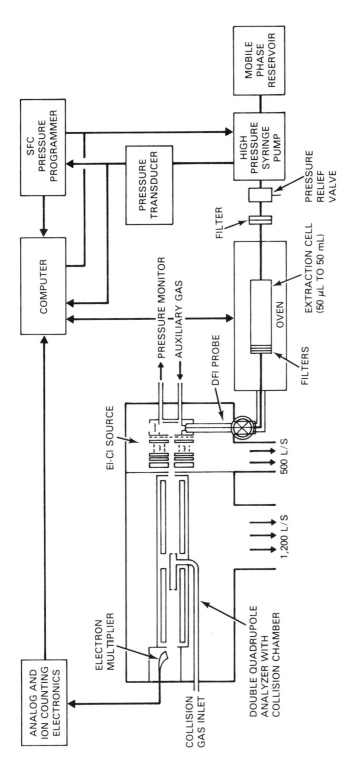

Figure 7.14 Schematic of the supercritical fluid chromatograph/mass spectrometer. (Smith and Udseth, 1983b; courtesy R. D. Smith; Reprinted from Ref. (*Sep. Sci. and Tech., 18*, No. 3, 245 (1983b) by courtesy of Marcel Dekker Inc.)

Supercritical Fluid Chromatography/Mass Spectrometry

The physical properties of a supercritical fluid lie between those of the gas and liquid states. Above but near the critical temperature, the fluid (or "dense gas") exhibits liquid-like properties as pressures are increased, with densities reaching several orders of magnitude higher than in the gas phase and with small changes in pressure resulting in large changes in density. The supercritical fluid has a greatly enhanced solvating power, so that thermally labile high molecular weight compounds can often be dissolved at low temperatures. The solvating power is a function of temperature, pressure, and fluid composition; therefore, it can provide selective extraction or chromatography under nonisobaric or nonisothermal conditions. While both HPLC/MS and supercritical fluid chromatography/mass spectrometry (SFC/MS) will resolve thermally labile and nonvolatile compounds, the latter has the advantage of simpler pressure programming in lieu of gradient elution; and the higher volatility of the mobile phase simplifies interfacing. The principles and instrumental aspect of supercritical fluid chromatography have been reviewed by Guow and Jentoft (1975), Peaden et al. (1982), and by Smith et al. (1984, 1985).

An SFC/MS interfacing system has been developed by Smith and co-

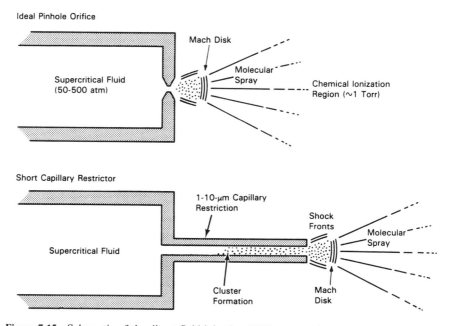

Figure 7.15 Schematic of the direct fluid injection (DFI) process for transfer of supercritical fluid solutes to the gas phase. (Smith et al., 1985; Reprinted with permission from Chromatography and Separation Chemistry, ACS Symp. Series, ed. S. Ahuja, Washington, D.C. (1985). Copyright 1985 American Chemical Society.)

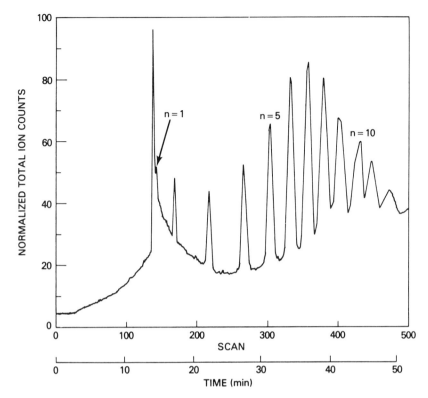

Figure 7.16 Total ion chromatogram of the first 12 oligomers of a polystyrene, with each having a molecular weight of 104 n + 58. (Smith et al., 1982; courtesy R. D. Smith; copyright Elsevier Science Publishers B.V.)

workers using either a capillary column, a small (< 50 ml) extraction cell Fig. 7.14), or direct fluid injection (Smith et al., 1982; Smith and Udseth, 1983a, 1983b). Pressure programming is used in all cases. The eluent is introduced through an isothermal, small-diameter, direct fluid injection probe into a dual electron-impact/CI QMS, with supercritical conditions maintained up to the 2–3 μm orifice (Fig. 7.15). In a typical measurement, nanogram quantities in a mixture of five polycyclic aromatic hydrocarbons in methylene chloride were analyzed on a 0.2 mm × 20 m long capillary column with a *n*-pentane mobile phase and pressure programming from 20–40 atm at 0.8 atm·min^{-1}. The total ion chromatogram obtained for a nonvolatile, high molecular weight polystyrene resolved into the first 12 styrene oligomers, with the same column and solvent when the pressure was programmed to 60 atm at 1.2 atm·min^{-1}, is shown in Figure 7.16. The molecular weight of each oligomer is 104 n + 58, so that the weight of the 12th oligomer is 1306 amu. With the extraction cell, supercritical fluid reactions or extraction processes can be determined under both nonisothermal and nonisobaric conditions. Pressure-dependent molecular weight distributions of

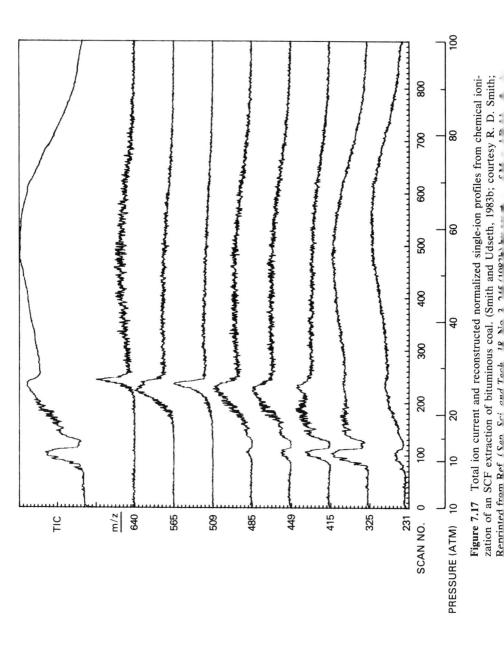

Figure 7.17 Total ion current and reconstructed normalized single-ion profiles from chemical ionization of an SCF extraction of bituminous coal. (Smith and Udseth, 1983b; courtesy R. D. Smith; Reprinted from Ref. (*Sep. Sci. and Tech.*, *18*, No. 2, 245 (1983))

extraction products are determined from scanning the appropriate mass range as the pressure is increased, while total and selected ion current profiles characterize fractionation processes. Figure 7.17 shows total and selected ion monitoring profiles of an extraction from a 40 mg sample of bituminous coal in a 75 μl extraction cell at 280°C with a 95% pentane-5% isopropranol mixture. The mass spectra were obtained by CI to reduce ionic fragmentation. The long tailing of the spectra above 50 atm indicates that a large amount of complex, higher molecular weight material is extracted in that region.

REFERENCES

Alexander, A. J. and P. Kebarle, *Anal. Chem. 58*, 471 (1986).

Apffel, J. A., U. A. Th. Brinkman, R. W. Frei, and E. A. I. M. Evers, *Anal. Chem., 55*, 2280 (1983).

Arpino, P. J., M. A. Baldwin, and F. W. McLafferty, *Biomed. Mass Spectrom., 1*, 80 (1974a).

Arpino, P. J., B. G. Dawkins, and F. W. McLafferty, *J. Chromatogr. Sci., 12*, 574 (1974b).

Arpino, P. J., and G. Guiochon, *Anal. Chem., 51*, 682A (1979).

Arpino, P. J., P. Krien, S. Vajta, and G. Devant, *J. Chromatogr., 203*, 117 (1981).

Arrendale, R. F., R. F. Severson, and D. T. Chortyk, *Anal. Chem., 56*, 1533 (1984).

Baldwin, M. A., and F. W. McLafferty, *Org. Mass Spectrom., 7*, 1353 (1973).

Blakley, C. R., M. J. McAdams, and M. L. Vestal, in *Advances in Mass Spectrometry*, Vol. 8A, ed. A. Quayle, Heyden and Son Ltd., London (1980a), p. 1616.

Blakley, C. R., J. J. Carmody, and M. L. Vestal, *Anal. Chem., 52*, 1636 (1980b).

Blakley, C. R., J. J. Carmody, and M. L. Vestal, *J. Am. Chem. Soc., 102*, 5931 (1980c).

Blakley, C. R., and M. L. Vestal, *Anal. Chem., 55*, 750 (1983).

Blumer, M., *Anal. Chem., 40*, 1590 (1968).

Bonelli, E. J., M. S. Story, and J. B. Knight, *Dynamic Mass Spectrom., 2*, 177 (1971).

Brunee, C., L. Jenchel, and K. Kronenberger, *Z. Anal. Chem., 189*, 50 (1962).

Brunee, C., H. J. Bultemann, and G. Kappus, Paper No. 46 at *17th Ann. Conf. on Mass Spectrom. and Allied Topics*, Dallas (1969).

Carroll, D. I., I. Ozidic, R. N. Stillwell, K. D. Haegele, and E. C. Horning, *Anal. Chem., 47*, 2369 (1975).

Chapman, J. R., *Anal. Proc. (London), 19*, 235 (1982).

Chapman, J. R., E. H. Harden, S. Evans, and L. E. Moore, *Int. J. Mass Spectrom. Ion Phys., 46*, 201 (1983).

Colby, B. N., in *Physical Methods in Modern Chemical Analysis*, Vol. 1, ed. T. Kuwana, Academic Press, New York (1978), p. 67.

Copet, A., and J. Evans, *Org. Mass Spectrom., 3*, 1437 (1970).

Cree, R., in *Pittsburgh Conf. on Anal. Chem. Appl. Spectrosc.* (1967).

Dedieu, M., C. Juin, P. J. Arpino, J. P. Bounine, and G. Guiochon, *J. Chromatogr. Rev., 251*, 203 (1982).

Dobberstein, P., E. Korte, G. Meyerhoff, and R. Pesch, *Int. J. Mass Spectrom. Ion Phys., 46*, 185 (1983).

Ewing, G. W., *Instrumental Methods of Chemical Analysis*, 4th ed., McGraw-Hill, New York (1973), Chapter 20.

Fenselau, C., *Appl. Spec.*, *28*, No. 4, 305 (1974).

Fenselau, C., in *Physical Methods in Modern Chemical Analysis*, Vol. 1, ed. T. Kuwana, Academic Press, New York (1978), p. 103.

Games, D. E., in *Soft Ionization Biological Mass Spectrometry*, ed. H. R. Morris, Heyden and Son Ltd., London (1980), p. 54.

Games, D. E., M. S. Lant, S. A. Westwood, M. J. Cocksedge, N. Evans, J. Williamson, and B. J. Woodhall, *Biomed. Mass Spectrom.*, *9*, 215 (1982).

Games, D. E., N. J. Alcock, L. Cobelli, C. Eckers, M. P. L. Games, A. Jones, N. S. Lant, M. A. McDowall, M. Rossiter, R. A. Smith, S. A. Westwood, and N. Y. Wong, *Int. J. Mass Spectrom. Ion Phys.*, *46*, 181 (1983).

Gohlke, R. S., *Anal. Chem.*, *31*, 535 (1959).

Grayson, M. A., and C. J. Wolf, *Anal. Chem.*, *39*, 1438 (1967).

Grayson, M. A., and C. J. Wolf, *Anal. Chem.*, *42*, 426 (1970).

Grayson, M. A., and R. L. Levey, *J. Chromatogr. Sci.*, *9*, 687 (1971).

Grayson, M. A., and J. J. Bellina, Jr., *Anal. Chem.*, *45*, 487 (1973).

Greenway, A. M., and C. F. Simpson, *J. Phys. E: Sci. Instru.*, *13*, 1131 (1980).

Grob, K., *J. High Resol. Chromatogr. and Chromatog. Comm.*, *2*, 599 (1979).

Gudzinowicz, B. J., M. J. Gudzinowicz, and H. F. Martin, *Fundamentals of Integrated GC/MS*, Vol. 7, Part 3, Marcel Dekker, New York (1977).

Guow, T. H., and R. E. Jentoft, in *Advances in Chromatography*, Vol. 13, eds. J. C. Giddings, E. Grushka, R. A. Keller, and J. Cazes, Marcel Dekker, New York (1975), p. 1.

Hardin, E. D., T. P. Fan, C. R. Blakley, and M. L. Vestal, *Anal. Chem.*, *56*, 2 (1984).

Hatch, F. W., and M. E. Parrish, *Anal. Chem.*, *50*, 1164 (1978).

Henion, J. D., and G. A. Maylin, *Biomed. Mass Spectrom.*, *7*, 115 (1980).

Henion, J. D., and T. Wachs, *Anal. Chem.*, *53*, 1963 (1981).

Henneberg, D., U. Henrichs, H. Husmann, and G. Schomburg, *J. Chromatogr.*, *167*, 139 (1978).

Heresch, F., E. R. Schmid, I. Fogy, and J. F. K. Huber, in *Advances in Mass Spectrometry*, Vol. 8A, ed. A. Quayle, Heyden and Son Ltd., London (1980), p. 1880.

Holmes, J. C., and F. A. Morrell, *Appl. Spectrom.*, *11*, 86 (1957).

Horning, E. C., M. G. Horning, D. I. Carroll, I. Szidic, and R. N. Stillwell, *Anal. Chem.*, *45*, 936 (1973).

Huber, J. F. K., A. M. VanUrk-Schoen, and G. B. Sieswerda, *Z. Anal. Chem.*, *264*, 257 (1973).

Junk, J. A., *Int. J. Mass Spectrom. Ion Phys.*, *8*, 1 (1972).

Kimble, B. J., in *High Performance Mass Spectrometry: Chemical Applications*, ed. M. L. Gross, ACS Symp. Series 70, American Chemical Society, Washington, D.C. (1978), p. 120.

Krick, T. P., H. Schlenk, and R. L. Glass, paper RP 29 at *26th Ann. Conf. on Mass Spectrom. and Allied Topics*, St. Louis (1978).

Krueger, P. M., and J. A. McCloskey, *Anal. Chem.*, *41*, 1930 (1969).

Kuster, T., and E. Wetzel, *Int. J. Mass Spectrom. Ion Phys.*, *46*, 173 (1983).

Lane, D. A., B. A. Thomson, A. M. Lovett, and N. M. Reid, in *Advances in Mass Spectrometry*, Vol. 8B, ed. A. Quayle, Heyden and Son Ltd., London (1980), p. 1480.

Lant, M. S., D. E. Games, S. A. Westwood, and B. J. Woodhall, *Int. J. Mass Spectrom. Ion Phys.*, *46*, 189 (1983).

Leclercq, P. A., G. J. Scherpenzeel, E. A. A. Vermeer, and C. A. Cramers, *J. Chromatogr.*, *241*, 61 (1982).

Levsen, E., and K. H. Schafer, *Int. J. Mass Spectrom. Ion Phys.*, *46*, 209 (1983).

Ligon, W. V., Jr., and R. J. May, *Anal. Chem.*, *52*, 901 (1980).

Llewellyn, P. M., and D. P. Littlejohn, in *Pittsburgh. Conf. on Anal. Chem. and Appl. Spectrosc.* (1966).

Mass Spectrometry Data Centre, *Eight Peak Index of Mass Spectra*, AWRE, Aldermaston, Reading, United Kingdom, 2nd Edition, 1974.

McFadden, W. H., *Techniques of Combined Gas Chromatography/Mass Spectrometry: Applications in Organic Analysis*, Wiley, New York (1973).

McFadden, W. H., *J. Chromatogr., Sci.*, *17*, 2 (1979).

McFadden, W. H., *Anal. Proc.* (*London*), *19*, 258 (1982).

McFadden, W. H., H. L. Schwartz, and S. Evans, *J. Chromatogr.*, *122*, 389 (1976).

Melera, A., *Advances in Mass Spectrometry*, Vol. 8B, ed. A. Quayle, Heyden and Son Ltd., London (1980), p. 1597.

Merritt, C., *Appl. Spec. Rev.*, *3*, 263 (1970).

Millard, B. J., *Quantitative Mass Spectrometry*, Heyden and Son Ltd., Philadelphia (1978).

Millington, D. S., D. A. Yorke, and P. Burns, in *Advances in Mass Spectrometry*, Vol. 8B, ed. A. Quayle, Heyden and Son Ltd., London (1980), 1819.

Mitchum, R. K., and W. A. Korfmacher, *Anal. Chem.*, *55*, 1458A (1983).

Okamura, J. P., and D. T. Sawyer, in *Physical Methods in Modern Chemical Analysis*, Vol. 1, ed. T. Kuwana, Academic Press, New York (1978), p. 2.

Peaden, P. A., J. C. Fjeldsted, M. L. Lee, S. R., Springston, and M. Novotny, *Anal. Chem.*, *54*, 1090 (1982).

Perry, J. A., *Introduction to Analytical Gas Chromatography: History, Principles, and Practice*, Marcel Dekker, New York (1981).

Rhyage, R., *Anal. Chem.*, *36*, 759 (1964).

Rhyage, R., *Arkivkemi, 26*, 305 (1967).

Rhyage, R., and S. Wikstrom, in *Mass Spectrometry: Techniques and Applications*, ed. G. W. A. Milne, Wiley-Interscience, New York (1971), p. 91.

Safe, S., and O. Hutzinger, *Mass Spectrometry of Pesticides and Pollutants*, CRC Press, Cleveland (1973), Chapter 3.

Schomburg, G., H. Husmann, and F. Weeke, *J. Chromatogr.*, *112*, 205 (1975).

Scott, R. P. W., C. G. Scott, M. Moroe, and B. Hess, Jr., *J. Chromatogr.*, *99*, 395 (1974).

Simpson, C. F., *Gas Chromatography—Mass Spectrometry Interfacial Systems, CRC Review of Analytical Chemistry*, CRC Press, Cleveland (1972), p. 1.

Smith, R. D., J. C. Fjeldsted, and M. L. Lee, *J. Chromatogr.*, *247*, 231 (1982).

Smith, R. D., and H. R. Udseth, *Anal. Chem.*, *55*, 2266 (1983a).

Smith, R. D., and H. R. Udseth, *Sep. Sci. and Tech.*, *18*, No. 3, 245 (1983b).

Smith, R. D., H. R. Udseth, and B. W. Wright, in *Conf. Am. Inst. Chem. Eng.*, San Francisco (November 1984).

Smith, R. D., B. W. Wright, and H. R. Udseth, in *Chromatography and Separation Chemistry*, ACS Symp. Series, ed. S. Ahuja, Washington, D. C. (1985).

Snyder, L. R., and J. J. Kirkland, *Introduction to Modern Liquid Chromatography*, 2nd ed., Wiley, New York (1979).

Stern, J. J., and B. Abraham, *Anal. Chem.*, *50*, 2161 (1978).

Szepsey, L., *Gas Chromatography*, Ilift Books Ltd., London (1970), p. 160.

Ten Noever de Brauw, M. C., *J. Chromatogr., 165*, 207 (1979).

Tsuchiya, M., and T. Taira, *Int. J. Mass Spectrom. Ion Phys., 34*, 351 (1980).

Vestal, M. L., *Int. J. Mass Spectrom. Ion Phys., 46*, 193 (1983).

Vestal, M. L., *Science, 226*, 275 (1984).

Voyksner, R. D., and J. T. Bursey, *Anal. Chem., 56*, 1582 (1984).

Voyksner, R. D., J. T. Bursey, and E. D. Pellizzari, *Anal. Chem., 56*, 1507 (1984).

Watson, J. T., and K. Biemann, *Anal. Chem., 36*, 1135 (1964).

Watson, J. T., *Introduction to Mass Spectrometry*, 2nd Ed., Raven Press, New York, (1985).

Willard, H. H., L. L. Merritt, J. A. Dean, and F. A. Settle, Jr., *Instrumental Methods of Analysis*, D. van Nostrand Co., New York (1981), p. 436.

ENGINEERING AND THE PHYSICAL SCIENCES

CHAPTER EIGHT

GEOCHEMISTRY AND GEOCHRONOLOGY

An ultimate goal of geochemistry is to unravel the detailed history of the elements as they have interacted in terrestrial and cosmic events and processes. These events and processes cannot be understood without the use of mass spectrometry to quantify the elements and their interaction at the atomic-molecular level. Classical studies of the major geospheres (atmosphere, hydrosphere, biosphere, crust, and mantle) have recently been extended to include interactions between the earth's surface and cosmic radiations, dynamic crustal-mantle phenomena that lead to earthquakes, soil-rock chemistry relevant to agriculture and prospecting, recycling of the ocean floor, and the concentration of some metals within living organisms. Furthermore, isotopic measurements provide the salient data for establishing the time scale of the solar system, for developing models of planetary evolution, and for building the foundations of cosmochronology. In a contemporary context, mass spectrometry also relates to such practical geoscience topics as ocean mining, earthquake prediction, climatology, energy production, and transport and rate phenomena that impact the global environment.

"COSMIC" ABUNDANCES OF THE ELEMENTS

The abundances of the elements in the solar system provide the data base for theories of nucleosynthesis and cosmochemistry. These abundances reflect primarily the chain of nuclear reactions by which the elements were synthesized; these values also relate to criteria of nuclear stability and isotopic abundances.

An interpretation of nucleosynthesis for all the heavy elements, from carbon to uranium, has been given by Fowler (1984), who shared the Nobel prize in physics for 1983. Table 8.1 lists estimates of the abundances of the elements in our solar system ("cosmic" abundances) relative to 10^6 silicon

TABLE 8.1 Relative Abundance of the Elements in the Solar System in Cosmic Abundance Units (Atoms per 10^6 Silicon Atoms)

Z	Element	Abundance	Z	Element	Abundance
1	H	4.8×10^{10}	44	Ru	1.83
2	He	3.9×10^9	45	Rh	3.3×10^{-1}
3	Li	16	46	Pd	1.33
4	Be	0.81	47	Ag	3.3×10^{-1}
5	B	6.2	48	Cd	1.2
6	C	1.7×10^7	49	In	1.0×10^{-1}
7	N	4.6×10^6	50	Sn	1.7
8	O	4.4×10^7	51	Sb	2.0×10^{-1}
9	F	2.5×10^3	52	Te	3.1
10	Ne	4.4×10^6	53	I	4.1×10^{-1}
11	Na	3.5×10^4	54	Xe	3.0
12	Mg	1.04×10^6	55	Cs	2.1×10^{-1}
13	Al	8.4×10^4	56,	Ba	5.0
14	Si	1.0×10^6	57	La	4.7×10^{-1}
15	P	8.1×10^3	58	Ce	1.38
16	S	8.0×10^5	59	Pr	1.9×10^{-1}
17	Cl	2.1×10^3	60	Nd	8.8×10^{-1}
18	Ar	3.4×10^5	61	Pm	—
19	K	2.1×10^3	62	Sm	2.8×10^{-1}
20	Ca	7.2×10^4	63	Eu	1.0×10^{-1}
21	Sc	3.5×10^1	64	Gd	4.3×10^{-1}
22	Ti	2.4×10^3	65	Tb	6.1×10^{-2}
23	V	5.9×10^2	66	Dy	4.5×10^{-1}
24	Cr	1.24×10^4	67	Ho	9.3×10^{-2}
25	Mn	6.2×10^3	68	Er	2.8×10^{-1}
26	Fe	2.5×10^5	69	Tm	4.1×10^{-2}
27	Co	1.9×10^3	70	Yb	2.2×10^{-1}
28	Ni	4.5×10^4	71	Lu	3.6×10^{-2}
29	Cu	4.2×10^2	72	Hf	3.1×10^{-1}
30	Zn	6.3×10^2	73	Ta	1.9×10^{-2}
31	Ga	2.8×10^1	74	W	1.6×10^{-1}
32	Ge	7.6×10^1	75	Re	5.9×10^{-2}
33	As	3.8	76	Os	8.6×10^{-1}
34	Se	2.7×10^2	77	Ir	9.6×10^{-1}
35	Br	5.4	78	Pt	1.4
36	Kr	2.5×10^1	79	Au	1.8×10^{-1}
37	Rb	4.1	80	Hg	6.0×10^{-1}
38	Sr	2.5×10^1	81	Tl	1.3×10^{-1}
39	Y	4.7	82	Pb	1.3
40	Zr	2.3×10^1	83	Bi	1.9×10^{-1}
41	Nb	9.0×10^{-1}	90	Th	4×10^{-2}
42	Mo	2.5	92	U	1×10^{-2}
43	Tc	—			

Source: Goles (1969); Faure (1977a).

atoms. Hydrogen and helium are the most abundant elements; lithium, beryllium, and boron are lower by many orders of magnitude. Elemental abundances beyond carbon tend to decrease exponentially with increasing atomic number, but elements of *even* atomic number are somewhat more abundant than their neighboring nuclei. These atomic abundances are also generally consistent with *mass* estimates for the elements based on the Big Bang model of the universe. Twenty-five percent of the mass of the universe is comprised of the single isotope ^{4}He, and this fraction is exactly what is predicted by modern cosmological theory (Schramm, 1983).

Most nuclides will also be characterized by several naturally occurring isotopes, but only a small number of these will be radioactive. These radioactive nuclides exist in nature due to one of the following reasons:

1. Some nuclei have not completely decayed since their nucleosynthesis because of their very long half-life (e.g., ^{40}K, ^{87}Rb, ^{235}U, etc.).
2. Some nuclei have very long-lived radioactive parents (e.g., ^{226}Ra, ^{230}Th, ^{234}U, etc.)
3. Some nuclei result from nuclear reactions occurring in nature (e.g., ^{10}Be, ^{14}C, ^{32}Si, etc.).

Nuclei that are continually being formed due to cosmic radiations, but have short half-lives, will exist only in trace amounts and require highly sensitive instrumentation (such as accelerator-mass spectrometry) for their detection.

TERRESTRIAL ABUNDANCES AND ISOTOPIC COMPOSITION

Terrestrial abundances of the elements in the earth's crust are based on extensive chemical geological data. Abundances of the elements for the earth as a whole are indirectly deduced primarily from the analysis of chondrites and meteorites. Figure 8.1 indicates the relative abundance of the more common elements in the earth's crust, and for the earth as a whole (Ahrens, 1965). It will be noted that oxygen (O) is the most common element in the crust; iron (Fe) is more abundant, however, for the entire earth. Approximately 90% of the earth is comprised of Fe, O, silicon (Si), and magnesium (Mg), and 99.9% is made up of only 15 of the elements. Thus, there exists a marked differentiation between our planetary body and stellar matter. Helium (He), an abundant stellar gas, is only a trace element on earth. It is available in quantity primarily in the United States (Kansas, Oklahoma, Texas, and New Mexico), where it is separated from natural gas. There are two naturally occurring isotopes of helium, ^{3}He and ^{4}He. Perhaps some of the ^{4}He is primordial, but most of it is assumed to be a product of alpha decay.

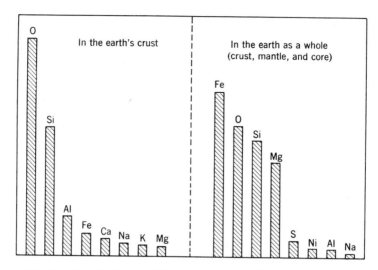

Figure 8.1   Relative abundances of the common terrestrial elements (Ahrens, 1965).

If we focus our attention on the isotopic composition of the terrestrial elements, we find that a chemical element will have almost the same isotopic abundance regardless of where it is found on earth. Also, if radioactive decay products are excluded, the variation in the percent abundance of the isotopes is very small. However, the "normal" isotopic abundance values per se are of interest to the nuclear geologist and the nuclear physicist alike. In particular, the relative abundances of the several isotopes of a given chemical element clearly indicate which nuclides are favored from a standpoint of nuclear stability. In other words, one would expect the more abundant isotopes to correspond to the most stable nuclear configurations of protons and neutrons. Consider the isotopic structure of tin (having the largest number of isotopes) and cerium. If we plot the percentage abundance of the isotopes of these elements according to the number of neutrons in the nucleus (A–Z), we obtain the graph of Figure 8.2. An even number of protons and neutrons is clearly favored over the even-odd proton-neutron species. The neutron number 82 for the ^{140}Ce isotope is also one of the "magic" numbers associated with a particularly stable nuclear configuration. Indeed, isotopic spectra of the naturally occurring nuclides have been important sources for developing modern theories of nuclear structure and for establishing limits of geological age. The isotopic invariance of widely dispersed matter (e.g., in the solar system) also suggests that such material had been exposed to a nuclear history comparable to that of the earth.

Thus, the general isotopic invariance of matter is an important phenomenon, because it establishes a primary baseline against which changes in isotopic ratios can be compared and evaluated for their significance. When very large isotopic deviations occur from normal abundances, we can expect

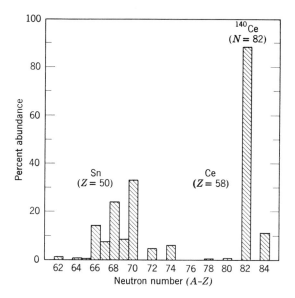

Figure 8.2 Isotopic abundance of tin and cerium, showing the nuclear stability of "even-even" nuclides.

that nuclear transmutations have occurred. An analogous situation takes place in a nuclear reactor in which high fluxes of neutrons breed many new nuclides, fission products, and literally hundreds of new isotopes in trace amounts. In nuclear geology, the problem is to extract information produced by the past "natural nuclear environment." We thus inquire as to the "natural" rather than artificially produced species that give rise to isotopic anomalies.

NATURALLY OCCURRING RADIONUCLIDES

Provided that the rate of disintegration of an unstable nuclide is relatively low, we should expect some residue of that substance in nature. Table 8.2 is a partial listing of some geochemically important nuclides, together with mass number, half-life, decay mode, and products. With the exception of the neutron, all of the nuclides listed in Table 8.2 can be detected by a mass spectrometer. By stretching a point, we could even consider that the neutron has been observed by mass spectrometry. The reason, in fact, is that the half-life of the neutron was determined by the ingenious coincident detection of its decay products—a beta particle and a proton, with a beta ray spectrometer and a proton spectrometer, respectively (Robson, 1955).

A glance at the half-lives shown in Table 8.2 suggests why the mass spectrometer is the ideal tool in geological studies. On a laboratory time scale,

TABLE 8.2 Natural Unstable Nuclides[a]

Nuclide	Mass No.	Half-Life	Decay Mode	Product
Neutron	1	12.6 min	β^-	^{1}H
Tritium	3	12.4 yr	β^-	^{3}He
Carbon	14	5730 yr	β^-	^{14}N
Potassium	40	1.25×10^9 yr	β^-; K-capture	^{40}Ca; ^{40}Ar
Vanadium	50	6×10^{15} yr	β^-; K-capture	^{50}Cr; ^{50}Ti
Rubidium	87	4.8×10^{10} yr	β^-	^{87}Sr
Lanthanum	138	1.1×10^{11} yr	β^-; K-capture	^{138}Ce; ^{138}Ba
Samarium	147	1.1×10^{11} yr	α	^{143}Nd
Lutetium	176	3.5×10^{10} yr	β^-	^{176}Hf
Rhenium	187	4.3×10^{10} yr	β^-	^{187}Os
Lead	204	1.4×10^{17} yr	α	^{200}Hg
Radium	226	1620 yr	α	^{222}Rn[b]
Thorium	232	1.39×10^{10} yr	α	^{228}Ra[c]
Uranium	234	2.5×10^5 yr	α	^{230}Th[b]
	235	7.1×10^8 yr	α	^{231}Th[d]
	238	4.5×10^9 yr	α	^{234}Th[b]

[a] Partial list.
[b] Final decay product: ^{206}Pb.
[c] Final decay product: ^{208}Pb.
[d] Final decay product: ^{207}Pb.

most of these nuclides are quite stable; but on a geological time scale, many of these nuclides have an appreciable decay rate. Table 8.3 lists some of the geochemically important nuclides that are produced by cosmic rays. Their abundance is often so low that detection and measurement require the ultimate in mass spectrometric sensitivity.

TABLE 8.3 Trace Cosmic Ray Produced Isotopes

Isotope	Half-Life (yr)
^{3}H	12.4
^{10}Be	1.6×10^6
^{14}C	5730
^{26}Al	7.2×10^5
^{32}Si	100
^{36}Cl	3.1×10^5
^{39}Ar	270
^{81}Kr	2.1×10^5
^{129}I	1.6×10^7

Source: Olsson (1981).

Tritium (^{3}H) has a half-life of 12.43 years (Unterweger et al., 1980). Natural tritium has been measured in snow deposited in Antarctica and elsewhere prior to the first arrival of bomb-produced tritium (1954). The data indicate mean deposition rates of 0.37 tritium atoms/cm^2/s over Antarctica—a mean global deposition rate of approximately 0.2 tritium atoms/cm^2/s—and that natural tritium results only from cosmic ray production (Jouzel et al., 1982). This corresponds to present tritium concentrations in pre-1950 snowfalls of only a few TU (1 TU is defined as a ^{3}H/H ratio of 10^{-18}). Such low concentrations require a high tritium enrichment prior to counting.

^{14}C is formed in the upper atmosphere by the reaction of ^{14}N with cosmic ray neutrons, that is, ^{14}N $+ n \rightarrow {}^{14}$C $+$ H, and it constitutes a very small part of the earth's carbon dioxide (CO_2) (approximately 1 part in 10^{12}). The CO_2 is in equilibrium with the large reservoirs of carbon in both the surface waters of the ocean and terrestrial life. When radiocarbon dating was first introduced in 1949, it was assumed that (1) the exchange of carbon within the atmosphere and other reservoirs was short compared to the 5730 year half-life of ^{14}C; and (2) the production rate of ^{14}C was invariant. This latter assumption is not strictly valid, because temporal variations have been noted in the radiocarbon content of atmospheric CO_2. The change in ^{14}C production is now attributed to changes in intensity of the earth's magnetic field and other parameters that control the amount of cosmic radiation reaching the upper atmosphere. Hence, a "calibration" to correct for such variations has been made on 1154 samples of dendrochronologically dated wood—principally bristlecone pine and giant sequoia with maximum ages of 8000 years—at radiocarbon laboratories of the universities of Arizona, Groningen, California at San Diego, Pennsylvania, and Yale (Klein et al., 1982). The cosmic ray correction is not large ($\pm 10\%$ over the past 50,000 years), but it is significant, and carbon production can be partially correlated with long-term magnetic field changes. Also, when maximum precision of a radiocarbon date is required, ^{13}C/^{12}C ratios are measured mass spectrometrically as a check on natural or laboratory fractionation effects (Ralph and Michael, 1974).

TRACE ISOTOPE DETECTION BY ACCELERATOR-MASS SPECTROMETRY

The detection of trace elements is important in many disciplines, but their measurement has special relevance in geoscience. Only recently, however, has it been possible to achieve a sensitivity exceeding 1 part per billion, thus making possible the detection of very long-lived radionuclides (Rucklidge et al., 1972). These trace isotopes are the result of spontaneous fission, neutron capture, and production by cosmic rays. Figure 8.3 shows their relative abundance in contrast to abundances of stable elements in the earth's crust (Rucklidge et al., 1981). A fundamental experimental problem in the

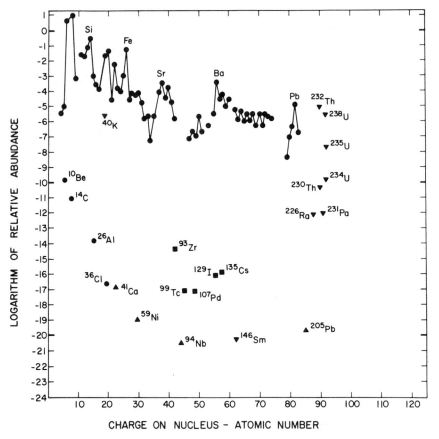

Figure 8.3 Relative abundance of stable and trace long-lived nuclides in the earth's crust. Nuclei shown as inverted triangles (▼) denote survivors from element formation prior to the birth of the solar system; upright triangles (▲) denote nuclides produced via neutron capture; squares (■) are ^{238}U fission products. (Rucklidge et al., 1981; *Nuclear Instruments and Methods*, *191*, 1 (1981), by permission of North-Holland Physics Publishing.)

past has been that radioactive isotopes that decay by beta particle emission produce isobars having very small mass differences, for example, ^{14}C and ^{14}N. While some isobars and interfering molecular species can, indeed, be resolved by high-resolution instruments, mass separation cannot be realized when the abundances of the species differ by many orders of magnitude.

The new alternative is essentially a charge-stripping and molecular dissociation, so that a subsequent mass spectrometric analysis of the ion beam requires instrumentation of only nominal resolution. Such an analyzing system is shown in Figure 4.22, in which a tandem accelerator is used for stripping electrons at high ion energies. For the case of ^{14}C and ^{14}N, the complete instability of N^- makes isobaric separation feasible at the ion source, and *measurable* isotope ratios for $^{14}C/^{12}C$ as low as 10^{-15} have been reported (Bennett et al., 1977). For the case of ^{10}Be and ^{10}B, substantial discrimination can be achieved because of the differential rate of energy loss (dE/dx); the range of the low Z isobar is greater in matter than that of the higher Z nuclide. Complete charge stripping has been used for the separation of $^{7}Be^{4+}$ and $^{7}Li^{3+}$ at a few MeV energies. ^{26}Al is a cosmic ray-produced isotope with a half-life of 7.2×10^5 years, and a separation of $^{26}Al^{13+}$ and $^{26}Mg^{12+}$ has been achieved for ions in the 100 MeV range (Raisbeck et al., 1979a, 1979b). In other accelerator-mass spectrometer experiments at Yale, $^{26}Al/^{27}Al$ ratios have been measured in the range of 1×10^{-12} to 1×10^{-9} atoms/atom, and the exposure age of an iron meteorite has been determined to be 200 ± 60 million years (Thomas et al., 1983). The separation of ^{36}Cl and ^{36}S has also been reported using dE/dx discrimination, and a ratio approaching 10^{-16} has been observed by Elmore et al. (1979) for $^{36}Cl/Cl$ using the University of Rochester tandem accelerator.

The inherent greater sensitivity of the accelerator-mass spectrometer approach in contrast to radioactive counting methods is made apparent by considering both practical counting rates and spurious background events. In radionuclide assays, the number of atoms present in a sample N_p is related to the disintegration rate dN/dt and half-life $T_{1/2}$ by the expression

$$N_p = \left(\frac{dN}{dt}\right)\left(\frac{T_{1/2}}{0.693}\right)$$

Consider ^{14}C and ^{26}Al with half-lives of 5730 years and 7.2×10^5 years, respectively. If $T_{1/2}$ is expressed in hours and dN/dt is measured in disintegrations per hour, then

$$N_p(^{14}C) = dN/h \times (7.2 \times 10^7)$$

$$N_p(^{26}Al) = dN/h \times (9.1 \times 10^9)$$

From these expressions, it is clear that for an 8 h counting period, only about 1 atom in 10^7 is detected for ^{14}C and 1 atom/10^9 atoms is detected for ^{26}Al. Furthermore, both the cosmic ray background and the naturally occurring radioactivity in the environment places a lower limit on meaningful counting rates and, hence, on the minimum number of radioactive atoms that can be detected.

The detection efficiency for ^{14}C is improved by a factor as high as 10^4 by accelerator-mass spectrometry, where approximately 1% of the ^{14}C atoms of a sample can be ionized, mass and charge analyzed, and counted. Even greater detection differentials, in principle, apply to nuclides of longer half-lives. Inasmuch as the accelerator-mass spectrometric analysis of microgram-to-milligram samples can be completed in a few hours, the problem of spurious background counts is also minimized.

TABLE 8.4 Radionuclides Assayed by Tandem Accelerator-Mass Spectrometry

Radionuclide	Stable Isotopes	Detection Limit[a]
^{10}Be	^{9}Be	7×10^{-15}
^{14}C	^{12,13}C	0.3×10^{-15}
^{26}Al	^{27}Al	10×10^{-15}
^{32}Si	$^{28-30}$Si	7×10^{-12}
^{36}Cl	^{35,37}Cl	0.2×10^{-15}
^{129}I	^{127}I	0.3×10^{-12}

Source: University of Rochester, Nuclear Structure Group

[a] Radioisotope-to-stable isotope ratio; that is, the limiting ratio due to background.

A partial list of the radionuclides that have been studied by the University of Rochester group is shown in Table 8.4, with reported sensitivities.

ION-LASER MICROPROBE ASSAY OF CRYSTALS

The microprobe has become a powerful tool for providing detailed spatial information of ores and minerals, and for performing microassays of single crystals. Rare gas distributions in minisized single grains has been of special interest in connection with meteorites that are enriched in solar wind-implanted elements (Kiko et al., 1979). In an early application of the ion microprobe to the study of galena, Brevard et al. (1978) reported oscillatory Pb isotope variations within a crystal; and subsequently, other studies have utilized ion microprobe spectrometry to delineate phenomena of deposition, mineralization, leaching, and "ore-fluid stratigraphy." The analysis of single zircons has also yielded a more detailed chronology than is evident in milligram and bulk samples.

In work reported by Hart et al. (1981), the variations of Pb isotope ratios within a single crystal of galena were measured with an ion microprobe spectrometer. The detected variation of almost 3% in the ^{208}Pb/^{206}Pb ratio across a 13 mm crystal is shown in Figure 8.4. The uniform concentric zonation of the Pb isotope ratio shows the crystal perimeter to be less radiogenic than the core, and this perturbation is potentially related to fluid circulation patterns in the mine from which the crystal was obtained. Analogous microprobe U-Pb studies have been made of single zircons separated from the Cambrian Potsdam sandstone of New York by Gaudette et al. (1981) in which four distinct population ages (1180, 1320, 2100, and 2700 m.y.) have been identified. These ages relate to sources in provinces of the Canadian Shield and in Grenville rocks of the Adirondack Mountains. In this same study, individual zircon grains were mounted in epoxy, coated with gold,

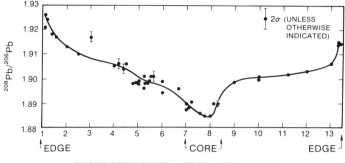

Figure 8.4 Ion microprobe assay of a single galena crystal, showing variations in the $^{208}Pb/$ ^{206}Pb ratio. (Hart et al., 1981; *Economic Geology*, *76*, 1873 (1981), by permission of the American Geophysical Union.)

and analyzed for their rare earth element composition, thus permitting additional interpretations to be made of continental crustal evolution.

The ion probe method has also been employed to assay Cu concentrations of mafic silicate minerals, including absolute abundances and high-resolution differential measurements. In a study of the Koluala igneous complex of the Solomon Islands (Hendry et al., 1981), the Cu contents of biotite (average, 880 ppm), amphibole (16 ppm), and magnetite (160 ppm) from barren intrusions were shown to be significantly greater than the Cu concentrations from unaltered mineralizing intrusions.

The ion microprobe method has had special relevance when analyses must be made in refractory inclusions of very small dimensions, as such inclusions are difficult to extract uncontaminated from their matrix. In a corundum-hibonite inclusion of the Murchison carbonaceous chondrite, Bar-Matthews et al. (1982) analyzed for ^{26}Mg, which was known to be the product of in situ radioactive decay of now extinct ^{26}Al. A primary ^{16}O beam of 2–3 nA and 2 keV was focused onto the surface of the sample with 3–8 μm diameter spatial resolution. By step-scanning the spectrometer magnet, secondary ions of ^{26}Mg, ^{25}Mg, and ^{24}Mg were monitored and compared with normalized $^{27}Al/^{24}Mg$ ratios. Even though the total magnesium content of the sample was in the range of 100 ppm, the ion microprobe (secondary ion mass spectrometry) generated sufficient ions to measure relative ion peaks and to correct for fractionation.

The measurement of diffusion rates of lithium in an alkali-basaltic melt has also been determined by ion microprobe techniques, using ^6Li as a tracer. Based on temperature measurements at 1300, 1350, and 1400°C, a tracer-diffusion relationship has been found to be (Lowry et al., 1981)

$$D_{Li} = 7.5 \times 10^{-2} \exp{(-27,600/RT)} \text{ cm}^2/\text{s}$$

This work complements previous studies that have utilized radioactive isotopes to investigate transport phenomena in crystalline and glassy solids. It further demonstrates the power and sensitivity of the ion microprobe (SIMS) technique for studies that involve the transport of major and minor elements in silicate melts and other regimes.

Currently, the laser probe is complementing the ion probe as another effective tool for directly assaying microfluid inclusions ($< 10 \mu m$) inside geological crystals. Elroy et al. (1983) have successfully employed the laser probe, in the reflection mode, to analyze minerals with a spatial resolution of 2 μm, and for sample volumes as low as 1.5×10^{-13} cm^3.

ISOTOPIC RATIOS OF THE NOBLE GASES

From a geological point of view, the noble gases are an especially interesting class of elements. They are chemically inert, so that if initially confined in some geological reservoir, they remain in their initial state. These gases are also produced in easily detectable quantities by natural nuclear reactions that have taken place over billions of years.

The low abundances of terrestrial rare gases confirm other evidence that the atmosphere is primarily due to the outgassing of the solid earth. Furthermore, isotopic variations in the rare gases indicate that outgassing is still proceeding from the earth's interior as well as the crust. More importantly, isotopic variations of helium, neon, argon, krypton, and xenon permit interpretations to be made relative to dating, nuclear processes, diffusion, the solubility of rare gases in silicate melts, mantle dynamics, etc. (Kyser and Rison, 1982a, 1982b).

Helium is found wherever alpha particles are a reaction product, as indicated in Table 8.5. Argon is produced by K-capture in ^{40}K. Krypton and xenon are fission product gases. Radon, the only naturally radioactive noble

TABLE 8.5   Helium Content of the Atmosphere, Natural Gas, and Minerals

Location	Material	He Content	He^3/He^4($\times 10^{-7}$)
Stamford, Conn.	Air	0.004%	12
Rattlesnake Field, Shiprock, N.M.	Natural gas	7.7%	0.5
Cliffside Field, Amarillo, Tex.	Natural gas	1.8%	1.5
Great Bear Lake, Canada	Pitchblende	—	0.3
Keystone, S.D.	Beryl	0.022 (cc/g)	1.2
Jokkmokk, Lapland (Sweden)	Beryl	0.023 (cc/g)	1.8
Cat Lake, Manitoba, Canada	Spodumene	0.01 (cc/g)	24
Edisonmine, S.D.	Spodumene	0.01 (cc/g)	120

Source: Faul (1954).

gas, has three isotopic species produced from the decay series of uranium and thorium.

We find two isotopes of helium in the earth's crust and in the atmosphere. The lightest isotope, ^{3}He, is thought to be generated in rocks (Hill, 1941) by the reactions:

$$^6Li(n,\alpha)^3He,$$

$$^3H(\beta^-)^3He$$

Production in the atmosphere (Libby, 1946) is believed to result from the three reactions

$$^{14}N(n,^{12}C)^3H$$

$$^{14}N(n,3\alpha)^3H$$

$$^3H(\beta^-)^3He$$

The upper atmosphere contains an adequate flux of neutrons from cosmic rays to explain the abundance of atmospheric ^{3}He. Cosmic ray neutrons, however, are highly attenuated below the surface of the earth. Hence, the generation of ^{3}He in minerals is ascribed to neutrons from the spontaneous fission of ^{238}U and an approximately equal number of neutrons arising from various (α,n) reactions on elements of low atomic number. An early substantiation of such a reaction was reported by Aldrich and Nier (1948), who discovered that spodumene ($LiAlSi_2O_6$) has a relatively high $^3He/^4He$ ratio. Their results and the analysis of other materials have been tabulated by Faul (1954). Faul also points out that 1 metric ton of granite containing 2 ppm of uranium and 10 ppm of thorium will produce approximately $0.22 + 0.29 = 0.51$ ml of helium per million years (at standard temperature and pressure). Such concentrations can be analyzed even for very small samples.

Data such as that above permit analyses to be made on the age of minerals, as helium is continuously being generated in the earth's crust. Both gross helium content and isotopic ratio measurements allow important interpretations to be made of nuclear and geological processes. In some minerals, the retention of helium is almost complete (e.g., in crystals of magnetite and zircon, diffusion coefficients are so small that there is a negligible helium loss even in a period of a billion years). However, in other instances, helium is lost by diffusion owing to radiation damage, to general migration to more stable crustal regions, or because the original helium content is contaminated from some other localized source. Hence, in practice, the basic data must be combined with other information to be of any significance.

In the case of the exceedingly high helium content of Rattlesnake Field, New Mexico ($\sim$8%), geologists have had difficulty ascribing this very rich helium source to radioactive products, because no significant radioactivity

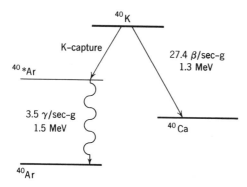

Figure 8.5   Decay scheme of ^{40}K.

has been observed in this region. Consequently, this large concentration has been tentatively ascribed to a migration from a reasonably remote source, with the helium finally becoming trapped, as in the cases of natural gas or petroleum reservoirs.

The distinct isotope ratio of mantle helium (^{3}He/^{4}He $\cong 10^{-5}$) makes it possible to detect this source in the presence of atmospheric helium (^{3}He/^{4}He $\cong 10^{-6}$) or helium produced from radioactive decay. For example, an extensive plume of water enriched with ^{3}He has been discovered, in the deep Pacific Ocean (2500 m depth), having a ^{3}He/^{4}He ratio 50% higher than in atmospheric helium. This measurement is consistent with other measurements in that a major input into the terrestrial helium inventory is due to the flux of primordial helium from the mantle into the oceans and, subsequently, into the atmosphere (Lupton and Craig, 1981). Other high ^{3}He/^{4}He ratios have been measured in volcanic gases, representing a deep mantle source; ratios as high as 23 times atmospheric have been reported in Iceland, and Craig et al. (1978) have measured ratio increases as high as 15 at Yellowstone Park. Recent measurements of helium in diamonds reveal ^{3}He/^{4}He ratios ranging from less than 10^{-7} to 3.2×10^{-4} (Ozima and Zashu, 1983). This latter value is higher than primordial ^{3}He/^{4}He ratios in meteorites, and it possibly suggests not only helium trapped soon after the formation of the earth, but the existence of nearly solar-type helium.

The abundance of argon in the atmosphere is much higher than that of helium, being slightly less than 1%. In natural gases, it is present to only about 0.1%. The isotopic abundance of argon in the air, as reported by Nier (1950), is ^{40}Ar, 99.6%; ^{36}Ar, 0.337%; and ^{38}Ar, 0.063%. ^{40}Ar is important in geological age determinations, because it is the radiogenic product of ^{40}K, according to the decay scheme indicated in Figure 8.5. This decay suggests why the ^{40}Ar isotope occurs in relatively high abundance; it also indicates that the argon content of potassium-bearing minerals permits the determination of mineral age. (The positron decay mode, 0.001%, is not shown).

Variations in the isotopic abundances of krypton and xenon are of interest because they can be related to the build-up of fission product gases in ura-

TABLE 8.6 Rare Gas Isotopic Ratios of Hawaiian Samples

Sample No.	^{3}He/^{4}He ($\times 10^{-6}$)	^{20}Ne/^{22}Ne	^{40}Ar/^{36}Ar	^{129}Xe/^{132}Xe
		Maui Island		
HA a, Cpx	47.8 ± 6.1	9.94 ± 0.38	311.2 ± 0.8	0.990 ± 0.007
HA a, Ol	51.5 ± 9.7	8.8 ± 1.3	417.6 ± 4.9	0.994 ± 0.017
HA b, Cpx	23.5 ± 4.8	9.9 ± 0.5	334.1 ± 1.9	1.010 ± 0.032
		Oahu Island		
SLC-45	10.5 ± 0.6	10.2 ± 0.4	7563 ± 66	1.016 ± 0.012
SLC-52	11.3 ± 0.7	9.64 ± 0.17	8054 ± 56	0.987 ± 0.032
SLC-57	11.0 ± 0.7	9.70 ± 0.19	4039 ± 50	1.002 ± 0.010
Air	1.4	9.81	295.5	0.983

Source: Kaneoka and Takaoka (1980).

nium- and thorium-bearing minerals. These fission products are generated by (1) neutron-induced fission; and (2) uranium that undergoes fission spontaneously. Spontaneous fission half-lives are very long, even compared with the age of the earth; half-lives for ^{232}Th, ^{235}U, and ^{238}U are 1.35×10^{18}, 1.87×10^{17}, and 8.07×10^{15} years, respectively (Segrè, 1952). Kaneoka and Takaoka (1980) have analyzed the rare gases in Hawaiian ultramafic nodules and phenocrysts in volcanic rocks (Table 8.6); this work demonstrates the dramatic range of isotopic ratios that relate to different source materials. The importance of dissolved noble gases in seawater has also been emphasized by Bieri and Koide (1972). Unlike oxygen, CO_2, or other gases that undergo chemical reactions in a body of water, noble gas concentration gradients can be correlated with evidence of mixing, temperature-solubility relationships, etc. in the oceans.

GEOCHEMISTRY OF CARBON AND SULFUR

Extensive mass spectrometric work has been undertaken to measure isotopic abundance variations of carbon and sulfur. Studies of the former have included specimens of carbonates, igneous rocks, diamonds, petroleum, natural gases, fossil woods, marine plants, marine invertebrates, and sediments. The isotopic studies of petroleums and their relation to other natural carbonaceous substances provide an insight into questions of petroleum origin and evolution. Studies with respect to sulfur have been of particular interest because of that element's wide distribution and abundance in the earth's crust and in meteorites.

Nier and Gulbransen (1939) were the first to report that ^{13}C/^{12}C ratios of sedimentary rock carbonates are 2–3% higher than those of natural organic matter, and the review by Silverman (1964) indicates the deviations, δ (in

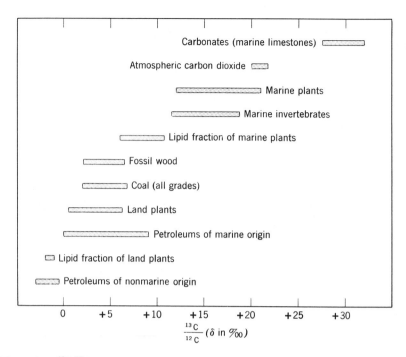

Figure 8.6 $^{13}C/^{12}C$ δ-value ranges in natural carbonaceous materials (Silverman, 1964).

parts per thousand), of $^{13}C/^{12}C$ ratios of natural carbonaceous materials (Fig. 8.6). The high ^{13}C content of carbonates in limestone is definitely attributed to a carbon isotopic exchange reaction. In marine and land plants, however, there is a preferential assimilation of $^{12}CO_2$ over $^{13}CO_2$ during photosynthesis, and there are other kinetic effects that are presumed to concentrate ^{12}C with respect to atmospheric CO_2. There exists a further line of demarcation between marine or land plants, in that CO_2 is provided and transported across cell walls in a different mode. In the former case CO_2 is supplied from a solution, whereas in the latter case CO_2 diffuses through stomata along a concentration gradient. McMullen and Thode (1963) thus concluded that a clear distinction can be made among limestone carbon, organic carbon of marine origin, and carbon of land plant origin.

Investigations relating to petroleum and oil formation products have been extensive. Chromatographic fractions of petroleum and of soluble and insoluble organic constituents of shale have been analyzed to detect isotopic changes that might be attributed to chemical composition. Such changes, if they occur, are exceedingly small. However, significant differences in $^{13}C/^{12}C$ ratios have been noted between hydrocarbon gas evolving from oil-bearing deposits and liquid petroleum. Fractionation results in a depletion of the ^{13}C in the gas phase. Petroleums of marine and nonmarine origin have lower $^{13}C/^{12}C$ ratios than some of the corresponding organic source materials. McMullen and Thode (1963) thus suggested that not all biogenic carbon is converted into petroleum.

Important mass spectral measurements of carbon also relate to dating. In 1977, it was suggested that radioisotope dating with ^{14}C and other nuclides could be achieved by using a cyclotron as a high-energy mass spectrometer, with sensitivities at the 10^{-16} level (Muller, 1977). Such a sensitivity was essentially demonstrated with the cyclotron, and subsequently with other accelerators. At present, ^{14}C dating by accelerator-mass spectrometry (see Table 8.4) is now complementing dating by beta particle counting. ^{14}C is formed by several nuclear reactions, but the dominant cosmic ray neutron process is:

$$^{1}n + {}^{14}N \rightarrow {}^{14}C + {}^{1}H$$

The ^{14}C atoms are incorporated into CO_2 molecules, which mix rapidly throughout the atmosphere and hydrosphere. $^{14}CO_2$ molecules then enter living plants by photosynthesis and by absorption through the roots. All carbon uptake ceases upon death of the plant; ^{14}C decays with a 5730 year half-life, and a $^{14}C/^{13}C$ ratio permits a calculation of a "carbon-14-date" of the specimen.

Isotopic differences in sulfur provide considerable insight into effects induced by bacterial action, in addition to fractionation arising from purely physical phenomena. Analyses have been made of (1) meteorites; (2) sulfides of igneous origin; (3) gases from volcanoes; (4) seawater sulfates of the Atlantic, Pacific, and Arctic oceans; (5) free sulfur produced in lake bottoms; (6) petroleum oils; and (7) the vast sulfur wells of Texas and Louisiana. The $^{32}S/^{34}S$ ratios of sulfates, sulfur, and sulfides from these latter two sources are shown in Figure 8.7, together with other data on naturally occurring sources.

The $^{32}S/^{34}S$ ratio in meteoritic samples is remarkably invariant, and it is assumed that this represents the ratio for terrestrial sulfur during the time the world was formed. For this reason, it is sometimes used as a standard. There is considerable evidence to indicate that the Texas and Louisiana sulfur deposits reflect the effects of living organisms. Bacterial reduction of sulfates are known to result in large fractionation effects; similar enrichments have been reported in the sulfur-producing lakes of Africa, where bacterial reduction of the sulfate has been analyzed (Slater, 1953). Thode et al. (1953) proposed that we can always expect a kinetic isotopic effect in the reduction of sulfate by bacterial action, with ^{34}S enrichment in the sulfate. This effect will occur in sediments of lakes and shallow ocean waters where bacterial action is considerable. A biological process is also ascribed to the fractionation of sulfur isotopes in sedimentary rocks, and the extent of this fractionation has suggested that isotopic ratios can reveal the biological activity

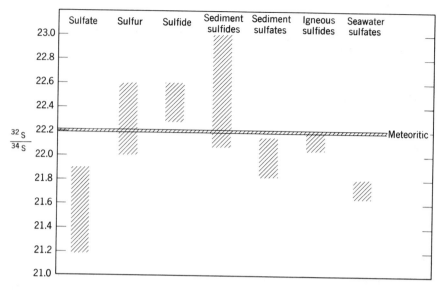

Figure 8.7 Variations in the $^{32}S/^{34}S$ ratios of naturally occurring deposits (McMullen and Thode, 1963).

at the time of deposition. In some instances, isotopic data can also provide a rough correlation of bacterial action with geological period.

Isotopic ratios of sulfur have also been measured in asphalts extracted from rocks in attempts to correlate age and depth of deposit. A biological theory of ^{34}S enrichment postulates that $(^{34}SO_4)^{2-}$ and $(^{32}SO_4)^{2-}$ are reduced at different rates, and that the H_2S produced is depleted in ^{34}S. The residual sulfate thus becomes enriched in the heavier isotope.

GEOLOGICAL AGE FROM URANIUM AND LEAD

Geological dating from uranium (U) and lead (Pb) presumes the following decay modes:

$$^{238}U \rightarrow 8\alpha \text{ particles} \rightarrow {}^{206}Pb$$

$$^{235}U \rightarrow 7\alpha \text{ particles} \rightarrow {}^{207}Pb$$

$$^{232}Th \rightarrow 6\alpha \text{ particles} \rightarrow {}^{208}Pb$$

Prior to Professor Nier's pioneering isotope measurements on many elements in the 1930s, a chemical lead-uranium assay had been employed for determining the age of a uranium mineral. The earlier work of Aston (1927) had clearly demonstrated that the atomic weight of an element reflected the relative abundances of its several isotopes, so that the constant atomic weight of common lead seemed to suggest an invariant isotopic composition. However, lead associated with uranium was found to have a "lighter" atomic weight, and this difference was correctly interpreted as resulting from the presence of radiogenic ^{206}Pb.

The advent of precision isotopic measurements led to two significant developments in the history of geochronology. First, these mass spectrometric measurements provided an accurate measure for determining the "common lead impurity" in samples of geological interest. Second, for relatively pure uranium-lead samples, "*a determination of the $^{207}Pb/^{206}Pb$ abundance ratio provided an independent method of measuring the age of a uranium mineral since ^{207}Pb and ^{206}Pb are the end products of the decay of ^{235}U and ^{238}U, respectively, and the decay rates of ^{235}U and ^{238}U are drastically different.*" (Nier, 1981). The early work on lead (Nier, 1938) was then complemented by measurements of the $^{238}U/^{235}U$ ratio (Nier, 1939), which was needed in the $^{207}Pb/^{206}Pb$ method for evaluating mineral age. The accepted value for the $^{238}U/^{235}U$ ratio is now 137.88.

From these and other studies, it is now known that a "constant" atomic weight of lead (207.21) does not mandate an invariant isotopic composition. Rather, all common lead samples can be viewed as mixtures of primordial and radiogenic lead. The primordial lead is comprised of isotopes ^{204}Pb, ^{206}Pb, ^{207}Pb, and ^{208}Pb. Quite fortuitously, these isotopic variations occur in such a manner as to have little effect on the atomic weight of the element. Thus, it is postulated that an early stage—when the earth was fluid—primordial lead, uranium, and thorium were uniformly distributed throughout the planet's mantle. Upon formation of the earth's crust, primordial lead was frozen into rocks together with various amounts of thorium and uranium. The U-Th-Pb composition of each locality was thus fixed. Subsequent thorium and uranium decay then continuously generated radiogenic lead. Therefore, it is presumed that if we knew the age and isotopic composition of a number of lead ores, we might compute the time when mixing ceased and the earth's crust became a solid.

It will be noted that isotopic measurements of lead are especially valuable in that several ages may be computed corresponding to the several decay modes. Ores and minerals that have been analyzed include pitchblende, uraninite, monazite, and zircon. The ages reported for a zircon from Ceylon are interesting, because they indicate exceptionally good agreement for a single specimen. Results are shown in Table 8.7. According to Nier (1966), the measurement of the $^{207}Pb/^{206}Pb$ ratio generally provides a more accurate age than uranium-lead measurements. The half-lives of the ^{238}U and ^{235}U leading to ^{206}Pb and ^{207}Pb are quite different (4.5×10^9 and 7.1×10^8 years, respectively), so that the lead abundance ratio directly yields the age of the mineral. Hence, in lead-lead mesurements, it is unnecessary to analyze for total amounts of uranium or lead in the sample.

Recent U-Pb calculations of the "age of the earth" range from 4.43–4.56

TABLE 8.7   Age Determination by
U-Th-Pb Measurements

Measurement	Calculated Age (yr)
$^{238}U/^{206}Pb$	5.40×10^8
$^{235}U/^{207}Pb$	5.44×10^8
$^{232}Th/^{208}Pb$	5.38×10^8
$^{207}Pb/^{206}Pb$	5.55×10^8

Source: Nier (1966).

billion years. The lowest age is based on an assay of oceanic sediment and the highest value is derived from analyses of whole rocks and associated sulfide deposits in Ontario, Canada. A critical review of Pb isotope systematics and its application to planetary evolution has been provided by Tera (1981). Based on old galena leads of different compositions, an age of 4.54 billion years is suggested for the age of the earth or the "age of the debris!"

Other recent measurements have been made of $^{238}U/^{235}U$ ratios in order to compare terrestrial rock and nine meteorites. However, this ratio appears to be the same within experimental error (Chen and Wasserburg, 1980) as the 137.88 National Bureau of Standards value reported by Shields (1960). Additional U, Th, and Pb measurements have focused on oceanic basalts, hydrothermal phenomena, and the vertical heterogeneity in the oceanic crust (Chen and Pallister, 1981). For example, uranium series ages have been reported for fossil corals from deposits of the southeastern Atlantic Coastal Plain (Cronin et al., 1981). This study involved isotopic measurements of $^{234}U/^{235}U$, $^{230}Th/^{232}Th$, and $^{230}Th/^{234}U$ to investigate coral age clusters at 440,000, 188,000, 120,000, 94,000, and 72,000 years ago. Most of these measurements involved alpha particle counting. Results demonstrated that a possible correlation could be made between deep sea isotope stages, coral terrace chronologies, and marine paleoclimates.

Many other U-Pb isotopic studies have been undertaken to determine the history of geological activity and sequences. An example is the U-Pb study on the timing and distribution of Mesozoic magmatic events in the Sierra Nevada batholithic complex, based on measurements of 133 zircon and 7 sphene separates from 82 samples of granitoid rocks. Figure 8.8 is a graphical relationship between various intrusion ages and isotopic ratio measurements (Chen and Moore, 1982).

Some interesting geological dating has also been derived from uranium mines and deep drill holes. The Midnite mine, near Spokane, Washington, is one of the few deposits in the United States producing uranium ore from crystalline rocks—having yielded 11 million pounds of U_3O_8 from three geological zones. Analyses of these ores define a $^{207}Pb/^{204}Pb$-$^{235}U/^{204}Pb$ isochron age of mineralization of 51 m.y. (Ludwig et al., 1981). This age coincides with a period of regional volcanic eruptions, shallow intrusive

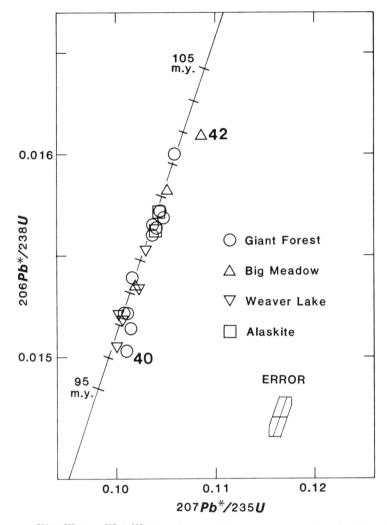

Figure 8.8 $^{206}Pb/^{238}U$ vs. $^{207}Pb/^{235}U$ for zircons and sphene separates from the Giant Forest-Alaskite sequence in the central Sierra Nevada batholith. (Chen and Moore, 1982; *J. Geophysical Research*, *87*, B6, 4761 (1982), by permission of the American Geophysical Union.)

activity, and faulting. U-Pb zircon ages of granite and gneiss cores from deep drill holes in Wyoming and southern Montana have also been consistent with a known period of late Archean metamorphism and igneous activity. Peterman (1981) has generated a concordia plot for zircon size fractions from this region, and the 2.88 b.y. intercept is comparable to ages of extensive plutonic rocks in the Bighorn Mountains.

An additional geochemical application for uranium isotopic analysis is a determination of the weathering rate of rocks. It can be shown that an important fraction of the dissolved lead in river water comes from the dissolution of minerals, so that uranium can be used as a natural tracer of the weathering process. In a specific study of the Preto River basin in Brazil (Moreira-Nordemann, 1980), uranium measurements were made of the content of rocks, soils, river water, and rain in a river basin area of approximately 850 km². The uranium concentration in the local rain (~0.01 μg/l) was 300 times less than the mean content of the river water, thus permitting a minimum correction to be attributed to this source. A detailed analysis of other parameters yielded a weathering rate for rocks in this basin of 100 tons/km²/yr, which corresponds to 0.04 mm/yr. An extrapolation of this rate is that one vertical meter of rock would require 25,000 years to be weathered under existing climatic conditions.

Isotopic measurements of lead are not limited, of course, to pure geochemical studies. For example, $^{206}Pb/^{207}Pb$ ratios have been used to identify natural and industrial soruces of lead in the coastal waters of British Columbia, Canada (Stukas and Wong, 1981). Such measurements also serve to support theoretical models of circulation patterns, etc., and to provide data for distinguishing between "old" lead ores that tend to have a lower content of ^{206}Pb and ^{207}Pb, in contrast to "modern lead" that may be enriched in these radiogenic nuclides.

THE RHENIUM-OSMIUM CHRONOMETER

The decay of the radioactive isotope ^{187}Re into the relatively stable ^{187}Os has a distinct advantage over the earlier uranium-thorium method for determining the age of the universe. The beta decay half-life is many billions of years greater, and the nuclear processes involved in the interior of stars are said to be better understood. The estimated rhenium-osmium (Re-Os) age of the universe is about 18×10^9 years, which is at least consistent with the astronomical age of the universe based on a Hubble constant for a stellar red shift of ~15 km/s per million light years (Cloud, 1980).

The Re-Os method for dating meteorites and the measurement of $^{187}Os/^{186}Os$ terrestrial samples were first reported by Herr et al. (1961), but this early work lacked adequate sensitivity and accuracy. In recent years, however, the decay constant, λ, of ^{187}Re has been measured to be about 1.6×10^{-11} yr^{-1}; and by measuring the $^{187}Os/^{186}Os$ ratios of well-documented samples, a valuable chronometer has evolved. In studies reported by Allègre and Luck (1980) using a mass spectrometer ion microprobe, osmium ratios were determined on a series of osmiridium grains whose source age had been carefully measured in independent analyses. These ratios vs. known ages are plotted in Figure 8.9. The relationship of this line to age is

$$^{187}Os/^{186}Os = (^{187}Os/^{186}Os)_0 + (^{187}Re/^{186}Os)_{mantle} (e^{\lambda_{Re}t} - 1)$$

Using $\lambda_{Re} = 1.61 \times 10^{-11}$ yr^{-1}, and since $(e^{\lambda_{Re}t} - 1)$ is approximately $\lambda_{Re}t$,

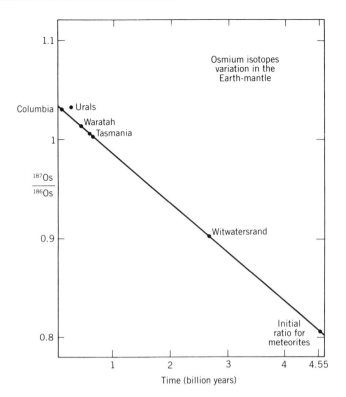

Figure 8.9 Variation of ^{187}Os/^{186}Os with age in several osmium-rich samples that are assumed to be derived from mantle materials. (Allègre and Luck, 1980; *Earth and Planetary Science Letters*, **48**, 148 (1980), by permission of Elsevier Science Publishers.)

a correlation can be made between the slope of the line and the (^{187}Re/^{186}Os) ratio of the mantle.

The linear approximation made by Allègre and Luck (1980) yields a ^{187}Re/^{186}Os value of 3.15, a value "which appears to be a constant in the solar system and quite independent of the various processes of planetary formation." As a consequence, these authors suggest that the age of osmium-rich samples coming from the mantle can be calculated by the expression:

$$T = 4.55 \times 10^9 - \frac{1}{\lambda_{Re}} \log \left[\frac{(^{187}Os/^{186}Os)_{sample}}{3.15} + 1 \right]$$

Subsequent studies by Luck and Allègre (1982), utilizing SIMS on iron meteorites and chondrites, have redetermined the rhenium decay constant to be 1.52×10^{-11} yr^{-1}. In this latter work, the ^{187}Re $- ^{187}$Os chronometer was further utilized for the assay of molybdenites and for differentiating between ore bodies coming from the crust or from the mantle.

^{40}Ar/^{39}Ar DATING OF MINERALS

The ^{40}Ar/^{39}Ar method of dating affords some advantage over the earlier K-Ar method, which assumed that a mineral at the time of its formation contained no argon and that radiogenically produced argon losses were negligible. In the ^{40}Ar/^{39}Ar age determination, potassium-bearing minerals are irradiated in a nuclear reactor and ^{39}Ar is generated by the (n, p) reaction with ^{39}K. The number of radiogenic ^{40}Ar atoms produced in the sample due to ^{40}K decay during its lifetime will be

$$^{40}\text{Ar}^* = \frac{\lambda_e}{\lambda}\,^{40}\text{K}\,(e^{\lambda t} - 1)$$

where λ_e and λ are the electron-capture and total decay constants, respectively, for ^{40}K. Argon isotopes may also be produced by neutron interactions with isotopes of calcium, potassium, and chlorine so that appropriate corrections must sometimes be made. With such corrections, it can be shown that if a sample has been closed to argon and potassium since the time of initial cooling and mineralization, an age can be calculated solely on the basis of the ^{40}Ar/^{39}Ar ratio of a neutron-irradiated sample (Faure, 1977b). Essentially, a stepwise heating of a mineral sample is employed to determine whether radiogenic argon was lost after initial cooling. This stepwise heating, degassing, and ^{40}Ar*/^{39}Ar measurement technique also permits a date determination that approaches, but may underestimate, an original date of cooling; a partial loss of radiogenic ^{40}Ar can then be interpreted in terms of metamorphism.

An improved method of laser stepwise heating has been reported on some separated micas, hornblende, and whole-rock slate. Specifically, York et al. (1981) have demonstrated that a 2 W argon ion laser can be used to melt individual grains of sample; the evolved gases are then leaked into the mass spectrometer for isotopic ratio analysis in the static mode. This laser fusion technique provides both an age microprobe capability and a convenient multistep heating cycle that compares well with more conventional radiofrequency (RF) fusion assays. Table 8.8 provides a comparison of isotopic data and laser heating of muscovite samples from the Kidd Creek mine near Timmins, Ontario, Canada.

Another example of the ^{40}Ar/^{39}Ar method employing neutron irradiation and incremental heating is the work reported by Baksi (1982) on samples taken from the Tudor Gabbro, Grenville Province, Ontario. Whole-rock samples were powdered to 10–30 mesh size, placed in aluminum capsules, and irradiated with an integrated fast neutron flux of $\sim 2 \times 10^{18}$ *nvt*. Subsequent to irradiation, the samples were RF induction heated to successively higher temperatures, each heating step being of ~ 30 min duration. The evolved gases were then analyzed in a mass spectrometer operated in the static mode. Figure 8.10 shows "age studies" based on incremental heating studies in

TABLE 8.8   RF Fusion vs. Laser-Heating
Technique Applied to Kidd Creek Muscovite
Specimens

Analysis	^{40}Ar*/^{39}Ar	Age (Ga)
RF fusion	578 ± 4	2.64 ± 0.01
	562 ± 8	2.60 ± 0.02
Laser	561 ± 7	2.60 ± 0.02
(single grain)	540 ± 47	2.55 ± 0.12
	550 ± 9	2.57 ± 0.02

Source: York et al. (1981).

which whole-rock age is correlated with ^{40}Ar/^{39}Ar ratios. Similar studies involving irradiated samples and mass spectrometric measurements have been reported by Harrison and McDougall (1982) to determine microcline ages and to reconstruct the temperature histories of various feldspars.

Additional salient isotopic ratio measurements of argon include the assay of diamonds. Diamond is one of the minerals found in kimberlite eruptions, and these eruptions provide one of the few routes for sampling the earth's upper mantle. Diamonds are also relatively inert and impermeable, so they can retain both fluids and solids that are entrapped during their crystallization. The recent work of Melton and Giardini (1980) relates to a 6.3 carat, well-crystallized stone derived from the Arkansas kimberlite eruption of known age. This stone was ultrasonically cleaned and placed in a diamond crusher, and the gas released upon crushing was passed directly into the inlet of a mass spectrometer. The argon isotopic abundances of the sample compared to those of atmospheric air are shown in Table 8.9.

The authors derive several conclusions from these measurements. First, the argon isotopic abundance in the mantle at the time and depth of crys-

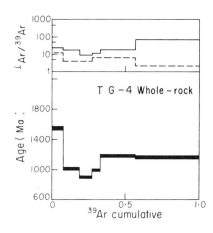

Figure 8.10   Age spectra for ^{40}Ar-^{39}Ar incremental heating of Tudor Gabbro whole-rock. In the upper section of plot, the solid line represents ^{37}Ar/^{39}Ar ratios and the dotted line ^{38}Ar/^{39}Ar ratios. (Baksi, 1982; *Geophys. J. R. Astr. Soc.*, 70, 545 (1982), by permission of Blackwell Scientific Publications, Ltd.)

TABLE 8.9   Isotopic Abundances of Argon Included in a 6.3 Carat Diamond from Arkansas, U.S.A.

Isotope	Included Argon	Atmospheric Argon
^{36}Ar	0.525 ± 0.011	0.339 ± 0.006
^{38}Ar	0.093 ± 0.007	0.064 ± 0.005
^{40}Ar	99.382 ± 0.011	99.597 ± 0.006
$^{40}Ar/^{36}Ar$	189.3	293.8

Source: Melton and Giardini (1980).

tallization is close to that of the earth's atmosphere at that time. Second, the $^{38}Ar/^{36}Ar$ ratio in the diamond is sufficiently close to that of the present atmosphere to indicate that the argon is of primordial origin and not enriched by radiogenic processes. Third, data suggest that the outgassing of the earth was "neither complete nor isotopically selective at the time and depth of diamond crystallization." Finally, the results are consistent with a diamond crystallization date of about 3.1 billion years.

RUBIDIUM-STRONTIUM DATING

Rubidium (Rb) is a relatively rare element with an abundance of about 0.035% by weight in igneous rocks in the earth's crust. It has two naturally occurring isotopes, ^{85}Rb and ^{87}Rb, with the latter being a long-lived beta emitter that decays to ^{87}Sr with a half-life of $\sim 5 \times 10^{10}$ years ($\lambda = 1.42 \times 10^{-11}$ yr^{-1}). Strontium (Sr) has four naturally occurring and stable nuclides, ^{84}Sr, ^{86}Sr, ^{87}Sr, and ^{88}Sr. The elemental abundance of nonradiogenic strontium in igneous rocks is only slightly less than that of rubidium. Nevertheless, by making careful mass spectral corrections, a clear differentiation can be made between the strontium resulting from rubidium decay and primeval strontium. Fortunately, also, a few minerals contain rubidium with much lower natural strontium concentrations than average—lepidolite, muscovite, and biotite—so that the build-up of radiogenic ^{87}Sr can be monitored with reasonable precision. The amount of this isotope in a "closed system" where no rubidium or strontium has been added or depleted during the lifetime t of the mineral will be

$$^{87}Sr = {}^{87}Sr_0 + {}^{87}Rb(e^{\lambda t} - 1)$$

where ^{87}Sr is the number of atoms at t years after mineralization, $^{87}Sr_0$ is the original number of ^{87}Sr atoms present at the time of mineralization, and λ is the decay constant in units of reciprocal years. The above equation can

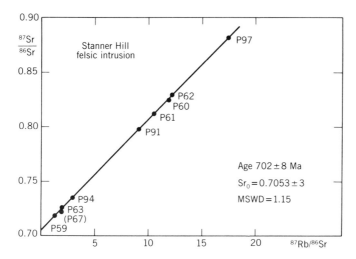

Figure 8.11   Rb-Sr isotopic data from late Precambrian-to-Cambrian igneous rocks in southern Britain, indicating a mineral age of 702 ± 8 million years. (Patchett et al., 1980; *J. Geological Society* (London) *137*, 649 (1980), by permission of Blackwell Scientific Publications, Ltd.)

also be written in terms of the actual mass spectrometric ratios that form the basis for age determination, namely

$$\frac{^{87}\text{Sr}}{^{86}\text{Sr}} = \left(\frac{^{87}\text{Sr}}{^{86}\text{Sr}}\right)_0 + \frac{^{87}\text{Rb}}{^{86}\text{Sr}} (e^{\lambda t} - 1)$$

This equation is of the general form $y = b + mx$, so that if $^{87}\text{Sr}/^{86}\text{Sr}$ and $^{87}\text{Rb}/^{86}\text{Sr}$ are the y and x parameters, respectively, $(^{87}\text{Sr}/^{86}\text{Sr})_0$ will be the y intercept (b) and $m = (e^{\lambda t} - 1)$ will represent the slope of the line. This line is designated as an "isochron" to indicate that all points on this line denote a common age t, and also the same initial $^{87}\text{Sr}/^{86}\text{Sr}$ ratio for specimens having undergone the same geological history.

This method of Rb-Sr determination by mass spectrometry was made practical with the advent of solid-sample analysis for these low-ionization potential elements, the application of isotopic dilution techniques, and the availability of separated and enriched isotopes. An example of the Rb-Sr isochron is shown in Figure 8.11, which represents age data for late Precambrian-to-Cambrian igneous rocks in southern Britain (Patchett et al., 1980). Mica samples were assayed in a 30 cm radius spectrometer using a mixed Rb-Sr spike and the isotopic dilution method. These samples were taken from a "Stanner-Hill" complex comprising outcrops of intrusive rocks that were essentially unmetamorphosed, but had been subjected to local shearing and hydrous alteration. A whole-rock isochron of 702 ± 8 million years is defined, with an initial $^{87}\text{Sr}/^{86}\text{Sr}$ ratio equal to 0.7053 ± 3. Rubidium-

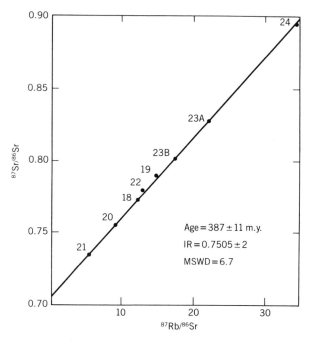

Figure 8.12 Rubidium-strontium isochron defined by eight whole-rock samples of Hallowell quartz monzonite, south-central Maine. (Dallmeyer and Van Breeman, 1981; *Contributions to Mineralogy and Petrology*, *78*, 61 (1981), by permission of Springer-Verlag.)

strontium ages have also been reported by Dallmeyer and Van Breeman (1981). Figure 8.12 shows their data for eight whole-rock samples from several localities within the Hallowell pluton in south-central Maine. The 387 m.y. isochron ages are well defined and date the time of intrusion and crystallization. Comparable Rb-Sr dating has been reported for granite complexes in Georgia (Ellwood et al., 1980), Nova Scotia (Clarke and Halliday, 1980), etc.

Rubidium-strontium measurements have also been applied to a systematic study of meteorites. Minster and Allègre (1981) utilized programmable mass spectrometers equipped with digital outputs and an on-line computer to assay whole-rock samples of meteorites. Reproducibility and accuracy were checked periodically by comparison with NBS standard 987 ($^{87}Sr/^{86}Sr$ = $0.71014 \pm 5 \times 10^{-5}$).

Rubidium-strontium isotope ratios have been used further to determine ages of hydrothermal alteration in the oceanic crust. Veins and fractures in the crust provide pathways by which carbonates and clay minerals may be altered by the penetration of the hydrothermal circulation of seawater. Hart and Staudigel (1978) have shown that by comparing strontium ratios in calcites with the known variations of seawater $^{87}Sr/^{86}Sr$ with time, the hydrothermal circulation is a salient process for no more than 5–10 m.y. after crust formation. In a related study (Richardson et al., 1980) involving deep sea drilling in the western Atlantic, measurements of strontium ratios of silicates and carbonates have been made to determine the vein mineral ages of old oceanic crust.

SAMARIUM-NEODYMIUM AND LUTETIUM-HAFNIUM SYSTEMATICS

The alpha decay of ^{147}Sm to ^{143}Nd provides an alternative method for dating both terrestrial and extraterrestrial rocks (DePaolo and Wasserburg, 1979; 1981); it is also relevant to studies of the earth's crustal growth and mantle evolution. Specificially, in oceanic crustal research, the samarium-neodymium (Sm-Nd) isotopic system has been shown to be relatively unaffected by seawater alteration in contrast to Rb-Sr and ^{18}O/^{16}O systems, which are sensitive to hydrothermal interactions (McCulloch et al., 1981). Furthermore, both Sm and Nd have low ionization potentials (5.63 eV and 5.49 eV, respectively), so that precision isotopic measurements can be made by surface-ionization mass spectrometry and isotopic dilution chemistry. It has also been pointed out by DePaolo (1980) that the Sm-Nd system is unique compared to other extended time-scale geochronometers, in that both "parent and daughter elements are refractory in their cosmochemical behavior; a consequence of this is that Sm-Nd fractionation during condensation of the solar nebula is expected to be small or zero". However, the fractionation that can occur in magmatic processes is large, so that the observed Sm-Nd can probably be attributed to the earth's internal differentiation and not to condensation phenomena. Hence, Sm-Nd measurements are becoming important in estimating crustal age, the extent of mantle differentiation, crustal recycling rates, and the development of theoretical transport models.

Samarium-neodymium measurements have been dominant in several other recent studies. Domenick and Basu (1982) have analyzed whole-rock samples in the Cortlandt Complex of southeastern New York to establish an intrusion age of 430 m.y. Their isotopic ratios were determined by using Sm and Nd phosphates that were loaded on a tantalum (Ta) ribbon in a triple-filament ion source of a 30 cm, 60° spectrometer. A comparison of U-Pb zircon and Sm-Nd mineral and whole-rock isochron ages has been reported by Nunes (1981), with an age agreement to within 0.4%. The Sm-Nd method has also been applied by Jahn et al. (1982) to date komatiites (MgO > 18%) of the Onverwacht Group (South Africa) that have been produced by partial melting of upper mantle materials. The Sm-Nd isochron in Figure 8.13 yields an age of 3.56 ± 0.24 billion years with an initial ^{143}Nd/^{144}Nd ratio of 0.50918 ± 23; this age is suggested as the time of initial Onverwacht volcanism. Combined Sm-Nd and Rb-Sr systematics have also been applied to document crustal displacements and the mixing of a mantle-derived

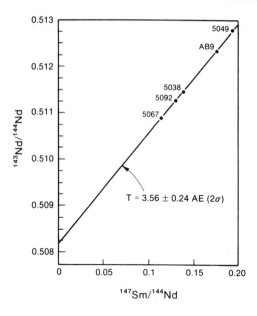

Figure 8.13 Sm-Nd isochron for the On-verwacht komatiites, South Africa. (Jahn et al., 1982; *Contributions to Mineralogy and Petrology*, *80*, 24 (1982), by permission of Springer-Verlag.)

magma in western Labrador, Canada (Zindler et al., 1981). In this study of a regional magmatic event, it is concluded that while the Rb-Sr systematics are somewhat ambiguous, the Sm-Nd analyses provide a reliable age estimate.

Collinear with specific ^{147}Sm-^{143}Nd measurements for interpreting lunar, meteoritic, and terrestrial evolution have been precise calibrations of standard Sm-Nd solutions by Wasserburg et al. (1981). The isotopic dilution method requires both precise mass spectrometric measurements and absolute gravimetric standards for Sm and Nd. Due to uncertainties with pure salts, these standards were derived from pure metallic specimens obtained from the Ames Laboratory. A mixed Sm-Nd normal solution was then prepared with a precisely known ^{147}Sm/^{144}Nd ratio that approximated average chondritic values. Aliquots of standard solutions have also been made available for interlaboratory comparison. In other measurements involving Sm and Nd, the concentrations of these two elements have been found to increase approximately linearly with depth in ocean waters; and it has been suggested that Nd may be useful as a tracer in paleo-oceanographic studies (Piepgras and Wasserburg, 1982).

The lutetium-hafnium (Lu-Hf) isotope method, involving the β-decay of ^{176}Lu to ^{176}Hf and the precision measurement of ^{176}Hf/^{177}Hf ratios, is a rather recent development (Patchett and Tatsumoto, 1980). A mass spectrometric difficulty has been in using the surface-ionization technique to provide stable ion beams of Hf, due to the refractory nature and high ionization potential (6.8 eV) of this element. However, a potential advantage cited by Patchett (1983) is the greater fractionation of Lu-Hf than Sm-Nd in planetary mag-

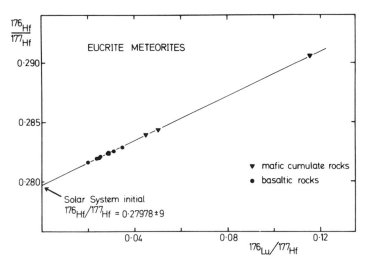

Figure 8.14 The first Lu-Hf isochron. The known 4.55 Gy igneous differentiation age of the eucrites yielded the decay constant λ, for ^{176}Lu of $1.94 \pm 0.07 \times 10^{-11}$ yr^{-1}. (Patchett, 1983; *Geochimica et Cosmochimica Acta*, **47**, 81 (1983), by permission of Pergamon Press Ltd.)

matic processes; and that, "in general, proportional variations of ^{176}Hf/^{177}Hf exceed those of ^{143}Nd/^{144}Nd by factors of 1.5–3 in terrestrial and lunar materials." Patchett (1983) has applied this Lu-Hf method to whole-rock eucrite meteorites. From the known 4.55×10^9 y differentiation age of the eucrites, he has obtained the decay constant λ for both ^{176}Lu of $1.94\ (\pm 0.07) \times 10^{-11}$ yr^{-1} and the solar system initial ^{176}Hf/^{177}Hf ratio indicated in Figure 8.14.

^{10}Be IN MARINE GEOCHEMISTRY

The recent development of accelerator-mass spectrometry has made possible many new studies in geochemistry. One area of research relates to variations in the ^{10}Be concentration in marine sediments as these might correlate to periods of geomagnetic reversal. A second and related study is focused on the growth rate of manganese nodules from the ocean floor, where such growth rates can be monitored by ^{10}Be profiling. ^{10}Be has a half-life of 1.6×10^6 years, being a nuclide produced by cosmic rays at a rate estimated at $\sim 1.5 \times 10^{-2}$ atoms/cm^2/s. (Somayajulu, 1977). The rationale for its use as a tracer is that this nuclide can attach to aerosols in the atmosphere, and it is rapidly (~ 1 year) carried to the earth's surface by precipitation and is subsequently incorporated in both the ocean and marine sediments.

The work of Raisbeck et al. (1979a) on the northern Pacific sediments in a search for a correlation of ^{10}Be concentrations during a geomagnetic field reversal assumes that the dipole intensity may decrease to a low level for

TABLE 8.10   Sedimentation Core Depth vs. Measured ^{10}Be Concentration

Sample Depth (cm)	^{10}Be Concentration (10^9 atoms/g)
977–978	3.66 ± 0.56
990–991	3.38 ± 0.51
995–996	3.66 ± 1.28
1000–1001	3.43 ± 0.60
1004–1006[a]	4.35 ± 0.80
1029–1031	3.62 ± 0.65
Average	3.68 ± 0.35

Source: Raisbeck et al. (1979).

[a] Location of calculated magnetic reversal.

a period of about 5000 years, during which time cosmogenic isotope production would be enhanced by a factor of 2 or 3 (O'Brien, 1979). Inasmuch as the reversal period is a small fraction of geological time and the production rate of ^{10}Be is so small, the detection of a variation is extremely difficult. Nevertheless, Raisbeck et al. (1979b) selected a marine sediment core, purported to have a uniform sedimentation rate (~1.1 cm/1000 y), and they examined a series of 5 cm core samples corresponding to ~5000 year intervals. The results of their assay showing ^{10}Be concentration vs. core depth are shown in Table 8.10. Clearly, the variation among samples is very small, and the authors do not claim a conclusive correlation; but it is interesting that the highest level of ^{10}Be does indeed correspond to a calculated magnetic reversal. In any event, this type of tracer study could only have been un-

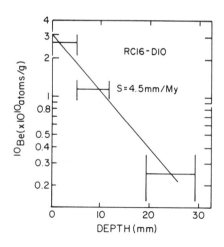

Figure 8.15   Depth profile for a manganese nodule from the Rio Grande Ridge in the Atlantic Ocean. The vertical axis is the ^{10}Be concentration per gram of the initial nodule material, and the concentrations represent ^{10}Be:^{9}Be ratios of 10^{-10} to 10^{-11} after addition of the pure ^{9}Be carrier. (Lanford et al., 1980; *Nuclear Instruments and Methods, 168,* 505 (1980), by permission of North-Holland Physics Publishing.)

dertaken with advanced instrumentation that is now extending the range and sensitivity of isotopic analysis.

The ^{10}Be measurements by Lanford et al. (1980) on manganese are another geochemical problem mandating the use of accelerator-mass spectrometry because of its inherent sensitivity. Manganese nodules "grow" on the ocean floor and are rich in manganese and other metals, and for this reason, they have attracted commercial as well as geophysical attention. Pertinent technical questions relate to growth mechanisms, growth rates, and why these nodules are found primarily on the ocean floor rather than being buried in sediments. ^{10}Be was selected as a chronological tracer for measuring nodule profiles because of its half-life, and because new accelerator techniques can bypass the stringent requirements of chemistry and very low-level β-decay counting. In a measurement of a nodule from the Rio Grande Ridge in the Atlantic Ocean, the ^{10}Be growth profile was determined using ^{9}Be as a carrier, applying the isotopic dilution method, and measuring the ^{10}Be/^{9}Be ratio. Figure 8.15 shows the results of the analysis.

The solid line indicates a nodule growth rate of 4.5 mm per million years and confirms the suspected very slow growth rate for these nodules in contrast to deep sea sediments. The authors indicate that there is no present answer to the question of why these nodules are not buried in the sediments that accumulate 1000 times faster than the nodule growth rate.

THE PLATINUM GROUP ELEMENTS

The detection of platinum and iridium at subparts per billion concentrations in rocks and minerals has been reported recently, using tandem accelerator-mass spectrometry (Rucklidge et al. 1982). A prime requirement for this type of assay is that the element can be transformed into negative ions—either directly by sputtering or by charge exchange. The efficiency of negative ion production by sputtering depends upon the work function of the emitting surface and electron affinity of the sample element. Pt and Ir have electron affinities of 2.13 eV and 1.6 eV, respectively, so that both elements can be ionized easily with a cesium (Cs) sputter source. However, conventional secondary ion mass spectrometry is not applicable at very high sensitivities because of molecular interferences at high mass numbers. Accelerator-mass spectrometry provides both sensitivity and freedom from spurious molecular species, resulting in a detection level for Pt and Ir of 10^{-11} g/g. This level can be achieved via neutron activation for Ir, but not for Pt, where a detection limit in the range of 10 ppb is usually reported because of the low neutron capture of Pt (Crocket and Teruta, 1977). The work of Rucklidge et al. (1982) on the Pt group elements in ultramafic rocks relates directly to the Ir anomaly at the cretaceous-tertiary (K/T) boundary, and also to the suggestion by Alvarez et al. (1980) that mass extinctions could be attributed to the impact of a large asteroid. Indeed, recent measurements confirm large abundances for Ir at the K/T boundary compared to concentrations existing in non-boundary shales (Brooks et al., 1984). Also, Ir is an element that is rare in terrestrial material, but it appears to be much more abundant in chondrites and other extraterrestrial matter.

Accelerator-mass spectrometric measurements of Pt and Ir in manganese nodules have also been important. An anomalous abundance of ^{193}Ir, ^{194}Pt, ^{195}Pt, and ^{189}Os at the K/T boundary in Denmark has also been established recently (Chew et al., 1984).

ANALYSES OF METEORITES

Isotopic anomalies in meteorites were detected as early as 1960 by Reynolds, who observed large enrichments of ^{129}Xe in the Richardson meteorite. This excess ^{129}Xe was later attributed to the in situ decay of ^{129}I. ^{3}He has been found in meteorites with an isotopic abundance as high as 25%, even though the atmospheric helium fraction is only $1.3 \times 10^{-4}\%$. Thus, similarities or differences in the isotopic composition of elements found in terrestrial minerals and meteorites permit conclusions to be made with respect to the history of mineral formation as well as the determination of age. In contrast to terrestrial samples that have undergone fractionation by various geological processes, carbonaceous chondrites are assumed to represent more accurately the relative abundances of the nonvolatile elements at the time of solar system formation.

Specifically, these meteorites are believed to be the condensed matter from a gaseous cloud of solar origin that has cooled to approximately 300 K. Elements that are gases at this temperature will be depleted from their abundance in the sun. However, the enrichment of lithium and various isotopic measurements provide the best clues concerning early thermonuclear reactions. Meteoritic analyses also yield essential data for estimating past and present intensities of cosmic radiations in interplanetary space. Nevertheless, isotopic anomalies in meteoritic inclusions continue to generate diverse interpretations depending upon the element, that is, interstellar dust that has been carried into the solar system, various nucleosynthesis products, fission products (in the case of Xe) from the in situ decay of a superheavy element, and other nuclear or mass fractionation processes.

As previously noted, there is also some evidence that a massive, chemically undifferentiated meteorite collided with the earth some 34 million years ago—producing "North American tektites" and leading to a dramatic change in climate with attendant mass extinction of life (Alvarez et al., 1982). This single event has generated substantial research relating to variations in elemental and isotopic abundances. The recent discovery of large numbers of Antarctic meteorites has also stimulated many other mass spectrometric measurements. ^{36}Cl($t_{1/2} = 3.1 \times 10^5$ years) has been measured in many of these meteorites by accelerator-mass spectrometry, and ^{10}Be/^{36}Cl ratio measurements have demonstrated a new method for dating ice (Nishiizumi et al., 1983). The isotopic composition and concentration of silver in iron meteorites has been studied by Kaiser and Wasserburg (1983), with some meteorites yielding $^{107}Ag/^{109}Ag$ values in the metal phase that markedly exceed the terrestrial value. Enrichments of ^{107}Ag ranging from 2–212% have been reported. It has been suggested that this excess ^{107}Ag is the result of decay of ^{107}Pd ($t_{1/2} = 6.5 \times 10^6$ years), a nuclide that is virtually extinct with an abundance of $^{107}Pd/^{108}Pd$ of only about 3×10^{-5}. A search has also been made in meteorites for isotopic anomalies in nickel (Ni), as Ni has five isotopes, and these are close to the iron group nuclei (Morand and Allègre, 1983; Shimamura and Lugmair, 1983). To date, however, no isotopic anomalies have been revealed.

The lithium isotopes are of considerable interest in meteorites, even though no significant differences in isotopic composition have been observed from that of terrestrial materials. According to theoretical models for the nucleosynthesis of lithium, approximately 90% of the ^7Li can be ascribed to the Big Bang model; the remaining ^7Li and virtually all the ^6Li is assumed to be produced by cosmic ray protons colliding with dust and gaseous matter throughout the galaxy. Rajan et al. (1980) have compared $^7Li/^6Li$ ratios for nine meteorites (including Allende) and four terrestrial samples and found no difference at the 1% level. While these measurements are not as precise as those made on oxygen, magnesium, neon, and other elements, they set important upper limits of 10^{19} to 10^{20} protons/cm^2 (with $E_p > 5$ MeV) for the excess irradiations possibly received by these meteorites compared to earth.

Nitrogen isotopic abundances of extraterrestrial material became important following the discovery that a large secular variation occurs in the $^{15}N/^{14}N$ ratio of solar wind trapped in lunar soils and breccias. Specifically, it has been shown that nitrogen implanted by the "modern" solar wind in mineral grains during the last ~50, m.y. is isotopically heavier (by about 120 ‰) than the earth's atmospheric nitrogen; however, in older lunar samples exposed to the solar wind several billion years ago, the nitrogen is substantially lighter ($\delta^{15}N = -210$ ‰) (Frick and Pepin, 1981). On the basis of this evidence in an ancient breccia that reflects a ^{15}N deficiency in ancient solar winds, it is suggested that nitrogen in the primitive solar nebula had a low $^{15}N/^{14}N$ ratio.

Noble gas analyses of the Murchison (C2) chondrite have been reported by many investigators. These measurements have included ratios of $^{20}Ne/^{22}Ne$, $^{21}Ne/^{22}Ne$, $^{86}Kr/^{82}Kr$, $^{120}Xe/^{132}Xe$, etc. Many of these have been assayed by a stepwise heating technique plus a detailed identification of individual components and host phases. At least some of these results strongly suggest that the noble gas components in Murchison are identical to their counterpart in Allende (Alaerts et al., 1980).

Measurements relative to models of nucleosynthesis and meteoritic age-dating also include $^{244}Pu/^{238}U$ ratios (Jones, 1982). Several studies had suggested that the excess of heavy Xe isotopes in some meteorites was due to the contribution of fission products from ^{244}Pu ($t_{1/2} = 82$ m.y.). This conclusion was reached, in part, because the fission Xe concentrations in some meteorites were too large to be attributed to either the spontaneous fission of ^{238}U or the neutron-induced fission of ^{235}U. A ^{244}Pu/^{238}U ratio measurement is very difficult, because plutonium (Pu) has no stable isotope and differences in meteoritic ^{244}Pu might be attributed to an age difference or to chemical processing.

Among other recent isotopic analyses of great interest has been that of the so-called "Allan Hills 81005" meteorite, whose oxygen isotope composition is identical to that of the moon and is unlike that of most meteorites (Kerr, 1983). Allan Hills (76° 45'S, 159° 00'E) is an Antarctic ablation region with many meteorites. It is suggested that this meteorite, which was ostensibly ejected and launched to earth by some massive impact on the lunar surface, is a more pristine moon rock sample than those collected by astronauts on the Apollo missions. Analyses performed on over 30 elements—the magnesium/iron ratio, the potassium/lanthanum ratio, and trace element abundances, etc.—all sufficiently matched other lunar specimens to suggest that this golf ball-sized meteorite, retrieved from the Antarctic ice in 1982, was indeed of lunar origin.

Additional recent measurements of meteorites include the Abee E4 enstatite chondrite, which fell in 1952 north of Edmonton, Canada (Marti, 1983; Bogard et al., 1983). Determinations of ^{40}Ar/^{39}Ar ratios and U-Th-Pb have provided data relating to thermal history and age. Oxygen isotopic measurements have also been made on various groups of stony meteorites (Clayton and Mayeda, 1983). In these latter measurements, it has been possible to identify two separate achondrite populations having distinct source reservoirs.

THE ALLENDE METEORITE

The Allende meteorite landed near Pueblito de Allende in Mexico during 1969. It is classified as a carbonaceous chondrite containing the most primitive material in the solar system, and whose isotopic anomalies have been reported for nitrogen, oxygen, magnesium, and many other elements. Thermal ionization mass spectrometry has been used for assaying Ca-Al-rich inclusions extracted from Allende samples, and ion/laser microprobe techniques have been extensively applied.

^{26}Al is one of the extinct nuclides that decays to ^{26}Mg with a half-life of 7.2×10^5 years. Hence, the existence of ^{26}Al in the early solar system must be detected via the mass spectrometric detection of ^{26}Mg. The work of Nishimura and Okano (1980) is one of the many significant measurements that have been reported in the Allende meteorite. In this microprobe work, an 8 keV primary oxygen beam of approximately 100 μm in diameter was fo-

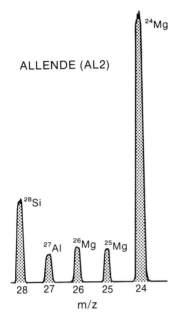

ALLENDE (AL2)

Figure 8.16 The Si-Al-Mg mass spectrum of a sample of the Allende carbonaceous chondrite obtained by ion microprobe analysis. (Nishimura and Okano, 1980; courtesy of H. Nishimura).

cused on the Allende meteorite with a current of ~3 × 10^{-7} A. ^{25}Mg/^{24}Mg, ^{26}Mg/^{24}Mg, and ^{27}Al/^{24}Mg ratios were determined across inclusion lines, and a mass spectrum is shown in Figure 8.16.

A detailed examination was made to detect interferences of molecular- and doubly charged ions, and a terrestrial laboratory standard was intermittently analyzed to normalize the data. The deviations (for sample inclu-

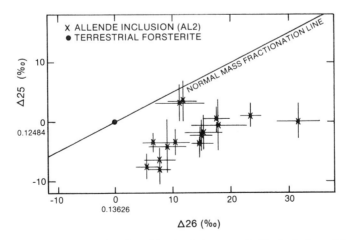

Figure 8.17 Deviation Δ_{25} as a function of Δ_{26} for dark gray and white inclusions of the Allende. (Nishimura and Okano, 1980; courtesy of H. Nishimura).

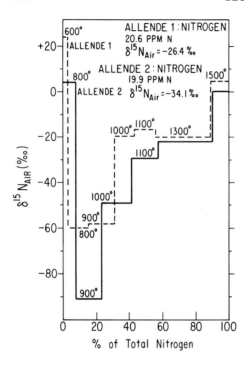

Figure 8.18 Isotopic composition of nitrogen from the stepwise pyrolysis of two Allende samples. The $-90\permil$ value for the 900° fraction of sample 2 is the lowest $^{15}N:^{14}N$ ratio observed in any meteorite; the excess ^{14}N is of possible nucleosynthetic origin. (Thiemens and Clayton, 1981; *Earth and Planetary Science Letters*, 55, 363 (1981), by permission of Elsevier Science Publishers.)

sions) Δ_{25} and Δ_{26} from a reference sample are shown as ordinate and abscissa, respectively, in Figure 8.17 and calculated according to the expression:

$$\Delta_n = \left[\frac{(^n Mg/^{24}Mg)_{meas}}{(^n Mg/^{24}Mg)_{ref}} - 1\right] \times 1000 \qquad (n = 25, 26)$$

where $(^{25}Mg/^{24}Mg)_{ref} = 0.12484$ and $(^{26}Mg/^{24}Mg)_{ref} = 0.13636$ are mean values of 20 runs for the terrestrial fosterite sample.

A ^{14}N-enriched component of presumed nuclear origin has been reported in the Allende meteorite (Thiemans and Clayton, 1981). In this work, two samples of Allende ($\sim$2 g each) were analyzed by stepwise heating in a vacuum with no chemical treatment prior to pyrolysis. Figure 8.18 shows the results of two samples where $\delta^{15}N$ is defined as

$$\delta^{15}N = \left[\frac{(^{15}N/^{14}N)_{sample}}{(^{15}N/^{14}N)_{standard}} - 1\right] \times 1000$$

The $^{15}N/^{14}N$ ratio, expressed as $\delta^{15}N$ (parts per thousand deviation from the terrestrial atmospheric standard), varies by more than $100\permil$ among components released at different temperatures, and the value of $-90\permil$ for the 900° fraction of sample 2 is the lowest $^{15}N/^{14}N$ ratio of any reported meteorite sample.

Additional measurements on the inclusions of the Allende meteorite have been those of I-Xe dating. If the ratio of ^{129}I/^{127}I produced in galactic nucleosynthesis is known, then I-Xe isochrons can yield age differences in the formation ages of meteoritic samples that contain radioactive ^{129}I. The short half-life of ^{129}I (17.2 m.y.) together with Xe isotopic data have yielded accurate relative age differences as small as 3.7 ± 1.5 m.y. in mineral formation for meteorites that formed approximately 4.5 billion years ago (Zaikowski, 1980).

Other dating measurements include ^{207}Pb/^{206}Pb ratios that yield a model age of 4.559 ± 0.015 billion years, which is an age consistent with the observation of normal uranium isotopic composition (Chen and Wasserburg, 1981). Tellurium is another element displaying an anomaly in the two mineral fractions that contain isotopically anomalous xenon; both elements have enhanced abundances of the neutron-rich and neutron-poor isotopes (Oliver et al., 1981). Refractory metal particles containing Os, Re, W, Mo, Ir, and Ru have also been formed in the calcium-aluminum-rich inclusions of the Allende meteorite. It is suggested that those particles are relatively unaltered primordial metal condensates that have been isolated from a nebula before the condensation of refractories was complete (Blander et al., 1980).

A search is continuing to detect isotopic anomalies in the Allende, namely, for argon (Villa et al., 1983), barium (Lewis et al., 1983), samarium (Lugmair et al., 1983), etc. These studies on Ca-Al-rich inclusions and carbon and chromite fractions represent state-of-the-art measurements in geochronology. They provide fundamental data relating to nucleosynthesis models and cosmic ray exposure ages. To date, however, the Allende meteorite has yielded little evidence to link it to a presolar history or an extinct superheavy element.

^{18}O/^{16}O RATIOS AND PALEOTEMPERATURES

In a remarkable paper entitled *The Thermodynamic Properties of Isotopic Substances* (1947) Professor Urey, discoverer of the second isotope of hydrogen, proposed that oxygen isotopes could be used to measure temperature. He showed that the thermodynamic properties of a system of molecules are mass-dependent, and that one might expect to observe fractionation if isotopic substitutions of low atomic weight elements were made in low molecular weight compounds. Thus, if Q_1 and Q_2 are the partition functions of isotopic molecules having molecular weights M_1 and M_2 and energy states E_1 and E_2, a partition function can be expressed by

$$\frac{Q_2}{Q_1} = \left(\frac{M_2}{M_1}\right)^{\frac{3}{2}} \frac{\sigma_1}{\sigma_2} \frac{\sum e^{-\frac{E_2}{kT}}}{\sum e^{-\frac{E_1}{kT}}}$$

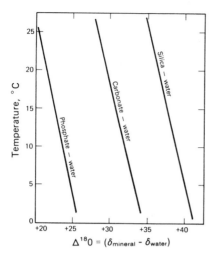

Figure 8.19 Temperature vs. $\Delta^{18}O$ for calcium carbonate, phosphate, and silica paleotemperature systems; $\Delta^{18}O$ is the difference between the δ value of the mineral and that of the water with which it is equilibrated (Faure, 1977c).

where σ_1 and σ_2 are symmetry constants and the sum is taken over all energy states. This mass-dependent equilibrium exchange relationship provides a classical basis for interpreting fractionation and temperature-dependent phenomena such as mineralization and hydrothermal change. In particular, Urey suggested a paleotemperature scale, based on the fact that the oxygen isotopic composition of calcium carbonate differs from that of water when the calcite precipitates from water under equilibrium conditions. In consequence, the enrichment of the ^{18}O isotope in marine calcite, compared to an assumed constant $^{18}O/^{16}O$ ratio for the seawater from which the calcite is equilibrated, provides a temperature scale. Other paleothermometers using ^{18}O are biogenic silica and biogenic phosphate (Faure, 1977c). Figure 8.19 shows the temperature dependence vs. ^{18}O enrichment for all three systems, where $\Delta^{18}O$ is the difference between the δ value of the mineral and that of the water with which it is equilibrated.

Isotopic compositions of oxygen and hydrogen are referenced to a standard mean ocean water (SMOW) sample of distilled water provided by the National Bureau of Standards. ^{18}O enrichments are then expressed as

$$\delta^{18}O = \left[\frac{(^{18}O/^{16}O)_{sample} - (^{18}O/^{16}O)_{SMOW}}{(^{18}O/^{16}O)_{SMOW}} \right] \times 1000$$

and have positive values; a negative $\delta^{18}O$ value thus implies a depletion of the heavy oxygen isotope relative to the sample.

Inasmuch as oxygen is the most abundant element in the earth's crust, geochemical-related studies of isotopic variation are many; for example, evaporation-condensation phenomena, the dating of ice cores, the mixing of surface and deep ocean waters, paleotemperature scales, geothermal waters, etc.

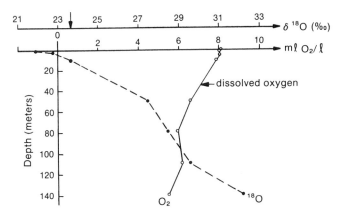

Figure 8.20 Increase in the ^{18}O content with depth for dissolved oxygen in the North Sea. The $^{18}O/^{16}O$ ratio of dissolved oxygen in the ocean is altered by the preferred consumption of ^{16}O in the respirative activity of living matter. (Förstel and Zielke, 1978; *Pure and Applied Geophysics*, *116*, 486 (1978), adapted by permission of H. Förstel.)

Typical of recent paleotemperature measurements is the investigation showing the ^{18}O enrichment of empty shells on the seabed compared to the shells of their living counterparts in surface waters (Duplessy and Blane, 1981). Two tropical planktonic species were examined in the Indian Ocean in a region where the ambient seawater temperature variations are small; the enrichment was attributed to a calcification process that extracted calcium carbonate from deeper and colder waters as the shells sank below the euphotic zone. Calculations indicated that one species with a shell length of 500 μm and an initial mass of 20 μg would increase to 27 μg through gametogenic calcification and then descend to a depth of 450 m in 12 h. Experimental isotopic measurements showed that an appreciable fraction of the calcification occurred with $\delta^{18}O$ values ranging from $+2-1.07$/mil, indicating a mean secretion temperature of 9–12.5°C or mean depths of 300–800 m.

The measurements of the $^{18}O/^{16}O$ ratio of dissolved oxygen in the North Sea (Förstel and Zielke, 1978) are representative of the recent work that has focused on dissolved oxygen in ocean waters. The purpose of this work was to elucidate biochemical and exchange phenomena at the air-water boundary. While the atmosphere is a reasonably well-mixed reservoir, ocean waters display stratification and distinct profiles as a function of depth. Figure 8.20 shows that with increasing depth, there is a decrease in the dissolved oxygen and a marked increase in the ^{18}O content. Dissolved oxygen originating from the atmospheric reservoir has an $^{18}O/^{16}O$ ratio of approximately 23/mil relative to SMOW, the seawater standard; and at the atmosphere-ocean surface interface, this $^{18}O/^{16}O$ ratio is evident. The ratio is a variable as a function of depth, however, partially due to the consumption of oxygen during respiration and because the lighter ^{16}O isotope is preferred in respirative activity. Other oxygen sources and sinks can also affect the amount of dissolved oxygen and the ^{18}O enrichment.

Additional oxygen isotope research has focused on characterizing minerals and rocks through a section of the oceanic crust. According to Gregory and Taylor (1981), such studies are important as they relate to (1) the effects of hydrothermal circulation at midocean ridges through geological time; and (2) the isotopic and chemical composition of the ocean crust that is recycled into the mantle at subduction zones. Variations in the $^{18}O/^{16}O$ ratio in shells of marine microfossils are greatly influenced by changes in global ice volume (Morley and Hays, 1981); such variations can thus furnish a set of global time lines in deep sea sediments. It is even suggested that the timing of major changes in oxygen isotope abundances in deep sea cores is strongly influenced by variations in the geometry of the earth's orbit. And since the variations in orbital motion can be accurately calculated, these authors suggest that "the opportunity exists for transforming this orbital chronology into a geological chronology."

THERMAL HISTORIES AND PLATE TECTONICS

Much of the climatic history of the past 100 million years lies buried in the huge reservoir of seafloor sediments and fossils that contain isotopic data on ocean surface temperatures, ocean current systems, and rates of biological activity. Because the mixing time of the oceans is short (a few hundred years) compared to major climatic periods, isotopic variations provide a reasonably well-defined paleochronology.

Isotopic measurements have also yielded ages of ice formation. For example, recently reported oxygen isotope compositions ($\delta^{18}O$) vs. age (or depth in a sediment core) reveal a distinct maximum about 29 million years ago, which suggests that a major ice sheet covered Antarctica at that time (Keigwin and Keller, 1984). Other $\delta^{18}O$ and $\delta^{13}C$ measurements in fossils from Lake Turkana in Kenya reveal climatic conditions at the time of deposition of these carbonates; the data are sufficiently detailed that they indicate changing climatic conditions, which range from a cool, wet climate prior to about 1.9 m.y. ago to a drier and warmer period about ca. 1.4 m.y. ago, followed by a cooler and moister climate at the end of the Pleistocene (Abell et al., 1982).

According to isotopic data acquired on a global scale, eight major periods of glaciation have occurred during the past 700,000 years. The beginning of the latest ice age was about 75,000 years ago, and at its maximum (~18,000 years ago), the continental ice sheet capping the northern hemisphere extended as far south as Long Island and Ohio. Oxygen isotope paleotemperatures figured prominently in interpreting the events of this period (Shackelton and Boersma, 1981). Deep sea cores from the northeastern Gulf of

Mexico identify a more recent period of rapid ice melting (coincident in time with a flood described in the writings of Plato) and a sea level rise about 9600 BC (Emiliani et al., 1975).

Oxygen and carbon isotopic measurements on mollusk and carbonate shells from sediment cores in Lake Erie have served to document climatic changes of the Great Lakes region and the evolution of Lake Erie since deglaciation (Fritz et al., 1975). In related studies of the glacial Norwegian Sea, oxygen isotope analyses have provided information on past ice cap volumes and sea temperatures (Thomas and Murphy, 1975). The isotopes of carbon and oxygen also provide records of seasonal temperature ranges and life history stages of tropical foraminifera shells that constitute an important component of present and ancient carbonate reefs. In waters off the Philippines and Bermuda, $\delta^{18}O$ variations have recorded the seasons, and $\delta^{13}C$ perturbations have reflected the metabolic responses of these organisms to seasonal temperature cycles (Wefer and Berger, 1980).

$^{18}O/^{16}O$ and D/H ratios have also been interpreted in terms of the three processes that link the hydrogen/oxygen isotopic composition to the climate: precipitation, evapotranspiration, and isotopic fractionation that accompany biochemical reactions in plants. Attempts have also been made to retrieve paleoclimatological information from tree rings; a recent study of a peat bog core in the Netherlands (Brenninkmeijer et al., 1982) has demonstrated the feasibility of establishing cellulose δD and $\delta^{18}O$ records for peat profiles.

$^{129}Xe/^{130}Xe$ measurements provide another important input for the internal thermal history of the earth for several reasons. First, xenon is a rare gas for which the ^{130}Xe isotope is relatively abundant, and changes in its abundance due to terrestrial nuclear processes are negligible. Second, exceedingly accurate mass spectrometric measurements can be made of the quantitites

$$\Delta_i = (^iXe/^{130}Xe)_s - (^iXe/^{130}Xe)_a$$

where the subscripts s and a refer to the sample and the atmosphere, respectively. The rationale is that ^{129}Xe from extinct ^{129}I ($T_{1/2} = 1.6 \times 10^7$ years) can establish limits on the thermal history of the upper mantle (Thomsen, 1980). Third, recent data by Smith and Reynolds (1981) indicating high $^{129}Xe/^{130}Xe$ ratios in CO_2 well gas in New Mexico are consistent with other thermal estimates. Specifically, this latter mass spectral assay was made in a newly built, 10 in. radius mass spectrometer in which no previous samples had been inserted that could cause contamination.

Oxygen isotope thermometry has been highly successful in elucidating the temperatures of coexisting minerals in lavas and mantle nodules (Kyser et al., 1981). According to these authors, isotopic fractionation between coexisting minerals can be used as a geothermometer provided that (1) the mineral assemblages do not show evidence of disequilibrium; (2) the minerals have retained their original isotopic ratios; and (3) the fractionations are known as a function of temperature. The per mil fractionation of ^{18}O between minerals A and B is given by

$$\Delta^{18}O\ (A\ -\ B)\ =\ \delta^{18}O_A\ -\ \delta^{18}O_B$$

where

$$\delta^{18}O_A\ =\ \left[\frac{(^{18}O/^{16}O)_A}{(^{18}O/^{16}O)_{V-\text{SMOW}}}\ -\ 1\right]\ \times\ 1000$$

and values are referenced to the V-SMOW standard.

$^{18}O/^{16}O$ ratios have also been important in studies of oxygen isotope exchange at an igneous-intrusive contact (Nagy and Parmentier, 1982). The $\delta^{18}O$ variation across the exchange zone provides evidence of diffusion, fluid flow, and other mechanisms of oxygen transport across the intrusive host rock contact.

Metamorphic temperatures have been determined by the fractionation of ^{13}C between graphite and calcite that coexist in a wide variety of metamorphosed rocks (Bohlen et al., 1980). In a specific investigation of marbles from the Grenville Province, Adirondack Mountains, New York, Valley and O'Neil (1981) have shown that a $^{13}C/^{12}C$ exchange takes place between graphite and calcite at temperatures as low as 300°C, and that this exchange may be used as a basis for geothermometry. In the temperature interval of 610–760°C, a fractionation of $\delta^{13}C$ between calcite and graphite, $\Delta(\text{Cc-Gr})$, is expressed as:

$$\Delta(\text{Cc-Gr})\ =\ -0.00748\ T\ +\ 8.68\qquad(T\ \text{in °C})$$

Further determinations of metamorphic temperatures in the Adirondack region have been reported by Bohlen et al. (1985), resulting in the isotherm map of Figure 8.21. This extensive work is of special significance, in that other independent thermometers corroborate data obtained from feldspars and oxides.

Finally, isotopic measurements play an important role in elucidating plate tectonics, that is, the dynamics of the rigid lithospheric plates ($\sim$100 km in thickness) that encompass the upper mantle and the overlying continental and oceanic crust. The movement of these plates, their intrusion into the asthenosphere, volcanic activity at midocean ridges and at island arcs, etc. are all processes that have been probed by mass spectrometric analysis (Faure, 1977d). A modern view suggests that the earth is characterized by mantle convection, and the thickness of the continental and oceanic lithospheres changes with time as pressure, temperature, and geochemical conditions are altered in the upper mantle (Tarling, 1980).

Specifically, trace element and isotopic data are primary sources for describing the evolution of the mantle and for providing evidence that oceanic

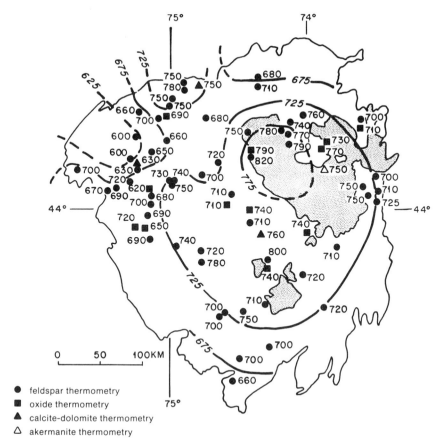

Figure 8.21 Outline map of the Precambrian terrane of upper New York state, showing peak metamorphic temperatures as determined by four independent mineral thermometers. (Bohlen et al., 1985; *J. Petrology*, 26, 971 (1985); courtesy of S. R. Bohlen.)

and continental basalts are mixtures of melts from ancient depleted and enriched reservoirs (Anderson, 1982). A detailed analysis of basaltic glass has been studied for Pb, Nd, and Sr isotope compositions and U, Pb, Sm, Nd, Rb and Sr concentrations (Cohen and O'Nions, 1982). Basalt glass has been identified as the most appropriate material for obtaining estimates of the primary (prealteration) chemistry of oceanic crust and for eliciting details of high- and low-temperature alteration processes. Nd and Sr isotopic studies have also been made of volcanic rocks from the Japanese islands for the purpose of examining the relationships between isotopic abundances and the general tectonic features of the western Pacific (Nohda and Wasserburg, 1981). This study suggests that the island arc of Japan is partly continental in nature, with an older continental basement admixed with young, rising oceanic arc magmas. Additional U-Pb measurements, in conjunction with

Sm-Nd and Rb-Sr systems, have played an important role in assays of mid-ocean ridge basalt (MORB) glasses from the Pacific, Indian, and Atlantic Oceans (Cohen and O'Nions, 1982).

Perhaps one of the most interesting recent ^{235}U measurements relating to plate tectonics and continental drift theory was the search for the "Oklo phenomenon" (Lima et al., 1982). Vestiges of a natural nuclear fission reactor discovered in the Gabon Republic on the coast of west Africa in 1972 were identified on the basis of isotopic anomalies of uranium found in the region. The postulation of a supercontinent some 1000 to 2000 million years ago, comprising Africa and South America, has now pointed to the desirability of searching for comparable isotopic anomalies on the coast of Brazil in the region where the two continents were purportedly joined.

The recent analysis by Lima et al. (1982) of Precambrian rocks on the Brazilian coast has provided no evidence of a ^{235}U depletion that would correlate with the Oklo nuclear reactor some two billion years ago. However, as the authors report, their isotopic measurements were based on neutron activation and gamma ray spectrometry, rather than on precision mass spectrometry; and an additional sampling of rocks has been recommended. In any event, it is clear that future isotopic measurements will play an important role in testing the validity of continental drift models.

PETROLEUM GEOLOGY

It is appropriate to conclude this chapter by citing the relevance of mass spectrometric measurements to the origin and location of petroleum deposits. Giardini and Melton (1983) have addressed this topic in some detail, and they suggested a quantitative theoretical model for the formation of petroleum. This model is a diffusion model that assumes that petroleum is a product of mantle outgassing, and that it formed from Fischer-Tropsch-type reactions of H_2, CO, and CO_2 in the upper 400 km of the earth. Their experimental measurements included analyses of solid inclusions in diamonds, quantitative mass spectrometric measurements of fluids trapped in diamonds, and appropriate isotopic measurements of entrapped argon that permitted an age of 3.1 billion years to be established as the age of diamond crystallization in the earth's upper mantle. This crystallization age has been recently confirmed as being correct by Richardson et al. (1984). From these data, Giardini et al. (1982) have described the nature of the upper 400 km of the earth since the age of diamond crystallization 3.1 billion years ago.

Their outgassing model and mantle-atmosphere transport calculations for N_2 that is outgassed to the atmosphere (4.9×10^{15} tons) are in agreement with the present-day atmospheric N_2 abundance (Melton and Giardini, 1982). Also, a calculated value for the quantity of outgassed ^{36}Ar is reasonably close to the experimentally measured present-day value of 2.5×10^{11} tons. ^{40}Ar/^{36}Ar-observed ratios are also consistent with this diffusion model, which can quantitatively account for the proven quantity of petroleum in crustal formations. Thus, petroleum accumulations (including the 78 "giant" fields) are quantitatively related to the H_2, CO, alcohols, and hydrocarbons that are constituents of the mantle-fluid compounds from which the atmosphere, hydrosphere, and surficial carbonaceous matter were evolved. The authors tend to discount the presence of trace nickel and vanadium porphyrins as a proof of a biological origin of petroleum. Rather, they suggest that traces of biologically derived compounds and fossils in "source rock" and petroleum reservoirs are "merely intrinsic to their sedimentary nature."

In any event, these investigators suggest that all of the petroleum that is known to exist can be connected to the outflow of H_2O, CO_2, N_2, and Ar from mantle outgassing, based on data from "diamond carriers" that crystallized 3.1 billion years ago. The model of Giardini and Melton (1983) accommodates the presence of modest quantities of biologically derived compounds, and it is not inconsistent with a large body of knowledge relating to sedimentary formations, plate tectonics, and the kinetics of mantle-to-surface transport.

Giardini and Melton (1982) have also obtained carbon isotopic data for CO_2 entrapped in diamonds. Their $\delta^{13}C$ value of $-35.2‰$, which falls in the range that is usually ascribed to biological activity, is found to be in general agreement with isotopic fractionation that would accompany the nonbiological reaction, CO, diamond, and CO_2.

The above-cited work is at variance with some generally accepted concepts of petroleum geology, but it does point to the continuing role of mass spectrometry in revealing the many facets of geochemistry and geophysics.

REFERENCES

Abell, P. I., S. M. Awramik, R. H. Osborne, and S. Tomellini, *Sediment. Geol., 32*, 1 (1982).

Ahrens, L. H., *Distribution of Elements in Our Planet*, McGraw-Hill, New York (1965) p. 20.

Alaerts, L., R. S. Lewis, J. Matsuda, and E. Anders, *Geochim. et Cosmochim. Acta, 44*, 189 (1980).

Aldrich, L. T., and A. O. C. Nier, *Phys. Rev., 74*, 1590 (1948).

Allègre, C. J., and J. Luck, *Earth and Planet. Sci. Lett., 48*, 148 (1980).

Alvarez, L. W., W. Alvarez, P. Asaro, and H. V. Michel, *Science, 208*, 1095 (1980).

Alvarez, W., F. Asaro, H. V. Michel, and L. W. Alvarez, *Science, 216*, 886 (1982).

Anderson, D. L., *Earth and Planet. Sci. Lett., 57*, 1 (1982).

Aston, F. W., *Nature, 120*, 224 (1927).

Baksi, A. K., *Geophys. J. R. Astr. Soc., 70*, 545 (1982).

Bar-Matthews, M., I. D. Hutcheon, G. L. MacPherson, and L. Grossman, *Geochim. et Cosmochim. Acta, 46*, 31 (1982).

Bennett, C. L., R. P. Beukens, M. R. Clover, H. E. Gove, L. R. Kilius, A. E. Litherland, and K. H. Purser, *Science, 198*, 5082 (1977).

Bieri, R. H., and M. Koide, *J. Geophys. Res., 77*, No. 9, 1667 (1972).

Blander, M., L. H. Fuchs, C. Horowitz, and R. Land, *Geochim. et Cosmochim. Acta, 44*, 217 (1980).

Bogard, D. D., D. M. Unruh, and M. Tatsumoto, *Earth and Planet. Sci. Lett., 62*, 132 (1983).

Bohlen, S. R., A. L. Boettcher, W. A. Dollase, and E. J. Essene, *Earth and Planet. Sci. Lett., 47*, 11 (1980).

Bohlen, S. R., J. W. Valley, and E. J. Essen, *J. Petrol. 26*, 971 (1985).

Brenninkmeijer, C. A. M., B. Van Geel, and W. G. Mook, *Earth and Planet. Sci. Lett., 61*, 283 (1982).

Brevard, O., N. Shimizu, and C. Allègre, *U.S. Geol. Survey, Open-File Rept. 78–701*, 49 (1978).

Brooks, R. R., J. D. Collen, J. Holzbecher, J. Lee, V. E. Neall, R. D. Reeves, D. E. Ryan, and X. Yang, *Science, 226*, 539 (1984).

Chen, J. H., and G. J. Wasserburg, *Geophys. Res. Lett., 7*, No. 4, 275 (1980).

Chen, J. H., and G. J. Wasserburg, *Earth and Planet. Sci. Lett., 52*, 1 (1981).

Chen, J. H., and J. S. Pallister, *J. Geophys. Res., 86*, B4, 2699 (1981).

Chen, J. H., and J. C. Moore, *J. Geophys. Res., 87*, B6, 4761 (1982).

Chew, S. H., T. J. L. Greenway, and K. W. Allen, *Nucl. Instr. and Methods in Phys. Res., B5*, 179 (1984).

Clarke, D. B., and A. N. Halliday, *Geochim. et Cosmochim. Acta, 44*, 1045 (1980).

Clayton, R. N., and T. K. Mayeda, *Earth and Planet. Sci. Lett., 62*, 1 (1983).

Cloud, P., *Am. Scientist, 68*, 381 (1980).

Cohen, R. S., and R. K. O'Nions, *J. Petrol. 23*, No. 3, 299 (1982).

Craig, H., J. E. Lupton, J. A. Whalen, and R. Poreda, *Geophys. Res. Lett., 5*, No. 11, 897 (1978).

Crocket, J. H., and Y. Teruta, *Can. J. Earth Sci., 14*, 777 (1977).

Cronin, T. M., B. J. Szabo, T. A. Ager, J. E. Hazel, and J. P. Owens, *Science, 211*, 233 (1981).

Dallmeyer, R. D., and O. Van Breeman, *Contrib. Min. Petrol., 78*, 61 (1981).

DePaolo, D. J., *Geochim. et Cosmochim. Acta, 44*, 1185 (1980).

DePaolo, D. J., and G.-J. Wasserburg, *Geochim. et Cosmochim. Acta, 43*, 615 (1979).

DePaolo, D. J., and G. J. Wasserburg, *Geochim. et Cosmochim. Acta, 45*, 1961 (1981).

Domenick, M. A., and A. R. Basu, *Contrib. Min. Petrol., 79*, 290 (1982).

Duplessy, J., and P. Blane, *Science, 213*, 1247 (1981).

Ellwood, B. B., J. A. Whitney, D. B. Wenner, D. Mose, and C. Amerigian, *J. Geophys. Res., 85*, B11, 6521 (1980).

Elmore, D., B. R. Fulton, M. R. Clover, J. R. Marsden, H. E. Gove, H. Naylor, K. H. Purser, L. R. Kilius, R. P. Beukens, and A. E. Litherland, *Nature, 277*, 22 (1979).

Elroy, J. F., M. Leleu, and E. Unsold, *Int. J. Mass Spectrom. Ion Phys., 47*, 39 (1983).

Emiliani, C., S. Gartner, B. Lidz, K. Eldridge, D. K. Elvey, T. C. Huang, J. J. Stipp, and M. F. Swanson, *Science, 189*, 1083 (1975).

Faul, H., *Nuclear Geology*, Wiley, New York (1954), p. 135.

Faure, G., *Principles of Isotope Geology*, Wiley, New York (1977a), p. 20.

Faure, G., *Principles of Isotope Geology*, Wiley, New York (1977b), p. 168.

Faure, G., *Principles of Isotope Geology*, Wiley, New York (1977c), p. 340.

Faure, G., *Principles of Isotope Geology*, Wiley, New York (1977d), p. 114.

Förstel, H., and H. Zielke, *Pure and Appl. Geophys., 116*, 486 (1978).

Fowler, W. A., *Science, 226*, 922 (1984).

Frick, U., and R. O. Pepin, *Earth and Planet. Sci. Lett., 56*, 64 (1981).

Fritz, P., T. W. Anderson, and C. F. M. Lewis, *Science, 190* 267 (1975).

Gaudette, H. E., A. Vitrac-Michard, and C. J. Allègre, *Earth and Planet. Sci. Lett., 54,* 248 (1981).

Giardini, A. A., and C. E. Melton, *J. Petrol. Geol. 4,* No. 4, 437 (1982).

Giardini, A. A., C. E. Melton, and R. S. Mitchell, *J. Petrol. Geol., 5,* 173 (1982).

Giardini, A. A., and C. E. Melton, *J. Petrol. Geol., 6,* No. 2, 117 (1983).

Goles, G. G., in *Handbook of Geochemistry,* Vol. 1, K. H. Wedepohl, New York, Springer-Verlag (1969), pp. 116–133.

Gregory, R. T., and H. P. Taylor, Jr., *J. Geophys. Res., 86,* B4, 2737 (1981).

Harrison, T. M., and I. McDougall, *Geochim. et Cosmochim. Acta, 46,* 1811 (1982).

Hart, S. R., and H. Staudigel, *Geophys. Res. Lett., 5,* 1009 (1978).

Hart, S. R., N. Shimizu, and D. A. Sverjensky, *Econ. Geol., 76,* 1873 (1981).

Hendry, D. A. F., A. R. Chivas, S. J. B. Reed, and J. V. P. Long, *Contrib. Min. Petrol., 78,* 404 (1981).

Herr, W., W. Hoffmeister, B. Hirt, J. Geiss, and F. G. Houtermans, *Z. Naturforch, 16,* 1053 (1961).

Hill, R. D., *Phys. Rev., 59,* 103 (1941).

Jahn, B., G. Gruau, and A. Y. Glikson, *Contrib. Min. Petrol., 80,* 24 (1982).

Jones, J. H., *Geochim. et Cosmochim. Acta, 46,* 1793 (1982).

Jouzel, J., L. Merlivat, D. Mazaudier, M. Pourchet, and C. Lorius, *Geophys. Res. Lett., 9,* No. 10, 1191 (1982).

Kaiser, T., and G. J. Wasserburg, *Geochim. et Cosmochim. Acta, 47,* 43 (1983).

Kaneoka, I., and N. Takaoka, *Science, 208,* 1366 (1980).

Keigwin, L., and G. Keller, *Geology, 12,* 16 (1984).

Kerr, R. A., *Science, 220,* 288 (1983).

Kiko, J., H. W. Muller, K. Buchler, S. Kalbitzer, T. Kirsten, and M. Warhout, *Int. J. Mass Spectrom. Ion Phys., 29,* 87 (1979).

Klein, J., J. C. Lerman, P. E. Damon, and E. K. Ralph, *Radiocarbon, 24,* No. 2, 103 (1982).

Kyser, T. K., J. R. O'Neil, and I. S. E. Carmichael, *Contrib. Min. Petrol., 77,* 11 (1981).

Kyser, T. K., and W. Rison, *J. Geophys. Res., 87,* B7, 5611 (1982a).

Kyser, T. K., and W. Rison, *J. Geophys. Res., 87,* 564 (1982b).

Lanford, W. A., P. D. Parker, K. Bauer, K. K. Turekian, J. K. Cochran, and S. Krishnaswami, *Nucl. Instr. and Methods, 168,* 505 (1980).

Lewis, R. S., E. Anders, T. Shimamura, and G. W. Lugmair, *Science, 222,* 1013 (1983).

Libby, W. F., *Phys. Rev., 69,* 671 (1946).

Lima, F. W., M. B. A. Vasconcellos, M. J. A. Armelin, R. Fulfaro, V. J. Fulfaro, and B. B. B. Neves, *J. Radioanalyt. Chem., 71,* No. 1–2, 311 (1982).

Lowry, R. K., S. J. B. Reed, J. Nolan, P. Henderson and J. V. P. Long, *Earth and Planet. Sci. Lett., 53,* 36 (1981).

Luck, J., and C. J. Allègre, *Earth and Planet. Sci. Lett., 61,* 291 (1982).

Ludwig, K. R., J. T. Nash, and C. W. Naeser, *Econ. Geol., 76,* 89 (1981).

Lugmair, G. W., T. Shimamura, R. S. Lewis, and E. Anders, *Science, 222,* 1015 (1983).

Lupton, J. E., and H. Craig, *Science, 214,* 13 (1981).

Marti, K., *Earth and Planet. Sci. Lett., 62,* 116 (1983).

McCulloch, M. T., R. T. Gregory, G. J. Wasserburg, and H. P. Taylor, Jr., *J. Geophys. Res., 86,* B4, 2721 (1981).

McMullen, C. C., and H. G. Thode, in *Mass Spectrometry*, ed. C. A. McDowell, McGraw-Hill, New York (1963), p. 422.

Melton, C. E., and A. A. Giardini, *Geophys. Res. Lett., 7*, No. 6, 461 (1980).

Melton, C. E., and A. A. Giardini, *Geophys. Res. Lett., 9*, No. 5, 579 (1982).

Minster, J. P., and C. J. Allègre, *Earth and Planet. Sci. Lett., 56*, 89 (1981).

Morand, P., and C. J. Allègre, *Earth and Planet. Sci. Lett., 63*, 167 (1983).

Moreira-Nordemann, L. M., *Geochim. et Cosmochim. Acta, 44*, 103 (1980).

Morley, J. L., and J. D. Hays, *Earth and Planet. Sci. Lett., 53*, 279 (1981).

Muller, R. A., *Science, 146*, 489 (1977).

Nagy, K. L., and E. M. Parmentier, *Earth and Planet. Sci. Lett., 59*, 1 (1982).

Nier, A. O., *J. Am. Chem. Soc., 60*, 1571 (1938).

Nier, A. O., *Phys. Rev., 55*, 150 (1939).

Nier, A. O., *Phys. Rev., 77*, 789 (1950).

Nier, A. O., *Am. Scientist, 54*, 368 (1966).

Nier, A. O., *Ann. Rev. Earth Planet, Sci., 9*, 1 (1981).

Nier, A. O., and E. A. Gulbransen, *J. Am. Chem. Soc., 61*, 697 (1939).

Nishiizumi, K., J. R. Arnold, D. Elmore, X. Ma, D. Newman, and H. E. Gove, *Earth and Planet. Sci. Lett., 62*, 407 (1983).

Nishimura, H., and J. Okano, in *Advances in Mass Spectrometry*, Vol. 8A, ed. A. Quayle, Heyden and Son Ltd, London (1980), p. 513.

Nohda, S., and G. J. Wasserburg, *Earth and Planet. Sci. Lett., 52*, 264 (1981).

Nunes, P. D., *Geochim. et Cosmochim. Acta, 45*, 1961 (1981).

O'Brien, K., *J. Geophys. Res., 84*, 423 (1979).

Oliver, L. L., R. V. Ballad, J. F. Richardson, and O. K. Manuel, *J. Inorg. Nucl. Chem., 43*, 2207 (1981).

Olsson, I. W., *Geologiska Föreningens i Stockholm Förhandlinger, 103*, No. 3, 522 (1981).

Ozima, M., and S. Zashu, *Science, 219*, 1067 (1983).

Patchett, P. J., and M. Tatsumoto, *Contrib. Min. Petrol., 75*, 263 (1980).

Patchett, P. J., N. H. Gale, R. Goodwin, and M. J. Humm, *J. Geol. Soc. (London), 137*, 649 (1980).

Patchett, P. J., *Geochim. et Cosmochim. Acta, 47*, 81 (1983).

Peterman, Z. E., *Geol. Soc. Am. Bull., 92*, 139 (1981).

Piepgras, D. J., and G. J. Wasserburg, *Science, 217*, 207 (1982).

Raisbeck, G. M., F. Yiou, M. Fruneau, J. M. Loiseaux, M. Lieuvin, J. C. Ravel, and J. D. Hays, *Geophys. Res. Lett., 6*, 717 (1979a).

Raisbeck, G. M., F. Yiou, and C. Stephan, *J. Phys. Lett., 40*, 241 (1979b).

Rajan, R. S., L. Brown, F. Tera, D. J. Whitford, *Earth and Planet. Sci. Lett., 51*, 41 (1980).

Ralph, E. K., and H. N. Michael, *Am. Scientist, 62*, 553 (1974).

Reynolds, J. H., *Phys. Rev. Lett., 4*, 8 (1960).

Richardson, S. H., S. R. Hart, and H. Staudigel, *J. Geophys. Res., 85*, B12, 7195 (1980).

Richardson, S. H., J. J. Gurney, A. J. Erlank, and J. W. Harris, *Nature, 310*, 198 (1984).

Robson, J. M., *Phys. Rev., 100*, 933 (1955).

Rucklidge, J. C., N. M. Evensen, M. P. Gorton, R. P. Beukens, L. R. Kilius, H. W. Lee, A. E. Litherland, D. Elmore, H. E. Gove, and K. H. Purser, *Nucl. Instru. and Methods, 103*, 437 (1972).

Rucklidge, J. C., N. M. Evensen, M. P. Gorton, R. P. Beukens, H. W. Lee, A. E. Litherland, D. Elmore, H. E. Gove, and K. H. Purser, *Nucl. Instr. and Methods, 191*, 1 (1981).

Rucklidge, J. C., M. P. Gorton, G. C. Wilson, L. R. Kilius, A. E. Litherland, D. Elmore, and H. E. Gove, *Can. Minerol.,20*, 111 (1982).

Schramm, D. N., *Phys. Today, 36*, No. 4, 27 (1983).

Segrè, E., *Phys. Rev., 86*, 21 (1952).

Shackelton, N., and A. Boersma, *J. Geol. Soc. (London), 138*, 153 (1981).

Shields, W. R., *Nat'l Bureau of Standards, Report No. 8*, Standards Meeting of the Advisory Comm. (1960).

Shimamura, T., and G. W. Lugmair; *Earth and Planet. Sci. Lett., 63*, 177 (1983).

Silverman, S. R., in *Isotopic and Cosmic Chemistry*, ed's. H. Craig, S. L. Miller, and G. J. Wasserburg, North Holland, Amsterdam (1964), p. 95.

Slater, N. B., *Trans. Roy. Soc. (London), A246*, 57 (1953).

Smith, S. P., and J. H. Reynolds, *Earth and Planet. Sci. Lett., 54*, 236 (1981).

Somayajulu, B. L. K., *Geochim. et Cosmochim. Acta, 41*, 909 (1977).

Stukas, V. J., and C. S. Wong, *Science, 211*, 1424 (1981).

Tarling, D. H., *J. Geol. Soc. (London), 137*, 459 (1980).

Tera, F., *Geochim. et Cosmochim. Acta, 45*, 1439 (1981).

Thiemans, M. H., and R. N. Clayton, *Earth and Planet. Sci. Lett., 55*, 363 (1981).

Thode, H. G., J. Macnamara, and W. H. Fleming, *Geochim. et Cosmochim. Acta, 3*, 253 (1953).

Thomas, G. J., Jr., and P. Murphy, *Science, 188*, 1208 (1975).

Thomas, J., P. Parker, G. Herzog, and D. Pal, *Nucl. Instru. and Methods, 211*, 511 (1983).

Thomsen, L., *J. Geophys. Res., 85*, 4374 (1980).

Unterweger, M. P., B. M. Coursey, F. J. Schima, and W. B. Mann, *Int. J. Appl. Rad. and Isot., 31*, No. 10, 611 (1980).

Urey, H. C., *J. Chem. Soc.*, 562 (1947).

Valley, J. W., and J. R. O'Neil, *Geochim. et Cosmochim. Acta, 45*, 411 (1981).

Villa, I. M., J. C. Huneke, and G. J. Wasserburg, *Earth and Planet. Sci. Lett., 63*, 1 (1983).

Wasserburg, G. J., S. B. Jacobsen, D. J. DePaolo, M. T. McCullock, and T. Wen, *Geochim. et Cosmochim. Acta, 45*, 2311 (1981).

Wefer, G., and W. H. Berger, *Science, 209*, 803 (1980).

York, D., C. M. Hall, Y. Yanase, J. A. Hanes, and W. J. Kenyon, *Geophys. Res. Lett., 8*, No. 11, 1136 (1981).

Zaikowski, A., *Earth and Planet. Sci. Lett., 47*, 211 (1980).

Zindler, A., S. R. Hart, and C. Brooks, *Earth and Planet. Sci. Lett., 54*, 217 (1981).

ATMOSPHERIC, LUNAR, AND PLANETARY MEASUREMENTS

The development of the modern mass spectrometer and the exploration of space are inseparably linked. In a recent review of instruments for measuring neutral species, Niemann and Kasprzak (1983) have pointed out that virtually all types of mass spectrometers (i.e., radiofrequency, time-of-flight [TOF], magnetic sector, and double focusing) have been used in flight and space applications. Except for a few special cases, such as the Viking GC/MS soil analyzer and the TOF to analyze dust particles in Halley's comet, the mass spectrometers have been specifically designed to determine the elemental and chemical composition of the atmosphere or to measure the mass and energy distribution of ionized gaseous species. Balsiger (1983) and Whalen (1983) have discussed the recent developments in energetic ion mass spectrometry (EIMS), particularly for energies below 100 keV. They have also described applications of EIMS in the characterization of the magnetosphere and in space plasma physics.

The instrument type, configuration, and its operating characteristics depend upon the anticipated gas composition and other related parameters. Weight and power allocations are often determined by mission objectives and vehicle constraints, which at times have limited the capabilities of the instrument. Nonperturbed representative sampling of the ambient atmosphere is usually the most critical factor for the mass spectrometer system regardless of the type used. The instruments discussed in the following sections are not meant to be a comprehensive review of flight mass spectrometry. Rather, they have been selected as important examples of applications and mass spectrometers that have been developed for determining the composition of the terrestrial and planetary atmospheres.

STRUCTURE OF THE EARTH'S ATMOSPHERE

The atmosphere is divided into a series of regions and subregions (Table 9.1) according to predominant physical and chemical properties, which also act to control the composition. In the near-earth atmosphere, the major divisions are based on physical properties, usually the mean variation of temperature with altitude. The boundaries of these regions, termed *pauses*, are not stationary, but vary in height according to latitude (being somewhat higher at the equator than at the poles), time of year, and solar activity.

TABLE 9.1 Typical Physical Properties of the Terrestrial Atmosphere

Regions	Altitude (km)	Temperature (K)	Pressure (torr)	No. Density (n /cm^3)	Mean Molecular Weight	Research Vehicles
	600	1000	2.1×10^{-10}	2.0×10^6	11.51	
	400	990	2.6×10^{-9}	2.0×10^8	15.98	
	300	976	1.4×10^{-8}	6.5×10^8	17.73	
	250	941	1.6×10^{-7}	1.9×10^9	19.19	
	200	854	8.4×10^{-7}	7.2×10^9	21.30	
	150	634	2.7×10^{-6}	4.9×10^{10}	24.10	
	140	520	4.6×10^{-6}	9.3×10^{10}	24.75	
	130	420	8.5×10^{-6}	2.0×10^{11}	25.44	
	120	355	2.0×10^{-5}	5.4×10^{11}	27.27	
	110	265	5.8×10^{-5}	2.1×10^{12}	27.90	
	100	210	2.4×10^{-4}	1.1×10^{13}	28.40	
	95	193	6.4×10^{-4}	3.2×10^{13}	28.60	
	90	176	1.4×10^{-3}	7.6×10^{13}	28.77	
	85	160	2.9×10^{-3}	1.9×10^{14}	28.88	
	80	177	7.9×10^{-3}	4.2×10^{14}	28.96	
	75	194	1.0×10^{-2}	9.6×10^{14}		
	70	211	4.4×10^{-2}	2.0×10^{15}		
	65	232	9.4×10^{-2}	3.9×10^{15}		
	60	253	1.9×10^{-1}	7.2×10^{15}		
	55	273	3.6×10^{-1}	1.3×10^{16}		
	50	274	6.6×10^{-1}	2.3×10^{16}		
	45	274	1.2×10^0	4.3×10^{16}		
	40	268	2.2×10^0	8.1×10^{16}		
	35	252	4.3×10^0	1.7×10^{17}		
	30	235	8.6×10^0	3.6×10^{17}		
	25	227	1.8×10^1	7.7×10^{17}		
	20	219	3.9×10^1	1.7×10^{18}		
	15	211	8.5×10^1	3.9×10^{18}		
	10	231	1.8×10^2	7.7×10^{18}		
	5	266	3.7×10^2	1.3×10^{19}		
	0	291	7.6×10^2	2.5×10^{19}	28.96	

Regions (left margin labels): Heterosphere (diffusive mixing), Homosphere (turbulent mixing), Ionosphere, Ozonosphere, Exosphere, Thermosphere, Mesosphere, Stratosphere, Troposphere

Research Vehicles (right margin labels): Satellite, Sounding rocket, Balloons, sounding rockets

TABLE 9.2 Composition of the Atmosphere[a]

Gaseous Species	Volumetric Mixing Ratio	Gaseous Species	Volumetric Mixing Ratio
Nitrogen (N_2)	7.8×10^{-1}	Helium (He)	5.24×10^{-6}
Oxygen (O_2)	2.09×10^{-1}	Krypton (Kr)	1.14×10^{-6}
Water vapor (H_2O)	4.00×10^{-2}	Hydrocarbons (HC)	1.01×10^{-6}
Argon (Ar)	9.34×10^{-3}	Hydrogen (H_2)	5.00×10^{-7}
Carbon dioxide (CO_2)	3.30×10^{-4}	Carbon monoxide (CO)	1.00×10^{-7}
Neon (Ne)	1.82×10^{-5}	Xenon (Xe)	8.70×10^{-8}

[a] The values for H_2O, CO_2, CO, and hydrocarbons will vary with local conditions, values given for N_2, O_2, and Ar are for dry air.

With the exception of water vapor, which becomes negligible at 10 km, the composition remains essentially constant to an altitude of 90–100 km (Table 9.2). This region includes the *troposphere, stratosphere, mesosphere*, and lower *thermosphere*, and is collectively called the *homosphere*. In the *troposphere* and *mesosphere*, an inverse temperature relationship with altitude results in considerable vertical mixing as the warm air moves upward to the cooler regions, and the composition remains constant despite the reduced pressure. The *stratosphere* is characterized by horizontal transport with little vertical mixing. Substantial quantities of aerosols and other particles can therefore persist in this region longer than 1 year. Above 100 km, in the *heterosphere*, radiatively induced dissociation occurs, molecular diffusion becomes significant, and the concentration of the lighter species increases by gravitational separation until at about 800 km helium and hydrogen predominate. The altitude for this transition to the lighter species is strongly influenced by solar activity. Temperatures at the *thermopause*, which marks the onset of an isothermal profile, can vary from 500 K to as much as 2000 K as the solar activity increases (Wallace and Hobbs, 1977), which is reflected in the differences in tabulated values of temperature and composition of the region. Above 400–500 km, in the *exosphere*, the mean free path is so long that the molecules move more or less independently and, hence, have a temperature of their own.

In addition to thermally controlled properties, important regions are also identified by chemical or ionization characteristics. The solar constant (the energy radiated by the sun) has a contemporary value of 1.37 ± 0.02 kW/m^2 (Evans, 1982). Atmospheric absorption excludes all wavelengths below 2900 Å from the earth's surface (Fig. 9.1). Below 2424 Å ultraviolet radiation photodissociates molecular oxygen into atoms, which oxidize other oxygen molecules in a reversible three-body collision to form ozone (O_3). This forms a layer called the *ozonosphere*, between about 20–50 km, in which the ozone

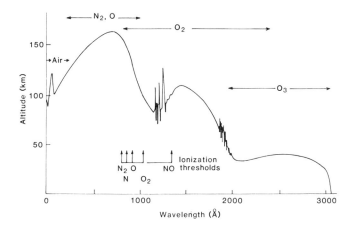

Figure 9.1 Absorption of solar radiation in the earth's atmosphere. (Evans, 1982; copyright 1982 by the AAAS).

levels are relatively high, having a maximum of about 10 ppm near 25 km (Fig. 9.2).

The population of charged species becomes significant in the *ionosphere*. The D region (60–90 km) contains both positive and negative ions (e.g., NO_3^-, CO_3^-, and CO_4^-), because the relatively high neutral gas density

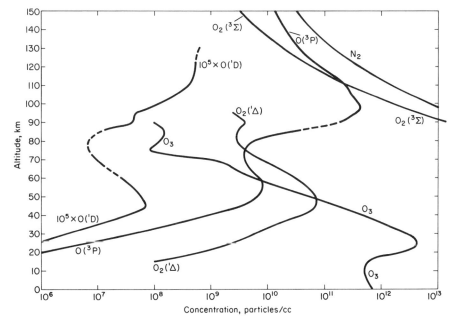

Figure 9.2 Semilog plots of typical nitrogen, oxygen, and ozone concentrations with altitude. (Heicklen, 1976; courtesy J. Heicklen, copyright Academic Press.)

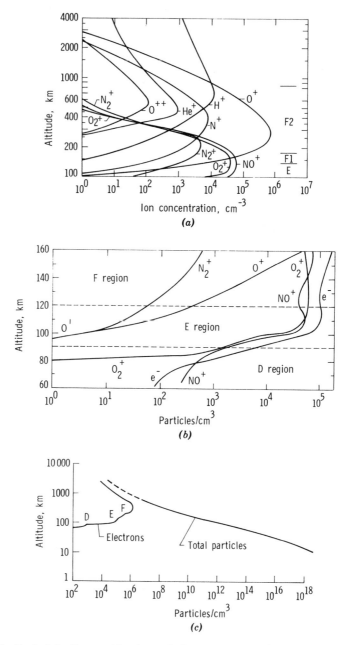

Figure 9.3 Typical daytime positive ion and electron concentrations in the ionosphere (*a*, Hansen and Carlson, 1977; *b*, Heicklen, 1976; *c*, Wallace and Hobbs, 1977; *a*, copyright National Academy of Sciences; *b,c*, copyright Academic Press.)

allows rapid three-body attachment of electrons to molecular oxygen forming O_2^-, which subsequently reacts with the neutral gases (Sechrist, 1977). Ionization results from galactic cosmic rays in the lower D region and from ultraviolet photoionization of nitric oxide and metastable oxygen in the upper D region. In the E (90–140 km) and F (above 140 km) regions, ionization of the major constituents by solar x-ray and extreme ultraviolet radiation predominates. Typical daytime ion and electron populations for the lower and upper ionosphere are shown in Figure 9.3. Molecular ions predominate in the F1 region and below, while above that in the F2 region atomic oxygen ions predominate. Since the ion population depends upon photoionization, it is highly dependent upon solar zenith angle, that is, time of day. As an example, the number of O_2^+ ions at 160 km may range from a daytime value of about 6×10^4 ions/cm^3 to 80 ions/cm^3 at night, so that large temporal variations may occur (Heicklen, 1976). The high abundance of the NO^+ ion in the F1 region was unanticipated prior to the first rocket-borne mass spectrometer measurements. Subsequent measurements with satellite mass spectrometers revealed that a corrrelation exists between the drift velocity and NO^+ enhancement (Brinton, 1974; Hansen and Carlson, 1977).

Beyond the upper reaches of the ionosphere, the mean free path becomes exceedingly long and the motions of the ions are controlled by the earth's magnetic field. This region is called the *magnetosphere* and its boundary is the *magnetopause*. Because of the complex interactions with the solar wind, the magnetosphere extends to about 10 earth radii (1 R_e = 6378 km or 3963 miles) on the sunward side and to more than 100 R_e on the downwind side, or past the orbit of the moon (Evans, 1982). The ionic composition of the *magnetosphere* and the radiation belts within are thought to result from (1) solar radiation (protons, electrons, ionized helium, etc.); and (2) from the transport of ions from the ionosphere, although the mechanisms are not yet fully understood.

Each of these regions of the atmosphere thus provides a different set of measurement conditions and problems. In the *troposphere* and *stratosphere*, mass spectrometric measurements of neutral species are made with instruments on balloons and sounding rockets. In the ionosphere and magnetosphere, charged energetic ions are also measured on rockets and satellites with spectrometers that are energy as well as mass-selective. The intervening region that is too high for balloons and too low for satellites (55–150 km) has thus far been explored primarily with sounding rockets and, to a lesser extent, with the reentering space shuttle. Consequently, less is known about the composition and atmospheric chemistry of this transitional region than at the altitudes where measurements are made over extended periods.

BALLOONS AND SOUNDING ROCKETS

The first measurements in the atmosphere were made with instruments carried aloft by pioneering eighteenth century balloonists. Because the modern balloon can reach the upper level of the stratosphere and remain aloft for extended periods, balloon-borne experiments continue to provide important data relative to atmospheric composition and chemistry. In a typical experiment, the mass spectrometer inlet is opened when the balloon reaches a desired altitude in the stratosphere, where it remains because of the minimal vertical movement. After several hours, a gas release valve is opened and the balloon descends, with the mass spectrometer then obtaining data as a function of altitude.

Mass spectrometric measurement of neutral composition and the measurement of positively and negatively charged ions have contributed significantly to the understanding of atmospheric chemistry. Mauersberger (1983) has described a magnetic sector mass spectrometer in which the neutral gas particles are formed into a molecular beam and traverse the ion source without wall interactions. A flag mechanism intercepts the neutral beam to measure instrument background; with the flag retracted, both background and ambient gases are measured, and the difference is the net signal. Using peak switching with 8 s integration times, mixing ratios of less than 10 ppb are detectable. In addition to measuring vertical profiles of the major constituents from 20–40 km, Mauersberger also verified an enrichment in stratospheric heavy ozone ($^{16}O_2{}^{18}O$) that was predicted by Cicerone and McCrumb (1980) since photodissociation of O_2 is the dominant source of O_3 and the photodissociation rate for $^{16}O^{18}O$ is significantly greater than for $^{16}O_2$ above 40 km. These data (Fig. 9.4) showed a pronounced increase in the $^{18}O^{16}O_2 : {}^{16}O_3$ ratio, with the maximum enhancement reaching 40% at 32 km.

Balloon-borne, cryogenically pumped quadrupole mass spectrometers

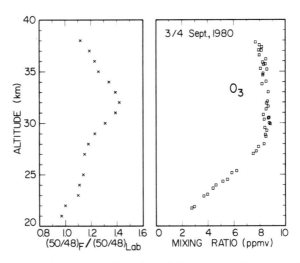

Figure 9.4 Stratospheric heavy ozone isotopic enrichment and mixing ratio. (Mauersberger, 1981; courtesy K. Mauersberger, copyright American Geophysical Union.)

TABLE 9.3   Positive Ions Measured at 35 km

Observed Peaks	Assigned Mass No.	Tentative Identification
18,20,22	19 ± 3	$H^+(H_2O)$
28	28 ± 2	$H^+ HCN$
34	33 ± 2	$H^+ CH_3OH$
37	37	$H^+(H_2O)_2$
42,44	42 ± 2	$H^+ CH_3CN$
50	51 ± 2	$H^+ CH_3OH(H_2O)$
55	55	$H^+(H_2O)_3$
60	60	$H^+ CH_3CN(H_2O)_2$
68	69 ± 2	$H^+ CH_3OH(H_2O)_2$
73	73	$H^+(H_2O)_4$
78	78	$H^+ CH_3CN(H_2O)_2$
82	82 ± 1	$H^+ Y(H_2O)_j$
86	87 ± 2	$H^+ CH_3OH(H_2O)$
91	91	$H^+(H_2O)_5$
96	96	$H^+ CH_3CN(H_2O)_3$
99	99 ± 1	$H^+ Y(H_2O)_{j+1}$

Source: Arijs et al. (1982).

(QMS) have been used to measure positive and negative ion clusters in the stratosphere (McCrumb and Arnold, 1981; Arnold et al., 1981a, 1981b; Arijs et al., 1982, 1983). Ions are sampled through a biased small-diameter orifice, which is dimensioned to maximize ion count rates while maintaining the internal pressure below 10^{-4} torr to minimize ion scattering. At the low-pressure side of the orifice, ions are focused into the mass filter with an electrostatic lens or extracted from the expanding gas beam by a homogeneous electric field between the orifice and a planar grid. In either case, dissociation of the ion clusters will occur, and the measured data must be corrected based on laboratory measurements.

Two groups of positive ions have been identified: (1) proton hydrates (PH) having the form $H^+(H_2O)_n$; and (2) nonproton hydrates (NPH) having the form $H^+ X_m(H_2O)_n$; and some of these are listed in Table 9.3. Arnold et al. (1978) proposed that NPH is formed by ion-molecule (proton transfer) reactions of PH with a trace gas having a mass of 41, a proton affinity greater than water (175 kcal/mol), and a number density on the order of 10^5 cm^{-3} (0.001 ppb) at an altitude of 36 km. Subsequent data have confirmed both these findings and that the NPH:PH abundance ratio decreases from about 90% at 25 km to 1% at 45 km. Several molecules have been proposed for X, including MgOH, but careful analysis of the data from higher resolution measurements suggest that X is most likely acetonitrile, CH_3CN (Arijs et al., 1983).

In a subsequent development, a QMS aboard a research aircraft was used

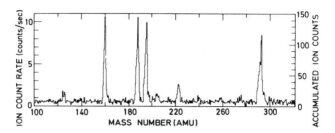

Figure 9.5 Typical nighttime spectrum for negative ions near 35 km altitude. (Arijs et al., 1981; copyright American Geophysical Union.)

to extend the measurement of ions to as low as 3.2 km in the upper troposphere (Arnold et al., 1984). Air sampled from above the boundary layer associated with the aircraft, which was flying at approximately 800 km/hr, passed into a low-pressure (10^{-6} torr) volume that was connected to the mass spectrometer by a small-diameter orifice. The data indicated the presence of massive cluster ions, with masses of approximately 540 amu being observed at 8.2 km, although the major ion species appeared below 240 amu. The concentration of heavy ions, which cannot be attributed to any single hydrated ion family, was observed to vary inversely as a function of altitude. They later measured trace levels of acetonitrile, acetone, and ammonia at parts-per-trillion levels with a method they called active chemical ionization mass spectrometry (Arnold and Hauck, 1985). Molecules possessing large proton affinities or gas phase basicities are detected by reactions with $H_3O^+(H_2O)$ ions formed in a glow-discharge 70 cm upstream of a sampling cone inlet to a quadrupole mass spectrometer.

Similar data for negative ions in the stratosphere were obtained by Arijs et al. (1981) using a microprocessor-controlled QMS with an electron multiplier detector operated in the pulse-counting mode. The instrument was programmed to measure neutral as well as positive and negative ions, and it was switched from high to low resolution during a flight. A typical negative ion scan is shown in Figure 9.5 and the peaks are identified in Table 9.4.

Trace levels of H_2SO_4 and HSO_3 vapors in the stratosphere lead to aerosol formation and control the composition of the stratospheric negative ions. The aerosols, thought to be droplets of a supercooled solution of 75% H_2SO_4 and 25% H_2O by weight, scatter solar radiation back to space, and therefore have the potential to affect the earth's climate. Stratospheric sulfur compounds originate primarily from carbonyl sulfide (COS) released at the earth's surface, which reaches the stratosphere by turbulent diffusion. It is subsequently photolyzed to form H_2SO_4 through a series of reactions that are not yet fully understood. A second major source of sulfur is large volcanic eruptions that periodically inject significant quantities of sulfur-bearing precursor gas—primarily SO_2—into the stratosphere. Arnold and his coworkers have described a method that they call passive chemi-ionization mass spectrometry (PACIMS), with which they have measured gas phase sulfuric acid

TABLE 9.4   Negative Ions Measured at 35 km

Mass No. (amu)	Identification	Abundance (%)
125	$NO_3^- \cdot HNO_3$	2
160	$HSO_4^- \cdot HNO_3$	16
178 ± 2	$HSO_4^- \cdot HNO_3 \cdot H_2O$	—
188	$NO_3^- \cdot 2HNO_3$	15
195	$HSO_4^- \cdot H_2SO_4$	14
206 ± 2	$NO_3^- \cdot 2HNO_3 \cdot H_2O$	—
213 ± 2	$HSO_4^- \cdot H_2SO_4 \cdot H_2O$	—
223 ± 1	$HSO_4^- \cdot 2HNO_3$	5
241 ± 2	$HSO_4^- \cdot 2HNO_3 \cdot H_2O$	—
258 ± 2	$HSO_4^- \cdot H_2SO_4 \cdot HNO_3$	—
276 ± 2	$HSO_4^- \cdot H_2SO_4 \cdot HNO_3 \cdot H_2O$	—
293 ± 1	$HSO_4^- \cdot 2H_2SO_4$	14
391 ± 2	$HSO_4^- \cdot 3H_2SO_4$	34

Source: Arijs et al. (1981).

in the stratosphere (Arnold and Fabian, 1980; Arnold et al., 1981a, 1981b, 1981c; Viggiano and Arnold, 1981). The method is particularly useful for the detection of sulfuric acid, since the ion processes are well known and the relevant kinetic data have been measured (Viggiano et al., 1980). The H_2SO_4 vapor reacts with $NO_3^- (HNO_3)_n$ ions to form $HSO_4^- (HNO_3)_n$, which then reacts to form ions of the general type $HSO_4^- (H_2SO_4)_x (HNO_3)_y$ that are eventually lost through ion-ion recombination. Data typical of that obtained in their experiments are shown in Figure 9.6, where the most distinct feature is the abrupt change in $[H_2SO_4]$ occurring at about 30 km. In other experiments, they noted a distinct increase in $[HSO_3 + H_2SO_4]$ in the 20–27 km region, which was attributed to the April 1982 eruption of El Chicon in Mexico and to other volcanic activity (Qui and Arnold, 1984). Hoffman and Cunningham (1985) have recently developed a balloon-borne system consisting of two cryogenically pumped magnetic sector mass spectrometers biased to measure positive and negative stratospheric ions, respectively. Each of the miniature magnetic sector instruments has a mass range of 700 amu with unit resolution at 500 amu, and a sensitivity sufficient to detect 1 ion/cm^3 for a 1 s accumulation time. The upper limit of the mass range can be extended to as high as 5000 amu by operating at very low acceleration voltages. In this mode, the unit resolution is reduced to about 50 amu; however, total concentrations and nominal mass distributions of the large cluster ions can be determined. The higher resolution and sensitivity of the magnetic sector at the higher masses, as compared to the previously flown QMS' are expected to greatly improve upon the quantitative analysis of the trace level ions and ion clusters.

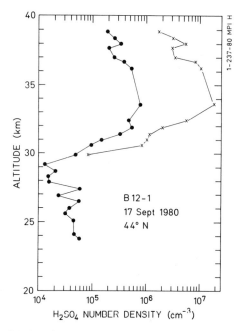

Figure 9.6 Corrected ($\times$) and uncorrected ($\bullet$) H_2SO_4 vapor concentration as a function of altitude. (Viggiano and Arnold, 1981; courtesy of A. Viggiano; copyright American Geophysical Union.)

In some instances, recoverable sampling systems transported on balloons or ejected from rockets have been used and, in fact, predate real-time in situ methods by several years. For example, Bieri et al. (1970) described an experiment in which 9.3 mol of air were collected over an altitude of 43–63 km in a liquid hydrogen-cooled sample volume through a supersonic diffuser inlet. After ejection and recovery of the volume, the concentrations and isotopic abundances of neon, argon, and krypton were determined by isotopic dilution mass spectrometry to be identical to those in dry surface air within limits of experimental error. This confirmed both an absence of gravitational effects at these altitudes and the homogeneous mixing of the atmosphere up to the lower mesosphere, in contrast to earlier measurements and predictions. More recently, Fabian et al. (1985) have measured the vertical distribution of Freon-22 ($CHClF_2$) in the stratosphere from 30 L-STP samples collected between 10–33 km altitude in a balloon-borne liquid neon (27.1 K) cryogenic sampler. Subsequently recovered and analyzed by GC/MS, the slowly decreasing mixing ratios with altitude indicated that Freon-22 may indeed become an important source of ClO_x in the upper stratosphere.

Above 45–50 km, data are obtained with mass spectrometers carried on sounding rockets. The sounding rocket presents several unique sampling problems for the mass spectrometer: (1) measurements are made while as-

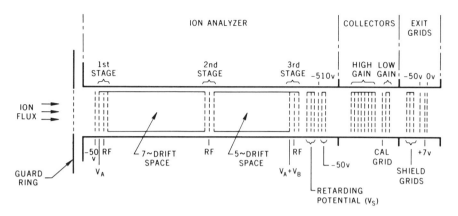

Figure 9.7 Typical Bennett-RF mass spectrometer for measuring ambient ion concentrations in spaceflight applications. (Brinton et al., 1973; courtesy M. W. Pharo; copyright American Geophysical Union.)

cending or descending, so that the experimental conditions may be continuously changing; (2) vehicle charging must be suppressed when measuring charged species; (3) vehicle outgassing must be accounted for; and (4) the sample inlet must be designed to obtain the sample at supersonic velocity.

Among the earliest in situ investigations of the ion and neutral composition were those of Johnson, Meadows, and Townsend using a modified Bennett-RF mass spectrometer (Fig. 9.7), which is an instrument also used extensively in the Soviet Union (Townsend, 1952; Johnson and Meadows, 1955). The instrument is well suited for upper atmospheric research because of its light weight, high-current efficiency, and large sampling area; and it is still being used today in some applications, most notably the Venera, Pioneer-Venus, and Galileo spaceprobes. Small magnetic sector and quadrupole mass spectrometers have largely replaced the RF spectrometer in the measurement of atmospheric composition. For example, Nier et al. (1964a, 1964b, 1971) measured the atmospheric composition above 100 km with small magnetic mass spectrometers, with which they were able to establish that a significant dissociation of molecular oxygen occurs. A classical experiment illustrative of these measurements was that of Narcisi and Bailey (1965), who measured the ion concentration in the D layer of the ionosphere using a small QMS. More recently, they have measured both positive and negative ion composition in the D and E regions during a solar eclipse on February 26, 1979 by using two cryogenically pumped quadrupole mass spectrometers in series, with the first being operated in the high-pass filter mode as a prefocus device and the second in either the mass analysis or high-pass mode (Narcisi et al., 1983). The gases were sampled through a shock-attaching conical sampler having a 1 mm diameter ion entrance orifice at the cone apex to avoid the thermodynamic break-up of weakly bound cluster ions that would occur if the shock were detached.

SATELLITES

At altitudes beyond 150 km, mass spectrometers carried on orbiting satellites have contributed significantly to the understanding of the composition and chemistry of the upper atmosphere. Time- and position-dependent variations have been observed over extended periods; and in the magnetosphere, the effects of geomagnetic activity on ionic composition have been detected (Hoffman, 1970).

Aside from vehicular constraints, the type of mass spectrometer used (e.g., scanning or nonscanning, single or double focusing, open or closed source, etc.) is determined by the measurements to be made and the dynamic range required. The major design difficulty lies in the inlet-ion source system, which must accurately produce ionized samples representative of the ambient atmosphere, and which may be required to operate with widely differing pressure, temperature, and composition during an elliptical orbit in which the apogee and perigee may differ by several hundred kilometers or more. Nier et al. (1973), for example, have pointed out that the variations in the number densities of oxygen, nitrogen, and helium are of the order of 10^8, 10^7, and 10^1, respectively, for an orbit with a 600 km apogee and 150 km perigee. The open ion source mass spectrometer flown on the Atmospheric Explorer satellites therefore used both an electrometer and an electron multiplier to provide a dynamic range of about 10^8 for the higher masses (Hayden et al., 1974).

In the lower thermosphere, neutral species are measured with mass spectrometers having open or closed ion sources, depending upon the information to be obtained. Open source instruments minimize gas-surface interactions of the ambient neutral beam. However, the measurements are somewhat difficult to accurately relate to atmospheric parameters because of both the apparent high velocity (typically 8.5 km/s) of the incoming gas and the difficulty in calibrating the instrument under simulated flight conditions. The closed source, in which the atmosphere is generally sampled through a knife-edged orifice into a spherical chamber, may be more easily calibrated. In the closed source, the gases reach thermal equilibrium with the source walls prior to ionization, so that the spread in ion energy is reduced, but recombinations or gas-surface interactions may be difficult to determine with precision. The composition and density of chemically inert gases can be estimated from measured conditions within the closed ion source chamber if the pressure drop across the orifice and the source pumping speed have been accurately characterized and the satellite velocity, angle of attack, and ambient gas temperature can be determined (Nier et al., 1973; Trinks and von Zahn, 1975). Spencer et al. (1973) showed that the ambient gas temperature could also be determined with a mass spectrometer from measurement of the velocity distribution of molecular nitrogen. The distribution is determined from the gas density in a spherical chamber coupled to the atmosphere through a knife-edged orifice. The temperature is computed from free diffusion of the gases between the atmosphere and the chamber as defined by kinetic theory. Measurement of reactive species, particularly atomic oxygen, is difficult with either source because the reaction mechanisms with and on the surface are not simple nor are they totally predictable or understood. After conditioning of the surfaces in the closed ion source, the atomic oxygen is measured after recombination as molecular oxygen at mass 32. Ambient molecular and atomic oxygen are indistinguishable. Above 200 km, however, the contribution from molecular oxygen becomes negligible and the number density of atomic oxygen can be determined with some confidence. Nier et al. (1974) have described a "fly-through" method for distinguishing gas particles that have not contacted the surface prior to ionization and those that have reached thermal equilibrium with the surface of the open ion source, based on the measurable energy differences between the two. An appropriate retarding potential applied to the draw-out and focusing grids of a double focusing, open source mass spectrometer exclude the thermally accommodated gas particles from the analyzer. The method can also be used to measure ambient molecular and atomic oxygen because of the apparent energy differences between the two in the spacecraft frame of reference. At a satellite velocity of 8.5 km/s, the gas particles have an equivalent energy of about 0.37 eV/amu; for example, atomic and molecular oxygen have energies of 5.9 eV and 11.8 eV, respectively. Philbrick et al. (1972) analyzed neutral species ionized by electron-impact using these same energy differences to obtain mass separation, which is a method later also used on some planetary probes. Data obtained with the open and closed source spectrometers are often complementary, and both are often included on the same mission (e.g., Nier et al., 1973; Pelz et al., 1973; Philbrick, 1974; Trinks and von Zahn, 1975).

Atmospheric models based on satellite drag and acceleration data alone do not accurately represent the global distribution of variations in density due to geomagnetic disturbances, but they are generally adequate for most applications involving orbital mechanics (Slowey, 1984). Measurements with mass spectrometers have led to more realistic, detailed, and accurate models of neutral composition. These models, such as the one developed by Jacchia et al. (1977) and the more recent revision of the mass spectrometric and incoherent scatter (MSIS) model of Hedin (1983), also include temporal and geopositional variations in temperature and density. The MSIS-83 model, for example, is comprised of data obtained from seven satellites, five incoherent scatter (IS) radars, and a comprehensive summary of rocket probe data taken above 85 km during a wide range of solar activities. Where temperatures were not measured directly, they were derived by the method of Spencer et al. (1973) from the measurement of nitrogen density. Representative annual and diurnal temperature and density profiles derived from MSIS-83 are shown in Figure 9.8. Forbes and Champion (1982) have also recently published a comprehensive review of theoretical and observed tidal oscillations of temperature, density, and composition in the thermosphere

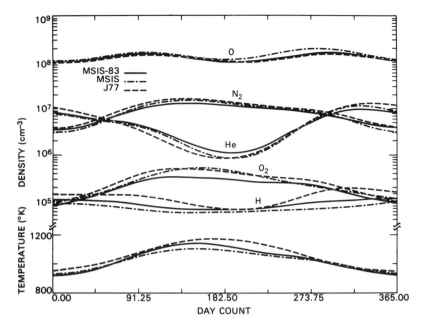

Figure 9.8 Average temperature and density variations at 400 km and 45° latitude. (Hedin, 1983; courtesy A. E. Hedin; published by the American Geophysical Union.)

above 100 km, and they have discussed the effects of wind-induced diffusion and other transport mechanisms. Order-of-magnitude higher levels of helium concentrations in the winter hemisphere ("winter helium bulge") were measured in situ with a closed source mass spectrometer on a satellite by Reber et al. (1971); and a "summer argon bulge" of similar magnitude was first observed by von Zahn et al. (1973) with a closed source monopole.

In the upper ionosphere and magnetosphere, charged species become dominant, and mass spectrometers of various types have been used in determining the ionic composition and its variations (e.g., Hoffman et al., 1973a, 1974; Brinton et al., 1973; Coplan et al., 1978; Chappell et al., 1981; Reasoner et al., 1982; Béghin et al., 1983). In the polar regions, atomic hydrogen ions (H^+) were observed to be streaming outward at velocities as high as 15 km/s, constituting a *polar wind*, as reported from ion mass spectrometer measurements in the ionosphere (Banks et al., 1968; Hoffman, 1970). Chappell et al. (1982) have also measured N^+ and N^{2+} in the magnetosphere, with the N^+ seen to be flowing from the northern polar cap to altitudes reaching 3 R_e. The existence of the polar wind as a major source of ions in the magnetosphere was recently confirmed by measurements with ion mass spectrometers on the Dynamic Explorer-A satellite (Chappell et al., 1981) and on the International Sun-Earth Explorer (ISEE-1) at distances as far as 23 R_e from the earth (Sharp et al., 1981). The ionic species in the

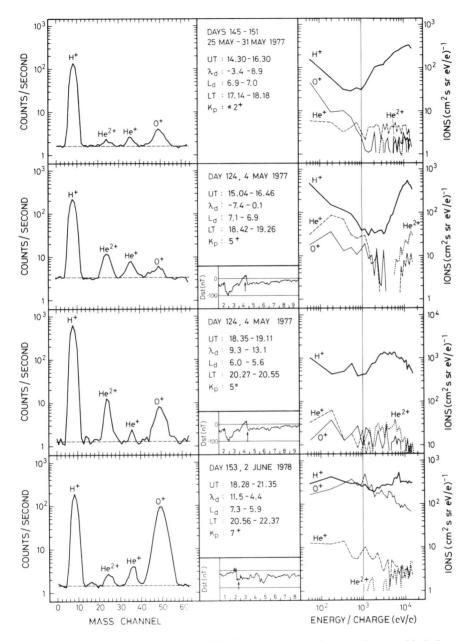

Figure 9.9 Ion concentrations measured in the magnetosphere in geostationary orbit during a progressively intense magnetic storm (Balsiger et al., 1980; courtesy H. Balsiger; copyright American Geophysical Union.)

magnetosphere are now thought to be source-specific, that is, most of the O^+ and He^+ originate from the terrestrial atmosphere, while the He^{2+} originates in the solar wind. The H^+ component, on the other hand, can originate in either region (Sharp et al., 1982). The ionic composition of the distant plasma sheet has also been determined to be dependent on magnetic activity (e.g., Kaye et al., 1981; Strangeway and Johnson, 1983). Balsiger et al. (1980) have reported that H^+ is normally the dominant species, but that heavier ions increase and may even become dominant in the outer magnetosphere during periods of magnetic storms (Fig. 9.9). The data also suggest that a source of He^+ and O_2^+ is the high-altitude thermal plasma.

LUNAR EXPLORATION

The Apollo lunar-exploration missions included mass spectrometers to measure atmospheric composition as part of both the Apollo 15 and 16 orbiter (LOSEP) and Apollo 17 surface (ALSEP) science experiment packages. In addition to determining the source of the lunar atmosphere, such measurements provided a means of studying large-scale, low-pressure transport processes in planetary exospheres uncomplicated by the hydrodynamic wind systems that limit such studies on earth.

The lunar-orbiter mass spectrometer (LOMS) measured concentrations in a 60 mile lunar orbit and in transearth coast, during which over 2000 h of data were obtained. The instrument was a small dual-collector, magnetic sector mass spectrometer in which second-order focusing was achieved by circular exit-field boundaries (Hoffman, 1972). The LOMS was mounted on a boom that was extended 7.5 m from the command and service module (CSM) to minimize the effects of spacecraft outgassing in the spectra. With the CSM oriented in the $-x$ direction, ambient species entered a scoop at ram velocity and were transported into a plenum that contained the electron-impact ionization source. Reversing orientation of the CSM to the $+x$ direction placed the inlet in the wake, thus excluding ambient species and providing a measure of instrument background. Spectra obtained in lunar orbit showed peaks occurring at every mass, with water vapor saturating the counting system. This would imply a large outgassing cloud co-orbiting at the same altitude as the CSM, however, a collisionless stable cloud in lunar orbit would require 10^5 tons of H_2O at 1 K to produce the measured levels. Measurements during transearth coast determined that molecules outgassing from the CSM formed a collisionless cloud 1.25 m from the surface, which led to the unexpected implication that orbital mechanics predominated in maintaining the high contaminant levels in lunar orbit. The measured species therefore resulted from vaporization of matter ejected from the spacecraft with small relative velocities that remained in nearby orbits of their own (Hoffman et al., 1972a). Neon thought to be of lunar

TABLE 9.5   Gas Concentrations and Solar Wind Fluxes on the Lunar Surface (mol·cm^{-3})

Parent Substance	Sunrise Concentration	Element	Solar Wind Flux[a] (cm^{-2}·s^{-1})
CH$_4$	$(1.2 \pm 0.4) \times 10^3$	C	1.6×10^6
NH$_3$	$\left(\begin{smallmatrix}4+4\\-2\end{smallmatrix}\right) \times 10^2$	N	2.8×10^5
H$_2$O	$(6 \pm 6) \times 10^2$	^{36}Ar	8×10^2
N$_2$	$(8 \pm 8) \times 10^2$		
CO	$\left(\begin{smallmatrix}0+16\\-0\end{smallmatrix}\right) \times 10^2$		
O$_2$	$(1 \pm 1) \times 10^3$		
CO$_2$	$(1.4 \pm 1.0) \times 10^3$		
^{36}Ar	$(1.6 \pm 0.4) \times 10^3$		

Source: Hoffman and Hodges (1975).
Instrument sensitivity; 1 count = 200 mol·cm^{-3}.
[a] Based on H flux of 3×10^8 cm^{-2}·s^{-1}.

origin was also measured in concentrations of 8000 atoms/cm^3 by the LOMS on Apollo 16.

The lunar-surface atmospheric composition experiment mass spectrometer (LACE) was also a magnetic sector instrument having a mass range of 1–110 amu. Portions of the mass range were simultaneously measured with three electron multiplier detectors with a sensitivity in which 1 count/s was equivalent to 200 molecules/cm^3 (Hoffman et al., 1972b). The LACE was deployed on the lunar surface with the entrance aperture oriented to intercept the downward flux of gases. Data were obtained mostly at night because of the relatively high (10^7 molecules/cm^3) daytime concentrations, which also included outgassing from the ALSEP site. The tenuous (5×10^5 molecules/cm^3) lunar atmosphere consists primarily of He, Ne, and Ar, which except for ^{40}Ar originates in the solar wind (Hoffman and Hodges, 1975). Noncondensable gases are distributed approximately as $T^{-5/2}$ in the lunar atmosphere; and since the daytime-to-nighttime temperature varies by a factor of 4, nighttime concentrations are about 30 times higher. A condensable gas, on the other hand (e.g., ^{40}Ar, CH$_4$, NH$_3$), shows distribution minima in both day and night and a maximum associated with the sunrise terminator following a predawn enhancement. Concentrations of condensable gases measured with the LACE are listed in Table 9.5, where the concentration of primordial ^{36}Ar is shown for comparison. The radiogenic ^{40}Ar exhibited a marked predawn enhancement peaking at about 1.8×10^4 molecules/cm^3 at 1 or 2° past the terminator crossing Fig. 9.10. The m/z 16 continued to increase past sunrise until masked by outgassing from the ALSEP site. Nevertheless, the data provide evidence of the existence of small amounts

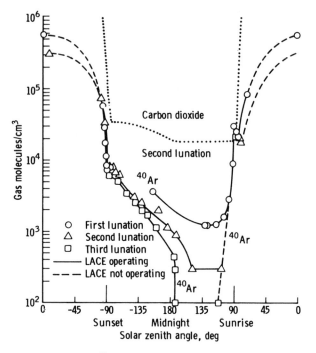

Figure 9.10 Diurnal variation of ^{40}Ar on the lunar surface. Daytime increases are due to out-gassing at the ALSEP site (Hoffman et al., 1973b).

of methane and ammonia and suggest that higher concentrations may exist at the poles.

PLANETARY EXPLORATION

The analysis of the atmospheric chemistry of the planets poses a severe set of operating conditions for the instrumentation. Despite this, important new scientific knowledge has been gained from the recent missions to Venus and Mars. Most future planetary missions will include an orbiting vehicle that may or may not have an on-board mass spectrometer and an entry vehicle or "sounder" that almost certainly will. The rapidly changing pressure, temperature, density, and composition of the atmosphere during descent present challenging engineering problems and make the high data acquisition rate of the mass spectrometer an important attribute. Real-time control of the instrumentation is impossible because of the long distances involved, and total automation of measurement sequencing and data transmission is necessary. Current technology thus requires that the entire experiment be totally defined substantially ahead of the launch date. In the future, however, the refinement of computer-based artificial intelligence methods should provide a means of modifying the measurement sequence based on real-time analysis of the data, thus allowing the mass spectrometer to search for previously unsuspected species.

Mars

The Viking mission to analyze both the atmosphere and composition of the Martian surface 223 million miles from earth was among the most successful of the unmanned planetary explorations. Two identical vehicles, each containing a double focusing mass spectrometer to analyze the upper atmosphere (UAMS) and a GC/MS to analyze the atmosphere on the surface and to examine the surface material itself, were soft-landed at different locations. The UAMS (Nier et al., 1972, 1977) was an electrically scanned Mattauch-Herzog double focusing mass spectrometer (Fig. 9.11) with an open electron-bombardment ion source oriented to allow ambient atmospheric gases to enter directly, thus aiding in the measurement of reactive species. A small titanium sublimation pump kept the pressure in the instrument to less than 10^{-8} torr during the two years that elapsed between launch and the encounter with the Martian atmosphere. The UAMS was designed to characterize the atmosphere along the trajectory of the descending Viking lander from 200 down to 100 km. A planar-retarding potential analyzer was also mounted to the aeroshell to provide supplementary measurements of atmospheric ions in the ionosphere.

Earlier data predicted a cold lower atmosphere composed primarily of

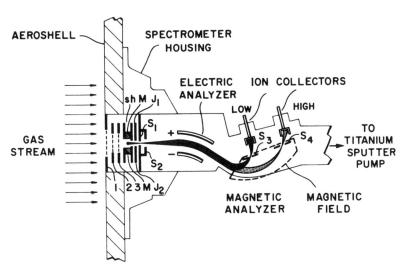

Figure 9.11 Double focusing mass spectrometer for terrestrial and planetary atmospheric studies. (Nier and McElroy, 1977; copyright American Geophysical Union.)

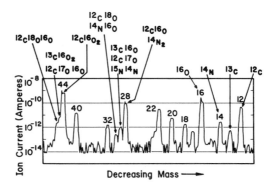

Figure 9.12 Spectrum of the Martian atmosphere at 125 km (Nier et al., 1976a, 1976b; courtesy A. O. Nier; copyright 1976 by the AAAS.)

CO_2, with a moderately dense ionosphere having a maximum density at approximately 130 km. A real-time spectrum obtained from the Viking 1 lander at an altitude of about 125 km is shown in Figure 9.12. These and other data show that the atmosphere is mixed to above 120 km, and that below 180 km, it is primarily CO_2 with low levels of N_2, Ar, O, O_2, CO, and NO (Nier et al., 1976a, 1976b; Nier and McElroy, 1977).

The isotopic ratio of ^{15}N to ^{14}N shows an increase of 60% over terrestrial levels, which may be attributed to the selective loss of the lighter ^{14}N isotope from an initially nitrogen-rich atmosphere, while carbon and oxygen isotopic ratios are essentially unchanged (McElroy et al., 1976). Data obtained from the planar-retarding potential analyzer at 130 km showed that the molecular ion O_2^+ was the major ionic constituent at this altitude, being about nine times more abundant than the CO_2^+ ion. This measurement indicates the importance of the reaction $CO_2^+ + O \rightarrow CO + O_2^+$ in converting the primary photoion CO_2^+ to the more stable O_2^+.

The GC/MS instrument (Biemann, 1974; Rushneck et al., 1978) was designed to determine atmospheric composition, particularly with respect to nitrogen, and to measure organic materials vaporized or pyrolized from the soil. The mass spectrometer was a Nier-Johnson double focusing instrument (Chapter 3) with a dual-filament closed ion source, an electron multiplier detector, and a separate detector to measure total ion current. Since telemetry transmission to earth required approximately 20 min, operation of the GC/MS system was under total control of the on-board computer once the command for initiating a data sequence was given. The system is shown schematically in Figure 9.13. Atmospheric samples were introduced directly into the mass spectrometer through a molecular leak having a conductance of approximately 2×10^{-4} ml/s. Soil samples were collected in a small scoop on the end of a retractable boom and introduced into a processor, where they were ground, sieved, and loaded into an oven. Vaporized or pyrolyzed gases were expanded into the carrier gas stream or purged into it with is-

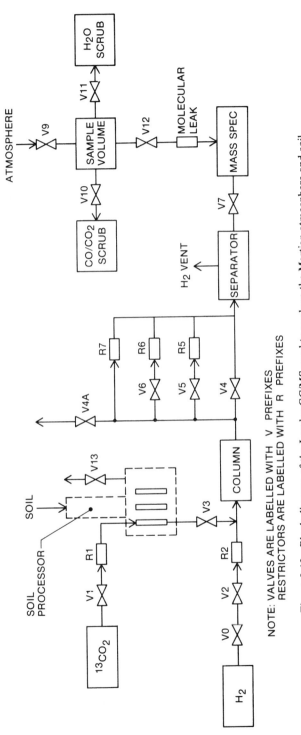

Figure 9.13 Block diagram of the Lander-GC/MS used to analyze the Martian atmosphere and soil. (Rushneck et al., 1978; courtesy D. R. Rushneck; copyright American Institute of Physics.)

NOTE: VALVES ARE LABELLED WITH V PREFIXES
RESTRICTORS ARE LABELLED WITH R PREFIXES

311

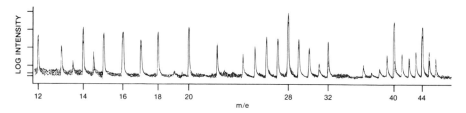

Figure 9.14 First mass spectrum of the atmosphere on the Martian surface. (Rushneck et al., 1978; copyright American Institute of Physics.)

otopically pure $^{13}CO_2$. The hydrogen carrier gas was removed from the stream with a palladium-alloy separator, and a five-stage effluent splitter activated by the mass spectrometer prevented overpressure of the ion source.

The first mass spectrum returned to earth from the Martian surface is shown in Figure 9.14 (Rushneck et al., 1978). Subsequent analyses (Owen and Biemann, 1976) showed an atmospheric composition on the surface of approximately 95% CO_2, 2–3% N_2, 1–2% Ar, 0.1–0.4% O_2, and trace levels of the rare gases. Measurement of the isotopic ratios of $^{13}C/^{12}C$ and $^{18}O/^{16}O$ based on the relative intensities of $^{12}C^{16}O_2$, $^{13}C^{16}O_2$, and $^{12}C^{16}O^{18}O$ showed no significant departure from terrestrial values, but the $^{15}N/^{14}N$ ratio showed a significant increase (Nier et al., 1976a). The ratio of ^{36}Ar to ^{40}Ar was approximately 1/10th the terrestrial value, while the $^{36}Ar:^{38}Ar$ ratio was unchanged. The low level of ^{36}Ar suggests that the total Martian atmosphere never exceeded its present value of 8 mbar by more than a factor of 10. Using an enrichment procedure to enhance the concentration of the rare gases, the presence of krypton and xenon in the atmosphere was confirmed. Furthermore, the isotopic abundance of ^{129}Xe was much higher than the ^{130}Xe and ^{131}Xe, in contrast to the nearly equal terrestrial values.

The objective of the soil analyses was to search for organic and volatile inorganic compounds in the surface material at the landing site. The samples were collected, processed, and heated to 50°C for 10 min before being raised to higher temperatures. The results showed no organic compounds even when the sample was heated as high as 500°C. However, in addition to CO_2, considerable quantities of water were evolved at 350 and 500°C, reaching as high as 1%, thus demonstrating a significant aqueous environment (Biemann et al., 1976).

Venus

The Venusian and terrestrial atmospheres are markedly different despite similarities in mass, size, and orbit of the two planets. Early remote observations of Venus determined that the major atmospheric constituent was

CO_2 and that the planet cloud cover contained large quantities of H_2SO_4. Sounder probes measured surface pressures of 90 bars (1323 psi) and temperatures of 700 K.

Venus has been explored both by Soviet (Venera) and American (Pioneer-Venus) spacecraft. The first five Venera descent probes provided basic information concerning the atmosphere; the second-generation probes included mass spectrometers and gas chromatographs that operated during parachute descent from 60 km and for about 1 h on the surface (Moroz, 1981). Venera 9 and 10 used a monopole mass spectrometer with a two-stage capillary divider venting into a pre-evacuated volume to reduce the pressure input to the ion source. Venera 11–14 mass spectrometers were Bennet-type RF analyzers. The samples were introduced through a rapidly pulsed, piezoelectrically driven metering valve into a chamber connected to the ion source by a 20 μm orifice. Gas flow through a tube past the input valve was maintained by a pressure differential created by the probe descent velocity.

The Pioneer-Venus spacecraft contained a total of 24 instruments, five of which were mass spectrometers. These were reviewed in a special issue of the *IEEE Transactions on Geoscience and Remote Sensing* (Vol. GE-18, No. 1, January 1980) and by Colin and Hunten (1977). An instrumented orbiter accumulated data at a nominal 200 km periapsis; and a bus containing one large and three small probes, each released to simultaneously enter the atmosphere, obtained data from about 1000 km down to the surface (Colin and Hall, 1977); none of the Pioneer-Venus probes were adapted for a soft landing as were those of Venera; however, one of the small probes continued to transmit from the surface for 67 min after impact.

The orbiter neutral mass spectrometer (ONMS); Fig. 9.15 measured number density distributions of the major neutral species and some of the isotopes over an altitude range of 150–300 km (Niemann et al., 1980a). Average vertical and horizontal resolutions were approximately 0.4 and 2.0 km, respectively. The ONMS was a hyperbolic-rod quadrupole mass filter having a mass range of 1–46 amu. The ONMS ion source included a retarding potential grid (RPG) biased to differentiate between ions created from neutrals that have or have not undergone collision with the walls according to the *fly-through* method described by Nier et al. (1974). Both free-streaming and surface-reflected gas particles were ionized; however, the kinetic energy of the reflected gas particles was lower, so that corresponding ions could be prevented from entering the analyzer by an appropriate potential placed between the RPG and a free drift space formed by the electron beam defining grids. Both the functional dependence of the free-stream ion current on mass and the RPG potential for the ONMS is shown in Figure 9.16, illustrating that energy discrimination could be obtained even for the lighter elements.

The bus neutral mass spectrometer (BNMS) was developed to measure number densities of neutral atmospheric constituents in the exosphere and thermosphere from approximately 1000 to 130 km (von Zahn and Mauers-

314

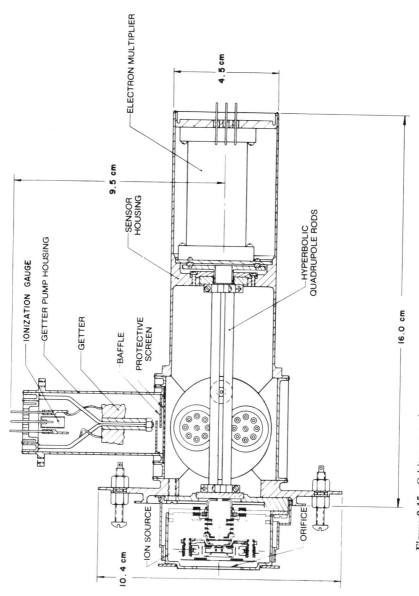

Figure 9.15 Orbiter neutral mass spectrometer. (Niemann et al., 1980a, 1980b; courtesy H. B. Niemann; © 1980 IEEE.)

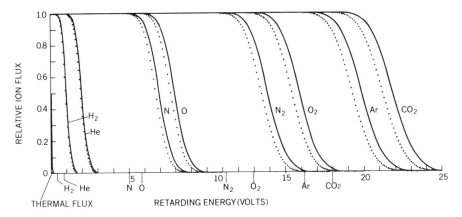

Figure 9.16 Dependence of the normalized ion current on the retarding potential at 0°. (Niemann et al., 1980a; courtesy H. B. Niemann; © 1980 IEEE.)

berger, 1978; Hoffmann et al., 1980). The BNMS (Fig. 9.17) was a Mattauch-Herzog double focusing instrument with dual-electron multiplier detectors. A separate electrometer measured total ion current extracted from the ion source at grid ET. Major design and operating constraints were imposed by the rapidly changing ambient pressure due to the 11 km/s entry velocity. The semiopen ion source was therefore configured to function at pressures as high as 10^{-2} torr, and the analyzer region was differentially pumped with small titanium sublimation and gettering pumps. The operating sequences gave preference to the lighter species and background at higher altitudes, and to the heavier species and isotopes at the lower altitudes. In-flight calibration was accomplished 2 days prior to entry by releasing approximately 700 μl of calibration gas at 10^{-3} torr into the analyzer.

The high entry velocity also results in a relatively high kinetic energy for the ambient neutral gas particles entering the ion source. Most of the neutral particles reach thermal equilibrium with the source walls, while the remainder are ionized at the original ram energy. This energy difference was again used to differentiate between the two by making use of the fly-through mode described by Nier et al. (1974).

Neutral composition of the lower atmosphere was measured with the large-probe neutral mass spectrometer (LNMS) from an altitude of approximately 62 km down to the Venusian surface (Hoffman et al., 1979, 1980a, 1980b). Other measurements included the isotopic composition of the rare gases and the composition of the cloud layer. The LNMS (Fig. 9.18) was a microprocessor-controlled, 90° magnetic sector with a mass range of 1–208 amu, a resolution $m/\Delta m$ of 400, and a sensitivity of 1 ppm relative to CO_2. Dual-electron multiplier detectors were used in the ion-counting mode. Ambient gases were sampled through a pair of passivated tantalum tubes that penetrated the spacecraft shell and extended beyond the boundary layer into

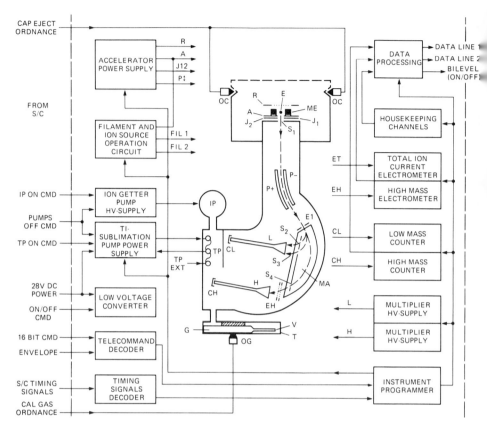

Figure 9.17   Bus neutral mass spectrometer. (Hoffmann et al., 1980; © 1980 IEEE.)

the free stream. Ion source pressure was maintained relatively constant, while the ambient pressure increased from 0.1 to 95 atm by a pneumatically activated gate valve that varied the pumping speed of the source getter pump from 1–2000 ml/s. In order to measure the isotopic ratio of the rare gases, high-altitude samples were collected in the isotope ratio measurement cell (IRMC), where the rare gas concentration was increased by removal of the active gases by adsorption and gettering.

Two identical ion mass spectrometers (IMS), one installed on the orbiter (OIMS) and the other on the bus (BIMS), provided continuous direct measurement of the composition of the ionosphere from the ionopause down to approximately 120 km, where the mean free path precluded further sampling (Taylor et al., 1980). The IMS were Bennett-type RF mass spectrometers similar to those used on Atmospheric Explorer (see Fig. 9.7), having a mass range of 1 (H^+) to 56 (Fe^+) amu and a 10^6 dynamic range in sensitivity. No ion source was required since only the charged species were to be measured, and the ambient ions entered directly through a sampling orifice and biased

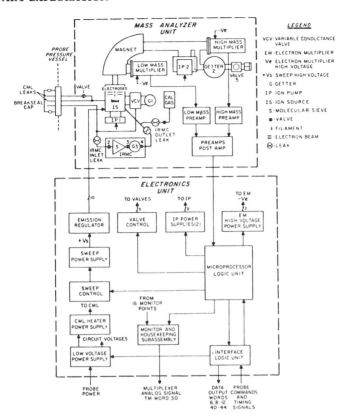

Figure 9.18 Large-probe neutral mass spectrometer. (Hoffman et al., 1980a; © 1980 IEEE.)

guard ring. Both the OIMS and BIMS were mounted with the analyzer axis parallel to the spin axis in order to maintain a low angle of attack and eliminate spin modulation of the ion currents. The sensitivity was such that ion concentrations as low as 5 ions/cm^3 and ion flow velocities as large as 9 km/s for O^+ could be detected.

The missions to Venus and the results of the measurements have been summarized by Hall (1980), Moroz (1981), Hunt and Moore (1982), Fimmel et al. (1983), and in special issues of *Science* (Vol. 203, 1979; Vol. 205, 1979) and the *Journal of Geophysical Research* (Vol. 85, A11, 1980). Scientific analyses continue to be made as the large quantities of data are further reduced, and Fimmel et al. (1983) have provided a bibliography of 269 publications issued through April 1982. In general, the observations supported earlier predictions that the predominant species was CO_2 (approximately 96%) and the minor one was N_2 (approximately 4%). However, several unexpected discoveries have led to much speculation concerning the evolution and atmospheric chemistry of the planet. For example, it was determined

TABLE 9.6 Volume Mixing and Isotopic Ratios of Gases Found in the Venus Atmosphere

Gas[a]	Venus Atmosphere Mixing Ratio (ppm)	Earth Atmosphere Mixing Ratio (ppm)
^{20}Ne	9^b	16
^{22}Ne	1^b	2
Total Ne	10^b	18
^{36}Ar	30_{-10}^{+20}	31
^{38}Ar	6^b	6
^{40}Ar	31^b	0.93%
Total Ar	$40-120^b$	0.93%
^{84}Kr	< 0.2	0.5
N_2	4 ± 2%	78%
O_2	< 30	21%
SO_2 55 km	< 10	
Below 24 km	< 300	
COS Above 24 km	< 3	
Below 20 km	< 500	
H_2S	3^b	
C_2H_6	2^b	
H_2O	< 1000	
Cl	< 10	
Hg	< 5	

Gas[a]	Venus Atmosphere Isotopic Ratio	Earth Atmosphere Isotopic Ratio
^{3}He/^{4}He	$< 3 \times 10^{-4}$	1.4×10^{-6}
^{22}Ne/^{20}Ne	0.07 ± 0.02	0.097
^{20}Ne/^{36}Ar	0.3 ± 0.2	0.58
^{38}Ar/^{36}Ar	0.18 ± 0.02	0.187
^{40}Ar/^{36}Ar	1.03 ± 0.04	296
^{13}C/^{12}C	$\leq 1.19 \times 10^{-2}$	1.11×10^{-2}
^{18}O/^{16}O	$2.0 \pm 0.1 \times 10^{-3}$	2.04×10^{-3}

Source: Hoffman et al. (1980b).

[a] In those rows where altitude is unspecified, result is average value from 24 km to surface.

[b] Deduced from ratio to ^{36}Ar.

that the isotopic ratio of the primordial argon (^{38}Ar:^{36}Ar) was almost identical to terrestrial values, while the radiogenic argon (^{40}Ar) abundance was markedly different (Table 9.6) Furthermore, absolute abundances of primordial argon and neon are greatly enhanced over terrestrial values, while those of xenon and krypton are more nearly equivalent (Donahue et al., 1981). Although present in small amounts, the coexistence of CO and O_2 is difficult to explain from a thermodynamic standpoint, as is the apparent low rate of photochemical production of oxygen.

Of particular interest is the formation of the sulfuric acid-bearing cloud layer, which had been predicated on a carbonyl sulfide (COS)-driven cycle similar to the terrestrial cycle responsible for the formation of the sulfate aerosols. However, both H_2S and SO_2 have been found in larger quantities at the higher altitudes (Fig. 9.19), leading to a proposed mechanism involving both of these compounds.

An analysis of the aerosols in the cloud layer between 63 and 47 km was also included in the Franco-Soviet VEGA-85 Venusian spaceprobe (Israel et al., 1984). Suspended particles accumulated on stainless steel fibers in a pyrolysis cell were subsequently heated to 400°C and analyzed over the range of 10–150 amu with the on-board mass spectrometer. Samples were taken both from the relatively stable middle cloud layer (64–53 km), which is most likely of photochemical origin, and from the lower layer (54 to 47 km) where the aerosols originate in the lower Venusian atmosphere.

Based on the ONMS data (Niemann et al., 1980b), Seiff and Kirk (1982) calculated the mean molecular weight in the mesosphere and thermosphere from the relationship $\bar{\mu} = \sum \mu_i n_i / \sum n_i$ where μ_i and n_i are the molecular weight and number density of the measured species (e.g., CO_2, CO, N_2, N, O, He). The marked temporal variations in the atmospheric composition resulted in two boundary profiles for noon and midnight (Fig. 9.20). Above 130 km, $\bar{\mu}$ diminishes rapidly as atomic oxygen on the dayside and helium on the nightside become dominant. The BNMS data (von Zahn et al., 1980) for 0830 illustrates the constancy of the dayside profile with solar zenith angle as the morning terminator is approached.

In the ionosphere, data taken with the IMS have shown that temporal variations, localized depletions of plasma density (holes), and electrodynamic perturbations occur (Nagy et al., 1979; Bauer and Taylor, 1981; Cravens et al., 1982; Luhmann et al., 1982; Taylor et al., 1983; Hartle and Taylor, 1983). In addition to the major ions (O_2^+, O^+, CO_2^+, He^+, and H^+), lesser concentrations of C^+, N^+, NO^+, CO^+, H_2^+, and N_2^+ were detected and their densities calculated (Fig. 9.21). Variations of the m/z 2^+ ion, as compared to the m/z 1^+ (H^+) measured with the OIMS in the chemical equilibrium region, identified D^+ as the dominant m/z 2^+ ion. This supports the suggestion that Venus initially contained a significant quantity of water (Donahue et al., 1982).

A mass-dependent day/night asymmetry has also been determined, that is, the composition as well as the plasma density depends upon local time. Typical variations of H^+, O^+, and O_2^+ are shown in Figure 9.22. Furthermore, measurements of the dayside ion composition exhibit a modulation corresponding to the 27 day variations in the solar flux. The compositional fluctuation is due mainly to changes in solar extreme ultraviolet (EUV) radiation and, to a much lesser extent, to exothermic temperature variations.

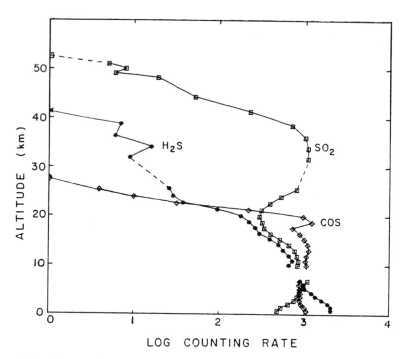

Figure 9.19 Concentrations of COS, SO₂, and H₂S in the Venusian atmosphere. (Hoffman et al., 1980; © 1980 IEEE.)

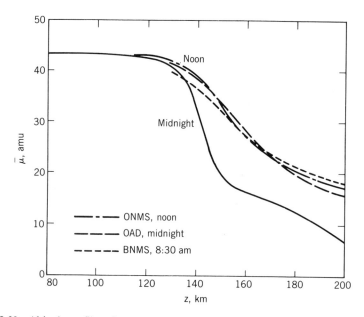

Figure 9.20 Altitude profiles of mean molecular weight in the Venusian atmosphere. (Seiff and Kirk, 1982; courtesy A. Seiff; copyright Academic Press.)

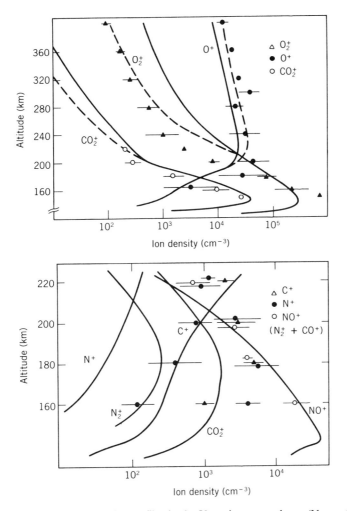

Figure 9.21 Calculated ion density profiles in the Venusian atmosphere. (Nagy et al., 1979; copyright 1979 by AAAS).

Jupiter

Unlike the Martian and Venusian atmospheres, which are thought to be the result of outgassing of rocky planets, the atmosphere of Jupiter is thought to be more star-like, being approximately 90% H_2 and 10% He. Also known to be present are trace amounts of Ne, Ar, Kr, Xe, H_2O, CH_4, NH_3, H_2S, C_2H_2, C_2H_6, PH_3, HCl, HF, CO, and GeH_4 (Niemann and Kasprzak, 1983). The Galileo-Jupiter spacecraft is the next planetary exploration mission that, in addition to an orbiter, will include an entry probe carrying a mass spectrometer (Casini, 1982). One of seven instruments to be carried on the probe,

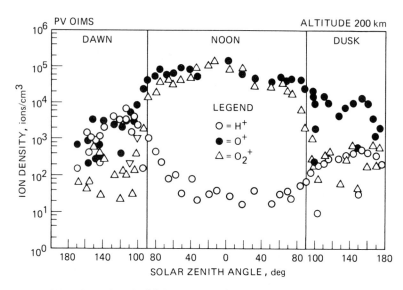

Figure 9.22 Mass-dependent day/night assymetry in the Venusian plasma density (Fimmel et al., 1983).

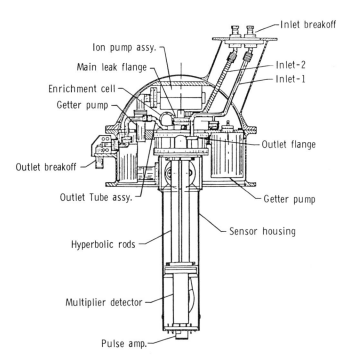

Figure 9.23 Galileo probe quadrupole mass spectrometer. (Niemann, 1985; courtesy H. B. Niemann.)

the Galileo-Probe mass spectrometer (GPMS) will measure atmospheric mixing and isotopic ratios over an ambient pressure and temperature range of 0.1–20 bars and 110–420 K, respectively. Total sampling time is expected to be about 60 min, during which the probe will pass through at least two cloud layers of H_2O and NH_3. Prominent masses in the range of 1–52 amu will be measured repeatedly, with periodic scans to 150 amu. Analyses will be made by direct sampling of the atmosphere and with enrichment- and noble gas-scrubbing cells.

The GPMS (Fig. 9.23) is a hyperbolic rod QMS with an electron-impact ionization source and an electron multiplier detector (Niemann, 1985). The ion source is an integral part of the sampling system to minimize gas-surface interactions, and it is designed to operate over an internal pressure range of 10^{-4} to 10^{-13} torr.

COMETS

The first in situ measurements in the coma of a comet were made by the ESA (Giotto), Soviet (Vega), and Japanese (Planet-A) spacecraft in an extraordinary mission to the comet Halley during March, 1986. The three missions were complementary in instrumentation (Table 9.7) and in measurement objectives; consequently, only the ESA (Giotto) mass spectrometers will be described. Mission objectives included the determination of the elemental and chemical composition of volatile components and dust particles, as well as the measurement of energetic ionized species (Reinhard, 1981). Data were to be taken over a 4 h period until time of closest approach or until the destruction of the spacecraft. Sampling of the atmosphere and, in fact, survival of the spacecraft were of great concern because of the density of ice and dust particles impacting at the spacecraft flyby velocity of 68 km/ s. At this velocity, impact energies are sufficient to totally vaporize (and partially ionize) the particles. Secondary ion emission from interaction of the energetic ions and neutrals also occur; however, these ions can be differentiated from those originating in the comet because of their large energy differences in the spacecraft frame of reference. All of the spacecraft successfully encountered the comet, the Planet-A and Vega passing within about 10^6 and 10^4 km of the nucleus respectively. Using navigational data obtained by Vega, the Giotto was targeted to a close encounter approximately 500 km on the sunward side to obtain measurements in both the outer and inner coma, which have a boundary at about 10^4 km from the nucleus. A large amount of scientific data were obtained, however, at the time of this writing, only the visual camera data have been sufficiently reduced to be released.

The particle impact analyzer (PIA, Fig. 9.24) mass spectrometer for determining the chemical and isotopic composition of individual dust particles (Kissel, 1981), was a modification of an instrument developed for the Helios micrometeoroid experiments (Friichtenicht and Grun, 1976). The PIA had

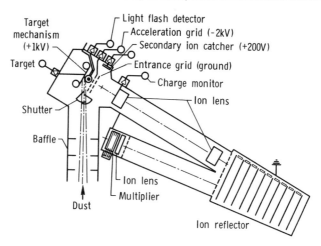

Figure 9.24 Particle impact ionizer to analyze cometary dust (Kissel, 1985).

a mass range of 1–100 amu, a resolution of $m/\Delta m = 200$, and a sensitivity sufficient to detect particles over a range of 3×10^{-16} to 5×10^{-10} g (0.1–10 μm in diameter) up to an impact rate of 100/s. Additionally, the sum of all masses in the range of 111–200 are measured as a single peak. Particles enter through an adjustable port and impact a target, producing an impact plasma consisting of electrons and ions of both the particle and target material, which rapidly expand to a collisionless state before recombination can occur. The free ions and electrons expand radially outward from the impact site with a velocity distribution characteristic of the original plasma temperature. The ions are accelerated into a folded 1 m flight tube, while electrons return to the target and a complete TOF spectrum is obtained. Compositional analysis of the particulates should provide important information concerning the origin and history of the comet. Measurement of the isotopic ratios of species such as ^{6}Li and ^{7}Li or ^{13}C and ^{12}C should provide insight into the age of the comet and into interstellar gas-grain chemistry.

The energetic ion mass spectrometer (EIMS, Fig. 9.25) for determining the chemical, elemental, and isotopic composition of ionized species in the coma (e.g., CO^+, H_2O^+, N_2^+, OH^+, CH^+, etc.) consisted of a high-energy range spectrometer (HERS) for measurements in the outer coma at distances greater than 10^4 km from the nucleus, and a high-intensity spectrometer (HIS) for the inner regions (Balsiger et al., 1981; Neugebauer et al., 1982). The HERS was to measure both ion abundance and three-dimensional velocity distributions in the region dominated by solar wind around the cometary ionosphere. Ions, but not dust particles, were deflected by an electrostatic mirror into the analyzer. Energizing the mirror switched the direction of sampling from the solar wind to the forward direction with an angular range of 30° (-1–29°; 45–75°), and a resolution of 3.5°. The ions

TABLE 9.7 Instrumentation and Measurements for the Halley's Comet Spacecraft

Instrument	Type	Measurements
	Giotto Payload	
Camera	Telescope, 960-mm focal length Detector, CCD 2.3° field-of-view, 4 filters Ultimate resolution ~30 m	Images of nucleus and inner coma
Neutral mass spectrometer	Dual sensors (mass and energy) Sensitivity to 10 cm^{-3} High resolution 1–35 amu Moderate resolution to 86 amu	Abundance of each molecular mass of neutral gas
Ion mass spectrometer	High intensity range, up to 10^4 ions cm^{-3} for energies to 1.5 keV High energy range, up to 8 keV for densities up to 10^2 ions cm^{-3}	Abundance, mass, and energy of ions
Ion cluster composition	Electrostatic analyzer 10 to >230 amu 10^{-3}–10^4 cm^{-3}	Mass analysis of positive ions and their hydrates
Dust mass spectrometer	Ionization-impact time-of-flight, $M/\Delta M$ ~150 for 1–110 amu	Abundance of chemical elements in the dust particles
Dust impact detector	Microphones for $M > 10^{-11}$ g Capacitor sensor for $M > 10^{-10}$ g Ionization sensor for $M > 10^{-16}$ g	Abundance and mass distribution function of coma dust particles
Optical probe experiment	Photopolarimeter 30-mm focal length, $f/2.5$ 8 filters	Scattering properties of dust and spatial densities of gas and dust
Magnetometer	Triaxial ring-core fluxgate on antenna mast	Ambient (DC) magnetic field
Plasma experiment	Fast ion sensor, all masses, 10 eV–20 keV Implanted ion sensor, six mass groups (1–45 amu), 100 eV–70 keV	Three-dimensional velocity distribution of positive ions
Energetic particles	Surface barrier detectors Electrons $\geq$ 30 keV Protons $\geq$ 100 keV Particles ($Z \geq 2$) $\geq$ 2.1 MeV	Energy spectrum and direction of energetic particles from the Sun, in Halley bow shock, and in ion tail

TABLE 9.7 (*Continued*)

Instrument	Type	Measurements
Electron distribution	Quadrispherical electrostatic analyzer 10 eV–30 keV 360° × 4° field-of-view 1.4° resolution $\Delta E/E \sim 0.1$ $\Delta T \sim 2$ s	Determine accurate angular distribution of suprathermal electrons at fast rate

Vega Payload

Instrument	Type	Measurements
Imaging (2)	Telescopes, 1200-mm and 200-mm focal length 0.4° × 0.6° and 3.5° × 5.2° field-of-view Detectors, CCDs with multiple filters Best resolution ~180 m	Structure of nucleus and coma
Spectrometer	Three-channel, 0.12–0.35 μm, 0.35–0.9 μm, and 0.9–2.0 μm Spatial resolution 3 × 6 arcmin in ultraviolet and visible, 6 × 60 arcmin in infrared, all 1° field-of-view Spectral resolution 5 Å in ultraviolet, 10 Å in visible, and 120 Å in infrared Polarimetry in visible	Abundances and distribution of atoms, molecules, and ions in coma
Infrared spectrometer	Three-channel, 4–8 μm, 8–16 μm, and 7–14 μm imaging Spatial resolution ~1 arcmin 1° field-of-view Detector, HgCdTe at 80 K	Nucleus properties Temperature, size distribution, and abundance of dust in coma Parent molecule spectra
Dust mass spectrometer	Ionization-impact time-of-flight, $M/\Delta M \sim 150$ for 1–110 amu	Composition from mass and number of component dust atoms
Dust counters (2)	Metal plates each with three piezoelectric sensors for $M \geq 10^{-10}$ g Plasma cloud counter for $M = 10^{-10}$–10^{-18} g	Number and mass of impacting particles
Neutral gas mass spectrometer	Time-of-flight	Number and mass of atoms and molecules in spacecraft vicinity
Ion mass spectrometer (2 of 5 sensors)	Electrostatic analyzer 1–150 amu, $M/\Delta M \sim 25$, energy 15–3500 eV Cylindrical retarding potential analyzer	Flux, mass, and velocity of cometary ions

326

TABLE 9.7 (*Continued*)

Instrument	Type	Measurements
Plasma mass spectrometer and electron analyzer (3 of 5 sensors)	Electrostatic analyzers (2) 3–5000 eV electrons 50–25,000 eV ions Cylindrical retarding potential analyzer	Flux and energy spectrum of solar wind electrons and ions
Energetic particle analyzer	Semiconductor detectors (2) Anticoincidence scintillator 175 keV–few MeV electrons 20 keV–~30 MeV nucleons	High-energy solar and galactic particles, particles accelerated near comet
Plasma wave detectors	Low frequency antenna with Faraday trap, 0.1–100 Hz High frequency antenna, 0–300 kHz Langmuir probe	Flux of high and low frequency electromagnetic waves propagating in cometary plasma
Magnetometers (2)	Fluxgate type on 5-m boom Accuracy 0.1 nT	Ambient (DC) magnetic field

Suisei Payload

Instrument	Type	Measurements
Ultraviolet camera	Ultraviolet telescope $2.5° \times 2.5°$ field-of-view Resolution 1 arcmin/pixel Intensified CCD detector, detection limit 1 kR at Lα	Abundance and distribution of hydrogen in coma by study of Lα (1216 Å)
Solar wind experiment	Spherical electrostatic energy analyzer $5° \times 60°$ field-of-view 30 eV–16 keV in 96 steps	Energy and direction of electrons and ions

Sakigake Payload

Instrument	Type	Measurements
Solar wind ions	Faraday cup, collection area 70 cm^2 at zero angle of attack	Ion temperature Ion density Bulk velocity
Magnetometer	Triaxial fluxgate of ring-core type on 2-m boom Sensitivity < 0.1 nT	Ambient (DC) magnetic field
Plasma wave probe	Dipole antenna, 10-m length, for electric field measurements Search coil, 10^5 turns, ferrite core for magnetic field, frequency range 70 Hz–200 kHz	Plasma waves Radio bursts Auroral radiation Bernstein waves

Source: The Halley Armada, JPL 400-278, Rev. 2, National Aeronautics and Space Administration, Jet Propulsion Laboratory, California Institute of Technology, Pasadena, CA (1985).

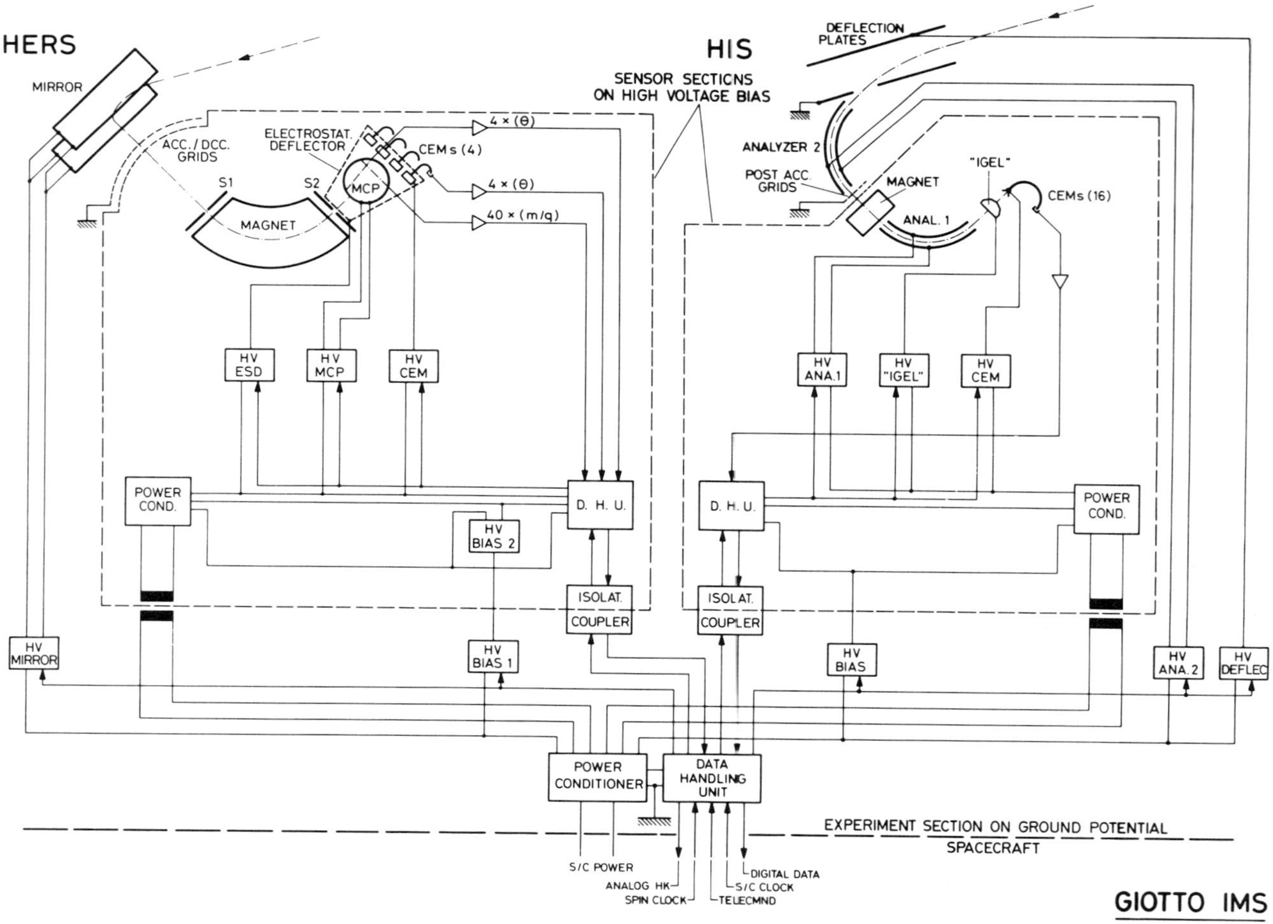

Figure 9.25 Block diagram of the high-energy range (HERS) and high-intensity (HIS) energetic ion

GIOTTO IMS

accelerated into an electrostatic field that separated the resolved ion beams as a function of energy-per-charge were detected on a two-dimensional, position-sensitive microchannel plate (MCP) detector (Liptak et al., 1984) those having a momentum-per-charge of 8 keV being distributed on the MCP plane dispersed in mass-to-charge on the ordinate and in angle-of-incidence on the abscissa. The angular coordinates were read directly from discrete anodes on the backplane of the MCP in eight sectors; and m/z were determined by 40 discrete anodes on a separate backplane located behind the MCP. Scanning was accomplished by changing the potential on the electrostatic analyzer.

The HIS measured ion composition, temperature, and angular distribution of the relatively cold cometary ion species over an angular range of $\pm 13.5°$ centered on spacecraft ram direction. Essentially a triple focusing mass spectrometer, the HIS consisted of a pair of quadraspherical analyzers (QA) separated by a magnetic sector. Energy-resolved ions from the leading analyzer (QA2) were accelerated through the magnetic field and the secondary analyzer (QA1). The mass- and energy-resolved ions were then detected by an array of 16 channeltron electron multipliers spatially separated to detect ions with a velocity spread of $\pm 10\%$.

The abundance and isotopic ratios of the gaseous constituents are to be obtained from the neutral mass spectrometer (NMS) data (Krankowsky et al., 1981). The primary measurement for a NMS is identification of neutral species; however, periodically switching off the ionizing electron beam permits measurement of the ionized gaseous species as well. The NMS system consisted of two separate spectrometers: (1) the E analyzer in which the mass of the energetic particle is calculated from the energy $E = 0.5 \, mV^2$, which is equivalent to $E = 25$ eV/amu for the comet approach velocity; and (2) the M analyzer, a double focusing configuration for measurement of the lower masses. Performance characteristics for the NMS are listed in Table 9.8. The E analyzer was a simple parallel plate device that provided differential energy analyses by focusing the ions on a plane at energy-dependent locations. This particular geometry also provides second-order focusing for the angular spread of gaseous particles at the entrance to the analyzer; and for the NMS, a half-angular width of $4.5°$ was acceptable without significant loss of resolution. The M analyzer was a parallel plate energy analyzer followed by a magnetic sector providing double focusing with unit resolution over a mass range of 1–36 amu for the constant velocity condition of 68 km/s. The focal plane detectors for both the E and M analyzers consisted of MCPs and linear arrays of discrete anode strips accumulating charges that are read out by fast-analog multiplexing at rates as high as 1 kHz to achieve a dynamic range of 10^6. Microprocessor control of MCP gain prevents saturation during periods of high flux, with the gain being quantitatively determined by measurement of total ion current from neutral species. To assure homogeneity of the gain, energy and mass scales are periodically shifted along the focal plane. The full mass and energy range are sampled contin-

TABLE 9.8   NMS Performance Characteristics

	M Analyzer	E Analyzer
Energy range		$20 \rightarrow 860$, $730 \rightarrow 2110$ eV
Mass range	$1 \rightarrow 36$ amu	$1 \rightarrow 35$, $30 \rightarrow 86$ amu
Resolution	0.25 amu	4.4/7.2 eV
Integration time		
Neutrals	1.4 s	0.70 s/0.70 s
Ions (encounter mode)	0.1 s	0.05 s/0.05 s
Ions (approach mode)	1.5 s	0.75 s/0.75 s
Repetition period		
Neutrals	1.5 s	1.5 s
Ions (encounter mode)	3.0 s	3.0 s
Ions (approach mode)	1.5 s	1.5 s
Density range		
Neutrals[a]	$10 \rightarrow 10^7$ cm^{-3}	$30 \rightarrow 10^7$ cm^{-3}
Ions (encounter mode)	$< 1 \rightarrow 10^5$ cm^{-3}	$< 1 \rightarrow 10^5$ cm^{-3}
Ions (approach mode)[a]	$3.10^{-5} \rightarrow 30$ cm^{-3}	$10^{-4} \rightarrow 50$ cm^{-3}

Source: Krankowski et al (1981).

[a] Density range upper limit can be extended when necessary by decreasing the gain of microchannel plates.

uously in order to detect all species without predisposition of their importance, and the integrated charge from each anode pixel element (256 for the M and 128 for the E) sequentially telemetered. Both analyzers made extensive use of the fly-through mode that effectively eliminates all background ions even at the lighter end of the mass spectrum. The density of neutral species was expected to become measurable at approximately 10^5 km from the nucleus; hence, during the approach from approximately 10^6 km to that distance, the NMS was operated mostly in the ion measurement mode, that is, with the electron beam turned off. When the particle count rate reached a preselected value, the operational mode was changed to preferentially measure the neutral species, and a series of programmed measurement sequences carried out.

REFERENCES

Arijs, E., D. Nevejans, P. Frederick, and J. Ingels, *Geophys. Res. Lett., 8,* 121 (1981).

Arijs, E., D. Nevejans, and J. Ingels, *J. Atmos. and Terres. Phys., 44,* 43 (1982).

Arijs, E., D. Nevejans, and J. Ingels, *Nature, 303,* 314 (1983).

Arnold, F., H. Bohringer, and G. Henschen, *Geophys. Res. Lett., 5,* 653 (1978).

Arnold, F., and R. Fabian, *Nature, 283,* 55 (1980).

Arnold, F., G. Henschen, and E. E. Ferguson, *Planet. Space Sci., 29,* 185 (1981a).

Arnold, F., R. Fabian, and W. Joos, *Geophys. Res. Lett., 8,* 293 (1981b).

Arnold, F., R. Fabian, E. E. Ferguson, and W. Joos, *Planet. Space Sci., 29,* 195 (1981c).

Arnold, F., H. Heitman, and K. Oberfrank, *Planet. Space Sci., 32,* 1567 (1984).

Arnold, F., and G. Hauck, *Nature, 315,* 307 (1985).

Balsiger, H., *Adv. Space Res., 2,* 3 (1983).

Balsiger, H., P. Eberhardt, J. Geiss, and D. T. Young, *J. Geophys. Res., 85,* A4, 1645 (1980).

Balsiger, H., J. Geiss, D. T. Young, H. Rosenbaur, R. Schwenn, W.-H. Ip, E. Urgstrup, M. Neugebauer, R. Goldstein, B. E. Goldstein, W. T. Huntress, E. G. Shelley, R. D. Sharp, R. G. Johnson, A. J. Lazarus, and H. S. Bridge, in *European Space Agency (ESA) SP-169,* eds. B. Battrick and J. Mort (1981), Paris, p. 93.

Banks, P. M., and T. E. Holzer, *J. Geophys. Res., 73,* 6846 (1968).

Bauer, S. J., and H. A. Taylor, *Geophys. Res. Lett., 8,* 840 (1981).

Béghin, C., J. J. Berthelier, R. Debrie, Y. I. Galperin, V. A. Gladyshev, N. I. Massevitch, and D. Roux, *Adv. Space Res., 2,* 61 (1983).

Biemann, K., *Orig. Life, 5,* 417 (1974).

Bieman, K., J. Oro, P. Toulmin, L. E. Orgel, A. O. Nier, D. M. Anderson, P. G. Simmonds, D. Flory, A. V. Diaz, D. R. Rushneck, and J. A. Biller, *Science, 194,* 72 (1976).

Bieri, R. H., M. Koide, E. A. Martell, and T. G. Scholz, *J. Geophys. Res., 75,* 6731 (1970).

Brinton, J. C., *In Situ Measurements of Plasma Drift Velocity and Enhanced NO^+ in the Auroral Electrojet by the Bennett Mass Spectrometer on AE-C,* NASA TMX-70821, Washington, D. C. (1974).

Brinton, J. C., L. R. Scott, M. W. Pharo, and J. T. Coulson, *Radio Sci., 8,* 323 (1973).

Casini, J. R., in *Space: Mankind's Fourth Environment,* ed. L. G. Napilitano, Pergamon Press, New York (1982), p. 335.

Chappell, C. R., S. A. Fields, C. R. Baugher, J. H. Hoffman, W. B. Hanson, W. W. Wright, H. D. Hammack, G. R. Carignan, and A. F. Nagy, *Space Sci. Instr., 5,* 477 (1981).

Chappell, C. R., R. C. Olsen, J. L. Green, J. F. E. Johnson, and J. H. Waite, *Geophys. Res. Lett., 9,* 937 (1982).

Cicerone, R. J., and J. L. McCrumb, *Geophys. Res. Lett., 7,* 251 (1980).

Colin, L., and C. F. Hall, *Space Sci. Rev., 20,* 283 (1977).

Colin, L., and D. M. Hunten, *Space Sci. Rev., 20,* 451 (1977).

Coplan, M. A., K. W. Olgilvie, P. A. Bochsler, and J. Geiss, *IEEE Trans. Geosci. Rem. Sens., GE-16,* 185 (1978).

Cravens, T. E., L. H. Brace, H. A. Taylor, C. T. Russell, W. L. Knudsen, K. L. Miller, A. Barnes, J. D. Mihalov, F. L. Scarf, S. J. Quenon, and A. F. Nagy, *Icarus, 51,* 271 (1982).

Donahue, T. M., J. H. Hoffman, and R. R. Hodges, *Geophys. Res. Lett., 8,* No. 5, 513 (1981).

Donahue, T. M., J. H. Hoffman, R. R. Hodges, and A. J. Watson, *Science, 216,* 630 (1982).

Evans, J. V., *Science, 216,* 467 (1982).

Fabian, P., R. Borchers, B. C. Krüger, S. Lal, and S. A. Penkett, *Geophys. Res. Lett., 12,* 1 (1985).

Fimmel, R. O., L. Colin, and E. Burgess, *Pioneer Venus,* NASA SP-461, Washington, D. C. (1983).

Forbes, J. M., and K. S. W. Champion, *Tides in the Thermosphere: A Review AFGL-TR-82-0264,* National Technical Information Service, Springfield, VA (1982).

Friichtenicht, J. F., and E. Grun, *COSPAR 19th Plenary Meeting,* Philadelphia (June 1976).

Hayden, J. L., A. O. Nier, J. B. French, N. M. Reid, and R. J. Duckett, *Int. J. Mass Spectrom. Ion Phys., 15,* 37 (1974).

Hall, C. F., in *Space Developments for the Future of Mankind,* ed. L. G. Napolitano, Pergamon Press, New York (1980), p. 309.

Hansen, W. B., and H. C. Carlson, in *Studies in Geophysics: The Upper Atmosphere and Magnetosphere*, National Academy of Sciences, Washington, D. C. (1977), p. 84.

Hartle, R. E., and H. A. Taylor, *Geophys. Res. Lett., 10,* 965 (1983).

Hedin, A. E., *J. Geophys. Res., 88,* 10170 (1983).

Heicklen, J., *Atmospheric Chemistry*, Academic Press, New York (1976), pp. 9, 126.

Hoffman, J. H., *Int. J. Mass Spectrom. Ion Phys., 4,* 315 (1970).

Hoffman, J. H., *Int. J. Mass Spectrom. Ion Phys., 8,* 403 (1972).

Hoffman, J. H., R. R. Hodges, and D. E. Evans, *Geochim. et Cosmochim. Acta, 3,* 2205 (1972a).

Hoffman, J. H., R. R. Hodges, F. S. Johnson, and D. E. Evans, *Geochim. et Cosmochim. Acta, 3,* 2865 (1972b).

Hoffman, J. H., W. B. Hanson, and C. R. Lippincott, *Radio Sci., 8,* 315 (1973a).

Hoffman, J. H., R. R. Hodges, F. S. Johnson, and D. E. Evans, in *Apollo 17 Preliminary Science Report*, NASA, Washington, DC (1973b), p. 17-1.

Hoffman, J. H., W. H. Dodson, C. R. Lippincott, and H. D. Hammack, *J. Geophys. Res., 79,* 4246 (1974).

Hoffman, J. H., and R. R. Hodges, *The Moon, 14,* 159 (1975).

Hoffman, J. H., R. R. Hodges, and K. D. Duerksen, *J. Vac. Sci. Tech., 16,* 692 (1979).

Hoffman, J. H., R. R. Hodges, Jr., W. W. Wright, V. A. Blevins, K. D. Duerksen, and L. D. Brooks, *IEEE Trans. Geosci. Rem. Sens.* GE-18, 80 (1980a).

Hoffman, J. H., R. R. Hodges, T. M. Donahue, and M. B. McElroy, *J. Geophys. Res., 85,* 7882 (1980b).

Hoffman, J. H., and A. J. Cunningham, Univ. of Texas at Dallas, private communication (1985).

Hoffmann, H. J., K. Pelka, U. vonZahn, D. Krankowsky, and D. Linkert, *IEEE Trans. Geosci. Rem. Sens., GE-18,* 122 (1980).

Hunt, G. E., and P. Moore, *The Planet Venus*, Faber and Faber, London (1982).

Israel, G., D. Imbault, C. Hantz, in *35th Congress Int. Astronautical Fed.*, Paper 214, Lausanne, Switzerland (October 1984).

Jacchia, L. G., J. W. Slowey, and U. von Zahn, *J. Geophys. Res., 82,* 684 (1977).

Johnson, C. Y., and E. B. Meadows, *J. Geophys. Res., 60,* 193 (1955).

Kaye, S. M., R. G. Johnson, R. D. Sharp, and E. G. Shelley, *J. Geophys. Res., 86,* A3, 1335 (1981).

Kissel, J., in *European Space Agency (ESA) SP-169*, eds. B. Battrick and J. Mort (1981), Paris, p. 53.

Kissel, J., Max-Plank Institut für Kernphysik, Heidleberg, private communication (1985).

Krankowsky, D., P. Lämmerzahl, P. Eberhardt, U. Herrmann, J. J. Berthelier, M. Sylvain, J. H. Hoffman, R. R. Hodges, U. von Zahn, J. U. Keller, M. Festou, in *European Space Agency (ESA) SP-169*, eds. B. Battrick and J. Mort (1981), Paris, p. 127.

Liptak, M., W. G. Sandie, E. G. Shelley, D. A. Simpson, and H. Rosenbauer, *IEEE Trans. Nucl. Sci., NS-31,* 780 (1984).

Luhmann, J. G., C. T. Russell, L. H. Brace, H. A. Taylor, W. C. Knudsen, F. L. Scarf, D. S. Colburn, and A. Barnes, *J. Geophys. Res., 87,* A11, 9205 (1982).

Mauersberger, K., *Geophys. Res. Lett., 8,* 935 (1981).

Mauersberger, K., *Adv. Space Res., 2,* 287 (1983).

McCrumb, J. L., and F. Arnold, *Nature, 294,* 136 (1981).

McElroy, M. B., Y. L. Yung, and A. O. Nier, *Science, 194,* 70 (1976).

Moroz, V. J., *Space Sci. Rev., 29,* 3 (1981).

Nagy, A. F., T. E. Cravens, R. H. Chen, H. A. Taylor, J. L. Brace, and H. C. Brinton, *Science*, *205*, 107 (1979).

Narcisi, R. S., and A. D. Bailey, *USAF Office of Aerospace Research*, Paper 82, Bedford, MA (1965).

Narcisi, R. S., A. Bailey, G. Federico, and L. Wlodyka, *J. Atmos. and Terres. Phys.*, *45*, 461 (1983).

Neugebauer, M., D. R. Clay, B. E. Goldstein, and R. Goldstein, *Rev. Sci. Instr.*, *53*, 277 (1982).

Neumann, G., *Dornier-Post (Engl. ed.)*, *2*, 75 (1983).

Niemann, H. B., J. R. Booth, J. E. Cooley, R. E. Hartle, W. T. Kasprzak, N. W. Spencer, G. H. Way, D. M. Hunten, and G. R. Carnignan, *IEEE Trans. Geosci. Rem. Sens.*, *GE-18*, 60 (1980a).

Niemann, H. B., W. T. Kasprzak, A. E. Hedin, D. M. Hunten, and N. W. Spencer, *J. Geophys. Res.*, *85*, A13, 7817 (1980b).

Niemann, H. B., and W. T. Kasprzak, *Adv. Space Res.*, *2*, 261 (1983).

Niemann, H., private communication, 1985.

Nier, A. O., in *AIAA Progress in Astronautics and Aeronautics—Rarified Gas Dynamics*, Vol. 51, ed. J. Potter (1977), p. 1255. American Institute of Astronautics and Aeronautics, New York.

Nier, A. O., J. H. Hoffman, C. Y. Johnson, and J. C. Holmes, *J. Geophys. Res.*, *69*, 979 (1964a).

Nier, A. O., J. H. Hoffman, C. Y. Johnson, and J. C. Holmes, *J. Geophys. Res.*, *69*, 4629 (1964b).

Nier, A. O., and J. L. Hayden, *Int. J. Mass Spectrom. Ion Phys.*, *6*, 339 (1971).

Nier, A. O., W. B. Hanson, M. B. McElroy, A. Seiff, and N. W. Spencer, *Icarus*, *16*, 74 (1972).

Nier, A. O., W. E. Potter, D. R. Hickman, and K. Mauersberger, *Radio Sci.*, *8*, 271 (1973).

Nier, A. O., W. E. Potter, D. C. Kayser, and R. G. Finstad, *Geophys. Res. Lett.*, *1*, 197 (1974).

Nier, A. O., M. B. McElroy, and Y. L. Yung, *Science*, *194*, 68 (1976a).

Nier, A. O., W. B. Hanson, A. Seiff, M. B. McElroy, N. W. Spencer, R. J. Duckett, T. C. D. Knight, and W. S. Cook, *Science*, *193*, 786 (1976b).

Nier, A. O., and M. B. McElroy, *J. Geophys. Res.*, *82*, 4341 (1977).

Owen, T., and K. Biemann, *Science*, *193*, 801 (1976).

Pelz, D. T., C. A. Reber, A. E. Hedin, and G. R. Carignan, *Radio Sci.*, *8*, 277 (1973).

Philbrick, C. R., *Space Res.*, *14*, 151 (1974).

Philbrick, C. R., R. S. Narcisi, D. W. Baker, E. Trzcinski, and M. E. Gardner, *15th Plenary Meeting COSPAR*, Madrid, Spain (1972).

Qui, S., and F. Arnold, *Planet Space Sci.*, *32*, 87 (1984).

Reasoner, D. L., C. R. Chappel, S. A. Fields, and W. J. Lewter, *Rev. Sci. Instr.*, *53*, 441 (1982).

Reber, C. A., D. N. Harpold, R. Horowitz, and A. E. Hedin, *J. Geophys. Res.*, *76*, 1845 (1971).

Reinhard, R., *ESA J.*, *5*, 273 (1981).

Rushneck, D. R., A. V. Diaz, D. W. Howath, J. Rampacek, K. W. Olson, W. D. Dencker, P. Smith, L. McDavid, A. Tomassian, M. Harris, K. Bulota, K. Biemann, A. L. LaFleur, J. E. Biller, and T. Owen, *Rev. Sci. Instr.*, *49*, 817 (1978).

Sechrist, C. F., in *Studies in Geophysics: The Upper Atmosphere and Magnetosphere*, National Academy of Sciences, Washington, D. C. (1977), p. 102.

Seiff, A., and D. B. Kirk, *Icarus*, *49*, 49 (1982).

Sharp, R. D., D. L. Carr, W. K. Peterson, and E. G. Shelley, *J. Geophys. Res.*, *86*, A6, 4639 (1981).

Sharp, R. D., W. Lennartsson, W. K. Peterson, and E. G. Shelley, *J. Geophys. Res., 87,* A12, 10420 (1982).

Slowey, J. W., *Dynamic Model of the Earth's Upper Atmosphere, NASA CR-3835,* National Technical Information Service, Springfield, VA (1984).

Spencer, N. W., H. B. Niemann, and G. R. Carignan, *Radio Sci., 8,* 287 (1973).

Strangeway, R. J., and R. G. Johnson, *J. Geophys. Res., 88,* A3, 2057 (1983).

Taylor, H. A., J. C. Brinton, T. C. G. Wagner, B. H. Blackwell, and G. R. Cordier, *IEEE Trans. Geosci. Rem. Sens., GE-18,* 44 (1980).

Taylor, H. A., J. M. Grebowsky, H. G. Mayr, H. B. Niemann, L. H. Brace, P. A. Cloutier, R. E. Daniell, and J. T. Coulson, *Adv. Space Res., 2,* 291 (1983).

Townsend, J. W., *Rev. Sci. Instr., 23,* 538 (1952).

Trinks, H., and U. von Zahn, *Rev. Sci. Instr., 46,* 213 (1975).

Viggiano, A. A., R. A. Perry, D. L. Albritton, E. E. Ferguson, and F. C. Fehsenfeld, *J. Geophys. Res., 85,* 4551 (1980).

Viggiano, A. A., and F. Arnold, *Geophys. Res. Lett., 8,* 583 (1981).

von Zahn, U., K. H. Fricke, and H. Trinks, *J. Geophys. Res., 78,* 7560 (1973).

von Zahn, U., and K. Mauersberger, *Rev. Sci. Instr., 49,* 1539 (1978).

von Zahn, U., K. H. Fricke, D. M. Hunten, D. Krankowsky, K. Mauersberger, and A. O. Nier, *J. Geophys. Res., 85,* A13, 7829 (1980).

Wallace, J. M., and P. V. Hobbs, *Atmospheric Science,* Academic Press, New York (1977), p. 17.

Whalen, B. A., *Adv. Space Res., 2,* 13 (1983).

CHAPTER TEN

METALS, GLASSES, CERAMICS, AND COMPOSITES

All engineering advances have been paced by the development of new materials. The availability of aluminum, titanium, stainless steel, super alloys, refractory metals, and graphite composites has made possible an aerospace industry that has produced subsonic and hypersonic aircraft and vehicles for space exploration. Every energy conversion system is also limited by the performance of materials, whether the system is a gas turbine, a windmill, or a fusion reactor. In the field of communications, electronic materials have ushered in a new era. The advent of new laser and ion implantation technologies is also having a profound impact on the variety of alloys and metallic glasses that can be used in critical mechanical components and in electric power transmission and distribution networks. In all of these applications, mass spectrometry is now complementing electron microscopy and x-ray analysis in characterizing the microstructures of solids (i.e., crystal and lattice defects, grain boundary impurities, precipitates, and surface inhomogeneities) that determine the macroscopic properties of materials.

ANALYSIS OF METALS AND ALLOYS

Spark source mass spectrometry continues to be a highly useful tool for determining the chemical composition of metals and alloys. When used in conjunction with standard reference materials (NBS standards), virtually all elements can be assayed. In work reported by Van Hoye et al. (1980), a double focusing instrument was used that incorporated a spherical electric sector, a radiofrequency spark source, and both photographic and electrical detection. Samples were analyzed with a spark source voltage of 30 kV, a pulse length of 20 μs, and a repetition frequency of 1 kHz. Nearly all elements had one or more interference-free isotopes; and in the case of zinc, the low abundance isotopes (^{67}Zn and ^{69}Zn) permitted electron multiplier response without saturation. Measurements quantified the relative sensitivity coefficients for P, Mg, Al, Si, Ti, Cr, Mn, Fe, Co, Ni, Cu, Zn, As, Cd, Sn, Sb, and Pb in iron, copper, aluminum, and zinc matrices. The influence of changing sparking conditions and electrode temperatures have also been reported, but the authors conclude that for accurate work, "the use of one or more reference materials of the same matrix remains mandatory."

In an iron matrix, some of the $3d$ elements such as V, Mn, Cr, Ni, and Mo are well known for greatly improving the mechanical properties of steels, whereas P, As, Sn, and Sb are embrittling and are undesirable impurities. Hence, a quantification of these elements remains important in quality steels. The determination of Se and Te in stainless steel, cast irons, and high-temperature alloys has also become important due to their synergistic effects with other elements—even at low concentrations. Kingston et al. (1984) have applied the isotopic dilution method and spark source spectrometry to assay these elements at the μg/g (ppm) and ng/g (ppb) levels, using standard reference materials developed by the National Bureau of Standards. In this work, the samples were spiked with ^{82}Se and ^{125}Te, and quantitation was achieved by measuring the ^{80}Se:^{82}Se and ^{130}Te:^{125}Te isotopic ratios.

The effect of impurities is of special concern with respect to the magnetic and electrical properties of certain alloys, as trace chemistry can affect crystal orientation, grain size, etc. For example, amounts of aluminum as low as 0.01 wt% degrade the magnetic properties of Fe-B-Si-C amorphous ribbons. Furthermore, secondary ion mass spectrometry (SIMS) of such ribbons showed the aluminum concentration to be increased in the surface regions to a depth of about 500 Å, with the increase at the surface of the ribbons being 10-fold (Fiedler et al., 1982). It is conjectured that this impurity concentration at the surface contributes to the formation of an oxide crystalline phase with a resultant decrease in permeability. In contrast, the addition of small amounts of tin and antimony, elements that are known to segregate to grain boundaries in iron, appear to have a beneficial magnetic effect (Fiedler, 1982).

A comparison has also been made by Van Craen et al. (1982) between spark source mass spectrometry (SSMS) and SIMS for the analysis of steel wires. NBS reference material No. 467 low-alloy steel was used for the reference specimen. SIMS results, for most elements, were within 25% of the concentrations determined by SSMS. Primary advantages of SIMS include (1) thin-layer, three-dimensional analysis; (2) detection of all elements including hydrogen; and (3) compound and isotope analysis. So-called "static" SIMS, in which the primary ion current density is less than about 10^{-9} A/cm^2 (total "dose" $\sim10^{12}$ to 10^{13}/cm^2), has also been successfully utilized for the assay of alloy surfaces (Toyokawa et al., 1981).

The preparation of reference standards for metals analysis by SIMS has been aided by ion implantation and has been carried out in a manner similar to the preparation of standards for semiconductors. Separated isotopes serve as dopants in metal matrices at concentrations and spatial distributions that permit quantitative depth-profiling assays. Thus, Pivin et al. (1982) have reported on matrices of Si, Al, Ni, NiO, and Cr_2O_3 and the implanted dopants ^{13}C, ^{29}Si, ^{34}S, and ^{89}Y for quantifying SIMS for specific application to corrosion studies. The preparation of such standards is essential in correcting for variations in secondary ion emission from metallic targets due to oxygen coverage, primary ion beam specie and energy, matrix effects, and other experimental parameters. A general review of problems relating to the quantification of SIMS for steels and other metals has been prepared by Morrison (1982).

Enriched isotopic dopants have also been utilized by Cristy and Condon (1979) for studying oxidation and corrosion. Their work included the use of enriched ^{62}Ni on the surface of a nickel substrate, exposing the substrate to reactants, and depth-profiling the $^{62}Ni/^{58}Ni$ ratio. In a related investigation, they observed water and oxygen reactions via isotopic labeling with ^{18}O and deuterium.

The analysis of bulk samples of metals and alloys by a laser microprobe time-of-flight (TOF) technique is a rather recent instrumental development. For thin specimens (films and powdered materials on thin supporting films), ions are usually extracted on the opposite side of the irradiated surface. For bulk samples, however, new laser TOF systems permit ions to be extracted from the same side as the incident laser beam (Heinen et al., 1983). Maximum power per pulse corresponds to an average power density of 10^{11} W/cm^2 of the irradiated area on the sample, with a spatial resolution approaching 2 μm. A single laser shot on an NBS standard stainless steel (No. 444) revealed Si, Mo, V, Co, Cu, and W at the 0.1% concentration range by weight. More importantly, this technique joins the ion microprobe in detecting inhomogeneous surface distributions of the chemical elements in various matrices.

In addition to general analyses of bulk specimens and surfaces, mass spectrometry is also being used to monitor reaction rates. In oxidizing atmospheres, the corrosion protection of some super alloys results from the formation of an adherent chromium oxide (Cr_2O_3) layer. However, this oxide layer can react with impurities to form low-melting and volatile chromium compounds that have been observed mass spectrometrically. Specifically, Hirayama and Lin (1979) have monitored the formation of these gas reaction products using Knudsen-cell and TOF spectrometry, in connection with materials research on gas turbines.

DIFFUSION IN METALS

Diffusion is a rate-determining parameter in many metallurgical processes, and it is of special concern when materials are subjected to high temperature regimes. Nondestructive methods such as Rutherford backscattering (RBS) spectroscopy and electron microprobe analysis at variable electron beam energies are among the several useful methods for measuring diffusion. However, the highest combined spatial resolution and sensitivity are achieved by either radioactive tracers or secondary ion mass spectrometry (SIMS). Lodding (1980) has cited the salient advantages of SIMS in comparison with other techniques for measuring diffusion:

1. Tracer concentration and penetration depth data are acquired simultaneously; within 1 h of sputtering time, penetration increments of $0.1–10$ μm can be analyzed.
2. Detection sensitivity for most elements is very high with stable isotopes.
3. Information on tracer depth in single crystals or amorphous solids can be obtained at very high resolution, from a few hundred angstroms to a single monolayer.
4. Very low diffusion coefficients ($\sim 10^{-16}$ cm$^2 \cdot$s^{-1} or lower) have been measured by SIMS.

The diffusion of metals in metals can also be studied mass spectrometrically using thermal ionization, spark probe, or laser probe. Surface diffusion and the diffusion of gaseous species may require special techniques; and in most instances, the availability of stable isotopes is exploited. Furthermore, the study of isotope effects, per se, is useful in differentiating between possible diffusion mechanisms, because in fine-grained polycrystalline solids, both lattice and grain boundary diffusion contribute to the effective bulk diffusion. In fact, atom transport along grain boundaries is the dominant diffusion mode in fine-grained specimens, and distinct diffusion coefficients for 316 stainless steels have been reported by Patil and Sharma (1982).

Diffusion phenomena on the surface layers of metals have been reviewed by Ehrlich (1982) for polycrystalline metals, single-crystal planes, and for both electronegative gases and metal atoms. The field ion microscope even permits the direct observation of individual atoms and their displacement at a surface; however, this technique is more appropriate for metal atoms, as electronegative gases are affected by the imaging electric fields.

Many investigators have reported on diffusion in aluminum and aluminum alloys. Vanderlinden and Gijbels (1980) have measured the diffusion of Li and Mg in aluminum over a large temperature range using the ion microprobe. The use of stable-enriched isotopes is especially useful, since no radioactive tracers of Li are available, and the only suitable radioactive isotope of Mg must be prepared via cyclotron or photon activation of ^{26}Mg or ^{30}Si, respectively. Isotope effects for the diffusion of lithium and magnesium impurities have also been determined by precise measurements of ^{6}Li:^{7}Li and ^{25}Mg:^{26}Mg ratios.

The rates of atomic transport vary over phenomenal limits. For example, the interstitial diffusion for carbon in *bcc* iron is such that an atom migrates about 10^{-5} cm in 10^7 s (~ 4 months) at room temperature (Cottrell, 1964).

This corresponds to a diffusion coefficient D, of approximately 10^{-17} $cm^2 \cdot s^{-1}$. In contrast, gold will diffuse through a 1.5 mm silicon wafer at 1000°C in about 20 min (Sprokel and Fairfield, 1965), corresponding to a diffusion coefficient of about 2×10^{-6} $cm^2 \cdot s^{-1}$.

The classical basis for computing the number of diffusing atoms in a volume element between two parallel planes x and $x + dx$ is given by

$$\frac{\partial c}{\partial t} = D\frac{\partial^2 c}{\partial x^2}$$

This is Fick's second law for the case when the diffusion coefficient does not depend upon the concentration c. In the general case, a variation of D with temperature should be expected to follow the exponential relationship $D = Ae^{-(E/RT)}$, where A is a constant and E is the "activation energy"; namely, the energy required to allow atoms to migrate over potential barriers that impede their motion. If the diffusion sample is semi-infinite (with a

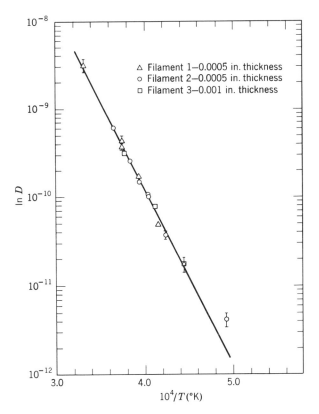

Figure 10.1 Diffusion coefficient of ^{235}U in polycrystalline tungsten. The slope of the graph indicates an activation energy (Schwegler and White, 1968).

thickness $\gg$ penetration depth), a solution to the above equation will have the form

$$c = \frac{Q}{2\sqrt{\pi Dt}} \exp\left(\frac{-x^2}{4Dt}\right)$$

Here c represents the concentration of diffusant atoms at a distance x, and Q is the surface density of original diffusant atoms deposited upon the sample. If a thin sample (not semi-infinite) is used, or initial conditions are different, a somewhat more complex expression is required.

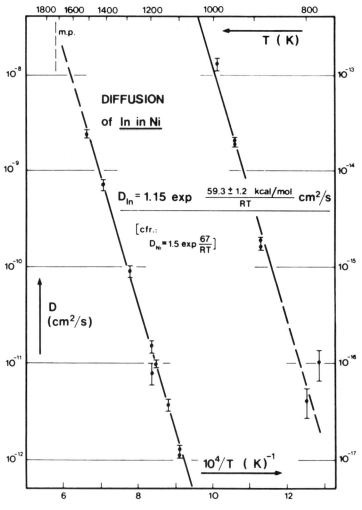

Figure 10.2 Diffusion of indium in single-crystal nickel as measured by SIMS. (Gust et al., 1980; Hintz, 1979; courtesy of A. Lodding.)

Two examples will be cited to demonstrate different measurement techniques for determining diffusion rates and diffusion coefficients. Schwegler and White (1968) measured the diffusion of uranium in tungsten by a surface-ionization method. A uranium-doped filament was heated to a temperature above the boiling point of uranium for the ambient pressure, and a small fraction of the uranium atoms were thermally ionized by the high work function tungsten surface, mass analyzed, and pulse counted at various filament temperatures. Data for the diffusion of ^{235}U in tungsten is shown in Figure 10.1.

SIMS has been used to measure the diffusion of indium in nickel single crystals over an exceedingly large temperature range. This work by Gust et al. (1980) and Hintz (1979) is shown in Figure 10.2, where D was measured between 0.44 T_m and 0.88 T_m (T_m = melting temperature).

As mentioned above, the SIMS technique has also been applied to determining the isotope effect in diffusion, and for identifying defect mechanisms affecting atomic mobilities. This type of measurement sometimes requires an accuracy of 1% in isotopic abundance ratios at the ppm level. Hehenkamp et al. (1979) have reported on the isotope ratios of Ge diffused into Cu, using the five stable Ge isotopes. Using a 5000 Å electrodeposited Ge film, the tracer concentrations ranged from $\sim$100 ppm next to the surface to $\sim$0.1 ppm at a 1 mm depth in the copper specimen.

Mass spectrometric measurements have also been made of the extent of

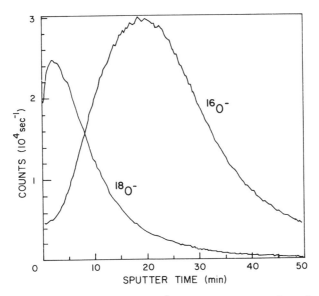

Figure 10.3 Negative SIMS profiles of a 470 Å chromium oxide film formed on Cr [100] at 700°C; a 330 Å film was produced in a $^{16}O_2$ environment followed by a 140 Å layer during exposure to $^{18}O_2$. (Mitchell et al., 1983; *J. Vacuum Science Technology*, *A1*, 991 (1981), by permission of the American Institute of Physics.)

oxygen diffusion in oxide scales formed on metals at high temperatures. Using a sequential oxidation in $^{16}O_2$ and $^{18}O_2$ followed by SIMS profiling, Mitchell et al. (1983) have performed analyses of thick oxide films on chromium. Figure 10.3 shows the $^{16}O^-$ and $^{18}O^-$ SIMS profiles of the 470 Å chromium oxide film formed on Cr [100] at 700°C in 5×10^{-3} torr oxygen. A 330 Å oxide film was produced in $^{16}O_2$ followed by 140 Å in $^{18}O_2$. The negative ion spectra were produced from a 4 keV primary beam of xenon ions.

HYDROGEN DIFFUSION IN ALUMINUM

Hydrogen-metal systems are of both practical and theoretical interest because dissolved hydrogen can be regarded as the simplest impurity. The solubility of hydrogen in aluminum is very low, about 1 ppm at atmospheric pressure near the melting point, and desorption experiments to measure diffusion are difficult because of surface effects. However, Hashimoto and Kino (1983) have successfully measured the diffusion of hydrogen in aluminum by the permeation method, using a quadrupole mass spectrometer (QMS) and an ultra-high vacuum system. Their aluminum specimens were made from zone-refined aluminum cold-rolled to thicknesses ranging from 50–500 μm and cut into discs 40 mm in diameter. These samples were then annealed in a vacuum at 400°C for 15–20 h prior to the permeation runs for hydrogen at various temperatures; the hydrogen permeation rate through the specimens was determined by monitoring the H_2^+ ion current. The diffusivity D of hydrogen in aluminum was then obtained from the time-temperature permeation data and the diffusion equation:

$$\frac{\partial c(x, t)}{\partial t} = D \frac{\partial^2 c(x, t)}{\partial x^2} \qquad \text{for } 0 < x < L$$

with the boundary conditions $c(0, t) = c_0$ and $c(L, t) = 0$ for all t, where (1) $c(x, t)$ is the fractional concentration of hydrogen as a function of position x and elapsed time t; (2) L is the specimen thickness; and (3) c_0 is the fractional concentration of hydrogen at the specimen surface. Calculated and experimentally determined permeation rates of hydrogen through a 95 μm thick specimen at several temperatures are shown in Figure 10.4, where the ion current is proportional to the permeation rate. Arrhenius plots of diffusivities obtained from the permeation data of Figure 10.4 and other aluminum specimens correspond to an activation energy of 0.61–0.62 eV.

From these and other data, Hashimoto and Kino (1983) have concluded that the diffusivity of hydrogen depends upon the relative concentrations of vacancies to the hydrogen concentration. At high temperatures, where the concentration of vacancies is higher than that of dissolved hydrogen, the

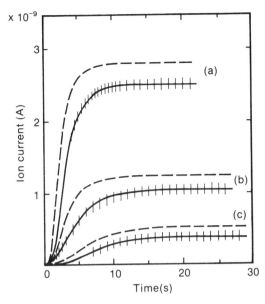

Figure 10.4 Calculated (dashed) and experimental (solid) curves showing the permeation rate of hydrogen through a 95 μm aluminum specimen at (*a*) 388°C, (*b*) 368°C, and (*c*) 348°C. (Hashimoto and Kino, 1983; *J. Physics*, F; *Met. Phys.*, *13*, 1157 (1983), by permission of the Institute of Physics, London.)

activation energy is near the migration energy of a vacancy in aluminum. At room temperature, however, where the hydrogen concentration greatly exceeds that of the vacancies, the diffusivity of hydrogen is much higher than that extrapolated from high temperatures.

DETECTION OF PHASE CHANGE

The suggestion was first made by Honig (1966) that inasmuch as diffusion measurements can be made mass spectrometrically over a large temperature range, it should also be possible to detect phase changes that occur in bulk material at high temperature by using a surface-ionization source. Iron, for example, will change its lattice from body-centered to face-centered cubic above 900°C, and zirconium is known to have an allotropic phase change at ~860°C. Hence, changes in the diffusion rate of impurity atoms could be monitored to detect the phase change. Accordingly, Schwegler and White (1968) used a polycrystalline zirconium ribbon, which contained approximately 50 ppm rubidium as an impurity, as a thermal ionization source. This ribbon filament was cycled in temperatures from 770–1030°C and subsequently down to 785°C. The diffusion coefficients of rubidium in zirconium

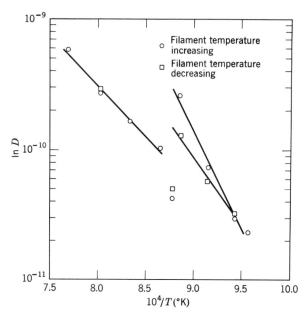

Figure 10.5 Mass spectrometric detection of the allotropic phase change in zirconium, using rubidium as the diffusing impurity (Schwegler and White, 1968).

at several temperatures were obtained by single-ion counting of the rubidium ions leaving the zirconium ribbon filament. The presence of the allotropic phase change was then clearly discernible from the discontinuity in the ln$D$ vs. inverse temperature plot of Figure 10.5. The activation energy for rubidium diffusion is considerably higher in the zirconium α phase than in the β phase. Also, this α phase activation energy seems to be larger before the filament has been subjected to the phase change than it is after cycling through the phase transition. This surface-ionization method is especially attractive for high-temperature phase change measurements of metals having high work functions and using diffusant atoms having first-ionization potentials below about 6 eV.

LASER SURFACE ALLOYING

Laser transformation hardening was first applied to cast iron for industrial applications. The initial laser heating raises the surface temperature of the metal, and the carbon exists in a solid solution with the iron in a soft austenitic phase. Subsequent rapid cooling alters the distribution of the carbon, and the surface layers are transformed into a hard martensitic phase. Laser surface alloying is an important cognate process that requires several forms of spectroscopy to provide detailed information on microstructure, depth profiles of the alloying elements, and impurity concentrations. SIMS and RBS are among the state-of-the-art techniques. Increased interest in metal surface alloying has been stimulated by the advent of high-powered lasers that can heat and melt thin layers of material (< 1—100 μm) at rates exceeding 10^8 °C/s. Rapid quenching after laser melting, together with the impregnation of alloying elements, can lead to the formation of unique alloys and to solute concentrations considerably higher than those predicted by equilibrium phase diagrams. These modified surfaces can be generated in a large variety of substrates, and concentration gradients can be tailored as a function of depth by means of appropriate laser programming. Another salient advantage of laser alloying is that surface layers can be heated with minimal heat transport into the bulk of the metal, thus permitting thermal conduction quenching to harden the surface without bulk heating and distortion. In addition to alloying, per se, laser heating offers the potential of (1) transformation hardening; (2) laser glazing; (3) cladding; and (4) laser melt/particle injection. The alloying or impregnated elements may be deposited on the host metal by electrodeposition, vapor deposition, sputtering, or as a powder. In the case of particle injection, wear-resisting particles of TiC and WC have been deposited as powders via a helium stream on surface layers of iron-, nickel-, titanium-, aluminum-, and copper-based alloys (Ayers et al., 1981). This latter process is expected to produce new metal/carbide combinations for cutting tools and abrasion-resistant surfaces.

Most recent laser alloying has been done with multikilowatt, continuous-wave CO_2 lasers, with special emphasis on producing stainless steel quality surfaces on ferrous alloy substrates by the addition of chromium, molybdenum, and nickel (Draper, 1982). RBS has been used to determine elemental impurity profiles (for depths $\leq 10,000$ Å) for a variety of dopants and laser irradiations. For many practical applications, the thickness of the alloyed region must extend to depths of 100 μm or greater, and quenching rates (10^5 to 10^6 °C/s) will be much lower than for shallow alloys. However, by controlling the laser power and dwell time, it is possible to manipulate microstructures and macroscopic physical properties (Narayan, 1980). In any event, the production of unique metal surfaces can be envisioned by this new technology—with surface-alloying concentrations in the range of 60% or greater. Strategic and noble metals can be conserved, "stainless steels" can be produced having electrochemical characteristics of 304 bulk alloys, and wear and friction properties can be tailored for specific applications.

ION IMPLANTATION OF METALS

Ion implantation has become one of the most important technologies in the electronics industry for doping semiconductors and in fabricating integrated circuits. This technique is now being applied by metallurgists for processing critical components and producing surfaces having unique properties. Over the long-term, ion implantation may also provide an option for greatly reducing the consumption of strategic materials. Potential advantages include the following:

1. Almost any element in the Periodic Table can be implanted.
2. In contrast to electroplating that produces a discrete coating, ion implantation changes the near-surface composition of the base metal to create a new alloy.
3. The alloyed region or complete component may contain 100–1000 times less strategic materials than required by conventional plating or other techniques.
4. The spectrum of alloys that can be created is not limited by classical thermodynamic parameters of diffusion and solubility.
5. Problems of bonding or adhesion are reduced, since alloying atoms are embedded within the crystal lattice of the substrate.
6. Surface properties can be optimized independently of the bulk properties, especially with respect to improved resistance to corrosion and wear.
7. The distributions of ion-implanted atoms in the substrate are usually well defined, and concentrations can be controlled and reproduced.
8. Ion implantation can be applied at low temperatures; and because there are no significant dimensional changes, the technique can be applied to components after their final stages of fabrication and finishing.
9. The technique also provides a means of depositing energy at a high rate into near-surface layers, so that the interdiffusion of a coating and a substrate can occur by radiation-enhanced diffusion.

Ion implantation technology is also unique in that the same generic particles (energetic ions) used to produce surface alloys are used for the post-implantation assay of dopant profiles by Rutherford backscattering (RBS) or secondary ion mass spectrometry (SIMS). The kinetic energy of the implanted ions can be varied over a wide range, but they are typically five orders of magnitude greater than the binding energy of lattice atoms of the substrate. Hence, during the very short time ($\sim 10^{-13}$ s) required for these implanted ions to be thermalized, their motion and detailed spatial distributions will be determined by atomic collision cascades (Sigmund, 1974). A final concentration profile will be conditioned by diffusion, if post-implantation annealing is applied; and a much greater penetration of a dopant element can be achieved by a subsequent bombardment by nitrogen.

The quantitative ion implantation of accurately known microgram-to-picogram quantities of a chemical element (usually a selected isotope) into the surface of a solid also can provide excellent reference standards not only for calibrating SIMS, but for other types of instrumental analysis (Gries, 1979).

Corrosion Resistance

The layers of highly protective compounds that form on many metals range in thickness from a few monolayers to about 10^{-6} cm, so that even very shallow ion implants can alter corrosion rates. In developing conventional corrosion-resistant alloys, the basic technical approach is to achieve passivity by incorporating elements that lower the passivation potential and critical current density. With ion implantation, a much greater latitude exists for effecting passivation, as solid solutions can be formed by incorporating elements that are normally insoluble under equilibrium conditions. Ion implantation also permits experimentation with unconventional alloying additives such as metalloids, and the kinetics of ion implantation lead to the removal of grain boundaries and the creation of amorphous layers (Clayton, 1981).

The implantation of 50 keV boron ions into Armco iron has been shown to enhance the passivating properties of iron oxides, and a drastic change in the surface morphology has been reported for 10^{16} B^+ ions/cm^2 implanted samples (Mazzoldi et al., 1982). Iron samples have also been implanted at 200 keV with Sn ions at doses between 5×10^{15} and 5×10^{16} ions/cm^2 (Baumvol, 1981); a reduction in the oxidation rate constant of up to a factor of 10 was noted in the temperature range of 300–500°C, with the surface being described as similar to that obtained with the more conventional method of tin-plating.

The implantation of aluminum and magnesium, which effectively inhibits the corrosion of copper, has also been studied by RBS. Thin films of copper were implanted with 60 keV $^{27}Al^+$ or $^{24}Mg^+$ ions from an isotope separator at a dose of 2×10^{17} ions/cm^2; then they were oxidized in air at 260°C (Svendsen and Børgesen, 1981). Compared with unimplanted specimens, these samples oxidized at a slower rate and, as shown in Table 10.1, the oxidation process stopped for the magnesium-implanted samples after 1 h. Apparently the implanted species form a protective layer and prevent further oxidation. Specifically, the unimplanted reference film was oxidized to Cu_2O at a rate of 2×10^{18} Cu atoms/cm^2 (4000 Å)/h, but the Mg-implanted corrosion was limited to a very shallow depth—less than ~7×10^{17} Cu atoms/cm^3 (~1500 Å). In the case of the aluminum implant, a thicker oxide layer was noted and tentatively attributed to the diffusion of copper through the implanted layer.

Specific examples of improved corrosion resistance by ion implantation include the treatment of main shaft bearings of turbo-jet engines. Pitting and corrosion occur when chlorides and water accumulate in the lubricants of aircraft in intermittent use, and the resultant fatigue spalling can lead to

TABLE 10.1 Total Oxygen Content vs. Oxidation
Times for Cu Films

Time (h)	Oxygen Content ($\times 10^{17}$ atoms/cm^2)		
	Unimplanted	Mg-Implant	Al-Implant
0.25	2.7	1.7	1.6
0.50	4.3	3.2	4.0
1.00	9.8	4.5	6.0–8.9
2.00	9.8	4.5	8.9
4.00	9.8	4.5	8.9

Source: (Svendsen and Børgesen (1981).

catastrophic failure (Wang et al., 1979). Steels that have been used for such bearings and that have been implanted with Cr, Cr + Mo, Cr + P, or Ta show a marked improvement relative to general corrosion and pitting corrosion (Clayton and Wang, 1983). Ion implantation has also been known to reduce hot gas erosion of nozzles in the oil-refining industry, and Preece and Kaufmann (1982) have observed a decrease in the cavitation erosion of boron-implanted copper and nickel.

Reduction in Wear Rates

Significant changes have been reported in the wear rates of steels ion-implanted with relatively high surface concentrations of nitrogen. The production of a hard and wear-resistant surface by ion implantation thus appears to provide an interesting alternative to expensive alloys when high mechanical strength is not required in the bulk material. The volumetric wear rate reduction of a typical steel as a function of nitrogen dose, implanted at 50 keV, has been reported by Dearnley and Hartley (1978) and Dearnley (1982). An improvement of > 30% at high doses is attributed, in part, to ion-induced dislocations in the steel that render it resistant to both adhesive and abrasive wear. Iron nitrides also form; but after annealing, the nitrogen is mobile, so that the effect of the ion implant is many times the depth of the initial penetration. It is this shallow kinetic implant of dopant atoms (which subsequently results in as much as a 100-fold effective penetration) that has generated considerable practical interest in metallurgical applications of implantation. In one test performed with 38NCD4 steel implanted with a dose of 2×10^{17} ions/cm^2 of nitrogen, the steady-state part of the wear curve (weight loss vs. time) was prolonged by a factor of 3, thus demonstrating that the mechanism responsible for severe wear can be delayed (Tosto, 1983). In this instance, the enhanced wear life was correlated with the surface precipitation hardening caused by the implantation. Improved wear rates for C^+ and B^+ implants in steel have also been reported by Harding (1977).

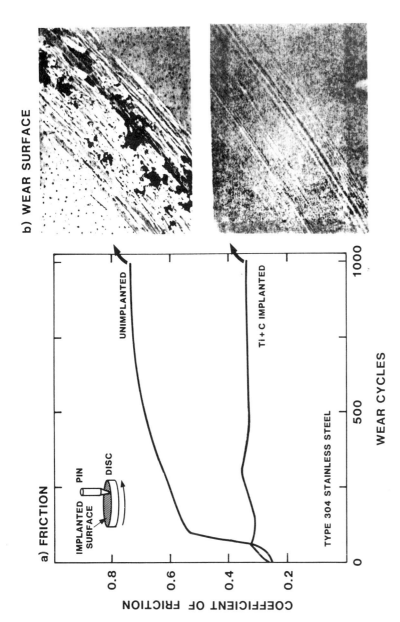

Figure 10.6 Coefficient of friction versus number of wear cycles with and without titanium and carbon implantation to form an amorphous iron (20 atom % Ti; 20 atom % C) layer in type 304 stainless steel. (Picraux and Pope, 1984; *Science*, 226, 615 (1984); courtesy of S. T. Picraux and by permission of the American Association for the Advancement of Science, copyright, 1984).

Reduced friction and wear have also been reported for carbon- and titanium-implanted layers in iron and stainless steel alloys (Singer and Jeffries, 1983; Nelson et al., 1983). Figure 10.6 shows the substantial reduction in friction for a titanium + carbon-implanted stainless steel (Picraux and Pope, 1984).

In some instances, the decreased wear rates of implanted specimens has been dramatic. The implantation of rare earth ions (yttrium) has resulted in reducing the wear rate of a steel component by a factor of 100, and extended lifetimes of cutting and forming tools, roll mills, ball bearings, valves, and molds used for the injection molding of plastics have been reported by Dearnley (1982).

Ion implantations have also produced marked increases in fatigue life. Hu et al. (1980) have tested AISI 1018 steel (0.18% C) fatigue specimens that were implanted with N_2^+ ions at 150 keV to a fluence of 1×10^{17} ions/cm^2. Fatigue results showed that both unimplanted and as-implanted samples failed at about 10^6 cycles, but samples implanted and "aged" at 100°C for 6 h (to simulate 4 months at room temperature) failed at about 10^8 cycles, that is, a two-order of magnitude improvement. The authors suggest that the results of this and other experiments emphasize the profound effect of near-surface layers on the macroscopic properties of metals. Frictional properties, as well as wear rates, have been studied for a variety of ion implanted species, and secondary ion mass analysis has been used to investigate the elemental concentrations that affect friction. In a study of Ni and Cu implanted in low carbon steel, Iwaki et al. (1981) have concluded that the number of implanted ions remaining in the surface layer (0–400 Å) is of primary importance.

Hirvonen (1980) has edited a compendium of papers that provides a general overview of ion implantation, and Picraux (1984) has evaluated both the advantages and limitations of ion implantation technology as it relates to several important metallurgical applications. Specific improvements in ion-implanted components have also been cited by Sioshansi (1984) in connection with drilling, orthopedic implants, bearings, extrusions, punching and stamping dies, and jet nozzles. In some of these industrial applications, the surface modification of materials is accomplished by ion beam mixing or plasma ion synthesis as well as by direct ion implantation processing.

ION BEAM DEPTH PROFILING

It is clear that ion beams provide the best means for probing the elemental composition of ion-implanted metals. Typical implantation depths are less than a few hundred nanometers for ions in the range of 10–200 keV, so that RBS or SIMS provide excellent near-surface resolution. In contrast to nuclear reaction analysis (e.g., $^{14}N[d,\alpha]^{12}C$ for nitrogen implants), these techniques apply to virtually all ion-implanted species. Furthermore, ion beam analysis techniques provide the quantitative standards against which other profiling methods, destructive or nondestructive, are compared (Hubler, 1981). The ion-implanted concentrations in metals are two to four orders of magnitude greater than in many semiconductor applications, so that the detection-counting rates of ion-implanted surface alloys are obtained with high signal-to-noise ratios.

Implanted species are initially dispersed on an atomic scale throughout the host lattice. In general, precipitates or clusters do not occur even at high doping levels unless the metals are heated to a temperature wherein the diffusion lengths $\sqrt{Dt}$ for all constituents are large compared to the lattice spacing. At high temperatures, the injected species can be expected to migrate, and a typical diffusion has been reported by Myers (1980). In this investigation, Cu atoms were initially implanted in Be, and a subsequent annealing at 673 K for 94 h shifted the distribution according to classical diffusion theory.

Campbell et al. (1980) have reported depth profiles for 25 keV Cr$^+$ and

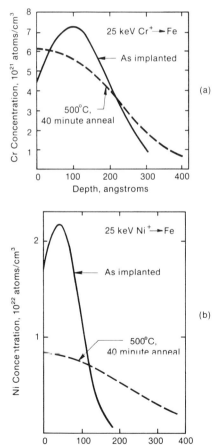

Figure 10.7 (a) Depth profiles of chromium-implanted iron, as implanted and after a 500°C, 40 min anneal. (b) Depth profiles of nickel-implanted iron, as implanted and after a 500°C, 40 min anneal (Campbell et al., 1980).

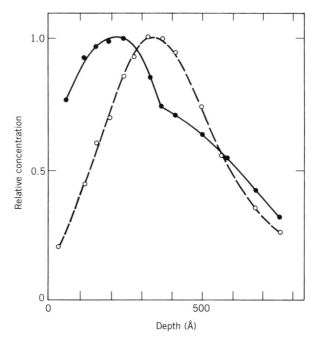

Figure 10.8 Concentration profiles of the double-ion implantation of Ni ($5 \times 10^{16}/cm^2$) and Cr ($1 \times 10^{17}/cm^2$) at 150 keV. Closed and open circles indicate the relative concentrations of chromium and nickel, respectively. (Okabe et al., 1981; *Nucl. Instr. Methods*, 182/183, 231 (1981), by permission of North-Holland Physics Publishing.)

Ni^+ in iron, and data for "as implanted" and after a 500°C, 40 min anneal are shown in Figures 10.7a and 10.7b. In this work, inert gas ion sputtering was supplemented by proton-induced x-ray emission analysis to generate elemental profiles for three different types of implanted alloys: Fe-Ni, Fe-Cr, and Fe-Al; fluences were greater than 10^{16} atoms/cm^2.

A secondary ion mass analysis has been used to determine the concentration profiles of chromium and nickel in pure iron, when both dopants were implanted in the same specimen at high doses. Figure 10.8 shows results of the double-ion implant reported by Okabe et al. (1981) that was carried out with 5×10^{16} Ni^+/cm^2 at 150 keV after performing a chromium implant with 1×10^{17} Cr^+/cm^2 at 150 keV. The authors conclude that the concentration profile of the first ion implant is shifted toward the surface due to the sputtering effect of the second implantation, lattice defects, and diffusion.

GRAIN BOUNDARIES AND MICROSTRUCTURE

The ion microprobe has become one of the most powerful techniques in metallurgical analysis, as it permits grain boundary chemistry to be distin-

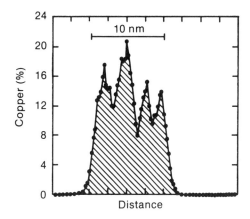

Figure 10.9 Atom probe analysis of a steel pressure vessel specimen that had undergone a 5 year neutron irradiation at 290°C, revealing the presence of small copper-rich regions. The ultra-high resolution of a field-ion microscope combined with a mass spectrometer make possible this type of assay. (Brenner and Miller, 1983; with permission from *J. Metals*, *35*, No. 3, 54 (1983), a publication of The Metallurgical Society of AIME, Warrendale, Pa.)

guished from bulk impurities. In some instances, order-of-magnitude differences exist in the elemental composition of the grain boundaries and the crystalline structure. The field ion microscope/mass spectrometer has also been used to elucidate grain boundary chemistry; and by the process of field evaporation, some degree of three-dimensional microstructure can be revealed. A distinction between phases and a spatial resolution of ultra-fine precipitates are also possible.

The practical significance of atom probe analysis can be seen from Figure 10.9, which is the result of several hundred thousand field-evaporated atoms being emitted from a random area scan of a nuclear reactor monitor test capsule (Brenner and Miller, 1983). The capsule had undergone neutron irradiation for 5 years at 290°C; the analysis was performed to locate the residual copper (0.17%) and phosphorus (0.011%) in the microstructure in order to explain the embrittling effect of these elements in steel pressure vessels. The high-resolution spatial profile revealed copper-rich regions as high as 20%, presumably formed as a result of the prolonged neutron irradiation at high temperature. Some phosphorus-rich zones were also detected. An analogous microstructural problem in steels is the chromium depletion adjacent to chromium-rich carbides precipitated at grain boundaries, where local chromium depletion zones can give rise to corrosion and the initiation of cracks (Loria, 1982). The segregation of phosphorus at grain boundaries in Ni-Cr and Ni-Cr-Mn steels has also been observed by Auger analysis (Clayton and Knott, 1982), as well as by mass spectrometric techniques.

Atom probe analysis has been used extensively for examining the grain boundary segregation of impurities and alloy elements in steels—relating to problems of temper embrittlement and the variation of ferrite compositions with aging. Wada et al. (1982) have reported depth profiles of Mn, Mo, Cr, P, C, and Si in 2.25 Cr-1 Mo water-quenched steel samples; and they also demonstrated that the microvolume assay by an atom probe was in good

TABLE 10.2   Comparison of an Atom Probe Assay of a 150 nm³ Volume to a Macrochemical Analysis of a 2.25 Cr–1 Mo Steel Sample[a]

	C	Si	Mn	P	Cr	Mo	(N_0)
Atom probe assay	0.91	0.54	0.48	0.065	2.11	0.43	[13,849]
Chemical analysis	0.79	0.65	0.56	0.077	2.39	0.59	

Source: Wada et al. (1982).

[a] N_0 is the total number of detected ions.

agreement with a macrovolume chemical analysis of a specially prepared sample. Table 10.2 shows a comparison of the atom probe assay from only 13,849 detected ions compared to the chemical analysis. The chemistry of regions between polycrystalline aggregation also influences tensile strength, yield, ductility, and hardness. In some instances, the yield strength of a material may change by as much as 1000% as a result of grain size and grain boundary occlusions. The variation of hardness, attributed by Westbrook and Aust (1963) to impurity concentrations of only 1 ppm in a Pb specimen, is shown in Figure 10.10. The role of grain boundaries on hardness, and the general nature of surfaces as they relate to friction and wear, are further reviewed by Buckley (1982). Figure 10.11 is a more detailed ion microprobe representation of the grain boundary chemistry of a specimen (Katz, 1984). This shows grain boundary Ni^-, C^-, H^-, and O^- images of a nickel-based alloy with a 150 μm image field. By raster scanning at high resolution, the spatial distribution of many elements can be assayed at the parts per million level.

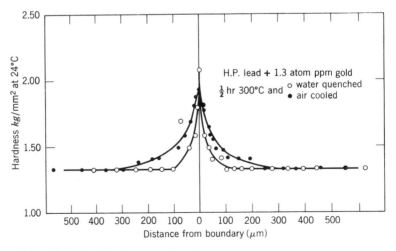

Figure 10.10 Variation of hardness, attributed to impurity concentrations near a grain boundary (Westbrook and Aust, 1963).

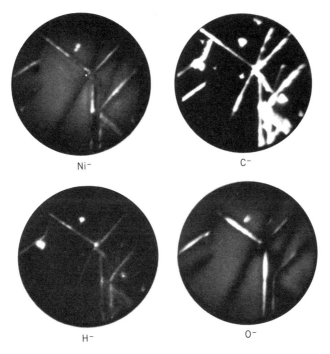

Figure 10.11 Grain boundary Ni^-, C^-, H^-, and O^- images of a nickel-based alloy, with a 150 μm image field (Katz, 1984).

Other materials analyzed for their microstructure by SIMS include tungsten lamp filaments that derive their unique high-temperature creep resistance and strength from the addition of small amounts of K, Al, and Si. For example, Pebler et al. (1975) used a mass spectrometer to image, for the first time, potassium-filled voids in annealed tungsten. The ion sputtering of these occlusions yielded data that permitted the measurement of their number density and spatial distribution, and it confirmed the persistence of these K bubbles throughout lamp filament lifetime.

MULTIELEMENT THIN FILMS AND ION BEAM SYNTHESIS

Thin films of cubic boron nitride have been formed at room temperature by the simultaneous evaporation of boron via electron-bombardment and nitrogen ion bombardment. Their deposition rates on a substrate were controlled by a thickness monitor for boron and by an ion current measurement for the 40 keV N_2 molecular ions (Satou and Fujimoto, 1983). The electric resistivities of samples prepared by various compositions and deposition rates are given in Table 10.3. The authors report that at a B/N composition

TABLE 10.3　Electrical Resistivity and Composition of Boron Nitride Films Prepared by Evaporation/Ion Implantation

Sample No.	Composition (B/N) Ratio	Thickness (Å)	Deposition Rate (Å/min)	Resistivity (Ω cm)
1	0.7	4000	38	
2	1.5	5500	65	4.2×10^{10}
3	1.8	5600	75	3.9×10^{10}
4	2.2	5400	74	3.2×10^{10}
5	2.5	6800	102	1.9×10^{10}
6	2.5	3000	42	1.9×10^{10}
7	2.7	5200	98	1.7×10^{10}

Source: Satou and Fujimoto (1983).

ratio of 2.5, a diffraction pattern and a lattice constant of 3.6 Å was observed corresponding to cubic boron nitride crystals. The production of such crystals, which are as hard as diamond, usually requires high temperatures, a catalyst, and pressures in the range of 10^4 to 10^5 atm.

Thus, in addition to using ion beams for implanting dopants in substrates, energetic ion beams have been employed to synthesize crystalline materials. Thin films containing cubic phase boron nitride have been synthesized using ions extracted from a borazine ($B_3N_3H_6$) plasma (Shanfield and Wolfson, 1983). Considerable effort has also been devoted to the synthesis of diamond by chemical vapor deposition, ion beam techniques, and plasma-induced vapor deposition. Microcrystalline diamond has even been formed on silicon and molybdenum substrates by vapor deposition from a gaseous mixture of methane and hydrogen (Matsumoto et al., 1982).

Another technique for producing multielement films is to use ion beams to induce "mixing" of two or more thin layers of elements that have been prepared by sequential vacuum deposition onto inert substrates such as SiO_2 or sapphire. Ion bombardment with inert gas ions (such as Xe^+) has then resulted in the formation of a homogenized solid solution of the two elements. This process of metastable phase production by ion beam mixing of very thin alternative layers has been termed "microalloying," and backscattering analyses with 2 MeV $^4He^+$ ions has been employed to determine the composition and depth of the alloy region (Tsaur et al., 1981).

The effect of heavy ion bombardment on RF-sputtered films of Ti, Mo, Ta, and W has been reported by Padmanabhan et al., (1982). Figure 10.12 shows the RBS spectrum from an RF-sputtered TiC film on Si. A 2 MeV $^4He^+$ beam from a Van de Graaff accelerator was used, and a surface barrier detector oriented at 160° monitored the backscattered ions. From this data, it is possible to monitor the thickness of the sputtered carbide films and their elemental composition. For the quantitative analysis of very thin (mono-

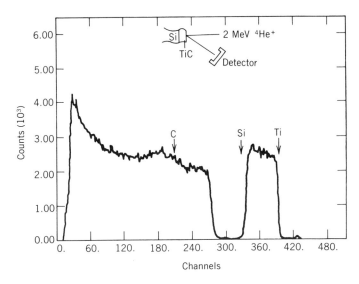

Figure 10.12   RBS spectrum of an RF-sputtered TiC on Si. (Padmanabhan et al., 1982; *J. Vacuum Science Technology*, *20*, 1406 (1982), and by permission of the American Institute of Physics.)

layer) films on metallic substrates, RBS is an especially powerful technique. For example, Frerichs et al. (1981) have analyzed electrochemically deposited layers of Pb on polycrystalline silver. The analysis was made with a $^4He^+$ ion beam of 0.5 MeV energy and resulted in a depth resolution of 22 Å (full width at half maximum [FWHM]) for both the Pb surface layer peak and the leading edge of the substrate spectrum. The area under the clearly resolved Pb peak corresponded to a total of only about 1000 counts, which were correlated to an actual number density of lead atoms. An overall review of RBS as a tool in surface composition has been prepared by Feldman (1982), who has also compared this technique to low-energy electron beam diffraction and Auger surface analysis.

Electrical properties of both metallic and insulating thin films are affected by the ion implantation of gaseous species. Changes in the conductivity of thin copper, aluminum, and bismuth films bombarded with N_2^+ and Ar^+ have been noted, and marked changes in the photoconductivity of N_2^+-implanted Bi_2O_3 has been reported (Ogale et al., 1982). The magnetic properties of sputtered Co-Pt alloy films have also been shown to be dependent upon both the platinum content and the argon pressure of the sputtering process (Naoe et al., 1983). These detailed analyses are of special relevance in achieving a high recording density in magnetic tapes and discs, where it is necessary to reduce the thickness of the recording medium and to increase the coercive force of the material.

Mass spectrometry is also employed for monitoring the many fabrication techniques for producing films required for heat insulation. In this case, the films must reflect broadband infrared heat radiation, but transmit light effectively in the visible region. Films of In_2O_3, SnO_2, and Cd_2SnO_4 are among the many transparent conducting oxides that have been investigated whose resistivity and optical properties are affected by oxygen pressure, doping, and substrate temperature (Dawar and Joshi, 1984).

Other surface diagnostics relate to electroplating and coating processes and to advanced vacuum plasma methods of deposition. In this latter category, a metallic powder is injected into a high-temperature plasma and molten particles are formed that solidify instantly upon impacting the surface of the object to be coated. This plasma method can produce superalloy coatings that are highly resistant to corrosion and thermal fatigue. The use of separated isotopes and SIMS profiling is also a powerful analytical method

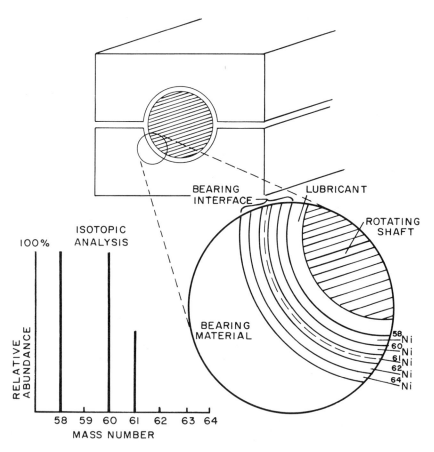

Figure 10.13 Schematic diagram of a proposed method for using thin films of separated isotopes to determine "on-line" bearing wear by monitoring isotopic concentrations in the bearing lubricant.

for studies relating to friction, lubrication, and wear, and the mass transfer of material between two interacting solids can be monitored at the subpicogram level. The schematic diagram of Figure 10.13 illustrates one method by which thin films of separated isotopes can be employed to determine bearing wear by monitoring trace isotopic concentrations of metal in the lubricant. Providing that the contacting surfaces are properly "coded," wear rates can be monitored quantitatively in situ by sampling the lubricant. This technique permits monitoring wear rates of submicron dimensions in "on-line" systems, where a knowledge of wear rates is crucial.

CATALYSTS

The phenomenon of *catalysis* provides the basis for many contemporary chemical and energy production technologies; the term was introduced by Berzelius in 1836 to describe the unique property of certain substances to "promote chemical change without being consumed during the reaction" (Spencer and Somorjai, 1983). At present, catalysts are as important to the petrochemical industry as semiconductors to the computer industry, and both technologies utilize the same surface science tools (David, 1984). Mass spectrometry is an especially powerful instrument for elucidating catalysis, as it can characterize molecular structures and metal clusters, identify "poisons," and measure reaction rates at a gas-solid interface. For example, automotive catalytic converters (containing only 2 g of platinum, palladium, and rhodium per car) have been developed with the aid of mass spectrometry to oxidize unburned hydrocarbons and carbon monoxide to water and CO_2. Other metallic solids with very large surface areas ($1–500$ m^2/g) are being studied for diverse applications. Photocatalysis, electrocatalysis, and artificial enzyme catalysis are additional areas of research where a detailed knowledge of surface chemistry and molecular topography require some type of mass spectral assay.

Mass spectrometry was also applied at an early date to study oxidation catalysts by isotopic exchange, as this method permits the study of the dynamics of the adsorption process. In a specific investigation of silver, Sandler and Hickam (1964) employed ^{18}O to perform a precise quantitative analysis of the different oxygen species in the gas phase and adsorbed phase during adsorption. This work indicated the importance of oxygen embedded below the surface in the oxidation activity of silver. In a later radiochromatographic, "pulse-dynamic" technique, Pscheidl et al. (1982) monitored impulses of known volumes of the gases or vapor phase components labeled with ^3H or ^{14}C. The degree to which reactants are chemisorbed upon a catalyst has also been observed by Drechsler et al. (1979) and Ertl (1982). Figure 10.14 shows the variation of the SIMS signal intensities of NH_3^+, NH_2^+, and NH^+ while continuously increasing the temperature of an ammonia-covered Fe [110] surface. The transition metals and interme-

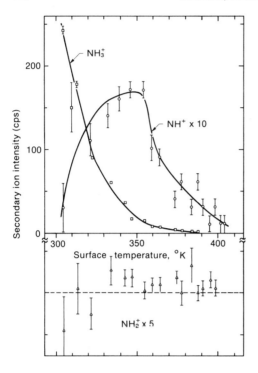

Figure 10.14 Secondary ion intensities of NH_3^+, NH_2^+, and NH^+ with increasing temperature of an ammonia-covered Fe [110] surface. (Drechsler et al., 1979; reprinted with permission from *Critical Reviews in Solid State and Materials Sciences*, *3*, 217 (1979). Copyright CRC Press.)

tallic compounds are of special interest because of their potential for hydrogen storage via metal hydrides, as well as catalysis; applications include heat pumps, cooling systems, transportation vehicles, and thermal storage (Buschow et al., 1982).

Nickel-copper alloys are also widely used as catalysts, and their preparation and role in chemical reactions have been reviewed by Khulbe and Mann (1982). The work function of clean Ni-Cu alloy surfaces, as reported by Takasu et al. (1974), is shown in Figure 10.15, but no simple relationship

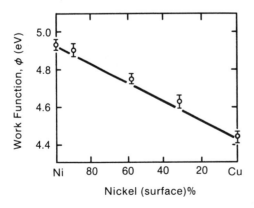

Figure 10.15 The work function of the well-defined surfaces of Ni-Cu alloys. (Khulbe and Mann, 1982; reprinted from *Catal. Rev. Sci. Eng.*, *24*, 311 (1982), by permission of Marcel Dekker, Inc.)

seems to exist between certain exchange reactions, catalytic activity, and purely electronic properties (i.e., work function, density of states at the Fermi level, etc.). However, progress has been substantial relative to a general understanding of metal surface regimes, catalytically active metal clusters (Muetterties, 1981), and metal-ligand interactions of the platinum metals—Ru, Os, Rh, Ir, Pd, and Pt (Davies and Hartley, 1981). The role of metal ions in biological catalysts is an additional field of intensive investigation (Williams, 1982), and "shape selectivity" has provided a new approach to catalyst design (Haggin, 1982).

Nickel complexes on oxide supports have been characterized by SIMS (Pierce and Walton, 1983), and molecular species in low concentrations have been observed for transition metal complexes supported on polymeric or oxidic materials. The catalytic decomposition of gaseous SO_3 by platinum and nine other metal or metal oxide powders has also been studied by monitoring gas composition (Brittain and Hildenbrand, 1983). Ion-scattering spectrometry (ISS) and secondary ion mass spectrometry (SIMS) have further been applied to the analysis of various natural and ion-exchanged synthetic zeolites (Suib et al., 1983). By using ISS and SIMS techniques simultaneously on a sample, a sensitivity as high as one part of a monolayer per million has been obtained. This combination also provides more detailed profiling information, as ISS indicates what is on the surface and SIMS characterizes what is leaving the surface. Finally, the use of multiple isotopes for following catalytic reaction is becoming increasingly effective. Otarod et al. (1983) have used ^2H, ^{13}C, and ^{18}O in studying the methanation of mixtures of carbon monoxide and hydrogen over nickel catalysts. Reactions involving five species—H_2, CH_4, H_2O, CO, and CO_2,—were observed with a quadrupole spectrometer to produce Figure 10.16, which allowed the investigators to compare dissociation rates with a theoretical model.

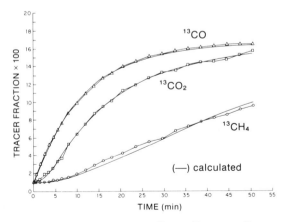

Figure 10.16   Use of ^{13}C as a tracer to observe $^{13}CH_4$, ^{13}CO, and $^{13}CO_2$ transients of methanation over a nickel catalyst. (Otarod et al., 1983; adapted from *J. Catalysis*, *84*, 156 (1983), by permission of J. Happel.)

SILICATE GLASSES

Differences in glasses can be characterized by the relative strength of their chemical bonds as measured by the energy gap between occupied and unoccupied electronic states. For oxide glasses, this gap is greater than 6 eV (ultraviolet regime), so that such glasses are transparent and colorless unless they contain impurities. Semiconductive glasses have energy gaps of ~1.5 eV and are colored yellow or red; in metallic glass, this energy gap is zero. And while glasses are generally regarded as having no crystalline structure, recent studies have revealed that amorphous materials can contain clusters, domains, or macromolecules in the range of 100 Å in diameter (Phillips, 1982). However, most practical glass investigations relate to bulk and surface chemistry, transport phenomena, and physical macroscopic properties.

SIMS composition profiles of borosilicate glasses reveal that the distribution of boron compounds in the near-surface regions is not always the same as the bulk composition. Differences in the surface and bulk chemistry can be attributed to mechanisms of volatilization, dissolution, or the addition of impurities during material processing. Using an ion beam probe, Malm and Riley (1982) reported that glass specimens prepared by conventional techniques (polishing, etc.) had a boron concentration increasing gradually to a constant level at 1–2 μm, but that impact-fractured surfaces reflected bulk composition and no change with ion probe depth. Increases in the amplitude of the Li^+, Na^+, and K^+ ion peaks were also observed. The authors suggest that these profiles indicate the existence of a diffuse boundary between the surface layer and the bulk composition. These profiles were taken at sputtering rates in the range of 0.1–0.4 μm/h and with a heated filament near the glass surface to provide charge neutralization of the specimens. Specifically, this work indicates the desirability of using freshly fractured samples in assaying bulk chemical composition, and that significant depth profile data can be obtained for glasses and other insulating materials.

Depth profiles within glass surface layers and coatings also provide data on the reactions of glasses and contacting materials. An in-depth resolution of a few nanometers has been reported by Bach and Baucke (1982) relative to interdiffusion-controlled reactions of thin oxide films. In this work, the profiling was done by controlled ion etching and the elemental analysis was simultaneously recorded by optical spectrometry, that is, the radiative deexcitations of the elements and molecules ejected from the ion-bombarded solid were detected by a grating monochrometer. Ion beam depth-profiling has also been used to study glass leaching—an ionic interdiffusion process that occurs when a silicate glass surface is in contact with water, wherein hydrogen (or hydronium H_3O^+) ions penetrate into the glass. This leaching or glass hydration forms a hydrated silica-rich surface layer where thickness is dependent on temperature and duration of the reaction

$$2H_2O + Na^+ \text{ (glass)} \rightarrow H_3O^+ \text{ (glass)} + NaOH.$$

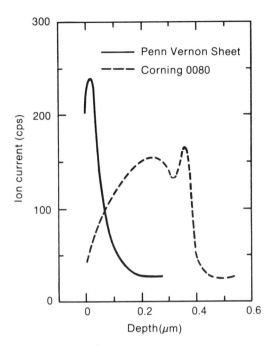

Figure 10.17 SIMS depth profiles of $^1H^+$ in two leached commercial glasses. (Houser et al., 1982; *Radiation Effects*, *64*, 103 (1982); courtesy of I. S. T. Tsong.)

Houser et al. (1982) monitored this interdiffusion process in commercial glasses and obtained SIMS depth profiles of $^{23}Na^+$ and $^1H^+$; results of the latter analysis are shown in Figure 10.17. The depth profiles suggest an interdiffusion mechanism in which sodium ions in the glass are depleted and replaced by H^+ or H_3O^+ ions from the solution. Variations in the surface chemistry of industrial float glass have also been reported by Borchardt et al. (1980). For surfaces that have been in contact with molten tin, bulk concentrations are not reached even at a depth of 2000 Å.

Enriched isotopes in glasses have been used to detect mass transport differences, since ionic transport is uniquely determined by the alkali cations. Also, isotopes provide the only means for distinguishing between the several parameters associated with different cations, that is, ion size, charge density, and activity coefficients. Using impedance measurement techniques, Van der Kouwe (1982) used commercial pyrex glass, in which sodium was replaced by ^6LiCl and ^7LiCl, and found that the lithium-electrolyzed borosilicate glass showed a mixed isotope effect. Specifically, at temperatures of 400, 450, 500, and 550°C, plots of log (resistivity) vs. lithium isotopic composition showed the expected deviation from linearity.

The self-diffusion coefficient of alkali ions in glass is often measured by radioactive isotope tracers (Kelley and Tomozawa, 1980). Such experiments involve the diffusion of a tracer into a sample and a determination of the radioactivity as a function of depth. However, stable isotopes and mass spectrometry provide an alternative method, although the mobility of alkali ions under SIMS ion bombardment may limit the spatial resolution (Katz, 1984).

A recent and practical mass spectrometric measurement relates to the analysis of bubbles in glass (Shick and Swarts, 1982). The gaseous content of bubbles enables the glass scientist to understand melting and refining reactions and to diagnose production problems. The assay of bubbles calls for a mass spectrometric analysis in the static mode (Chapter 4), with appropriate analyzer bake-out and calibration standards. Bubbles having diameters as small as 0.22 mm in diameter have been assayed for species that include N_2, O_2, CO_2, H_2O, SO_2, and Ar.

METALLIC GLASSES

A glass is a material that does not crystallize during solidification from the liquid state; hence, glasses are amorphous solids. In the case of metallic glasses, in contrast to silicate glasses, very rapid cooling rates ($>10^5$ °C/s) are required to prevent crystallization. Pure amorphous metals can only be prepared by vapor deposition on very cold (liquid helium cooled) substrates, and these films are relatively unstable and crystallize at approximately 40 K. Alloy metallic glasses, however, can be prepared by several techniques, and this new class of materials is envisioned for a number of practical applications. The synthesis of the first metallic glass, an alloy of gold and silicon, was first accomplished at the California Institute of Technology in 1959 (Klement et al., 1960), but within the last decade, glassy alloys of many metals are being studied for their unique electric, magnetic, mechanical, and corrosion-resistant properties. Cooling rates achieved by the so-called "splat-quenching" or "melt-spinning" methods now range from 10^5 to 10^8 °C/s. Furthermore, pulsed lasers have been used to investigate the rapid melting and resolidification for surface layers in semiconductors and metals; for example, nanosecond and picosecond pulses have produced amorphous layers with cooling rates of 10^{10} to 10^{13} °C/s (Lin and Spaepen, 1982).

Most metallic glasses contain at least two kinds of atoms of different atomic size and as many as seven species in some instances (Duwez, 1983). Probably the most important use of these materials over the short-term will be in utility distribution and power transformers, which are magnetized and demagnetized 120 times per second and dissipate ~30 billion kWh/yr of energy as heat in the United States. It has been estimated that this loss could be reduced by about 50–60% if present-technology silicon-steel cores were replaced by metallic glasses having low hysteresis losses. The use of metallic glasses as windings in high-efficiency electric motors is already being tested in laboratories. The corrosion resistance of metallic glasses results from another feature of glasses; namely, the absence of grain boundaries where corrosion is often initiated in conventional alloys. Furthermore, this unique corrosion resistance is retained at high temperatures, so that metallic glasses are of potential interest for geothermal turbine blades, electrochemical electrodes, and corrosive chemical piping. For example, the Allied Corporation's "METGLAS"® 2826A (Fe_{32}, Ni_{36}, Cr_{14}, P_{12}, and B_6) has a corrosion rate two orders of magnitude lower than No. 316 stainless steel in concentrated hydrochloric acid.

The use of mass spectrometry and other spectrometric techniques in analyzing metallic glasses relates to basic thermodynamic and kinetic studies of amorphous intermetallic phases, chemical compositions, activation energies of alloys, and surface analysis. Practical glassy metal alloys have been grouped into three categories by Raskin and Davis (1981):

1. Metal-metalloid alloys: gold, iron, nickel, cobalt palladium, and/or platinum, and from 10–30 atom% of boron, phosphorus, carbon, and silicon.

2. Transition metal alloys: zirconium, niobium, and/or tantalum, and from 30–60 atom% of iron, cobalt, nickel, and copper.

3. Alloys containing metals such as magnesium, calcium, and beryllium from group IIA of the Periodic Table.

The ferromagnetic transition metals Fe, Co, and Ni appear to be ideal hosts for the formation of various amorphous alloys. Cobalt is monoisotopic ^{59}Co, but Fe and Ni have four and five isotopes, respectively, so that the isotopic dilution technique can be used for the assay of these elements, plus most of the other metals; exceptions include ^{9}Be, ^{27}Al, ^{93}Nb, and ^{197}Au. The addition of small quantities of both Mo and Si (2–4 atom%) have been shown to have an effect on the crystallization kinetics of Fe-B-Si metallic glasses (Ramanan and Fish, 1982), and the soft magnetic properties of these glasses are also affected by post-fabrication heat treatments.

A metallic glass of the type Fe_{40}, Ni_{40}, and B_{20} has been analyzed for chemical inhomogeneity after rapid quenching by atom probe, field ion microscopy (Piller and Haasen, 1982). This study revealed that this alloy decomposed into two amorphous phases, and boron concentration profiles indicated that the local boron content substantially affected the stability of the glass with respect to decomposition and embrittlement.

Glasses that have been developed for high-frequency applications ($f <$ 50 kHz) appear to require the introduction of a small amount of crystallinity by suitable annealing. In a study of Fe_{80}, B_{12}, and Si_8 metallic glasses, crystalline phases observed were α(Fe, Si) and Fe_3B, and the growth rate of these particles appeared to be diffusion-controlled (Chang and Marti, 1983). The growth process of the particles followed the equation

$$r = \alpha(Dt)^{0.5}$$

where r is the size of the particles, α is a dimensionless parameter determined by the composition, D is the volume diffusion coefficient, and t is the annealing time.

A rather specialized projected use for metallic glasses is in the fabrication of Au-based spheres and microballoons (100 μm) for inertially confined fusion (ICF) targets. An alloy of composition $Au_{55}Pb_{22.5}Sb_{22.5}$ has been used because of its low melting point ($\sim$260°C) and high average Z number (73). These microspheres exhibit a high degree of sphericity and surface smoothness, and they are expected to deform homogeneously under the high strain rates involved in an ICF implosion (Johnson and Lee, 1983). Finally, in addition to the use of esoteric compounds for metallic glasses, iron-rare earth compounds are providing a new generation of permanent magnets. These rare earth-iron-boron and rare earth-cobalt materials possess an energy product (a composite parameter relating to both the strength of the magnet and its coercivity) greater than six times that of widely used ceramic or ferrite materials (Robinson, 1984).

CERAMICS

In comparison to metals, ceramics are more resistant to high temperatures, oxidation, and corrosion, so that despite their brittleness, ceramics will be increasingly used in gas turbines, stationary power generators, and remotely powered aircraft. Among the more widely used ceramics are compounds that include silicates, oxides, carbides, nitrides, and borides. Diesel engines made substantially of silicon nitride (Si_3N_4) can run without water cooling and survive at temperatures up to 1200°C, which is above the operating range of even advanced metallic alloys. In consequence, some ceramic engines will be built without radiators, fans, or water pumps, with fewer parts and appreciable weight savings. Major automobile manufacturers have already reported on tests of ceramics for automotive turbines (McLean, 1982); and a low-inertia turbocharging Si_3N_4 rotor, which operates at up to 150,000 rpm and weighs 60% less than its metal counterpart, has already been successfully tested (Robb, 1982). Under U.S. Department of Energy sponsorship, several manufacturers are also engaged in the development of a lightweight 100 hp advanced turbine engine that has a turbine inlet temperature of $\sim$1300°C. A technical objective is to have such an engine power a 3000 pound vehicle with a combined city/highway mileage of 43 miles/gal. A siliconized silicon carbide has also been investigated for applications in both solar converters and heat exchangers in aluminum and steel mills because of its high temperature limit, low thermal expansion coefficient, and high thermal conductivity. Furthermore, an increased use of ceramics is occurring in the electronics industry, where multilayer ceramic capacitors are desirable because of their small size, high capacitance, electric stability, and low cost (Jang, 1983).

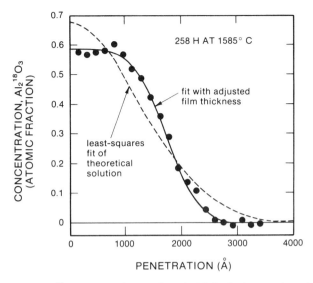

Figure 10.18 Corrected ^{18}O concentration gradients in Al_2O_3 single crystals as determined with an ion probe (Reed and Wuensch, 1980).

Mass spectrometric measurements of ceramics include (1) diffusion, (2) composition profiles, and (3) Knudsen-cell spectrometry at high temperatures. Ion probe measurements of oxygen self-diffusion in single-crystal Al_2O_3 at temperatures of 1585–1840°C have been reported by Reed and Wuensch (1980) using ^{18}O as a tracer. The stable isotope was introduced from an initial film of $Al_2^{18}O_3$ of approximately 1000 Å in thickness. Concentration gradients were then determined by measuring the $^{18}O/(^{16}O + {}^{18}O)$ isotope fraction obtained as the isotope-coated surface was sputtered by bombardment with a focused 5 keV beam of Ar^+ ions. The single-crystal diffusion specimens of Al_2O_3 ($1 \times 1 \times 0.2$ cm) were cut from a single-crystal boule; and the crystal orientation, normal to the $(1\overline{1}02)$ plane, was determined by using a back-reflection diffraction technique. Diffusion annealing of the samples was programmed in a vacuum furnace that maintained pressure at $< 10^{-5}$ torr. Corrected ^{18}O concentration gradients for various annealing times and temperatures are shown in Figure 10.18. The data correlate to a diffusion coefficient $D_0 = 6.4 \times 10^5$ cm^{2}s^{-1}, with an activation energy of 188 ± 7 kcal/mol.

Diffusion measurements of ^{59}Fe in both undoped and magnesium-doped Al_2O_3 have been studied at temperatures ranging from 1480–1600°C (Lloyd and Bowen, 1981). In this investigation, spark source mass spectrometry was used for elemental impurity determinations (Na, Mg, Si, S, Cl, K, Ca, Fe, and Ga) of the sample, and diffusion data was obtained via standard radioactivity techniques. Results showed that the relative silicon concentration had a greater effect on the diffusivity than the magnesium when both impurities were present at the ppm level.

An alternative method for determining the oxygen diffusion coefficient for oxides with an ^{18}O tracer has been reported by Komatsu et al. (1982). The method consists of monitoring the oxygen isotope exchange reaction between the ^{18}O gas and solid phases of the specimen, that is

$$M^{16}O(s) + {}^{18}O_2(g) \rightarrow M^{18}O(s) + {}^{16}O_2(g) + {}^{16}O^{18}O(g)$$

where M is a cation with an assumed valence of $2+$. Instead of using a mass spectrometer, however, the exchange reaction was followed by an accurate measurement of a weight change that was continuously recorded from a microbalance.

Other important diffusion measurements include the determination of diffusion profiles at glass-ceramic bonding interfaces. The bonding involves a diffusion zone of mixed components, and microprobe profiles permit a detailed examination to be made of the interdiffusion zone. In the case of Al_2O_3 bonded to soda-lime glass, the diffusion coefficients of Al were deduced and mass flow was also monitored at various temperatures by observing the motion of Pt particles placed at the glass-alumina interface (Willis and Doremus, 1980).

High-temperature Knudsen-cell spectrometry becomes a powerful tool in studying the thermodynamic properties of ceramics, because volatile species are often more abundant. Each partial pressure P_i of a gaseous molecule i leads to the activity of its component through the instrumental sensitivity factor S_i: $P_i S_i = I_i^+ T$, where I_i^+ are the ionic intensities and T is the temperature of the Knudsen cell (Granier et al., 1982). The activity of any species can then be related to the observed ionic intensities I_i^+ and to measurements from an appropriate reference cell. Granier et al. (1982) developed the ternary phase diagram for Ti-O-N and measured the thermodynamic activities of titanium and titanium monoxide with a multiple Knudsen-cell mass spectrometer, operating up to 1800 K.

The activities of CaO and Al_2O_3 in lime-alumina melts have also been studied by Knudsen-cell mass spectrometry (Allibert et al., 1981). In this investigation, six effusion cells were drilled in a molybdenum block. Two contained the reference mixtures, a CaO-saturated melt and pure Al_2O_3, with the remaining cells containing homogeneous melts of different composition. The gaseous species effusing from the melts were thus compared directly to those emerging from the reference cells, with all being ionized and analyzed under identical conditions. The use of oxides for high-temperature structural materials has also stimulated diffusion measurements of gases, including inert gases, in sintered oxides. Popov and Kupryazhkin (1981) have reported on the diffusion of ^4He in densely sintered corundum (Al_2O_3, 98%; TiO_2, 1.0%; SiO_2, 0.5%; plus minor constituents) over a temperature range of 400–850°C. Diffusion coefficients were obtained by (1) initial sample bake-out in vacuum; (2) saturation of the sample with helium for 6 h; and (3) desorption of the gas with temperature programming and mass spectrometric measurements.

The ion implantation of ceramics, as with metals, has resulted in surface modifications that have been analyzed by RBS. Both Al_2O_3 and SiC have been ion-implanted with metal ions at reasonably high doses. In some metallic ion implants of alumina, an amorphous surface layer was generated of increased hardness, wear/erosion resistance, and resistance to chipping and crack propagation (McHargue and Yust, 1984). Within the last decade, ceramic materials have also become important in photovoltaic devices and in biomaterials applications. Specifically, large photovoltages have been observed in lead titanate ($PbTiO_3$)-based ceramics, and the photovoltage has been observed to increase by a factor of 5 with MnO_2 doping (Uchino et al., 1982). With respect to bioceramics, materials such as alumina are now being used for hip joint prostheses, dental implants, and in instances where metals are of questionable toxicity or compatibility. In Japan, single-crystal sapphire (Al_2O_3) has been used for dental implants and as bone screws for over 10 years, with more than 50,000 cases of clinical documentation (Fisher, 1983a). Porous ceramics are also being tested for the controlled release of antibiotics. Ceramic materials are also important for state-of-the-art packaging of integrated circuits and other electronic devices (Fisher, 1983b), and their use in capacitors is discussed in Chapter 11.

COMPOSITES

Almost all naturally occurring objects are composites. In the field of agriculture alone, virtually all plants, trees, fruits, nuts, etc. have protective coatings, skins, and exterior layers that are markedly different from their interior. Furthermore, from calculations on the positions of veins in plant leaves, it even appears that nature has distributed the reinforcing members in a pattern to ensure the greatest photoexposure for a given size and shape (Baker, 1981). In an analogous fashion, the mechanical parameters of metals and polymers can be altered not only by reinforcing fibers, but by the introduction of inert particulates and gases. Thus, a commercial composite usually consists of two or more homogeneous constituents bonded together to provide properties unattainable by any single constituent. On this basis, composites include paper, plywood and metallic laminates, pneumatic tires, carbide cutting tools, and fiber or particulate-reinforced glasses, metals, and polymers. An inherent advantage of laminated composites, having well-defined boundaries between adjacent layers, is the potential for optimizing the properties of each component. The impact strength and other properties of high carbon/mild steel laminates is an example (Kum et al., 1983).

With respect to fiber/polymer composites, mass spectrometry is directly

TABLE 10.4   Density of Commercial Fibers (g/cm³)

Kevlar 49	Graphite	Boron	Glass	SiC	Al_2O_3	W
1.44	1.6–2.0	2.68	2.5	2.8	3.95	19.3

Source: Chawla (1983).

or indirectly involved in the many phases of research and product development, that is:

1. Basic studies of polymer molecular structure.
2. Quality control of raw materials.
3. SIMS assay of fiber surfaces and coatings.
4. Diagnostics of fiber-matrix interface bonding.
5. GC/MS monitoring of the curing cycle.
6. Surface analysis of fractured test specimens.

Factors affecting interfacial bonds have been reviewed by Jean-Baptiste (1981) and include (1) effect of the interfacial area; (2) wetting of the reinforcing fibers by the matrix; (3) effect of impurities, including water; and (4) the number and nature of the chemical/physical bonds. Surface analyses of fibers and particulates have been extensive in order to answer questions relating to bonding. Frequently, nonwettability between oxide particles and metals results in a segregation of metal and ceramic phases, so that coatings have been applied to ceramic particles, for example, noble metals, or copper on titania (TiO_2) particles. Coatings have also been extensively applied to laminates, bonded sandwiches, and clad-bonded materials to prevent delamination and to increase shear strength (Johnson and Ghosh, 1981).

In addition to classical mechanical tests, some composites have been examined in relationship to their extensive use in space. In space, materials are subject to hard vacuum, thermal cycling plus ultraviolet, electron, and proton bombardment. Vacuum causes outgassing and migration of low molecular weight components from the matrix; and each time satellites circle the earth and pass from sunshine to shadow, they experience a temperature cycle of several hundred degrees Fahrenheit. Dimensional stability is also important, and Frisch (1983) has noted that the next generation of communication satellites operating at 20–30 GHz will have to retain their dimensional integrity and shape within millimeters.

Most reinforcing fibers used in composites are of low density; among the practical fibers listed in Table 10.4, only tungsten has a high density. There are three types of Kevlar®fiber produced and designated by DuPont: Kevlar, Kevlar 29, and Kevlar 49 (Miner, 1982). Kevlar is primarily used in tires and rubber products. Kevlar 29 is utilized for ropes, cables, gaskets, etc.

Kevlar 49 is an advanced reinforcement fiber for polymeric composites in the aerospace industry and in the many products where a high strength-to-weight ratio is a prerequisite. Kevlar 49 has twice the modulus and a 40% lower density than glass, two-thirds the modulus of graphite fiber, and a tensile strength exceeding 200,000 lb/in.[2]. Pound for pound, it is five times stronger than steel. Many tests have also been made of other hybrid materials for aircraft, such as ARALL (aramid-reinforced aluminum laminate). This material is made by the adhesive bonding of a number of thin aluminum alloy sheets and aramid layers, and it possesses several properties that are superior to monolithic aluminum alloys (Vogelesung, 1983).

Important carbon composites include tires, because the carbon-rubber composite gives enhanced wear resistance over pure rubber. The polymer PVC (polyvinylchloride) is a long-chain molecule with a carbon backbone; when carbon black and PVC are milled together, an entirely new structure with different electric properties is formed (Sickel et al., 1982). In some instances, two or more composites are also used. Advanced wing structures are now being built consisting of layers of graphite and boron fiber-reinforced polymers, and the multiple layers are directionally tailored to resist flexing and to control the load distribution (Chawla, 1983).

Silicon carbide fibers or whiskers have been successfully incorporated into aluminum alloys, as well as some magnesium alloys. Whisker diameters vary from 0.2–1 μm and lengths vary up to 50 μm. A distinguishing feature of silicon carbide-reinforced aluminum composites is that these materials can be shaped from billets using conventional metal-working techniques. Silicon carbide/aluminum composite billets have been forged into sheets, rods, tubes, crosses, and even a pressure vessel cover (Divecha et al., 1981). Alumina in both single-crystal and polycrystalline form is used in metal composites, and these fibers are attractive in high-temperature applications. Aluminum matrices containing a high wt% of Al_2O_3 particles have exhibited greatly improved friction and wear properties. Specifically, the composite of alloy 2024 plus 20 wt% of 142 μm-sized Al_2O_3 particles showed a wear rate approximately two orders of magnitude less than that of the matrix alloy alone under identical test conditions (Hosking et al., 1982). High-temperature features of tungsten composites are also important in turbine blades currently being designed for operating temperatures that exceed those for superalloys by 150°C.

The use of both metallic and nonmetallic particles has given rise to the improved performance of other materials. A 60-fold increase in the toughness of glass has been observed with the addition of aluminum particles. A strong interfacial bonding and the nearly equal elastic and thermal properties of the glass-partly oxidized aluminum composite are factors contributing to the increased strength (Kristic et al., 1981). In continuous filament-reinforced composites, the ductility is usually reduced by the high-modulus, low-elongation filament; and since most high-modulus filaments are ceramics (e.g., B, Al_2O_3, or SiC) or ceramic-like, their elongation-to-failure never exceeds

0.5–1%. However, in short-fiber composites, some plastic flow occurs at the fiber or whisker ends.

Depth profiling has been used in conjunction with Auger electron spectroscopy to analyze the surfaces of coated and uncoated fibers in both polymer and metal matrices. The ion beam continuously mills atom layer by atom layer, while a high-energy-focused electron beam causes Auger transitions—thus providing a continuous chemical analysis of each newly exposed surface. This technique has been used for the in situ analysis of boron fiber/titanium alloy metal matrix composites (Clough, 1981), where the fiber/matrix interface region revealed boron, carbon, sulfur, oxygen, titanium, potassium, and sodium.

Chromatography-mass spectrometry is important in monitoring the actual fabrication and processing of composites. In contrast to metal structures that are fabricated essentially by mechanical techniques, composites are formed by both mechanical and chemical processes, requiring the controlled hardening of an organic matrix and the quality control of raw materials. Commercial composites are also characterized by a variety of auxiliary catalysts, secondary resins, diluents, and volatiles, etc.; hence, a chemical assay of the actual curing cycle, transition temperatures, and reaction rates is essential. Figure 10.19 is a high-performance liquid chromatogram (HPLC), reported by May (1983), of a typical TGMDA/DDS (tetraglycidylmethylenedianiline/diaminodiphenylsulfone) formulation in which peaks define the resin, curing agent, diluent, and reacted material. In the aircraft industry, this type of computer-based monitoring is a prerequisite for the control of heating rates, the onset of hardening, and general control of product quality.

The depth profiling of polymeric films on substrates has also been demonstrated with dynamic SIMS. Figure 10.20 is a depth profile by Katz (1984)

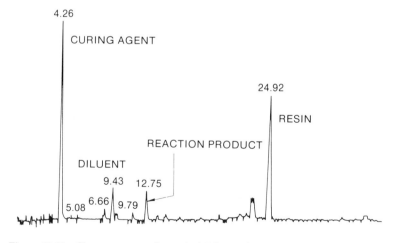

Figure 10.19 Chromatogram of a typical TGMDA/DDS formulation (May, 1983).

of a specimen in which the deuterium was tagged to a 2-hydroxy-4-dode-cyloxy-benzophenone contained in a thin polyalkylmethacrylate (PAMA) that was flow-coated onto the surface of a polycarbonate sheet. The samples were then heated at 130°C for 0, 2, 5, and 30 min, and the subsequent SIMS analysis permitted measurement of the diffusion rates of large molecules through a polycarbonate matrix (Valenty et al., 1984).

Finally, one of the most interesting recent developments has been that of polymer composites consisting of both rigid and flexible molecules. A very small fraction of the rigid molecules (~5%) shows a marked reinforcing effect on the matrix polymers when the rigid component forms a quasi-three dimensional lattice. According to Takayanagi (1983), this new concept will permit composites to be designed at the molecular level; properties of these molecular composites should compare well with macrofiber-reinforced plastics.

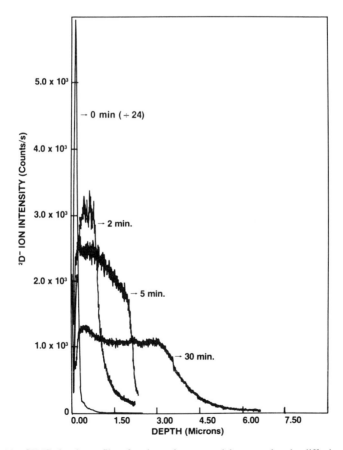

Figure 10.20 SIMS depth profile of a deuterium-tagged large molecule diffusing through a polycarbonate matrix (Katz, 1984; Valenty et al., 1984).

ION EMISSION ACCOMPANYING THE FRACTURE OF POLYMERS

Acoustic emission is now a widely used phenomenon for detecting the failure of structural materials. Only recently, however, have analogous studies focused on particle and photon emissions that accompany and follow the fracture of nonmetals. Collectively called "fractoemissions" because the propagation of a crack is a prerequisite for their generation, electrons, positive and negative ions, neutral molecules, and photons have been detected by several investigators; and the term stress-induced mass spectroscopy is sometimes applied to the emission of molecular species. Specifically, SO_2 has been detected in the fracture of an epoxy resin (Wolf et al., 1981), and the mass-to-charge ratio (m/q) of positive ion emission (PIE) has been reported from Kevlar 49®, E glass fibers, and aluminum oxide coatings (Dickinson et al., 1982). Typically, the ions are emitted in bursts of less than 1 μs duration and in coincidence with crack growth or fracture; observation of the m/q values also suggests that both ions and neutrals are atomic or molecular fragments resulting from fracture.

The experimental TOF apparatus of Dickinson et al. (1982) is shown in Figure 10.21. Two channel-electron multipliers (CEM) are positioned on opposite sides of the sample that is undergoing fracture, and an electron detector (CEM-EE) is positioned within 1 cm of the specimen and with a 300 V bias, so that electron transit time (~2 ns) is very short compared to ion TOF intervals. For the distance parameters in Figure 10.21 of $d_1 = d_3$

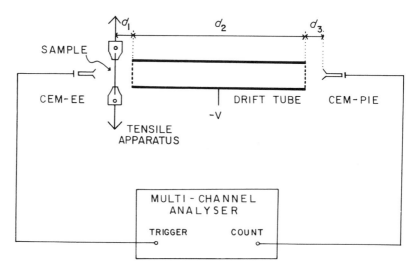

Figure 10.21 Schematic of time-of-flight apparatus applied to the positive ion emission (PIE) accompanying fracture. The distances are $d_1 = d_3 = 1$ cm and $d_2 = 25$ cm. (Dickinson et al., 1982; *J. Materials Science*, *17*, 3173 (1982); courtesy of J. T. Dickinson.)

$= 1$ cm, $d_2 = 25$ cm, and a drift tube potential V, the ion transit time T in microseconds is given approximately by

$$T = 20 \left(\frac{m/q}{V}\right)^{\frac{1}{2}}$$

The emitted ions have an energy distribution and the leading edge of the ion burst spreads out in time, as indicated in Figure 10.22, for accelerating voltages of 100, 500, and 1000 V. Nevertheless, this experimental approach is highly interesting, and the authors report m/q values for Kevlar, E glass, and aluminum oxide coatings of 60 ± 20, 48 ± 12, and 17 ± 6, respectively. They also suggest that (1) these positive ion distributions are indeed a fractoemission product; (2) PIE is intimately related to the production of high-energy sites on the surface by cleavage and bond breaking; and (3) high-resolution quadrupole instrumentation should be able to characterize the

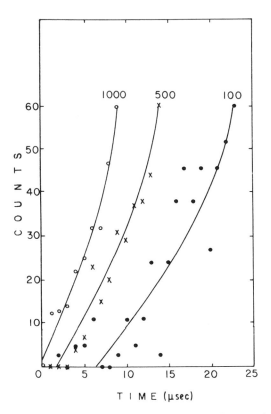

Figure 10.22 Leading edge of PIE-TOF distributions for Kevlar fibers. (Dickinson et al., 1982; *J. Materials Science*, *17*, 3173 (1982); courtesy of J. T. Dickinson.)

many parameters and reactions that relate to the macroscopic phenomenon of fracture.

Materials currently being studied with this new technique include single fibers, fiber/epoxy strands, particulate-filled epoxy, and multi-ply fiber/epoxy systems (Dickinson et al., 1985).

REFERENCES

Allibert, M., C. Chatillon, K. T. Jacob, and R. Lourtau, *J. Am. Ceramic Soc.*, *64*, No. 5, 307 (1981).

Ayers, J. D., R. J. Schaefer, and W. P. Robey, *J. Metals*, *33*, 819 (1981).

Bach, H., and F. G. K. Baucke, *J. Am. Ceramic Soc.*, *65*, No. 11, 527 (1982).

Baker, W. O., *Science*, *211*, 359 (1981).

Baumvol, I. J. R., *J. Appl. Phys.*, *52*, No. 7, 4583 (1981).

Borchardt, G., H. Scherrer, S. Weber, and S. Sherrer, *Int. J. Mass Spectrom. Ion Phys.*, *34*, 361 (1980).

Brenner, S. S., and M. K. Miller, *J. Metals*, *35*, No. 3, 54 (1983).

Brittain, R. D., and D. L. Hildenbrand, *J. Phys. Chem.*, *87*, 3713 (1983).

Buckley, D. H., *Progr. Surface Sci.*, *12*, 2 (1982).

Buschow, K. H. J., P. C. P. Bouten, and A. R. Miedema, *Rep. Prog. Phys.*, *45*, No. 9, 937 (1982).

Campbell, A. B., III, B. D. Sartwell, and P. B. Needham, Jr., *J. Appl. Phys.*, *51*, No. 1, 283 (1980).

Chang, C. F., and J. Marti, *J. Materials Sci.*, *18*, 2297 (1983).

Chawla, K. K., *J. Metals*, March 82 (1983).

Clayton, C. R., *Nucl. Instr. and Methods*, *182/183*, 865 (1981).

Clayton, C. R., and Y. F. Wang, *IEEE Trans. Nucl. Sci.*, *NS-30*, No. 2, 1752 (1983).

Clayton, J. Q., and J. F. Knott, *Metal Sci.*, *16*, 145 (1982).

Clough, S. P., *J. Metals*, *33*, No. 4, 12 (1981).

Cottrell, A. H., *The Mechanical Properties of Matter*, Wiley, New York (1964), p. 201.

Cristy, S. S., and J. B. Condon, in *Secondary Ion Mass Spectrometry, SIMS II*, ed. A. Benninghoven, C. A. Evans, R. A. Powell, R. Shimizu, and H. A. Storms, Springer-Verlag, Berlin (1979), p. 151.

David, E. E., Jr., *Am. Soc. Mass Spectrom.*, San Antonio, TX Keynote address, (May 29, 1984).

Davies, J. A., and F. R. Hartley, *Chem. Rev.*, *81*, 79 (1981).

Dawar, A. L., and J. C. Joshi, *J. Materials Sci.*, *19*, 1 (1984).

Dearnley, G., *J. Metals*, *34*, No. 9, 18 (1982).

Dearnley, G., and N. E. W. Hartley, *Thin Solid Films*, *54*, 215 (1978).

Dickinson, J. T., L. C. Jenson, and M. K. Park, *J. Materials Sci.*, *17*, 3173 (1982).

Dickinson, J. T., A. Jahan-Latibari, and L. C. Jensen, in *Molecular Characterization of Composite Interfaces*, eds. H. Ishida and E. Kumar, Plenum, New York (1985), pp. 111–131.

Divecha, A. P., S. G. Fishman, and S. D. Karmarkar, *J. Metals*, *33*, No. 9, 12 (1981).

Draper, C. W., *J. Metals*, *34*, No. 6, 24 (1982).

Drechsler, M., H. Hoinkes, H. Kaarmann, H. Wilsch, G. Ertl, and M. Weiss, *Appl. Surf. Sci.*, *3*, 217 (1979).

Duwez, P., *J. Vac. Sci. Technol.*, *B1 No. 2*, 218 (1983).

Ehrlich, G., *Crit. Rev. Solid State Material Sci.*, *10*, No. 4, 391 (1982).

Ertl, G., *Crit. Rev. Solid State Material Sci.*, *10*, No. 4, 349 (1982).

Feldman, L. C., *Appl. Surf. Sci.*, *13*, 211 (1982).

Fiedler, H. C., *J. Magnetism Magnetic Mat.*, *26*, 22 (1982).

Fiedler, H. C., J. D. Livingston, and S. C. Huang, *J. Magnetism Magnetic Mat.*, *26*, 157 (1982).

Fisher, G., *Am. Ceramic Soc. Bull.*, *62*, No. 11, 1237 (1983a).

Fisher, G., *Ceramic Ind.*, *121*, No. 1, 22 (1983b).

Frerichs, H. P., S. Kalbitzer, F. J. Demond, B. L. Roth, and D. M. Kolb, *Surface Sci.*, *105*, L 271 (1981).

Frisch, B., *Astronaut. Aeron.*, *May*, 33 (1983).

Granier, B., C. Chatillon, and M. Allibert, *J. Am. Ceramic Soc.*, *65*, No. 10, 465 (1982).

Gries, W. H., *Int. J. Mass Spectrom. Ion Phys.*, *30*, 97 (1979).

Gust, W., M. B. Hintz, A. Lodding, H. Odelius, and B. Predel, in *Advances in Mass Spectrometry*, Vol. 8A, ed. A. Quayle, Heyden & Son, Ltd., London (1980), p. 475.

Haggin, J., *Chem. Eng. News, December 13*, 9 (1982).

Harding, S. E., M. Sc. Thesis, Brighton Polytechnic. (1977).

Hashimoto, E., and T. Kino, *J. Phys. F; Met. Phys.*, *13*, 1157 (1983).

Hehenkamp, T. L., A. Lodding, H. Odelius, and V. Schlett, *Acta Metall.*, *27*, 829 (1979).

Heinen, H. J., S. Meier, H. Vogt, and R. Wechsung, *Int. J. Mass Spectrom. Ion Phys.*, *47*, 19 (1983).

Hintz, M. B., Thesis, University of Stuttgart (1979).

Hirayama, C., and C. Y. Lin, *NBS Special Publ.*, No. 561, 1539 (1979).

Hirvonen, J. K., ed., *Treatise on Materials Science and Technology, Ion Implantation*, Vol. 18, Academic Press, New York (1980).

Honig, R. E., private communication (1966).

Hosking, F. M., F. F. Portillo, R. Wunderlin, and R. Mechrabian, *J. Materials Sci.*, *17*, 477 (1982).

Houser, C. A., I. S. T. Tsong, W. B. White, A. L. Wintenberg, P. D. Miller, and C. D. Moak, *Radiation Effects*, *64*, 103 (1982).

Hu, W. W., H. Herman, C. R. Clayton, R. A. Kant, J. K. Hirvonen, and R. K. MacCrone, in *Implantation Metallurgy*, ed. C. M. Preece and J. K. Hirvonen, Metallurgical Society of AIME, Warrendale, NY (1980), p. 92.

Hubler, G. K., *Nuclear Instr. and Methods in Phys. Res.*, *191*, 101 (1981).

Iwaki, M., H. Hayashi, A. Kohno, and K. Yoshida, *Jpn. J. Appl. Phys.*, *20*, No. 1, 31 (1981).

Jang, S., *Am. Ceramic Soc. Bull.*, *62*, No. 2, 216 (1983).

Jean-Baptiste, D., *Pure and Appl. Chem.*, *53*, 2223 (1981).

Johnson, W., and S. K. Ghosh, *J. Materials Sci.*, *16*, 285 (1981).

Johnson, W. L., and M. C. Lee, *J. Vac. Sci. Technol.*, *A1*, No. 3, 1568 (1983).

Katz, W., private communication (1984).

Kelley, J. E., and M. Tomozawa, *J. Am. Ceramic Soc.*, *63*, No. 7–8, 478 (1980).

Khulbe, K. C., and R. S. Mann, *Catal. Rev. Sci. Eng.*, *24*, No. 3, 311 (1982).

Kingston, H. M., P. J. Paulsen, and G. Lambert, *Appl. Spectroscopy*, *38*, No. 3, 385 (1984).

Klement, W., R. H. Willens, and P. Duwez, *Nature*, *187*, 869 (1960).

Komatsu, W., Y. Ikuma, M. Kato, and K. Uematsu, *J. Am. Ceramic Soc.*, *65*, No. 12, C211 (1982).

Kristic, V. V., P. S. Nickolson, and R. G. Hoagland, *J. Am. Ceramic Soc.*, *64*, No. 9, 499 (1981).

Kum, D. W., T. Oyama, J. Wadsworth, and O. D. Sherby, *J. Mech. Phys.*, *31*, No. 2, 173 (1983).

Lin, C., and F. Spaepen, *Appl. Phys. Lett.*, *41*, No. 8, 721 (1982).

Lloyd, I. K., and H. K. Bowen, *J. Am. Ceramic Soc.*, *64*, No. 12, 744 (1981).

Lodding, A., in *Advances in Mass Spectrometry*, Vol. 8A, ed. A. Quayle, Heyden & Son Ltd., London (1980), p. 471.

Loria, E. A., *J. Metals*, *34*, No. 9, 16 (1982).

Malm, D. L., and J. E. Riley, Jr., *J. Electrochem. Soc.*, *128*, No. 8, 1819 (1982).

Matsumoto, S., Y. Sato, M. Kamo, and N. Setaka, *Jpn. J. Appl. Phys.*, *21*, No. 4, L183 (1982).

May, C. A., *Pure and Appl. Chem.*, *55*, No. 5, 811 (1983).

Mazzoldi, P., S. LoRusso, and P. L. Bonora, *Radiation Effects*, *63*, 17 (1982).

McHargue, C. J., and C. S. Yust, *J. Am. Ceramic Soc.*, *67*, No. 2, 117 (1984).

McLean, A. F., *Am. Ceramic Soc. Bull.*, *61*, No. 8, 861 (1982).

Miner, L. M., *Astronaut. Aeron.*, *June*, 33 (1982).

Mitchell, D. F., R. J. Hussey, and M. J. Graham, *J. Vac. Sci. Technol.*, A1, No. 2, 1006 (1983).

Morrison, G. H., in *Secondary Ion Mass Spectrometry*, SIMS III, eds. A. Benninghoven, J. Giber, L. László, M. Riedel, and H. W. Werner, Springer-Verlag, Berlin (1982), p. 244.

Muetterties, E. L., *Catal. Rev. Sci. Eng.*, *23*, No. 1/2, 69 (1981).

Myers, S. M., *J. Vac. Sci. Technol.*, *17*, No. 1, 310 (1980).

Naoe, M., Y. Hoshi, S. Yamanaka, and M. Kume, *Jpn. J. Appl. Phys.*, *22*, No. 10, Part 1, 1519 (1983).

Narayan, J., *J. Metals*, *32*, No. 6, 15 (1980).

Nelson, G. C., L. E. Pope, and F. G. Yost, *J. Vac. Sci. Technol.*, A1, No. 2, 486 (1983).

Ogale, S. B., S. V. Ghaisas, A. S. Ogale, M. R. Bhiday, A. S. Nigavekar, and V. N. Bhoraskar, *Radiation Effects*, *63*, 73 (1982).

Okabe, Y., M. Iwaki, S. Namba, and K. Yoshida, *Nucl. Instr. and Methods*, *182/183*, 231 (1981).

Otarod, M., S. Ozawa, F. Yin, M. Chew, H. Y. Cheh, and J. Happel, *J. Catalysis*, *84*, 156 (1983).

Padmanabhan, K. R., J. Chevallier, and G. Sørensen, *J. Vac. Sci. Technol.*, *20*, No. 4, 1406 (1982).

Patil, R. V., and B. D. Sharma, *Metal Sci.*, *16*, 389 (1982).

Pebler, A., G. G. Sweeney, and P. M. Castle, *Metallurg. Trans.*, *6A*, 991 (1975).

Phillips, J. C., *Physics Today*, *February*, 27 (1982).

Picraux, S. T., *Physics Today*, *November*, 38 (1984).

Picraux, S. T., and L. E. Pope, *Science*, *226*, 615 (1984).

Pierce, J. L., and R. A. Walton, *J. Catalysis*, *81*, 375 (1983).

Piller, J., and P. Haasen, *Acta Metall.*, *30*, 1 (1982).

Pivin, J. C., D. Loison, C. Roques-Carmes, J. Chaumont, A. M. Huber, and G. Morillotin, in *Secondary Ion Mass Spectrometry. SIMS III*, ed. A. Benninghoven, J. Giber, L. László, M. Riedel, and H. W. Werner, Springer-Verlag, Berlin (1982), p. 244.

Popov, E. V., and A. Y. Kupryazhkin, *Soviet Phys. Tech. Phys.*, *26*, No. 5, 604 (1981).

Preece, C. M., and E. N. Kaufmann, *Corrosion Sci.*, *32*, No. 4, 267 (1982).

Pscheidl, H., I. Kiricsi, and K. Varga, *J. Chromatogr.*, 243, 51 (1982).

Ramanan, V. R. V., and G. E. Fish, *J. Appl. Phys.*, 53, No. 3, 2273 (1982).

Raskin, D., and L. A. Davis, *IEEE Spectrum*, *November*, 28 (1981).

Reed, D. J., and B. J. Wuensch, *J. Am. Ceramic Soc.*, 63, No. 1–2, 88 (1980).

Robb, S., *Am. Ceramic Soc. Bull.*, 61, No. 5, 556 (1982).

Robinson, A. L., *Science*, 223, 920 (1984).

Sandler, Y. L., and W. M. Hickam, *Proc. Third Int. Congr. on Catalysis*, Amsterdam (July 1964), p. 227.

Satou, M., and F. Fujimoto, *Jpn. J. Appl. Phys.*, 22, No. 3, L171 (1983).

Schwegler, E. C., and F. A. White, *Int. J. Mass Spectrom. Ion Phys.*, 1, 191 (1968).

Shanfield, S., and R. Wolfson, *J. Vac. Sci. Technol.*, A1, No. 2, 323 (1983).

Shick, R. L., and E. L. Swarts, *J. Am. Ceramic Soc.*, 65, No. 12, 594 (1982).

Sickel, E. K., J. I. Gittleman, and P. Sheng, *J. Electronic Mat.*, 11, No. 4, 699 (1982).

Sigmund, P., *Appl. Phys. Lett.*, 25, 169 (1974).

Singer, I. L., and R. A. Jeffries, *Appl. Phys. Lett.*, 43, No. 10, 925 (1983).

Sioshansi, P., *Thin Solid Films*, 118, 61 (1984).

Spencer, N. D., and G. A. Somorjai, *Inst. of Phys. on Prog. in Phys.*, 46, No. 1, 1 (1983).

Sprokel, G. J., and J. M. Fairfield, *J. Electrochem. Soc.*, 112, 200 (1965).

Suib, S. L., D. F. Coughlin, F. A. Otter, and L. F. Conopask, *J. Catalysis*, 84, 410 (1983).

Svendsen, L. G., and P. Børgesen, *Nucl. Instr. and Methods*, 191, 141 (1981).

Takasu, Y., H. Konno, and T. Yamashina, *Surf. Sci.*, 45, 321 (1974).

Takayanagi, M., *Pure and Appl. Chem.*, 55, No. 5, 819 (1983).

Tosto, S., *J. Materials Sci.*, 18, 899 (1983).

Toyokawa, F., K. Furuza, and T. Kikuchi, *Surface Sci.*, 110, 329 (1981).

Tsaur, B. Y., S. S. Lau, and J. W. Mayer, *Nucl. Instr. and Methods*, 182/183, 67 (1981).

Uchino, K., Y. Miyazawa, and S. Nomura, *Jpn. J. Appl. Phys.*, 21, No. 12, Part 1, 1671 (1982).

Valenty, S. J., J. J. Chera, D. R. Olson, K. K. Webb, G. A. Smith, and W. Katz, *J. Am. Chem. Soc.*, 106, 6155 (1984).

Van Craen, M., J. Verlinden, R. Gijbels, and F. Adams, *Talanta*, 29, 773 (1982).

Van der Kouwe, E. T., *J. Electrochem. Soc.*, 129, No. 11, 2617 (1982).

Vanderlinden, J., and R. Gijbels, in *Advances in Mass Spectrometry*, Vol. 8A, A. Quayle, Heyden & Son Ltd., London (1980), p. 486.

Van Hoye, E., R. Gijbels, and F. Adams, in *Advances in Mass Spectrometry, Vol. 8A*, ed. A. Quayle, Heyden & Son Ltd., London 1980, p. 357.

Vogelesung, L. B., *Ind. Eng. Chem. Prod. Res. Dev.*, 22, 492 (1983).

Wada, M., K. Hosoi, and O. Nishikawa, *Acta Metall.*, 30, 1013 (1982).

Wang, Y. F., C. R. Clayton, G. K. Hubler, W. H. Lucke, and J. K. Hirvonen, *Thin Solid Films*, 63, 11 (1979).

Westbrook, J. H., and K. T. Aust, *Acta Metall.*, 11, 1151 (1963).

Williams, R. J. P., *Pure and Appl. Chem.*, 54, No. 10, 1889 (1982).

Willis, R. P., and R. H. Doremus, *J. Am. Ceramic Soc.*, 63, No. 5–6, 352 (1980).

Wolf, C. J., D. L. Fanter, and M. A. Grayson, *ONR Final Report MDCQO743* (July 21, 1981).

CHAPTER ELEVEN

ELECTRONIC MATERIALS AND DEVICES

Electronic materials have now joined structural materials as essential building blocks of contemporary civilization. In some instances, they provide new options for reducing the overall demand for metals and high-energy fuels. In a more general context, electronic materials furnish the design engineer with ingredients for effecting vast technical, economic, and social change. Telecommunications, industrial robotics, automated intelligence systems, and advanced medical diagnostics exemplify the new era that began with the announcement of the transistor by Bell Laboratories in 1947. At present, a broad spectrum of electronic materials includes semiconductors, superconductors, superionic conductors, lasers, piezoelectrics, magnetics, and optical fibers. From these have been derived new primary sensors, amplifiers, light-emitting diodes, liquid crystal displays, batteries, microwave generators, memories, signal processors, and large-scale integrated circuits. The electronics industry also makes use of most of the chemical elements in the Periodic Table, although a single element, silicon (~25% of the earth's crust), provides the basis for both semiconductor devices and fiber optics. However, some elements affect electronic transport phenomena at exceedingly small concentrations, and it is this *impurity-sensitive* property of electronic materials that has made mass spectrometry an indispensable tool in their analysis.

SEMICONDUCTORS AND IMPURITY ANALYSIS

Unlike metals, the electric properties of semiconducting specimens can vary over many orders of magnitude if small concentrations of foreign atoms or "defects" are introduced. As an example, the addition of one boron atom to 10^5 atoms of silicon increases the conductivity of silicon by a factor of 1000. On the other hand, the very high conductivity of copper ($\sim 10^{-6}$ Ω-

cm) is substantially independent of trace impurities, as conduction results from the exceedingly large number ($\sim 10^{23}$ cm^{-3}) of free and mobile electrons that are available for charge transfer. In insulators, most electrons are tightly bound, and perhaps fewer than 10^{10} electrons/cm^3 will contribute to conduction under the influence of an electric field. Semiconductors range between these extremes; the actual number of charge carriers varies widely, depending upon crystal structure, temperature, and impurities.

The two primary classes of semiconductors are (1) *intrinsic* and (2) *extrinsic*. The first class includes "impurity-free" solids such as silicon, germanium, selenium, and tellurium. Compounds generally classed as intrinsic include PbTe, ZnO, and Cu$_2$O. The purity of an element will determine the conductivity or resistivity of a specimen at any given temperature. The intrinsic resistivity of silicon exceeds 10^5 Ω-cm, but only material of the highest purity approaches this value. For intrinsic or impurity-free crystals, an elevated temperature or the exposure of the semiconductor to radiation provides a mechanism for conduction. Thermal excitation, for example, will generate some electrons with sufficient energy to free them from an interatomic bond or valence state, thereby making them available for charge transport.

The conducting properties of extrinsic or impurity semiconductors are primarily determined by the type and number of impurities in the crystal lattice. Impurities that provide extra electrons are termed "donors" or n-type. Electron-deficient impurities provide positive carriers and are termed "acceptors" or p-type. In a germanium crystal, for example, donor impurities include group V elements such as arsenic, antimony, or phosphorus, and acceptor impurities include group III elements such as indium, gallium, or boron. Figure 11.1 suggests one important feature of impurity semiconductors: the "energy levels" introduced through doping are often very close to either the conduction or valence band. In many instances, this energy gap is so small that room temperature is sufficient to "ionize" the impurity center, and thus to (1) excite an electron into the conduction band or (2) have a group III atom contribute a positive "hole" in the valence band. The

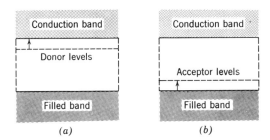

Figure 11.1   (*a*) Donor levels introduced by impurities in n-type material. (*b*) Acceptor levels introduced in a p-type semiconductor.

presence of certain impurities at even a few parts per billion is known to provide such levels, and these markedly affect the electrical properties of semiconductors. Impurities also affect the electron and hole mobilities in silicon, and a relationship for mobility as a function of impurity concentration and temperature has been developed for atom concentrations up to 10^{20} cm^{-3} and for temperatures up to 500 K (Arora et al., 1982).

Impurities in electroluminescent devices and phosphors also have a sizeable effect. For example, in very pure ZnS, as little as 0.0000001% copper activator is sufficient to give green emission. In this case, the activator impurity for light emission is analogous to the donor or acceptor impurities that are responsible for electron-hole charge transport in semiconductors. Yet trace quantities of iron or nickel ($\sim$0.00001%) significantly decrease the phosphor efficiency. Light-emitting diodes (LEDs) are known to degrade upon aging, and at least a portion of the deterioration of light output is attributed to a decrease in effective nitrogen concentration—specifically that part of the nitrogen concentration upon which radiative recombination takes place (Albrecht, 1983). Copper and other transition metals are also suspected as contaminants responsible for the degradation of LEDs fabricated with GaAsP (Damestani and Forbes, 1981).

Impurities are also known to degrade photovoltaic device performance, and specific impurities have been identified in the fabrication of hydrogenated, amorphous silicon (a-Si:H) thin films by glow-discharge deposition. Volatile impurities in silane (SiH$_4$) and disilane (Si$_2$H$_6$) gas at concentrations as low as 20 ppm have been shown to affect photoconductivity and short-circuit current (Corderman and Vanier, 1983). Thus, on-line mass spectrometric monitoring is a prerequisite for many plasma, sputtering, and vapor deposition techniques. Bulk impurities are of special concern in buried channel circuitry, such as charge-coupled devices, as some impurities (e.g., iron) affect device characteristics—including thermal leakage currents, charge transfer efficiency, and noise (McNutt and Meyer, 1982).

Carbon, in addition to oxygen, is a primary impurity in electronic-grade silicon that is used in the fabrication of solid-state control devices. Typical concentrations of this element are about 10^{16} cm^{-3}; concentrations exceeding about 5×10^{16} cm^{-3} in float-zoned silicon can lead to the formation of process-induced defects in the manufacture of power rectifiers and thyristors (Kolbesen and Mühlbauer, 1982). With increasing carbon concentrations, in which some local carbon concentrations may surpass solubility limits, precipitates can form that result in both the formation of extended defects and a deterioration of the electric performance of these devices.

The spatial distribution of impurities is also of paramount importance, and secondary ion mass spectrometry (SIMS) has made possible impurity profiles in three dimensions with very high resolution. It is also interesting that the first major semiconducting detector, (i.e., the photographic emulsion) has been the object of three-dimensional assay. In a recent analysis of the spatial distribution of iodide in Ag (Br, I) crystals, SIMS measurements have shown that the surface iodide concentration is roughly four times that of the interior, and that most of the enrichment occurs within the first 800 Å of the surface (Furman et al., 1980). This distribution, as well as the percentage of silver iodide in a silver bromide matrix, is related to both grain size and photographic response.

In addition to the obvious need for mass spectrometry to analyze impurity profiles in semiconductor devices produced by ion implantation and diffusion, Mashovets (1982) has cited the most probable causes of change in semiconductor properties resulting from heat treatment or quenching:

1. Formation of intrinsic defects—Frenkel pairs, multivacancy complexes, clusters of interstitial atoms, etc.
2. Interaction of intrinsic defects with impurity atoms, and formation of electrically active complexes.
3. Precipitation of impurities from the solid solution.
4. Transfer of impurities from clusters into the solid solution.
5. Diffusion of impurities from the surface of the sample.
6. Movement of intrinsic, interstitial, or impurity atoms to the surface.

In conjunction with other types of spectroscopy, mass spectrometry is now elucidating all such phenomena in solid-state microstructures.

IMPURITY PROFILING BY SIMS

Impurity profiles can be determined in semiconductors by several techniques. Junction capacitance-voltage (C-V) measurements are applicable for measuring selected impurity profiles in silicon, and both Hall effect measurements and neutron activation analysis are suitable techniques in some instances. However, secondary ion mass spectrometry (SIMS) has become the dominant analytical tool for determining impurity profiles for virtually all dopants—providing both very high sensitivity and excellent spatial resolution. There are, indeed, limits to the resolution that can be obtained by SIMS due to the energetics of the sputtering process. Kinetic energy transfer from bombarding primary ions can produce differential sputtering rates, changes in sample surface morphology, atomic redistribution or mixing, macroscopic heating, and radiation damage. These phenomena, plus the interaction of primary ions with residual gases, will cause perturbations in the observed secondary mass spectrum from a semiconductor material or device. Nevertheless, SIMS represents the most versatile technique for acquiring both surface and extended depth profiles for all atomic species. Typical impurity profiling of semiconductor materials is achieved with primary ions of 3–20 keV, highly focused beams, raster scanning, and beam current controls for programming sputtering rates. Commercial instruments also provide dual primary ion sources (e.g., oxygen and cesium), for enhancing the ionization and yield of the SIMS spectra. Table 11.1 indicates detection limits for selected impurities in GaAs as reported by Clegg (1982), using O_2^+ as the primary ion beam. Comparable sensitivities can be obtained for electronegative ions sputtered by using Cs^+ ion bombardment.

The literature relating to SIMS profiling is extensive, and only a few representative papers can be cited. The electric properties and atom depth

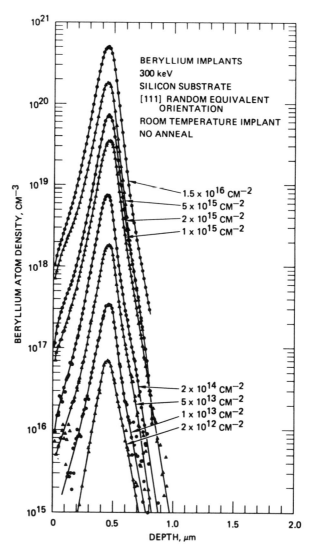

Figure 11.2 SIMS depth distributions as a function of ion fluence for 300 keV Be implanted into Si. (Wilson and Comas, 1982; *J. Appl. Phys.*, *53*, 3003 (1982), by permission of the American Institute of Physics.)

TABLE 11.1   Detection Sensitivity in SIMS Depth
Profiling for Selected Elements in GaAs

Element	Ion	ppm (atomic)	atoms (cm^{-3})
Be	$^9\text{Be}^+$	0.002	8×10^{13}
B	$^{11}\text{B}^+$	0.01	5×10^{14}
Al	$^{27}\text{Al}^+$	0.01	4×10^{14}
Si	$^{28}\text{Si}^+$	0.01	4×10^{14}
Cr	$^{52}\text{Cr}^+$	0.001	4×10^{13}
Mn	$^{55}\text{Mn}^+$	0.001	4×10^{13}
Fe	$^{56}\text{Fe}^+$	0.002	7×10^{13}
Cu	$^{62}\text{Cu}^+$	0.02	1×10^{15}
Sn	$^{120}\text{Sn}^+$	0.1	5×10^{15}

Source: Clegg (1982).

distributions of Be implanted into Si have been studied as a function of
implant fluence by Wilson and Comas (1982). Their SIMS depth distributions
for 300 keV Be implants are shown in Figure 11.2, and the results have been
compared with differential C-V profiling for the lower fluences. The ion
implantation of ^{75}As at 50 keV in single-crystal silicon has been measured
by SIMS when the impurity atoms were implanted through an SiO_2 layer
(Yen, 1981). Figure 11.3 shows the doping profile after 1×10^{16} ^{75}As atoms
were implanted through 230 Å of SiO_2 and annealed for 70 min at 1000°C.

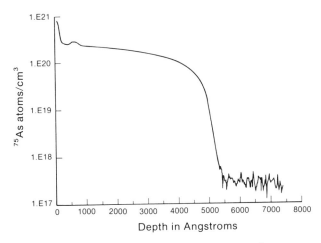

Figure 11.3   SIMS profile of 1×10^{16} ^{75}As implanted through 230 Å SiO_2 and annealed for 70
min at 1000°C under dry nitrogen flow, with ~200 Å SiO_2 remaining on the surface. (Yen, 1981;
J. Vacuum Science & Technology, 18, 895 (1981), by permission of the American Institute of
Physics.)

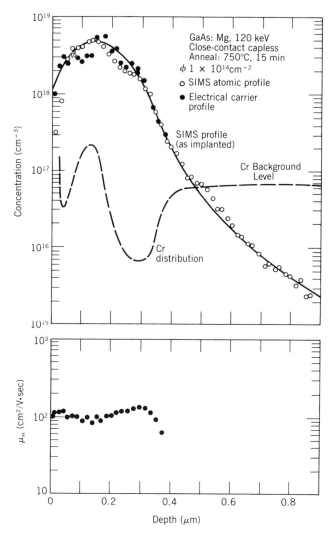

Figure 11.4 Atomic and electrical carrier concentrations determined by SIMS and Hall measurements for a 120 keV Mg implant in GaAs. (Yeo et al., 1982; *J. Appl. Phys.*, *53*, 6148 (1982), by permission of the American Institute of Physics.)

In a related study, SIMS profiles have been used to observe As dopant concentrations modified by laser-induced diffusion (Kiang et al., 1982). SIMS depth profiling has also been extensively utilized to examine differences between (1) "as-implanted," (2) thermally annealed, and (3) laser-annealed distributions (Christie and White, 1980). Furthermore, it has been important in examining the diffusion of impurities and possible precipitation phenomena at high doses (Simondet et al., 1980).

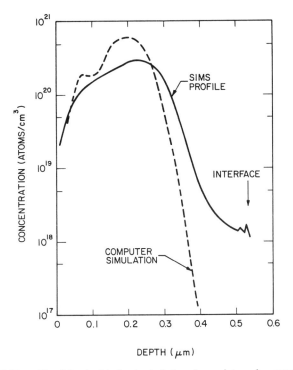

CONCENTRATION (ATOMS/cm³)

SIMS PROFILE

INTERFACE

COMPUTER SIMULATION

DEPTH (μm)

Figure 11.5 SIMS profile of the double implant of phosphorus into polycrystalline silicon: 170 keV at 7.5×10^{15} cm^{-2} plus 70 keV at 1×10^{15} cm^{-2}. (Smeltzer, 1982; *Appl. Phys. Lett., 41,* 649 (1982), by permission of the American Institute of Physics.)

A comparative study has been made between SIMS profiles and electric carrier profiles obtained via Hall measurements for Mg-implanted GaAs (Yeo et al., 1982). The Mg implants shown in Figure 11.4 do not indicate significant diffusion during the 750°C, 15 min anneal. Other SIMS measurements, however, have revealed that Be is a rapid diffusant in semi-insulating InP ($T \geq 700$°C; t = 15–30 min), and a redistribution of the compensating impurity (Fe or Cr) has also been observed (Oberstar et al., 1981).

SIMS profiling has also been employed to characterize doping profiles generated by chemical vapor deposition and subsequent diffusion, and to analyze the effect of various irradiations. Comparisons have also been made of computer simulations based on theoretical models to experimental implants. Figure 11.5 shows data for a double implantation of phosphorus in silicon: 170 keV at 7.5×10^{15} cm^{-2} plus 70 keV at 1×10^{15} cm^{-2} (Smeltzer, 1982).

Recently, impurity profiling has become an even more challenging task as investigators are experimenting with (1) multiple implantation, (2) co-implantation, and (3) implantation of complementary species, especially in compound semiconductors. N-type layers in GaAs with high free-electron

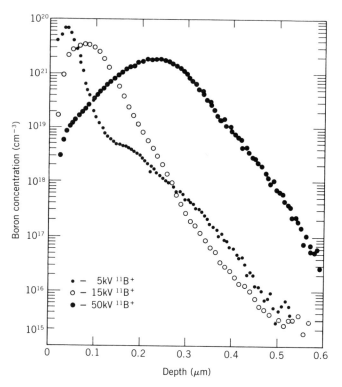

Figure 11.6 SIMS depth profiles of 5, 15, and 50 keV B implants at 3×10^{15} cm^{-2}. (Shepherd et al., 1983; *J. Vacuum Science & Technology*, *A1*, 991 (1983), by permission of the American Institute of Physics.)

concentrations have been produced by the multiple implantation of Se, Si, or Ge followed by a subsequent "complementary" species implant of P, As, or Ga (Stoneham et al., 1980). The latter implants were introduced as a means of minimizing the effect of the first dopant on the stoichiometry of the implanted region. Multiple simultaneous depth profiles and three-dimensional image profiles have been reported by Patkin and Morrison (1982) for a metal-oxide-semiconductor (MOS) integrated circuit. Using a digital data acquisition system, these investigators were able to visually observe ion images during the course of the SIMS profile for Na$^+$, Al$^+$, Si$^+$, and SiO$^+$.

SIMS measurements have also been made of very shallow implantation profiles; and a comparison has been made between calculated distributions of B and As in silicon, and experimental data has been obtained by (1) ion microscope (Cameca 3-f); (2) quadrupole SIMS; and (3) Rutherford backscattering (Shepherd et al., 1983). Ion microscope profiles of 5, 15, and 50 keV B implants at 3×10^{15} cm^{-2} are shown in Figure 11.6, and there is reasonable agreement between calculations and other experimental data. The authors suggest that the pronounced shoulder for the 5 keV implant at

~0.15 μm might be attributed to channeling during the implantation process. Clearly, as semiconductor device dimensions are reduced, the ability to monitor very shallow profiles becomes more important.

ION IMPLANTATION OF DEVICES

The concept of "forming semiconductive devices by ionic bombardment" was first proposed and successfully demonstrated by Shockley (1954); for this work, he was awarded U.S. Patent 2,787,564. There are several reasons why ion implantation has now become such an integral part of semiconductor technology. First, dopant atoms can be introduced into a wafer at the highest degree of chemical purity. Only a mass-resolved ion beam can restrict the introduction of dopant atoms to a single species. Second, impurity concentrations can be controlled over many orders of magnitude and be accurately monitored. Inasmuch as each ion possesses a positive or negative charge, the quantity of dopant can be monitored by sensitive current measurements. Finally, a high degree of spatial control can be achieved by ion beam programming by varying electric and magnetic fields, collimators, and ion source intensities. Very uniform depositions or steep gradients are possible; and by using stepping motors to control wafer orientation and throughput, the fabrication of devices has become both a quality-controlled and highly efficient process. The fact that semiconductor devices require only small quantities of dopants to affect their electrical properties is, of course, a primary reason for the successful use of ion implantation as a "high-dose," high-throughput, manufacturing technique. Subsequent to ion implantation, it is often necessary to anneal the sample to remove ion displacement damage and to "activate" the ions in the host lattice, but this can be easily accomplished by laser irradiations of appropriate wavelengths or by standard thermal techniques.

During the last decade, ion beam currents in practical ion implanters have increased by three orders of magnitude; currents are now as high as 10 mA over a large kinetic energy range, and silicon wafers have been rotated on target assemblies at 1000 rpm (Hanley, 1981). Electric fields disperse ions of mass M, energy E, and charge q according to E/q ratios; and magnetic fields disperse ions in the dimension ME/q^2, so that magnetic deflection is mass-dependent. Dopants such as Be, B, Mg, Ar, Cr, Fe, As, and Sb are among the many species that have been implanted into various specimens by magnetic, electric, and hybrid implantation systems. Silicon, using Si^+ or Si^{2+} ion beams, has also been implanted into semi-insulating GaAs single crystals at energies of 50–600 keV to form n-type layers of GaAs field-effect transistors (FETs) and GaAs integrated circuits (Onuma et al., 1981). In this work, the ion beam was incident to the crystal surface at an angle of 8° in order to minimize channeling, and the implanted carrier concentration profiles were approximated by a Gaussian distribution.

In some instances, ion implantation is carried out at very low kinetic energies (150–1000 eV) in connection with the molecular beam production of Si epitaxial films. This doping technique can be implemented without degrading the crystallographic or electrical quality of the films (Sugiura, 1980), and such films can be doped n- or p-type over a wide range of resistivities. Abrupt changes in the doping level of multilayer films can also be achieved with independent control of Si epitaxial growth and ion implant currents. A practical technique for rapidly alternating boron and arsenic doping in silicon molecular beam epitaxy (MBE) has been reported by Schwartz et al. (1982) that is expected to be useful in the further development of high-speed bipolar junction transistors.

Controlled impurity doping is now routine for producing a wide variety of devices. For example, doping crystalline silicon with lithium is the method for fabricating lithium-drifted detectors for nuclear spectroscopy. In such devices, lithium is an interstitial donor at 0.033 eV below the conduction edge in silicon (Beyer et al., 1979). Silicon planar varactor diodes, of large capacitance variation ratios, have been fabricated by ion implanting phosphorus into n-type silicon. These devices are now used for UHF/VHF TV tuner circuits (Gupta et al., 1982). Phosphorus-doped silicon and lithium-doped germanium infrared photoconductive detectors have been produced for low-photon background applications (Fishbach et al., 1981). Diode lasers have been fabricated and characterized in which the p-n junction is formed by causing type conversion in a p-type PbTe substrate using proton bombardment (Staudte and Bryant, 1983). In photoferroelectric storage devices, the intrinsic (near-ultraviolet) photosensitivity has been increased about four orders of magnitude by co-implanting Ar, Ne, and He ions into the surface exposed to image light (Land and Peercy, 1982). For the measurement of magnetic fields, GaAs Hall devices have been produced by multiple ion implantation steps; these devices possess a high sensitivity and linearity, a small temperature dependence, and a low residual voltage (Pattenpaul et al., 1981).

Ion bombardment is also an attractive alternative to conventional techniques for fabricating either ohmic or rectifying contacts on semiconductor circuitry (England and Rhoderick, 1981). The semiconductor is bombarded by low-energy ions through a mask, and is then placed in an ordinary commercial plating solution and illuminated by light whose photon energy exceeds the band gap of the material. The metal then plates on the part of the surface that has been bombarded. Advantages include (1) excellent adhesion; (2) no external electric connection is required as in electroplating; (3) no beam heater is required as with evaporation for refractory metals; and (4) metal deposition is highly spatially resolved. Bryant and Staudte (1981) have also implanted silver into PbTe diode lasers to produce improved low-resistance and longer life ohmic contacts.

Implantation techniques have also been widely applied in both basic and applied semiconductor technologies to compensate donor or acceptor pro-

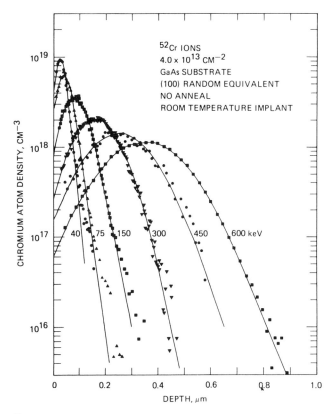

Figure 11.7 ^{52}Cr atom depth distributions in [100] GaAs; ion implantation fluence was 4 × 10^{13} cm^{-2} at energies ranging from 40–600 keV. (Wilson, 1981; *J. Appl. Phys.*, *52*, 3954 (1981), by permission of the American Institute of Physics.)

files at selected depths and to produce semi-insulating barriers in microcircuitry. Chromium, for example, has been introduced into GaAs during epitaxial growth to render the GaAs semi-insulating because of its deep acceptor levels and compensation for donor and acceptor impurities (Wilson, 1981). Figure 11.7 shows unannealed depth distributions for ^{52}Cr implanted into crystalline [100] GaAs at energies from 40–600 keV and at a fluence of 4 × 10^{13} cm^{-2}. Depth distribution profiles were measured by SIMS using an oxygen ion beam.

Overall, ion implantation has distinct advantages over diffusion technology in semiconductor device fabrication, some of which have been cited above. The technique requires only low-temperature processing, and there is little lateral spreading of the dopant, so that stray capacitance is minimized. Very shallow junctions can also be fabricated, and low-temperature processing permits the use of organic resist patterns for selective implantation.

In addition to using ion beams for impurity doping and device fabrication, per se, energetic ion beams are now used to form metastable compounds, silicides, amorphous layers, and solid solutions. The interface between metal films and silicon can be modified by using the ion beam to deposit large amounts of energy in this highly localized region, and thus to induce metal-semiconductor reactions. Ar, Kr, or Xe ions are often used in such studies at energies in the range of 100–300 keV and at doses of 10^{15} ions/cm^2. The rapid deposition and dissipation of beam energy at the metal-semiconductor interface results in a rapidly "quenched" zone that produces compounds (e.g., Ni_2Si, Pd_2Si, Pt_2Si, Co_2Si, $FeSi$, and $CrSi$) that have been subsequently identified by RBS (Mayer et al., 1981).

ION BEAM MILLING AND ETCHING

Mass spectrometry is essential in monitoring several important processes in semiconductor device fabrication, namely, (1) ion milling, (2) plasma etching, and (3) ion beam-assisted maskless etching.

Ion milling is the process in which a collimated beam of ions is incident upon a substrate to remove material by physical sputtering. Ions produced in a low-pressure (10^{-1} to 10^{-2} Pa; 1 torr = 133.32 Pa) discharge or plasma source are extracted and accelerated toward the sample with sufficient energy to remove surface atoms. Ion milling permits (1) the repeatable etching of line patterns as small as 0.2 μm; (2) the production of 10:1 depth-to-width ratio channels without undercutting; and (3) the removal of virtually any material to precise depths and geometries by proper selection of ion beam energy, angle of incidence, and masking techniques (Kowalski and Rangelow, 1983). Ion milling is also important in surface cleaning for subsequent epitaxy or for the controlled final thinning of specimens for transmission electron microscopy. For example, I^+ ion milling has been used for such a processing of InP and InSb, and the results compared with Ar^+ and Xe^+ ions (Chew and Cullis, 1984). Other semiconductor devices use multilayer films of SiO_2, Si_3N_4, Al, Cu, Ti, Pt, Au, etc., and these materials can be milled at rates as high as 1000 Å/min. In all ion-milling production systems, a mass spectrometer is required to monitor the primary gases plus background gases that arise from desorption, diffusion, and electron ion bombardment.

Plasma etching involves a glow discharge to produce chemically reactive species from relatively inert gases. The subsequent interaction of these reactive gases on certain solids gives rise to volatile compounds that are removed by the vacuum pumping system. This *reactive ion etching* is a complex mechanism, but it is believed that the dissociative process in a CF_4 plasma produces primarily CF_3^+, CF_3, F, CF_3^-, and F^-. Technical parameters include feed and effluent gas composition, RF frequencies, plasma etching pressure, ion energy distributions, ratio of ionic-to-neutral plasma species, maximization of etch in the direction perpendicular to the wafer, chemical selectivity, and plasma uniformity over the wafer surface, etc. Some of these parameters cannot be determined without the direct monitoring of the impinging, adsorbed, sputtered, and desorbed species during the etching process (Smith and Bruce, 1982).

Coburn and Winters (1981) have prepared a detailed paper on many aspects of plasma etching and identified the many atomic and molecular species that are tabulated in Table 11.2. These data were obtained with a quadrupole mass spectrometer, and a dramatic decrease is noted in the abundance of reactive species when the RF power (discharge) is turned off. In similar studies of fluorocarbon plasmas, Smolinsky et al. (1982) have reported the etching rates of Si and SiO_2. In addition, the energy dependence of ion-enhanced etching has been examined, and the usable ion energy range for broad-beam ion etching has been extended down to tens of electron volts (Harper et al., 1981).

Asakawa and Sugata (1983) have measured the reactive ion etching of GaAs with a very pure Cl_2 plasma flux, and they found the relationships shown in Figure 11.8. The increase of the sputter yield with increasing gas pressure at a constant extraction voltage suggests that the etching is significantly enhanced by the number density of the neutral radicals. It is also noted that the reactive sputter yield is much larger than that due to pure physical sputtering. In obtaining these data, a mass spectrometer monitoring of the plasma revealed that ions other than Cl^+, Cl_2^{2+}, and H^+ were negligible, and that ions undesirable for clean etching (C^+, O_2^+, and H_2O) were

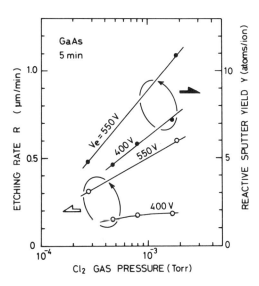

Figure 11.8 Etching rates (open circles) and reactive sputtering yields (solid circles) as a function of Cl_2 gas pressure. (Asakawa and Sugata, 1983; *Jpn. J. Appl. Phys., 22,* L653 (1983), by permission.)

TABLE 11.2 Ionic Species Observed in a Plasma Etch System with CF_4, H_2, O_2, and a Si Target

Mass Number (amu)	Ion	Probable Parent	Si Target CF_4 Gas	Si Target CF_4-22% H_2 Gas	Si Target CF_4-18% O_2 Gas	SiO_2 Target CF_4 Gas	Discharge Off CF_4 Gas
1	H^+	—	0.063	1.860	0.039	0.200	0.011
2	H_2^+	H_2	0.008	0.252	0.006	0.023	0.009
16	O^+	—	0.025	0.074	1.70	1.02	0.015
18	H_2O^+	H_2O	0.155	0.309	0.151	0.562	0.129
20	HF^-	HF	0.204	12.6	0.692	1.05	0.026
28	CO^+	CO	1.91	2.69	21.4	16.6	0.513
32	O_2^+	O_2	a	a	6.61	2.29	0.155
38	F_2^+	CF_4, F_2	0.056	0.068	0.093	0.107	0.076
44	CO_2^+	CO_2	0.035	0.129	7.94	7.24	0.007
47	COF^+	COF_2	b	b	4.68	4.27	<0.001
51	CHF_2^+	CHF_3	0.513	2.51	c	c	c
66	COF_2^+	COF_2	0.191	0.162	2.04	2.24	<0.001
69	CF_3^+	CF_4	100	100	100	100	100
85	SiF_3^+	SiF_4	14.8	11.7	15.8	15.1	<0.001
100	$C_2F_4^+$	—	0.257	0.282	0.013	0.009	<0.001
119	$C_2F_5^+$	C_2F_6	4.17	0.589	0.191	0.100	0.001
131	$C_3F_5^+$	—	0.418	0.053	0.011	0.010	0.001
169	$C_3F_7^+$	C_3F_8	0.182	0.018	0.010	0.008	<0.001

Source: Coburn and Winters (1981).

Note: Spectra were recorded on a quadrupole mass spectrometer. Ionizing electron energy = 70 eV. Fragmentation pattern for CF_4 gas: CF_3^+ = 100, CF_2^+ = 9.55; CF^+ = 3.31, F^+ = 2.24, C^+ = 1.95, CF_2^{2+} = 1.58, CF_3^{2+} = 0.65, F_2^+ = 0.07. (Of these fragment peaks, only F_2^+ was influenced by the discharge.) Glow discharge parameters: pressure = 2.7 pA (20 mTorr), CF_4 flow rate = 8.4 sccm, target electrode area = 182 cm^2, 13.56 MHz RF power ~200 W, interelectrode spacing = 6 cm, vacuum system volume ~100 l.

a Obscured by $^{13}CF^+$ but < 0.06.
b Obscured by SiF^+ but < 0.8.
c Obscured by $^{13}CF_2^+$ but < 0.1.

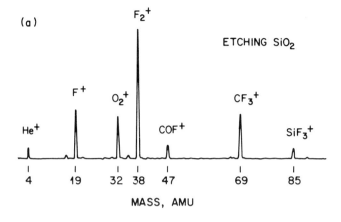

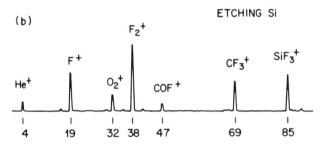

Figure 11.9 Mass spectra showing the composition of extracted ions when an He-F_2 discharge is etching (a) SiO_2 and (b) Si. (Vasile, 1980; *J. Appl. Phys.*, *51*, 2510 (1980), by permission of the American Institute of Physics.)

below the detection limit. Hayashi et al. (1982) have used photoionization mass spectrometry to identify organic compounds and numerous C_mF_n radicals that are generated in reactive ion beam etching of SiO_2. The mass spectra of these radicals is shown to vary dramatically, depending upon various feed gases, pressures, and other parameters. Vasile (1980) has also used direct ion sampling to investigate the etching of both SiO_2 and Si in an 80% helium/20% fluorine plasma; the spectra of Figure 11.9 show the difference in the composition of the extracted ions when the He-F_2 discharge is etching SiO_2 (Fig. 11.9a) and Si (Fig. 11.9b).

Mass spectrometry is almost always used in conjunction with ion beam-assisted maskless etching. Such ion beams furnish a high-energy deposition rate, and energetic ions undergo very little small-angle scattering. The schematic arrangement shown in Figure 20.1 by Gamo et al. (1982), using a 50 keV-focused Au^+ ion beam on GaAs in chlorine ambient gas, is typical of this technique.

THIN FILMS AND SEMICONDUCTOR INTERFACES

A detailed characterization of thin films, interfacial layers, and epitaxial growth is substantially aided by both SIMS and RBS. The term epitaxy is derived from the Greek *epi*, meaning outer or attached to, and *taxis*, denoting arrangement or order. Thus, epitaxy is the growth of a thin crystalline layer upon a crystal substrate, where the lattice arrangement of the growth layer replicates that of the substrate. Such epitaxial layers can be produced by liquid phase, vapor phase, or molecular beam techniques. Phillips et al. (1983) have studied the growth of BaF_2 insulating films on InP [100] by RBS using 1.6 MeV $^4He^+$ ions. For this particular system, they have found that the epitaxial quality is only a weak function of lattice mismatch, depending more strongly on growth temperature and substrate orientation. Figure 11.10 shows channeling and random RBS spectra of a 5000 Å BaF_2 film on InP [100]. The ratio of the backscattered ion yield from the BaF_2 layer in the aligned normal direction to the random direction (X_{min}) is very low, showing excellent epitaxy (see also Chapter 12).

SIMS analysis has been extensively for measuring very low concentrations of impurities in epitaxial layers. Profiles of oxygen concentrations for GaAs samples whose epitaxial layers were formed by vapor phase expitaxy have been measured by Albert (1982). High oxygen concentrations between the GaAs substrates and the silicon- and chromium-doped layers were observed to be sharply defined. The reported analytical limits for this SIMS investigation were: (1) minimum thickness needed for analysis, 0.05 μm; (2) depth resolution, approximately 100 Å; and (3) sensitivity, 0.01–0.02 ppm (atomic). Specific SIMS measurements have also been reported for monitoring of complex-layered structures. For example, Slusser and Slattery (1981) have monitored the deposition of 20 nm of Si_3N_4 and 300 nm of

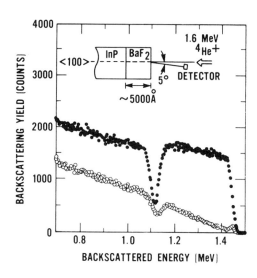

Figure 11.10 Channeling and random RBS spectra of a 5000 Å epitaxial BaF_2 film on In [100]. (Phillips et al., 1983; *J. Vacuum Science & Technology, A1,* 1006 (1983), by permission of the American Institute of Physics.)

polysilicon on top of a boron-doped silicon substrate wafer, which was subsequently implanted by a 40 keV ^{11}B ion beam.

Molecular beam epitaxy (MBE) is a specialized type of evaporation/deposition in which atomic and molecular beams impinge upon a heated substrate under ultra-high vacuum conditions. A molecular beam is produced from a Knudsen cell, and a portion of the desorbed flux is detected by a mass spectrometer (Foxon, 1981). With this and related techniques, it is now possible to produce layered structures of thicknesses of 1–100 nm and to add controlled amounts of electrically active dopants. It is questionable if this method will ever be widely used for thin film fabrication, but it is an excellent tool for monitoring the growth of binary compounds. Furthermore, by employing modulated beam techniques, thin films can be grown, doped, and characterized beyond the limits imposed by other schemes.

An important byproduct of laser annealing is the technology for forming crystalline silicon on amorphous substrates. By using high laser intensities that scan over small-grain polycrystalline films, it is now possible to produce regions of molten silicon that recrystallize into single-crystal islands. This new type of crystal growth or *lateral epitaxy* is becoming important in semiconductor manufacturing (Poate and Brown, 1982). In addition, various metallic films (Pt, Pd, Ni, Mo, and Nb) have been found to react with silicon under pulsed laser irradiation to form metal silicides. Since silicide interfaces are important for reliable Shottky barriers, ohmic contacts, etc., they have been studied extensively by SIMS and RBS. For example, studies have been made of the contact resistance between heavily doped p^+ and n^+ shallow junctions and an Al-0.9% Si alloy, and the dopant concentration was determined by SIMS (Cohen et al., 1982). Another important interface problem is the aluminum-silicon interconnect. Aluminum is the metal of choice because of its low electrical resistivity; it makes low-resistance contacts to both p and n regions, and it adheres well to the dielectric layers commonly used in device fabrication, that is, silicon dioxide and silicon nitride. However, under the influence of electric fields, silicon or aluminum electromigration can occur that may potentially cause device failure. Hence, very thin (1500 Å) films of tungsten are being tested as stabilizing barriers to inhibit aluminum-silicide reactions (Gargini, 1983). RBS is also a powerful technique for studying thin films, for example, the amorphous-crystal interface (Si-SiO$_2$), the crystal-crystal interface (Si-Ni silicides), and the liquid-solid interface—such as the laser-induced crystallization of silicon. Poate (1981) has discussed the Si-SiO$_2$ interface in detail and summarized recent developments relating to silicide thin film analysis. Figure 11.11 shows the result of an RBS experiment using 2 MeV ^{4}He$^+$ ions to probe a 5000 Å SiO$_2$ structure on a silicon crystal. The interface is sharply defined, and the technique has now been sufficiently developed so that the physical structure of the interface can be resolved within two to three monolayers.

The importance of SIMS in monitoring platinum silicide formation and the redistribution of boron resulting from thermal processing is dramatically

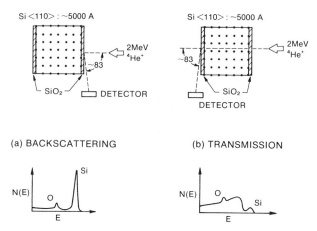

Figure 11.11 Schematic of Rutherford backscattering analysis of a SiO₂ film on silicon. (Cheung et al., 1979; Poate, 1981; *Nucl. Instr, Methods, 191,* 39 (1981), by permission of North-Holland Physics Publishing.)

shown in Figure 11.12. This depth profiling by Wei et al. (1983) shows distributions representing (*a*) as-deposited sample; (*b*) 400°C anneal for 5 min; (*c*) 400°C anneal for 30 min; and (*d*) 600°C anneal for 30 min. No other technique is able to monitor both silicide growth and dopant redistributions at low concentrations and with high spatial resolution.

Rutherford backscattering has also been employed for determining thin film densities since the densities of thin films may differ from bulk densities, and the RBS measurements can be made nondestructively (Simons et al., 1982). Diffusion is also an important parameter in thin metallic films, including the effect of annealing ambients. Diffusion rates in films are known to be affected by the presence of gases, and the effect of oxygen and hydrogen has been studied on the diffusion of Si, Cr, Co, Cu, and GaAs in gold and silver. The diffusion of Al in Pt films has also been measured in the temperature range of 200–600°C, including the effect of oxygen (Chang, 1981). Thin films of InAs deposited on mica substrates have been characterized for magnetoresistance and electron mobility, and marked changes have been observed depending upon evaporation techniques. The lower vapor pressure of arsenic resulted in a film containing an excess of indium, a lower mobility, and a higher magnetoresistance (Okimura, 1980).

Both SIMS and RBS are also important techniques for optimizing integrated circuit components. For example, very high-resistance load resistors are required to limit power consumption in static random access memory (RAM) circuits. Hence, gigaohm-range polycrystalline silicon resistors have been fabricated into integrated circuitry, using chemical vapor deposit techniques and patterned plasma etching to define resistor lengths (Mahan et al., 1983).

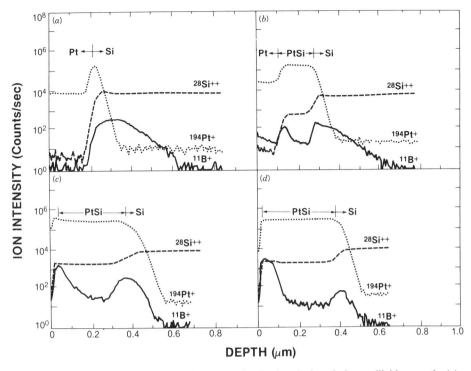

Figure 11.12 SIMS depth profiles of boron redistribution during platinum silicide growth: (*a*) as-deposit sample, (*b*) 400°C anneal for 5 min, (*c*) 400°C anneal for 30 min, and (*d*) 600°C anneal for 30 min. (Wei et al., 1983; courtesy of W. Katz.)

DIFFUSION MEASUREMENTS

Diffusion is a salient parameter in semiconductor technology with respect to the controlled introduction of dopants and the exclusion of undesired contaminants. Mass spectrometry has inherent advantages for investigating this parameter, inasmuch as it has a lateral spatial resolution on the order of 1 μm, a depth resolution of a few tens of angstroms, and virtually all elements are amenable to detection at trace levels. Diffusion profiles and diffusion rates have been measured in optoelectronic devices, optical waveguides, p-n junction diodes, integrated circuits, etc.; and diffusion measurements are crucial in monitoring thermal, laser, and electron beam annealing. Interdiffusion is also important at metal-semiconductor interfaces, especially in relatively unreactive systems where no interface layer is formed to retard atomic transport.

The diffusion of phosphorus in silicon has been studied extensively in both single-crystal and large-grained or relatively thick films of polysilicon. A recent study by Losee et al. (1983) is also of special interest, as it points

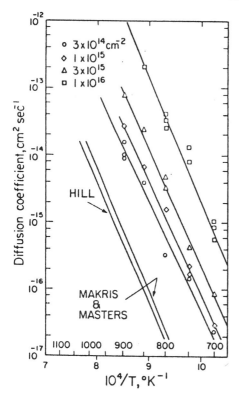

Figure 11.13 Effective diffusion constants for phosphorus in polysilicon vs. inverse temperature. Each line is labeled by the implant dose, and data for single-crystal silicon is shown for comparison. (Losee et al., 1983; *J. Electrochem. Soc., 83*, 597 (1983), by permission of the Electrochemical Society.)

out differences that can occur between fine-grained and large-grained films. Figure 11.13 also indicates that diffusion constants for phosphorus in polysilicon are dose-dependent. SIMS measurements combined with electric measurements have also been used to measure ion-implanted selenium in GaAs (Lidow et al., 1980). Results of this study suggest that the annealing behavior of Se in GaAs can be predicted for concentrations less than solid solubility. Furthermore, SIMS and RBS measurements have been made of high-dose ^{75}As implants in silicon, and the diffusion of this dopant has been observed in connection with continuous-wave laser and electron beam anneals (Gibbons and Sigmon, 1981).

Surface diffusion, in excess of that expected from thermally activated diffusion alone, has been observed for ion-impacted surfaces of silicon and metallic thin films (Robinson and Rossnagel, 1982). This enhanced diffusion due to Ar$^+$ bombardment has been observed to increase as ion beam current density is increased, and the effect is attributed to changes in surface topography resulting from multiple-ion impacts.

The simultaneous diffusion of zinc and cadmium into InAs has been reported by Horikoshi et al. (1981), and the effective diffusion coefficient indicated in Figure 11.14 shows a comparison with the separate diffusants. The result suggests that the simultaneous diffusion of Zn and Cd will produce

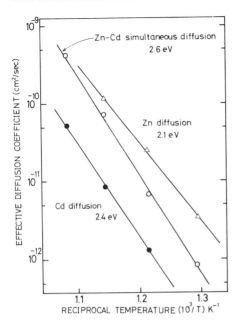

Figure 11.14 Diffusion constants for Cd and Zn, as separate and simultaneous dopants in InAs. (Horikoshi et al., 1981; *Jpn. J. Appl. Phys.*, **20**, 437 (1981), by permission.)

a more abrupt junction than the separate diffusion of either dopant at fairly high concentrations. SIMS studies of semi-insulating InP have been made following high fluence implants of Mg and Si. Following annealing, the Mg depletion zone was observed to occur in the region of maximum implant damage (Oberstar et al., 1981a, 1981b).

The effect of annealing ambients on the diffusion of various species has also been reported in metal thin films, and the effect of oxygen on the diffusion of Al in Pt has been determined by RBS (Chang, 1981). Diffusion at metal-semiconductor interfaces is also of special concern, as metal films are utilized in solid-state technology either as primary metallizations or as diffusion barriers. A study by Doyle et al. (1982) of the diffusion of Au showed the diffusion coefficient of this element in polycrystalline NiNb at 400°C to be 1.6×10^{-15} cm^{2}s^{-1}, but it is only $\sim 10^{-21}$ cm^{2}s^{-1} in amorphous NiNb at the same temperature. This result supports the general concept of using amorphous films as effective diffusion barriers in solid-state device applications.

The diffusion of Ga in intrinsic silicon has been measured between 700–1100°C, and diffusion profiles have been observed by both SIMS and neutron activation (Handoss et al., 1980). In this ion microprobe analysis, a fine beam of Ar ions was used to erode the Si samples, and the Ga profiles were obtained by monitoring the ^{69}Ga secondary ions. The use of capacitance-transient spectroscopy for studying trace contamination in semiconductor devices should be mentioned, as this is an important ancillary tool that can be used in conjunction with SIMS (Benton and Kimerling, 1982). This technique measures electrically active defects in the depletion region of a semi-conductor junction at concentrations as low as $\geq 10^{10}$ cm^{-3}. A salient feature of this technique is that it is effective in examining finished devices as well as in monitoring individual manufacturing steps.

VERY LARGE SCALE INTEGRATED (VLSI) MICROCIRCUITS

A generally recognized goal for very large-scale integrated circuits (VLSICs) is to produce devices with 100 K gates or memory bits per circuit, or devices with geometries less than 1 μm. An equally challenging objective for very high-speed integrated circuits (VHSICs) is to produce devices with 1 μm features that have "an equivalent gate clock frequency product exceeding 5×10^{11} gate Hz/cm^2 and a minimum clock rate of 25 MHz" (McGuire et al., 1983). Also, superfast transistors with switching times in the range of 10^{-12} s are being developed with doped aluminum gallium arsenide in a field effect structure. These new devices exploit the phenomenon that extra-high electron mobility is possible if electrons can be transferred from a doped layer to an adjacent undoped layer (Morkoc and Solomon, 1984). The essential role of mass spectrometry in the further development of such circuits becomes apparent from the following considerations:

1. The quality of the silicon starting material is critical. A National Research Council report estimates that VLSICs will require material whose variation in resistivity is no more than 1% from point to point in the wafer.

2. Thermal donors, introduced by oxygen, can change the resistivity of silicon during processing—which can shift threshold voltages of transistors and affect drive currents.

3. Some oxide films will have dimensions of only a few, hundred angstroms, and these must have no defects or pinholes.

4. Leakage currents caused by subsurface defects that may be tolerated in individual devices may give rise to unacceptably high leakage currents in VLSI circuits.

5. Passivating overcoat layers must be defect-free in order to prevent corrosion or the oxidation of thin film resistors.

6. Metallic interconnects such as aluminum tend to dissolve silicon; and although some solubility is desirable for ensuring good electric contact, this type of reaction can degrade circuit performance.

7. Diffusion and interdiffusion phenomena will have to be known and controlled over the operating temperature regimes for which the microcircuitry is designed.

8. Impurity doping profiles and n- or p-type regions must have sharply

TABLE 11.3 Qualitative Effect of Shallow vs. Deep Implants on
VLSI Transistor Parameters

Device Parameter	Shallow Implant	Deep Implant
Subthreshold voltage	−	+
Threshold voltage	−	+
Body effect	+	−
Avalanche	+	−
Parasitic capacitance	+	−
Breakdown voltage	−	+

Source: Risch et al. (1982).

defined boundaries. Also, the redistribution of dopants profiles, subsequent to laser or electron beam annealing, must be analyzed with
high resolution and sensitivity.

9. Multiple masks (possibly up to 15) must be registered within 1 part
    in 10^5.

10. In plasma and reactive ion-etching processes, the purity of gaseous
    constituents is crucial.

11. Hybrid electro-optical chips, a new generation of VLSICs, will im-
    pose even further demands upon the quality control of materials and
    processing technology.

Only a limited number of investigations can be cited relative to these areas
from among those being undertaken by industrial research and development
groups worldwide. In order to optimize the performance of VLSI transistors,
many processing techniques have been explored. With respect to dopant
profiles, multiple implants (i.e., shallow low-dose implants followed by
higher dose-deep implants) have been investigated (Risch et al., 1982). Table
11.3 compares the advantageous (+) and negative (−) effects associated
with shallow and deep implants on 1 μm MOSFET (metal-oxide semiconductor field-effect transistor) structures. SIMS measurements are also being
used to measure single shallow dopant profiles and to monitor those distributions for different annealing times (Ramsey, 1983). The depth resolution
of SIMS is graphically illustrated in the profiles obtained by Katz (1984) of
19 alternating layers of 29 Å (C) and 21 Å (Nb) on Si (Fig. 11.15). The analysis
of this multilayered structure was achieved using a primary beam of O_2^+ at
9.5 keV.

The use of SIMS or RBS is also required for examining the redistribution
of impurity dopants after laser or electron beam annealing. In this type of
semiconductor annealing, high densities of electron-hole pairs are created
and form a high-temperature plasma that transfers energy to the vibrational
modes of the lattice atoms. The time scale of this process is very short

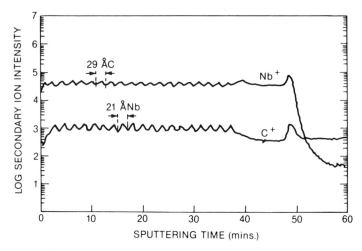

Figure 11.15 SIMS depth profile of VLSI circuitry, showing 19 alternate layers of 29 Å (C) and 21 Å (Nb) on Si. (Katz, 1984; courtesy of W. Katz.)

($<10^{-10}$ s in silicon), and with focused scanning laser or electron beams, temperatures close to the melting point of silicon can be reached in about 10 μs; with pulsed beams, melting point temperatures can be attained in about 20 ns (Brown, 1981). This ultra-fast and localized melting (1) allows the growth of single-crystal material on amorphous substrates; (2) anneals semiconductor damage and defects; (3) limits the dopant distribution profile; and (4) increases the potential for fabricating three-dimensional IC configurations.

VLSI requirements have imposed more stringent demands on the etching process, as wet chemical and conventional plasma etching techniques lack the resolution to transfer 1 μm dimension patterns into films that make up the device (Ephrath, 1982). In VLSI-etching processes, dry etching is carried out in a barrel or tunnel etcher and RF power is applied to external coils or capacitor plates. Wafers are loaded along the center axis, at pressures of 0.5–1 torr. In reactive ion-etching systems, wafers are loaded on an RF electrode in a system similar to a sputtering reactor, and reactive gases are maintained at pressures between 0.01–0.1 torr. Accompanying this new etchant technology, mass spectrometric methods have been devised for monitoring the purity of gases in device manufacturing (i.e., silane, phosphine, arsine, diborane, boron trichloride, etc.) inasmuch as these gases are required for sputtering, plasma etching, chemical vapor deposition (CVD), etc. Furthermore, a methodology has been developed for an accurate determination of hermetically sealed-in contaminants (Lin and Burden, 1978). In a recent investigation, mass spectral analyses of microvolumes as small as 40 μL have been obtained for certain compounds, with sensitivities in the range of 0.05–10 ppm (Schubert and Augis, 1982).

Low-energy, ion-scattering spectrometry (ISS) has also been used to study the formation and properties of silicides. Yabuuchi et al. (1982) have used Ne$^+$ ions at 500 eV to examine one of the most widely used silicide systems, Pd/Si(111), and to determine the geometrical configuration of Pd and Si surface atoms in conjunction with other forms of spectroscopy.

Collinear with the technology of integrating active components of VLSIC has been research to achieve integration and miniaturization of *passive* components, such as capacitors. The equation for the capacitance of a parallel plate capacitor is $C_A = \epsilon_0\epsilon_r/d$, where C_A is the capacitance per unit area, ϵ_0 is the permittivity of free space, ϵ_r is the relative permittivity or dielectric constant, and d is the interelectrode spacing. Thus, to achieve a high capacitance per unit volume, the dielectric constant must be very high and the interelectrode spacing must be reduced to the micron range. High dielectric constant ceramics (1000–15,000) are usually modifications of barium titanate, $BaTiO_3$; recent work has been reported with dopants of Cr, Mn, Fe, Co, or Ni (Hagemann et al., 1983–1984). The maximum field strength ratings for most ceramics is in the range of 2–3 V/μm, so that for normal voltage ratings of ~50 V, a 25 μm spacing would appear to suffice in an idealized device. In practice, such dimensions are difficult to attain with conventional technology, so that "ceramic barrier-layer" capacitors and "ceramic multilayer capacitors" have been proposed (Hagemann et al., 1983–1984).

"SOFT ERRORS" IN COMPUTER MEMORIES

So-called "soft errors" are sporadic losses of bits of information without any permanent change in the circuits, for example, due to noise spikes and ionizing radiation (Wallmark, 1982). One soft error that places an inherent limit on VLSICs relates to spurious signals that are generated by alpha particles. These alpha particles can be emitted from trace impurities of uranium or thorium in the packaging materials that enclose an integrated circuit chip. Another source of ionizing radiation are cosmic ray neutrons that produce alpha particles during their passage through the silicon transistors, per se. In the case of a ^{238}U impurity, the alpha particle range is on the order of about 25 μm and the alpha particle energy is ~4.2 MeV. If all of this energy is expended in silicon, the total number of electron-hole pairs generated is $4.2 \times 10^6/3.6$ or $>10^6$ electronic charges, as in silicon 3.6 eV is required to produce one electron-hole pair. Even if only a fraction of this ionization is collected by nearby junctions (in 10–50 ns), it may be sufficient to charge or discharge an MOS transistor-capacitor, and thus change the state of a memory cell. This uranium impurity problem was first identified by Intel Corporation engineers in RAMs being developed for their 16 K RAM. And while it was possible to redesign chips to account for such a spurious signal, this engineering solution imposed the penalty of using a fraction of the memory chip to store a suitable correction code. The alternative has been to

TABLE 11.4 Naturally Occurring Alpha Emitters of Concern in Semiconductor Memories

Isotope	Natural Fractional Abundance (%)	Half-Life (yr)	Alpha Energy (MeV)	Specific Activity ($\alpha/g - s$)
^{147}Sm	15	1.0×10^{11}	2.2	1.2×10^2
^{190}Pt	0.01	6.0×10^{11}	3.2	1.4×10^{-2}
^{232}Th	100	1.4×10^{10}	4.0	4.1×10^3
^{235}U	0.7	7.0×10^8	4.4	5.5×10^2
^{238}U	99	4.5×10^9	4.2	1.2×10^4

Source: Bouldin, (1981).

carefully screen virtually all plastics, ceramics, metals, and other materials that house memory chips for trace quantities of naturally occurring radionuclides.

The typical hermetically sealed computer memory chip is packaged in a ceramic/glass or epoxy-formed module. The silicon chip itself, of course, must be of the highest purity together with all the reagents that are associated with its fabrication. The maximum allowable alpha particle activity for materials in proximity to the active surface of a semiconductor memory chip is ~0.001 alpha particle/cm²/h. Trace levels of uranium and thorium in either silicon or packaging materials can be assayed by (1) mass spectrometry and the isotopic dilution method; (2) alpha particle counting; and (3) neutron activation analysis (NAA) with fission track counting. If mass spectrometry is used, ^{233}U can be used as a "spike," and an ether extraction can be employed to concentrate the uranium prior to analysis. Mass spectrometry provides the most unambiguous information relative to uranium and thorium, as isotopic abundance data can distinguish between naturally occurring radionuclides, enriched uranium, or transuranic contamination from other sources. However, neutron activation, with fission track counting that uses high-purity synthetic-fused silica as the detector, has been reported with a lower limit for uranium of 0.02 ppb (Riley, 1982). In order to attain this sensitivity, neutron doses were in the range of 10^{17} to 10^{19} n/cm². Actual measured concentrations of uranium in a variety of materials were 0.1–1 ppm for ceramics and glass, 1.5 ppb for silicone rubber, and 1.4 ppm for an imprinting material. An advantage of neutron activation is the high degree of spatial information that can be obtained from a wide variety of materials. A limitation is the vulnerability of the sample material to radiation degradation or some process during high-intensity irradiation that would alter sample/detector geometry.

Table 11.4 is a list of a few naturally occurring alpha emitters with long half-lives. Based on specific activities of these nuclides, Bouldin (1981) suggests that "if one part per million is a detrimental level of uranium contamination, roughly that level of thorium, one part in ten thousand of samarium, and pure platinum would be of equivalent concern" (neglecting the alpha energy differentials). In addition to these nuclides, which can occur in ceramics, Al_2O_3 substrates, and metal films, alpha particles can emanate from ^{210} Po—the daughter of ^{210}Pb—from lead/tin solder, or from lead impurities in copper.

OPTICAL WAVEGUIDES

In an optical waveguide system, a light signal is launched via a laser or light emitting diode and propagated through an optical fiber by total internal reflection. In 1982, Bell Laboratories reported the transmission, without repeaters, of digital signals of 274 megabits per second at 1.31 μm over single-mode fibers 101 km in length. British Telecom Research Laboratories and Nippon Telegraph and Telephone Company also reported high bit rate optical transmissions, virtually error free, over comparable distances. By the following year, the Bell system had installed approximately 200,000 km of fiber and 10,000 regenerators or repeaters for transmitting and receiving data, voice, and video signals. At present, optical fibers provide a broadband communication link between major East and West Coast cities of the United States and local networks worldwide; and in 1988, a transatlantic cable is scheduled for service capable of transmitting simultaneously 40,000 phone conversations. Thus, fiber optics has merged with its related technologies of satellite communications and integrated circuits in ushering in a global telecommunication system.

The potential advantages of using carrier frequencies of some 10^{15} Hz were clearly recognized in the 1960s with the advent of commercial lasers. Having such carrier frequencies transmitted through a practical waveguide, however, posed materials problems analogous to those of semiconductors. Specifically, the two salient problems were (1) the attenuation of the optical signal; and (2) group velocity dispersion—both as a function of wavelength. The attenuation limits the practical distance for signal transmission without a repeater, whereas the dispersion $dn/d\lambda$ limits the bandwidth and the retention of distortion-free waveshape (see Fig. 11.16). It can be shown that distortion-free transmission can be achieved (in fused silica fibers at 1.3 μm); also, the recompression of optical pulses broadened by their passage through optical fibers at other frequencies has been realized by ingenious means (Nakatsuka and Grischkowsky, 1982).

The basic attenuation problem in optical fibers is twofold. First, the waveguide must be fabricated with an index of refraction that varies radially from the maximum index at its center; and only with an appropriate gradient index can an optical ray be transmitted sinusoidally without attenuation at the fiber surface. In the graded index (multimode) fiber, the core has an

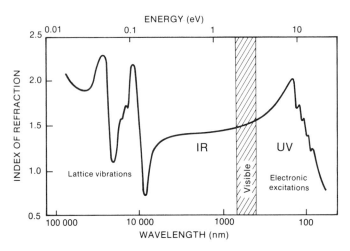

Figure 11.16 Index of refraction as a function of wavelength for a colorless glass. (Adapted from Nassau, 1983.)

index of refraction that decreases parabolically from the center toward the cladding, so that off-axis rays near the cladding travel faster than in the center (Miller, 1984). This causes modal dispersion. Although several techniques have been used for manufacturing gradient index materials, a CVD method is preferred for both telecommunication fibers whose diameters are on the order of 100 μm and an index change Δn, approximating 0.01. In the CVD process, glass of a given index is deposited on either the inside or outside of a fuzed quartz tube; each successive deposited layer has a slightly different chemical composition and index. Gaseous molecules used in the CVD process include silicon chlorides and doping agents, that is, SiCl, $PoCl_3$, BCl_3, $SbCl_5$, etc. (Mitchell, 1982). Subsequent to vapor deposition, the fiber is drawn, the layers become smaller than the wavelength of light, and the index profile appears to be continuous. Precise index control has been achieved with an inside vapor deposition process by Bell Laboratories (U.S. Patent 4,217,027) and an outside vapor deposition technology developed by the Corning Glass Works.

The second attenuation problem relates to the control of impurities at every stage of optical fiber manufacture, that is, raw materials, glass-making, chemical vapor deposition, and fiber-drawing. Desirable impurities must be introduced to generate the inhomogeneous index of refraction, but deleterious chemical species must be removed to obtain high optical transmissions. For example, measurements on CsBr infrared fibers indicate that an SO_4^{2-} ion impurity of only 2 ppm corresponds to an absorption loss of 3–8 dB/m at the 10.6 μm CO_2 laser frequency (Mimura et al., 1982). For some elements, impurities must be controlled at the parts per billion level. Limits

TABLE 11.5   Impurity Levels (ppb) Introducing 1 dB/km Absorption Losses at 800 nm in Various Glasses

Impurity	Na-Ca-Si	SiO$_2$	Na-B-Si
Co	0.2	40	100
Cr	2.1	2	20
Cu	5.1	100	1(II)[a], 10(I)[a]
Fe	2.1	20	1(II)[a], 1000(III)[a]
Mn	10.1	50	100
Ni	2.1	40	10
V	10.1	1	25

Source: Mitchell (1982).

[a] (I), (II), and (III) are oxidation states of impurities.

for certain transition elements in glass of various compositions are given by Mitchell (1982) in Table 11.5.

Studies of impurity absorption losses have also been made for BaF$_2$-GdF$_3$-ZrF$_4$ glass, as this material is a promising candidate for optical fibers in the 3–4 μm wavelength region. The rare earth elements have characteristic absorption bands in this infrared region, so that any contamination by these

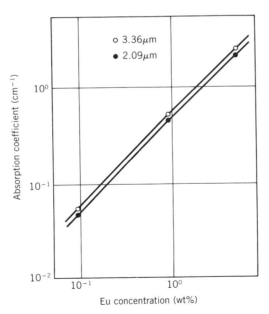

Figure 11.17   Relationship between peak absorption coefficients (cm^{-1}) and Eu concentration in wt%. (Ohishi et al., 1981; *Jpn. J. Appl. Phys. 20*, L191 (1981), by permission.)

TABLE 11.6 Rare Earth Ion Concentrations (ppb) Resulting in a 10^{-3} dB/km Absorption in Fluoride Glasses

	Wavelength (μm)			
Ion	2.5	3.0	3.5	4.0
Ce^{3+}	—	0.13	0.02	0.02
Pr^{3+}	0.57	2.52	0.22	0.05
Nd^{3+}	0.05	1.57	—	1.47
Sm^{3+}	0.39	0.18	0.85	0.14
Eu^{3+}	0.74	0.32	0.07	0.54
Tb^{3+}	—	0.07	0.88	—
Dy^{3+}	1.34	0.12	0.99	4.56

Source: Ohishi et al. (1981).

elements poses a problem in fabricating very low-loss fluoride optical fibers. Ohishi et al. (1981) have shown that peak absorption coefficients at 2.09 μm and 3.36 μm linearly increase in magnitude with the Eu concentration (Fig. 11.17), and this linear relationship appears to be confirmed for a number of other rare earth elements. Specifically, their data in Table 11.6 show some rare earth ion concentrations, in parts per billion, that would give 10^{-3} dB/ km absorption loss.

Thus, as in the case of semiconductors, the importance of impurity control for optical fibers has stimulated improved methods for producing high-purity reagents and for pin-pointing the specific role of certain dopants. Figure 11.18 shows the general shape of the absorption versus wavelength curve in high-purity optical waveguides, where the losses are approaching the Rayleigh scattering limit ($\sim 1/\lambda^4$) in the region of 1.3 and 1.5 μm (Lines, 1984).

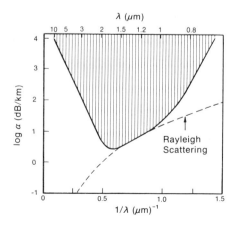

Figure 11.18 Logarithmic plot of attenuation α as a function of reciprocal wavelength $1/\lambda$ in silica. Small numbers of trapped OH^- radicals can contribute to increased losses at 1.4–2.8 μm, and Rayleigh scattering imposes a fundamental limit to transmission in glasses. (Adapted from Lines, 1984.)

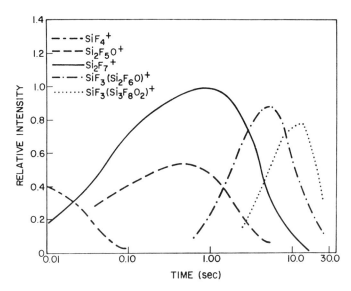

Figure 11.19 Time variation of selected ions produced from impure SiF_4 in Fourier transform mass spectrometer; pressure (SiF_4) $\approx 2 \times 10^{-7}$ torr. (Reents et al., 1985; reprinted with permission from *Anal. Chem., 57*, 104 (1985). Copyright 1985 American Chemical Society.)

In actual chemical vapor deposition fibers, an increased absorption has been attributed to small numbers of hydroxyl (OH) radicals trapped in the silica during manufacture (Miller, 1984). High-sensitivity mass spectrometry and the isotopic dilution method, in conjunction with other spectroscopic techniques, have been crucial in providing other qualitative and quantitative measurements. These analyses have also been carried over into a new generation of buried waveguides for microcircuitry, and to gradient index fiber arrays for copying machines. Commercial optical fiber cables are currently available with attenuations as low as 35 dB/100 km (including splicing), and very low-loss materials (<0.1 dB/km) are becoming available in the infrared (1–5 μm). In this region, Rayleigh scattering effects are reduced, and infrared applications now include repeaterless communications, control circuitry, focal plane imaging, pyrometry, and the transmission of laser power for surgery, cutting, drilling, and welding.

Among the several non-oxide glasses that are being explored are the "heavy metal fluoride glasses"; these vitreous materials may contain large amounts of ZrF_4 or HfF_4, and include minor components such as alkaline earths, rare earths, etc. Theoretical studies of these new classes of optical materials project absorption losses of 10–100 times lower than in the best silicate glasses, if virtually all impurities can be removed. Other glasses include a chalcogenide material designed for the mid-infrared (6–11 μm), consisting of three metals: germanium, antimony, and selenium. This glass has been produced commercially by Texas Instruments and fabricated into

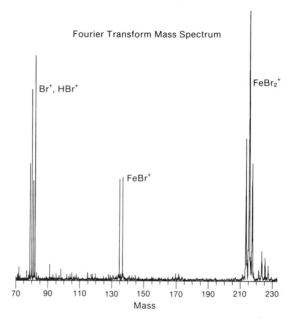

Figure 11.20   Positive ion Fourier transform mass spectrum of $FeBr_2$. (Gohlke, 1984; courtesy of the Eastman Kodak Co.)

fibers by Galileo Electro-Optics for applications in thermal sensing, control and medical systems, and spectrometry.

The advent of Fourier transform mass spectrometry (FTMS) is providing an important additional means for impurity identification in SiF_4, which has been considered as a fluorine dopant in advanced optical waveguides. Reents and Mujsce (1984) and Reents et al. (1985) have demonstrated that FTMS permits both the detection of impurities in SiF_4 and ion-molecule reactions that cannot be observed by conventional spectrometry. Exact mass measurements were made on the product ions (SiF^+, SiF_2^+, SiF_3^+, and SiF_4^+) from chemical ionization, using SiF_4^+ itself as the reagent gas. Other ions were then identified relative to this established mass scale. Quantification was achieved by monitoring specific ion-molecule reactions involving impurity molecules without the use of reference compounds, that is, observing the formation rate of an ion to the concentration of the corresponding neutral molecule. Figure 11.19 is a spectrum obtained by Reents et al. (1985) that shows the time variations of intensities for selected ions and demonstrates the potential of FTMS in observing ion-molecule kinetics.

Other important applications for FTMS include photosensitive materials and compounds of low vapor pressure. Figure 11.20 shows the positive ion Fourier transform mass spectrum of $FeBr_2$ obtained by Gohlke (1984).

OPTOELECTRONIC DEVICES

The response of an optoelectronic device is dependent upon many parameters, including photon wavelength, energy band structure of the semiconductor, number and type of impurities, etc. Clearly, many types of spectroscopy, including mass spectrometry, have been essential to the development of the following transducers and detectors:

1. Injection lasers.
2. Avalanche photodiodes.
3. Photovoltaic detectors.
4. Light emitting diodes.
5. Fiber optic magnetic field sensors.
6. Nonemissive liquid crystals.
7. Thin film electroluminescent and electrochromic displays.
8. Cathodoluminescent sensors.
9. Charge-coupled arrays.
10. Subnanosecond switches.

Specifically, mass spectrometry is a prerequisite tool in basic research relating to the physics and chemistry of all these transducers; it is also an important adjunct to actual device fabrication.

In connection with semiconductor optical amplifiers, ion beam technology is involved in (1) controlling the doping levels of diodes; and (2) the deposition of ultrathin films for obtaining large amplifier bandwidths, reducing coupling losses, and suppressing noise (Kobayashi and Kimura, 1984). Only a few selected examples, however, of mass spectrometric-related techniques, ion implantation, or use of separated isotopes will be cited.

High-radiance, surface-emitting LEDs are attractive sources for fiber optics communications as well as for displays. The light power of heterostructures of InGaAsP/InP LEDs can be several milliwatts, and such devices can now be butt-coupled to 50 μm core, graded index fibers (Borsuk, 1983). Light emitting diodes, injection lasers, photocathodes, and avalanche photodiodes have all been fabricated using gallium-indium arsenide ($Ga_{1-x}In_xAs$), as the semiconductor material in the 1-2 μm wavelength region (Whiteley and Ghandhi, 1982). Fabrication techniques require "electronic-grade" reactant gases, partial pressure measurements, "in situ" etching, and controlled epitaxial growth on appropriate substrates.

Thin film electroluminescent devices are multilayer structures that provide high brightness and contrast. ZnS:Mn, which produces a yellow luminescence, has been studied extensively, and ZnS has been doped with rare earth fluorides such as TbF_3 to produce a green luminescence (Howard, 1981). The distinguishing feature of an alternating current electroluminescent

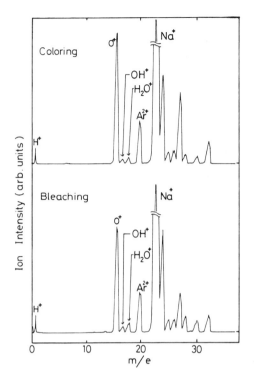

Figure 11.21 SIMS spectra of (*a*) colored and (*b*) bleached WO_3 thin films. (Muramatsu et al., 1982; *Jpn. J. Appl. Phys., 21*, L73 (1982), by permission.)

device is the presence of a true insulating layer that conducts no current and an active layer that responds to *changes* in applied external voltage (Smith, 1981).

SIMS analysis has been employed by Muramatsu et al. (1982) to identify the mechanism of electrochromic coloring in vacuum-deposited amorphous WO_3 thin films. In one instance, SIMS revealed that the chemical composition could not account for the difference between colored and bleached WO_3 films (Fig. 11.21). Rather, an ultraviolet photoelectron spectroscopy plot of intensity vs. electron kinetic energy (Fig. 11.22) is interpreted as a change of electronic state in the valence band, and by the effect of interactions between tungsten $5d$ electrons and hydrogen atoms. A theoretical model thus suggests that the coloring mechanism is due to the transfer of hydrogen from a passive site to an active site within the films.

Cathodoluminescence emission spectra of zinc telluride single crystals implanted with Zn, Ar, and Al has been reported by Bryant and Verity (1981). Samples were excited using 200 μA/cm² of 10 keV electrons, and time-resolved emission spectra were obtained by pulsing the electron beam at a frequency of 10 kHz and with a pulse width of 50 μs.

Recent investigations of luminescent materials also include the use of enriched stable isotopes. Energy levels of impurity atoms in a solid, and therefore the energy of the corresponding optical transitions, are influenced

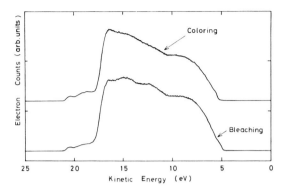

Figure 11.22 Ultraviolet photoelectron spectroscopy of colored and bleached WO_3 thin films having the same chemical composition. (Muramatsu et al., 1982; *Jpn. J. Appl. Phys.*, *21*, L73 (1982), by permission.)

by vibrational effects as well as the properties of the static host lattice. Furthermore, inasmuch as the vibrational energy depends markedly on the mass of the vibrating atoms, the optical spectrum should reveal an isotope effect. This isotopic effect is small, but it has been observed by Broser et al. (1981), who have diffused separated nickel isotopes ($M = 58, 60, 61, 62,$ and 64) into single-crystal ZnO, ZnS, CdS, and CdSe in a series of experiments. At low doping concentrations, an isotope splitting was clearly detected; and these data are providing a clearer understanding of electron-phonon interactions relating to impurities in solids.

Liquid crystals are of particular interest to both chemists and physicists, as they occupy a position intermediate between those of crystalline solids and isotropic liquids. They play an important role as display elements—in watches, calculators, thermometers, etc., and they produce good contrast with low power consumption. One prototype storage oscilloscope has a power consumption of less than 0.5 W. Liquid crystals are also known to exist in living organisms, where they appear to control biological processes (Brown and Crooker, 1983). The basic physics and display applications of nonemissive liquid crystals have been reviewed by Shanks (1982). In the absence of an applied electric field, a thin film of liquid crystal (6–50 μm thick) may be held in alignment by physical or chemical treatment of the transparent and conductive surfaces that contain the film. An electrode system and suitably applied voltages (2–3 V) can then change the optical properties of the film—due to birefringence, color, or optical scattering.

Optoelectronic switches made from high-resistivity semiconductor materials have found use in streak cameras, laser circuits, etc. For high-voltage switching, very pure (type IIa) diamond provides subnanosecond-switching capabilities; it can conduct very large pulsed currents and withstand large *DC* bias voltages. Diamond has a band gap of 5.48 eV. Less than $10^{15}/cm^3$ of its nitrogen impurity concentration is electrically active (having an ionization energy of 4 eV); carrier mobilities are higher than those of silicon and germanium, and the thermal conductivity of pure (type IIa) diamond is more than five times that of copper at room temperature (Bharadwaj et al., 1983). Ultraviolet excitation is obviously required to generate electron-hole pairs for conduction, but switching efficiencies up to 50% have been obtained at DC bias voltages as high as 8 kV.

Finally, the advent of monolithic-integrated optoelectronic circuits, which incorporate optical and electronic elements on the same substrate, provide potential advantages of smaller size and higher reliability compared to conventional hybrid circuitry (Bar-Chaim et al., 1982). Injection lasers can be fabricated on GaAs substrates, and the high electron mobility in GaAs contributes to the performance of GaAlAs heterojunction transistors. Frequencies of 4 GHz have already been reported for heterostructure bipolar transistors, and projected frequencies range to 10 GHz. In integrated optoelectronic circuits, parasitic inductances and capacitances are also reduced, and photodiodes and phototransistors can be matched relative to wavelength. Mass spectrometry can also be expected to contribute further to the development of new materials for all-optical processing, wherein one light beam is used to control a second light beam by having the refractive index n or absorption α changed by the propagation of light in the medium. In such a system, the material response time is critical, that is, the speed with which n or α can be changed by the control pulse and the speed with which the material recovers so that it may respond to a second control signal (Glass, 1984).

SUPERCONDUCTORS

In 1908, Kamerlingh Onnes, at the University of Leiden in Holland, liquified helium for the first time, and in 1911, he observed that the electrical resistance of a rod of frozen mercury dropped to zero as its temperature reached 4 K. Subsequently, the superconducting state was observed at very low temperatures in various metals, alloys, and intermetallic compounds. One of the earliest ideas for exploiting this superconducting state was to build electromagnets, but early attempts failed because the superconductivity was quenched in various metals by magnetic fields of only a few hundred gauss. A theoretical explanation of superconductivity was not forthcoming until Bardeen, Cooper, and Schrieffer (1957) postulated that conduction electrons, with equal and opposite moments and spins, form pairs ("Cooper pairs") that condense into a superconducting state. Superconductivity is then assumed to occur with the highly coordinated motion of these electron pairs, with each electron in a pair moving with the same speed but in opposite directions. Furthermore, it was not until the early 1960s that materials were found that could sustain both high-current densities and magnetic fields of several hundred thousand gauss, thereby making feasible the contemporary

TABLE 11.7   Superconducting Transition
Temperature T_c, and Critical Magnetic
Field H_{c2} of Several Metals and
Compounds

Material	T_c (K)	H_{c2} (tesla)[a]
Cd	0.52	—
Zn	0.85	—
Al	1.18	—
Pb	7.20	—
Nb	9.25	—
Nb_3Sn	18.0	29.6
Nb_3Al	18.7	32.7
Nb_3Ga	20.2	34.1
Nb_3Ge	23.2	37.1
V_3Si	17.0	34.0
V_3Ga	14.8	34.9

Source: Hulm and Matthias (1980).
[a]1 tesla = 10^4 G.

design of commercial electric generators and giant magnets for fusion experiments.

Today, mass spectrometry is important for monitoring the processing of sputter-deposited superconducting films and for characterizing the micro-structure of such films. For example, the superconducting performance of diffusion-processed Nb_3Sn is substantially influenced by its microstructure and oxygen content (Jergel et al. 1982), and films of Nb_3Ge of the highest critical current J_c have grain sizes of only ~1000 Å (Suzuki et al., 1983). Mass-resolved ion beams have also been employed for producing super-conducting alloys. In experiments by Leiberich et al. (1981), a supercon-ducting transition temperature T_c of 12.3 K was observed in Pd implanted with Cu ions and electrolytically charged with hydrogen. The singly charged Cu ions were implanted on a 38 μm thick Pd foil at an energy of 100 keV and at a doping concentration of 8×10^{16} atoms/cm². Other Cu implants in Pd foils have yielded even higher critical temperatures—with onsets as high as 16.6 K and showing a strong correlation between the implant doping concentration and T_c (Standish et al., 1983). This work is collinear with recent materials research that is focused on binary compounds of the type included in Table 11.7. In contrast to pure metals, they have a relatively high critical temperature T_c, and they remain superconducting up to very high field strengths (H_{c2}) where H_{c2} relates to a so-called type II superconductor.

Specific criteria for high T_c values and other important properties of su-perconductors have been summarized by Hulm and Matthias (1980). For materials primarily based on the transition metals, a high T_c relates to the

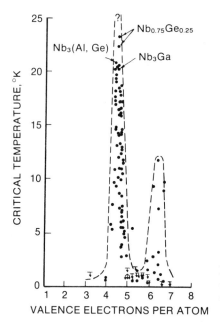

Figure 11.23 Critical temperature as a function of electron to atom ratio, e/a, of superconducting compounds. (Roberts, 1974; Hulm and Matthias, 1980.)

average electron concentration per atom, e/a. This e/a value is determined by counting all electrons outside a filled shell for a single element, but an arithmetic average has to be evaluated for a compound. As indicated in Figure 11.23, certain values of e/a are highly favored for high T_c temperatures—maxima occurring at 4.7 and 6.4. This rough selection rule appears to be valid for elements, their solid solutions, and compounds. Superconducting properties are also affected by crystal defects, and an increase in T_c has been observed in rhenium films irradiated with 350 keV nitrogen ions and 700 keV doubly charged argon ions (Haq and Meyer, 1982).

The technological impact of superconductors is already profound. With the advent of new materials that can have a resistance of zero while simultaneously maintaining current densities of 1 million A/cm², applications are increasing. Mass spectrometry, per se, has already made use of superconducting systems in the context of advanced cyclotron resonance systems (Chapter 3), and superconducting magnets will be utilized further with high-voltage ion accelerators and high-magnetic field analyzers. Molecules in the mass range of 100,000 amu and greater will also be amenable to analysis. In the field of biochemistry, nuclear magnetic resonance (NMR) imaging systems are now commercially available; and in electrical engineering, superconductors are the key ingredient for generators that are now being built by General Electric and Westinghouse. An even more dramatic use of superconductors is being explored at the University of Wisconsin for electric energy storage; instead of pumped hydro or battery storage, giant coils can, in principle, store many megawatt-hours of energy for utility load leveling and emergency power. This system is expected to achieve an efficiency of 90%, and its fast response time (<100 ms) makes it especially attractive for stabilizing electric power grids (Hassenzahl, 1983).

With respect to conducting polymers, "the chances of finding superconductors having transition temperatures greater than their inorganic counterparts seems remote" (Greene and Street, 1984). However, a greater understanding of the basic physics and chemistry of the conduction process in organics will require mass spectrometry in some form.

The Isotope Effect

Although the isotope effect was not reported until 1950, Kamerlingh Onnes almost discovered this phenomena some 30 years earlier, when he attempted to find a difference in the critical temperatures of ordinary lead and in lead specimens associated with the radioactive decay of uranium. However, his measurements were not sufficiently precise to resolve small differences in the transition temperature. The fact that T_c is mass-dependent for some materials was firmly established by two independent investigations (Maxwell, 1950; Reynolds et al., 1950) in which several mercury samples of different isotopic compositions were tested. Experimentally, it was observed that the critical superconducting temperature for mercury isotopes satisfies the relation

$$T_c M^a = \text{constant}$$

where M is the isotopic mass, T_c is the critical temperature, and $a \approx \frac{1}{2}$. This formulation was determined by using an external magnetic field for which there is a threshold H_c, which can quench the superconducting state according to the expression (Lynton, 1962)

$$H_c \approx H_0 \left[1 - \left(\frac{T}{T_c} \right)^2 \right]$$

This critical or threshold magnetic field H_c, plotted against T_c for several enriched isotopic samples of mercury, is shown in Figure 11.24.

The mass dependence upon the critical or transition temperature has now been investigated both theoretically and experimentally for many elements. The general relationship

$$T_c \propto 1/\sqrt{M}$$

is applicable to nontransition elements, but no such effect has been found for the transition elements. Hence, it has been postulated that a different

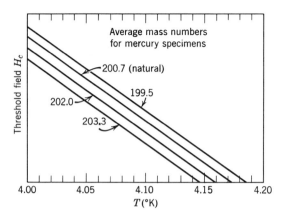

Figure 11.24 Dependence of the superconducting temperature on isotopic mass. (Lynton, 1962.)

mechanism governs superconductivity in the transition elements (Nicholson and Coffman, 1981).

From a practical point of view, this isotopc cffcct could be utilized as a cryogenic temperature monitor (White, 1977). By using enriched isotopes in thin film circuitry that would function as on-off gates at varying critical temperatures, a unique digital device coud be engineered to complement more conventional cryogenic sensors, namely, resistance thermometers, semiconductor junctions, and thermocouples.

JOSEPHSON JUNCTIONS

Josephson junctions represent a unique application of semiconducting materials. The concept of this transducer was first hypothesized by Brian Josephson in 1962 while he was a graduate student at Cambridge University. These junctions are somewhat analogous to transistors as in their simplest form they consist of two superconducting films between which is a thin (~40 Å) oxide layer that can be penetrated by superconducting electrons via quantum mechanical tunneling. By means of a control voltage, the impedances of this layer can be changed from zero to a high resistance, and by 1990, these junctions may be essential elements in a new generation of ultra-fast computers. The potential use of Josephson junctions in a supercomputer is derived from stringent specifications that include (1) nanosecond cycle times, (2) high packing densities in logic and memory circuits, and (3) low power dissipation. In principle, Josephson junctions not only provide greater switching speed (~10^{-12} s) with integrated circuitry of smaller dimensions than with silicon technology, but power dissipation can be reduced by a factor of 1000. In practice, working tunnel junctions have already been produced with a junction area as small as 1×10^{-10} cm^2 using advanced electron beam technology (Hu, 1981).

Research and development on Josephson junctions makes use of many types of spectrometry and spectroscopy, inasmuch as the technical problems include semiconductor and superconducting materials, advanced electron beam lithography, oxide films, dielectrics, control of grain size and bonding, contraction and expansion in thermal cycling, etc. Thus, at Bell Laboratories, IBM, and elsewhere, basic studies are in progress relating to superconducting alloys and insulating oxides; equally intensive work is ongoing relative to fabrication techniques and process control. In a fabrication step relating to an insulating oxide film, one objective is to control thickness within a monolayer. Hence, problems of surface physics, impurity control, diffusion, etc. directly relate to mass spectrometric measurements and ion beam diagnostics. δ-Phase niobium nitride, NbN, is among the more attractive materials envisioned as a base layer for Josephson devices because of its high critical temperature ($T_c = 17$ K), good mechanical properties, relative immunity to radiation damage, and ease of production by RF-reactive sputtering and subsequent thermal annealing (Cuculo et al., 1984).

Josephson junctions can also be utilized as sensitive detectors of magnetic fields and microwave radiation. Specifically, the current-voltage (I-V) curve of a junction irradiated by a microwave signal of frequency f will exhibit voltage increments ΔV directly proportional to the relation $\Delta V = (h/2e) f$, where h is Planck's constant and e is the electronic charge. The correlation is so precise that this transducer is now used to define the standard volt. A matrix array of junctions has also been considered as a real-time imaging system for millimeter wavelength (thermal) radiation. As an alternative to radar, it would have the advantage of being a passive system to "see" through clouds and fog and be independent of interference from spurious reflected signals.

SUPERIONIC CONDUCTORS

Superionic conductors are materials in which one or more ionic species within a host lattice undergo transport by a liquid-like diffusion. The mobile ions are also characterized by an activation energy (~0.1–0.3 eV) that is one order of magnitude lower than for ions in a typical ionic crystal. Thus, these solid electrolytes possess a high *ionic* conductivity, even though they have a high impedance to the flow of electrons.

In general, the conductivity σ will be equal to the sum of the ionic and electronic contributions, that is,

$$\sigma = \sum_i = \sigma_i + \sigma_e = \sum_i n_i \mu_i (z_i e) + n \mu_e e$$

For materials where the electronic conductivity is negligible,

$$\sigma = \sigma_i = n_i\mu_i(z_ie)$$

where n_i, μ_i, and (z_ie) are concentration, mobility, and charge, respectively, of the ionic species. The ions are able to move through the solid via defects such as interstitial sites and vacancies, so that superionic materials are characterized by a high degree of disorder (Hooper, 1978). Potential applications for materials that exhibit high ionic and low electronic conductivities include batteries, fuel cells, oxygen sensors, ion-selective membranes, capacitors, solid-state ion pumps, and electrochemical devices (Linford and Hackwood, 1981). Furthermore, certain mixed conductors (i.e., materials having appreciable charge transport by both ions and electrons) are attractive candidates as catalysts, solid-state electrodes, and electrochromic displays. (Huggins, 1981; Beni, 1981). The form of these electrolytes encompasses inorganic crystalline solids, glasses, and polymers.

Research on various solid electrolytes for high-energy density batteries has been stimulated by their applicability to electric vehicle propulsion, load leveling, central power stations, and space modules. Such batteries can produce the same electric energy as lead-acid types with about one-third of the weight.

Beta-alumina is one of the superionic conductors of interest for battery application. It is a layered structure in which spinel-type blocks composed of Al^{3+} and O^{2-} are separated by a loosely packed plane containing M (M = Na, K, Rb, Ag, and Tl) and O^{2-}. This loosely packed structure allows for the easy movement of ions and resultant high conductivity.

In a sodium-sulfur battery that uses ceramic sodium beta-alumina ($Na_2Al_{22}O_{34}$) as its electrolyte (Bates et al., 1982), sodium ions diffuse through the electrolyte and react with polysulfide ions to form Na_2S_x, where x has values from 3–5. Electric energy is expended in the discharge reaction, which transports electrons from the negative to positive electrode through an external circuit; the reactions are:

Anode	$2Na \rightarrow 2Na^+ + 2e^-$
Cathode	$2e^- + S_x \rightarrow S_x^{2-}$
Net	$2Na + S_x \rightarrow Na_2S_x$

The cell is recharged by applying an external voltage to the electrodes to reverse the electron current. The discovery of very high sodium ion mobility in sodium beta-alumina by Weber and Kummer at the Ford Motor Company initially stimulated a high degree of interest in this system (Kummer, 1972).

In principle, mass spectrometry can be utilized to measure ion mobility indirectly by measuring diffusion. Specifically, McKee (1981) has shown that the Nernst-Einstein equation can be generalized for a high-defect concen-

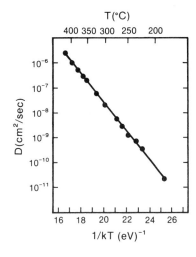

Figure 11.25 Arrhenius graph of the diffusion of Rb$^+$ in AgCl, 184–447°C. (Laskar and Cardegna, 1981; *Solid State Ionics, 5*, 465 (1981), by permission of North-Holland Physics Publishing.)

tration solid so as to relate mobility or conductivity to the self-diffusion coefficient, namely,

$$\frac{\mu_i}{D_i} = \frac{q_i}{kT}$$

where μ_i is the mobility of the charge carriers, q_i is the charge, D_i is the self-diffusion coefficient, k is the Boltzmann constant, and T is the temperature. This relationship assumes that the diffusing particles interact strongly and that the mobility is concentration-dependent.

Tracer diffusion studies have been undertaken using neutron activation (Dejus et al., 1981), but it is clear that stable isotopes could also be employed to determine diffusion rates in a number of fast ionic conductors. The diffusion measurement of ^{86}Rb in AgCl, by Laskar and Cardegna (1981) and shown in Figure 11.25, is a typical example of the type of radiotracer study that could be undertaken via stable vs. radioactive nuclides. The use of ^{18}O as a tracer has also been reported for measuring the self-diffusion coefficients of oxygen ions in single-crystal thoria (Ando and Oishi, 1981). A mass spectrometer has also been used in combination with a solid-state electrochemical cell to determine the oxygen ion transport number in yttria-doped zirconia and thoria (Alcock and Butler, 1981).

Superionic materials research of a more general nature has included the utilization of SIMS (Miyamura et al., 1981) and RBS. In this latter category, the displacement of ions from symmetric positions in layer-type electrolytes has been probed, using high-energy (~1 MeV) He$^+$ ions. For ions that are being *channeled* along the layers, there will be some backscattering from the host lattice and mobile ions, and the energy loss of the incident ions will be related to the mass and position of these scattering ions. If the ion beam is slightly divergent from the channeling direction, other data can be obtained relating to the displacement of ions from high symmetry sites. Studies of the ionic lattices of Ag and $K\beta$-Al_2O_3 have utilized this ion probe technique (Linford and Hackwood, 1981). Roth et al. (1983) have employed back-scattering and channeling to study beta-alumina crystals doped with Mn by ion implantation, and with Li and Mg from melts. Recently, superionic solids with Cu^+ ions as the charge carrier have also been investigated (Lal and Srivastava, 1984).

REFERENCES

Albert, P., *Pure and Appl. Chem., 54*, No. 4, 689 (1982).

Albrecht, H., *IEEE Trans. Electr. Devices, Ed-30*, No. 4, 499 (1983).

Alcock, C. B., and J. Butler, *Solid State Ionics, 3/4*, 499 (1981).

Ando, K., and Y. Oishi, *Solid State Ionics, 3/4*, 473 (1981).

Arora, N. D., J. R. Hauser, and D. J. Roulston, *IEEE Trans. Electr. Devices, Ed-29*, No. 2, 292 (1982).

Asakawa, K., and S. Sugata, *Jpn. J. Appl. Phys., 22*, No. 10, L653 (1983).

Bar-Chaim, N., S. Margalit, A. Yariv, and I. Ury, *IEEE Trans. Electr. Devices, Ed-29*, No. 9, 1372 (1982).

Bardeen, J., L. N. Cooper, and J. R. Schrieffer, *Phys. Rev., 108*, 1175 (1957).

Bates, J. B., J. C. Wang, and N. J. Dudney, *Physics Today, July*, 46 (1982).

Beni, G., *Solid State Ionics, 3/4*, 157 (1981).

Benton, J. L., and L. C. Kimerling, *J. Electrochem. Soc., 129*, No. 9, 2098 (1982).

Beyer, W., R. Fischer, and H. Wagner, *J. Electronic Mat., 8*, No. 2, 127 (1979).

Bharadwaj, P. K., R. F. Code, H. M. van Driel, and E. Walentynowicz, *Appl. Phys. Lett., 43*, No. 2, 207 (1983).

Borsuk, J. A., *IEEE Trans. Electr. Devices, Ed-30*, No. 4, 296 (1983).

Bouldin, D. P., *J. Electronic Mat., 10*, No. 4, 747 (1981).

Broser, I., R. Germer, A. Schröder, E. Birkicht, and R. Broser, *J. Luminescence, 24/25*, 225 (1981).

Brown, G. H., and P. P. Crooker, *Chem. Eng. News, January 31*, 24 (1983).

Brown, W. L., *IEEE Spectrum, April*, 50 (1981).

Bryant, F. J., and D. M. Staudte, *Solid-State Electr., 24*, 675 (1981).

Bryant, F. J., and D. Verity, *J. Luminescence, 22*, 171 (1981).

Chang, C., *J. Appl. Phys., 52*, No. 7, 4620 (1981).

Cheung, N. W., L. C. Feldman, P. J. Silverman, and I. Stensfaard, *Appl. Phys. Lett., 35*, 859 (1979).

Chew, N. G., and A. G. Cullis, *Appl. Phys. Lett., 44*, No. 1, 142 (1984).

Christie, W. H., and C. W. White, *Surface Sci., 100*, 43 (1980).

Clegg, J. B., in *Secondary Ion Mass Spectrometry III*, ed. A. Benninghoven, J. Giber, J. László, M. Riedel, and H. W. Werner, Springer-Verlag, Berlin (1982), p. 308.

Coburn, J. W., and H. F. Winters, *CRC Crit. Rev. Solid State Material Sci., 10*, No. 2, 119–141 (1981).

Cohen, S. S., G. Gildenblat, M. Ghezzo, and D. M. Brown, *J. Electrochem. Soc., 129*, No. 6, 1335 (1982).

Corderman, R. R., and P. E. Vanier, *J. Appl. Phys., 54*, No. 7, 3987 (1983).

Cuculo, A. M., L. Maritato, A. Saggese, and R. Vaglio, *Cryogenics, 24*, No. 1, 45 (1984).

Damestani, A., and L. Forbes, *J. Electronic Mat., 10*, No. 5, 879 (1981).

Dejus, R., K. Sköld, and B. Granéli, *Solid State Ionics, 3/4*, 499 (1981).

Doyle, B. L., P. S. Peercy, J. D. Wiley, J. H. Perepezko, and J. E. Nordman, *J. Appl. Phys., 53*, No. 9, 6188 (1982).

England, A. A., and E. H. Rhoderick, *Solid State Electr., 24*, 337 (1981).

Ephrath, L. M., *J. Electrochem. Soc., 129*, No. 3, 62C (1982).

Fishbach, J. U., V. Brunsmann, O. Frenzl, H. Egle, and O. Meyer, *J. Phys., 21*, 207 (1981).

Foxon, C. T., *CRC Rev. Solid State Material Sci., 10*, No. 3, 235 (1981).

Furman, B. K., G. H. Morrison, V. I. Saunders, and Y. T. Tan, *J. Appl. Phys., 51*, No. 10, 5342 (1980).

Gamo, K., Y. Ochiai, and S. Namba, *Jpn. J. Appl. Phys., 21*, No. 12, Part 2, L792 (1982).

Gargini, P. A., *Ind. Res. and Dev., March*, 141 (1983).

Gibbons, J. F., and T. W. Sigmon, *Nucl. Instr. and Methods, 191*, 115 (1981).

Glass, A. M., *Science, 226*, 657 (1984).

Gohlke, R. S., private communication (1984).

Greene, R. L., and G. B. Street, *Science, 226*, 651 (1984).

Gupta, R. P., A. K. Roy, W. S. Khokle, and A. Jain, *Radiation Effects, 63*, 61 (1982).

Hagemann, H. J., D. Hennings, and R. Wernicke, *Philips Tech. Rev., 41*, No. 3, 89 (1983–1984).

Handoss, Sl, F. Bénière, W. Gaumeau, and A. Rupert, *J. Appl. Phys., 51*, No. 11, 5833 (1980).

Hanley, P. R., *Nucl. Instr. and Methods, 189*, 227 (1981).

Haq, A. ul, and O. Meyer, *J. Low Temp. Phys. 49, 1/2*, 151 (1982).

Harper, J. M. E., J. J. Cuomo, P. A. Leary, G. M. Summa, H. R. Kaufman, and F. J. Bresnock, *J. Electrochem. Soc., 128*, 1077 (1981).

Hassenzahl, W. V., *Proc. IEEE, 71*, No. 9, 1089 (1983).

Hayashi, T., M. Miyamura, and S. Komiya, *Jpn. J. Appl. Phys., 21*, No. 12, L755 (1982).

Hooper, A., *Contemp. Phys., 19*, No. 2, 147 (1978).

Horikoshi, Y., H. Saito, and Y. Takanashi, *Jpn. J. Appl. Phys., 20*, No. 2, 437 (1981).

Howard, W. E., *J. Luminescence, 24/25*, 835 (1981).

Hu, E. L., *Am. Scientist, 69*, 517 (1981).

Huggins, R. A., *Solid State Ionics, 5*, 15 (1981).

Hulm, J. K., and B. T. Matthias, *Science, 208*, 881 (1980).

Jergel, M., T. Melisek, H. Allarova, D. I. Synak, J. Neuschl, and J. Ivan, *J. Materials Sci., 17*, 1579 (1982).

Katz, W., private communication (1984).

Kiang, Y. C., J. R. Moulic, W. Chu, and A. C. Yen, *IBM J. Res. Develop., 26*, No. 2, 171 (1982).

Kobayashi, S., and T. Kimura, *IEEE Spectrum, May*, 26 (1984).

Kolbesen, B. O., and A. Mühlbauer, *Solid State Electr., 25*, No. 8, 759 (1982).

Kowalski, Z. W., and I. W. Rangelow, *J. Materials Sci., 18*, 741 (1983).

Kummer, J. T., *Progress in Solid State Chemistry*, v. 7, eds., H. Reiss and J. O. McCaldin, Pergamon, Oxford (1972).

Lal, H. B., and O. P. Srivastava, *J. Materials Sci., 19*, 303 (1984).

Land, C. E., and P. S. Peercy, *Ferroelectrics, 45*, 25 (1982).

Laskar, A. L., and P. Cardegna, *Solid State Ionics, 5*, 465 (1981).

Leiberich, A., W. Scholz, W. J. Standish, and C. G. Homan, *Phys. Lett., 87A*, No. 1/2, 57 (1981).

Lidow, A., J. F. Gibbons, V. R. Deline, and C. A. Evans, Jr., *J. Appl. Phys., 51*, No. 8, 4130 (1980).

Lin, K., and J. D. Burden, *J. Vac. Sci. Technol., 15*, No. 2, 373 (1978).

Lines, M. E., *Science, 226*, 663 (1984).

Linford, R. G., and S. Hackwood, *Chem. Rev., 81*, 327 (1981).

Losee, D. L., J. P. Lavine, E. A. Trabka, S. T. Lee, and C. Jarman, *J. Electrochem. Soc., 83*, 597 (1983).

Lynton, E. A., *Superconductivity*, Wiley, New York (1962), p. 132.

Mahan, J. E., D. S. Newman, and M. R. Gulett, *IEEE Trans. Electr. Devices, Ed-30*, No. 1, 45 (1983).

Mashovets, T. V., *Soviet Phys. Semicond., 16*, No. 1, 1 (1982).

Maxwell, E., *Phys. Rev., 78*, 477 (1950).

Mayer, J. W., B. Y. Tsaver, S. S. Lau, and L. S. Hung, *Nucl. Instr. and Methods, 182/183*, 1 (1981).

McGuire, G. E., L. B. Church, D. L. Jones, K. K. Smith, and D. T. Tuenge, *J. Vac. Sci. Technol., A1*, No. 2, 732 (1983).

McKee, R. A., *Solid State Ionics, 5*, 133 (1981).

McNutt, M. J., and W. E. Meyer, *J. Electrochem. Soc., 128*, No. 4, 892 (1982).

Miller, S. E., *Am. Scientist, 72*, 66 (1984).

Mimura, Y., Y. Okamura, and C. Ota, *J. Appl. Phys., 63*, No. 8, 5491 (1982).

Mitchell, J. W., *Pure Appl. Chem., 54*, No. 4, 819 (1982).

Miyamura, M., S. Tomura, A. Imai, and S. Inomata, *Solid State Ionics, 3/4*, 149 (1981).

Morkoc, H., and P. M. Solomon, *IEEE Spectrum, February*, 28 (1984).

Muramatsu, H., T. Itoh, A. Watnabe, and K. Hara, *Jpn. J. Appl., Phys., 21*, No. 2, L73 (1982).

Nakatsuka, H., and D. Grischkowsky, *Opt. Lett., 6*, No. 1, 13 (1982).

Nassau, K., *The Physics and Chemistry of Color*, Wiley, New York (1983).

Nicholson, J. E., and K. G. Coffman, *J. Low Temp. Phys., 45*, 5/6, 429 (1981).

Oberstar, J. D., B. G. Streetman, J. E. Baker, and P. Williams, *J. Electrochem. Soc., Solid State Sci. Tech., 129*, 1320 (1981a); *129*, 1312 (1981b).

Ohishi, Y., S. Mitachi, S. Shibata, and T. Manabe, *Jpn. J. Appl. Phys., 20*, No. 3, L191 (1981).

Okimura, H., *J. Electronic Mat., 9*, No. 6, 989 (1980).

Onuma, T., T. Hirao, and T. Sugawa, *J. Appl. Phys., 52*, No. 10, 6128 (1981).

Patkin, A. J., and G. H. Morrison, *Anal. Chem., 54*, 2 (1982).

Pattenpaul, E., J. Huber, H. Weidlich, W. Flossmann, and V. von Borcke, *Solid State Electr., 24*, No. 8, 781 (1981).

Phillips, J. M., L. C. Feldman, J. M. Gibson, and M. L. McDonald, *J. Vac. Sci. Technol., B1*, No. 2, 246 (1983).

Poate, J. M., *Nucl. Instr. Methods, 191*, 39 (1981).

Poate, J. M., and W. L. Brown, *Physics Today, June*, 24 (1982).

Ramsey, J. W., *J. Vac. Sci. Technol., A1*, No. 2, 721 (1983).

Reents, W. D., Jr., and A. M. Mujsce *Int. J. Mass Spectrom. and Ion Processes, 59*, 65 (1984).

Reents, W. D., Jr., D. L. Wood, and A. M. Mujsce, *Anal. Chem., 57*, 104 (1985).

Reynolds, C. A., B. Serin, W. H. Wright, and L. B. Nesbitt, *Phys. Rev., 78*, 487 (1950).

Riley, J. E., Jr., *J. Radioanalyt. Chem., 72*, No. 1–2, 89 (1982).

Risch, L., C. Werner, W. Müller, and A. W. Wieder, *IEEE Trans. Electr. Devices, ED-29*, No. 4, 601 (1982).

Roberts, B. W., *Nat. Bur. Stand., U.S. Tech. Note No. 825*, 7 (1974).

Robinson, R. S., and S. M. Rossnagel, *J. Vac. Sci. Technol., 21*, No. 3, 790 (1982).

Roth, W. L., R. E. Benenson, C. Ji, L. Wielunski, and B. Dunn, *Solid State Ionics, 9/10*, 1459 (1983).

Schubert, R., and J. A. Augis, *J. Vac. Sci. Technol., A1*, No. 2, 248 (1982).

Schwartz, R. G., J. H. McFee, A. M. Voshchenkov, S. N. Finegan, and Y. Ota, *Appl. Phys. Lett., 40*, No. 3, 239 (1982).

Shanks, I. A., *Contemp. Phys., 23*, No. 1, 65 (1982).

Shepherd, F. R., W. H. Robinson, J. D. Brown, and B. F. Phillips, *J. Vac. Sci. Technol., A1*, No. 2, 991 (1983).

Shockley, W. B., U.S. Patent 2,787,564.

Simondet, F., C. Venger, and E. M. Martin, *Appl. Phys., 23*, 21 (1980).

Simons, D. G., C. R. Crowe, and M. D. Brown, *J. Appl. Phys., 53*, No. 5, 3900 (1982).

Slusser, G. J., and J. S. Slattery, *J. Vac. Sci. Technol., 18*, No. 2, 301 (1981).

Smeltzer, R. K., *Appl. Phys. Lett., 41*, No. 9, 649 (1982).

Smith, D. H., *J. Luminescence, 23*, 209 (1981).

Smith, D. L., and R. H. Bruce, *J. Electrochem. Soc., 129*, No. 9, 2045 (1982).

Smolinsky, G., T. M. Mayer, and E. A. Truesdale, *J. Electrochem. Soc., 129*, No. 8, 1770 (1982).

Standish, W. J., A. Leiberich, W. Scholz, and C. G. Homan, in *Electronic Structure and Properties of Hydrogen in Metals*, ed. P. Jena and C. B. Satterthwaite, Plenum, New York (1983).

Staudte, D. M., and F. J. Bryant, *Solid State Electr., 26*, No. 2 (1983).

Stoneham, E. B., G. A. Patterson, and J. M. Gladstone, *J. Electronic Mat., 9*, No. 2, 371 (1980).

Sugiura, H., *J. Appl. Phys., 51*, No. 5, 2630 (1980).

Suzuki, M., H. Ouchi, and T. Anayama, *Jpn. J. Appl. Phys., 22*, No. 5, Part 2, L307 (1983).

Vasile, M. J., *J. Appl. Phys., 51*, No. 5, 2510 (1980).

Wallmark, J. T., *IEEE Trans. Electr. Devices, ED-29*, No. 3, 451 (1982).

Wei, C. Y., W. Katz, and G. Smith, *Thin Solid Films, 104*, 215 (1983).

White, F. A., *Superconducting Thermometer for Cryogenics*, NASA Langley Research Center Report LAR-12055, NASA Tech. Brief, Summer (1977), p. 215.

Whiteley, J. S., and S. K. Ghandhi, *J. Electrochem. Soc., 129*, 383 (1982).

Wilson, R. G., *J. Appl. Phys., 52*, No. 6, 3954 (1981).

Wilson, R. G., and J. Comas, *J. Appl. Phys., 53*, No. 4, 3003 (1982).

Yabuuchi, Y., F. Shoji, K. Oura, T. Hanawa, Y. Kishikawa, and S. Okada, *Jpn. J. Appl. Phys., 21*, No. 12, Part 2, L752 (1982).

Yen, A. C., *J. Vac. Sci. Technol., 18*, No. 3, 895 (1981).

Yeo, Y. K., Y. S. Park, F. L. Pedrotti, and B. D. Choe, *J. Appl. Phys., 53*, No. 9, 6148 (1982).

ELECTROPHYSICS

Mass spectrometry is currently utilized in virtually all areas of research that can be circumscribed by the term *electrophysics*, for example, (1) the effect of impurities on thermionic emission; (2) materials research in semiconductors, thermoelectrics, and superconductors; (3) ionization and discharge phenomena; (4) fluorescence and gaseous dielectrics; (5) temperatures in rarefied gases and shock waves, etc. Mass spectrometry is also an important tool in analyzing *plasmas*—a nomenclature borrowed from the medical glossary by Nobel Prize winner Irving Langmuir in the early 1920s. In these and other studies, electrons, photons, and primary ions function as probes or excitation sources, and "secondary" electrons, photons, and ions are the diagnostic particles. Furthermore, the range of primary ion energies now extends from a few electron volts to approximately 100 MeV, and the resulting mass spectrum extends to almost 50,000 amu. Contemporary electrophysics-mass spectral investigations also include channeling in single crystals, ion clusters, and high charge state ionization produced by multiphoton interactions.

LASER AND RF ISOTOPE SEPARATION

Considerable attention has been focused on the development of laser isotope separation (LIS) technology based upon selective multistep photoionization of atomic beams. Current research and development expenditures in this field are at the $100 million level in the United States alone (Hitz, 1984). Among the light elements, there exists a growing market for ^2D, ^6Li, ^{10}B, ^{13}C, ^{15}N, ^{17}O, ^{18}O, and ^{33}S as nonradioactive tracers in analytical and medical applications. Based upon energy consumption, there also exists an economic incentive for enriching ^{235}U for fission fuel in light-water nuclear reactors. In contrast to other enrichment schemes, the laser separation process is highly selective, there is no minimum facility size, and the theoretical energy consumption is low. An early process for uranium separation involved a two-step photoionization in which xenon and krypton lasers irradiated a stream of vapor from a heated uranium sample. The xenon laser selectively excited ^{235}U atoms to a higher energy level, and a subsequent krypton laser pulse provided the final energy for ionization. The positive ions were then collected on a negatively biased collector, and the enrichment was assayed by mass spectrometry. Among recent investigations, uranium isotope separation has been achieved in UF_6 gas via the simultaneous irradiation from CF_4 and CO_2 lasers, using various gas pressures and laser fluences (Koren et al., 1982).

Laser isotope separations of deuterium and tritium have also received considerable attention in both fission and fusion technologies. Extremely high single-step enrichments have been measured via the CO_2 laser photolysis of trifluoromethane near 10.3 μm by irradiating a prepared mixture of $^{12}CDF_3/^{13}CHF_3$ and by determining the ^{13}C content in the product C_2F_4 (Herman and Marling, 1980). The tetrafluoroethene product mass $101:100$ ratio ranged from 9–65%, depending upon laser wavelength and fluence, fluoromethane pressure, and initial mixture D/H ratio. With respect to tritium, a practical use of laser isotopic separation relates to the removal, rather than the enrichment, of this radioisotope from waste waters of various nuclear facilities. Using a CO_2 laser, Takeuchi et al. (1982) have reported on the infrared, multiphoton dissociation of tritiated trifluoromethane that has a selective absorption band at 9.4 μm. This technique appears to be especially attractive for the removal of selected trace impurities at the pàrts per million level, where the optical energy from lasers can be effectively utilized.

Other laser isotope separation studies include many gaseous and metallic species. Sulfur-33 isotope enrichment has been reported by the isotopically selective, two-step photodissociation of COS, with chemical scavenging of S atoms by O_2 and NO_2 (Zittel and Darnton, 1982). The enrichment was monitored by a quadrupole mass spectrometer (QMS); the sample beam was physically chopped at 800 Hz, and phase-locked detection was employed to eliminate signal contributions from background gases.

Time-of-flight (TOF) mass spectrometry has been used to detect the multiphoton ionization of transition metal complexes such as $Mn_2(CO)_{10}$ and $Re_2(CO)_{10}$ (Lichtin et al., 1982). A key feature of this work was the ability to vary the time between the laser pulse and the application of the ion drawout pulse, which determined the "zero" of the TOF mass analysis. The vapor components of PbTe and PbSe have also been ionized and dissociated by various resonant multiphoton processes and monitored by mass spectrometry (Martin, 1982). The molecular beam was produced by effusion from a resistively heated quartz cell. A lead atom requires 7.4 eV for ionization, but the highest photon energy used in this work was 3.4 eV. Hence, at least three photons were needed for ionization from the ground state.

The laser isotope separation method utilizing selective photochemical reactions is of practical importance, since it requires only a single exciting light source, as opposed to multiple photoexcitation optics. To achieve separation, however, it is important to conserve some degree of selectivity in the chemical processes that follow the photoexcitation. Suzuki et al. (1981) have achieved some enrichment of the ^{37}Cl isotope using an Ar ion laser-induced reaction of Cl_2 with C_2Cl_4; and they reported on the selectivity factor as a function of laser power and gas pressure. A QMS was used for measuring the reaction of Cl_2 with C_2Cl_4 and for analyzing recombination mechanisms.

It should be noted that in addition to utilizing high-power lasers to separate isotopes, enriched isotopes in lasers, per se, can be used to increase laser efficiency and to generate new frequencies. A continuous-wave, isotopically enriched CO_2 laser ($^{13}C^{16}O_2$) has been used to pump a far-infrared (FIR) laser optically, producing a new FIR-lasing wavelength with high efficiency and very low CO_2 power threshold (Wood et al., 1980). A high-efficiency, low-threshold $^{15}NH_3$ FIR laser emission has also been reported by these investigators, who point out that enriched isotopes provide a valuable and practical option for pumping FIR lasers.

The separation of isotopes by RF or microwave radiation makes use of the fact that various isotopes possess different nuclear magnetic moments and spins. Potential advantages of supersonic molecular beam-RF spectroscopy have been cited by Amirav and Even (1980):

1. Isotopic spectral differences are much greater in the RF region than in the optical region.
2. Lifetimes of excited states are longer, reducing the power required for excitation.
3. The RF molecular spectrum is easily tunable by external magnetic (or electric) fields.
4. The combination of molecular beam magnetic resonance and supersonic molecular beam techniques can be used for isotope separation on a large scale.

This isotope separation method has been demonstrated for oxygen, and it is applicable to many molecular species.

IONS FROM LASER-PRODUCED PLASMAS

Laser-generated plasmas from solid targets can serve as a copious source of ions for accelerator-based physics research. This type of ion source has some unique advantages over conventional ion sources: (1) no differential vacuum pumping is required; (2) a large number of ions can be produced per laser pulse; (3) ions can be produced in many charge states; (4) a variety of solid targets allow a wide range of ion species to be generated; (5) short laser pulse times permit the acquisition of TOF data as well as electromagnetic analysis; (6) highly directional plasma plumes can be oriented along the ion accelerator axis; (7) little recombination occurs from an expanding plasma in vacuum; and (8) the ion optics for such a source is not complex.

Hughes et al. (1980) have reported on such an ion source, using a Q-switched Nd:Yag laser to generate 20 MW bursts of 1.06 μm radiation on solid targets. Laser power densities in the range of 10^{11} W/cm^2 on carbon, aluminum, copper, and lead produced thermalized plasma plumes that drifted to a 15 kV-gridded electrode system, where the ions were extracted, accelerated, and subsequently electrostatically focused. This extraction-acceleration system permitted TOF information to be acquired in the plasma drift region, so that ion temperatures and mass flow velocities could be monitored. The ions were then magnetically analyzed to determine their charge state. The laser was capable of generating 15 ns pulses at a maximum repetition rate of 50 pps, and the target discs were rotated within the vacuum chamber by an electrically isolated stepping motor. The detection system following the analyzing magnet consisted of an array of Faraday cup detectors spatially oriented to accept ions of different charge states. Figure 12.1 shows a reproduction of ion oscillograph traces from the Faraday cups for single-laser bursts incident on a solid aluminum target. Charge state distributions $Z = 1$ to $Z = 5$ are sharply defined on a microsecond time scale. Table 12.1 shows additional data on the charge state and ion temperatures for plasmas produced from carbon, aluminum, and copper with the same laser. The ion temperatures were measured using TOF data and were compared to Maxwell-Boltzmann distributions. These comparisons indicated that thermal equilibrium was established in the laser plume in times that were short compared with the ion flight times to the extractor grid.

Diagnostic studies on laser-induced plasmas expanding into a vacuum have been complemented by measurements at low pressures. At about 10^{-4} torr,

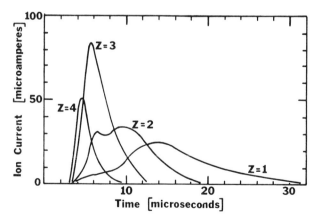

Figure 12.1 Multiple-charge states produced from 10^{11} W/cm^2 laser burst on an aluminum target. (Hughes et al., 1980; *J. Appl. Phys.*, *51*, 4088 (1980), by permission of the American Institute of Physics.)

TABLE 12.1   Charge States and Ion Temperatures from Laser-Generated Plasmas

Target	Ion Energy (keV)	Ion Charge	Ionization Energy (eV)	Ion Temperature (eV)
Carbon	15	1+	10.6	17
	30	2+	26.1	81
	45	3+	50.4	237
	60	4+	67.6	135
	75	5+	374.6	148
Aluminum	15	1+	5.7	55
	30	2+	18.7	34
	45	3+	29.2	228
	60	4+	116.2	—
	75	5+	161.5	—
	90	6+	206.7	—
	105	7+	252.0	—
Copper	15	1+	6.7	147
	30	2+	21.3	154
	45	3+	48.4	128
	60	4+	75.6	172
	75	5+	102.7	108

Source: Hughes et al. (1980).

charge exchange occurs, and the cross-sections for this process between residual gas molecules and plasma ions in high charge states are expected to be high ($> 10^{15}$ cm^2). Clement et al. (1980) have made measurements at this pressure and observed ion energy distributions attributed to this phenomenon, using a unique TOF/electrostatic analyzer that recorded the distributions of all the ion species from a single laser pulse.

CHANNELING IN SINGLE CRYSTALS

Channeling can be loosely defined as the enhanced penetration of energetic ions along preferred directions in a crystal structure. A somewhat analogous phenomenon has been observed with respect to interstitial diffusion, where large directional differences in diffusivity have been noted for atoms in a host lattice. An example of a channeling path, the $\langle 110 \rangle$ direction in a diamond lattice can be noted in Figure 12.2. For an ion to undergo channeling in a crystal, it must be collinear to a crystalline axis within a "critical angle" ψ_c that is related to several ion-target parameters identified in the schematic diagram of Figure 12.3. Basically, the ions are "steered" according to the theory first proposed by Linhard (1965), in which a channeled particle con-

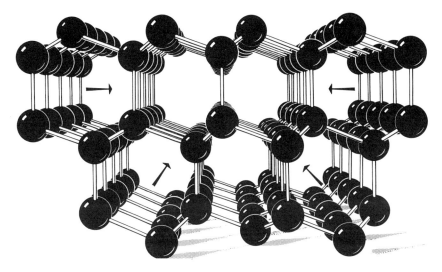

Figure 12.2 View of ⟨110⟩ direction in a diamond lattice, showing paths of easiest ion penetration.

serves its transverse energy $E_\perp$ during each encounter with an atomic row or plane. Thus, each channeled trajectory can be characterized by equating the transverse kinetic energy $E\psi^2$ in the midchannel region to the transverse potential at the distance of closest approach of the incident particle to the row r_{min}. A relationship between this distance of closest approach, incident angle ψ, and ion-target parameters has the form:

$$\psi^2 = \left(\frac{Z_1 Z_2 e^2}{Ed}\right) \log\left[\left(\frac{Ca}{r_{min}}\right)^2\right] + 1$$

where $Z_1 e$ is the nuclear charge of the incident ion, $(Z_2 e/d)$ is the mean

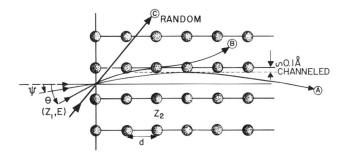

Figure 12.3 Schematic diagram of the channeling process of an energetic ion (A) within a crystal.

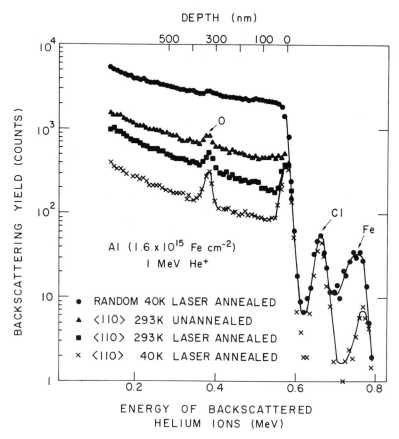

Figure 12.4 Channeled vs. random backscattered $^4He^+$ ions for an Al crystal that had been implanted with ^{56}Fe to a fluence of 1.6×10^{15} ions/cm^2. (Swanson et al., 1983; *Nucl. Instr. Methods, 218,* 643 (1983), by permission of North-Holland Physics Publishing.)

nuclear charge density along the aligned target atom row of interatomic lattice distance d, E is the incident ion energy, $C(\sim \sqrt{3})$ is an numerical constant, and a is the Thomas-Fermi screening length parameter (Davies, 1983). The lattice spacings along the $\langle 100 \rangle$ direction in several crystals are Si, $d = 5.43$ Å; Ge, $d = 5.66$ Å; and W, $d = 3.16$ Å; and the critical angles for 1 MeV He$^+$ ions have been experimentally observed to be Si, $\psi_c = 0.63°$; Ge, $\psi_c = 0.80°$; and W, $\psi_c = 1.97°$ (Feldman et al., 1982).

In practical channeling experiments, a collimated ion beam impinges upon the crystal, which is oriented by means of a goniometer, and a solid-state detector is employed that is sensitive to the energy of the backscattered ions. If the incident ion is within $\sim 1°$ of a crystalline axis, the ion can be channeled or steered by the coulomb fields of the rows and planes of atoms that comprise the lattice. These lattice atoms execute thermal motion about their equilibrium positions with displacements on the order of 0.1Å and with velocities on the order of 10^5 cm/s. The incident ion, however, sees a stationary or "frozen lattice" with atoms being only slightly displaced from their average positions (Feldman et al., 1982). Thus, if an impurity or displaced atom resides within a channeling path, the incident ions can be blocked and backscattered; and the backscattered angle and energy of these ions can provide information that relates to the depth at which an ion impurity scattering event occurred within the crystal. Consequently, ion channeling is useful in determining both site locations of implanted atoms in semiconductors and irradiation-produced defects, and in studying the structure of metal surfaces (Howe and Davies, 1980; Swanson and Howe, 1983).

Figure 12.4 shows channeling data obtained by Swanson et al. (1983) that illustrate the effect of excimer laser annealing on the backscattering spectra of 1 MeV He^+ ions from an Fe-implanted Al crystal. The spectra taken for the $\langle 110 \rangle$ alignment are compared with the "random" spectrum obtained by rotating the crystal away from the channeled orientation. The single crystal was originally implanted with ^{56}Fe at 40 keV in a fluence of 1.6×10^{15} ions/cm^2, and then it was excimer laser annealed (~ 6 J/cm^2). The oxygen peak is attributed to a surface oxide layer of ~ 6 nm, and the Cl peak is a result of electropolishing of the crystal in a perchloric acid solution. Recent ion channeling experiments have also been performed on *superlattices* formed by periodic alternate layers of two different materials. In an investigation of InGaAs/GaAs strained-layer superlattices, Picraux et al. (1983) demonstrated that the channeling technique could detect a lattice mismatch of $\sim 0.2\%$ and lattice atom distortions of ~ 0.01 Å.

SECONDARY ION EMISSION

Secondary ion emission is important in basic studies relating to the chemistry and physics of surfaces and to an understanding of ionization, sputtering, and cluster formation. It is also important in the analytical applications of secondary ion mass spectrometry (SIMS), where a complex spectrum can represent either an asset or a complication. When particles are emitted from a surface as a result of momentum transfer from a collision cascade induced by an incident particle, the process is called *physical sputtering*. An overview of the many mechanisms relating to physical sputtering has been presented by Williams (1982). However, if the incident particle or radiation (ion, electron, or photon) induces a chemical reaction that leads to the subsequent desorption of particles, the reaction is termed *chemical sputtering* (Winters, 1982). This latter process occurs for molecules such as CH_4, CF_4, etc., whose binding energy to many surfaces is small.

Energy Distributions and Yields

At low primary ion energies, secondary ion emission results from a small number of near-surface collisions. For a primary particle of mass m_1 and

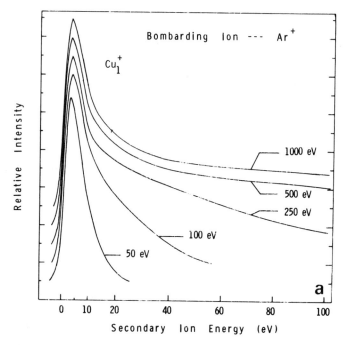

Figure 12.5 Secondary Cu^+ ion energy distributions for Ar^+ ions of 50, 100, 250, 500, and 1000 eV incident upon a polycrystalline copper target. (Hart and Cooper, 1980; *Surface Science*, *94*, 105 (1980), by permission of North-Holland Physics Publishing.)

kinetic energy E_1, in collision with a stationary particle of mass m_2, the energy transferred to mass m_2 is given by:

$$E_2 = \frac{4\,m_1 m_2}{(m_1 + m_2)^2}\,E_1\,\sin^2\frac{\theta}{2}$$

where θ is the scattering angle of m_1 in the center of a mass system. This classical kinematics is dominant for primary ions below about 200 eV. Hart and Cooper (1980) have used He^+, Ne^+, and Ar^+ ions with energies E_1 between 50–1000 eV to bombard a polycrystalline Cu target at an angle of 45° to the surface normal; and they measured the energies E_2 of the Cu^+ ions sputtered in a direction of 90° from the primary ion beam. Figure 12.5 shows their data for Ar^+ at several primary ion energies, with the most probable secondary or sputtered ion energy being 4 eV. This most probable energy of ~4 eV is also noted for He^+ and Ne^+, although the widths of the energy distributions are somewhat greater. However, the average emission energies $\overline{E}_2$ of the sputtered Cu^+ ions (calculated by integrating over the energy distributions) are shown in Figure 12.6. These graphs indicate that the average secondary ion energy increases with increasing bombarding ion en-

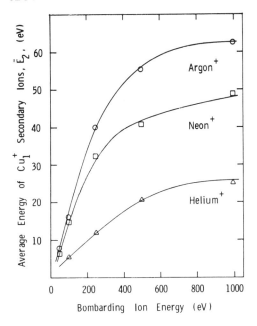

Figure 12.6 Average energy of Cu^+ secondary ions vs. bombarding ion energy for Ar^+, Ne^+, and He^+. (Hart and Cooper, 1980; *Surface Science*, *94*, 105 (1980), by permission of North-Holland Physics Publishing.)

ergy, but asymptotically approaches a limit in the range of 800–1000 eV. In this energy range, there are a sufficiently large number of collisions so that the energies of the secondaries are becoming "randomized" and the emitted ions have "forgotten" the energy imparted to the collision cascade by the incident ion. At higher kinetic energies, the primary ion can penetrate deeper into the target lattice, and collision sequences may give rise to the emission of a number of secondary ions or neutral particles.

From 1 keV to about 10 keV, the number of sputtered atoms per ion is usually observed to be a monotonically increasing function of energy; but the fraction of sputtered atoms that leave the surface as charged particles is small (10^{-2} to 10^{-4}) compared to the number of neutral atoms or molecules. Furthermore, the actual fraction will be highly dependent upon the particular surface, and the ionic species can vary over wide limits as indicated in Table 12.2. In this work by McHugh and Sheffield (1964), 7.4 keV $^{202}Hg^+$ ions at a beam intensity of 2×10^{-9} A were incident on a tantalum target maintained at 750°C. The pressure near the tantalum specimen was 9×10^{-7} torr, so that the primary ion beam was smaller than the arrival rate of foreign atoms. It was inferred, however, that at this elevated temperature, the metallic ions were produced from the bulk solid rather than from the fragmentation of surface oxides. At higher energies, Almen and Bruce (1962) have supplemented mass spectrometric investigations by measurements in an electromagnetic isotope separator in which the primary ion current density was 10–300 μA/cm². Targets were commercial metal foils, and the sputtering ratio (in atoms per bombarding ion) was determined by

TABLE 12.2   Secondary Ion Species from 7.4 keV $^{202}Hg^+$ Ions on a Tantalum Surface

Mass No.	Secondary Ion	Relative Intensity
181	Ta^+	9250
197	TaO^+	10,000
213	TaO_2^+	5600
362	Ta_2^+	790
378	Ta_2O^+	1000
394	$Ta_2O_2^+$	2100
410	$Ta_2O_2^+$	500
426	$Ta_2O_4^+$	380
442	$Ta_2O_6^+$	290
543	Ta_3^+	51
559	Ta_3O^+	48
575	$Ta_3O_2^+$	170
591	$Ta_3O_3^+$	80
607	$Ta_2O_4^+$	75
623	$Ta_3O_5^+$	20
639	$Ta_2O_7^+$	12
655	$Ta_3O_7^+$	3
~780	Ta_4O_{2-6+}	1

Source: McHugh and Sheffield (1964).

measuring (1) the weight loss of the bombarded target, (2) the primary ion beam current, and (3) the time of the bombardment. Figure 12.7 shows the atom per ion yield of Ne^+, Ar^+, Kr^+, and Xe^+ for a copper target. Their data fit the general analytical expression of Rol (1960), who assumed that the sputtering yield S is proportional to the energy dissipated by the im-

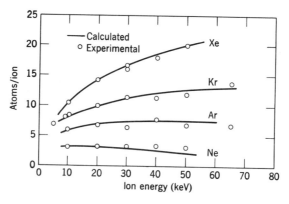

Figure 12.7   Sputtering yield for copper as a function of ion energy for several noble gases (Almen and Bruce, 1962).

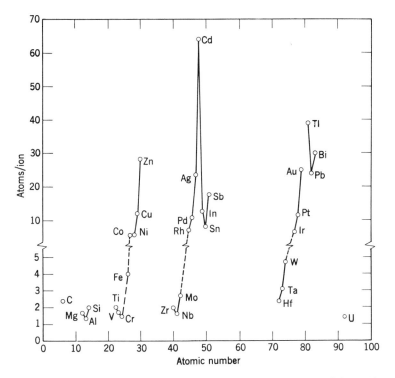

Figure 12.8 Sputtering yields for different target materials from 45 keV Kr$^+$ ions (Almen and Bruce, 1962).

pinging particle in the first few atomic layers, that is,

$$S = K \frac{E_1}{\lambda} \frac{m_1 m_2}{(m_1 + m_2)^2} \text{ atoms/ion}$$

where E_1 is the energy of the incident ion, λ is a calculated mean free path, m_1 and m_2 are the mass numbers of the ion and target atoms, respectively, and K is a constant for a certain combination of ion and target materials. Sputtering yields for many metallic targets bombarded with 45 keV Kr$^+$ ions are shown in Figure 12.8.

Other recent investigations include (1) the low energy Ar$^+$ ion sputtering of liquid and solid indium (Hurst and Cooper, 1982); (2) atomic and oxide ion yields from titanium by Ne$^+$ bombardment; (3) F$^+$ sputtering from a fluorinated silicon surface by Ar$^+$ and O$_2^+$ (Williams, 1981); (4) energy distributions of sputtered Nb and Cu atoms from amorphous targets bombarded by Ar$^+$ (Meyer et al., 1981); (5) the emission of Cu$^+$, Cu$_2^+$, and Mo$^+$ ions from polycrystalline copper and molybdenum bombarded by Zn$^+$ (Chakraborty and Dey, 1981); (6) the sputtering of single-crystal Si and polycrystalline Ag, Ca, Ni, Ti, and Al, produced by Ar^+ impact (Okajima, 1980); and (7) the O_2^+-induced emission of positive ions from five binary alloys: Cr-Ni, Fe-Ni, Cu-Ni, Pd-Ni, and Ag-Pd (Yu and Reuter, 1981). In this latter study, a significant enhancement or reduction of ionization probabilities due to the presence of the second alloy component was observed. Other investigators have examined the sputtering of Si under the simultaneous exposure of the surface to an etching reaction, with XeF_2, and Ar^+ ion bombardment (Haring et al., 1982). Vasile (1984) has measured the emission probability $S^+(v)$ of secondary ions emitted (normal) from polycrystalline Cr, Ag, Cu, and Zr and observed a strong dependence upon the velocity of the secondary ion. Plots of $\log_{10} S^+(v)$ vs. $1/v$ show linearity in the energy range $4 \leq E \leq 30$ eV with slopes of the order of $(2-3) \times 10^6$cm/s. This work utilized the unique ion energy analyzer cited in Chapter 3 (Siegel and Vasile, 1981). It also required accurate data on the transmission of this analyzer, as the number of detected secondary ions at any energy is a convolution of analyzer transmission and the kinetic energy of the ions. Vasile (1984) measured the absolute transmission of the analyzer vs. ion energy over the energy range $3 \leq E \leq 55$ eV, using a thermal source of K^+ and Rb^+ ions for calibration.

Cluster Ions and Organic Molecules

The emission of polyatomic secondary ions from alkali metal salts has also been an active area of research employing SIMS, and clusters have been observed over a wide range of masses. Figure 12.9 shows the ultra-high mass spectral data of CsI cluster ions for $n = 1$–70 ($m/z > 18,000$) induced by Xe^+ ion bombardment (Colton et al., 1982; Campana et al., 1984). The authors suggest that the anomalous ion intensities at $n = 13$–15, 22–24, 37–39, and 62–64 can be attributed to the emission of ions with discrete geometrical structures. Specifically, the observed ion intensity enhancements at $n = 13, 22, 37$, and 62 are assigned to stable $3 \times 3 \times 3, 3 \times 3 \times 5, 3 \times 5 \times 5$, and $5 \times 5 \times 5$ cubic-like structures, respectively. The relatively low intensity of cluster ions following stable n values is interpreted as indicating "a barrier to either their formation or to the addition of one or two more salt molecules to the symmetrically (and energetically) stable structures."

Additional investigations of CsI cluster ions have been reported by Baldwin et al. (1983) using fast atom bombardment and by Katakuse et al. (1984) using a 260 cm radius spectrometer and a Xe ion bombardment source. Iodide cluster ions up to $m/z = 90,000$ were generated in the latter study and resolved mass spectra were obtained up to 31,000 with new "magic" numbers being identified as 72, 87, 102, 122, 144 ± 3, 172 ± 3, and 194 ± 3. Similar phenomena have been observed for sodium halide salts. These and other SIMS studies of alkali halide clusters show that (1) the cluster intensity decreases several orders of magnitude as n (cluster size) increases,

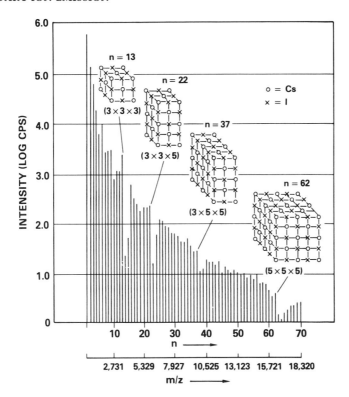

Figure 12.9 Cluster ion spectrum produced by 4.7 keV Xe^+ ions on CsI. Spectral evidence of stable "cubic-like" structures is apparent at n = 13, 22, 37, and 62. (Colton et al., 1982; Campana et al., 1984; *J. Vacuum Science & Technology, 20,* 421 (1982), by permission of American Institute of Physics.)

and (2) the secondary ion intensity distribution can be related to several properties of the halides (Barlak et al., 1983).

Cluster ion emission studies have also included observations on metals and rare gas solids. The latter are of interest, as they provide the simplest examples of "molecular solids." Figure 12.10 is the positive SIMS spectrum of a solid xenon matrix (at 18 K) from primary Ar^+ ions of 4 keV (Orth et al., 1981). In silicon cluster emissions, it has been suggested that Si cluster ions can be classified into two groups: (1) cluster ions produced by rare gas bombardment in high vacuum (representing kinetic emission) and (2) clusters produced by direct oxygen bombardment, resulting in a predominantly chemical emission process (Richter and Trapp, 1981). Gibbs et al. (1982) have observed ion clusters from CO chemisorbed on an Ni (001) surface bombarded by Ar^+. For this specific system, ion yields have been measured at various angles and energies, and results suggest that the ionization probability of the ejecting particle is isotropic and only weakly dependent upon particle velocity.

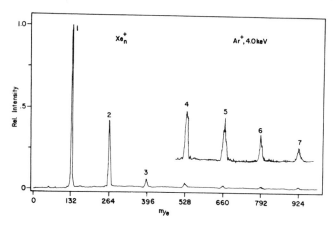

Figure 12.10 SIMS spectrum of a solid xenon matrix at 18 K produced by Ar⁺ ions at 4 keV. (Orth et al., 1981; reprinted with permission from *J. Am. Chem. Soc., 103*, 6026 (1981) copyright 1981 American Chemical Society.)

The ejection of large organic molecules by energetic ions or neutrals is also attributed to collision cascades, which transfer sufficient energy to surface molecules to overcome surface binding forces. Magee (1983) reviewed the salient aspects of organic sputtering from both solid and liquid matrices and identified the following relevant phenomena:

1. Liquid organic specimens will have diffusion coefficients on the order of 10^{-5} cm²/s (molecular velocities of ~30 μm/s), so that the uppermost 100 Å of a liquid film can be renewed every 3×10^{-4} s.

2. The charge on the bombarding particles does not affect the sputtering process, although fast-atom bombardment (FAB) results in less sample charging than ion bombardment.

3. The time scale of the collision cascade in a liquid is comparable to that in a solid (10^{-13} to 10^{-12} s).

4. The energetic particle bombardment is not a "soft ionization" method, but rather it is a mode of energy transfer producing a "graded fragmentation" capable of producing atomic species to very large, intact molecules.

5. The secondary ion energy distribution of large molecules is quite small (~1 eV); the random motions of many atoms in the collision cascade "push" a large molecule from the surface.

6. The degree of fragmentation of secondary molecular ions is affected by the primary ion angle of incidence. Normal incidence results in greater surface damage and a higher degree of molecular fragmentation.

NEUTRAL AND IONIC CLUSTERS

Clusters are formed by many processes in addition to ion impacts on surfaces. These small atomic aggregates also have properties sufficiently different from other forms of matter that they have been called the "fifth state of matter." The synthesis and characterization of this new class of molecules is relevant to basic studies of nucleation, the electronic structure of solids, and transition metal chemistry. Technological interest in clusters encompasses many fields, for example, magnetics, superconductors, solar energy converters, solid-state devices, catalysis, etc. Certain types of metal clusters (iron-sulfur clusters such as Fe_4S_4) are also found at the active sites of many enzymes. Two general methods are used to generate metallic clusters in a beam. In the first, metallic vapor passes through a nozzle and undergoes an isentropic expansion with a decrease in temperature and pressure so that nucleation can occur. Mercury aggregates as high as $n = 12$ have been observed with this technique (Cabaud et al., 1981). A second is a so-called "seed beam technique" in which condensation takes place through collisions with a rare gas.

Mass spectrometric measurements of ion clusters have been reported by many investigators. The laser production of supersonic metal cluster beams and the TOF mass spectral characterization of aluminum clusters have been reported by Dietz et al. (1981). Homogeneous clusters of Ar, Kr, and Xe have been observed in adiabatic, high Mach number expansions through small, diverging Laval nozzle sources (Kim and Stein, 1982). Ion clusters have been monitored in He-CO and Ar-CO glow discharges, and a strong dependence of cluster concentration upon gas temperature has been found by Kaufman et al. (1980). A supersonic molecular beam together with a pulsed laser and TOF spectrometer have been utilized to study the photo- and autoionization, two-photon ionization, and fragmentation processes of alkali clusters as a function of laser wavelength; alkali clusters M_n ($n < 21$) have been observed (Delacrétaz et al., 1982). Lithium clusters have been formed by evaporating lithium in argon at a pressure of 10^{-4} to 10^{-2} torr, and the mass numbers of these clusters containing up to 15 atoms and their abundances were measured by a quadrupole mass analyzer (Kimoto et al., 1980). The condensation of rare gases has been studied in free gas expansions, together with measurements of XeF_6 dimer concentrations and the effects of carrier gases (Vasile and Stevie, 1981). Microclusters M_x of Sb, Bi, and Pb have been generated and analyzed by TOF spectrometry, with M_x up to $x = 650$ being detected and mass peaks $x = 100$ (Sb), $x = 21$ (Bi), and $x = 17$ (Pb) being resolved (Mühlbach et al., 1981).

The existence of the stable lithium cluster Li_4 has been observed by Wu (1983) as one of the species in the effusate from a heated molybdenum Knudsen cell containing the lithium sample. In order to minimize the interference contributed by impurities such as nitrogen, oxygen, and hydrogen in the

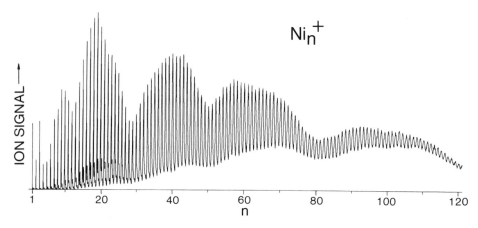

Figure 12.11 TOF mass spectrum of nickel clusters produced by a pulsed laser on an isotopically pure nickel target. The substructure around $n = 20$ is attributed to oxides. (Riley, 1985; courtesy of S. J. Riley.)

TOF measurements, various compositions containing enriched ^{6}Li were employed.

Other cluster spectra have been produced for such relatively refractory metals as Al, Cr, Ni, Cu, and Ag using laser evaporation. Riley et al. (1982) have used a laser evaporation combined with a liquid nitrogen-cooled quench cell for condensation and cluster growth to produce TOF spectra of copper and the nickel cluster spectrum shown in Figure 12.11.

Cluster ions provide insights relative to surfaces—that is, phases in alloys, dissociative and molecular adsorption, oxidation, and sputtered ions (emitted within three lattice constants) contain information on the short range order of the surface. Garrett and MacDonald (1981) have examined cluster ion emission induced by 50 keV Ar^+ bombardment on five niobium-vanadium alloys, and they observed clusters up to Nb_6^+. A theoretical interpretation of the size of clusters (for 10^4 atoms or less and $R < 50$ Å) has also been given by Bachelet et al. (1983) in relation to electronic structure and the density of states at the Fermi level.

Martin (1983) has observed sulfur-rich clusters of varying composition from As_2S_3 vapor quenched in He gas. These clusters varied from arsenic-rich $As_{2n}S_{2n}$ to sulfur-rich $As_{2n}S_{4n+2}$ having masses extending to 1400 amu. It is suggested that such distributions are important not only to indicate the relative stability of clusters, but to indicate the local composition and structure of glasses.

Pulsed laser vaporization has been employed to generate both metal and carbon clusters. The carbon clusters have shown "magic" numbers in their abundances, "with a period of 4 for C_n when n is less than 30 and a period of 2 when n is greater than 40" (Cardillo, 1985). The small cluster periodicity is attributed to the stabilities of unsaturated rings.

ELECTRON EMISSION FROM ION IMPACT

Two mechanisms contribute to ion-induced electron emission, depending on whether the primary source of energy required to release electrons from the surface is provided by (1) the neutralization energy of the incident ion or (2) by its kinetic energy. The first mode of release is via potential energy, and, being exothermic, it can presumably occur even at zero kinetic energy (Hagstrum, 1956). The latter mechanism predominates above a particle velocity threshold in the range of $(0.4-2) \times 10^7$ cm/s for clean surfaces (Ferrón et

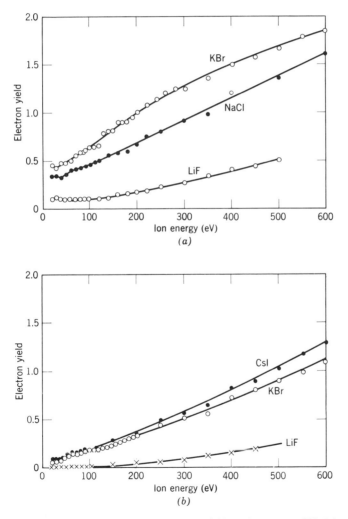

Figure 12.12 (*a*) Dependence of secondary electron yield on the energy of He$^+$ ions on KBr, NaCl, and LiF. (*b*) Dependence of secondary electron yield on the energy of Ar$^+$ ions on CsI, KBr, and LiF (Kondrashev and Petrov, 1965).

al., 1981). This kinetic emission of secondary electrons from solids is the result of a number of scattering and energy transfer processes. For ions with energies in the keV range or higher, a cascade of recoiling atoms may be generated, which in turn produces excited electrons. A fraction of these electrons can escape from a 5–20 Å surface layer of a metal and a somewhat deeper layer in the case of semiconductors and insulators (Schou, 1980).

Secondary yields have been measured as a function of incident ion species, ion energy, amorphous or single-crystal surfaces, crystal orientation, and other primary particle and surface parameters. Ewing (1966) used protons in the energy range of 50–225 keV to study the yield of electrons from a (100) tungsten crystal surface and found a maximum of 1.51 ± 0.03 electrons per proton at an energy of 125 keV. This yield represented an 11% decrease from a polycrystalline surface under otherwise comparable conditions. For this measurement, the tungsten surface was cleaned by flash heating in a vacuum environment of 2×10^{-10} torr. Typical of other early work on single crystals was that of Kondrashev and Petrov (1965), who measured secondary electron yields for alkali halide single crystals bombarded by helium and argon ions having energies in the range of 20–600 eV. Figure 12.12 shows electron yields for He^+ ions on KBr, NaCl, and LiF, and for Ar^+ ions on CsI, KBr, and LiF.

At low ion energies (<1 keV), the secondary electron emission from metal surfaces is primarily potential emission, and it is not a strong function of the energy of the incident ions. Yamauchi and Shimizu (1983) have shown that this potential emission is only about 0.07–0.08 for argon ions on alu-

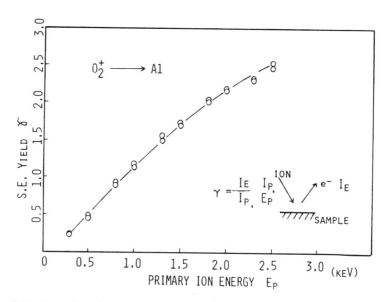

Figure 12.13 Secondary electron yield vs. ion energy for O_2^+ ions impacting a clean aluminum target. (Yamauchi and Shimizu, 1983; *Jpn. J. Appl. Phys.* 22, L227 (1983), by permission.)

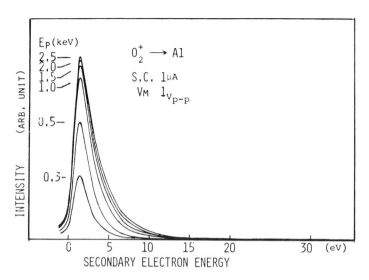

Figure 12.14 Energy distributions of secondary electrons for O_2^+ ions, at various energies, impacting a clean aluminum target. (Yamauchi and Shimizu, 1983; *Jpn. J. Appl. Phys. 22*, L227 (1983), by permission.)

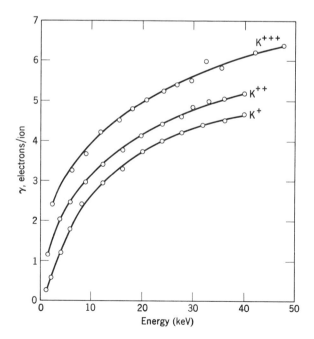

Figure 12.15 Secondary electron yield as a function of primary ion energy and charge state for K^+ ions incident upon a Pt surface (Flahs, 1965).

minum. However, above 1 keV, where kinetic emission takes place, the bombardment of aluminum with oxygen ions results in an order of magnitude increase in secondary electron yield compared to argon. Figure 12.13 shows this secondary yield for an aluminum target as a function of O_2^+ ion energy, and Figure 12.14 shows the corresponding secondary electron energy distributions for various O_2^+ ion energies on the same aluminum target.

Somewhat typical of the monotonically increasing yield of secondaries with primary ion energy are the data of Flahs (1965) in Figure 12.15. Results are shown for singly, doubly, and triply charged potassium ions on platinum. Also, in this case, the Pt surface is gas-covered rather than clean. In the case of single crystals, crystal orientation and the phenomenon of channeling will affect the secondary electron yield significantly. A few specialized studies have focused on a possible isotope effect for ion electron emission in connection with electron multiplier response. In order to compensate for mass discrimination when these multipliers are employed in the DC mode, isotope ratios are often corrected by a factor proportional to the inverse square root of the isotope masses. However, the work by Fehn (1974, 1976) on ions of 15 elements ranging from Mg to Pb on three dynode materials (CuBe, Al, and Ni) indicates that in addition to the general dependence on ion velocity, there is a periodic dependence of electron yield on atomic number. This additional correction is not large, and its periodicity is attributed to the influence of the electronic shell structure of the ions.

ELECTRICAL DISCHARGES AND HIGH TEMPERATURE VAPORS

Studies of plasmas, electrical discharges, and high-temperature vapors have been stimulated primarily by their many applications in high technology, that is, microelectronics, fusion, oxidation processes, chemical synthesis, lamp research, etc. At present, high-temperature chemistry and physics are also contributing to less exotic industries. For example, pilot scale plants are now being designed utilizing electrical discharges to break down toxic wastes into their constituent atoms. These atoms can then be combined into simple molecules that are either trapped by scrubbers or flared to the atmosphere. Combustion engineering is also now receiving input from mass spectral data relating to flames and chemical intermediates.

The pioneering investigation of Shahin (1966) on corona discharges utilized the quadrupole arrangement shown in Figure 12.16. The spectrometer was operated at a frequency of 1.87 MHz with a variable AC voltage that allowed scanning the mass range up to 250. In order to maintain the discharge tube at high pressure, an aperture of 30 μm in diameter separated the discharge and accelerating region, and high-speed vacuum pumps were employed at both the accelerating and analyzing sections. Studies of corona discharges were made in air, nitrogen, oxygen, and water vapor mixtures. Results pointed up the importance of trace quantities of water vapor in ni-

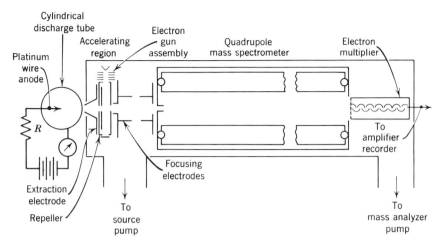

Figure 12.16 Quadrupole spectrometer system for the analysis of ions from a discharge tube (Shahin, 1966).

trogen and/or oxygen when these gases are subjected to corona; and the predominance of $(H_2O)_nH^+$ ion clusters in the mass spectra revealed these groups as positive charge carriers. Values of n as high as 8 were reported at water concentrations of 0.65 mol% at atmospheric pressure. In a related study, Dawson and Tickner (1966) analyzed a DC-negative glow discharge to detect H_2-D_2 exchange by an ionic chain reaction. This H_2-D_2 exchange in the discharge was measured by sampling the product gas downstream from the discharge region. In addition, small percentages of rare gases were added in order to assess their role in radiation-induced reactions. Trends that were noted with the addition of small quantities of krypton are shown in Figure 12.17.

The ongoing development of electric discharge lamps has generated further mass spectrometric studies on arc phenomena and the thermodynamic and electrical properties of vapors. Quadrupole mass spectrometers have been used by several investigators to measure concentrations of various ionic species present in the positive column of DC discharges. Coborn and Kay (1971) have developed a theoretical model for the attenuation of an ion beam sampled from glow discharges. For the Knudsen region, their expression for the ion current I_i is:

$$I_i = I_{i0} \exp \left[-\sigma n_0 \left(\frac{U}{U + S} \right) z \right]$$

where I_{i0} is the ion current at $z = 0$ (the orifice), σ is the cross-section for ion-neutral collisions, U is the orifice conductance, z is the distance between the sampling orifice and the quadrupole, S is the pumping speed downstream

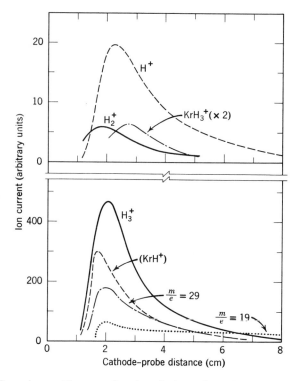

Figure 12.17 Ions observed in a negative-glow discharge for a (H_2 + 7.5% Kr) mixture at 0.3 torr (Dawson and Tickner, 1966).

from the orifice and the quadrupole, and n_0 is the density of neutrals in the discharge ($n_0 = 3.24 \times 10^{16}\ P$ at 298 K, where P is the pressure, torr). El-Kelish and Hattori (1983) have verified the validity of this model and determined a calibration constant for a quadrupole mass filter for neon discharges at pressures up to 8 torr.

Tin (II) halides, because of their continuum radiation in the visible spectrum, have been studied with Knudsen-cell techniques (Hirayama and Kleinosky, 1981); positive ion spectral measurements are shown in Figure 12.18. Sn^+, SnI^+, and SnI_2^+ determinations are given in the usual semilog plot of i^+T vs. $1/T$, where i^+ is the ion intensity and T is the absolute temperature. The data also indicate that the ions all originate from a single parent, that is, SnI_2, and an enthalpy of sublimation of 34 ± 3 kcal/mol is derived for the reaction $SnI_2(c) \rightarrow SnI_2(g)$ at 533 K. Other thermodynamic mass spectral studies have been made of selenium (Grimley et al., 1982) and refractory-mixed oxide compounds that contain alkaline earths (Hirayama et al., 1980). Spectra have also been reported of positive ions in spark discharges in SF_6 in the pressure range 13–67 kPa. Very complex ion mass spectra were pro-

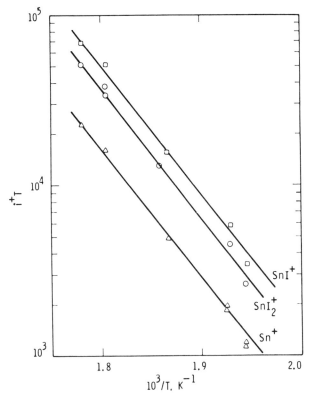

Figure 12.18   Semilog plot of i^+ T vs. $1/T$ for positive ions from SnI$_2$. (Hirayama and Kleinosky, 1981; *Thermochim. Acta, 47,* 355 (1981), by permission of Elsevier Science Publishers.)

duced when the SF$_6$ contained only a few parts per million of either N$_2$ or O$_2$ (Frees et al., 1981).

Other discharge-related studies have included spark source mass spectrometry (SSMS) of very high resolution that has been compared with Knudsen-cell data. Specifically, Cornides and Gál (1981) have suggested that the mechanism by which molecular complexes are formed in the vapor phase can be deduced by comparing the ionic species of a spark plasma to ions produced from the high-temperature evaporation of solids. By using stable isotopes, they have obtained experimental evidence elucidating a formation mechanism for complexes of Ga$_x$O$_y$ generated in a spark ion source with Ga$_2$O$_3$/graphite electrodes. Table 12.3 shows the very high resolution of their data, which include species not previously detected.

Mass spectrometry also provides the only practical means for measuring the species, yields, degree of ionization, ratio of ions to neutrals, and energy distributions of electrode products emitted by DC arcs in a vacuum ambient. In such arcs, current passes solely by means of the vaporized electrode

TABLE 12.3 Mass Measurements for Ions Formed in a Spark Plasma

Mass No.	Ga_xO_y	Mass Value Calculated	Measured
85	^{69}GaO	84.92060	84.92080
87	^{71}GaO	86.91976	86.92199
101	$^{69}GaO_2$	100.91551	100.91593
103	$^{71}GaO_2$	102.91467	102.91781
154	$^{69}GaO_2O$	153.84627	153.84683
156	$^{69}Ga^{71}GaO$	155.84543	155.84733
158	$^{71}Ga_2O$	157.84459	157.84557
170	$^{69}Ga_2O_2$	169.84119	169.84043
172	$^{69}Ga^{71}GaO_2$	171.84035	171.84081
174	$^{71}Ga_2O_2$	173.83951	173.83979

Source: Cornides and Gál (1981).

material produced by the arc itself. Davis and Miller (1969) used a double focusing magnetic analyzer to measure the radial ion fluxes emitted from several metal vapor arcs to obtain the data shown in Table 12.4. These and other data indicate that ions can be formed having energies substantially greater than the anode potential, but that the average energy of neutral atoms formed is only 1 or 2 eV. A partial explanation for the energetic ions is that

TABLE 12.4 Fractional Distribution of Radial Ion Flux From 100 A DC Metal Vapor Arcs in Vacuum

Element	Arc Voltage[a] (V)	Fraction of given Ion (%) Degree of Ionization: I	II	III	IV	V	VI
Ca	13	53	47	T^b	—	—	—
C	16	96	4	0	—	—	—
Ag	16½	65	34	1	0	—	—
Ni	18½	48	48	3	*T*	—	—
Al	20	49	44	7	0	—	—
Cu	20½	30	54	15	0.4	*T*	—
Zr	21½	14	60	21	5	*T*	—
Ta[c]	24	13	35	28	13	10	0.3
Mo	25½	16	69	13	1.5	—	—

Source: Davis and Miller (1969).

[a] ±1 V.

[b] Present, but ≤ 0.1%.

[c] Fractions at 140 A.

just above the cathode, where the densities of electrons and neutral atoms are very high, an atom can be ionized by a series of inelastic collisions in which the time between collisions is less than the lifetime of the excited atom. Table 12.4 also confirms the general rule that electrode elements with higher arc voltages tend to generate a larger fraction of ions in a higher charge state.

ION MOBILITIES IN GASES

A knowledge of the mobilities of ions is important to a quantitative understanding of ion-molecule reactions, laboratory and naturally occurring plasmas, discharge phenomena, and ionospheric chemistry. Furthermore, a new tool in chemical analysis, the "plasma chromatograph," is simply an ion mobility spectrometer in which the size and nature of a molecule, together with instrumental and drift gas parameters, determine the transit time. The apparatus for measuring ion mobilities described by Johnsen et al. (1982) is interesting inasmuch as a QMS is used to select ions from a source, and a second quadrupole is employed in connection with ion detection. A mass-selected pulse of ions ($\sim 10^5$ ions) is injected into the drift region with an energy of about 10 eV and at a repetition rate of ~ 1 kHz. Subsequent to thermalization through collisions with neutral gas atoms, the ions traverse the drift tube with a drift velocity corresponding to their ionic mobility, applied electric field, and gas density.

The reduced ion mobility μ_0 is defined by the relation

$$v_d = \mu_0 N_0 E / N$$

where v_d is the ion drift velocity in the direction of the applied electric field E through a gas of density N, and N_0 is the number density of gas molecules at STP. For a given ion-gas combination, only the E/N ratio and the gas temperature T determine v_d, that is, $\mu_0 = \mu_0 (E/N, T)$. It is then customary to plot reduced mobility μ_0 as a function of E/N in Townsends (1 Td = 10^{-17} Vcm^{-2}) at a fixed temperature.

It has been possible to differentiate between mobilities of ground state and metastable O^+ and O_2^+ ions in helium and neon; although in many instances, the ion mobilities of ground state and metastable ions are nearly identical (Johnsen et al., 1982). However, mobility measurements, in general, agree with theoretical calculations of ion-neutral interaction potentials. Mobilities with this drift tube technique have been reported for (1) K^+ ions in Ar, Kr, and Xe (Lamm et al., 1981); (2) Ne^+ and Xe^+ ions in argon (Jones et al., 1981); (3) O^+, O^{+*}, and O_2^{2+} in He and Ar (Fhadil et al., 1982); (4) cluster chloride ions in Ar, Kr, and Xe (Jówko and Armstrong, 1981); (5) SF_6^- ions in He; and (6) F^-, Cl^-, Br^-, and I^- in Ar, Xe, H_2, N_2, CO, and CH_4 (Fujii and Meisels, 1981). Experimental measurements have also been

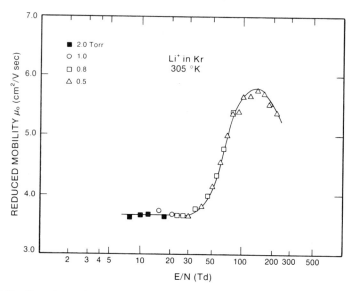

Figure 12.19 Experimental mobilities for Li$^+$ ions in a Kr-filled drift tube at several pressures. (Takebe et al., 1982; *J. Chem. Phys.*, **76**, 2672 (1982), by permission of American Institute of Physics.)

made of the mobility of Li$^+$ ions in Ar, Kr, and Xe, using a simple drift tube (but without a mass spectrometer) and a thermionic emitter (Takebe et al., 1982). Figure 12.19 shows mobility data acquired for Li$^+$ in Kr over an E/N range of 8–216 Td at pressures of 0.5–2 torr. The zero field-reduced mobility μ_0 was found to be 3.65 cm^2/Vs by averaging the results over the E/N range of 8–20 Td. Overall, these measurements and appropriate calculations compare favorably with longitudinal diffusion coefficients obtained for Li$^+$ ions in noble gases by other methods.

TEMPERATURE MEASUREMENT OF RAREFIED GASES

At gas pressures in the range of 1 atm or greater, there are several techniques, including acoustics, for measuring temperature. At very low pressures, however, most ordinary thermometers are poor sensors because of their large heat capacity. Therefore, TOF mass spectrometry provides an interesting alternative at very low pressures ($\sim 10^{-5}$ torr or lower), because it can accurately determine the velocity distribution of molecules. Furthermore, this type of temperature measurement is useful in both laboratory and upper atmospheric investigations.

In contrast to conventional TOF spectrometry, Matsumura (1980) has successfully employed a metastable time-of-flight (MTOF) technique for determining the temperatures of extremely rarefied Ar and N_2 gases at temperatures ranging from 300–500 K. The basic concept is to operate an electron gun in a pulsed mode to generate atoms or molecules in metastable states. These neutral, but excited, particles then proceed along a flight path and are detected by a so-called Auger surface detector and a 20-stage electron multiplier. With an Auger detector, if the energy difference of the metastable state of the incident particle exceeds the work function of the cathode, there is some probability that secondary electrons will be ejected from the cathode surface. In the MTOF apparatus, it is also necessary to deflect charged particles out of the flight path, as ions would greatly distort the assumed Maxwellian velocity distribution of the neutrals. In addition, it is essential to have a flight path whose length is appropriate for the metastable lifetimes of the gaseous species (e.g., 160 µs for N_2), and to acquire data in a low enough pressure regime so that the mean free path of the molecules is much greater than the source-to-detector distance. If this latter condition is not satisfied, gas scattering will occur, producing an effect that may be interpreted as radiative decay.

In the experiments of Matsumura (1980), the essential conditions were fulfilled: (1) the neutral species emerged from an orifice of an oven, and (2) the velocities of the gas molecules had a Maxwellian distribution. Hence, gas temperatures could be calculated from basic kinetic theory, and the maximum of the distribution function is observed at a time

$$t_{max} = \sqrt{\frac{\beta}{2}} \cdot L = \sqrt{\frac{2}{\pi}} \cdot \frac{L}{\bar{v}}$$

where $\beta = M/2kT$, M is the mass of the particle, k is the Boltzmann constant, T is the absolute temperature, L is the distance between the modulated source and the detector, and $\bar{v}$ is the mean thermal velocity. On the basis of calculations and experimental measurements using a multichannel counting system, temperature measurements for both Ar and N_2 were determined to within ±3 K.

TEMPERATURE MEASUREMENTS IN SHOCK WAVES

Direct temperature measurements of gases in shock waves have been made by optical spectroscopic techniques, and these observed temperatures have been in reasonable agreement with incident shock velocities over a limited temperature range. However, mass spectrometry has also been employed for temperature measurements, using the TOF method to determine molecular velocities. In a study by Teshima and Takahashi (1983), reflected shock wave temperatures were measured by sampling a small volume of shock-heated gas through a nozzle that connected a vacuum chamber equipped

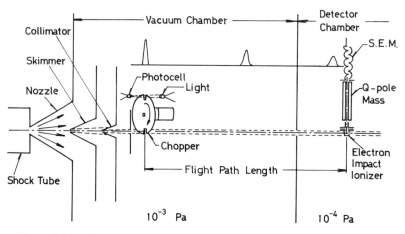

Figure 12.20 Schematic diagram of time-of-flight apparatus for measuring temperatures of shock-heated gas. (Teshima and Takahashi, 1983; *Jpn. J. Appl. Phys.*, 22, Pt. 1, 1742 (1983), by permission.)

with a mechanical chopper, an electron impact ionizer, and quadrupole analyzer, as shown in Figure 12.20. These investigators then calculated the effective stagnation temperature, which is expressed by the energy conservation equation:

$$T_{0\text{eff}} = \frac{1}{5}\frac{m}{k}u^2 + T$$

where m is the atomic mass, k is the Boltzmann constant, u is the measured flow velocity, and T is the static gas temperature obtained from the TOF measurements. Referring to the schematic of Figure 12.20, the shock tube had a diameter of 128 mm with a 2.5 m driver section and a 4.7 m driven section—connected to the convergent-divergent nozzle with a 3.2 mm throat diameter. Hydrogen was used as a driver gas to yield shock waves in argon (0.13–6.5 kPa), with shock Mach numbers of 2.5–7.1. Polyester diaphragms were burst mechanically by the compressed hydrogen at slightly below their spontaneous bursting pressure. Part of the expanding gas emerging from the nozzle passed through a skimmer and collimator and was chopped by a slotted disc rotated at 500–7000 rpm.

Basically, the measurement of shock temperature in this experiment was made by using a photocell signal to indicate the "zero" for each TOF signal of the 104 cm flight path. The slotted chopper disc (19 cm in diameter with 24 slits on its periphery) chopped the molecular beam emerging from the nozzle into segments that were subsequently detected at the end of the flight path via the ionizer, quadrupole, and secondary electron multiplier detector. The observed TOF signals were then fitted to a Maxwellian velocity distribution to yield the flow velocity and static gas temperature. A temperature as high as 10,500 K was observed by means of this measurement technique. This specific shock wave experiment is also typical of the increasing number of studies in which dynamic mass spectrometers are being employed to provide detailed molecular velocity distributions associated with transient phenomena and high-temperature regimes.

HIGH-CHARGE STATE MULTIPHOTON IONIZATION

With the advent of lasers that can produce ultra-short pulses at gigawatt power levels, ionization phenomena are being observed of atoms in very high charge states. These charge state distributions provide a means for determining the cross-section for multiphoton energy transfer and for observing nonlinear interactions in atoms and molecules. TOF mass spectrometry has been employed by Boyer et al. (1984) for quantifying charge states produced by picosecond ArF* radiation (93 nm) at intensities up to 10^{15} W/cm^2. In contrast to multiphoton ionization by lasers for isotope separations, this study required low pressures, with molecular densities in the interaction volume being $10^{12}/cm^3$. A 1 m TOF spectrometer was attached to a stainless steel laser irradiation chamber, and sample gases included He, Ne, Ar, Kr, Xe, and UF_6. The entire system was baked and evacuated to a background pressure of $<10^{-8}$ torr to prevent spurious, singly ionized species from interfering with spectra of the sample gases.

Multiphoton ionization experiments with this apparatus were studied of the general physical process:

$$N\gamma + X \rightarrow X^{q+} + qe^-$$

where N is the number of photons, γ is the photon energy, X is the irradiated species, and q is the multiplicity of electronic charge. The TOF spectrum of uranium revealed clearly resolved positive ions up to the charge state U^{10+}. This spectrum was remarkable in that it represented a multiphoton energy transfer of at least 633 eV, corresponding to values of approximately 99 quanta to produce the U^{10+} ion from the neutral atom and ~133 eV (21 quanta) to remove the 10th electron (Carlson et al., 1970). Thus, TOF spectrometry appears to be an effective tool for studying charge state distributions, where instrumentation of only modest resolution, $m/\Delta m \sim 100$, is adequate to identify various ionized states in both low Z and high Z species.

REFERENCES

Almen, O., and G. Bruce, *Trans. Eighth Nat. Symposium on Vacuum Technol.*, Pergamon, New York (1962), p. 245.

Amirav, A., and U. Even, *J. Appl. Phys., 51,* No. 1, 1 (1980).

Bachelet, G. B., F. Bassani, M. Bourg, and A. Julg, *J. Phys. C.: Solid State Phys., 16,* 4305 (1983).

Baldwin, M. A., C. J. Proctor, I. J. Amster, and F. W. McLafferty, *Int. J. Mass Spectrom. and Ion Processes, 54,* 97 (1983).

Barlak, T. M., J. E. Campana, J. R. Wyatt, and R. J. Colton, *J. Phys. Chem., 87,* 3441 (1983).

Boyer, K., H. Egger, T. S. Luk, H. Pummer, and C. K. Rhodes, *J. Opt. Soc. Am., B1,* No. 1, 3 (1984).

Cabaud, B., A. Hoareau, and P. Melinon, *J. Phys. D: Appl. Phys., 13,* 1831 (1981).

Campana, J. E., R. J. Colton, J. R. Wyatt, R. H. Bateman, and B. N. Green, *Appl. Spectroscopy, 38,* No. 3, 430 (1984).

Cardillo, M. J., *Physics Today, 38,* No. 1, S-16 (1985).

Carlson, T. A., C. W. Nestor, Jr., N. Wasserman, and J. D. McDowell, *Atomic Data, 2,* 63 (1970).

Chakraborty, P., and S. D. Dey, *J. Appl. Phys., 52,* No. 11, 7002 (1981).

Clement, R. M., R. A. Davies, H. T. Miles, and S. K. Sethhuraman, *J. Phys. D: Appl. Phys., 13,* 1643 (1980).

Coborn, J. W., and E. Kay, *J. Vac. Sci. Technol., 8,* 738 (1971).

Colton, R. J., T. M. Barlak, J. R. Wyatt, J. J. DeCorpo, and J. E. Campana, *J. Vac. Sci. Technol., 20,* No. 3, 421 (1982).

Cornides, I., and T. Gál, *High Temp. Sci., 14,* 71 (1981).

Davies, J. A., *Physica Scripta, 28,* 294 (1983).

Davis, W. D., and H. C. Miller, *J. Appl. Phys., 40,* 2212 (1969).

Dawson, P. H., and A. W. Tickner, *J. Chem. Phys., 45,* 4330 (1966).

Delacrétaz, E., J. D. Ganière, R. Monot, and L. Wöste, *Appl. Phys., B29,* 55 (1982).

Dietz, T. G., M. A. Duncan, D. E. Powers, and R. E. Smalley, *J. Chem. Phys., 74,* No. 11, 6511 (1981).

El-Kelish, S., and S. Hattori, *Jpn. J. Appl. Phys., 22,* No. 10, Part 1, 1577 (1983).

Ewing, R. I., *Am. Phys. Soc. Bull., 11,* 816 (1966).

Fehn, U., *Int. J. Mass Spectrom., 15,* 391 (1974).

Fehn, U., *Int. J. Mass Spectrom., 21,* 1 (1976).

Feldman, L. C., J. W. Mayer, and S. T. Picraux, *Materials Analysis by Ion Channeling,* Academic press, New York (1982), p. 6.

Ferrón, J., E. V. Alonso, R. A. Baragiola, and A. Oliva-Florio, *Phys. Rev., B24,* No. 8, 4412 (1981).

Fhadil, H. A., D. Mathur, and J. B. Hasted, *J. Phys. B: Atom. Molec. Phys., 15,* 1443 (1982).

Flahs, I. P., *J. Technol. Phys. (S.S.S.R.), 25,* 2463 (1965).

Frees, L. C., I. Sauers, H. W. Ellis, and L. G. Christophorou, *Chem. Phys. Lett., 81,* No. 3, 528 (1981).

Fujii, T., and G. G. Meisels, *J. Chem. Phys., 75,* No. 10, 5067 (1981).

Garrett, R. F., and R. J. MacDonald, *Nucl. Instr. and Methods, 191,* 308 (1981).

Gibbs, R. A., S. P. Holland, K. E. Foley, B. J. Garrison, and N. Winograd, *J. Chem. Phys., 76,* No. 1, 684 (1982).

Grimley, R. T., Q. G. Grindstaff, T. A. DeMercurio, and J. A. Forsman, *J. Phys. Chem., 86,* 976 (1982).

Hagstrum, H. D., *Phys. Rev., 104,* 672 (1956).

Haring, R. A., A. Haring, F. W. Saris, and A. E. deVries, *Appl. Phys. Lett., 41,* No. 2, 174 (1982).

Hart, R. G., and C. B. Cooper, *Surf. Sci., 94,* 105 (1980).

Herman, I. P., and J. B. Marling, *J. Chem. Phys., 72,* No. 1, 516 (1980).

Hirayama, C., R. L. Kleinosky, and R. S. Bhalla, *Thermochim. Acta, 39,* 187 (1980).

Hirayama, C., and R. L. Kleinosky, *Thermochim. Acta, 47,* 355 (1981).

Hitz, C. B., *Laser Appl., 3,* 51 (1984).

Howe, L. M., and J. A. Davies, in *Site Characterization and Aggregation of Implanted Atoms in Materials,* ed. A. Perez and R. Coossement, Plenum, New York (1980), p. 241.

Hughes, R. H., R. J. Anderson, C. K. Manka, M. R. Carruth, L. G. Gray, and J. P. Rosenfeld, *J. Appl. Phys., 51,* No. 8, 4088 (1980).

Hurst, B. L., and C. B. Cooper, *J. Appl. Phys., 53,* No. 9, 6372 (1982).

Johnsen, R., M. A. Biondi, and M. Hayashi, *J. Chem. Phys., 77,* No. 5, 2545 (1982).

Jones, T. T. C., J. D. C. Jones, and K. Birkinshaw, *Chem. Phys. Lett., 82,* No. 2, 377 (1981).

Jówko, A., and D. A. Armstrong, *J. Chem. Phys., 76,* No. 12, 6120 (1981).

Katakuse, I., H. Nakabushi, T. Ichihara, T. Sakurai, T. Matsuo, and H. Matsuda, *Int. J. Mass Spectrom. and Ion Processes, 57,* 239 (1984).

Kaufman, Y., P. Avivi, F. Dothan, H. Keren, and J. Malinowitz, *J. Chem. Phys., 72,* No. 4, 2606 (1980).

Kim, S. S., and G. D. Stein, *J. Colloid and Interface Sci., 87,* No. 1, 180 (1982).

Kimoto, K., I. Nishida, H. Takahashi, and H. Kato, *Jpn. J. Appl. Phys., 19,* No. 10, 1821 (1980).

Kondrashev, A. I., and N. N. Petrov, *Soviet Phys. Solid State, 7,* 1255 (1965).

Koren, G., Y. Gertner, and U. Shreter, *Appl. Phys. Lett., 41,* No. 5, 397 (1982).

Lamm, D. R., M. G. Thackston, F. L. Eisele, H. W. Ellis, J. R. Twist, W. M. Pope, I. R. Gatland, and E. W. McDaniel, *J. Chem. Phys., 74,* No. 5, 3042 (1981).

Lichtin, D. A., R. B. Bernstein, and V. Vaida, *J. Am. Chem. Soc., 104,* 1830 (1982).

Linhard, J., *Kgl. Danske Vidensk. Selsk Mat. Fys. Medd, 34,* No. 14 (1965).

Magee, C. W., *Int. J. Mass Spectrom. Ion Phys., 49,* 211 (1983).

Martin, T. P., *J. Chem. Phys., 77,* No. 7, 3815 (1982).

Martin, T. P., *Solid State Comm., 47,* No. 2, 111 (1983).

Matsumura, S., *Jpn. J. Appl. Phys., 19,* No. 7, 1205 (1980).

McHugh, J. A., and J. C. Sheffield, *J. Appl. Phys., 35,* 512 (1964).

Meyer, K., I. K. Schuller, and C. M. Falco, *J. Appl. Phys., 52,* No. 9, 5803 (1981).

Mühlbach, J., E. Recknagel, and K. Sattler, *Surf. Sci., 106,* 188 (1981).

Okajima, Y., *J. Appl. Phys., 51,* No. 1, 715 (1980).

Orth, R. G., H. T. Jonkman, D. H. Powell, and J. Michl, *J. Am. Chem. Soc., 103,* 6026 (1981).

Picraux, S. T., L. R. Dawson, G. C. Osbourn, and W. K. Chu, *Appl. Phys. Lett., 43,* No. 10, 930 (1983).

Richter, C. E., and M. Trapp, *Int. J. Mass Spectrom., 38,* 21 (1981).

Riley, S. J., E. K. Parks, C. R. Mao, L. G. Pobo, and S. Wexler, *J. Phys. Chem., 86,* 3911 (1982).

Riley, S. J., private communication (1985).

Rol, P. K., Ph.D. Thesis, Amsterdam (1960).

Schou, J., *Phys. Rev., B22,* No. 5–6, 2141 (1980).

Shahin, M. M., *J. Chem. Phys., 45,* 2260 (1966).

Siegel, M. W., and M. J. Vasile, *Rev. Sci. Instr., 52,* 1603 (1981).

Suzuki, K., P. H. Kim, and S. Namba, *Jpn. J. Appl. Phys., 20,* No. 4, 753 (1981).

Swanson, M. L., and L. M. Howe, *Nucl. Instr. Methods, 218,* 613 (1983).

Swanson, M. L., L. M. Howe, A. F. Quenneville, and J. A. Nilson, *Nucl. Instr. Methods, 218,* 643 (1983).

Takebe, M., Y. Satoh, K. Linuma, and K. Seto, *J. Chem. Phys., 76,* No. 5, 2672 (1982).

Takeuchi, K., I. Inoue, R. Nakane, Y. Makide, S. Kato, and T. Tominaga, *J. Chem. Phys., 76,* No. 1, 398 (1982).

Teshima, K., and N. Takahashi, *Jpn. J. Appl. Phys., 22,* No. 11, Part 1, 1742 (1983).

Vasile, M. J., *Phys. Rev., B29,* No. 7, 3785 (1984).

Vasile, M. J., and F. A. Stevie, *J. Chem. Phys., 75,* No. 5, 2399 (1981).

Williams, P., *Phys. Rev., B23,* No. 11, 6187 (1981).

Williams, P., *Appl. Surf. Sci., 1;3,* 241 (1982).

Winters, H. F., *Radiation Effects, 64,* 79 (1982).

Wood, R. A., B. W. Davis, A. Vass, and C. R. Pidgeon, *Opt. Lett., 5,* No. 4, 153 (1980).

Wu, C. H., *J. Phys. Chem., 87,* 1534 (1983).

Yamauchi, Y., and R. Shimizu, *Jpn. J. Appl. Phys., 22,* No. 4, L227 (1983).

Yu, M. L., and W. Reuter, *J. Appl. Phys., 52,* No. 3, 1478 (1981).

Zittel, P. F., and L. A. Darnton, *J. Chem. Phys., 77,* No. 7, 3464 (1982).

CHAPTER THIRTEEN

ENERGY SYSTEM DIAGNOSTICS: SOLAR, FOSSIL, FISSION, FUSION

Many analytical instruments are required to characterize the detailed chemistry and physics of various energy conversion cycles. Materials in some of these cycles must be operational at near-absolute zero or contain plasmas at temperatures $>10^7$ K; other materials must possess unique properties for energy conversion, energy storage (e.g., metal hydrides), or heat transfer. Mass spectrometry assumes an especially important diagnostic role because it can elucidate energy transmutations occurring at the nuclear, atomic, and molecular level, and can translate microscopically observed phenomena into overall system parameters. The identification of combustion products of fossil fuels and a determination of temperatures in plasmas are two examples. Isotopic measurements, in particular, yield additional information on reaction kinetics and transport phenomena, thereby complementing data obtained from other methods of analysis.

SOLAR CELLS

The conversion of sunlight to electric energy was demonstrated on a reasonably large scale when *Solar One*, a 10 megawatt-electric (MWe) plant at Barstow, California became operational in 1982. An array of 1818 heliostats, each having a reflecting area of approximately 39 m^2, focused solar energy on a central receiver; this thermal energy was converted to electric power via heat exchangers and generators (Krieth and Meyer, 1983).

Solar cells, however, have been available for generating electric power for almost three decades, but their most important use has been in space-

and telecommunication-related applications. Their potential future market is enormous, because in many regions of the world, the radiant solar energy incident upon a typical house is adequate to supply its electric demands—provided that an efficient conversion and storage system can be engineered. Furthermore, most developing countries are located in the sunbelt, where decentralized generating stations are needed for agriculture and village power. Photovoltaic power may even find residential applications in industrial countries if costs of $2–3 per installed watt in the system level can be reached (Van Overstraeten et al., 1982). The concept of placing a large, photovoltaic solar array in geosynchronous orbit represents another option. Despite serious practical problems, the attractiveness of this scheme is that a satellite solar cell would receive the high solar radiation flux beyond the earth's atmosphere; this *solar constant* is ~1.37 kW/m^2 (Duncan et al., 1982). A solar satellite in a geosynchronous orbit of 35,800 km (22,000 miles) above the equator would always remain above the same point on the earth's surface, and would be able to beam microwave power to a line-of-sight receiving station capable of converting microwave power to DC power. A terrestrial-based system would receive only about 7% as much radiant energy as the satellite; in addition to atmospheric absorption and scattering, the solar energy will vary with geographical location, the season, and the time of day. For example, the average solar energy received in 1 day on a horizontal surface is 1.1 kWh/m^2 (an average flux of 46 W/m^2) in Seattle in January and approximately 9 kWh/m^2 (375 W/m^2) in the Mojave Desert in July (Williams, 1975).

In contemporary photovoltaic technology, the engineering approaches that show the most promise are (1) tandem, amorphous silicon thin films, (2) crystalline silicon ribbons, and (3) high-concentration systems that utilize high-efficiency cells.

Photovoltaics

The photovoltaic effect was first reported by Becquerel in 1839, but it was not until the early 1940s that photovoltaic action was discovered at a p-n junction within a silicon crystal. Shortly thereafter, other p-n junction devices were fabricated using PbS, Ge, and other compound materials. By 1954, the conversion efficiency of some solar cells reached about 6%. At present, silicon and selenium are the two most important single-element materials for energy conversion, but several binary compounds are being explored, that is, gallium arsenide (GaAs), cadmium telluride (CdTe), cadmium sulfide (CdS), copper sulfide (Cu_2S), plus higher order inorganic compounds.

Mass spectrometry and mass-resolved ion beams are now directly or indirectly utilized for solar cell research in the following areas:

1. Impurity identification and control.
2. Mobility and diffusion measurements.

3. Ion implantation.

4. Secondary ion mass spectrometry (SIMS) measurements of impurity profiles.

5. Thin film antireflection coatings.

6. Glow discharge and chemical vapor deposition technology.

7. Photochemical reactions.

8. Multilayered solar-selective coatings.

9. Thin film amorphous silicon cells.

10. Electrochemical and high-temperature thermodynamic measurements.

The most common form of solar cell has a thin collecting junction at its front surface. This junction may be a diffused or ion-implanted type, and may have configurations as exemplified in the schematic diagram of Figure 13.1, that is, (1) PN or NP cell, (2) metal insulator semiconductor (MIS) structure, or (3) metal insulator n-type p-type (MINP). Photons absorbed in the cell will generate minority carriers that diffuse to the junction, where they are accelerated through the built-in potential field to produce a photocurrent. In such front-surface, single-crystal configurations, the cell must be sufficiently thin, so that charge carriers produced near the surface can reach the junction and be collected. A typical silicon photovoltaic cell will develop 0.5 V and generate about 0.1 W/cm^2. The energy band gap E_g in pure silicon is 1.120 eV at 300 K, so that the wavelength threshold for producing electron-hole pairs is

$$\lambda = \frac{12399 \text{ Å}}{E_g(\text{eV})} \sim 11{,}000 \text{ Å} (1.1 \ \mu\text{m})$$

Thus, a silicon cell will derive power from the infrared as well as the visible

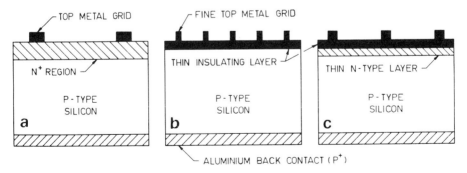

Figure 13.1 Solar cell types. (*a*) PN or NP, (*b*) MIS (metal insulator silicon), and (*c*) MINP (metal insulator n-type p-type). (Blakers et al., 1981; *Nucl. Instr. Methods, 191*, 51 (1981), by permission of North-Holland Physics Publishing.)

and ultraviolet portions of the spectrum. Quantitatively, the minority charge carrier generation will depend upon the spectral distribution of the incident radiation; and in a typical single-crystal cell, a thickness of 200 μm is required for a 95% absorption of the solar spectrum.

The electron and hole mobilities in pure silicon are given by

$$\mu_e = 1360 \ (T/300)^{-2.42} \ cm^2/V\text{-}s$$

and

$$\mu_h = 495 \ (T/300)^{-2.20} \ cm^2/V\text{-}s$$

but these values are substantially decreased by impurity scattering in heavily doped materials. However, solar cell performance is more likely to be affected by the diffusion rates of minority carriers to the collecting junction than by mobility values, per se (Hall, 1981). The diffusion coefficient D is given by the Einstein relation $D = kT\mu/q$, so at 300 K in pure silicon, the coefficients for electrons and holes are

$$D_e = 35.1 \ (T/300)^{-1.42} \ cm^2/s$$
$$D_h = 12.7 \ (T/300)^{-1.20} \ cm^2/s$$

Furthermore, the effective current generated by a solar cell is determined by the minority charge carrier lifetime τ and diffusion length L, whose relation to the diffusion coefficient are expressed by $L = \sqrt{D\tau}$.

All of these parameters (mobilities, diffusion rates, and lifetime) are influenced by impurity concentrations, so that mass spectrometry is crucial to both the chemistry and solid-state physics of solar devices.

Single-crystal solar cells have been fabricated by many techniques, including both mass-resolved and mass-unresolved ion implantation. The latter scheme has been successfully demonstrated for producing solar cells of large area, with the radiation damage from atomic and molecular beams being annealed by an on-line laser melting of the implanted area (Muller and Siffert, 1981). Ions are produced via a glow discharge in a gas containing the dopant (B_2H_6, BF_3, or BCl_3 for boron and PH_3 or PF_5 for phosphorus) at a pressure of 10^{-2} to 10^{-3} torr. These ions are then extracted from the plasma and accelerated directly onto the cell elements through potentials up to 50 kV and current densities up to 1 mA/cm². The laser annealing is very fast and is accomplished with short ruby laser pulses having 25 ns half-power width at 0.69 μm and power densities in the range of 20–50 MW/cm². SIMS profiles of implanted specimens before and after annealing then provide detailed data on various ion implantation-laser annealing distributions. Muller and Siffert (1982) suggest that this continuous on-line fabrication technique for large area wafers should be able to compete with production rates of silicon ribbons.

Other factors contributing to the widespread use of ion implantation in solar cell developments include (1) accurate dopant control; (2) uniformity of deposition; (3) cleanliness and purity of chemical processing; (4) high yields; and (5) feasibility of automatic fabrication. In some newer solar cell structures such as the MINP shown in Figure 13.1, it has also been possible to produce relatively high voltage cells in wafers with very low dopant levels. Blakers et al. (1981) have reported phosphorous ion doping at 20 keV with dopant implant levels as low as $10^{13}/cm^2$, which has yielded open circuit voltages of 661 mV. This is a higher output voltage than for conventional ion-implanted solar cells, and with implant doses 200 times lower; this low-dose implant value should also translate into a significant reduction in fabrication cost. Recent work in solar cells has also included "direct gap" materials such as gallium arsenide and gallium indium arsenide, whose projected efficiencies may ultimately approach 30%. Their cost is substantially higher than silicon cells, but they would be used in conjunction with low-cost concentrator systems. These cells would also be fabricated of multiple layers of materials in order to capture both the long and short wavelengths of the solar spectrum.

The amorphous silicon cell is one alternative to the single-crystal type. Its glass-like composition is neither pure nor ordered, and fabrication techniques include both chemical vapor deposition and gaseous discharge. With the latter scheme, silane gas (SiH_4) is heated and subjected to an electric discharge, causing the gas to decompose into silicon-hydrogen molecules and ions—with a subsequent deposition on a substrate. In related amorphous silicon research, the DC-reactive sputtering of silicon in an argon-hydrogen discharge has been studied, using a glow discharge mass spectrometry method in which ionic species arriving in the substrate plane are unambiguously identified (Tardy et al., 1981). Both atomic (Ar^+, Si^+) and molecular ions (SiH^+, SiH_3^+, H_3^+, AH^+) have been observed, and the relative intensities of the three Si^+ (28, 29, 30) species were found to be in good agreement with their isotopic abundances. It is deduced that the reaction between hydrogen and the silicon target leads to the formation of SiH molecules, which are sputtered and ionized in the discharge, and that various plasma parameters influence the nature of the deposited amorphous silicon films. Figure 13.2 is a typical schematic of apparatus for obtaining mass spectra from a multipole discharge where thin films of hydrogenated amorphous silicon are produced from silane and disilane plasma composition (Perrin et al., 1984).

Secondary mass spectrometry (SIMS) has been used to characterize hydrogenated amorphous silicon (a-Si:H), and Magee and Carlson (1980) have specifically employed SIMS (1) to detect all elements from hydrogen to uranium; (2) to obtain depth profile data with 100 Å resolution; and (3) to determine quantitative impurity concentrations with the aid of ion-implanted standards. In a related study, Carlson and Magee (1978) have also measured the diffusion of deuterium in hydrogenated amorphous silicon. Figure 13.3 shows their measured diffusion coefficients plotted as a function of $1000/T$,

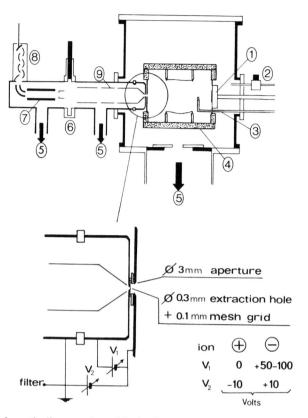

Figure 13.2 Schematic diagram of a multipole plasma reactor, ion extraction system, and mass spectrometer: (*1*) substrate; (*2*) regulated gas inlet; (*3*) Langmuir probe; (*4*) multipolar magnetic wall; (*5*) diffusion pump port; (*6*) valve; (*7*) quadrupole mass filter; (*8*) electron multiplier; (*9*) cylindrical electrostatic lenses. (Perrin et al., 1984; *Int. J. Mass Spectrom. 57*, 249 (1984), by permission of Elsevier Science Publishers.)

corresponding to an activation energy of $E_a = 1.53 \pm 0.15$ eV. Based on this mass spectrometric diffusion data, these investigators conclude that device degradation due to the out-diffusion of hydrogen should not occur for 10^4 years at 100°C.

Recently, Fourier transform mass spectrometry (FTMS) has been employed to analyze high-purity silicon tetrafluoride (SiF_4) with respect to detrimental impurities (Reents et al., 1985). Identification was determined by making exact mass measurements on the product ions from chemical ionization, using SiF_4 itself as the reagent gas; and quantification was achieved by observing ion-molecule reactions without the use of reference compounds. This same technique has been used for monitoring impurities in the fluorine doping of optical waveguides (Chapter 11).

Some of the present-day optimism for amorphous silicon cells stems from

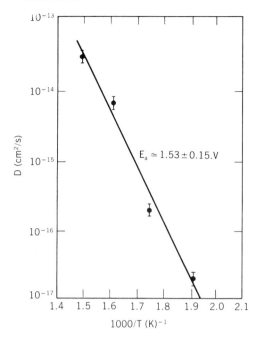

Figure 13.3 Diffusion coefficient of deuterium in amorphous hydrogenated silicon. (Carlson and Magee, 1978; *Appl. Phys. Lett., 33*, 81 (1978), by permission of American Institute of Physics.)

the concept of a *tandem configuration*, with two or more thin contiguous layers of material responding to different portions of the solar spectrum (DeMeo and Taylor, 1984). Whereas single-junction amorphous silicon cells have conversion efficiencies of only about 10%, tandem cells are expected to have ratings in the 16% range.

Ion implantation technology has also been used to fabricate titanium nitride films as an antireflection coating on silicon single crystals. Specifically, N_2^+ ions, at dopant levels of $3.4 \times 10^{17}/cm^2$, have been implanted in 600 Å thick titanium layers to yield a TiN layer of ~650 Å; this corresponds to a minimum in the reflectivity curve at 0.55 μm according to the "quarter wavelength" criterion ($nt = \lambda/4$), if the refractive index is $n = 2.3$ (Armigliato et al., 1982). These films are reported to exhibit better electric and optical properties than obtained by either evaporation or sputtering processes.

Electrochemical Cells

In addition to SIMS and other mass spectrometric assays of silicon cells, low-energy (helium) ion-scattering spectrometry (ISS) has been used for photoelectrochemical cell research. In such cells, light impinging upon a semiconductor generates electrons and holes that are transferred to an electrolyte that can react to produce hydrogen. In order to maximize this charge transfer to the solution and to increase cell efficiency, the recombination of electrons and holes at the photocathode surface must be suppressed. Recent work at Bell Laboratories (Heller, 1982; Lewerenz et al., 1982) has demonstrated that thin layers of rhodium or rhenium ($\sim$10 Å) can be used to produce "islands" of catalytic activity that cause reaction rates to proceed thousands of times greater than at the semiconductor-electrolyte interface, thereby affecting the electrochemical potential at the surface and enhancing cell efficiency. Surface characterization and depth profiles utilized a 1.5 keV helium probe ion beam focused to a 1 mm spot for both sputtering and analysis. The sputtering rate was estimated at 2 Å/min and provided detailed data on the major elements and oxygen in the near-surface region.

In other general areas of solar research, mass spectrometry has been utilized extensively. For example, the composition of electrodeposited black chrome-absorbing films has been analyzed by SIMS. Both positive and negative SIMS have provided detailed data on surface and bulk species, for example, O^-, OH^-, Cr^+, H^+, CrO^+, $Cr(OH)^+$, etc. (Smith et al., 1982b). A high intensity of OH^- ions in the SIMS spectrum suggests that the nonmetallic component of black chrome is predominantly a hydroxide rather than an oxide. Ion beams have also been employed to produce "microtextured" surfaces having very high absorption coefficients in the visible and near-infrared portions of the electromagnetic spectrum. This texturing is effected by sputtering with a low-energy (500 eV), broad-beam ion source (Ar^+, 1 mA/cm^2), while simultaneously adding impurities to "seed" the surface (Rossnagel and Robinson, 1982). Various micron-sized structures (cones, pyramids, ridges, etc.) are then formed by surface diffusion and clustering. Absorption coefficients for such microtextured structures on W, Cu, and Ni have been measured in the range of 0.93–0.97 for wavelengths from 0.4–2.2 μm.

FOSSIL POWER GENERATION

Historically, the first commercial mass spectrometers in the United States were built to assay fossil fuels. Initial interest focused on the determination of very small quantities of hydrocarbon impurities in soil gases, relative to oil prospecting. Subsequently, attention was directed to the analysis of gasoline in refining streams, and in 1943 the Consolidated Engineering Corporation in Pasadena sold its first commercial mass spectrometer to the Atlantic Refining Company. At present, mass spectrometry is not only the dominant instrumentation of the petrochemical industry, but both chemical and isotopic measurements have been extended to coals. The natural process of coal formation evolves in successive steps from plant debris to peat, lignite, sub-bituminous coals, and bituminous coals, to anthracitic coal. Chemically, the carbon content increases from approximately 40% in immature coals to 80% in high-grade anthracite, and mass spectrometry is utilized to identify their molecular structure. Very recently, $^{18}O:^{16}O$ isotope ratios have been used to study the formation and chemical origin of these coals (Dunbar and Wilson, 1983), so that mass spectrometry now encompasses measurements of fossil fuel precursors (cellulose, lignin, and plant resins) to the many byproducts of combustion and the assay of trace elements in fly ash.

Synthetic Gaseous Fuels

The reaction of carbon-containing solids with hydrogen can proceed according to the relation

$$C + 2H_2 = CH_4 \quad (\Delta H^\circ_{298} = -17.9 \text{ kcal·mol}^{-1})$$

and this exothermic process occurs in many gasifiers. However, the rate of CH_4 formation is low, so that a catalyst is required in commercial systems that are designed to produce hydrocarbons in large quantities. In one recent investigation, both Auger spectroscopy and mass spectrometry were employed to study the hydrogenation of graphite in the presence of nickel and tungsten, as these two transition metals have different d-electron concentrations, and the stabilities of the metal carbides are different (Holliday et al., 1975). The mass spectrometric measurements of the graphite-metal system were monitored with a quadrupole at hydrogen partial pressures ranging from 1×10^{-8} to 5×10^{-6} torr (Bliznakov et al., 1983). With pure graphite, only a small amount of CH_4 was detected at temperatures higher than 850 K. In the presence of Ni and W, CH_4 was formed at temperatures down to 750 K, and it increased with increasing surface metal concentrations. Mass spectrometric observations were also made of the amount of CH_4 liberated for fixed concentrations of Ni and W, but for increasing partial pressures of hydrogen. Results show a monotonically increasing amount of CH_4 formed with increasing H_2 partial pressure. This work is typical of many catalytic reactions that can be monitored by mass spectrometry for relating the basic electronic and adsorptive properties of solids to rates of methane formation.

Other recently reported applications of mass spectrometry relating to fossil fuels include both a totally computer-controlled triple quadrupole for structure elucidation (Wong et al., 1983) and a computerized capillary GC-MS quadrupole system for characterizing organic compounds in leachates from surface-retorted oil shales (Pereira and Rostad, 1983).

A considerable stimulus to the use of synthetic fuels was the successful completion of the first U.S. commercial-sized (100 MW-e) coal gasification plant at the Southern California Edison Company (1984). This plant produces a medium BTU gas consisting primarily of carbon monoxide and hydrogen. After further processing and conversion, this clean synfuel is fed to a combustion turbine that drives an electric generator. This development represents a milestone for clean-coal technology, and further basic studies in coal gasification provide some additional challenges for mass spectrometry.

Trace Elements in Coals and Petroleum

Most of the elemental concentrations in coals are lower than their average concentrations in the earth's crust, as shown in Table 13.1. Only four elements, on the average, are present in coals in concentrations significantly greater than their crustal abundance (Valković, 1983). Trace elements from coal have been studied extensively because of environmental concerns and levels in the food chain. In recent years, trace element monitoring has also been critically related to the production of semiconductor materials and devices.

Concentrations of many of these elements vary over an exceedingly wide range of values. For example, aluminum concentrations in Nebraska coals have been found to range from 0.4–13% (Burchett, 1977), and magnesium concentrations in U.S. coals are reported to be highest for lignites (0.31%), followed by sub-bituminous coal (0.18%) and bituminous coal (0.08%) (Swanson et al., 1976). Vanadium concentrations have also been reported as low as 0.03 ppm and as high as 51,000 ppm, and uranium concentrations in some samples of western coals have been reported to range as high as 50–100 ppm (Facer, 1979). The mean mercury level in some British coals has been reported to vary from 0.13 ppm to ~2 ppm, and as high as 3 ppm in other parts of the world. As a result, the release of mercury from coal is in the range of about 0.14–2.72×10^9 g/yr—depending upon coal concentrations and the amount of mercury trapped by precipitators before release from chimney stacks (Airey, 1982).

Trace elements are also of interest in crude petroleum, fuel, and lubricating oils (Hofstader et al., 1976). In addition to environmental concerns, some metallic elements such as Al, As, Fe, Ni, and V are major poisons that affect catalysts used in refining and processing. In one study by Ndiokwere (1983) of Nigerian crude petroleum, Al, As, Sb, and V were found to be in the range of 0.9–2.5, 0.18–0.6, 0.04–0.42, and 0.1–0.64 μg/g, respectively. The arsenic concentration, a serious poison for catalysts because of its affinity for Pt, was observed to be substantially higher than its concentration range of 0.005–0.077 μg/g in oils from Venezuela, Libya, Iraq, and Kuwait. Impurity deposits in gas turbines from fuels containing sodium and vanadium have also been the object of detailed studies relating to condensation chemistry and corrosion (Luthra and Spacil, 1982).

Combustion

"We possess a highly detailed knowledge of the products resulting from the fission of a single ^{235}U atom, but we have only a limited understanding of what happens when coal burns" (Dondes, 1983). This succinct appraisal admits to the much greater difficulty of detecting the chemical intermediates of the combustion process than the highly energetic fission products and radiations emitted from a fissioning nucleus. Combustion diagnostics not

TABLE 13.1   Average Concentrations of Selected Elements in Coals

Element	Concentration (%) U.S.	Worldwide
Sulfur	3.0	2.0
Silicon	2.6	2.8
Aluminum	1.4	1.0
Calcium	0.54	1.0
Magnesium	0.12	0.02
Potassium	0.18	0.01
Iron	1.6	1.0

Element (Trace)	Concentration (ppm) U.S.	Worldwide
Antimony	1.1	3.0
Arsenic	15	5.0
Barium	150	500
Beryllium	2.0	3
Boron	50	75
Cadmium	1.3	—
Chromium	15	10
Chlorine	207	1000
Cobalt	7	5
Copper	19	15
Germanium	0.71	5
Lead	16	25
Lithium	20	65
Manganese	100	50
Mercury	0.18	0.012
Molybdenum	3	5
Nickel	15	15
Rubidium	2.90	100
Selenium	4.1	3
Silver	0.20	0.50
Strontium	100	500
Titanium	800	500
Thorium	1.9	—
Tin	1.6	—
Tungsten	2.5	—
Uranium	1.6	1.0
Vanadium	20	25
Zinc	39	50

Source: U.S. National Committee for Geochemistry (1980).

only require the time-resolved sampling of reactant products at relatively high pressures and temperatures, but the sampling must be done in a non-perturbing mode. Thus, tunable lasers are well suited for providing high-resolution absorption measurements of flames. Specifically, in the 3–30 μm wavelength regime, tunable diode lasers are now available for the measurement of all infrared-active combustion species; for example, such lasers have been used to measure NO at 5.25 μm (Falcone et al., 1983). In addition, continuous tuning of laser radiations can be programmed over many individual absorption lines at repetition rates up to 10 kHz, so that absorption profiles can be obtained that are highly time-resolved.

A reasonable number of combustion-related studies have been focused on spark-ignited automobile engines, using various techniques; but most of these have been focused on measuring concentrations of exhaust hydrocarbon species as a function of engine-operating variables (Kaiser et al., 1983). Gas chromatography has also been used recently to assay CO, CO_2, H_2, Ar, O_2, CH_4, and N_2 in an investigation of the effect of moisture on pulverized coal and combustion emissions (Asay et al., 1983). However, only recently has time-resolved, molecular beam mass spectrometry been applied directly to detect transient combustion phenomena (Sloane and Ratcliffe, 1983). In this investigation, some trade-off was made in operating pressure and the combustion flame was somewhat perturbed, but time-resolved data (0–25 ms) was obtained for several species on a stoichiometric CH_4-O_2-Ar mixture in a combustion bomb. Figure 13.4 is a schematic diagram of the apparatus that included a combustion sampling system, vacuum pumps, a spark ignition discharge in a combustion bomb, an electron-bombardment axial-type source, a quadrupole mass spectrometer (QMS), and a channel electron multiplier. The combustion bomb is located within a stainless steel diffusion-pumped chamber, a typical initial bomb pressure is 32 kPa (240 torr), and the pressure in the chamber housing is ~7 × 10^{-6} kPa. Ignition of the mixture is triggered by discharging a capacitor charged to ~5 kV across a 0.2 cm gap between two electrodes. Subsequent to the capacitor discharge, a flame propagates across the bomb toward a quartz-sampling cone that has a 0.15 mm orifice in the top. The gases that pass through this sampling cone are then collimated into a molecular beam and enter the ionizing region of the QMS. According to the authors, the number of repeated ignitions to produce acceptable data varies from 10–20 for major components to about 500 for a radical such as OH. Partial pressure profiles for H_2, OH, CH_4, CO, and H_2CO were measured within a 0–25 ms transient, but no H or CH_3 species were observed. The authors suggest, however, that the H atom partial pressure should be comparable to the OH partial pressure, but that the low sensitivity of the spectrometer to light gases may have precluded the detection of this radical.

In another combustion study, the concentration of the OH radical in a stoichiometric methane-air flame at atmospheric pressure was measured by

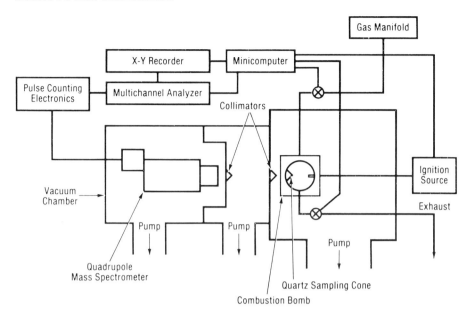

Figure 13.4 Schematic diagram of an apparatus for obtaining time-resolved molecular beam mass spectra of combustion phenomena. (Sloane and Ratcliffe, 1983; *Combustion Sci. & Technology, 33,* 65 (1983), courtesy of T. M. Sloane.)

both molecular beam mass spectrometry and laser absorption spectroscopy (Cattolica et al., 1982).

Other methane-oxygen flame studies by Goodings and Karellas (1984) have utilized a mass spectrometric technique that provides for the simultaneous and correlated determination of ion densities and flame temperatures. However, the in situ assay of reactive intermediates in flames and other combustion cycles remains generally unresolved due to the problems of sampling, and accounting for the complex phenomena of ionization cross-sections, charge exchange, fragmentation, and recombination.

Fly Ash

There are a number of on-going studies for utilizing the millions of tons of fly ash that are produced annually from coal-fired power plants. The bulk of fly ash is collected by wet scrubbers and electrostatic precipitators, but some 1–5% of the particles escape with the stack gases to the atmosphere. One feasibility study suggests that fly ash can be utilized to facilitate the reclamation of acidic soil. Another alternative is to recover the constituent metals of the ash, inasmuch as aluminum, iron, manganese, calcium, titanium, and magnesium are all contained in this residue. The soil reclamation concept is supported by Pennsylvania strip mine data, in which fly ash soil

TABLE 13.2   Representative Composition of Power Plant Ash

Constituent	%
SiO_2	45.7
Al_2O_3	26.0
Fe_3O_4	17.1
CaO	3.8
SO_3	2.6
K_2O	1.5
TiO_2	1.2
MgO	1.2
Na_2O	0.6
P_2O_5	0.3

Source: Morrison (1976).

cover resulted in beneficial changes in soil pH, grain size, and moisture. Similar beneficial results have been reported in Arkansas experimental areas when fly ash has been substituted for agricultural lime (Coombs, 1982).

The recovery of certain metals from ash is attractive on the basis of environmental, self-sufficiency, and strategic planning considerations (Calzonetti and Elmes, 1981). Projected annual ash from U.S. coal-burning power plants alone is in the 100 million ton range, so that the potential recovery of several of the metals listed in Table 13.2 is attractive. The United States now imports approximately 85% of its total alumina requirements, and ferric oxide has magnetic properties that make it useful in several industrial processes. Titanium and manganese are strategically important, and uranium, a trace element, has its obvious use in the energy cycle. Thus, even though the recovery of these metals may not be economically feasible at present, such a recovery program in the United States could eventually lead to a substantially reduced volume of utility waste and a decreased dependence upon overseas sources of aluminum and essential metals. It is also expected that mass spectrometry will continue to be utilized in the research phases of such a program and in monitoring product yields and process control.

Other mass spectral studies of fly ash have been stimulated, in part, by the increased disposal of municipal refuse by incineration. By burning municipal waste together with coal in utility power plants, it is possible to generate substantial amounts of usable energy as well as to dispose of the waste. However, the electrostatically precipitated fly ash formed during incineration contains a variety of polynuclear aromatic hydrocarbons (PAHs), polychlorinated dibenzo-p-dioxins (PCCDs), plus other organic and inorganic compounds. The distribution of these organic compounds, adsorbed on size-fractionated incinerator fly ash particles, has been reported by Clem-

TABLE 13.3 Concentrations (ng/g) of Phthalates, *n*-Hydrocarbons, and PAHs Extracted From Different Size Fly Ash Samples

	Average Particle Size (μm)					Light Ash (> 850)	Agglomerate Particles (> 850)	Totals (ng/g)
	30	80	125	200	550			
Diethyl phthalate	240	95	150	650	1200	3200	750	6300
Dioctyl phthalate	690	510	510	680	910	2800	330	6400
Biphenyl	43	35	33	160	240	700	55	1300
Fluorene	24	3.8	3.5	—	69	—	—	100
Anthracene[a]	16	4.0	2.9	2.9	—	—	—	26
Fluoranthene	76	51	68	97	120	150	7.9	570
Pyrene	44	16	13	18	9.1	—	—	100
Normal alkanes	14,000	7900	5200	5400	11,000	5500	900	50,000
Totals	15,000	8600	5900	7000	14,000	12,000	2000	65,000

Source: Clement and Karasek (1982).

[a] Concentrations based on fluoranthene response with relative response factor of unity.

ent and Karasek (1982), using GC-MS analysis. Results of their SIMS analyses of extracts of different-sized fractions are given in Table 13.3, which indicates that the smallest particles contain a greater concentration of the higher molecular weight PAHs and that larger particles have a greater concentration of low molecular weight PAHs. Such data are important in the broad range of studies that attempt to correlate concentrations of both organic compounds and trace elements for particulates of respirable size to those with diameters in excess of 800 μm.

Additional environmental studies have focused on the behavior of fly ash metals in aqueous systems and the potential use of such metals as isotopic tracers. Trace metals such as Cu, Pb, Zn, Cd, Cr, and Ni can be leached from coal ash by percolating water when large quantities of waste are buried in landfills. Subsequently, this leachate or contaminated water can enter local ground water systems. Furthermore, experimental measurements and mathematical models indicate that the mobilization of trace metals from fly ash is primarily a surface desorption phenomenon (Chapelle, 1980).

The use of strontium as an isotopic tracer to monitor wind-transported fly ash distribution is based on several factors. First, precision $^{87}Sr/^{86}Sr$ isotopic measurements are of interest, since fly ash is strongly enriched with strontium. Second, many large coal-fired power stations in the western United States are located in desert sites, and strontium appears to be a chemical tracer appropriate for monitoring desert soils and vegetation; specifically, the desert plant brittlebush (*Encelia farinosa*) is a Sr accumulator. Finally, recent data indicates that brittlebush isotopically equilibrates with desert soils having fly ash components as low as 0.25% by weight (Hurst and Davis, 1981). These conclusions have been reached, in part, from the assay of samples collected in the Mojave Desert in the vicinity of Southern California Edison's coal-fired facility.

Water Chemistry of Steam Turbines

The condensate/steam cycle transports thermal energy generated by fossil combustion (or nuclear fuels) to an electric generator via steam turbines.

TABLE 13.4 Turbine Steam Purity Recommendations

Sodium	< 5	ppb
Oxygen	< 10	ppb
Chlorides	< 5	ppb
Sulfates	< 5	ppb
Silica	< 10	ppb
Copper	< 2	ppb
Iron	< 20	ppb

Source: Jonas (1978).

TABLE 13.5 Purity of Typical Waters

Water Type	Impurity Level
Ocean	30,000 ppm
Lake	1000 ppm
Drinking	100 ppm
Turbine (Unacceptable)	50 ppb
Turbine (Acceptable)	< 20 ppb

Source: Hickam (1982).

Water chemistry is crucial in this system, in which millions of pounds per hour of steam/condensate are reprocessed with pressure and temperature differentials exceeding 2000 psi and 900°F, respectively. Failure to maintain control of impurities or essential chemical additives results in accelerated corrosion rates of blades, discs, shrouds, piping, bellows, and other turbine components. In fact, some impurity limits for steam/condensate are orders of magnitude lower than for drinking water. Recommended limits for steam purity of modern power plant turbines in normal operation are listed in Table 13.4.

Turbines are especially vulnerable to impurities because steam is traveling at near-supersonic velocities and is interacting with large masses rotating at 1800 rpm; under such conditions, stress corrosion cracking can result. In a large steam generator, where some 10 million pounds (4.5×10^6 kg) of feedwater circulates through the system each hour, a 10–20 ppb iron impurity translates to about 1 ton/yr of iron oxide depositing in the steam generator. Such an impurity not only promotes cracking, but it inhibits heat transfer and restricts flow in the generator tubes (Hopkinson, 1982).

Conductivity measurements are useful for measuring the total conductivity of all impurity ions in the steam or water environment. However, ion chromatography or mass spectrometric assays are required for identifying specific impurities at the parts per billion level (Hickam et al., 1981). As a point of comparison, Table 13.5 lists acceptable turbine purity vs. that of other typical waters.

FISSION AND NUCLEAR PHYSICS

It is difficult to conceive of atomic and nuclear physics or of contemporary nuclear engineering without mass spectrometry. It is true that many nuclear phenomena are observable by means of alpha, beta, and gamma ray spectroscopy, and by a variety of other techniques. However, the establishment of a precise mass scale, the positive identification of nuclides that result from the fission of the heavy elements, the measurement of nuclear cross-sections, etc. are fundamental to nuclear and reactor physics—and mass spectrometry has provided many of the important data relating to these and other measurements. Mass spectrometry also plays an important role in monitoring the many isotope-sensitive parameters of nuclear reactors utilized for electric power generation, for example, fuel inventory, nuclear burn-up, and environmental assays of transuranic nuclides.

The Mass-Energy Scale

The mass scale in nuclear physics is as important as the mass scale in organic chemistry. Mass measurements indicate the *binding energy* of the particles comprising the nucleus, the stability of a nuclide, and the energy or q-value of a nuclear reaction. The Einstein relation $E = c^2 \Delta M$ predicts the q value precisely. If the reaction is one of simple beta decay, the q value is the difference between parent and daughter masses. In the case of alpha decay, it is the mass difference between the original nuclide and the daughter atom plus a helium atom (^4He). For positron decay, in order that the q value include all the energy released in the transition, $2m_0c^2$ must be added to the beta end-point energy to account for annihilation radiation.

A measure of the stability of a given nucleus of mass M is conventionally expressed in two forms:

$$\text{packing fraction } f = \frac{(M - A)}{A},$$

$$\frac{\text{binding energy}}{\text{nucleon}} = \left(\frac{Z^1H + (A - Z)n - {}_Z^A M}{A} \right) c^2$$

where A, Z, ^1H, and n denote mass number, atomic number, hydrogen mass, and neutron mass, respectively. A small packing fraction reflects high stability. Likewise, a high binding energy per nucleon suggests a stable system. The overall stability pattern for the elements is shown in Figure 13.5; this figure also makes plausible the fact that fission energy can be obtained only from heavy nuclei and fusion energy from a few light nuclei. However, if the general binding energy vs. mass data is examined very carefully, there is evidence of "fine structure" in the plot of Figure 13.5 and a distinction exists between nuclides of *odd* or *even* mass numbers.

Figure 13.6 is a mass doublet typical of the precision that has been obtained by Professor Nier (1966) and coworkers at the University of Minnesota with a 16 in. double focusing spectrometer. Each of the compound nuclei has a mass slightly greater than that of a ^{238}U atom (238.05077). The masses of all naturally occurring nuclides are now known based on $^{12}C = 12.000000$, together with the kinetic energies of various radioactive species. Collectively, these data permit calculations to be made of the energetics of

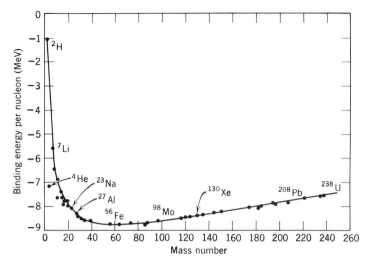

Figure 13.5 Binding energy of nuclei as a function of mass number.

nuclear fission, fusion, and other nuclear reactions. The mass-energy conversion constant is now given as:

$$1 \ \mu/1 \ eV = 931.5012 \times 10^6 \ (\pm 0.3 \ ppm)$$

as calculated from a recent evaluation of mass measurements and other constants (Cohen and Wapstra, 1983). Table 13.6 also lists atomic masses $M(\mu)$ and $M - A$ (keV) values of the neutron and several light nuclei.

It is interesting to note that mass spectrometry was also used to establish an absolute energy scale for high-energy alpha particles that are emitted spontaneously from radionuclides or that result from nuclear reactions. Relative kinetic energies of energetic particles can be measured by magnetic

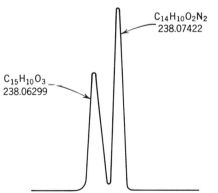

$C_{14}H_{10}O_2N_2$
238.07422

$C_{15}H_{10}O_3$
238.06299

Figure 13.6 Mass doublet as measured in the Minnesota 16-in. double focusing instrument (Nier, 1966).

TABLE 13.6 Atomic Mass and Energies of Several Light Nuclei

Nuclide	Mass (μ)	$M - A$ (keV)
n	1.008664907 (14)	8071.371 (13)[a]
^{1}H	1.007825032 (14)	7289.027 (10)
^{2}D	2.014101775 (22)	13,135.821 (21)
^{3}T	3.016049265 (34)	14,949.910 (32)
^{3}He	3.016020297 (34)	14,931.309 (32)
^{4}He	4.002603231 (49)	2424.913 (46)
^{13}C	13.003354822 (17)	3125.021 (16)

Source: Cohen and Wapstra (1983).

[a] Digits in parentheses represent the uncertainty (1σ) in the final digits of the quantity.

analysis, as evidenced in the high resolution alpha particle spectra of ^{210}Po and ^{244}Cm in Figure 13.7, or by gaseous or solid-state ionization detectors. However, the normalization of this energy scale is achieved by directly comparing positive ions and alpha particles in a common magnetic field trajectory, where both the mass and accelerating potential of the ion are known precisely (White et al., 1958a). Consider the schematic diagram in Figure 13.8 (see also Fig. 3.3), which indicates the 180°, 76 cm radius trajectory of an alpha particle, that is, a doubly charged ^{4}He ion. Assume also that a positive ion is accelerated through a precisely determined potential

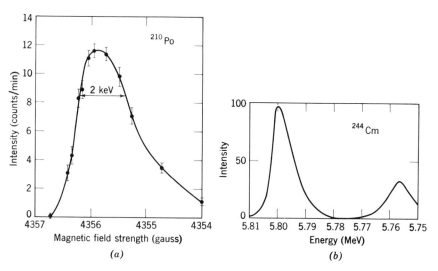

Figure 13.7 Alpha particle spectra of ^{210}Po and ^{244}Cm obtained with a 180°, 76 cm radius magnetic analyzer (White et al., 1958).

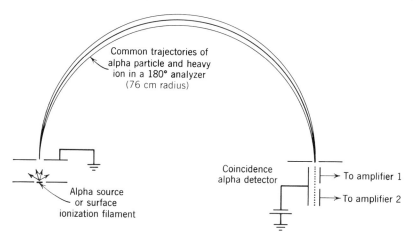

Figure 13.8 Schematic diagram of a mass spectrometric method for measuring the absolute kinetic energies of alpha particles (White et al., 1958a).

V that will cause the ion to have a path coincident with the alpha particles originating from the same source slit. If B is the magnetic field intensity, r is the radius of curvature, e is the electronic charge, v is the velocity, m is the mass, and subscripts 1 and 2 refer to the singly charged ion and alpha particle, respectively, then

$$Br = \frac{m_1 v_1}{e_1} = \frac{m_2 v_2}{e_2}$$

Applying the usual relativistic correction to the alpha particle of rest mass m_0, and kinetic energy E, it can also be shown that

$$\frac{m_1 v_1 e_2}{m_2 e_1} = c\left[1 - \frac{m_0^2 c^4}{(E + m_0 c)^2}\right]^{\frac{1}{2}}$$

Now, c is the velocity of light, $e_2 = 2e$, $m = m_0 + E/c^2$, and $v_1 = (2 e_1 V/ m_1)^{\frac{1}{2}}$. An expression can then be obtained for the kinetic energy of the alpha particle of the form

$$E = \frac{4 m_1 e_1 V}{m_0}\left(1 - \frac{E}{2 m_0 c^2} + \frac{E^2}{4 m_0^2 c^4} - \cdots\right)$$

Thus, the energy of an alpha particle can be determined without data on the magnetic field; only the mass of the ion and its accelerating potential need be known.

For the case of a ^{210}Po alpha particle energy determination, a ^{175}Lu ion

$(m_1 = 174.995)$ was accelerated to traverse a common trajectory to that of the alpha particle. The accelerating potential corresponding to this trajectory was measured by a precision potentiometer that was referenced to an NBS standard cell (1.01906 absolute V at 25°C). A calculated value of the alpha particle was found to be 5.3054 MeV. Thus, the normalization of an MeV energy scale was experimentally measured in terms of the known masses of a heavy ion and the alpha particle and a 1 V standard cell.

Somewhat analogous measurements have recently been reported by Lennard et al. (1983), in which ions of known mass and charge state were used with a time-of-flight (TOF) technique to calibrate a large analyzing magnet. Using energetic ions over a mass range from 1–208 amu, magnetic field intensities were calibrated over a range of 0.4–12.8 kG. In related measurements, this TOF technique is reported as being appropriate for determining the kinetic energies of heavy ions within a few keV at 1 MeV.

Neutron Cross Sections and Half-Lives

The first mass spectrometric observation of a change in isotopic composition caused by neutron irradiation was reported by Dempster (1947). After irradiation of a cadmium sample, he noted that the abundance of ^{113}Cd, normally 12.26%, had decreased to 1.6% whereas the abundance of ^{114}Cd, normally 28.86%, had increased to 39.5%. Dempster's experiment furnished conclusive proof that the isotope responsible for the high absorption of thermal neutrons in cadmium was ^{113}Cd, and subsequent work by his group revealed that other isotopes having exceedingly large absorption cross-sections included ^{149}Sm, ^{155}Gd, and ^{157}Gd. For many nuclides having a reasonably long half-life, neutron cross-sections can be obtained by simply comparing the known or experimentally measured isotopic ratios prior to irradiation to the post-irradiation ratios.

Unless isotopically enriched samples are available, all of the isotopes of a material will be undergoing nuclear transmutation. However, a three-nuclide transmutation expression will usually suffice in any specific experiment. Thus, consider an element with three isotopes designated as $^{(A-1)}X$, ^AX, and $^{(A+1)}X$, where $(A-1)$, A, and $(A+1)$ are the mass numbers. Let σ_a, σ_b, and σ_c represent the respective effective neutron absorption cross-sections of these nuclides. Under neutron irradiation, a nuclear transmutation will take place,

$$^{(A-1)}X \xrightarrow{\sigma_a} {^AX} \xrightarrow{\sigma_b} {^{(A+1)}X} \xrightarrow{\sigma_c} {^{(A+2)}X},$$

where $^{(A+2)}X$ is the daughter product of $^{(A+1)}X$.

If we let N_a, N_b, and N_c represent the respective number of atoms of isotopes $(A-1)$, A, and $(A+1)$ that exist at any time during the neutron irradiation, then the nuclear transmutations of these isotopes can be represented by the set of differential equations

$$\frac{dN_a}{dt} = -N_a\sigma_a\phi,$$

$$\frac{dN_b}{dt} = N_a\sigma_a\phi - N_b\sigma_b\phi,$$

$$\frac{dN_c}{dt} = N_b\sigma_b\phi - N_c\sigma_c\phi,$$

where ϕ is the neutron flux.

The initial conditions, at $t = 0$, are

$$N_a = N_{ao}; \quad N_b = N_{bo}, \quad \text{and} \quad N_c = N_{co}.$$

The solutions to these equations are then

$$N_a = N_{ao}e^{-\sigma_a\phi t},$$

$$N_b = \frac{\sigma_a}{\sigma_b - \sigma_a} N_{ao}(e^{-\sigma_a\phi t} - e^{-\sigma_b\phi t}) + N_{bo}e^{-\sigma_b\phi t},$$

$$N_c = N_{ao}\sigma_a\sigma_b\left[\frac{e^{-\sigma_a\phi t}}{(\sigma_b - \sigma_a)(\sigma_c - \sigma_a)} + \frac{e^{-\sigma_b\phi t}}{(\sigma_a - \sigma_b)(\sigma_c - \sigma_b)}\right.$$
$$\left. + \frac{e^{-\sigma_c\phi t}}{(\sigma_b - \sigma_c)(\sigma_a - \sigma_c)}\right] + N_{bo}\frac{\sigma_b}{(\sigma_c - \sigma_b)}[e^{-\sigma_b\phi t} - e^{-\sigma_c\phi t}] + N_{co}e^{-\sigma_c\phi t}.$$

The above are the general nuclear transmutation equations for three isotopes. In mass spectrometry, however, the usual measurable quantity is an isotopic ratio, rather than any absolute determination of the number of atoms of a single nuclide. It is thus convenient to define

$$R \equiv \frac{N_c}{N_b}; \quad R_o \equiv \frac{N_{co}}{N_{bo}}; \quad R'_o \equiv \frac{N_{ao}}{N_{bo}}.$$

It will be noted that R_o and R'_o are the known or experimentally measured isotopic ratios prior to neutron irradiation, and R is evaluated in terms of preirradiation isotopic ratios, effective neutron cross section, and neutron "fluence" or time-integrated neutron flux, ϕt, where t is the total irradiation time. If the time-integrated neutron flux and any two effective neutron cross sections are known, the third cross section can be determined.

There is one case that permits the use of a simple explicit expression for a neutron cross section (White, 1968). It is for the condition that

$$\sigma_a \ll \sigma_b, \quad \sigma_c \ll \sigma_b, \quad \text{and} \quad R_o' \simeq 0,$$

which, in turn, leads to the simple relationship that is sometimes referred to as the "burn-up equation":

$$\sigma_b = \frac{1}{\phi t} \ln \left(\frac{1 + R}{1 + R_o} \right) .$$

This expression has been used in the mass spectrometric measurement of neutron capture cross sections of ^{149}Sm and ^{150}Sm by Aitken and Cornish (1961).

For the case in which only a single isotope is present in the irradiated sample, and $\sigma_b \gg \sigma_c$,

$$\sigma_b - \frac{1}{\phi t} \ln (1 + R).$$

Forman and White (1967) used a four-stage mass spectrometer to produce an exceedingly high degree of enrichment for ^{147}Sm and also to measure its

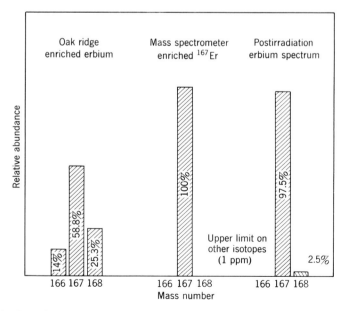

Figure 13.9 Isotopic changes in erbium produced by neutron irradiation (Su and White, 1967).

cross-section; a similar technique was used by Su and White (1967) to measure the high thermal neutron cross-section of ^{167}Er. In this latter investigation, a rhenium ribbon was placed in the position normally occupied by the electron multiplier detector, and $\sim 10^{-8}$g of ^{167}Er was collected by "ion implantation" and subsequently exposed to a neutron flux. These experiments demonstrated the potential of ion implantation in nuclear research; the four-stage mass spectrometer permitted a postirradiation assay to be made of this very small sample, as depicted in Figure 13.9—from which a thermal neutron cross-section was calculated. In connection with this particular mass spectrometric technique, it might be noted that (1) high isotopic purity affords greater sensitivity and precision, (2) "self-shielding" corrections are negligible when small samples are used, and (3) no sample chemistry is required that might introduce impurities.

Half-lives are generally measured by observing the exponential decay of a sample, using radiotracer methods. However, mass spectrometry has also been used for half-life determinations by either (1) detecting the decay of the parent nuclide or (2) monitoring the growth of a daughter product. The latter is usually a more sensitive method for relatively long-lived species. As an example, the half-life of ^{137}Cs has been measured by observing the rate of growth of stable ^{137}Ba from a known quantity of ^{137}Cs. Using the isotopic dilution method, a value of 29.2 $\pm$ 0.3 years was determined by Rider et al. (1963). Figure 13.10 shows their experimental measurements from both single-stage and two-stage magnetic spectrometers. Precision mass spectrometry has also been employed recently to remeasure the half-lives of certain nuclides of interest in cosmochemical studies. For example, the

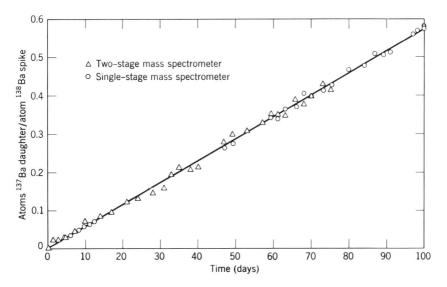

Figure 13.10   Growth of stable ^{137}Ba from the decay of ^{137}Cs (Rider et al., 1963).

half-life of ^{202}Pb has been redetermined and found to be $5.25 \pm 0.28 \times 10^4$ years (Nagai et al., 1981), which is six times shorter than a previously reported value. The ^{202}Pb was produced by proton bombardment of natural thallium, and the number of ^{202}Pb atoms was determined by isotope dilution mass spectrometry. The absolute activity of ^{202}Pb was then determined by gamma ray counting of the daughter nuclide ^{202}Tl, using a Ge(Li) detector.

Even very short half-lives (< 1 s) have been measured by mass spectrometry. The technique involves an "on-line" mass spectrometer in which (1) new nuclides are produced by energetic charged particles, (2) the reaction products are promptly ionized, and (3) the short-lived daughter products are accelerated, mass analyzed, and their decay noted after they impinge on a mass spectrometer electron multiplier. For example, Klapish and Bernas (1965) used a high-energy beam of charged particles (150 MeV protons from the Orsay synchrocyclotron) directed on a "target" that consisted of a series of thin carbon foils. Lithium isotopes were produced by the reaction

$$p + {}^{12}C \rightarrow {}^{6}Li, {}^{7}Li, {}^{8}Li, {}^{9}Li$$

The target was maintained at approximately 1600°C so that the lithium atoms diffused to the surface, and a reasonable fraction left this target as ions. Using this technique, the decay of ^{8}Li to ^{8}Be was detected on the first dynode of an electron multiplier by a 2-α particle decay, and half-lives of 0.8 s for ^{8}Li and 0.17 s for ^{9}Li were reported. This general on-line approach is very useful for half-life and nuclear reaction measurements if the daughter products can be ionized almost instantaneously after their formation.

With modern accelerator-mass spectrometer systems (Chapter 20), even shorter lifetimes and a greater range of reaction products can be observed.

Fission Yields

Although the relative abundances of most nuclides produced by neutron-induced fission are now well known, interest continues in the systematics of *odd-even* charge distributions and measurements of very short-lived fission products. Specifically, isotope separators operating on-line with research reactors that produce high thermal neutron fluxes on fissile targets make possible (1) isotopic separations in the microsecond regime and (2) measurements of very low-fission yields.

Fission theory predicts that the charge distribution in an isobaric decay chain is a Gaussian function modified by a fine structure attributed to both an odd-even proton effect and a smaller odd-even neutron perturbation (Amiel and Feldstein, 1975). The fractional-independent yield (FIY) is characterized by this Gaussian having a width parameter c, the charge of the nuclide Z, the most probable charge for a given mass chain Z_p, and the odd-even factor, f_{oe}, that is,

$$\text{FIY} = f_{oe} \frac{1}{\sqrt{\pi c}} \exp\left[-\frac{(Z - Z_p)^2}{c} \right]$$

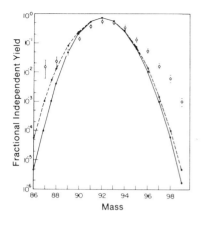

Figure 13.11 Isotopic fission yield distribution for rubidium. Solid and broken lines indicate theoretical values; Φ = experimentally determined relative abundances. (Schmid et al., 1981; *J. Inorganic Nucl. Chem. 43*, 867 (1981), by permission.)

In recent studies to verify this theory, independent fission yields of ^{235}U have been made for $^{92-99}$Rb, $^{94-100}$Sr, $^{142-148}$Cs, and $^{143-149}$Ba. The isotopic yield distribution for Rb is shown in Figure 13.11, together with a theoretical curve (dotted line) that accounts for prompt neutron emission (Shmid et al., 1981). In acquiring this data, a ^{235}U target enriched to 93% was exposed to a thermal neutron flux of $\sim 5 \times 10^8$ $n/cm^2/s$; and a surface-ionization source operating at 2000 K was used to ionize and selectively discriminate against contaminants and interfering isobars. Isotopes collected at appropriate mass positions by the on-line isotope separator were then beta-counted with silicon surface barrier detectors.

Assay of Nuclear Fuel

Uranium burn-up in a thermal reactor core is determined by changes in the isotopic abundances of the nuclear fuel. As a reactor core undergoes a continued high-flux operation, the number of available fissionable atoms is depleted. If the burn-up is large ($\sim 10\%$), it is convenient to assume that the capture of thermal neutrons by ^{238}U is negligible compared with the number of ^{235}U atoms that undergo fission or absorb a neutron to form ^{236}U. This assumption is valid as the thermal capture cross-section of ^{238}U is about 3 barns, but the ^{235}U fission and capture cross-sections are about 580 and 100 barns, respectively. An illustration of a typical burn-up calculation has been given by Dietz (1964), using the mass spectrometric measurement of four uranium isotopes. Consider the preirradiated and burn-up sample described in Table 13.7. Because the ^{238}U is assumed to remain constant, each of the isotopes in the burn-up sample may be multiplied by (5.47/13.30) = 0.4113. The normalized percent abundances can then be determined, that is, 41.13 is the total number of uranium atoms left per 100 uranium atoms prior to the irradiation. The number of ^{235}U atoms that fissioned (percent of fission burn-up) is (100–41.13) = 58.87; the number of ^{235}U atoms that capture

TABLE 13.7   Typical Isotopic Abundances in ^{235}U Burn-up

Preirradiated Sample		Burn-up Sample		Normalized Burn-up Sample	
Isotope	Atom (%)	Isotope	Atom (%)	Isotope	Atom (%)
^{234}U	1.16	^{234}U	1.96	^{234}U	0.81
^{235}U	92.84	^{235}U	53.04	^{235}U	21.81
^{236}U	0.53	^{236}U	31.70	^{236}U	13.04
^{238}U	5.47	^{238}U	13.30	^{238}U	5.47
Total	100.00		100.00		41.13

Source: Dietz (1964).

neutrons to form ^{236}U (percent of capture burn-up) is (13.04–0.53) = 12.51 atoms per 100 original uranium atoms.

For very low burn-up or accurate capture-to-fission measurements, high-abundance mass spectrometry is required. The spectrum of uranium in Figure 13.12, taken with a three-stage mass spectrometer (White et al., 1958b) shows all isotopes completely resolved and with virtually no contributions from adjacent mass numbers. This type of mass spectrometric assay provides a measurement of high or low burn-up in fuel; it will also be noted that the method takes into account a significant amount of neutron capture by ^{234}U to form new ^{235}U atoms.

The simultaneous measurement of all the major actinides in the fuel cycle

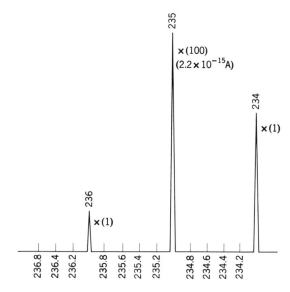

Figure 13.12   Uranium spectrum of nuclear fuel after low-level "burn-up" (White et al., 1958b).

(Th, U, and Pu) is now possible, and this technique provides an enhanced precision in the overall assay of nuclear fuels (Koch and Steinert, 1980). Interference occurs only with the isobaric nuclides ^{238}Pu and ^{238}U. However, Pu and U are emitted at slightly different temperatures in a surface-ionization source, and ^{238}Pu can be determined independently by alpha spectrometry. As for ^{236}Pu and ^{232}U, these latter nuclides occur in such low abundances that they have little effect on the accuracy of the ^{232}Th and ^{236}U analyses.

Substantial efforts have been devoted to special ion source techniques for determining the isotopic abundances of the transuranics—for nuclear fuel burn-up, spent fuel assays, fuel inventory, nuclear safeguard studies, and environmental measurements. Surface ionization of samples is the preferred method, because nanogram amounts of sample are often adequate. The so-called "resin bead" technique has become useful for loading samples, and results can be compared with NBS standards (Smith et al., 1982a). Carburized rhenium ribbons have also been widely used for plutonium; and for particulate samples of PuO_2, as many as 5% of the Pu atoms in a particle have been detected as mass-analyzed ions (Gordon, 1983–1984). An interlaboratory comparison of isotopic measurements for U and Pu (nanogram amounts) indicates that major isotopes can be determined to about 0.3%, using resin beads and thermal ionization mass spectrometry (Fassett and Kelly, 1984). A triple-filament thermal ionization source has also been used for Th and U assays of thorium-uranium dioxide fuels (Green et al., 1983).

Nuclear Reactor Materials

Virtually all the materials used in nuclear reactors must meet stringent criteria in terms of both their nuclear properties and their engineering suitability. Major categories of reactor materials are:

1. Nuclear fuels.
2. Burnable "poisons."
3. Moderators and control rods.
4. Coolants.
5. Structural materials.

The assay of the fissionable nuclides is of primary concern, as indicated in the previous section; and the simultaneous measurement of several elements is now being explored via surface-ionization mass spectrometry. The rather small spread in the ionization energies, listed in Table 13.8, indicates why a multielement analysis is feasible from a single sample, provided suitable isotopic spikes and normalizations are employed. The lower, experimentally observed ionization potential of Pu compared to U also confirms results of many other investigators, namely, Pu is easier to ionize than U by surface-ionization techniques.

TABLE 13.8 Ionization Potentials of Uranium and Transuranic Elements

Element	Experimental	Calculated
Uranium	6.09	6.05
Plutonium	6.03	—
Americium	5.96	5.99
Curium	6.16	6.09
Californium	6.20	6.41

Source: Chetverikov et al. (1981).

Clearly, all reactor core loadings must be analyzed for their isotopic composition. The addition of so-called "burnable poisons" also involves mass spectrometry because such elements can readily alter reactor reactivity and permit extended reactor life. The net effect of employing a "burnable poison" is to compensate for loss of reactivity caused by fuel depletion. The rare earth elements are especially effective as reactor "poisons." They have a multiplicity of isotopes, and these elements can easily be analyzed by surface-ionization mass spectrometry. Samarium has a high thermal cross-section, and it is relatively inexpensive and abundant. Gadolinium and europium also provide considerable latitude in reactor control. Europium has two naturally occurring isotopes, ^{151}Eu and ^{153}Eu; the light isotope has a very large resonance at about 0.5 eV. Dysprosium has a continuous chain of high cross-section isotopes, and it has the advantage of short-lived radioactive daughters.

The importance of isotopic control via "burnable poisons" is illustrated in Figure 13.13, where the europium isotopic abundance is plotted as a function of total neutron absorptions per initial europium atom (Stevens, 1958). The dramatic change in isotopic ratios and the "burn-out" of ^{151}Eu illustrate the importance of isotopic composition in reactor control materials and the general dependence upon mass spectrometry in reactor design and operational control.

The measurement of ^{10}B/^{11}B ratios is of even more general interest, as it is necessary to know accurately the "boron inventory" of a reactor for dynamically controlling neutron flux. Aluminum oxide/boron carbide composite pellets are used as neutron shims (control rods) and contain 1.5–4% B that may or may not be enriched in ^{10}B—the high-neutron, cross-section isotope. During reactor operations, the ^{10}B/^{11}B ratio will change as ^{10}B is depleted via an n, α reaction to lithium (^{10}B + $^{1}n \rightarrow$ ^{4}He + ^{7}Li). By using SIMS for this type of analysis, isotopic data is obtained for both boron and lithium with a minimum of chemistry. The precision of SIMS isotopic analysis for natural boron samples is about 0.5%, and for natural lithium, about 1%. Other recent SIMS measurements of borosilicate glasses for ^{10}B/^{11}B ratios have also been reported as they relate to the storage of uranium (Chris-

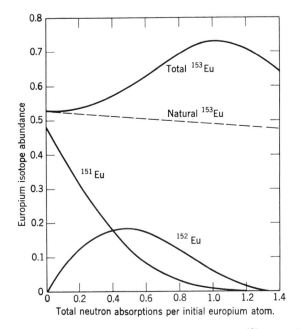

Figure 13.13 Isotopic changes induced by neutron absorption in [151]Eu and [153]Eu when this element was used as a "burnable poison" in controlling reactor reactivity (Stevens, 1958).

tie et al., 1981). In certain instances, solutions of this fissionable element are in vessels that are monitored via borosilicate glass rings. These rings are sampled periodically and analyzed for [10]B content to verify that no significant neutron-producing event has occurred in the storage tanks due to an unsafe storage configuration from the standpoint of nuclear criticality.

When [10]B/[11]B ratios are measured by surface ionization, a conventional technique has been to analyze $Na_2BO_2^+$ ions from borax and to observe peaks at masses 88 and 89. An alternative method, recently reported by Duchateau and DeBièvre (1983), is to use the negative BO_2^- ion. The authors claim both an ion emission that is less affected by impurities in the sample and an improvement in sensitivity by a factor of 50.

With respect to reactor coolants, in both light- and heavy-water (Canadian-deuterium/uranium-CANDU) reactors, the water chemistry must be carefully controlled if corrosion and other adverse effects are to be avoided in the piping, the steam generator, and all components of the heat transfer circuit. A deterioration in water quality can also lead to the potential failure of the fuel element cladding, the release of fission products, and the accumulation of deposits on surfaces, with a subsequent development of hot spots. In fast reactors that use liquid sodium as a coolant, mass spectrometry has been employed to quantify virtually all trace elements. Specifically, Hickam and Berkey (1972) not only utilized spark source spectrometry to assay reactor-grade sodium at the parts per million level, they also demonstrated an on-line technique for monitoring this low melting point and chemically reactive material.

Structural materials in nuclear reactors must meet specifications imposed by the unique environments to which they are exposed. The use of substandard materials not only leads to economic penalties imposed by forced shutdowns, but also to component failures that can lead to the initiation of accidents. Thus, design and operational considerations must be given to irradiation-enhanced corrosion, void swelling, irradiation creep, embrittlement, phase stability, and atomic dislocations of interstitials induced by nuclear collisions. In some instances, collisions between neutrons and the nuclei in structural alloys can give rise to the formation of microbubbles, which can then act as nucleation sites for voids. For example, nickel is a major component of austenitic stainless steels, and ^{58}Ni can produce helium via an (n, α) reaction. Hence, structural materials as well as reactor core materials must be evaluated for their isotopic as well as chemical, mechanical, thermal, and metallurgical properties. Some trace constituents, such as niobium, that have been added to steels to inhibit cracking are also of future concern, as ^{94}Nb has a half-life of 20,300 years and emits energetic gamma rays. A 1980 report by Battelle Pacific Northwest Laboratories even suggests that the decay of ^{94}Nb could dominate the radiation dose rate of irradiated stainless steel for 70 years after reactor shutdown.

Surveillance of Radioactive Wastes

The generation of high-level nuclear wastes in the United States originated with the start-up of the Clinton and Hanford reactors in 1943–1944. In 1957, civilian nuclear wastes began to appear. At present, the radwaste scenario must address the disposal of low-level biomedical radioactive wastes, industrial and research wastes, the spent fuel of nuclear power plants, and the substantial byproducts of military and defense programs.

It is generally agreed that any radwaste burial system must deal with public acceptance as well as technical feasibility, and that continuous monitoring must be maintained in order to insure both containment and isolation (Hammond, 1979). Fission products such as ^{90}Sr or ^{137}Cs, with half-lives of about 30 years, pose no major problem in reactor high-level wastes after a few centuries. These and other longer-lived fission products will, nevertheless, dominate the radioactivity and heat production during the so-called "thermal period" of the first 1000 years after a repository is sealed. However, the alpha-emitting transuranic actinides such as ^{237}Np, ^{239}Pu, ^{241}Am, and ^{246}Cm have half-lives ranging from 500 years to 2 million years.

The escape of radionuclides to the biosphere can occur through (1) exposure of the geological formation that contains the radwaste to the surface; or (2) the transport of the waste by ground water to a river, lake, or other mode of entry into the biosphere. Conceivably, human activities (surface

TABLE 13.9   Typical Composition of SYNROC and Its Constituent Minerals (wt%)

	"Hollandite"	Zirconolite	Perovskite	Bulk SYNROC
TiO_2	71.0	50.3	57.8	60.3
ZrO_2	0.2	30.5	0.2	10.8
Al_2O_3	12.9	2.5	1.2	6.3
CaO	0.4	16.8	40.6	16.2
BaO	16.0	—	—	6.4
Total	100.5	100.1	99.8	100.1

Source: Ringwood (1982).

explosions, etc.) or geological activity (crustal uplift, earthquakes, etc.) could lead to the exposure of buried waste; hence, most burial proposals call for a multiple-barrier approach that makes use of both natural and man-made barriers. The transport of radionuclides through ground water, host formations, etc. is considered to be a somewhat more complex problem that is dependent upon the sorptive properties of the media, its porosity and permeability, and other geohydrological and geochemical parameters (Skinner and Walker, 1982).

Thus, for at least two decades, the nuclear industry has been exploring borosilicate glasses for immobilizing high-level wastes, so that radwaste in a solid matrix might withstand leaching by ground water or dispersal in various burial environments. Research has included both ion bombardment to evaluate radiation damage effects on leaching rates (Primak, 1982) and SIMS studies of borosilicate glasses (McIntyre et al., 1980). In this latter work, quantitative measurements were made of minor glass constituents, the transport of fission product ions, and depth profiles of elemental distributions of fractured specimens exposed to aqueous leaching. More recently, an Australian group has proposed an even more stable material for immobilizing radioactive wastes, that is, a titanate ceramic called SYNROC (synthetic rock) consisting of the three minerals listed in Table 13.9. These minerals are similar to those in nature that have demonstrated their capacity to lock up radioactive nuclides (^{238}U, ^{232}Th, ^{40}K, and ^{87}Rb) for millions of years. Specifically, a mass spectrometric study has been made of naturally occurring zirconolites and perovskites for uranium and lead, using the isotopic dilution method. The study revealed that zirconolites could immobilize uranium, thorium, and their decay products even under extreme conditions of radiative damage ($>10^{19}$ α/g) (Oversby, 1981). Other general attributes of SYNROC include:

1. Ability to accept 20% by weight of fission products with minimal phase change.
2. Stability at high temperatures and pressures.

3. Leach rates for many elements that are lower (by orders of magnitude) than for borosilicate glasses.

Overall, this titanate-based, polyphase, crystalline ceramic is rated as the leading candidate for immobilizing radwaste prior to emplacement in a mined cavern at a Federal repository (Bernadzikowski et al., 1983).

Geological formations that have been evaluated as potential burial sites include salt domes and deep-granite cores. Among the relevant parameters pertinent to these sites are permeability and fluid transport properties, since the water transport of dissolved oxides of radioisotopes represents one of the most probable escape paths (Trimmer, 1982). In the salt domes bordering the northern coast of the Gulf of Mexico, geochemical measurements have focused on the origin of water that is external to the domes or incorporated in the salt. Using mass spectrometry to determine the $^{18}O:^{16}O$ and D:H ratios in exploratory holes and mines, it has been possible to distinguish between air condensate, sea and mine waters, and surface waters of meteoric origin; and at least one study of ^{18}O enrichments suggests that certain mine waters can be identified with formation water that has been trapped within the salt for 10–13 million years (Knauth et al., 1980).

The migration of radionuclides in granite and bedrock has been studied extensively, because even dense rocks such as granite can be porous and contain microfissures. Hence, the radionuclides can migrate through porous rock and fissures by diffusion. Furthermore, for an actual disposal site to be validated, diffusion measurements must be made not only for the transport times and distances that can be observed in the laboratory, but diffusion data on the migration of nuclides must be supported by geological evidence. A comprehensive assessment of this problem has been reported by the Earth Science Division, Lawrence Berkeley Laboratory (Neretnieks, 1980), which takes into account both the concentration and half-lives of radionuclides in spent fuel and their migration distance in granitic rock. The distance calculations in Table 13.10 correspond to values where the concentration c of the radionuclide has been attenuated by a factor of 10^{-9}. The criterion in the analysis is that if the nuclide at no time reaches the "biosphere" with a concentration 10^{-9} times that in the repository, it has decayed to insignificance. From Table 13.10, the only four nuclides that do not decay "totally" within 350 m are ^{135}Cs, ^{129}I, ^{99}Tc, and ^{238}U.

Recent geophysical studies also suggest that deep-granite burial should provide a corrosion-resistant containment for long-term waste encapsulation. For example, mass spectrometric studies have indicated that lead has been retained in zircons extracted from deep (960–4310 m) granite cores where the ambient temperature increases from 105 to 313°C (Gentry et al., 1982). The retention of lead derived from U and Th was determined from lead isotopic ratios and helium concentrations resulting from alpha decay. Even from zircon samples extracted at the greatest depths (and highest temperatures), the Pb was relatively immobile. Such measurements that relate

TABLE 13.10   Radionuclides in Spent Nuclear Fuel and Their Potential Migration in Granitic Rock

Nuclide	Initial Amount per ton U (Ci/ton)	Half-life (yr)	Distance in Rock (m) Beyond Which $c < C_0 \times 10^{-9}$
^{137}Cs	1.1×10^5	30	4.1
^{90}Sr	7.6×10^4	28	7.8
^{241}Am	7.8×10^2	458	0.7
^{243}Am	2.1×10^1	7370	2.9
^{239}Pu	3.2×10^2	24,400	53
^{240}Pu	4.9×10^2	6580	28
^{241}Pu	1.1×10^5	13.2	1.3
^{242}Pu	1.4×10^0	379,000	210
^{244}Cm	2.0×10^3	17.6	0.2
^{99}Tc	1.4×10^1	210,000	380
^{237}Np	3.3×10^{-1}	2.1×10^6	250
^{135}Cs	2.5×10^1	3.0×10^6	1300
^{129}I	3.8×10^3	17×10^6	565,000
^{226}Ra	1.1	1600	10.6
^{229}Th	8.5	7300	10.3
^{238}U	—	4.5×10^9	36,200

Source: Neretnieks (1980).

to actual long-term geophysical phenomena reinforce the view that deep-granite storage may effectively immobilize radwaste and withstand its corrosion or dissolution from invasive aqueous solutions.

Finally, in any radwaste surveillance scenario, mass spectrometry will be called upon to be the "expert witness"—as the most sensitive technique for monitoring the potential transport of all long-lived actinides and fission products, and for differentiating between naturally occurring radioisotopes and all other sources of radwaste.

FUSION RESEARCH AND ENGINEERING

Fusion of the light elements is responsible for the 3×10^{26} W of power continually being radiated by the sun. Furthermore, almost all terrestrial energy sources (solar, fossil, biomass, hydro, etc.) are derived from this nuclear power plant that has an internal temperature of some $15\text{–}20 \times 10^6$ K. In order to build a terrestrial power plant operating at a comparable temperature, however, many engineering and technological problems must be resolved. A magnetic confinement system must address problems of first-wall sputtering, heat transfer, radiation damage, the retention of radioactive gases, and the remote assembly and disassembly of complex components.

Laser-induced fusion must deal with the economic production of millions of fuel pellets and the short lifetimes of high-power lasers. An appraisal of fusion by Edward Teller in a lecture at Stanford University in 1974 is still valid: "it won't help us with this energy crisis or the next one; maybe it will help us with the one after that." Nevertheless, the oceans contain enough deuterium to provide power generation for millions of years; and tritium, the other basic fuel in fusion reactors, is bred in the reactor and recycled. There is even some optimism in the plasma physics community based on progress during the last two decades (Furth, 1985):

1. Confinement times in the tokamak have increased from a few milliseconds to 0.8 s.
2. The product of plasma density and confinement time has increased from 10^{10} cm^{-3}·s to 8×10^{13} cm^{-3}·s.
3. Vastly increased ion temperatures have been attained, from a few hundred eV to 7 keV ($\sim 7 \times 10^7$ K).

The Lawson Criterion and Plasma Temperatures

The generation of energy by controlled thermonuclear fusion requires that a high-temperature plasma of fusionable nuclei be confined sufficiently long enough so that fusion reactions can occur. In order to produce energy substantially greater than that delivered to the plasma, the ion density n and the time of confinement τ must exceed critical values. For a deuterium-tritium plasma at a temperature of close to 10^8 K, the product $n\tau$ must approximate 10^{14}. The relationship $n\tau \cong 10^{14}$, where n is the number of nuclei/cm^3 and τ is in seconds, is known as Lawson's criterion. In laser fusion, the density is high and τ is very short; in magnetic confinement, a lower density demands much longer confinement times. Also, in laser or other inertial confinement schemes that utilize particle beams, the objective is to uniformly focus energy upon a small spherical pellet so as to compress it by the ablation of surface material and the resultant spherical implosion. This heating and compression can yield high densities, and it is convenient to express the Lawson criterion by the expression $\rho r = 0.3$ g/cm^2, where ρ is the mass density of the compressed core of the pellet and r is the radius. This alternative product relationship results from the substitutions: (1) $n \to \rho/\overline{A} \times 6 \times 10^{23}$, (2) $\tau \to 10^{-8} r$, and (3) $10^{14} \to 6 \times 10^{14}$. The first substitution expresses number density in terms of mass density, Avagadro's number, and the average atomic mass $\overline{A}$, which is 2.5 for a 50/50 deuterium-tritium plasma. The second substitution approximates the "disassembly" time of the core by equating it to the core radius divided by the approximate velocity of sound, 10^8 cm/s at 10^8 K. The final relation is derived from an estimate in which only about one-sixth of the laser or particle beam energy is transferred to the core; the remainder is dissipated in the blow-off plasma that provides the compression (Mallozzi and Epstein, 1976).

The Saha-Langmuir equation is important in surface-ionization mass spectrometry; it is also important in relating the amount of ionization to plasma temperatures. For a gas in thermal equilibrium, this relationship, expressed in the mks system, has the form

$$\frac{n_i}{n_n} \approx 2.4 \times 10^{21} \frac{T^{3/2}}{n_i} e^{-U_i/kT}$$

where n_i and n_n are, respectively, the density (number/m^3) of ionized atoms and of neutral atoms, T is the gas temperature in K, k is Boltzmann's constant, and U_i is the ionization potential of the gas (Chen, 1984). Thus, the amount of ionization will remain small until U is only a few times the product kT. The range of kT values in plasmas is enormous, that is, from 0.1 to 10^6 eV, and it is estimated that 99% of the universe exists in the plasma state.

Plasmas in controlled thermonuclear fusion involve deuterium and tritium atoms in the following principal reactions:

$$D + D \rightarrow {}^3He + n + 3.2 \text{ MeV}$$

$$D + D \rightarrow T + p + 4.0 \text{ MeV}$$

$$D + T \rightarrow {}^4He + n + 17.6 \text{ MeV}$$

The cross-sections for the fusion of these nuclides are appreciable only for energies above 5 keV, so that it is necessary to create plasmas where the thermal energies are 10 keV ($\sim10^8$ K) or greater.

Mass spectrometry can furnish data that complements other forms of spectroscopy relating to ionization and charge exchange; it can also provide data on temperature distributions via the direct measurement of ion velocities. For example, Takeuchi et al. (1983a; 1983b) have measured the ion temperature of an ohmically heated *JET*-tokamak using a neutral beam scattering system, where a continuous beam of 15 keV hydrogen atoms was used with a scattering angle of 4°. Assuming Maxwellian target particles with a velocity that is small compared to the incident particles, the ion temperature T_i is:

$$T_i \approx \left(\frac{\Delta E}{4\,\theta}\right)^2 \frac{m_2}{E \ln 2m_1}$$

where E is the energy of the incident particles of mass m_1 and ΔE is the half-maximum width of the energy spectrum of the scattered particles of mass m_2 and scattering angle θ in the laboratory system. Higher energy neutral beam scattering systems are envisioned for measuring higher plasma temperatures.

A multichannel, mass-separated neutral particle analyzer for simultane-

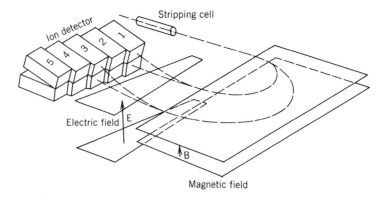

Figure 13.14   Schematic of E/B-type, mass-separated neutral particle energy analyzer. (Takeuchi et al., 1983b; *Jpn. J. Appl. Phys.*, **22**, Pt. 1, 1709 (1983), by permission.)

ous measurements of hydrogen and deuterium atoms emitted from the *JET-2* tokamak has also been described by Takeuchi et al. (1983b). Figure 13.14 is a schematic of the analyzer, which consists of a stripping cell, a bending magnet, deflection plates, and ion detectors. The analyzing magnetic field and electrostatic field are superposed, and the apparatus provides energy and mass separation in the energy range of 0.4–9 keV.

Assay of Deuterium-Tritium Fuel

Gases used in fusion reactors must be analyzed for both isotopic composition and chemical purity. The tokamak fusion test reactor (TFTR) at the Princeton Plasma Physics Laboratory (PPPL) will be the first magnetically confined fusion reactor to utilize deuterium-tritium fuel, and specifications have already been determined for the prerequisite high-purity gases. Experimental measurements have also been made of the response of a quadrupole mass spectrometer to the various species of hydrogen and helium impurities in a specialized tritium mass spectrometry laboratory (Ellefson, 1980).

Tritium purity specifications are designed to minimize plasma radiation losses from nonhydrogenic species (Jensen, 1978). For species of atomic number Z, impurities are limited to the following concentrations: $Z \leq 2$ (2 atom%), $2 < Z \leq 10$ (0.1 atom%), and $Z > 10$ (0.01 atom%). Ions of specific concern in the mass range from $m/q = 2$–6 are shown in Table 13.11 together with the required resolution ($m/\Delta m$) for the separation of ion pairs of interest (Ellefson et al., 1981). Specifically, ^{3}He is a natural decay product of tritium, HT is an anticipated contaminant during storage and handling of the fusion fuel, and DT is present from the original isotopic separation of tritium. Other impurities of higher mass number that might occur in trace quantities in high-purity tritium are N_2, CO, Ar, CO_2, CT_4, HTO, and T_2O.

In addition to monitoring the initial purity of tritium, remote-controlled

TABLE 13.11 Ion Species in the m/q Range of 2–6 and the Required Resolution for the Separation of Ion Pairs

m/q	Ion Species[a]	Resolution (Pair)
2	D^+, (H_2^+)	1302 (D – H_2)
3	$^3He^+ + T^+$, (H, D^+), (H_3^+)	512 (^3He – HD)
4	($^4He^+$), HT^+, (D_2^+), (H_2D^+)	189 (^4He – HT)
		930 (HT – D_2)
5	DT^+, (H_2T^+), (D_2H^+)	856 (DT – D_2H)
6	T_2^+, (D_3^+)	591 (T_2 – D_3)

Source: Ellefson et al. (1981).

[a] The ion species in parentheses are not expected in high-purity tritium.

mass spectrometry is required to monitor the complex tritium storage and delivery system (TSDS) through which tritium is delivered to the plasma torus. This system includes three injection assemblies, a 100 ft long tube, and the tritium generators (uranium getters). When tritium is required in the torus, the uranium tritide within a generator will be heated to 400°C, releasing T_2 gas to the delivery manifold. The three generators have individual secondary containment vessels, and each has a rated tritium capacity equal to twice the allowed inventory of 25,000 Ci (Gill et al., 1983). The entire TSDS has multiple sampling points in the tritium manifold, stringent specifications as to gaseous leak rates (no leak $>1 \times 10^{-9}$ std·cm³/s) in the vacuum system lines, and secondary containment manifolds; and the system has a seismic testing requirement that mandates leak tightness during the most intense anticipated earthquake (Richter, ~5.7).

Breeder Materials and Lithium Isotopes

Lithium, in some form, is the only material suitable for breeding tritium in a commercial fusion reactor. At low neutron energies, the major tritium (and thermal) reaction is associated with the ^6Li isotope, that is,

$$^6Li + n \rightarrow {}^3T + {}^4He + 4.76 \text{ MeV}$$

As seen in Figure 13.15, the thermal neutron cross-section is very high (3000 barns); but at the high neutron energies (14 MeV) that will exist near the first wall of a fusion reactor, the ^6Li capture cross-section decreases to less than 0.1 barn. Only if the fusion neutrons are moderated by collisions with the first wall and blanket materials will the (n, ^6Li) reaction produce ^3T at an acceptable rate. At energies greater than 5 MeV, however, ^7Li has a neutron cross-section of about 1 barn, so that tritium can be generated by the reaction

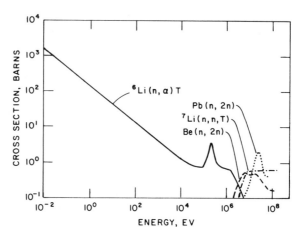

Figure 13.15 Neutron cross-section of ^6Li as a function of energy, and nuclear reactions of other nuclides above 1 MeV. (Johnson et al., 1981; Argonne National Laboratory).

$$^7Li + n \rightarrow {}^3T + {}^4He + n$$

The lower energy neutron that is produced in this reaction is also available for capture by ^6Li (Johnson et al., 1981).

Clearly, mass spectral data is essential for monitoring ^6Li:^7Li ratios and other nuclear reactions. Spectrometry is also required for evaluating isotopic enrichment schemes; for example, Amirav and Even (1982) have obtained separation factors of up to 30 in a supersonic, molecular beam magnetic resonance spectrometer. For practical breeder materials, mass spectrometry has also been essential for acquiring thermochemical data of lithium compounds relating to blanket operating temperatures, tritium retention and release rates, and blanket coolants. Thus, studies have been made of many lithium compounds, such as Li_2O, lithium aluminates, lithium silicates, Li_2S, Li_3N, Li_2C_2, LiAl, LiPb, etc. Vaporization data of $Li_2TiO_3(s)$ have been investigated by the mass spectrometric effusion method over a broad temperature regime (1180–1628 K), and the atomization energies for LiO(g) and $Li_3O(g)$ have been determined from measurements of the partial pressures of the vapor species (Nakagawa et al., 1982). Lithium oxide is also a serious candidate as a breeder material due to its high lithium atom density and potential for breeding without a neutron multiplier.

First Wall Interactions

Plasma wall interactions will place limits on the design and performance of all magnetic fusion devices, for example, the TFTR, the Joint European Torus (JET), the Japanese tokamak (JT-60), and the international tokamak reactor study (INTOR). Limiters, divertor components, and the first wall

TABLE 13.12   Threshold Energy (in eV) for
Sputtering of Materials by H, D, and ^{4}He

Target	H	D	^{4}He
Al	53	34	20.5
Be	27.5	24	33
C	9.9	11	16
Fe	64	40	35
Mo	164	86	39
Ni	47	32.5	20
Si	24.5	17.5	14
Ta	460	235	100
Ti	43.5	—	22
V	76	—	27
W	400	175	100

Source: Roth et al. (1979); Brossa et al. (1981).

are subject to plasma and neutral atom bombardment at rates as high as $\sim 10^{23}$ particles/s, and particle energies range from tens of eV to 2–5 keV depending on plasma boundary conditions. Neutral beam injection can also bombard special components with 100–200 keV particles (Conn, 1981). For example, the INTOR divertor/limiter will absorb loads on the order of 8 × 10^7 W and 3 × 10^{23} ions/s (Heifetz et al., 1983). A limiter lifetime as short as 1 year is deemed acceptable, but the vacuum vessel wall must have a life span of ~ 20 years and a limited erosion rate. The actual net erosion will be primarily via physical sputtering, but it will also be affected by chemical erosion, redeposition of wall material, and the surface desorption of hydrogen.

Impurities that are sputtered from the first wall can contribute to both lower plasma temperatures and plasma instabilities. If the sputtered species are secondary ions rather than neutrals, and are of low kinetic energy, then the influx of impurities from the boundaries may be reduced; some ions may even be redeposited on structural surfaces (Krauss and Gruen, 1981). Thus, some mass spectrometric measurements have been made of the positive charge fractions of H, D, and He backscattered from solid surfaces (Bhattacharya et al., 1980). Other SIMS "ion impact desorption" studies have been made of first-wall impurities that can influence plasma temperatures and fuel recycling rates (Bastasz, 1983).

Tantalum, used as a first-wall coating, has been studied because it has a high melting point, a low vapor pressure, good mechanical properties at high temperature, and a low sputtering yield. The threshold energy (in eV) for tantalum sputtering is given in Table 13.12 for several ion species and in comparison with several other materials (Brossa et al., 1981; Roth et al., 1979). Clearly, at low ion temperatures, tantalum is attractive even though it is a high Z material.

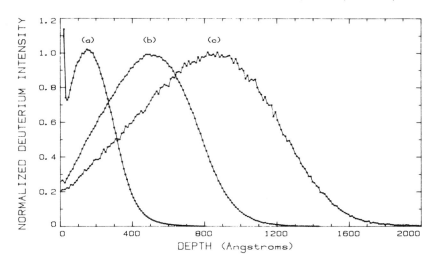

Figure 13.16   Normalized depth profiles of deuterium implanted into a Si sample at (*a*) 750 eV, (*b*) 2500 eV, and (*c*) 5000 eV; fluences were $1 \times 10^{17}/cm^2$, $2 \times 10^{16}/cm^2$, and $2 \times 10^{15}/cm^2$, respectively. (Magee, 1983; *J. Vacuum Sci. & Technology, A1*, 901 (1983), by permission of American Institute of Physics.)

An additional first-wall problem relates to the retention and reemission of hydrogen that impacts the surface. Some energetic atoms that are implanted will remain mobile and diffuse back to the surface, where they may recombine to form molecules and reenter the vacuum vessel. Hence, implantation phenomena are important to the understanding of fuel recycling in tokamaks and for estimating the tritium inventory of vacuum walls in fusion reactors (Thomas et al., 1982). By employing SIMS to measure the depth profiles of first-wall and limiter materials exposed to tokamak discharges, Magee (1983) has also shown that it is possible to determine the flux and energy of the hydrogen isotopes. Meaningful measurements, however, require (1) good depth resolution; and (2) a background pressure in the range of 10^{-10} torr. Figure 13.16 shows the excellent depth resolution of a test Si sample implanted with D at 750, 2500, and 5000 eV obtained with a primary ion beam of low energy (5 keV) and high mass (^{40}Ar or ^{133}Cs). The primary ion beam also struck the sample at a high angle of incidence (60° with respect to the surface normal). SIMS depth profiles of H and D have also been made in an Si sample that had been exposed to several high-power discharges near the wall of the PLT (Princeton Large Torus); Magee (1983) reports that the D profile corresponds to a plasma temperature of ~300 eV (~3 × 10⁶ K) at the vacuum wall. High-resolution depth profiles of hydrogen and deuterium implanted in carbon have also been reported by Wampler and Magee (1981). This experimental work relates to a theoretical model for calculating particle deposition profiles for H, D, and T for any energy or distribution of energies in materials exposed to plasmas.

Other "first-wall" studies have focused on (1) radiation damage, (2) surface texturing, and (3) particle reflection coefficients from compound targets. Neutrons are less effective than charged particles in creating defects. However, a neutron of energy E_n and mass M_1 passing through a medium of mass M_2 will occasionally collide (mean free path ~ 1 cm) with a lattice atom, and will impart to it a kinetic energy up to a maximum given by

$$E_{\max} = \frac{4\, M_1 M_2}{(M_1 + M_2)^2}\; ; E \sim \frac{4\, M_1}{M_2}\, E_n \qquad \text{for } M_2 \gg M_1$$

For fast-fusion neutrons of 14 MeV, this means that a primary knock-on can transfer up to approximately 2 MeV to a light atom such as Al, or 0.2 MeV to a heavy atom such as U. Thus, substantial radiation damage will result from neutron-induced collisions for all practical metals and alloys.

Relative to surface effects, type 304 stainless steel substrates with a variety of machined surface structures (micro-V groove patterns) have been exposed to low-energy (250–1000 eV) hydrogen and argon ion beams. This surface texturing has been found to significantly reduce the effective ion erosion rate in contrast to flat surfaces, and a similar improvement has been noted with hydrogen plasmas bombarding honeycombed molybdenum surfaces (Panitz and Sharp, 1980). The reflection of keV light ions from compound targets such as WO_3, TiC, and TiB_2 has also been studied using computer simulations and in experimental measurements (Jackson and Eckstein, 1982; Morita and Tabata, 1984).

Laser Fusion Ion Spectrometry

One of the most challenging mass spectrometric problems is to characterize the mass, velocities, and charge states of ions produced in laser-induced fusion. Various diagnostic instruments have utilized multiple Faraday cup charge collectors, TOF techniques, and electrostatic and magnetic analyzers. Charge collectors yield qualitative data on ion velocities based on TOF information, provided that the ion flight time is substantially longer than the ion generation time. This assumption is usually valid. Electrostatic analyzers can yield more detailed data by measuring z/m ratios and velocity distributions simultaneously (Oron and Paiss, 1973). For a fusion target (ion source) to detector distance L and ion velocity v, the ion transit time is L/v; and by filtering through the electrostatic analyzer, ions are additionally selected according to $v^2(z/m)$. Hence, the time of arrival of an ion is proportional to $(m/z)^{\frac{1}{2}}$, and (m/z) species can be correlated with laser pulse-collector time measurements. If a sufficient number of collectors are employed in the detection plane, it is possible to reconstruct the velocity spectrum of each (z/m) state. This method requires many collectors, however, and a continuous measurement of (z/m) states and energy distribution is not feasible.

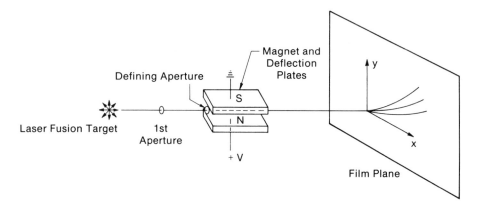

Figure 13.17   Schematic of electrostation/magnetic analyzer for analyzing ion species produced by laser-induced fusion. (Sarraf et al., 1983; *Phys. Rev.*, A27, 2110 (1983), by permission of American Institute of Physics.)

Magnetic spectrometers for fusion diagnostics have also been constructed—some being equipped with solid-state ion track detectors where the density of tracks can be related to a velocity distribution. The distribution, however, is provided for only a single-ion species. Thus, the group at the Laboratory for Laser Energetics, University of Rochester has applied the classical "Thomson parabola" analyzer for fusion diagnostics because of its ability to simultaneously provide energy distribution and charge-to-mass ratio data for a broad spectrum of ion species. Optional detectors are channeltron electron multiplier arrays or a solid-state track device (Sarraf et al., 1983).

Figure 13.17 is a schematic of this "Thomson parabola" scheme. A pulsed laser beam impinges upon a target and a plasma is generated. Ions pass through a defining aperture and through the combined electrostatic/magnetic analyzer, where two mild steel plates are connected to, but electrically insulated from, an electromagnet. The electric field is collinear with the magnetic field, and it is produced by biasing one plate relative to the opposite grounded one. The vacuum housing is aluminum and operates at a pressure of $\sim 10^{-7}$ torr to eliminate charge exchange phenomena. Overall, system design parameters were selected to observe ions with velocities between 10^8 and 10^9 cm/s. In this "parabolic" type of ion spectrometry, there is clearly no focusing, resolution is dominated by the geometrical divergence of the ion beam, and ion species in the detector plane have x and y coordinates that are dependent upon their velocity, charge-to-mass ratios, and electric and magnetic field strengths. The final analysis is complex, but an ion intensity plot in velocity space generated by a computer for hydrogen and oxygen ions in various charge states is shown in Figure 13.18.

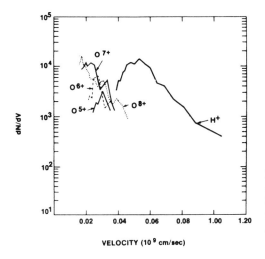

Figure 13.18 Computer-generated ion density plot of the charge state and velocities of ions in a laser-produced plasma. (Sarraf et al., 1983; *Phys. Rev., A27*, 2110 (1983), by permission of American Institute of Physics.)

REFERENCES

Airey, D., *Sci. Total Environ. 25*, 19 (1982).

Aitken, K. L., and F. W. Cornish, *J. Inorg. Nucl. Chem., 17*, 6 (1961).

Amiel, S., and H. Feldstein, *Phys. Rev., C11*, 845 (1975).

Amirav, A., and V. Even, *Chem. Phys. Lett., 86*, No. 4, 331 (1982).

Armigliato, A., G. Celotti, A. Garulli, S. Guerri, R. Lotti, R. Ostaja, and C. Summonte, *Appl. Phys. Lett., 41*, No. 5, 446 (1982).

Asay, B. W., L. D. Smoot, and P. O. Hedman, *Comb. Sci. and Technol., 35*, 15 (1983).

Bastasz, R., *IEEE Trans. Nucl. Sci., NS-30*, No. 2, 1183 (1983).

Bernadzikowski, T. A., J. S. Allender, J. A. Stone, D. E. Gordon, T. H. Gould, Jr., and C. F. Westberry, III, *Am. Ceramic Soc. Bull., 62*, No. 12, 1364 (1983).

Bhattacharya, R. S., W. Eckstein, and H. Verbeek, *Surf. Sci., 93*, 563 (1980).

Blakers, A. W., H. A. Green, and M. R. Willison, *Nucl. Instr. and Methods, 191*, 51 (1981).

Bliznakov, G. M., M. P. Kiskinova, and L. N. Surnev, *J. Catalysis, 81*, 1 (1983).

Brossa, F., G. Piatti, and M. Bardy, *J. Nucl. Mat., 103/104*, 261 (1981).

Burchett, R. R., *Coal Resources of Nebraska*, Report NP 23879, University of Nebraska, Lincoln (1977).

Calzonetti, P. J., and G. A. Elmes, *Geo. J., 3*, Suppl. Iss., 59 (1981).

Carlson, D. E., and C. W. Magee, *Appl. Phys. Lett., 33*, No. 1, 81 (1978).

Cattolica, R. J., S. Yoon, and E. L. Knuth, *Comb. Sci. and Technol., 28*, 225 (1982).

Chapelle, F. H., *Environ. Geol., 3*, 117 (1980).

Chen, F. F., *Plasma Physics and Controlled Fusion*, Vol. 1, Plenum, New York (1984), p. 1.

Chetverikov, A. P., V. Y. Gabeskiriya, and V. V. Puchkov, *Soviet Phys. Technol. Phys., 26*, No. 1, 73 (1981).

Christie, W. H., R. E. Eby, R. J. Warmack, and L. Landau, "Secondary-Ion Mass Spectrometry: Some Applications in the Analysis of Nuclear Materials." in *25th Conf. on Anal. Chem. and Nucl. Technol.*, Gatlinburg, TN (October 6, 1981).

Clement, R. E., and F. W. Karasek, *J. Chromatogr. 234*, 395 (1982).

Cohen, E. R., and A. H. Wapstra, *Nucl. Instr. Methods, 211*, 153 (1983).

Conn, R. W., *J. Nucl. Mat., 103/104*, 7 (1981).

Coombs, M. M., *Power Eng., December*, 42 (1982).

DeMeo, E. A., and R. W. Taylor, *Science, 224*, 245 (1984).

Dempster, A. J., *Phys. Rev., 73*, 829 (1947).

Dietz, A., *KAPL Report M-MS-4*, 1964. Knolls Atomic Power Laboratory, Schenectady, N.Y.

Dondes, S., personal communication (1983).

Duchateau, N. L., and P. DeBièvre, *Int. J. Mass Spectrom. Ion Processes, 54*, 289 (1983).

Dunbar, J., and A. T. Wilson, *Geochim. Cosmochim. Acta, 47*, 1541 (1983).

Duncan, C. H., R. C. Willson, J. M. Kendall, R. G. Harrison, and J. R. Hickey, *Solar Energy, 28*, No. 5, 385 (1982).

Ellefson, R. E., in *Tritium Technology in Fission, Fusion, and Isotopic Applications, Proc. Am. Nucl. Soc. Topical Meeting*, CONF-800427, 240–244, Dayton, OH (April 1980).

Ellefson, R. E., W. E. Moddeman, and H. F. Dylla, *J. Vac. Sci. Technol., 18*, No. 3, 1062 (1981).

Facer, J. F., *Uranium in Coal*, U.S. Department of Energy, Report GJBX-56(79). Grand Junction Office, CO (May 1979).

Falcone, P. K., R. K. Hanson, and C. H. Kruger, *Comb. Sci. Technol., 35*, 81 (1983).

Fassett, J. D., and W. R. Kelly, *Anal. Chem., 56*, 550 (1984).

Forman, L., and F. A. White, *Nucl. Sci. Eng., 28*, 139 (1967).

Furth, H. P., *Physics Today, 38*, No. 3, 53 (1985).

Gentry, R. V., G. L. Glish, and E. H. McBay, *Geophys. Res. Lett., 9*, No. 10, 1129 (1982).

Gill, J. T., B. E. Anderson, R. A. Watkins, and C. W. Pierce, *J. Vac. Sci. Technol., A1*, No. 2, 856 (1983).

Goodings, J. M., and N. S. Karellas, *Int. J. Mass Spectrom. Ion Processes, 62*, 199 (1984).

Gordon, R. L., *Int. J. Mass Spectrom. Ion Processes, 55*, 31 (1983–1984).

Green, L. W., N. L. Elliot, and T. H. Longhurst, *Anal. Chem., 55*, 2394 (1983).

Hall, R. N., *Silicon Photovoltaic Cells*, General Electric Technical Information Series Report No. 80CRD274 (1981).

Hammond, R. P., *Am. Scientist, 67*, 146 (1979).

Heifetz, D., J. Schmidt, M. Ulrickson, and D. Post, *J. Vac. Sci. Technol., A1*, No. 2, 911 (1983).

Heller, A., *Solar Energy, 29*, No. 2, 153 (1982).

Hickam, W. M., private communication (1982).

Hickam, W. M., and E. Berkey, in *Trace Analysis by Mass Spectrometry*, Academic Press, New York (1972), p. 373.

Hickam, W. M., S. H. Peterson, D. F. Pensenstadier, J. C. Bellows, "Steam Purity Analysis Instrumentation," in *Instr. Soc. Am. Conf.*, Anaheim, CA (October 1981).

Hofstader, R. A., O. I. Milner, and J. H. Runnels, *Analysis of Petroleum for Trace Metals*, American Chemical Society, Washington, D. C. (1976).

Holliday, A. K., G. Hughes, and S. M. Walker, *The Chemistry of Carbon*, Pergamon, New York (1975), p. 1203.

Hopkinson, J., *EPRI J1, October*, 22 (1982).

Hurst, R. W., and T. E. Davis, *Environ. Geol., 3*, 363 (1981).

Jackson, D. P., and W. Eckstein, *Nucl. Instr. Methods, 194*, 671 (1982).

Jensen, R. V., D. E. Post, and D. L. Jasby, *Nucl. Sci. Eng., 65*, 292 (1978).

Johnson, C. E., R. G. Clemmer, and G. W. Hollenburg, *J. Nucl. Mat., 103/104*, 547 (1981).

Jonas, O., *Turbine Steam Purity*, Southeastern Electric Exchange, Washington, D. C. (1978).

Kaiser, E. W., W. G. Rothschild, and G. A. Lavoie, *Comb. Sci. Technol., 32*, 245 (1983).

Klapish, R., and R. Bernas, *Nucl. Instr. Methods, 38*, 291 (1965).

Knauth, L. P., M. B. Kumar, and J. D. Martinez, *J. Geophys. Res., 85*, B9, 4863 (1980).

Koch, L., and D. Steinert, *Advances in Mass Spectrometry*, Vol. 8A, ed. A. Quayle, Heyden & Son Ltd., London (1980), p. 330.

Krauss, A. K., and D. M. Gruen, *J. Nucl. Mat., 103/104*, 239 (1981).

Krieth, F., and R. T. Meyer, *Am. Scientist, 71*, 598 (1983).

Lennard, W. N., D. Phillips, and R. Hill, *IEEE Trans. Nucl. Sci., NS-30*, No. 2, 1532 (1983).

Lewerenz, H. J., D. E. Aspnes, B. Miller, D. L. Malm, and A. Heller, *J. Am. Chem. Soc., 104*, 3325, (1982).

Luthra, K. L., and H. S. Spacil, *J. Electrochem. Soc., 129*, No. 3, 649 (1982).

Magee, C. W., *J. Vac. Sci. Technol., A1*, No. 2, 901 (1983).

Magee, C. W., and D. E. Carlson, *Solar Cells, 2*, 365 (1980).

Mallozzi, P. J., and H. M. Epstein, *Res. Dev., February*, (1976) p. 30.

McIntyre, N. S., G. G. Strathdee, and B. F. Phillips, *Surf. Sci., 100*, 71 (1980).

Morita, K., and T. Tabata, *J. Appl. Phys., 53*, No. 3, 776 (1984).

Morrison, R. E., in *Proc. Fourth Int. Ash Utilization. Symp.*, St. Louis, MO (March 24, 1976).

Muller, J. C., and P. Siffert, *Nucl. Instr. and Methods, 189*, 205 (1981); *Radiation Effects, 63*, 8 (1982).

Nagai, H., O. Nitoh, and M. Honda, *Radiochim. Acta, 29*, 169 (1981).

Nakagawa, H., M. Asano, and K. Kubo, *J. Nucl. Mat., 110*, 158 (1982).

Ndiokwere, Ch. L., *Radiochem. Radioanal. Lett., 59*, No. 4, 201 (1983).

Neretnieks, I., *J. Geophys. Res., 85*, B8, 4379 (1980).

Nier, A. O., *Am. Scientist, 54*, 377 (1966).

Oron, M., and Y. Paiss, *Rev. Sci. Instr., 44*, 1293 (1973).

Oversby, M., and A. E. Ringwood, *Radwaste Mange., 1*, 289 (1981).

Panitz, J. K. G., and D. J. Sharp, *J. Vac. Sci. Technol., 17*, No. 1, 282 (1980).

Pereira, W. E., and C. E. Rostad, *Int. J. Environ. Anal. Chem., 15*, 73 (1983).

Perrin, J., A. Lloret, G. DeRosny, and J. P. M. Schmitt, *Int. J. Mass Spectrom. and Ion Processes, 57*, 249 (1984).

Primak, W., *Nucl. Sci. Eng., 80*, 689 (1982).

Reents, W. D., Jr., D. L. Wood, and A. M. Mujsee, *Anal. Chem., 70*, 202 (1985).

Rider, B. F., J. P. Peterson, and C. P. Ruiz, *Nucl. Sci. Eng., 15*, 284 (1963).

Ringwood, A. E., *Am. Scientist, 70*, 202 (1982).

Rossnagel, S. M., and R. S. Robinson, *J. Vac. Sci. Technol., 20*, No. 3, 336 (1982).

Roth, J., J. Bohdansky, and W. Ottenberger, *IPP Report*, IPP 9/26 (1979).

Sarraf, S. P., G. A. Williams, and L. M. Goldman, *Phys. Rev., A27*, No. 4, 2110 (1983).

Shmid, M., Y. Nir-el, G. Engler, and S. Amiel, *J. Inorg. Nucl. Chem., 43*, 867 (1981).

Skinner, J., and C. A. Walker, *Am. Scientist, 70*, 180 (1982).

Sloane, T. M., and J. W. Ratcliffe, *Comb. Sci. Technol., 33*, 65 (1983).

Smith, D. H., R. L. Walker, and J. A. Carter, *Anal. Chem., 54*, No. 7, 827A (1982a).

Smith, G. B., C. Zajac, and A. Ignatiev, *Surf. Sci., 114*, 614 (1982b).

Stevens, H. E., *Nucl. Sci. Eng., 4*, 73 (1958).

Su, C. S., and F. A. White, *IEEE Trans. Nucl. Sci., 15*, 284 (1967).

Swanson, V. E., J. H. Medlin, J. H. Hatch, J. R. Coleman, S. L. Wood, G. H. Wood, Jr., S. D. Woodruff, and R. T. Hildebrand, *U. S. Geol. Survey*, Report 76-468, Reston, VA, (1976), p. 503.

Takeuchi, H., T. Matsuda, Y. Miura, T. Nishitani, M. Shiho, C. Konagai, H. Kimura, and H. Maeda, *Jpn. J. Appl. Phys.*, *22*, No. 11, Part 1, 1717 (1983a).

Takeuchi, H., T. Matsuda, Y. Miura, M. Shiko, H. Maeda, K. Hashimoto, and K. Hayashi, *Jpn. J. Appl. Phys.*, *22*, No. 11, Part 1, 1709 (1983b).

Tardy, J., J. M. Poitevin, and G. Lemperière, *J. Phys. D: Appl. Phys.*, *14*, 339 (1981).

Thomas, E. W., I. G. Petrov, and M. Braun, *J. Appl. Phys.*, *53*, No. 9, 6365 (1982).

Trimmer, D., *Rev. Sci. Instr.*, *53*, No. 8, 1246 (1982).

U. S. National Committee for Geochemistry, *Trace Element Geochemistry of Coal Resource Development Related to Environmental Quality and Health*, National Academy Press, Washington, D. C. (1980).

Valković, V., *Trace Elements in Coal*, Vol. I, CRC Press (1983), p. 70.

Van Overstraeten, R., R. Mertens, and J. Nijs, *Reports Progr. Phys.*, *45*, No. 10, 1041 (1982).

Wampler, W. R., S. T. Picraux, S. A. Cohen, H. F. Dylla, G. M. McCraken, S. M. Rossnagel, and C. W. Magee, *J. Nucl. Mat.*, *85/86*, 983 (1979).

Wampler, W. R., and C. W. Magee, *J. Nucl. Mat.*, *103/104*, 509 (1981).

White, F. A., *Mass Spectrometry in Science and Technology*, Wiley, New York (1968), p. 126.

White, F. A., F. M. Rourke, J. C. Shuman, and J. R. Huizenga, *Phys. Rev.*, *109*, 437 (1958a).

White, F. A., J. C. Sheffield, and F. M. Rourke, *Appl. Spectroscopy*, *12*, No. 2, 46 (1958b).

Williams, J. R., *Astronaut. Aeron.*, *November*, *24*, 46 (1975).

Wong, C. M., R. W. Crawford, V. C. Barton, H. R. Brand, K. W. Neufeld, and J. E. Bowman, *Rev. Sci. Instr.*, *54*, No. 8, 996 (1983).

CHAPTER FOURTEEN

ON-LINE MONITORING AND PROCESS CONTROL

Process monitoring and control is a systematic method for maintaining the parameters essential to a process within required and often narrow limits. While usually related to industry, the principles apply to virtually all human activities. In industrial or chemical engineering, the function is to monitor the conversion of raw materials or reactants into an end-product in a manner that maximizes yield and minimizes waste or the formation of undesirable byproducts. The processes are generally continuous, although closed or batch processes are also monitored to determine when an optimum end-point has been reached. As an example of the latter, a quadrupole mass spectrometer (QMS) has been used in connection with a complex batch process for producing brake-lining material; an increased profitability in this case resulting from improved economic control rather than from increased yields (Shaw, 1984).

The effectiveness of mass spectrometry for on-line monitoring depends upon the characteristics of the stream. McKeown (1981) has classified process streams into several categories according to the number of constituents to be monitored and the information required:

(1) Processes involving a single but critical component in a low concentration present in a matrix of feedstock and byproducts. The component is typically the product being manufactured; hence, the goal is to maximize the yield while minimizing cost and the production of other compounds. Examples of this category are the production of ammonia, ethylene dichloride, ethylene oxide, and methyl alcohol.

(2) Processes using multiple feedstocks to produce more than one product, in which product mix and the consumption of feedstock are to be optimized according to market demand. An example is the process in which the feedstocks are air, ammonia, and propylene and the products are acrylonitrile, hydrogen cyanide, and acrylic acid.

(3)   Processes using a single high-purity component with only trace contaminants present in which it is necessary to maintain purity of the feedstock. In the manufacture of polyethylene from polymer-grade ethylene (C_2H_4) for example, trace levels of acetylene, saturated aliphatics, carbon dioxide, carbon monoxide, hydrogen, oxygen, or water vapor can poison the catalyst, cause an explosion, or otherwise affect product quality.

(4)   Processes in which reactions in a liquid phase must be inferred from measurements in the accompanying gas phase. Examples are the conversion of pig iron into steel in a basic oxygen furnace, the manufacture of antibiotics or beverages by fermentation, and coal gasification. Measurements in this category are often difficult because of adverse operating conditions, problems with obtaining representative samples, and the rapidity with which the concentrations change.

ON-LINE MASS SPECTROMETERS

The first commercial use of mass spectrometry outside of the research laboratory was for process control in the petroleum industry, beginning with the delivery of an instrument to the Atlantic Richfield Refining Company in 1943 (Meyerson, 1984). These instruments were essentially adaptations of the laboratory spectrometers, which limited on-line applications because of the frequent calibrations and highly trained personnel that were required for effective operation. The analytical requirements of the space program subsequently encouraged the development of small, reliable, high-performance mass spectrometers that were designed to continuously perform specific tasks for extended periods without operator involvement, and that required infrequent or automatic calibration and correction for electronic drift. This, in turn, led to the development of industrial mass spectrometers configured specifically for process applications. These are generally under microprocessor control, with automated data acquisition and reduction and with limited allowable operator interaction once analytical procedures are established. Samples are continuously introduced into the ion source, and stream selection is provided with automatically actuated switching valves.

These commercial instruments are often QMSs because of their capability to rapidly record a number of peaks (or to scan) over the mass range of interest, and both magnetic sector and time-of-flight (TOF) instruments are also used effectively. Data may be recorded as analog signals if monitoring the intensity of one or more resolved peaks is sufficient to control the process. If quantitative analysis of a mixture is required, the data may be entered into the computer in a matrix of peak intensities and sensitivity coefficients to correct for interferences and to automatically perform the analysis (Chapter 6). In practice, the number of compounds that can be determined by this method is limited by the difficulty in selecting peaks providing sufficient leverage for accurate analysis and in the inverting of large matrices. For most applications, identification of only a few species at low or intermediate molecular weight is adequate, and 10 × 10 or smaller matrices are sufficient. Larger matrices can be used if necessary, however, and a prototype QMS system has been described in which the matrix size can be varied from 1 × 1 to 40 × 40 to analyze trace levels of hydrocarbons in the air, such as might occur in stack effluents from industrial processes (Judson et al., 1978; Wood and Yeager, 1980). Although a limited number of the prototypes were manufactured, the computer methods and other salient features were subsequently incorporated into several commercially available systems.

The rapid scanning instruments have the advantage in that reprogramming to monitor other masses or to reconfigure the matrix is a straightforward procedure. When monitoring a limited number of components in a stream, the multiple collector, nonscanning magnetic anayzer with a permanent magnet has the advantage in that all signals simultaneously and continuously arrive at the detectors. This maximizes sensitivity and response, and it makes drifts in sensitivity due to ion source fluctuations readily correctable (Savin et al., 1983). Adams (1980) and Whistler and Schaefer (1983) have described the instrument shown in Figure 14.1, which measures up to eight species, electronically adjusts ion source drift, and corrects for peak interference at the masses being monitored. The resolution of the multiple collector instrument and the number of species that can be monitored are determined by the length of the focal plane and the size of the individual detectors. Development of the microchannel-plate/CCD-hybrid ion detector (Chapter 5)

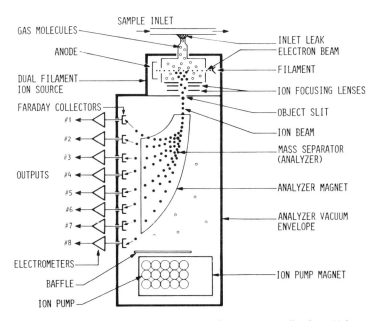

Figure 14.1 Multiple collector mass spectrometer for process applications (Adams, 1980).

may well represent the next major improvement in the simultaneous monitoring of mass-resolved ion beams; but for now, magnetically or electrostatically scanned instruments are used when high resolution is required.

In some cases, the complexity of the sample is beyond the capability of the mass spectrometer alone and requires the use of a gas chromatographic inlet. Such multicomponent samples are found in the beverage and flavor industries, and also in the fragrance industry, which requires the identification of animal excretions and essential oils in natural and synthetic complex mixtures. While these analyses are not carried out on-line in the usual sense in order to automatically adjust flows, rapid analysis of a large number of samples is required. Green (1983) has described an automated system using quartz capillary chromatography coupled with a high-performance QMS, in which 60 samples can be sequentially introduced and analyzed. The entire process from sample introduction through identification and quantization of the resolved component is computer controlled. Identification of the unknown species is accomplished by comparison to specially compiled libraries of mass spectra relevant to the application, although the larger data bases discussed in Chapter 6 can also be used.

CHEMICAL APPLICATIONS

The use of on-line monitoring and process control is best exemplified by the chemical industry, which annually produces more than $150 billion in sales of chemicals and chemical-based products. To be economically productive, a single chemical plant will produce 10^6 to 10^9 pounds of product per year— a typical base petrochemical plant, for example, consumes 1.1×10^7 barrels of naphtha (a mix of paraffins, naphthenes, and aromatics) to produce 3.5×10^9 pounds of ethylene and coproducts (Bailey, 1983). The quantitative flow diagram in Figure 14.2 for manufacturing nitric acid by the oxidation of ammonia illustrates the large quantities of feedstocks, intermediate products, and final products that occur in even a simple multistep manufacturing process (Peters and Timmerhaus, 1968).

Because of these large quantities, a small percentage increase in efficiency will have a significant effect on output. Major cost reductions are realized by maintaining optimized (stoichiometric) ratios of feedstocks, substituting lower cost-starting materials, or reducing the number of reaction steps. In an example discussed by Bailey (1983), acetic acid and acetic anhydride were initially produced as coproducts in a three-step synthesis from ethylene: (1) acid hydrolysis to ethanol; (2) catalytic dehydrogenation to acetaldehyde; and (3) direct liquid phase oxidation,

(ethylene)   (ethanol)   (acetaldehyde) (anhydride)   (acetic acid)

$$CH_2CH_2 \rightarrow C_2H_5OH \rightarrow CH_3CHO \rightarrow (C_2H_3O)_2O + CH_3COOH$$

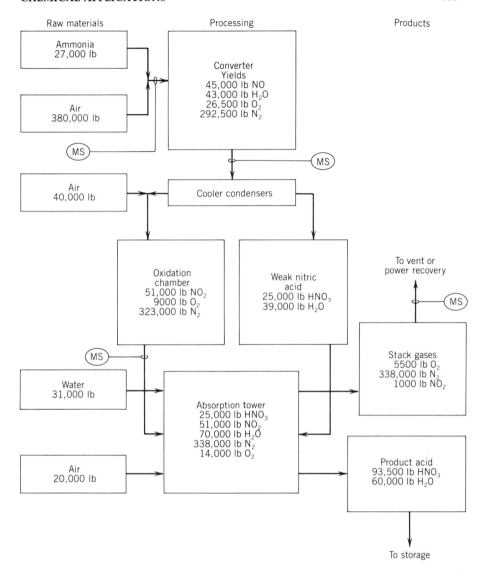

Figure 14.2 Quantitative flow diagram for the manufacture of nitric acid by the oxidation of ammonia. (Adapted from Peters and Timmerhaus, 1968; copyright McGraw Hill.)

This process was eventually replaced by direct synthesis from methanol and carbon monoxide to produce acetic acid essentially without coproducts (e.g. $CH_3OH + CO \rightarrow CH_3COOH$), with an attendant eightfold increase in production over the original method. The methanol itself is often made from *synthesis gas* (carbon monoxide and hydrogen) obtained from coal, that is, $CO + 2H_2 \rightarrow CH_3OH$. The acetic acid process has therefore been simplified, so that only purity of feedstock and stoichiometry are controlled to optimize production.

Processing Petroleum Fuels

The major constituents of petroleum are carbon (83–87%) and hydrogen (11–15%), with small amounts of combined sulfur, nitrogen, oxygen, and inorganic salts present as impurities. To be transformed into useful products, the petroleum is first cleaned of impurities (desalted), distilled into fractions as shown in Table 14.1, and further restructured by catalytic cracking or reforming. If a fraction containing a large number of hydrocarbon compounds, such as gasoline, heating oil, or lubricants, is extracted from the distillation column, it is called a *refined product*; if it results from further processing of a refined (or gaseous) product and contains only one or two hydrocarbon compounds, it is called a *petrochemical* (Hoffman, 1983).

An example of a typical single-stage distillation process is shown schematically in Figure 14.3. The crude oil feedstock is preheated by the outgoing stream before entering the furnace. The components are separated in the column according to their boiling points and are removed as side streams at appropriate levels, with those having the lowest boiling point exiting from the top. The side streams may be further fractionated in small columns, or

TABLE 14.1 Petroleum Distillation Fractions and Boiling Ranges (°F)

Product	Approximate Boiling Range (°F)
Natural gasoline	30–180
Light distillates	
Gasoline	80–380
Naphthas	200–450
Jet fuels	180–450
Kerosene	350–550
Light heating oils	400–600
Intermediate distillates	
Gas oils	480–750
Diesel oils	380–650
Heavy fuel oil	550–800
Heavy distillates	
Lubricating oils	600–1000
Waxes	Above 625
Residues	
Lubricating oils	Above 900
Asphalt	Above 900
Residium	Above 900

Source: Adapted from Galli (1974).

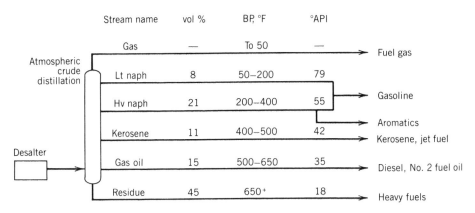

Figure 14.3 Distillation of desalted crude oil. (Adapted from Hoffman, 1983; copyright van Nostrand Reinhold Co., Inc.)

strippers, and additional stages of distillation are often used. Light and heavy naphtha fractions are combined to produce gasoline and aromatic feed-stocks, with the fraction ratio adjusted to optimize volatility of the product. Quality of the product is further enhanced by increasing the octane number, defined as the percent composition of iso-octane (2,2,4-trimethylpentane), in a mixture with normal heptane that matches the knock (spontaneous ig-nition) in a test engine using the fuel being rated; that is, an octane rating of 90 is equivalent to 90% iso-octane and 10% normal heptane in the test mixture. Octane enhancement is principally accomplished by catalytic re-forming, which converts paraffins or naphthenes to their aromatic homolog. The process is highly efficient, and the reformed mixture contains a much higher percent composition of the aromatic compounds than did the original.

Most crude contains at best about 20% or less of the naphtha fractions necessary to directly formulate gasoline. Increased production of light frac-tions, with an attendant increase in octane rating, is achieved through cat-alytic cracking—sometimes in the presence of hydrogen, which fractures large molecules into two or more smaller ones. Production yields are also increased through polymerization of olefinic (C_nH_{2n}) compounds, which link several molecules to form a compound within the gasoline boiling range; or, by the alkylation (addition of an alkyl group) of paraffinic or olefinic hy-drocarbons. The latter process typically reacts isobutane with a light olefin, for example, propylene or butylene, to form a product with a high concen-tration of 2,2,4-trimethylpentane. Jet fuel, kerosene, and heating oils are obtained from the middle range distillates, while lubricants are produced by vacuum distillation of high boiling point (>600°F) fractions. Modern fuels and lubricants, therefore, are highly complex mixtures of blends derived from carefully monitored and controlled multiple-step processes. The for-mulation of these mixtures are determined by the properties required of the end-product, while the steps involved are dictated by the composition of the crude feedstock.

Catalytic Processing of Petrochemicals

Most commercial processes involve noncatalytic steam cracking of light hydrocarbons, as in the production of ethylene, propylene, and some higher olefins and dienes, or the use of metallic catalysts to initiate and maintain the reaction. The reactions may be complex and involve large molecules, for example, 90% of the primary feedstocks for chemical production originate in crude oil, while 90% of the crude oil is converted into fuel (Edmonds, 1981). Alternatively, the reaction may involve a simple process with low molecular weight compounds, such as in the production of ammonia (NH_3) by the catalytically induced reaction at elevated temperature and pressure of nitrogen from air and hydrogen from methane.

The control of a less complicated process becomes important when, as in the case of ammonia, the product has economic significance. In 1984, the United States alone produced approximately 36 billion pounds of ammonia at a market value of $3 billion, with about 80% used in agriculture and the remainder in the manufacture of feed additives, polymers for fibers and fabricated plastics, and explosives for mining operations.

The ammonia-manufacturing flow diagram in Figure 14.4 illustrates an integrated catalytic process to which mass spectrometry is applied for mon-

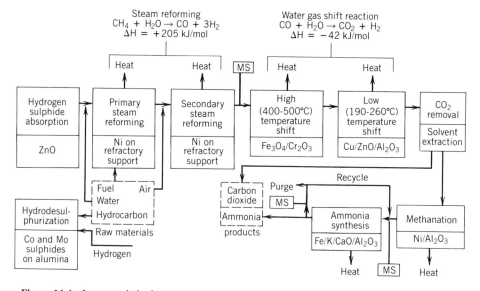

Figure 14.4 Integrated single-stream synthesis of ammonia. (Adapted from Pearce and Twigg, 1981; Leonard Hill, Glasgow and London, with permission.)

itoring and control. In this kind of process, intermediate products are continuously produced and not stored; hence, the system efficiency is determined by each of the in-line catalytic reactions. Feedstocks are produced (1) in the primary and secondary *steam-reforming* processes in which superheated steam mixes with methane in the presence of a catalyst to form hydrogen and carbon monoxide in the endothermic reaction $CH_4 + H_2O \rightarrow CO + 3H_2$; and (2) in the *shift reactions* that produce more hydrogen through oxidation of CO in the exothermic reaction $CO + H_2O \rightarrow CO_2 + H_2$. Air is introduced into the stream to form a stoichiometric hydrogen-nitrogen mixture of 3:1, and ammonia is synthesized through the exothermic reaction $N_2 + 3H_2 \rightarrow 2NH_3$. Gas leaving the reactor is cooled to remove the ammonia, and unreacted nitrogen and hydrogen are recirculated along with additional feedstock, with conversion of 15–39% per pass being typical. Removal of sulfur prior to the steam reformers and of CO_2 prior to the NH_3 converter is critical to protect the catalyst—a 15% nickel catalyst is poisoned at 775°C when containing only $5 \times 10^{-3}\%$ sulfur (Green and Li, 1983)— and residual methane and argon must be removed to maximize the ammonia synthesis. As simple as the single-stream process is, control of the hydrogen:nitrogen ratio to optimum values for stream and catalyst conditions will result in significantly reduced operating costs and increased output. Pearce and Twigg (1981) illustrate this by pointing out that in a large plant, a 0.1% increase in the efficiency of the low-temperature carbon monoxide shift stage can result in an additional 350 metric tons/yr of ammonia at minimal additional cost. In one such application, computer control of a synthesis process with a multiple collector, magnetic sector mass spectrometer resulted in an increased production of 2 tons/day of ammonia with no increase in feedstock, resulting in a recovery of the cost of the instrument in only 130 days (Bourg, 1982).

One of the more important petroleum-based petrochemical feedstocks is ethylene, which is used in the manufacture of polyethylene and polyvinyl chloride polymers, ethylene oxide and glycol, and ethyl alcohol. In the production of ethylene, the feedstocks (generally ethane and propane) are mixed with steam to increase the olefin yields and are cracked in a high-temperature (1200–1600°F) tubular pyrolysis furnace, followed by quenching to separate and purify the components. A typical ethylene plant includes a number of furnaces operating in parallel, and yields are optimized by controlling both steam-to-gas ratios and furnace temperatures based on measured reaction products. Overcracking produces an excess of methane and hydrogen, and it leads to an excessive rate of coke formation, while undercracking results in high recycle rates and increased production of less important byproducts.

In a typical operation using an ethane feedstock, optimum ethylene yields were obtained with a conversion index of 65.0, (Adams, 1984), where the conversion index is defined as

$$C = \frac{\text{wt\% feedstock (inlet)} - \text{wt\% feedstock (outlet)}}{\text{wt fraction feedstock (inlet)}}$$

Standard deviations (SD) from the optimized value of C were typically on the order of 4%, with the plant under manual control. Operating under computer control, with feedstock and furnace effluents monitored by gas chromatography, reduced the SD to about 1.3% and increased yields by 2%. With the mass spectrometer furnishing input to the computer, the high speed of the analysis enabled the control loop to further reduce the SD to only 0.5%, maintaining output at near maximum levels.

Fermentation Chemistry

Industrial fermentation is the process in which the activities of micro-organisms lead to industrially significant products. Most often thought of in terms of alcoholic beverages, the major output of the process actually involves other products. Bacteria, viruses, yeasts, filamentous molds, enzymes, and actinomycetes are used to produce a wide variety of industrial and medicinal chemicals, many of which are otherwise unobtainable and which are large molecules with complex structural and stereochemical characteristics. The process is either anaerobic (without oxygen) or aerobic (requiring large amounts of oxygen from the air), depending upon the physiological requirements of the particular organism. Examples of important fermentation products are the amino and organic acids, enzymes, fragrances, flavors, liquid fuels, solvents, antibiotics, steroids, vitamins, plant growth regulators, insecticides, and microbial polysaccharides.

Control of the fermentation process is primarily achieved through monitoring the gaseous products associated with the reaction. Accurate analysis of exhaust gases is particularly important for mycelial culture fermentations, which constitute the majority of all commercial aerobic fermentations. Buckland and Fastert (1981), for example, have detailed the control of cefoxitin and thienamycin fermentation by directly interfacing the fermenters and reactor vessels to mass spectrometers. Generally speaking, only O_2, N_2, CO_2, and components that might be detrimental to the organism need to be monitored. Because of the rapid response of the mass spectrometer in measuring these gases, a large number of fermenters can be simultaneously controlled with suitable stream-switching methods. The addition of both nutrients and air during the fermentation are controlled, and an analysis to determine the presence of bacterial contamination is made prior to switching the seed tank into the production rate fermenter. The cessation of product synthesis is indicated by a corresponding cessation in the production of CO_2, as shown in Figure 14.5. The optimum time for harvesting can therefore be readily determined from the mass spectrometer data.

Volatile components in either the liquid or gas phase can be monitored for on-line control using a silicone rubber membrane separator. The device is similar to that developed by Llewellyn and Littlejohn for the GC/MS interface (Chapter 7), and it is directly immersed in the liquid medium (Heinzle and Bournes, 1983; Kallos and Mahle, 1983). Weaver (1982) has de-

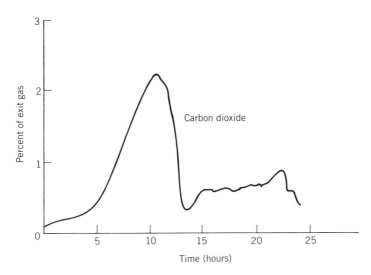

Figure 14.5 Carbon dioxide evolution during a yeast fermentation. (Adapted from Weaver, 1982; courtesy J. C. Weaver; copyright John Wiley & Sons.)

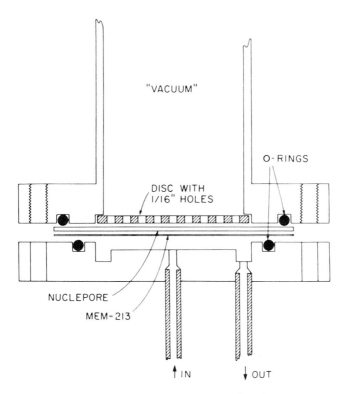

Figure 14.6 Silicon rubber membrane interface for measuring volatile species in the gas or liquid phase. (Weaver, 1982; courtesy J. C. Weaver; copyright John Wiley & Sons.)

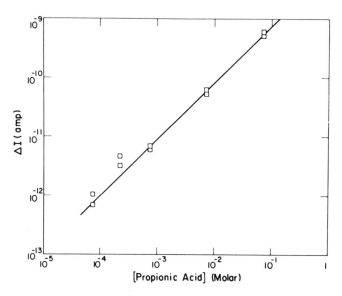

Figure 14.7 Response curve of dissolved propionic acid with the interface acidified to pH 1. (Weaver, 1982; courtesy J. C. Weaver; copyright John Wiley & Sons.)

scribed the device shown in Figure 14.6 in which the fluid or gaseous stream enters at the center and flows radially to an outer annulus and exit port, while the volatile species permeate a polymeric membrane and diffuse to the mass spectrometer ion source. Measurements of dissolved ethanol made by monitoring the intensity of m/z 31 and of DHO in aqueous solution by monitoring m/z 19 as a function of concentration showed that the interface is linear with concentration over four orders of magnitude. Volatile acids and bases can also be analyzed by controlling the acidity of the sample at the interface (Weaver and Abrams, 1979). A linear response curve of ion current intensity at m/z 74 as a function of the concentration of dissolved propionic acid (CH_3CH_2COOH), an esterifying agent used in the manufacture of solvents, fruit flavors, and perfume bases, is shown in Figure 14.7. In this measurement, formic acid ($HCOOH$) was used to acidify the samples containing the propionic acid to pH 1, that is, to be highly acidic. Dissolved formic, acetic (CH_3COOH), acrylic (CH_2CHCO_2H), and pyruvic ($CH_3COCOOH$) acids have been measured using this technique, as have the volatile bases, alcohols, and aldehydes.

   Quantitative analysis of gases and liquids in an aqueous solution is also used to provide both process and quality control in the food and brewing industries. The sample is continuously introduced into the mass spectrometer through a membrane inlet immersed in a stirred tank or flowing stream. The liquid medium is maintained at constant temperature to minimize variations in permeation rate of the sample through the membrane, and in some

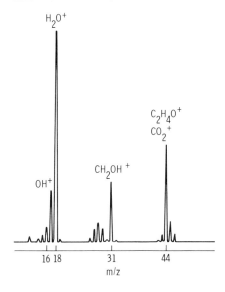

Figure 14.8 Mass spectrum of a strong beer obtained from the aqueous solution with a silicone rubber membrane inlet and an on-line quadrupole mass spectrometer (Balzers, Inc.).

cases, quantitative measurements can be made to within $\pm 0.1\%$. A mass spectrum of a strong beer obtained with an on-line QMS is shown in Figure 14.8, where it is seen that the primary peaks result from water (m/z 18), carbon dioxide (m/z 44), and the characteristic CH_3O fragment from ethanol at m/z 31. When compared to the spectrum of a normal beer, the strong beer showed an increase in the ethanol-to-water peak ratio of 27%, so that mass spectrometry provides a viable method of monitoring the brewing process while measuring only a few significant mass peaks.

In addition to regulating the fermentation, analyses of the unit processes and reaction products are essential to scaling up from the developmental system in the laboratory to the full-scale production facility (Humphrey and Lee, 1983). Fermentation unit processes (e.g., the reaction mechanisms by which the reactants are converted into products) include oxidations, reductions, transformations, polymerizations, hydrolyses, and biosyntheses. Scaling up these processes has become less empirical due to improvements in gas analysis methods, which enable a more reasoned approach. By comparing the experimental observations with known metabolic pathways, environmental conditions for the production processes can be defined, along with the instrumentation and measurements necessary to maintain and control the process. The fermentation control methods are generally based on achieving oxygen transfer rates comparable to those that are optimal in the laboratory, which are typically on the order of 1.5–4.0 g/l/h. Under steady-state conditions, the volumetric oxygen transfer rate equals the oxygen uptake rate, which is measured by monitoring m/z 32 at the air input and output points. If the CO_2 evolution rate is similarly measured, the respiratory quotient can be readily calculated.

INDUSTRIAL APPLICATIONS

On-line monitoring with the mass spectrometer is most useful for those industrial processes in which a gaseous or vaporized constituent is, in itself, indicative of the state of the process or the degree of completion. These are often, but not always, closed or batch processes, and are typified by (1) the curing and processing of rubber and other polymeric materials; (2) the hydrolysis or distillation of wood to obtain sugar, liquid fuels, turpentine, pine oil, and rosin; (3) the production and purification of the industrial gases such as nitrogen, oxygen, carbon dioxide, acetylene, and the rare gases; and (4) the high-temperature production of iron or steel. Equally important is the use of the mass spectrometer in an off-line analysis to determine the composition and quality of the product—in itself a form of process control—and to participate in the development of new manufacturing procedures and products. It is beyond the scope of this chapter to discuss all of the industrial processes to which the mass spectrometer may be applied; however, the processes described below are typical.

Polymeric Curing

In one recent year, approximately 6 million metric tons of natural and synthetic rubber and 7 million metric tons of plastics and resins were processed, mixed into various formulations, and transformed into useful products. Polymerization reactions are difficult to follow with physical measurements alone, particularly when new materials or processes are being developed. In many cases, however, volatile monomers and other gases indicative of the reactions are evolved during the polymerization, especially when elevated temperatures and condensation-type reactions are involved during which a nonpolymeric molecule is released as a byproduct. The evolution of volatile components can also occur with low-temperature reactions, however, as demonstrated by the familiar odor of vinegar associated with the curing of room temperature vulcanizing (RTV)-type synthetic rubbers. The mass spectrometer provides a highly sensitive and versatile method both for monitoring the polymerization process and for characterizing the finished polymeric material. The major difficulty of the measurement is to assure that the sample introduced into the instrument is indicative of the process and has not undergone further reaction in the sample collection and inlet system.

As an example of a continuously monitored process, phenol evolved during a polymerization of a mixture of phenolic microspheres, phenol formaldehyde prepolymer, and polyamide powder were monitored to determine the extent of cure during dielectrically heated molding of an ablative heat shield material for manned-spacecraft applications (Aikens et al., 1975). Heating past the fully cured state resulted in severe internal charring of the material, while insufficient heating produced a molding that was structurally

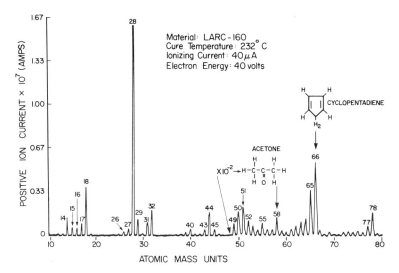

Figure 14.9 Gases evolved during autoclave curing of a condensation polymer at 232°C. (Smith, 1983; courtesy A. C. Smith; copyright SAMPE.)

unacceptable. A mass scan of the gases evolved during the process identified a significant evolution of phenol, and the subsequent measurements were made by monitoring m/z 94. From the phenol evolution data, it was possible to predict when polymerization was essentially complete and when charring of the material would occur.

Off-line analyses of evolved volatiles obtained at intervals during polymerization are also useful in identifying components that can be used to characterize the process or that can subsequently be selected for on-line monitoring. Smith (1983) has discussed off-line measurements of this type obtained at several increasing temperatures during autoclave curing of a condensation-type polymer, using a QMS with a differentially pumped inlet. The material evolved water and ethanol during the reaction, both of which were measured, as seen in the spectrum of Figure 14.9. The cyclopentadiene (C_5H_6) at m/z 66 resulted from a chemical rearrangement of the starting materials, while the peak at m/z 78 was from methyl substitution of the same compound; the acetone (C_3H_6O) resulted from a solvent used in preparation of the monomer.

Laser-induced vaporization of the polymer surface or thermal decomposition of the bulk material with mass spectrometric analysis of the evolved components have become standard methods for analyzing polymers. Lincoln (1983), for example, has measured the gases evolved during *in situ* laser heating of a carbon-phenolic material in a vaporization chamber directly attached to a TOF mass spectrometer. In an earlier work, Covington et al. (1977) showed that laser heating produces a quasisteady, supersonic, free-jet expansion of the vapor products to form an isentropic flow with stream-

TABLE 14.2   Mole Fractions of Vapors Evolved During
Laser Heating of a Carbon-Phenolic Polymer Composite

	Pulsed (58 kW/cm^2)	Continuous (270 W/cm^2)
CH_4	0	0.12
H_2O	0.10	0.48
C_2H_2	0.026	0
C_2H_4	0.021	0.05
CO	0.065	0.14
C_3	0.69	0
CO_2	0	0.030
C_4H_2	0.088	0
Toluene	0	0.023
Phenol	0	0.031
Cresols	0	0.028
Xylenols	0	0.012
C_3H	0.02	0

Source: Lincoln (1983).

lines appearing to radiate from a point source. Furthermore, the rapid gas-dynamic expansion freezes the components in a collisionless regime, assuring that the measured species are representative of the gases at the surface. A comparison of measurements obtained with rapid high-energy heating with a 58 kW/cm^2 pulsed laser and with a lower energy, 270 W/cm^2 continuous-wave laser are listed in Table 14.2. The longer low-energy exposure produces products corresponding to pyrolysis of the polymer, while species released at the higher heating rate are typical of polymer vaporization, thereby providing information on the final state of the polymerized material.

Blast and Basic Furnace Operation

Elemental iron, the principle constituent of cast iron and steel, is also used both as a catalyst and in the formulation of ferric and ferrous chemicals. Commercially important ores contain iron in the form of oxides, for example, hematite (Fe_2O_3), magnetite (Fe_3O_4), ilmenite ($FeTiO_3$), limonite (FeO[OH]), and siderite ($FeCO_3$). The iron is extracted from the ore by high-temperature reduction with carbon (coke), carbon monoxide, hydrogen, or gaseous hydrocarbons. A number of processes are used in reducing the iron ore; however, the majority of all iron is extracted in a blast furnace, which is a tall refractory-lined vessel. The feedstock, consisting of iron ore, coke, and limestone for fluxing the gangue (mineral aggregates), is introduced continuously into the top of the furnace, and hot (1600°C) air from a blast stove is injected into the bottom. The rising gases burn the coke to reduce and melt the charge as it settles towards the bottom of the furnace, where the slag and liquid iron are periodically removed.

Efficient operation of the blast furnace is achieved by maintaining optimum carbon monoxide-to-carbon dioxide ratios. The multiple-collector mass spectrometer in Figure 14.1, as well as several quadrupole- and TOF-based systems, provide on-line monitoring of these and other gases and are used by the furnace operator to adjust air or coke supplies to obtain maximum yields. The gaseous combustion products contain about 28% carbon monoxide, having a calorific content of approximately 90 BTU/ft^3, along with a lesser concentration of hydrogen. This gas is taken from the top of the furnace, filtered to remove dust, and used as a fuel for the hot gas blast stoves. Total calorific content is determined and maintained by monitoring the CO and H$_2$. The top gas is also monitored for the presence of oxygen to assure that an explosive gas mixture does not occur.

An important process for the production of high-grade steels is the basic-oxygen-furnace, which is capable of reaching yield rates as high as 270 metric tons/h. In this process, the molten pig iron is refined by blowing oxygen at high pressure and supersonic velocity onto the surface of the melt through a water-cooled lance, thus oxidizing the carbon and impurities. The addition of lime to the oxygen stream assists in the formation of slag, especially when the iron has a high (0.3%) level of phosphorous. Operating efficiency of the basic oxygen furnace is increased by analyzing the gases above the melt, in that the extent of the conversion and, hence, the required oxygen flow time can be determined by mass flow measurements and calculations of the degree of decarburization. Both magnetic sector spectrometers and QMSs are used in this application to measure N$_2$ (0–79%), CO (0–80%), CO$_2$ (0–60%), O$_2$ (0–21%), and Ar (0–2%) over the ranges shown. H$_2$ is also monitored to detect explosive mixtures, which can be a common occurrence in basic-oxygen-furnace operations.

Fluidized Bed Combustion

The electric power generating industry annually consumes more than 400 million metric tons of coal, which is about 65% of the total produced in the United States. Therefore, large economic benefits are realized through the improvement or development of methods that increase the thermal efficiency of the combustion process. One such method is the fluidized bed combustor, in which a pulverized mixture of coal and limestone is burned while held in suspension by a high-pressure stream of air or oxygen. Sulfur is removed by reaction with the limestone, production of oxides of nitrogen and volatilization of sodium and potassium are reduced because of the lower (1500–1600°F) fuel bed temperatures, thermal efficiencies approach 50%, and heat transfer rates as high as 100 BTU/(h-ft^2-°F) are realized.

Operational control of the fluidized bed combustor is accomplished by real-time analysis of the combustion products for SO_x, NO_x, H_2O, CO_2, CO, O_2, and CH_4; and control of emissions requires the measurement of trace (10^{-9} g/m^3) levels of polycyclic aromatic hydrocarbons and their heterocyclic analogs (Johnson et al., 1982). Because of the complexity of the combustion products, the most promising approach for this application is atmospheric pressure ionization, triple-quadrupole mass spectrometry (API/MS/MS/MS) (Chapter 4). In this method, a regulated portion of the flue gas is filtered, dried, and passed through a transfer line maintained at 250°C into the ion source, where it is ionized at atmospheric pressure either by a corona discharge or by a nickel-63 beta emitter, and the ionized gas is expanded through a small diameter orifice into the high-vacuum region. An ion of interest selected by the first quadrupole is injected into the second, which is operating in the RF only mode, where it undergoes collision with a neutral gas independently injected into the region. The parent ion and collision-induced daughter ions are then mass analyzed in the third quadrupole.

When individual gas monitors are used, current practice requires additional measurements in which the polycyclic aromatic hydrocarbons are collected on an adsorbing material for later analysis by GC/MS or HPLC/MS (Chapter 7). On-line monitoring with API/MS/MS/MS avoids the need for collection of the adsorbent, minimizing reactions that may occur during sampling and eliminating the necessity for use of a multiplicity of measurement methods to determine the predominant combustion products.

CONTROLLED SPACECRAFT AND SUBMARINE ATMOSPHERES

The limits of unaided excursions from the terrestrial surface range from an underwater depth of 100 ft to an altitude of 18,000 ft, while breathing apparatus with suitable gas mixtures extend these limits to 300 and 40,000 ft, respectively (Mills, 1971; Robinson, 1973). Beyond these extremes, enclosed pressurized vessels with atmospheric monitoring and real-time control to maintain a life-supporting atmosphere are necessary. These systems are classified as *open, semiclosed,* or *biologically closed,* depending upon the regenerative processes used (e.g. Spurlock and Modell, 1979). The nonregenerative *open* system provides basic necessities as expendables, and might consist of an oxygen supply, adsorbents for removing carbon dioxide and water, minimally interactive temperature, humidity, and waste control, and enough food and water to accomplish the mission. In the *semiclosed* system, some regenerative or recycling processes are used. For example, electric power for the Apollo command and service module was generated by three 1.42 kW fuel cells, which combined hydrogen and oxygen to concurrently produce 1.5 lb/h of potable water under normal load (Sharpe, 1969); while water removed from the spacecraft atmosphere was recycled into the temperature control system. Operational lifetimes can also be extended if O_2 is recovered from CO_2 removal systems. This can be accomplished by reaction with an alkali metal superoxide (e.g., $4CO_2 + 4NaO_2 + 2H_2O \rightarrow 4NaHCO_3 + 3O_2$), which is a process used in Soviet spacecraft.

The fully regenerative, closed environment is one in which all wastes or products from functional units of the ecological system are transformed into self-supporting reactants, that is, food, water, and atmosphere. This self-sufficient system is not thermodynamically isolated, however, since an external input of energy is always required (Krauss, 1979). Other than the earth itself, the fully *closed biological* system does not exist. Conservation of resources is of concern in long-term spaceflights or space colonization, and this has stimulated research in closed system technology in both the United States and the Soviet Union. The open systems consume material, while the regenerative systems consume only power (McConnaughy, 1958), so that multiple-function regenerative systems will be used whenever possible.

Spacecraft

The environmental control system (ECS) for a high-altitude research craft must (1) maintain pressures commensurate with the composition of the cabin atmosphere; (2) control temperature and humidity; (3) protect against acceleration, noise, radiation, and the build-up of toxic compounds; (4) manage and dispose of wastes; (5) maintain physiological and psychological well-being of the crew; and (6) provide the metabolically required atmospheric composition. The complexity of the ECS is determined by the number of personnel, length and purpose of the mission, and weight, volume, and power allocations. The extent of on-board monitoring is, in turn, influenced by these same parameters and also by the number and kind of gases used in the spacecraft atmosphere. On-board gas analyzers can monitor the ECS performance and can provide data to the crew or to the ground control stations by telemetry. In missions of short duration, measurements of the relatively high-abundance primary constituents (O_2, N_2, CO_2, and H_2O) are sufficient, reducing demands on instrument sensitivity. Single-gas analyzers are thus most expedient for measuring the primary constituents on short flights; but for longer missions, mass spectrometers are advantageous, since both primary and trace constituents can also be measured (Smylie and Reumont, 1964). Lehotsky (1973), for example, described a 3 cm radius, 90° magnetic sector mass spectrometer with multiple collectors for monitoring the primary constituents of the spacecraft atmosphere at m/z 18, 28, 32, and 44, having a generalized configuration based on the prototype instrument shown in Figure 14.10. In Lehotsky's version additional detectors were also included for m/z 2 (H_2) and for monitoring the sum of all ion currents between m/z 50 and m/z 120 as a measure of total organic contamination.

The first enclosed spacecraft were, in fact, pressurized gondolas for high-altitude balloon flights, beginning with that designed by Augusté Picard in 1931 (e.g., Engle and Lott, 1979). Oxygen or alternative mixtures of oxygen, helium, and nitrogen were supplied and monitored, carbon dioxide and water

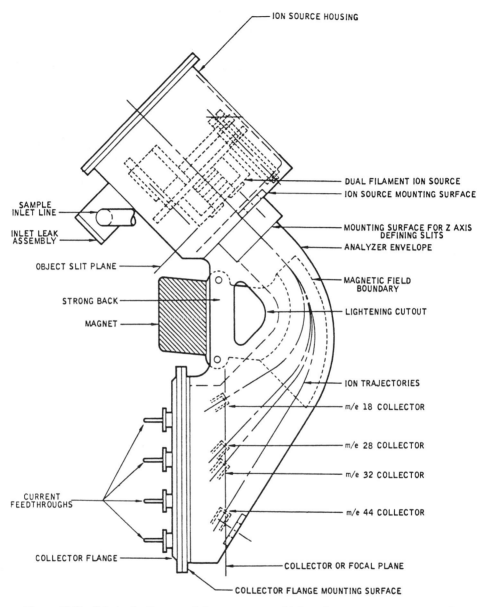

ION SOURCE HOUSING

DUAL FILAMENT ION SOURCE
ION SOURCE MOUNTING SURFACE

SAMPLE INLET LINE

INLET LEAK ASSEMBLY

MOUNTING SURFACE FOR Z AXIS DEFINING SLITS
ANALYZER ENVELOPE

OBJECT SLIT PLANE

MAGNETIC FIELD BOUNDARY

STRONG BACK

LIGHTENING CUTOUT

MAGNET

ION TRAJECTORIES

m/e 18 COLLECTOR

m/e 28 COLLECTOR

m/e 32 COLLECTOR

CURRENT FEEDTHROUGHS

m/e 44 COLLECTOR

COLLECTOR FLANGE

COLLECTOR OR FOCAL PLANE

COLLECTOR FLANGE MOUNTING SURFACE

Figure 14.10 Schematic diagram of the prototype multiple collector mass spectrometer for monitoring spacecraft atmospheres (NASA files).

were removed by adsorbents, cabin pressures were maintained at safe levels, and pressure suits were subsequently worn as back-up at very high altitudes. The early manned space capsules were similar but more sophisticated systems, with cabin atmospheres also maintained in an expendable manner and spacesuits worn as back-up for the ECS for the duration of the flight.

Mission goals and the ability to place heavy spacecraft in orbit led the Soviet Union to develop a two-gas system approximating atmospheric composition and pressure. Cabin pressures are maintained at 765 ± 10 mm, temperature at 20 ± 6°C, humidity at 54 ± 3%, and oxygen levels at 23 ± 2%. Gases supplied from tanks circulate through an air regeneration system containing an alkali metal superoxide that removes CO_2, releases O_2, and removes excess water and odors. Both O_2 and CO_2 are monitored and controlled along with the other parameters listed above (Johnson, 1980).

The two-gas atmospheric pressure system requires little acclimation by the crew. It does require continuous monitoring and control to maintain the proper atmosphere, and the hazard of acute dysbarism (decompression sickness) or ebullism (vaporization of body fluids) exists in the event of cabin decompression resulting from meteorite impact or from system failure.

Weight restrictions and a goal of lunar exploration led to a single-gas, low-pressure system for the U.S. Mercury, Gemini, and Apollo spacecraft, which operated at an equivalent pressure altitude of 27,000 ft with 100% oxygen at a pressure of 258 mm Hg (Baker, 1981). Oxygen was supplied from pressurized tanks, and CO_2 (produced at a rate of about 400 ml/min per astronaut) was removed by reaction with lithium hydroxide. Odors and other contaminants were adsorbed on activated charcoal, which was subsequently removed for post-flight analysis by GC/MS. The low-pressure oxygen system is simple and reliable in that the possibility of a cabin leak is reduced and only a single gas needs to be controlled; however, it also has hazards that must be addressed. In addition to monitoring oxygen concentrations, the maintenance of cabin total pressure is critical. With 100% oxygen, blood oxygenation levels increase by 0.3% for each 100 mm increase in pressure, until at a pressure altitude of about 12,000 ft saturation occurs with an onset of hypoxia or oxygen toxicity (Engle and Lott, 1979). There is also an extreme fire hazard that is reduced, but not eliminated, with flameproof materials but is minimized in the zero gravity of orbital altitude (Sharpe, 1969). A modified two-gas system used in later U.S. missions had an initial atmosphere at launch of approximately 60% nitrogen and 40% oxygen. After launch, nitrogen was purged from the atmosphere at a controlled rate and was continuously monitored until an approximate 100% oxygen level was obtained when the spacecraft was in orbit.

To minimize toxic compounds in the atmosphere, materials used in the spacecraft cabin are selected based on measured outgassing characteristics and on toxicological studies of the identified constituents. Post-flight analyses of extracts from charcoal adsorbents removed from the ECS, using MS, GC/MS, GC, and IR methods, are performed to quantitatively assay toxic

TABLE 14.3   Contaminants Extracted Post-Flight From the Mercury A6 (Friendship 7) Charcoal Adsorbent

Contaminant	Minimum Concentration (ppm)
Vinylidene chloride	3
Benzene	3
Vinyl chloride	2
Methyl chloroform	0.5
Methylene chloride	0.4
p-Dioxane	0.3
Unidentified	0.3
Cyclohexane	0.2
Toluene	0.2
Methyl alcohol	0.2
Ethyl alochol	0.05
Trichlorofluoromethane	0.05

Source: Saalfeld and Saunders (1979).

species evolved during the mission. Table 14.3 lists contaminants identified in the charcoal adsorbent material taken from the Mercury A6 (Friendship 7) spacecraft following the three-orbit flight, which lasted about 5 h. Post-analyses following the much longer Apollo missions identified about 50 compounds, including both hydrocarbons and halocarbons; however, in all cases, concentrations were too low to be of toxicological significance even when considering synergistic effects (Berry, 1970).

Orbital Space Stations

The U.S. Skylab, the largest manned orbiting spacecraft to date, was constructed from the upper stage of a Saturn 5 rocket and had a controlled working volume of approximately 12,000 ft^3. Structural constraints and the requirement for docking with the Apollo spacecraft dictated that a total system pressure of 259 $\pm$ 10 torr be retained. The duration of the mission precluded the use of 100% oxygen, and a controlled mixture of approximately 78% O_2 and 22% N_2 was substituted. CO_2 was maintained at an average concentration of 5.5 mm with molecular sieve material (Linde 5A and 13X), odor and contaminants were removed with activated charcoal, and humidity was maintained below a dewpoint of 286 K by condensing heat exchangers and the Linde 13X molecular sieve beds (Hopson et al., 1976). The Skylab gas control instruments monitored ECS performance and analyzed for contaminants that were of concern not only for crew safety, but also for possible effects on optical surfaces of on-board scientific instruments (Disher, 1975). The ECS data were also telemetered to ground control and support centers, from which mission control could send commands to remotely reconfigure the ECS or to assist the crew in correcting anomalies.

Smaller than Skylab, the Soviet Salyut space station had a volume of appoximately 3500 ft^3 with one Soyuz transport attached. Like all other Soviet spacecraft, it was designed to maintain terrestrial atmosphere and conditions. The ECS removed contaminants, odors, and particulates with filters, liberated O_2 while chemically removing CO_2, and regenerated water from atmospheric moisture condensate. Atmospheric parameters were maintained within assigned limits of 0.9 torr CO_2, 160–280 torr O_2, and a total pressure of 760–960 torr by manual connection and disconnection of adsorbers and regenerators as directed by readings from the gas analyzers (Balayan, 1983).

Despite the regeneration of water (and the liberation of O_2 by the Salyut system), both the U.S. and Soviet space stations were at best semiclosed systems. For Skylab, sufficient gases and other consumables were carried aloft at launch to provide for the manned missions of 144 (later extended to 177) days, plus a comfortable safety margin. Salyut, on the other hand, was regularly resupplied with atmospheric gases and other necessities for missions lasting months at a time. Initially resupplied from manned Soyuz spacecraft, resupply was later accomplished with an unmanned expendable transport (Progress) capable of carrying 2.3 metric tons and of automatically docking with the Salyut. Instrumentation for the ECS has therefore been commensurate with an expendable system, but it will of necessity become more extensive and complex as regenerative systems become fully developed and the duration of missions and crew size increase.

The space transportation system (STS), or space shuttle, is the first U.S. spacecraft designed to operate with a standard atmosphere and 760 torr pressure. In addition, because of the complexity of the system and the diversity of the planned missions, constraints on materials used in construction and furnishings are less stringent from an outgassing standpoint than for prior manned spacecraft. Therefore, there is an increased possibility of contaminants occurring at physiologically significant levels from material outgassing, fluid system leaks, and metabolic byproducts. These contaminants include a large number of alcohols, aldehydes, aliphatics, aromatics, ketones, esters, heterocyclics, halocarbons, and inorganic compounds. Analyzing for this many species at low concentrations precludes single-gas analyzers and is beyond the analytical capability of small mass spectrometers alone, requiring in addition a gas chromatographic inlet. An automated GC/MS developed to periodically sample and analyze the atmosphere in a spacecraft (Dencker, 1975; Anonymous, Perkin-Elmer Corp., 1981) typifies the instrumentation required for this application. Based on the Viking-lander GC/MS (Chapter 9), the inlet consists of a dual-column chromatograph (Fig. 14.11) and includes an internal gas supply for periodic recalibration of the system. One column separates carbon monoxide from the major atmospheric constituents while the second separates organic compounds, and a two-

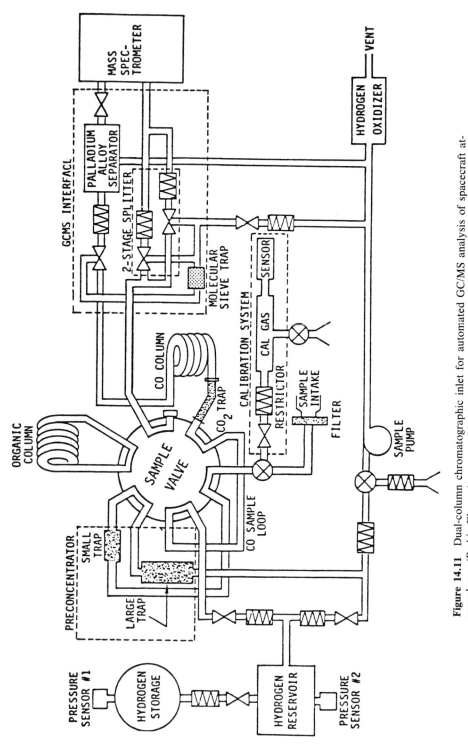

Figure 14.11 Dual-column chromatographic inlet for automated GC/MS analysis of spacecraft atmospheres (Perkin-Elmer Aerospace Division).

stage, temperature programmable preconcentrating trap is used to increase sensitivity. The GC is interfaced to the MS both with an effluent divider to extend the dynamic range for the organic compounds and with a silver-palladium electrochemical separator to remove the hydrogen carrier gas when analyzing carbon monoxide. The Nier-Johnson mass spectrometer is scanned from m/z 25–250 as the GC peaks elute, with a total cycle time for the anaysis of carbon monoxide and the hydrocarbons of about 2 hours.

Biologically Closed Systems

The logistics of providing periodic resupply during extended space travel (e.g., manned missions to Mars) or for space or planetary colonization involving large crews will require a closed regenerative life support system to maintain atmospheric composition, provide food, and remove wastes. Such a system maintains individual organisms in balance with each other, and therefore is a closed but very limited ecosystem. Botkin et al. (1979) suggest that the system will ultimately consist of interconnected compartments having specialized functions with controlled and measured flows between them; for example, an oxygen-consuming compartment will be connected to an oxygen producer, and so forth. Removal of CO_2 and the generation of O_2 and food will result from photosynthesis—the direct conversion of electromagnetic energy into chemical energy. The most likely supporting organisms to function in the life support system are bacteria (some of which may not be photosynthetic), higher plants, and (most promising) some forms of algae, which in photosynthesis absorb CO_2, release O_2, and produce a highly nutritious cellular product (Krauss, 1979).

   Such a system must be continuously monitored and carefully managed to maintain the state of one organism and its activity with respect to all others, and to detect and prevent events that might lead to the extinction of any of the species. In a larger, fully developed ecosystem such as the earth, a catastrophe occurring to one species is generally offset by increased opportunities and activities of others. In the limited system, this is not the case. The production and accumulation of toxins from outgassing of materials and from crew or photosynthetic activity is indeed serious, since it can lead to the loss of a critical organism in the regenerative cycle. On-line and redundant instrumentation to detect, identify, and isolate toxins at trace levels and effect their removal will therefore be mandatory, with the most likely instrumentation for this monitoring being the computer-controlled, gas chromatograph/high-resolution mass spectrometer.

Submarines

It was not until the development of nuclear power that the true submarines appeared—their predecessors having, in fact, been surface vessels capable of operating submerged for periods limited by the capacity of the batteries

534 ON-LINE MONITORING AND PROCESS CONTROL

rather than by the purity of the atmosphere (Cross, 1959). The ability to submerge almost indefinitely required concurrent development of atmospheric control and monitoring systems, along with the identification of a multitude of contaminants and an assessment of their physiological effects.

In contrast to the rigidly controlled spacecraft atmosphere, the modern submarine environment is exceedingly complex, and it is believed to consist of more than 10,000 species (Saalfeld and Saunders, 1979). These compounds arise from (1) lubricants and machinery operations; (2) out-gassing of batteries, materials, and housekeeping supplies; (3) byproducts of chemical or catalytic contaminant removal systems; and (4) crew activities such as cooking or smoking. Most of these are not harmful, but some are toxic even at extremely low levels. An example is dichloroacetylene (C_2Cl_2), a highly toxic, spontaneously explosive compound that is stabilized in air by other contaminants, and which may be incapacitating after 7–8 days exposure to concentrations of a few parts per billion. Dichloroacetylene is not released into the atmosphere through outgassing, but is generated when trichloroethylene (C_2HCl_3, a commonly used degreasing agent) reacts with alkaline materials, such as lithium hydroxide (LiOH), that are used to remove carbon dioxide. Synergistic effects may also occur, and some contaminants may be strong pulmonary or ocular irritants.

Analysis of all or most of the compounds in the submarine atmosphere is an immense undertaking and is generally not required in day-to-day monitoring. Most of the species are hydrocarbons, and classification by types (i.e., aliphatics or aromatics) is sufficient. Halogenated hydrocarbons are also a concern because of undesirable physiological effects and the possibility of decomposition to form toxic vapors. Saalfeld et al. (1971) identified 56 contaminants found in the atmosphere of the U.S.S. Hammerhead, including saturated and unsaturated aliphatics, C_6–C_{14} aromatics, ketones, acetates, alcohols, nitriles, sulfides, chlorides, aldehydes, oxides, silanes, methylsilicones, and halocarbons. Typical analyses by hydrocarbon type indicate that about 30% are aromatics, and measured levels for representative species are listed in Table 14.4.

The submarine atmosphere passes through a number of purifiers, one of which is an activated charcoal bed. This bed is replaced at periodic intervals, at which time samples of the adsorbent can be sealed in bags for later analysis. The results represent an integrated qualitative analysis of the total atmosphere over the sampling time. Adsorption samples have also been obtained at specific locations with small, hermetically sealed canisters containing a few grams of activated charcoal (Saalfeld et al., 1971) or by adsorbent-filled personnel sampling tubes (DeCorpo et al., 1976). The contaminants are desorbed by heating or extracted with solvents to be analyzed by appropriate methods, with the most effective method being GC/MS (Chapter 7). Quantitative accuracy is limited because of differing adsorption-desorption efficiencies for the various constituents, and because some may desorb at such high temperatures that chemical reaction or alteration occurs.

TABLE 14.4   Analyses of Submarine Atmospheres by Hydrocarbon Type

Vessel	Volume Fraction of Hydrocarbons		
	Saturated	Unsaturated	Aromatic
Abraham Lincoln	0.631	0.073	0.296
George Washington	0.656	0.063	0.282
Snook	0.668	0.079	0.252
Robert E. Lee	0.736	0.034	0.231
Nautilus	0.704	0.059	0.237
Patrick Henry	0.565	0.065	0.370
Average	0.660	0.062	0.270

Source: Saalfeld and Saunders (1979).

All nuclear-powered submarines in the U.S. fleet are equipped with on-line analyzers to provide real-time measurement of the most important atmospheric constituents. Initially (1) oxygen was measured by paramagnetic analysis methods; (2) carbon monoxide by dual-isotope, fluorescent, non-dispersive infrared sensors; (3) hydrogen by thermal conductivity; and (4) hydrocarbons by gas chromatography. With the exception of the infrared analyzer, which was retained to monitor CO, all of these were replaced with a single, nonscanning, magnetic deflection mass spectrometer with multiple detectors (Ruecker, 1973; Saalfeld and Wyatt, 1976; Saalfeld and Saunders, 1979) that simultaneously measures m/z 2 (H_2), 18 (H_2O), 28 (N_2), 32 (O_2), 44 (CO_2), 85 (freon-12, CCl_2F_2), 101 (freon-11, CCl_3F), and 135 (freon-114, $C_2Cl_2F_4$). Monitoring of these fluorinated hydrocarbons is important, because they can be converted into highly toxic and corrosive compounds in the catalytic burners used to remove hydrocarbon contamination from the atmosphere. The Central Atmospheric Monitoring System (CAMS-I) mass spectrometer was the result of a joint Navy/NASA program to develop a highly reliable automated monitoring system. The instrument was based on the earlier but smaller version developed for use on manned spaceflights (Bicksler, 1971), and it was the predecessor of the process monitoring and control instrument shown in Figure 14.1. Prior to becoming fully operational, two prototype CAMS-I instruments were operated on submarines for approximately 25,000 h without a major failure, demonstrating that the reliability of the mass spectrometer was adequate for this application. Gases are continuously pumped to the system from various locations in the submarine and are sequentially analyzed by connecting the mass spectrometer to each gas transport line through a heated inlet leak valve. The constituents are simultaneously measured; and if the concentration for any constituent is outside of a specified tolerance, an alarm is initiated.

As noted earlier, there is a physical limit to the number of separate collectors that can be placed on the focal plane of a small, homogeneous field magnetic mass spectrometer. In order to measure additional trace level contaminants in air that have been determined to be physiologically important and to provide for an investigative survey of all compounds with the mass range of the instrument, a scanning mass spectrometer with a dynamic sensitivity range of 10^7 is required (Wyatt et al., 1976; Saalfeld and Saunders, 1979). Long-term stability and moderately high resolution, in this case, are more important than rapid scanning; hence, the CAMS-II mass spectrometer (Fig. 14.12) is a small, computer-controlled, magnetically scanned Nier-Johnson double focusing instrument (Cason and Koslin, 1981; Wyatt, 1984). The system is similar to that previously developed for spaceflight applications (Dencker, 1975); however, active control of the submarine atmosphere precludes the use of a GC inlet, and gases are directly sampled through a flow rate-controlled variable leak valve. The instrument has a mass range of 2–300 amu, and it is programmed to automatically step through the sequence of m/z listed in Table 14.5. Additional constituents may be monitored through simple modification of the computer software. The major constituents are measured with a Faraday cup using a dual-range linear electrometer, and the lesser constituents are measured with a channeltron electron multiplier operated in a pulse-counting mode. This minimizes temporal gain changes and provides direct digital input to the data processor. To compensate for long-term drift in the gain of either detector, the isotopic ratio of oxygen obtained by measuring m/z 32 with the Faraday cup and m/z 34 with the electron multiplier is compared to a known value.

In addition to the continuously monitored constituents, investigative survey scans over the entire mass range are periodically obtained and stored for subsequent analysis. While this information is not used as a normal monitoring and control function, the data are analyzed to identify other physiologically important contaminants related to operation of the vessel and the crew activities.

Deep-diving research habitats have high-pressure, nonstandard atmospheres equalling the pressures at the depth at which the research is being carried out. Under these conditions, personnel are able to leave and reenter the vessel without experiencing the effects of dysbarism, can remain at great depths for extended periods, and are able to return to the surface much faster than when working in a pressure suit and in a standard atmosphere. For example, four Navy divers lived and worked in Sealab-I for 11 days at a depth of 192 ft in an atmosphere of 4% O_2, 16% N_2, and 80% He at a pressure of 110 psi (Saalfeld and Saunders, 1979). The relatively short habitation time limits the need for continuous on-board monitoring and control, but studies of the atmosphere are necessary to identify and remove sources of contamination, because the lessened oxygen pressure and the relatively small internal volume amplify the effects of trace level contaminants. Saunders et al. (1967) identified 38 hydrocarbon contaminants in Sealab-II by GC/MS of desorbates from activated charcoal. Of these, three cycloalkanes, which occurred in significant concentrations, were eventually traced to a

TABLE 14.5 Gaseous Constituents, Ranges, and Sensitivities for the CAMS-II Mass Spectrometer

Compound	Monitored m/z	Range	Typical Range	Relative Sensitivity[a]
N_2	28	0–1000 torr	500–700 torr	1.00
$^{32}O_2$[b]	32	0–300 torr	155–165 torr	0.83
$^{34}O_2$[b]	34	0–1200 mtorr	620–660 mtorr	0.83
CO_2	44	0–70 torr	5–25 torr	1.30
H_2O	18	0–30 torr	—	0.60
H_2	2	0–40 torr	—	0.11
^{40}Ar[b]	40	0–7 torr	5–7 torr	1.50
^{36}Ar[b]	36	0–25 mtorr	18–25 mtorr	1.50
Aliphatic hydrocarbons	43, 57, 71, 99, 113	0–100 mtorr	—	2.10
Aromatic hydrocarbons	91, 105, 119, 133, 147	0–5 mtorr	—	2.60
Freon-11	101	0–1000 mtorr	—	1.70
Freon-12	85	0–1000 mtorr	—	1.50
Freon-114	135	0–1000 mtorr	—	0.83
Trichlorethylene	130	0–100 mtorr	—	1.00
Benzene	78	0–5 mtorr	—	2.30

Source: Cason and Koslin (1981).

[a] Relative sensitivity is the efficiency of ionization compared to N_2.

[b] For cross-calibration purposes.

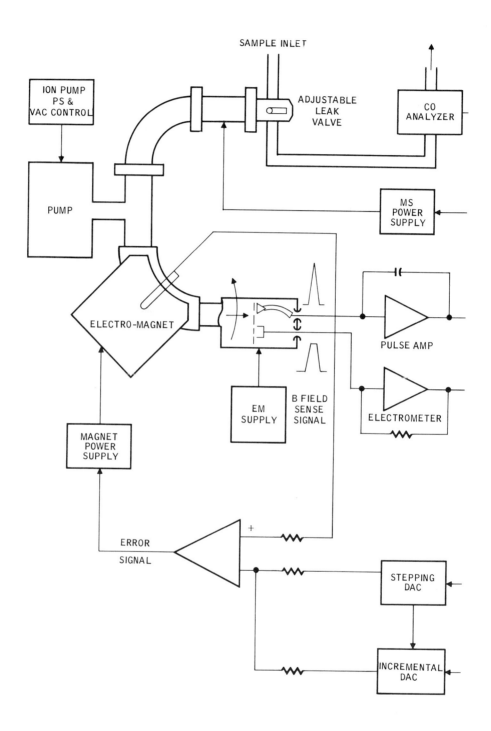

Figure 14.12 Functional diagram of the CAMS-II central atmosphere monitoring system for submarine applications. (Cason and Koslin, 1981; courtesy M. E. Koslin; copyright SAMPE.)

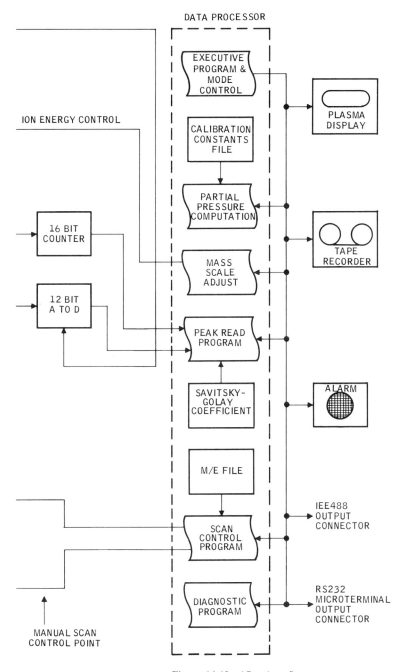

Figure 14.12 (*Continued*)

cork adhesive, and on Sealab-I, a source of heavy hydrocarbon contamination was determined to be oil-treated dust filters.

SPECIALIZED MEASUREMENTS

In addition to its capability for on-line monitoring, mass spectrometry is an important investigative method for quality control and for other industrial applications. Because of the sensitivity and speed of response, quantitative results can usually be obtained from exceedingly small samples, and because of its qualitative specificity, information concerning the mechanism or state of the process is obtained.

Gaseous Inclusions

Quality control of mass-produced enclosed devices, such as small gas-filled lamps or hermetically sealed microcircuitry, requires testing for leaks as well as determining composition or changes in composition of fill gases during the expected life of the product. Additionally, a measurement of gaseous inclusions when the encapsulating material is glass is useful in identifying and eliminating the source of these inclusions from the glass production process. These measurements require sample preparation and testing at elevated temperatures under clean high-vacuum conditions and instrumentation that has high sensitivity and rapid response over a wide dynamic range. Koprio et al. (1983) have tested such devices using a computer-controlled QMS system that includes a small, heated, gas-sampling and device-piercing chamber evacuated by a 40 l/s turbomolecular pump. Gases released in the chamber are pumped through a closed axial ion source in the mass spectrometer by a 330 l/s turbomolecular pump. Time-dependent, preselected ion currents are monitored by peak switching, and unknown peaks are detected by mass scanning. After testing the device for leaks, the sample is opened by a mechanically activated, bellows-sealed metal pin that pierces the container, releasing gas into the sample chamber. Detection limits are on the order of 1 ppm for N_2, O_2, and CO_2 and 10 ppm for H_2, H_2O, SO_2, etc.

Trace Levels of Volatiles in Polymers

The outgassing of moisture or other volatile species, such as unreacted monomers, plasticizer residue, or antistatic coatings into sealed containers, may result in contamination of the contents or failure of encapsulated electronic components. These effects may result even when the volatiles are present in trace levels and may often lead to degradation of the polymer itself. A careful analysis of the materials under the conditions to which they will be

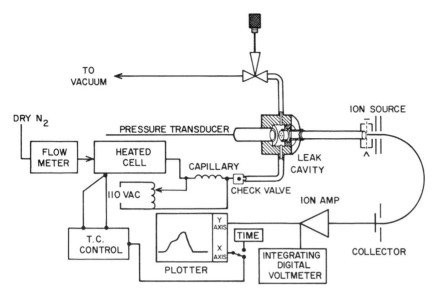

Figure 14.13 Schematic diagram of the system for the quantitative measurement of volatiles desorbed from solids (Wood et al., 1975).

exposed is therefore required to identify the potential contaminants and the rate of their evolution.

To illustrate, the presence of trace amounts of adsorbed moisture may degrade performance or cause failure in hermetically sealed microelectronic circuitry due to dendritic growth between gold-plated electrodes. Because water is very difficult to eliminate from any system, the measurement at trace levels is not trivial, even with a mass spectrometer. Wood et al. (1975) have described a method with which 250–280 mg samples of a Teflon-FEP copolymer suspected of being the source of moisture resulting in failure of hermetically sealed microcircuitry for the Viking-Mars probe were examined and found to contain 2.7–4.1 μg of water (10–16 ppm by weight), as opposed to 2.3 μg (8 ppm) for a 276 mg sample of Teflon alone. The polymer samples were placed in a heated and outgassed Pyrex cell. Rather than allowing the vapor to diffuse to the ion source in vacuum, the evolved water was transported through a heated capillary directly into a heated inlet (Fig. 14.13) by a stream of superdry nitrogen in order to maximize instrument sensitivity and response. The leak cavity, which has essentially zero volume, contains a flush diaphragm pressure transducer and a gold foil molecular leak through which a quantitatively representative amount of sample passes directly into the ion source of the mass spectrometer.

The peak intensity at m/z 18 with the cell at 150°C was recorded as a function of time, and the areas were integrated to obtain the concentration of water in each sample. Measurement precision at the parts per million level was calculated to be better than 15%, with a detection limit of 300 ng for water. Instrument sensitivity for water was determined by the total dehydration of weighed milligram samples of disodium tartrate dihydrate, which contain 15.56% water by weight. Accuracy of the method was verified by analyzing similar quantities of barium chloride dihydrate containing 14.75% water by weight. In addition to measuring total water content in the sample, the procedure also provides the capability for generating temperature-dependent curves indicative of desorption from the surface and from different sites in the bulk, when the cell is operated in a temperature-programmed mode rather than an isothermal mode.

When larger samples are available, the outgassing studies are generally performed by heating the sample in a vacuum chamber connected to the mass spectrometer ion source to which the gases diffuse, and a number of such studies have been reported in the literature. While less quantitative than the first method described because of the difficulty in obtaining an accurate calibration and the adsorption of condensable species in the sample transport system, this technique provides an efficient means of identifying the evolved species and of calculating diffusion rates through the material. As an example, Grotheer and Smith (1983) described an apparatus to be used in the measurement of diffusion coefficients and evolution rates of volatile or absorbed components in polymeric materials by using TOF mass spectrometry. Total condensed volatile components are determined using a standard ASTM weighing method, and some limitations on accuracy of the measurement are discussed. Rigby (1984) has pointed out that corrosion in hermetically sealed electronic packages may also be enhanced by potentially ionic species such as CO_2, HCl, or organic halides, and that active surface layers may be influenced by oxidizing or reducing atmospheres. He described a computer-controlled QMS system and the analytical procedures used to assure quantitative transfer of the sample gas for quality control and device failure assessment in the manufacture of highly reliable devices. Hermeticity is determined by immersing the package in an inert tracer gas (such as helium) at elevated pressure. Measurement of gases effusing from the package is nondestructive, but has a detection limit of approximately 10^{-8} atm-ml-s^{-1}, corresponding to a 50% exchange of ambient gas in a 0.1 ml package in 6 weeks. Measurement of infused tracer along with the other gaseous constituents by penetrating the package with a needle-sampling device extends the detection limit to less than 10^{-10} atm-ml-s^{-1}. The procedure, of course, is destructive, but it is necessary for the quantitative analysis of the package atmosphere. Measurement of volatiles released when the polymer is stressed or abraded in vacuum at room temperature is also an emerging technique that eliminates interferences in the data from pyrolyzates formed during pyrolysis of the material. The stress-induced events release nanogram amounts of organic vapors in short (<1 s) bursts, which are then analyzed by TOF mass spectrometry. These and other mass spectrometric methods for the characterization of polymers have recently been discussed in a comprehensive review by Shulten and Lattimer (1984).

High Temperature Corrosion and Oxidation

In quantitative studies of high-temperature corrosive or oxidative atmospheres, it is essential that on-line analysis and real-time control of the reactants be provided. These experiments include oxidation, carburization, reforming, combustion, petrochemical processes, coal gasification and liquefaction, and studies of gas-cooled reactors (Kane and Goodell, 1982). At high temperatures, the gas composition is determined by the starting mixture, catalytic effects, and molecular rearrangement. Hydrogen and carbon dioxide, for example, are the principal high-temperature products of a mixture of hydrogen, water, and methane. Reactions may be carried out at measurable rates by mixing limited amounts of reactants in an inert carrier or by saturation of a primary component to obtain high-temperature products. In this manner, oxidation and exchange mechanisms on crystalline and amorphous quartz at temperatures reaching 1400 K have been measured with a 1–2% mixture of $^{18}O_2$ or $^{18}O_2 + {}^{16}O_2$ in a neon carrier flowing over the surface of the quartz (Batten et al., 1985); and the effects of oxidation potential on the carburization of alloys in petrochemical streams have been studied with alcohol-saturated hydrogen to provide the required $H_2/CH_4/CO$ ratios (Kane, 1981).

A unique nondestructive method for measuring corrosion inside a closed vessel, such as a stainless steel fuel tank, monitors catalytically or chemically induced changes in the composition of a probe gas (Aerospace Corp., 1983). In these tests, hydrazine (N_2H_4) vapor was catalytically decomposed at an accelerated rate by the iron oxide to form N_2 and H_2, resulting in a measurable change in hydrazine partial pressure in the test tank with the smallest detectable rust patch having an area of approximately 1 mm^2. Because of this sensitivity, about 1 mol of liquid (32 g) would provide sufficient vapor to test a tank having a volume of 1.5 m^3 (50 ft^3). About 3 h are required to detect the corrosion, with typical run times on the order of 24 h to obtain statistical accuracy.

Earthquake Prediction

Evidence that earthquakes may sometimes be preceded by geochemical signals was reported about 1970 in the Soviet Union and Japan, and some Chinese geochemical data have been correlated with several strong earthquakes. In the United States, variations in radon concentration have been the most widely studied parameter in ground water and in soil gas near active faults (King, 1980).

Other chemical, radioisotope, and mass spectrometric measurements have included helium, carbon dioxide, nitrogen, argon, methane, sodium, fluorine, radium, ^{210}Pb, and isotope ratios of D/H, ^{13}C/^{12}C, ^{18}O/^{16}O, and ^{234}U/^{238}U. The rationale for such measurements is that earthquake-related geophysical phenomena, such as strain and crustal movement, will induce small chemical changes within the fault zone.

Radon (^{222}Rn) is an inert and water-soluble gas that is produced as an intermediate decay product of ^{238}U, which is widely dispersed throughout the earth's crust. Because of its inertness, radon concentrations in ground water are seasonably independent of fundamental hydrological parameters such as water composition and temperature, but tectonic perturbations that induce microfractures of rock or alter flow rates of pore fluids can sometimes be sensed by radon monitoring. Thus, locally produced radon serves as a tracer gas to detect changes in flow rates of subsurface fluids that may occur in association with earthquakes (Mogro-Campero et al., 1980). The effect of stress on tensile cracks and radon emanation from uranium-bearing granitic rock has also been observed in laboratory studies (Holub and Brady, 1981). In addition, both solid-state nuclear track detectors and real-time instrumentation have been employed to detect at least 91 radon anomalies associated with 46 different earthquakes worldwide (Hauksson, 1981). In general, the size of the region where radon anomalies are detected appears to scale with the earthquake magnitude; and radon data, per se, seems to correlate with a region, but it fails to locate a specific fault.

Variations in the stable isotope ratios of hydrogen and oxygen in ground waters of seismically active regions of California have also been investigated. Figure 14.14 shows changes of deuterium vs. δ^{18}O for water samples col-

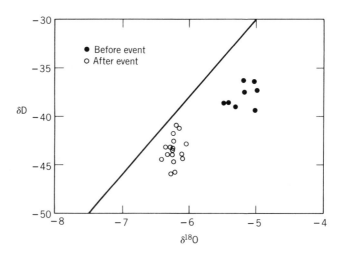

Figure 14.14 Deuterium vs. δ^{18}O for the water samples collected before and after the isotopic anomaly related to the largest and closest earthquake at the Mission Farm Campground near San Juan Bautista, CA. The straight line shown is the "meteoric water line." (O'Neil and King, 1981; copyright American Geophysical Union.)

lected before and after the largest and closest earthquake at the Mission Farm Campground along the San Andreas Fault (O'Neil and King, 1981). The potential utility of uranium isotopes as fluid phase earthquake precursors has also been assessed by monitoring water from 24 wells and springs associated with southern California fault zones (Finkel, 1981). It was anticipated that changes in the flow rate of water and the exposure of fresh rock surfaces might influence $^{234}U/^{238}U$ ratios in fault zone ground water. ^{234}U goes into solution more readily than its parent ^{238}U, either because of nuclear recoil or the differential chemical state of the two species. Data correlating to the $M = 6.6$ Imperial Valley earthquake of October 1979 correlate with changes in uranium concentrations of approximately 60 and activity ratios of a factor of 3, as observed about 70 km from the epicenter.

Variations in soil-gas helium concentrations have also been monitored in order to determine whether helium might serve as a possible geochemical parameter for predicting earthquakes. Helium, as well as radon, is generated by the radioactive decay of uranium and thorium; helium is stable, and it can be detected mass spectrometrically at the parts per billion level. Limitations on measuring significant variations in soils, however, include the atmospheric background of 5240 ppb He and perturbations due to precipitation, vegetation, and diurnal and bacterial life cycles (Reimer, 1980). Nevertheless, small changes in helium concentrations in ground water have been noted to precede earthquake activity, and decreases in the helium concentration of soil gas have been observed to precede six of eight recent central California earthquakes (Reimer, 1981). For this latter study, 10 monitoring stations were established near Hollister, California and along the San Andreas Fault. The data reveal a quasisinusoidal seasonal variation reflecting soil moisture content, together with small changes that tend to correlate with times of earthquake activity.

Furthermore, the mass spectrometric studies of Giardini et al. (1976) at least seem to indicate that a functional relationship exists between the volume of occluded gas released from various rocks and levels of stress. In this work, the rock samples were placed in a crusher that was directly connected to the inlet of a mass spectrometer after an initial sample heating and outgassing to 1×10^{-9} torr. Each rock released gas with a characteristic composition. Figure 14.15 shows the release of methane from samples of granodiorite and gneiss as a function of applied pressure. Other released gases were H_2, N_2, CO, O_2, and CO_2. Hence, it is reasonable to assume that some form of mass spectrometry will be included at major earthquake monitoring stations in the future.

Acoustic Mass Spectrometry of Natural Gas

One of the most challenging on-line measurements relates to the vast pipeline system that transmits natural gas to millions of industrial and residential customers throughout the United States. This system is analogous to the

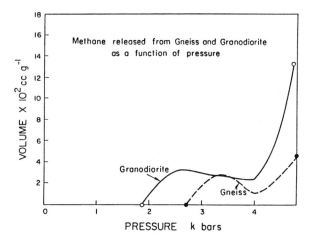

Figure 14.15 Mass spectrometrically observed emission of methane from rocks subjected to stress. (Giardini et al., 1976; copyright American Geophysical Union.)

interstate electrical power grid in the sense that energy is distributed over great distances and from many energy sources. Unlike electrical power, however, that is transmitted with a high degree of voltage, current, and frequency control, pipeline natural gas is a mixture of various gaseous components having quite different thermal or BTU contents. The gas will be principally methane, with low levels of ethane, propane, and butane—plus trace levels of carbon dioxide and nitrogen noncombustibles. The actual gas content will depend on the source, that is, Texas, Louisiana, Oklahoma, Mexico, Canada, etc., or liquid natural gas (LNG) supplied by tanker from abroad. The heat from natural gas combustion is the result of energy release during formation of CO_2 in the presence of excess oxygen. The greater the number of hydrocarbon bonds, the proportionally greater the thermal content. This proportionality is evident in the molecular weight-energy relationship shown in Table 14.6.

Thermal or BTU standards, of course, are established by gas calorimetry, but the blending and mixing of gases from diverse sources calls for on-line measurements for purposes of equitable customer billing, and for thermal control in some industrial processes. Among the several on-line potential measurements for determining BTU content, an acoustic approach appears to be the most simple, nonintrusive, and economical. Also, because the measurement is essentially mass-dependent, it can be properly identified as one form of mass spectrometry.

Acoustic mass spectrometry is based on the classical kinetic theory that states that the "average" velocity u of a molecule can be expressed as

$$u = \left(\frac{3\,kT}{M}\right)^{\frac{1}{2}}$$

TABLE 14.6 Natural Gas Constituent Data

Gas Constituent	Chemical Formula	Molecular Weight (M)	Thermal Content (BTU/ft³)	Bond Thermal Content (BTU/ft³/n bonds)	Thermal Content (per g)
Methane	CH_4	16	1012	253.0	50.20
Ethane	C_2H_6	30	1772	253.1	46.88
Propane	C_3H_8	44	2523	252.3	45.51
N-Butane	C_4H_{10}	58	3260	250.8	44.61
Pentane	C_5H_{12}	72	4009	250.6	44.19

Source: Watson and White, 1982.

where k is the Boltzmann constant, T is the temperature in degrees Kelvin, and M is the mass of the molecule. An alternative expression for an *ideal* gas ($PV = nRT$) that relates sound wave velocities to gas parameters is

$$c = \left(\frac{\gamma RT}{M}\right)^{\frac{1}{2}}$$

where c is the velocity of sound, R is the gas constant, T is the Kelvin temperature, M is the molecular weight of the gas, and γ is the ratio of the specific heat of the gas at constant pressure to its specific heat at constant volume (White, 1975). From this equation, the speed of sound in an *ideal* gas depends only on the type of gas and temperature, and it is independent of changes in pressure. For pipeline pressures (up to 300 psi), this pressure invariance is important; also, the temperature in underground piping can easily be monitored by other sensors.

There is nothing unique about relating gas densities to sound velocity, but the BTU on-line application differs from standard laboratory measurements. A recent feasibility study has provided measurements that show an excellent correlation to calculated sound velocities for pure hydrocarbon gases and various gas mixtures (Watson and White, 1982). The measurements also indicated that with appropriate refinements in pulsed sound velocity measurements, BTU content could be determined with a precision that would satisfy most commercial concerns, and that methane-ethane constituent changes could be monitored at sensitivity levels comparable to gas chromatography.

For laboratory measurements, of course, gas chromatography will continue to be a preferred method of energy analysis, as it can identify noncombustibles as well as combustibles (Villalobos, 1982). However, acoustics provides a real option for certain gaseous mixtures, and in the case of helium, it has even been applied to the isotopic assay of ^{3}He/^{4}He ratios (Fraser, 1972).

REFERENCES

Adams, V. H., in *Proc. Inst. Soc. Am. Conf.*, Houston (1980), p. 1.

Adams, V. H., Perkin-Elmer Corp., private communication (1984).

Aerospace Corp., *Industr. Res. Develop., December*, 47, (1983).

Aikens, D. A., G. M. Wood, and B. T. Upchurch, *On-Line Mass Spectrometric Monitoring of the Polymerization of a Phenolic Resin Based Material*, NASA TN D-8075, Washington, DC (1975).

Anonymous, Perkin-Elmer Corp., *Operations Manual, NASA Contract NAS9-15432*, Johnson Space Center, Houston, Texas (1981).

Bailey, F. E., in *Riegal's Handbook of Industrial Chemistry*, Vol. 8, ed. J. A. Kent, Van Nostrand Reinhold, New York (1983), p. 1.

Baker, D., *History of Manned Space Flight*, Crown Publishing Co., New York (1981).

Balayan, in *SALYUT: Soviet Steps Toward A Permanent Human Presence in Space—A Technical Memorandum*, OTA-TM-STI-14, U.S. Congress, Office of Technology Assessment, Washington, DC (1983), Appendix A, p. 51.

Batten, C. E., B. T. Upchurch, G. M. Wood, R. T. Swann, R. F. Hoyt, and G. J. Allen, in *Abs. 189th Natl. Mtg. Am. Chem. Soc.*, Miami (1985), Paper 103.

Berry, C. A., in *Recent Advances in Aerospace Medicine*, ed. D. F. Busby, D. Reidel, Dordrecht-Holland (1970), p. 3.

Bicksler, B. F., *Undersea Atmospheric Analysis*, NASA CR-111980 National Aeronautics and Space Administration, Washington, DC (1971).

Botkin, D. B., S. Golubic, B. Maguire, B. Moore, H. J. Morowitz, and L. B. Slobodkin, in *Life Sciences and Space Research*, Vol. 17, ed. R. Holmquist, Pergamon Press, Elmsford, NY (1979), p. 3.

Bourg, W., in *Proc. 28th Ann. ISA Anal. Instr. Symp.*, Baton Rouge (1982), p. 99.

Buckland, B. C., and H. Fastert, in *Soc. Chem. Ind. 3rd Ann. Conf. on Comput. Appl. in Fermentation Technol.*, Manchester, UK (August 1981).

Cason, R. M., and M. E. Koslin, *ASME Paper 80-ENAs-35, Intersoc. Conf. on Environ. Sys.*, San Diego (July 1981).

Covington, M. A., G. N. Liu, and K. A. Lincoln, *AIAA J.*, 5, 1174 (1977).

Cross, W., *Challengers of the Deep*, Wm. Sloan Assoc., New York (1959).

DeCorpo, J. J., F. E. Saalfeld, J. R. Wyatt, F. W. Williams, and H. G. Eaton, in *24th Ann. Conf. Mass Spectrom. and Allied Topics*, San Diego (1976), p. 124.

Dencker, W., *AAS Paper 75-259, Joint Meeting on Space Shuttle Missions of the '80s*, Denver (August 1975).

Disher, J. H., in *Skylab and Pioneer Report, 36, Science and Technology*, eds. P. H. Bolger and P. B. Richards, AAS Publications, Tarzana, CA (1975), p. 1.

Edmonds, T., in *Catalytic and Chemical Processes*, eds. R. Pearce and W. R. Patterson, Wiley, New York (1981), p. 88.

Engle, E., and A. S. Lott, *Man in Flight: Biomedical Achievements in Aerospace*, Leeward Publications, Annapolis, MD (1979), pp. 15, 317.

Finkel, R. C., *Geophys. Res. Lett.*, 8, No. 5, 453 (1981).

Fraser, J. C., *Rev. Sci. Instr.*, 43, No. 11, 1692 (1972).

Galli, A. F., in *Riegel's Handbook of Industrial Chemistry*, 7th ed., ed. J. A. Kent, Van Nostrand Reinhold, New York (1974), p. 407.

Giardini, A. A., G. V. Subbarayudu, and C. E. Melton, *Geophys. Res. Lett., 3*, No. 6, p. 355 (1976).

Green, D. T., *Int. J. Mass Spectrom. Ion Phys., 48*, p. 43 (1983).

Green, R. V., and H. Li, in *Riegel's Handbook of Industrial Chemistry*, 8th ed., ed. J. A. Kent, Van Nostrand Reinhold, New York (1983), p. 143.

Grotheer, E. W., and C. H. Smith, *Ann. Tech. Comp. Soc. Plastics Eng,*, Chicago (May 1983).

Hauksson, E., *J. Geophys. Res., 86*, B10, 9397 (1981).

Heinzel, E., and J. R. Bournes, *Chem. Rundschace, 50*, 3 (1983); *Balzers Techn. Note BG 800 004*, (8304) Fürstentum Liechtenstein.

Hoffman, H. L., in *Riegel's Handbook of Industrial Chemistry*, 8th ed., ed. J. A. Kent, Van Nostrand Reinhold, New York (1983), p. 488.

Holub, R. F., and B. T. Brady, *J. Geophys. Res., 86*, B3, 1776 (1981).

Hopson, G. D., J. W. Littles, and W. C. Patterson, in *Space Stations Present and Future*, eds. L. G. Napolitano, P. Contensou, and W. F. Hilton, Pergamon Press, New York (1976), p. 223.

Humphrey, A. E., and S. E. Lee, in *Riegel's Handbook of Industrial Chemistry*, 8th ed., ed. J. A. Kent, Van Nostrand Reinhold, New York (1983), p. 631.

Johnson, I., K. M. Myles, and D. C. Fee, *Report No. ANL/FE-82-10*, Argonne National Laboratory, Argonne, IL (1982).

Johnson, N. L., *Handbook of Soviet Manned Space Flight*, Vol. 48, Science and Technology Series, Advances in Astronautical Sciences, Unevelt, San Diego (1980) p. 17.

Judson, C. M., J. L. Lawrence, C. S. Josias, and R. A. Suchter, paper FA5 in *26th Ann. Conf. on Mass Spectrom. and Allied Topics*, St. Louis (1978), p. 466.

Kallos, J., and N. H. Mahle, *Anal. Chem., 55*, 813 (1983).

Kane, R. H., *Corrosion, 37*, No. 4, 187 (1981).

Kane, R. H., and P. D. Goodell, *J. Test. and Eval., 10*, No. 6, 286 (1982).

King, C. Y., *J. Geophys. Res., 85*, B6, 3051 (1980).

Koprio, J. A., H. Gaug, and H. Eppler, *Int. J. Mass Spectrom. Ion Phys., 48*, 23 (1983).

Krauss, R. W., in *Life Sciences and Space Research*, Vol. 17, ed. R. Holmquist, Pergamon Press, Elmsford, NY (1979), p. 13.

Lehotsky, R. B., in *21st Ann. Conf. on Mass Spectrom. and Allied Topics*, San Francisco (1973), p. 403.

Lincoln, K. A., *AIAA J., 21*, 1204 (1983).

McConnaughey, W. E., in *Man's Dependence on the Earthly Atmosphere*, ed. K. E. Schaefer, Macmillan, New York (1958), p. 390.

McKeown, M., *In Tech.*, 45 (August 1981).

Meyerson, S., in *Retrospective Lectures, 32nd Ann. Conf. on Mass Spectrom. and Allied Topics*, San Antonio (May 1984), p. 65.

Mills, S., in *Encyclopedia of Sports Science and Medicine*, eds. L. A. Larson and D. E. Herrmann, Macmillan, New York (1971), p. 881.

Mogro-Campero, A., R. L. Fleischer, and R. S. Likes, *J. Geophys. Res., 5*, B6, 3053 (1980).

O'Neil, J. R., and C. King, *Geophys. Res. Lett., 8*, No. 5, 429 (1981).

Pearce, R., and M. V. Twigg, in *Catalytic and Chemical Processes*, eds. R. Pearce and W. R. Patterson, Leonard Hill, Glasgow, London (1981), p. 23.

Peters, M. S., and K. D. Timmerhaus, *Plant Design and Economics for Chemical Engineers*, McGraw-Hill, New York (1968), p. 23.

Reimer, G. M., *J. Geophys. Res.*, *5*, 3107 (1980).

Reimer, G. M., *Geophys. Res. Lett.*, *8*, No. 5, 433 (1981).

Rigby, L. J., *Int. J. Mass Spectrom. Ion Processes, 60*, 251 (1984).

Robinson, D. H., *The Dangerous Sky: A History of Aviation Medicine*, University of Washington Press, Seattle (1973), p. 10.

Ruecker, M. R., *ASME Paper 73-ENAs-9, Intersoc. Conf. on Environ. Sys.*, San Diego (July 1973).

Saalfeld, F. W., F. W. Williams, and R. A. Saunders, *Am. Lab., 8* (July 1971).

Saalfeld, F. E., and J. R. Wyatt, *NRL's Central Atmosphere Monitoring Program*, NRL MR-3432, Naval Research Laboratory, Washington, DC (1976).

Saalfeld, F. W., and R. A. Saunders, in *Mass Spectrometry*, eds. C. Merritt, Jr., and C. N. McEwen, Marcel Dekker, New York (1979), p. 119.

Saunders, R. A., M. E. Umstead, W. D. Smith, and R. H. Gammon, in *Atmos. Trace Contaminant Patterns of Sea Lab II, Proc. 3rd Ann. Conf. on Atmos. Contam. of Confined Spaces*, AMRL-TR-67-200 (1967).

Savin, O. R., V. F. Shkurdoda, and V. I. Simmowsky, *Int. J. Mass Spectrom. Ion Phys., 48*, 63 (1983).

Sharpe, M. R., *Living in Space: The Astronaut and His Environment*, Doubleday Science Series, Doubleday, Garden City, NY (1969), p. 105.

Shaw, G., Extranuclear Corp, private communication (1984).

Shulten, H. R., and R. P. Lattimer, *Mass Spectrom. Rev., 3*, 231 (1984).

Smith, A. C., *SAMPE Quart., 14*, 1 (1983).

Smyle, R. E., and M. R. Reumont, "Life Support Systems," in *Manned Spacecraft: Engineering Design and Operation*, eds. P. E. Purser, M. A. Faget, and N. F. Smith, Fairchild, New York (1964).

Spurlock, J. M., and M. Modell, in *Life Sciences and Space Research*, Vol. 17, ed. R. Holmsquist, Pergamon Press, New York (1979), p. 27.

Villalobos, R., *ISA Transact., 21*, No. 2, 93 (1982).

Watson, J. W., and F. A. White, *Oil and Gas J., 80*, No. 14, 217 (1982).

Weaver, J. C., and J. H. Abrams, *Rev. Sci. Instr., 50*, 478 (1979).

Weaver, J. C., in *Noninvasive Probes of Tissue Metabolism*, ed. J. S. Cohen, Wiley, New York (1982), p. 25.

Whistler, W. J., and K. Schaefer, *Int. J. Mass Spectrom. Ion Phys., 48*, 159 (1983).

White, F. A., *Our Acoustic Environment*, Wiley, New York (1975), p. 28.

Wood, G. M., B. T. Upchurch, and D. B. Hughes, in *23rd Conf. on Mass Spectrom. and Allied Topics*, Houston (1975), p. 213.

Wood, G. M., and P. R. Yeager, in *Environmental Impact of Coal Utilization*, eds. A. Deepak and J. J. Singh, Academic Press, New York (1980), p. 187.

Wyatt, J. R., J. J. DeCorpo, and F. E. Saalfeld, in *24th Ann. Conf. on Mass Spectrom. and Allied Topics*, San Diego (1976), p. 111.

Wyatt, J. R., *CAMS-II Technical Evaluation, Phase I*, NRL MR-5309, Naval Research Laboratory, Washington, DC (1984).

PART THREE

ENVIRONMENTAL MEASUREMENTS AND THE LIFE SCIENCES

CHAPTER FIFTEEN

AIR AND WATER MONITORING

In conjunction with other methods of chemical analysis, mass spectrometry is currently (1) monitoring the overall biosphere; (2) elucidating the global and local cycles of volatile hydrocarbons; (3) providing detailed data on transport and degradation; (4) detecting trace metals in the air, oceans, ground water, and soils; and (5) distinguishing between naturally occurring compounds and pollution from anthropogenic sources. Approximately 13 million tons of C_6-C_8 aromatics and about 3 million tons of naphthalene and higher alkylated, volatile aromatic hydrocarbons enter the environment annually from the oil industry and automobile exhausts (Merian, 1982). Other major inputs to the environment are from the combustion of coal and industrial processes. For the assay of organic pollutants in many environmental matrices, tandem mass spectrometry with collision-activated dissociation is becoming increasingly important. This analytical approach is direct and rapid, and organics can be characterized by both molecular weight and functional group (McLafferty, 1981; Hunt et al., 1983). Tandem mass spectrometry is also extending the range of trace metals that can be quantified in air and water samples. In most instances, even radionuclides can now be detected with a sensitivity that exceeds that of conventional radiation instrumentation.

ATMOSPHERIC AEROSOLS

Aerosols have been categorized by Bowen (1979) into five major groups, namely:

1. Salt particles evaporated from seawater.
2. Soil and volcanic particles containing substantial concentrations of S, Al, Si, and Ti.
3. Organics evaporated from forest vegetation ($\sim 4 \times 10^{10}$ kg/yr).

4. Soot particles, carbon, plus elements such as As, Cr, Cu, Pb, V, and Zn.

5. Industrial particulates.

These tropospheric aerosols vary in size from 5 nm to 20 μm, although the bulk of their mass is associated with particles in the 0.1–20 μm range, and it is the larger particles that give rise to haze and fog.

The substantial difference in the abundance of selected elements in "ocean air" over the northern and southern hemispheres has been reported by Egorov et al. (1971) (Table 15.1). These data not only indicate the lesser amount of pollution in the southern hemisphere, but the differences of elemental concentrations.

In general, all elements exist in higher concentrations in urban atmospheres, and many elements in aerosols reflect specific localized conditions, that is, types of soils, industrial activity, etc. Seasonal variations in the concentrations of some elements (As, Br, Hg, Sb, Se, V, and Zn) have also been noted due to the increased burning of coal in winter (Salmon et al., 1977).

Many studies have been made of sulfur compounds in soils, freshwater systems, and the atmosphere, and a general review of sulfur in the environment has been prepared by Brown (1982). This element exists in the earth's crust at an average concentration of approximately 1%. Also, because it is an essential nutrient for all life forms, it differs from some other pollutants. Thus, sulfur concentrations in soils can range from "deficient" to "toxic," depending upon the life form, the sulfur compound, etc. Sulfates occur in all natural waters, but the maximum allowable concentration in drinking water has been targeted at 400 ppm by the World Health Organi-

TABLE 15.1 Elemental Composition of Oceanic Air from the Two Hemispheres[a]

Element	Northern Hemisphere (ng/m^3)	Southern Hemisphere (ng/m^3)
Al	121	12
Cd	2	0.14
Cr	7.2	0.23
Cu	12	12
Fe	180	7
Mn	7.9	0.24
Ni	2.9	0.35
Sn	3.5	0.27
Pb	4.4	1

Source: Egorov et al. (1971).

[a] Mean values from numerous measurements at sea.

zation. The combustion of fossil fuels is, of course, a source of atmospheric pollution, and sulfur is also present in many fungicides and insecticides.

In many environmental investigations, sulfur levels are so high that the sensitivity of mass spectrometry is not required. However, isotopic measurements have been made to distinguish between possible sources of this element. For example, in Utah, Grey and Jensen (1972) demonstrated that various sources of atmospheric sulfur could be distinguished from coking, refining, and smelting operations—based on slight variations in $^{34}S/^{32}S$ ratios. A new technique for analyzing nitrate anions in natural precipitation is also included in the last section of this chapter.

The importance of mass spectrometry in analyzing "organic priority pollutants" as identified by the Environmental Protection Agency (EPA) has been well documented. With chromatographic instrumentation alone, 12 separate methods are required with different GC detectors. Utilizing GC/MS, however, all compounds may be analyzed using only three separate methods, with the mass spectrometer functioning as the common detector (Stafford et al., 1983).

TECHNIQUES FOR MONITORING AEROSOLS AND PARTICULATES

Measurements relating to atmospheric particulates are important in several disciplines. Studies of crustal dust, volcanic ash, and ocean-generated aerosols are of interest to both the geochemist and the meteorologist. Particulates and aerosols from anthropogenic sources also directly or indirectly relate to basic studies in physical chemistry, combustion, corrosion, and medical research, as well as to environmental science. Clearly, many analytical tools are required to investigate mechanisms of the formation and dispersion of particles and their interaction with photons, atmospheric constituents, surfaces, and living organisms. However, mass spectrometry is important not only in determining macroscopic concentrations of various species, but also in determining the chemistry of single particles—and even the heterogeneous distribution of the chemical elements within a single particle. This latter factor is important, since the surface and interior compositions are often different and the surface layers will affect catalytic activity. For example, in coal fly ash, the surface region chemistry (within ~300 Å of the surface) is quite different from that of the bulk. Table 15.2 shows the differentiation when the surface was analyzed by the ion microprobe and the bulk was analyzed by spark source mass spectrometry (Linton, 1979).

The work of Davis (1977) in using a surface-ionization method for analyzing atmospheric particulates merits special mention. Figure 15.1 is a schematic of the apparatus. Particles in the atmosphere pass through a small collimating capillary and impinge directly on a heated rhenium (Re) ribbon in a mass spectrometer ion source. Oxygen in the air raises the work function

**TABLE 15.2   Comparison of the Surface and Bulk
Elemental Composition of Coal Fly Ash**

Element	Surface Region (~300 Å)[a]	Bulk Concentration[a]
S	130,000	7100
Fe	100,000	92,000
Pb	12,000	620
Mg	7500	12,000
Ti	3500	4700
Cr	3000	310
Mn	1300	380
Tl	800	30
V	700	380

Source: Linton (1979).

[a] Concentrations in parts per million by weight.

of rhenium to about 7.2 eV at 1000 K, so that elements with ionization potentials as high as 8 eV can be ionized and mass analyzed in a 2 in. radius-of-curvature magnetic sector instrument. As each individual particle impacts the ribbon, a burst of ions is produced, and the number of ions in the burst is a measure of the number of atoms of that element in the particle. If the analysis of single particles is not required, the temperature of the ribbon filament can be lowered until the desired degree of signal integration is obtained and the average ion current can be monitored. For some elements,

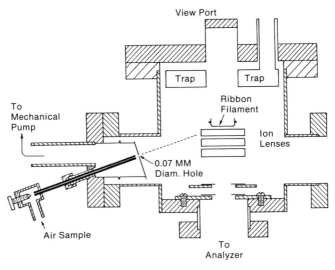

Figure 15.1 Surface-ionization method for analyzing aerosols and particulates (Davis, 1977).

TABLE 15.3   Concentrations of Several Elements in Air as Measured by a
Surface Ionization of Particulates

Element	I.P. (eV)	Re Temp. (°C)	Percent Ionized	Background $(ng/m^3)^a$	Laboratory Air (ng/m^3)
Cs	3.9	600	100	0.002	2
Rb	4.2	600	100	0.0008	0.05
K	4.3	600	100	0.0004	0.5–200
Na	5.1	600	100	0.0004	0.4–40
Li	5.4	840	100	0.0001	0.1–6
Sr	5.7	1500	100	0.02	10
U	6.1	1070	50	0.003	<0.003–0.7
Cr	6.8	1130	50	0.005	0.05
Pb	7.4	910	3	0.6	60–600
Cu	7.7	1000	0.1	1	<1–15

Source: Davis (1977).
^{a} Background refers to filtered air.

very low concentrations can also be measured by turning off the ionizing ribbon filament, collecting particles on the surface for a short interval, and then rapidly reheating the rhenium ribbon to obtain a burst of ions. In favorable cases, concentrations of elements as low as 10^{-13} g/m^3 can be detected with this direct impact technique, and measurements have been made for Li, Na, K, Rb, Cs, Sr, U, Pb, Cr, and Cu.

Table 15.3 shows the concentration of these elements in laboratory air as measured with this technique, and also for "background" or filtered air. Figure 15.2 is a continuous recording of the Pb$^+$ signal from atmospheric air near a laboratory parking lot. The graph clearly shows the departure times of employees shortly after 4:30 PM and 5:00 PM, based on the Pb emissions from their automobiles. Many organic compounds can also be detected by this surface-ionization method, although many spectra are too complex to unambiguously identify individual components. Figure 15.3 shows the mass spectrum produced from air in which a single cigarette has been ignited compared to normal air; nicotine is noted at the mass 161 peak.

Sinha et al. (1982) have applied this same technique to the assay of various aerosols. In this work, the authors generated particle beams of dioctyl-phthalate, glutaric acid, adipic acid, ammonium sulfate, and amino acids by expansions of these aerosols through a capillary nozzle, and they measured particle densities with a laser-type optical counter. Relative intensities of different mass fragments from ammonium sulfate particles are shown in Figure 15.4. The authors also report that 1 in 10^5 molecules from the vaporization can be detected, and that the method has the sensitivity to identify a 0.02 μm diameter single-component particle.

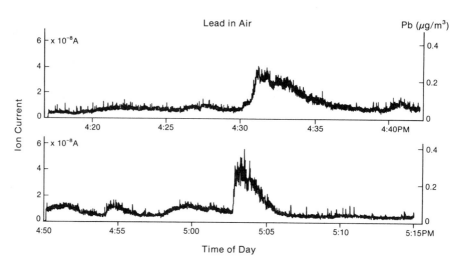

Figure 15.2 Continuous recording of atmospheric Pb concentrations showing an increase caused by automobiles (Davis, 1977).

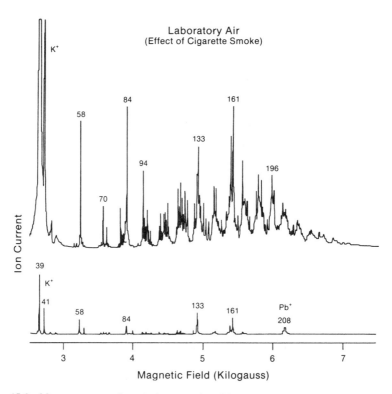

Figure 15.3 Mass spectrum of particulates produced from cigarette smoke (upper graph) compared to normal laboratory air (lower graph) (Davis, 1977).

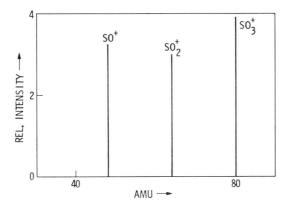

Figure 15.4 Relative intensities of different mass fragments from ammonium sulfate particles. (Sinha et al., 1982; *J. Colloid and Interface Science, 87,* 140 (1982), courtesy of M. P. Sinha.)

AIRBORNE METALLIC ELEMENTS

Mercury is the only metallic element that is found in ambient air in elemental form and in the gaseous state. Data obtained in the lower troposphere over the northern Atlantic show average values of 2.8 ng/m^3 in the gaseous state and $\leq$ 0.3 ng/m^3 in aerosols (Seiler et al., 1980). However, virtually every metal in the Periodic Table can be found in the atmosphere in the form of compounds, aerosols, and particulates. Coal-burning power plants are a major source of airborne mercury in the United States, Europe, and northern Asia, as Hg concentrations in coal have been observed to range from 0.18–3.3 ppm, with an average value of approximately 0.5 ppm being recorded for many coals (Pearson, 1981).

The EPA-sponsored assay of airborne particulates taken in six urban locations (Cincinnati, Denver, St. Louis, Washington, D.C., Chicago, and Philadelphia) has provided useful information on the overall range of metal concentrations for selected elements. Table 15.4 shows some of the data that was acquired during 10 day integrated sampling periods at these various sites (Thompson, 1979). These measurements were obtained using a variety of analytical techniques that included spark source mass spectrometry of samples collected from high-volume filtering—in contrast to the techniques described in the previous section. The large spread in values is a result of inputs from many sources. For example, a high concentration of vanadium is to be expected in the environs of coal-burning power plants even though this metal is a trace element, and lead concentrations can be expected to reflect certain industrial activities. Lead is an element whose airborne concentration has been extensively monitored because it is closely correlated with industrial activities and automobile emissions. Lead levels vary from about 0.1 μg/m^3 in remote rural areas to about 10 μg/m^3 near freeways where

TABLE 15.4   Overall Ranges of Metal
Concentrations (ng/m³) in Airborne
Particulates

Element	Low	High	Typical
Be	0.1	0.6	0.2
C	28,000	110,000	50,000
Si	4000	45,000	10,000
K	600	10,000	1000
V	1	300	30
Ni	2	100	20
Fe	1500	9000	4000
As	2	100	—
Sn	6	200	50
Pb	300	5000	—
S	3000	11,000	5000

Source: Thompson (1979).

automotive traffic is dense. In contrast, concentrations in the polar regions
as low as 0.05–0.2 ng/m³ have been recorded. This element is also sufficiently
related to health concerns that sample analyses have been made according
to particle size, and electrostatic precipitators have been used for collecting
very fine particulates.

Cadmium is a potentially toxic element. It has a low melting point (321°C),
and it is closely associated with lead processing. Average airborne concentrations in 1969 for major U.S. cities (in ng/m³) have been reported in New
York City, 14; Chicago, 15; Los Angeles, 6; Philadelphia, 15; and Detroit,
12 (Thompson, 1979). A high value of ~100 ng/m³ has been measured in El
Paso, Texas, where there is substantial ore-processing activity.

Nickel, chromium, and cadmium are important metals in common use,
and their local concentrations correlate with various types of industrial activity. All have multiple isotopes and all are amenable to trace assay by the
isotopic dilution method. The first ionization potentials of Ni and Cr are
sufficiently low that surface-ionization techniques can be applied to these
two elements. Global emissions of nickel to the atmosphere proceed at a
rate of ~8.5 × 10⁶ kg·yr⁻¹ from natural sources (windblown dust, volcanoes,
vegetation, etc.) and ~43 × 10⁶ kg·yr⁻¹ from industrial and other activities,
such as combustion, mining, steel production, etc. Residual and fuel oil
combustion constitute the most important single source, accounting for more
than 50% of all emissions to the atmosphere. Representative concentrations
of this element, plus a range of their measured concentrations in the environment, are given in Table 15.5 (Schmidt and Andren, 1980).

Measurements of chromium concentrations by the National Air Surveillance Network (NASN) at selected urban sites show concentrations ranging

TABLE 15.5   Concentrations of Nickel in the Environment

	Concentrations	Range of Values
Atmosphere		
Rural	10 ng·m^{-3}	6–20
Urban	20 ng·m^{-3}	10–60
Lithosphere		
Agricultural soil	30 μg·g^{-1}	4–230
Hydrosphere		
Fresh water	10 μg·l^{-1}	< 10–960
Marine	0.3 μg·l^{-1}	0.1–0.5
Biosphere		
Land plants	0.5 μg·g^{-1}	0.01–3
Foods	0.5 μg·g^{-1}	0.01–3

Source: Schmidt and Andren (1980).

from approximately 10–100 ng/m^3. Five year average data (1965–1969) are reported (in ng/m^3) as: Baltimore, MD, 69; Jersey City, NJ, 32; Pittsburgh, PA, 32; Charleston, WV, 29; Youngstown, OH, 26; and Newark, NJ, 16.

Platinum exists in the atmosphere in only trace amounts (0.02–0.1 ng/cm^3), and some mass spectrometric measurements have been made to detect its emission from the catalytic converters of automobiles. However, there is no evidence that this is a significant source; and, in general, platinum concentrations are attributed to geochemical abundances or highly localized industrial sources.

Selenium is an essential element in nutrition, but it is also potentially toxic even at low levels. For mammals, the lethal dose is 0.004 mg/kg of body weight. Furthermore, because of its volatility at the high temperature of stack gases (from coal combustion, municipal incinerators, etc.), it enters the vapor phase and subsequently condenses on particulates (Kut and Sarikaya, 1981). Selenium has six isotopes, so the isotopic dilution method can be applied for trace analysis. Its ionization potential is quite high (9.75 eV), so it is not a good candidate for surface ionization, but other ion sources can be employed.

The measurement of atmospheric concentrations of 14 rare earth elements by spark source mass spectrometry has been reported in the Osaka area of Japan by Sugimae (1980). These rare earths are used extensively as cracking catalysts in the petroleum industry, in the manufacture of glass as components, colorants, etc., and in other high-technology applications. The reported concentrations of the elements as a fraction of the total collected aerosols range from a high of > 30 μg/g for La and Ce to less than 0.1 μg/g for Eu, Tb, Ho, Er, Tm, Yb, and Lu. The atmospheric concentrations of these elements were also observed to be dependent upon the area's geology—reflecting the abundance of the crustal materials near the sampling site.

Even though the concentration of most metallic elements in the atmosphere is usually small, for some metals the total emissions from the combustion of fossil fuels may be greater or of the same order of magnitude as the worldwide industrial production of the refined product. For example, Vouk and Piver (1983) have pointed out that this is true for arsenic, cadmium, selenium, and vanadium; and in the United Kingdom alone, about 2000 metric tons of germanium are discharged in stack gases from coal-burning sources, whereas the total world production is on the order of only ~100 tons/yr. The recent increase in wood burning has also given rise to higher levels of gaseous emissions and particulates. In some residential areas, approximately half of the respirable particulates are attributed to wood burning. It should be pointed out that mass spectrometry (as well as radioactive techniques) can be applied to the $^{14}C/^{12}C$ measurements of carbon particulates as well as metals. Specifically, living material such as wood has a higher $^{14}C/^{12}C$ ratio than fossil fuels, so that by determining the ^{14}C content of the ambient particulate, it is possible to identify the "contemporary carbon" fraction (Dasch, 1982).

An additional technique for monitoring trace metals over an extended period relates to tree ring diagnostics. The annual uptake of metals in trees is now being used to study (1) climate; (2) pollution; (3) the relationship between pollution and growth; and (4) the possible relationship between growth and acid rain (Baes and McLaughlin, 1984).

LEAD IN THE ATMOSPHERE AND HYDROSPHERE

Lead is characterized as one of the "ancient metals," reflecting human exposure to this element for thousands of years. It has been produced since about 2500 BC, when technologies for smelting lead sulfide ores and extracting silver from lead were developed in Southwest Asia. Present-day atmospheric concentrations of ~10 ng/mg^3 are now largely attributed to industrial sources (Shirahata et al., 1980). Only a small fraction of atmospheric lead is directly traced to volcanic eruptions, etc. In the hydrosphere, of course, much of the lead is naturally occurring, and both hydrospheric and atmospheric concentrations are listed in Table 15.6. It can now be measured at exceedingly low concentrations, and in some instances, the relative abundances of three isotopes, ^{204}Pb, ^{206}Pb, and ^{207}Pb, have revealed the source (Rabinowitz and Wetherill, 1972).

Recent lead concentration profiles have been determined by isotope dilution mass spectrometry in surface water and deep water samples in the Pacific Ocean between Hawaii and California and near the California coast (Schaule and Patterson, 1981). This work shows that dissolved lead concentrations are about 10-fold higher in surface and thermocline waters than

TABLE 15.6 Lead Concentrations in the Atmosphere and Hydrosphere

Atmosphere levels	
Nonurban sites near cities	200 ng/m^3
Rural areas	100 ng/m^3
Remote areas	20 ng/m^3
North Atlantic Ocean	10 ng/m^3
North Pacific and Indian Oceans	5 ng/m^3
Hydrosphere levels	
Rainwater	34 µg/l
Rivers (U.S.A.)	7 µg/l
Tap water	4 µg/l
Ocean near shore (surface)	0.16 µg/l
Near shore (deep layers)	0.03 µg/l
Open ocean	0.06 µg/l
Lithosphere levels	
Soil (agricultural)	50 µg/g
Rocks	22 µg/g
Sediments fresh water	25 µg/g
Ocean (open)	148 µg/g
Ocean (near shore)	39 µg/g

Source: Clarkson et al. (1983); O'Brien (1979).

in deep waters (Fig. 15.5), and are as low as 1 ng/kg below 3500 m depths. The study also indicates that lead is supplied to the ocean surface largely from the atmosphere at the rate of about 60 ng/cm^2/yr due to anthropogenic inputs such as smelters and the combustion of leaded gasoline. Concentration profiles in the north Atlantic Ocean are expected to be greater by a factor of 2–3 due to even higher atmospheric inputs from industrial sources.

Measurements of lead in ancient Arctic ice and polar snow also indicate that virtually all of the present day ~300-fold excess of Pb above natural levels can be attributed to industrial and vehicular emissions to the atmosphere (Ng and Patterson, 1981). Over an extended period, additives in the form of lead alkyls have been used by refiners to increase the octane of gasoline. However, commencing in 1986, the Environmental Protection Agency in the United States has limited lead additives to 0.1 g/gal. This represents a reduction of greater than an order of magnitude from previous limits, and even more stringent restrictions are anticipated in future years. In Canada, lead will be eliminated from gasoline by the end of 1992.

VOLCANIC ASH

"The largest and deadliest volcanic eruption in recorded history was the explosion of Mount Tambora (8°S, 118°E) on the island of Sumbawa, In-

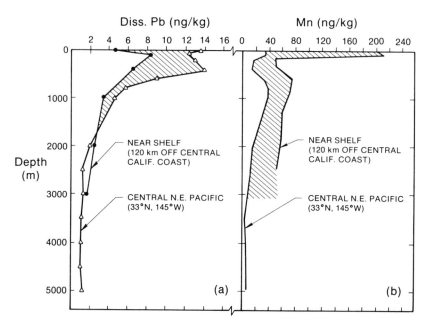

Figure 15.5 Vertical concentration profiles of (*a*) dissolved lead and (*b*) manganese at the same stations in the central and northeast Pacific. (Schaule and Patterson, 1981; *Earth and Planetary Science Lett., 54*, 97 (1981), by permission of Elsevier Science Publishers.)

donesia, on 10 and 11 April, 1915'' (Stothers, 1984). Some 88,000 people perished during the discharge of > 150 km³ of molten rock and ash, with the latter having a range of 1300 km. By comparison, the May 18, 1980 *Mount St. Helens* eruption released only ~4 km³ of rock and ash into the atmosphere, but some material reached an altitude of 25 km, so that over 1 million people in the northwestern United States were temporarily exposed to elevated concentrations of particulates. The majority of the ash particles were of some respiratory concern, that is, having an average "diameter" of ~1.7 μm (Vallyathan et al., 1983). The mineral form and chemical composition of these particles have been reported in some detail. Other large eruptions, such as that of the *Agung* in Bali in 1963 and of *Vulcan Fuego* in Guatemala in 1974, have injected particles into the troposphere and stratosphere, where their long lifetime (months to years in the stratosphere) can affect the earth's radiation balance. Many gaseous species are also emitted during major volcanic activity (e.g., SO_2, H_2S, COS, CS_2, HCl, HF, HBr, CO, CO_2, Hg vapor, organic compounds, etc.), but their average annual emission rates are small compared to global fluxes from other sources (Cadle, 1980). In the

$\longrightarrow$

Figure 15.6 Reconstructed mass chromatogram (*m/z* 57) characteristic of alkanes in volcanic ash (Pereira et al., 1982).

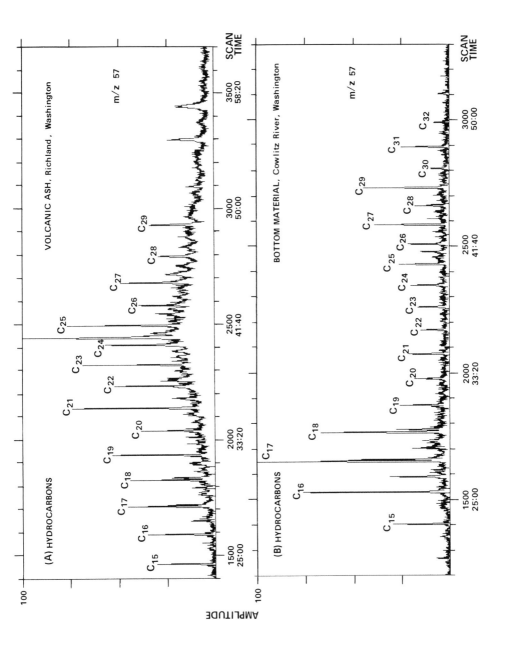

TABLE 15.7   Enrichments of Elements in Airborne Particles from Kilauea Volcano, Hawaii

Degree of Enrichment	Element
1–3	Na, Mg, Al, Cl, Ca, Sc, Ti, V, Mn, Fe, Co, Ga, Br, I, Cs, La, Ce, Sm, Eu, Yb, Hf, Ta
3–10	Sb, Ag, K, Rb, Au
10–50	S, Cu, Zn, As
> 50	Se, Hg, Cd, In, Ir

Source: Zoller et al. (1983).

case of the Mt. St. Helens eruption, SO_2 and H_2S were the dominant sulfur gas species, but the dominant sulfur gas in the background stratosphere, carbonyl sulfide (COS), showed only a slight increase (Pollack, 1981).

Substantial quantities of organic material were found condensed in the particulates of the ash cloud. Some of these organics were from the forests, vegetation, and soils that were completely pyrolyzed and vaporized in the immediate vicinity of the volcanic eruption. Gas chromatography/mass spectrometry analysis of these samples revealed compounds that included n-alkanes, fatty acids, dicarboxylic acids, aromatic acids and aldehydes, phenols, resin acids, terpenes, etc. (Pereira et al., 1982). The compound identifications were made by computerized matching of mass spectra with the National Bureau of Standards (NBS) library of approximately 32,000 compounds, published spectra, and fragmentation patterns. A reconstructed mass chromatogram (m/z 57) of alkanes is shown in Figure 15.6. This investigation also included high-resolution mass spectrometry of samples taken from nearby Spirit Lake.

Other measurements of volcanic ash have been monitored by atmospheric sampling stations that collect particles in global circulation for subsequent analysis by neutron activation. For example, the 1983 eruption of the *Kilauea* volcano on the island of Hawaii produced some strikingly large concentrations of trace elements, suggesting that Hawaiian volcanoes may be fed by magma from very deep in the earth, and possibly from the mantle (Zoller et al., 1983). Table 15.7 shows the degree of enrichments observed compared to "normal" atmospheric particulates. The authors also report a ratio of iridium to aluminum as high as 17,000 times its value in Hawaiian basalt.

NATURALLY OCCURRING RADIONUCLIDES IN THE ENVIRONMENT

Primary sources of airborne radioactivity are from the earth's crust and the combustion of fossil fuels. The earth's crust contains about 4 ppm of uranium, and thorium has an abundance of about 12 g/1 million g of rock. About 100 mineral species contain high concentrations of uranium (1% or greater), and uranium is widely dispersed in granites, metamorphic rocks, lignite, monazite sands, and phosphate deposits. Uranium is also a naturally occurring element in seawater; an estimated 4×10^9 tons exist in the oceans, with concentrations ranging from 0.0003–0.003 mg/l (Cothern and Lappenbusch, 1983). At its natural isotopic composition of ^{238}U (99.27%), ^{235}U (0.72%), and ^{234}U (0.006%), 1 μg of uranium has an activity of 0.67 pCi, with most of the alpha particle activity emanating from ^{238}U and ^{234}U in roughly equal amounts.

Nuclear radiation instrumentation is usually used to detect radionuclides, but mass spectrometry is often an even more sensitive technique and it provides unambiguous isotopic data. This is especially true for uranium and plutonium assays, where the presence of uranium can obscure the detection of plutonium. In the case of soil analysis, the relatively high natural uranium content (0.1–2 ppm) will also usually exceed any trace amounts of plutonium by several orders of magnitude. In addition, isotopic abundance measurements of ^{234}U, ^{235}U, and ^{238}U permit a distinction to be made between natural and other sources of uranium in both the atmosphere and the aquatic environment.

A recent review of the current capabilities of mass spectrometry for assaying low-level environmental radionuclides has been made by Halverson (1984). This review contains specific sensitivity comparisons for radionuclide-counting methods vs. mass spectrometric methods. Also, advances in laser resonance ionization sources (Chapter 2) indicate that presently available laser systems can ionize any element except helium and neon.

Local industrial activities, such as major construction, will be reflected in increased levels of particulates that contain radionuclides. For example, air samples taken during the rebuilding of the state capitol complex at Albany, New York revealed both high, total suspended particulates and corresponding higher levels of uranium—approximately 0.5 ng/m^3 of uranium as measured by both mass spectrometry and fission track analysis (McEachern et al., 1971). This concentration was attributed to uranium in cements, soils, etc. In contrast, total airborne uranium in the environs of a nuclear fuel-processing facility within the same state was assayed to be lower, but the $^{235}U/^{238}U$ ratio showed an enrichment of the ^{235}U isotope, thus indicating some contribution from the nuclear site. In both instances, however, the uranium concentrations were substantially below contemporary standards.

The mining of phosphate rocks for the production of phosphate products as well as uranium mining is another local source of uranium and its decay products. Approximately 120 million metric tons of phosphate ore are mined in the United States each year, and phosphate deposits contain 10–100 times the concentration of uranium in the normal terrestrial environment (EPA, 1979). Coal is a significant source of airborne uranium, because trace quan-

TABLE 15.8   Radionuclide Concentrations of Various Coals
and Coal Ash

	Coal (ppm)		Ash (ppm)	
	^{232}Th	^{238}U	^{232}Th	^{238}U
Anthracite	5.4	1.5	32	9
Eastern bituminous	5.0	1.9	99	38
Western bituminous	5.0	1.9	42	16
Lignite	6.3	2.3	57	22

Source: EPA (1979).

tities of each of the three natural radioactive series plus ^{40}K are found in all coals. A fraction of the combustion products are released to the atmosphere, depending upon the mineral content of the coal, furnace design, emission control systems, and ash disposal. The radionuclide concentrations, in parts per million, of various coals mined in the United States, and ash, are shown in Table 15.8.

The behavior and transport of the naturally occurring radionuclides ^{40}K, ^{210}Pb, ^{226}Ra, ^{228}Th, ^{235}U, and ^{238}U in coal-fired power plants have been reviewed in some detail by Coles et al. (1978). Their scenario for combustion and emission suggests that the uranium dispersed in the coal as uranite becomes volatile as the UO_3 species, and it continues along with the gases, Pb, and the fly ash. Subsequently, uranium and then lead preferentially condense on the finer fly ash particles because of their higher surface-to-volume ratio. Electrostatic precipitators will collect most of the solids, but the finer particles that continue up the stack with the gases are very enriched in ^{210}Pb relative to coal, and are moderately enriched in ^{235}U and ^{238}U relative to coal.

While considerable attention has been focused on airborne radioactivity, the annual human uptake of naturally occurring uranium is largely derived from the aquatic environment. On the average, drinking water contributes about 85% and food about 15% (Cothern and Lappenbusch, 1983). The average radioactivity due to naturally occurring uranium in drinking water in the United States is about 2 pCi/l, so that uranium normally in the atmosphere is a minor contributor to the human ingestion of this element. A comprehensive assessment of the production and transport of transuranic elements that are released to the atmosphere, soils, and aquatic environment from various nuclear-related facilities and programs is available from the U.S. Department of Energy (Hanson, 1980).

DIESEL EMISSIONS

The characterization of diesel particulates is important to basic studies of combustion and to problems of environmental concern. Health risks associated with diesel emissions relate to their small sizes and sorptive surface properties, plus the fact that diesel engines emit 50–80 times more particles than a catalyst-equipped spark ignition engine (Springer and Baines, 1977). Changes in the character of diesel exhaust particulates have also been observed as the air/fuel ratio is varied. For ratios below 20:1 the particulates are primarily carbonaceous agglomerates. Increases in the air/fuel ratio lead to an increase in the volatile organic content and to a decrease in particle size. The specific optical extinction (defined as the coefficient per unit mass concentration of smoke) also decreases from over $10 \text{ m}^2 \cdot \text{g}^{-1}$ at air/fuel ratios below 20 to less than $5 \text{ m}^2 \cdot \text{g}^{-1}$ for air/fuel ratios above 30 (Roessler and Faxvog, 1981). The diameters of a large fraction of these particles is less than 1 μm, which results in the particles having atmospheric residence times on the order of days or weeks; these small diameter particles also tend to deposit in human lungs.

Yergey et al. (1982) have used chromatographic/mass spectrometric techniques to study the formation of the particle-adsorbed, polycyclic aromatic compounds that have exhibited mutagenic activity in some bacterial assays. The authors have clarified some aspects of the combustion process by using an engine with a synthetic lubricant and by utilizing controlled air or argon/oxygen mixtures as oxidant systems. Their findings suggest that some polycyclic aromatic compounds found in diesel particles are inherent products of the diffusion-controlled combustion process, and that they are independent of the structure and chain length of the hydrocarbon fuel. Specifically, they reported that the presence of nitropyrene in only the air oxidant samples indicates that this compound is formed as a secondary process and is not dependent upon fuel-bound or lubricant-bound nitrogen.

Overall, diesel emissions have been analyzed using liquid chromatography, gas chromatography, gas chromatography/mass spectrometry, and MS/MS techniques. In one investigation of nitrated, polycyclic aromatic hydrocarbons, the analysis was made with a triple-stage quadrupole system (Schuetzle et al., 1982). The first quadrupole was set to transmit the parent ion, the second quadrupole functioned as a collision cell, and the third quadrupole was scanned repetitively to collect daughter ion spectra. In this study, one of the nitro-PAH compounds identified was 1-nitropyrene (1-NP). Spectra were obtained using methane reagent gas at 0.4 torr. Figure 15.7a and 15.7b show a composite average of 60 scans from a 46 ng standard of 1-NP, together with a 54 μg sample spectrum of unfractionated NI-1 diesel particulate extract produced from 130 scans. The authors state that the m/z abundance ratio for 231/238 for the sample and standard, respectively, indicates the possibility of another ion at $m/z = 248$, in addition to the $C_{16}H_{10}NO_2$ ion from 1-NP. Therefore, they conclude that the daughter ion m/z 231 may be preferred for quantitation. In any event, this and similar investigations have pointed out the advantages of MS/MS techniques for rapid and semiquantitative assays of isomer groups of nitro-PAH in diesel emissions and other complex compounds.

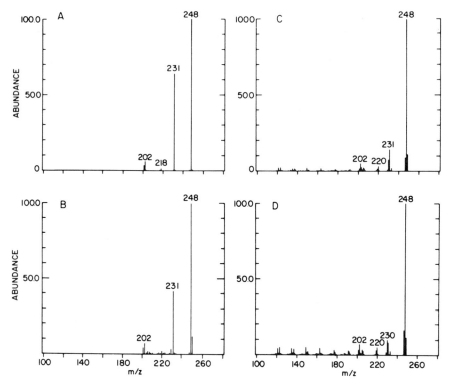

Figure 15.7 Three-stage quadrupole spectra for m/z 248 obtained from the positive methane chemical ionization of (A) 46 ng of 1-NP standard, and (B) 54 μg of unfractionated NI-1 diesel particulate extract. Spectra shown in (C) and (D) are composite averages of other samples revealing additional compounds. (Schuetzle et al., 1982; reprinted with permission from *Anal. Chem., 54,* 265 (1982), copyright 1982 American Chemical Society.)

PHOTOIONIZATION OF AIR CONTAMINANTS

In addition to standard mass spectrometric techniques for analyzing air contaminants, a photoionization method has been demonstrated that can extend the detection limit for some organic molecules. Many contaminants have ionization potentials below 12 eV in contrast to the higher ionization potentials of the primary air constituents. Therefore, a higher signal-to-noise can sometimes be achieved if a means of selective ionization is employed. Surface ionization is useful for metal particulates and chemical ionization is widely employed, but photons can be generated that have a narrow energy band and which permit the identification of contaminant molecules at concentrations well below 1 ppm.

Lasers are highly selective, and Chen et al. (1983) have also successfully demonstrated the use of an argon resonance lamp that emitted photons with an energy of 11.6 and 11.8 eV with an intensity of 3×10^{12} photons/s. In

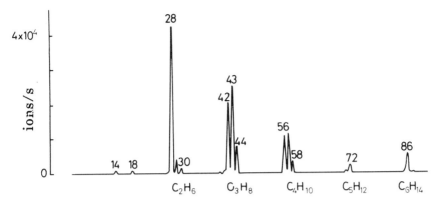

Figure 15.8 Photoionization (11.8 eV) mass spectrum of a 15 ppm alkane mixture using an argon resonance lamp and a quadrupole mass filter. (Chen et al., 1983; *Int. J. Mass Spectrom.*, *51*, 207 (1983), by permission of Elsevier Science Publishers.)

their apparatus, the light beam passes through the ion source region in which the sample gas is admitted at a pressure of 10^{-2} torr. The photoionized molecules are then extracted and accelerated by a three-electrode lens and pass into a quadrupole mass filter. Pumping of the mass analyzer is effected by a 1000 l/s cryopump that produces a "clean" vacuum substantially free of hydrocarbons that would interfere with the analysis.

Figure 15.8 is a photoionization mass spectrum of a mixture of six alkanes (methane to hexane) at 15 ppm concentration in nitrogen, and the authors report that the extrapolated detection limit for four of these alkanes is about 10 ppb. The appearance of the nitrogen peak (mass 28) is due to the fact that a small fraction of the argon lamp photons ($< 1\%$) have an energy higher than 12 eV. These investigators also suggest that the detection limit of a standard quadrupole mass spectrometer (QMS) "for most air contaminants can be improved by a factor of a hundred" by selective photoionization; and that for the analysis of gas mixtures, this method provides a simpler spectrum than that generated by conventional electron-impact ionization.

TOXIC SUBSTANCES

The Toxic Substances Control Act (P.L. 94-469) was signed into law by President Ford in October 1976 in an attempt to limit human exposure to chemicals that are known to be carcinogenic. The role of mass spectrometry in identifying these toxic substances has been important with respect to both (1) isotopic ratio measurements; and (2) the elucidation of molecular structures. The first have been useful in tracing the input of industrial chemicals into lake sediments and the aquatic environment. For example, Pb and Cd isotope ratios in sediment cores in southern Lake Huron have been corre-

TABLE 15.9   Molecular Formula, Weight, and Isomers of PCDD's

Chlorinated dibenzo-p-dioxin	Molecular Formula	Molecular Wt	No. of Isomers
Monochloro (MCDD)	$C_{12}H_7ClO_2$	218.0133	2
Dichloro (DCDD)	$C_{12}H_6Cl_2O_2$	251.9744	10
Trichloro (T₃CDD)	$C_{12}H_5Cl_3O_2$	286.2865	14
Tetrachloro (TCDD)	$C_{12}H_4Cl_4O_2$	319.8965	22
Pentachloro (P₅CDD)	$C_{12}H_3Cl_5O_2$	353.8577	14
Hexachloro (H₆CDD)	$C_{12}H_2Cl_6O_2$	387.9592	10
Heptachloro (H₇CDD)	$C_{12}HCl_7O_2$	421.7799	2
Octachloro (OCDD)	$C_{12}Cl_8O_2$	455.7410	1

Source: Karasek and Onuska (1982).

lated with depositions of dioxins and dibenzofurans over a period of several decades (Hites and Czuczwa, 1983). With respect to the latter, chromatography/mass spectrometry provides the most unambiguous fingerprint of the chemical compound—in soils or any part of the environment.

Studies on dioxins have focused on postulated DNA damage to cells, thereby causing mutations in a possible first step of a process leading to cancer. Both dioxins and dibenzofurans are known to concentrate in the fat of laboratory animals, so that fatty tissue in humans has also been subjected to analysis. Using high-resolution mass spectrometry, one study showed 2,3,7,8-TCDD (tetrachlorodibenzo-p-dioxin) concentrations ranging from 3–99 parts per trillion, with mean levels in control subjects being 5.7 ± 3.1 ppt and suspected exposed subjects measuring 8.3 ± 7 ppt (Gross, 1983).

PCDDs and PACs

The polychlorinated dibenzo-p-dioxins (PCDDs) are compounds formed by chlorine atom substitution on the dibenzo-p-dioxin nucleus. Table 15.9 lists the 75 possible isomers. Both PCDDs and polycyclic aromatic compounds (PACs) are formed in various combustion processes, and members of both groups are currently under study because of their toxic properties in animals. Chromatography and low-resolution mass spectrometry can be used for screening purposes, but high-resolution mass spectrometry is required to eliminate interference from other chlorinated species and to provide an acceptable signal-to-noise ratio (SNR). High-resolution gas chromatography is required to differentiate structural isomers. Historically, PCDDs were detected in the fly ash of municipal incinerators in Switzerland (Buser et al., 1978). Subsequently, Dow Chemical Company scientists reported that many burning processes were responsible for PCDDs being released to the environment. PACs also enter the environment from many combustion sources, for example, power plants, vehicular exhausts, etc.

TABLE 15.10   Chlorine Content of Fuels

Fuel	Chlorine (ppm)
Coal	1300 (mean)
Refuse	2500 (average)
Paper	300–600
Leaded gasoline	300
Unleaded gasoline	1–6

Source: Bumb et al. (1980).

Trace quantities of TCDD will also be generated from the combustion of wood, because one of the components of wood is lignin, a phenolic material, and the natural chlorine content of wood is in the range of 15–75 ppm. The chlorine content of coal is even higher (Table 15.10), and there are ample ring compounds to serve as precursors of dioxins, whose formation appears to be maximized at rather low combustion temperatures ($\sim$750–1000°C). For this reason, municipal incinerators may produce more dioxins from available carbon, oxygen, hydrogen, and chlorine than industrial burners operating in a higher temperature regime.

Dioxins are quickly degraded by sunlight or artificial ultraviolet light. However, the radiation must be able to penetrate the dioxin molecules, and hydrogen for the reaction must be available from some organic donor. Hence, dioxins (from pesticides or effluents) spread thinly and exposed to the sun will usually degrade and diminish to undetectable concentrations within a few days.

Furthermore, analyses by Bumb et al. (1980) of soils, dusts, wastewater, and effluents from mufflers, fireplaces, cigarettes, etc. suggest that some chlorinated dioxins are ubiquitous and occur during combustion at concentrations as low as $10^{-10}\%$. In one specific test, these investigators used 2 ng of ^{13}C-labeled 2,3,7,8-TCDD as an internal control. Overall, they conclude that chlorinated dioxins are generated in trace quantities from the combustion of most types of organic material, and that combustion processes must be more than 99.9% efficient to ensure reduction of the concentration of dioxins from 1 ppm to 1 ppb. The toxicity of the compounds is usually isomer-dependent, and the toxicity of the 2,3,7,8-TCDD is much greater than that of the other isomers. This particular isomer has also been identified with the pesticide 2,4,5-T. Since the tetrachlorodioxins appear in 22 different isomeric forms, it has been necessary to employ high-resolution mass spectrometry in combination with multiple capillary column chromatography to achieve total differentiation. Figure 15.9 is a mass fragmentogram (*m/z* 320) showing the elution of all 22 TCDD isomers as reported by Buser and Rappe (1980). For quantitative measurements of PCDDs, internal standards have been applied using isotopically labeled compounds. Gross et al. (1981) have used [37]Cl-labeled 2,3,7,8-TCDD and Mitchum et al. (1982) have used [13]C-

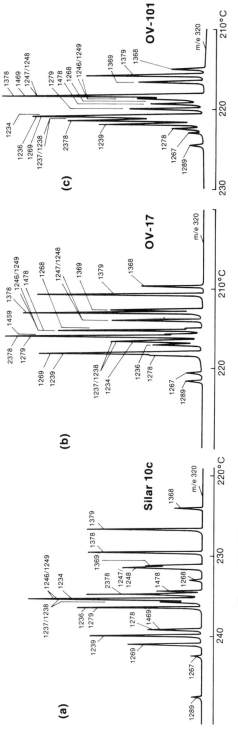

Figure 15.9 Mass fragmentograms (*m/z* 320) of a sample showing the elution of all 22 TCDD isomers on (*a*) 55 m Silar 10c, (*b*) 50 m OV-17, and (*c*) 50 m OV-101 HRGC columns. (Buser and Rappe, 1980; reprinted with permission from *Anal. Chem.*, **52**, 2257 (1980), copyright 1980 American Chemical Society.)

**TABLE 15.11   Estimated Detection
Limits for Chlorinated Dioxins Using a
GC/MS/MS System**[a]

Congener	Detection Limit (pg)
mono	0.02
di	0.08
tri	0.11
tetra	0.06
penta	0.10
hexa	0.23
hepta	0.56
octa	1.7

Source: Sakuma et al. (1983).

[a] Based on 3σ, single-daughter ion.

labeled compounds. The detection of femtogram amounts of samples of 1,2,3,4-TCDD and 2,3,7,8-TCDD by GC/MS/MS techniques has also been reported by Shushan et al. (1983). Table 15.11 shows estimated detection limits for dioxins, as reported by Sakuma et al. (1983), using a GC/MS/MS system. A daughter ion spectrum of 2,3,7,8-TCDD is also shown in Figure 15.10a, using this same system. Figure 15.10b shows a direct analysis of TCDDs in a mixture of chlorinated pesticides.

The analysis procedures for PACs are comparable to those for PCDDs, and they have been reviewed by Freudenthal (1982). Schuetzle et al. (1982) have also used a triple-stage quadrupole for screening nitrated polycyclic aromatic hydrocarbons from diesel exhausts. Samples are introduced directly into the spectrometer without chromatography. The nitrated polycyclic aromatic hydrocarbon ions lose their OH in the collision cell, and the third quadrupole is adjusted to detect a mass 17 amu lower than that of the first quadrupole.

PCDFs

Polychlorinated biphenyls (PCBs) are industrial compounds that are now widely distributed in the environment. Approximately 1.3 billion pounds were made in the United States between 1929–1977, primarily for use in electrical equipment such as transformers and capacitors. All electrical-grade PCBs were nonflammable and high boiling, and they possessed high chemical and thermal stability. This stability was an important parameter that originally led to their use, but it was also a factor that led to their being discontinued due to their persistence in the environment. Government-sponsored studies have revealed no conclusive evidence of above-normal mortality rates or diseases for occupationally exposed workers to transformer-

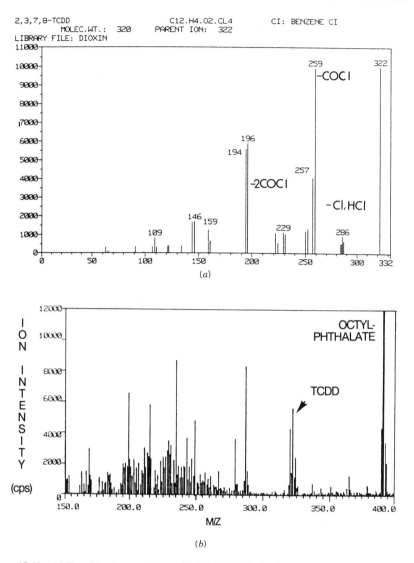

Figure 15.10 (*a*) Daughter ion spectrum of 2,3,7,8-TCDD obtained with a GC/MS/MS system. (*b*) Direct analysis of TCDDs in a mixture of chlorinated pesticides. (Sakuma et al., 1983; courtesy of Sciex, Inc.)

type PCBs. However, partially burnt PCBs can generate polychlorinated dibenzofurans, or PCDFs, which are highly toxic in some animal species. These PCDFs have been detected in PCBs after 1 week at 500°F and after 1 s at 950°F. At much higher temperatures, PCBs are transformed completely to CO_2, H_2O, and HCl (Brown, 1981). The highly publicized incident that occurred in 1968 in Japan is now attributed to PCDFs. In this episode, some

TABLE 15.12 PCB Airborne Concentrations in
Selected Areas

Environment	PCB Concentration ($\mu g/m^3$)
Air over open ocean	0.001
Urban air	0.01–1
PCB plant or dump site	1–10
Transformer repair area	10–100
Capacitor manufacturing area	10–1000

Source: Brown (1981).

1100 people developed a severe form of chloracne after eating a rice oil ("yusho") that was contaminated with a Japanese PCB containing 48% chlorine, and that was partially pyrolyzed. In contrast, human uptake studies of PCBs, but without conversion to PCDFs, have generally revealed no significant differences between exposed and control populations. In one study by the Michigan Department of Public Health, an adult group that had consumed PCB-contaminated Great Lakes fish was examined. Even for those who had eaten ~26 pounds of fish per year, no symptoms characteristic of PCB/PCDF toxicity were observed (Humphrey, 1980).

Nevertheless, because of their chemical stability and inertness to bio-degradation, environmental measurements will continue to be made of their airborne and aquatic concentrations. Table 15.12 contains order-of-magnitude concentrations, reported by Brown (1981), for various environments that have been assayed. For air, the values in $\mu g/m^3$ are roughly equivalent to parts per billion levels.

The in situ measurement of PCBs in ambient air has also been reported by Thomson et al. (1980) using a mobile, atmospheric pressure chemical ionization instrument. Air is drawn into the system through a glass line, and contaminants are adsorbed on an "integrator that consists of a conical coil of Chromel wire that is coated with a thin layer of OV-17 (a silicone based GC liquid phase). The sample line is isolated from the ionization region by a valve, and a flow of C_6H_6 is maintained during the corona discharge. Subsequently, the integrator is moved forward, the ionization region is isolated from the carrier line, and the collected sample is desorbed from the wire by resistive heating. The cycle is repeated in order to obtain average concentrations of PCBs in the sample air stream." The response of the system is reported to be linear for concentrations up to 200 ng/m^3 for each particular PCB homolog. In practice, the authors report that measurements can be obtained every 3–4 min with this adsorption/desorption device.

The chemical ionization method chosen for this system was $C_6H_6^+$ (I.P. = 9.25 eV), as this species provided an adequate specificity and sufficient sensitivity to detect the monochloro-to the hexachloro-biphenyls in air at ng/m^3 concentrations.

Trace levels of PCBs have also been measured in marine sediments by Lewis and Jamieson (1983). In their GC/MS work, the authors used negative chemical ionization (NCI) and employed several reagent gas mixtures: CH_4/O_2 (5:1), Ar/O_2 (5:1), CH_4/H_2O (1:2), and CH_4. Freeze-dried sediment samples were extracted with hexane; the extract was subsequently transferred, reduced in volume, and chromatographed. Mass spectra were then obtained for "Aroclor 1260," and interesting differences were observed with the various mixtures. The authors also conclude that the $(M-19)^-$ ions are well suited for quantifying specific PCB compounds, since "under oxidizing conditions—there is little interference due to fragments from more highly chlorinated congeners."

It should also be noted that selected chlorinated biphenyls have been studied by collision-induced dissociation, and fragmentation spectra have been correlated with PCB isomers. In one investigation, Hass et al. (1979) used a reversed Nier-Johnson-type instrument, with a collision cell between the magnetic and electric sectors, to show that the loss of a single chlorine atom is associated with a wide range of kinetic releases, but that a correlation could be made with a single-reaction mechanism.

GROUND WATER QUALITY CONTROL

Ground water, primarily from underground aquifers, provides 25% of all fresh water used in the United States, and it accounts for approximately 50% of the drinking water. Ground water also constitutes approximately 4% of the water in the overall hydrologic cycle, second only to the oceans and seas, which account for about 94%. Furthermore, the stored volume of ground water exceeds that of fresh surface waters in all streams, rivers, and lakes (Pye and Patrick, 1983). It is preferred for public water supplies because it can often be distributed without chemical treatment. If treatment is required to remove iron, manganese, or ammonia, the procedure is simple and inexpensive. In contrast, surface water from lakes and rivers almost always requires the filtering of suspended matter, plus chemical additives to remove pathogenic organisms (Huisman, 1981). Ground water is derived primarily from rainwater or from lakes and rivers having a water level above the ground water table in an aquifer below. For aquifers that are far below the surface and are reasonably distant from rivers, ground water quality may be conditioned by local agricultural and industrial activity and by discharges from miscellaneous sources. When municipal dumps are percolated by rainfall, chemicals may be dissolved and transported to the water table; and in coastal areas, the migration of seawater to an aquifer is of concern. In addition, because ground water moves slowly and mixing rates are low, it is possible for contaminants to remain localized and to exist in higher concentrations than in some surface waters.

The role of mass spectrometry in ground water quality relates to its sensitivity and selectivity in determining (1) chemical species, (2) transport, and (3) conversion. Also, as pointed out by Engelen (1981), a clear distinction is essential in terms of which *element* of quality is meant in a dynamic ground water system, that is, heavy metals, anions, cations, synthetic and natural compounds, organic and inorganic solutes, dissolved gases, and biological, chemical, and physical quality. The simplest case is where quality along a flow line depends only upon transport, in contrast to conversion phenomena such as ion exchange, etc. This type of transport can be detected by a tritium tracer, which due to its decay serves as a clock that moves through the system, or by means of enriched stable isotopes of metals with periodic monitoring. Representative of ground water research utilizing the tritium tracer is the 1975–1980 study on the distribution of solutes from agricultural land in some major British aquifers (Oakes et al., 1981). A relationship was found between the concentration of certain solutes, especially nitrates, and farming practice; and in some cases, these nitrate concentrations in water supply sources have exceeded the World Health Organization limit of 11.3 mg/l nitrate nitrogen. Nitrate profiles (often > 20 mg/l of nitrogen) are shown as a function of depth in the interstitial water beneath a farming area at Kent, England (Fig. 15.11a); and a corresponding tritium profile is shown in Figure 15.11b. It is not completely clear whether the nitrate and tritium profiles reflect downward migration only or include hydrogeological factors such as striations and zones of high and low permeability.

In contrast to the diffuse source of pollution from the application of agrichemicals, many other major and persistent pollutants are discharged from "point" sources, for example, mines, industrial waste sites, textile, paper, or petrochemical plants, power-generating stations, accidental leaks from storage tanks and pipelines, etc. From these and other sources, a plume of contaminated water, characterized by low-dissolved oxygen and high chloride, sulfate, and ammonia, will extend in the direction of ground water; and the rate of pollution attenuation will be dependent on a number of geochemical and hydrogeological factors (Young, 1981). One survey of 50 industrial sites in the United States indicated 43 examples of some contamination (Miller et al., 1977); and in northeastern England, traces of hexavalent chromium were detected from an ore sludge disposal at a distance of 250 m (Barber et al., 1981).

Compounds that have been observed in ground waters in the Netherlands and whose transportability half-lives have been estimated include hydrocarbons, ketones, and halogenated hydrocarbons (Zoeteman, 1981). Furthermore, this study suggests that among the hundreds of compounds that can be detected in contaminated ground water, the relatively low molecular weight and well-known chemicals constitute the greatest problems.

For a rapid, qualitative analysis of heavy metals in water samples, the

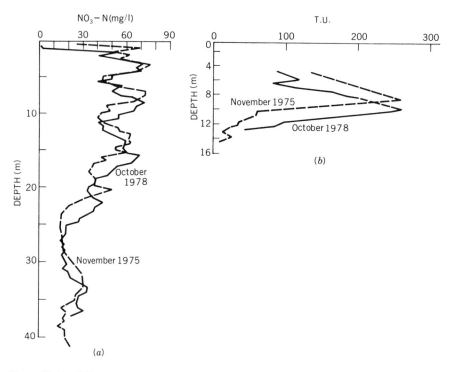

Figure 15.11 Effects of farming practice on aquifers. (*a*) Unsaturated zone pore-water nitrate profiles beneath farmland. (*b*) Unsaturated zone pore-water tritium profiles. (Oakes et al., 1981; *Science of the Total Environment, 21,* 17 (1981), by permission of Elsevier Science Publishers.)

inductively coupled plasma (ICP) ion source provides some specific advantages (Gray and Date, 1983). General background levels are low, and many elements can be determined with detection limits below 1 ng/ml. The ICP source is also relatively free from matrix and interelement interferences. (Figure 15.12 shows a quadrupole mass spectrum by Gray and Data (1985) of a sample containing W, Au, Hg, Pb, Bi, Th, and U, each at a concentration of 100 ng/ml and where the peak count rate for $^{209}Bi^+$ is 9.6×10^4/s.

AROMATIC HYDROCARBONS IN LAKE SEDIMENTS

In a continuing program to assess the environmental impact of fossil fuel combustion, analyses have been made of sediments in lakes for polynuclear aromatic hydrocarbons (PAH). These sediments provide a memory of pollution, and remote lakes can reflect the long-distance atmospheric transport of anthropogenic pollutants. One typical investigation by Tan and Heit (1981) has focused on two lakes in New York State's Adirondack Park in an area that is forested and remote, and receives negligible pollution from local

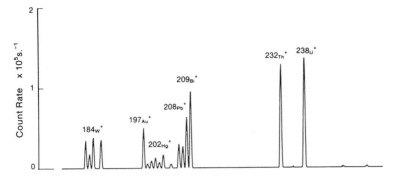

Figure 15.12 Spectrum of water sample containing W, Au, Hg, Pb, Bi, Th, and U, each at a concentration of 100 ng/ml (Gray and Date, 1985).

sources. Both Sagamore Lake and Woods Lake are located on the western slope of the Adirondack Mountains, and both lakes are oligotrophic and acidic throughout the year. Two cores were taken from each of the lakes; the core samples had a water content ranging from 81–92% by weight and had siliceous diatom skeletons. High-resolution gas chromatography/mass

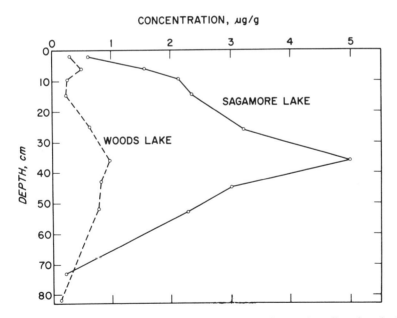

Figure 15.13 Concentration of perylene in lake sediments with depth attributed to the long-distance atmospheric transport of fossil fuel combustion byproducts. (Tan and Heit, 1981; reprinted with permission from *Geochimica et Cosmochimica Acta*, **45**, 2267 (1981), copyright 1981 Pergamon Press, Ltd.)

**TABLE 15.13 Observed Detection Sensitivities for Selected
Polynuclear Aromatic Hydrocarbon Compounds in a Standard
Solution by HPLC/MS**

Compound	Molecular Wt	Concentration (ng/μl)
Naphthalene	128	11.3
Acenaphthylene	152	9.6
Acenaphthene	154	10.5
Fluorine	166	2.5
Phenanthrene	178	2.5
Anthracene	178	1.7
Fluoranthene	202	5.1
Pyrene	202	4.9

Source: Quillian et al. (1984).

spectrometry was then used to determine indicated PAH composition and distribution. With the exception of one species, perylene, all the parental PAH decreased rapidly in concentration with depth. The concentration of perylene showed an increase to a maximum at approximately 35 cm, and then it decreased monitonically at greater depths (Fig. 15.13). The possibility of combustion products from forest fires was ruled out, as the last major forest fire in this specific Adirondack locale occurred about 80 years ago; no significant PAH concentrations were found at sediment depths corresponding to such a date. While a complete understanding of the environmental chemistry of perylene is lacking, this and other studies have made it possible to establish a correlation between sediment hydrocarbons, biogenically converted species, and remote pollutant inputs.

Other analytical studies relating to marine sediments have focused on increasing the sensitivity for PAH detection and developing suitable reference standards. For example, Quilliam et al. (1984) have compared several techniques and determined that high-performance liquid chromatography (HPLC) interfaced with mass spectrometry yields good reproducibility, and detection limits for PAH compounds are listed in Table 15.13.

POLLUTION SOURCE IDENTIFICATION

A number of studies have indicated that chromatography combined with mass spectrometry can be effectively used to identify pollution sources. Furthermore, variations in the isotopic abundances of carbon, oxygen, sulfur, and lead have helped in elucidating the origin of certain pollutants. Ogata and Miyake (1980) have determined that organic sulfur compounds usually found in crude oil can be used as a marker of oil pollution in shellfish and fish. Their mass spectra and chromatograms of short-necked clam extract revealed the presence of alkyl-(from C_1–C_{10}) benzothiophenes, dibenzo-thiophene, alkyl-(C_1–C_8) dibenzothiophenes, naphthalene, and alkyl-(C_1–C_6) naphthalene. Other polyaromatic compounds were also detected in other extracts that the authors reported as markers of oil pollution.

Gas chromatography and selective chemical ionization mass spectrometry have been reported by Burke et al. (1982) in identifying crude oils from worldwide sources. Their objectives were to devise a rapid method for iden-tifying a spilled oil from a short list of possible candidates and to reduce the mass spectral ambiguities produced by weathering processes such as evap-oration, solution, and photolysis. In their study, crude oils were sampled from the four geographical regions that represented the main oil-producing countries. Their experimental technique included the use of N_2O in the ion source through a separate inlet where it mixed with the oil/hexane vapor for 30 s. This gaseous mixture was then bombarded to form OH^- ions, which subsequently ionized the aromatic fraction of the crude oils. By a judicious selection of 30 ion peaks, they were able to reduce their crude oil "finger-print" to a "30 dimensional plot" that could be rapidly compared to stan-dards. On this basis, the authors report that their method can provide a positive identification of the geographical area of origin of the oil, and that the identification of individual fields can be realized in a majority of cases.

For highly weathered petroleum residues, an effort has been made to correlate isotopic data with various petroleum sources. Specifically, carbon and sulfur isotope ratios and total sulfur content have been evaluated by Hartman and Hammond (1981) for correlating beach tars depositing near Los Angeles with their probable sources. Their analytical data show that tanker crudes imported into the area differ significantly from California oils. Furthermore, on the basis of $\delta^{13}C$, %S, and $\delta^{34}S$ measurements, tars col-lected from Santa Monica bay beaches could be correlated with various petroleum natural oil seeps.

Measurements of $^{206}Pb/^{204}Pb$ ratios obtained in the Helsinki area in Fin-land also showed that it was possible to evaluate the contribution of motor traffic to the lead content of environmental samples (Jaakkola et al., 1983). A triple-filament thermal ionization source was used in this analysis, with the samples being loaded as $Pb(NO_3)_2$. Table 15.14 gives representative data

TABLE 15.14 Average Isotope Lead Ratios Observed in Finland from Various Environmental Sources

Source	$^{206}Pb/^{204}Pb$	$^{206}Pb/^{207}Pb$	$^{206}Pb/^{208}Pb$
Lichen	17.7	1.146	0.475
Air filter	17.5	1.133	0.462
Incinerator ash	18.4	1.156	0.471
Gasoline	16.47	1.072	0.461

Source: Jaakkola et al. (1983).

obtained by these investigators for lichen, air filter samples, incinerator ash, and high octane gasoline. Lichen obtains its mineral content from air, rain-water, and snow, so it should be expected to represent the average "lead fallout" for a period of years. The substantially different isotopic content of gasoline permits a distinction to be made of the lead contribution from this source.

In addition, GC/MS assays have been used to locate industrial discharges by matching spectra of specific types of compounds with sample ion profiles (Carter, 1982). By using computerized mass spectrometry as a detector in capillary column chromatography, the possibility of identifying specific sources of pollution is greatly enhanced; and an archival storage of computerized data permits a convenient reexamination of spectral information from various sources.

ATMOSPHERIC TRACERS

Gaseous and particulate tracers can be injected into the atmosphere to investigate transport and dispersion processes. Important uses for atmospheric tracer technology include basic atmospheric dispersion studies, the long-range transport of aerosols, acid rain, meteorology, and environmental modeling and analysis. For studies that require measurements hundreds or even thousands of kilometers from the release point, constraints on the tracer and its detection are severe:

1. The tracer should be inert and nontoxic.
2. The tracer must be measurable at the part per trillion (10^{-12}) level or lower.
3. If possible, multiple tracers should be used to minimize ambiguities and interferences.
4. Labeling a compound with a stable isotope is desirable, but costs may be prohibitive.
5. The mass spectrometer detector should effectively have a "zero" background and count single atoms or molecules.

In addition, it is desirable that tracer gases for global scientific experiments be excluded from commercial use, or their usefulness will be quickly obscured.

The development of isotopically labeled methane has been reported by Cowan et al. (1976) and Fowler (1979), and the properties and atmospheric lifetimes of perfluorocarbon tracers have been reviewed by Lovelock and Ferber (1982). CF_4 has an estimated residence time greater than 10^4 years, and other perfluorocarbons have probable lifetimes of 100 years. Table 15.15 contains a few tracer compounds from a more complete list prepared by

TABLE 15.15 Tracer Compounds and Background Concentrations

Compound	Molecular Wt	Atmospheric Background ($\times 10^{-15}$)
Permanent		
SF_6	146	600
C_7F_{14} (Perfluoromethyl cyclohexane)	350	2
C_8F_{16} (Perfluorodimethyl cyclohexane)	400	26
$C_{10}F_{18}$ (Perfluorodecalin)	462	2
CF_3SF_5	196	1
Temporary		
$^{12}CD_4$	20	< 1.5
$^{13}CD_4$	21	0.008
C_6F_6	186	10

Source: Lovelock and Ferber (1982).

Lovelock and Ferber (1982) that includes data on the atmospheric background (by volume) and molecular weight. "Permanent" denotes an estimated atmospheric residence time exceeding 100 years, while temporary denotes a lifetime less than 50 years. Sulfur hexafluoride (SF_6) is an excellent potential tracer, but it has been widely used in industry, so the background level is high. Lovelock and Ferber (1982) also point out that if "a detection limit of one part in 10^{19} is ultimately possible, then this concentration requires the release of only a few tens of kilograms to label the entire atmosphere!"

Shields (1982) has made mass spectrometric measurements of heavy methane ($^{13}CD_4$) in a "600 km experiment" in which 84 g of this tracer and perfluorocarbons were released at the National Severe Storms Laboratory in Norman, Oklahoma. Since the level of $^{13}CD_4$ naturally occurring in air is below 10^{-21} (vs. 1.5 ppm for normal methane [$^{12}CH_4$]), the signal-to-noise level for this tracer is very high. Cryogenically collected samples were recovered in Nebraska and Missouri on a 600 m arc from the release point. One measurement at Maryville, Missouri taken 9 h later showed a concentration of 5×10^{-17} (wt. fraction).

Other measurements of natural and anthropogenic trace gases have been made in connection with global studies of the atmosphere both below and above the boundary layer. Rasmussen et al. (1982) have reported on a number of atmospheric trace gases in the lower atmosphere (0–4 km) of the Southern Hemisphere, including CCl_3F, CCl_2F_2, $CHClF_2$, C_2Cl_3F, CCl_4, C_2Cl_4, CH_3I, $CHCl_3$, CO, CH_3Cl, CH_4, N_2O, and COS. In these measurements, concentrations were detected at the parts per trillion level. Their observations for CH_4 in particular were consistent with other measurements showing an increase of methane in the Southern Hemisphere at a rate of 1.2 ± 0.5%/yr.

The long-range transport of aerosols from the European continent and England has been monitored in Norway for many years. Airborne aerosols collected by high-volume samplers on glass fiber filters and analyzed by gas chromatography have provided quantitative data for 20 different PAH compounds, with concentrations ranging from ~100–6,000 ng/1000 m³ (Lunde and Bjorseth, 1977).

Instrumentation particularly designed to study trace gases in geological surveys, plumes arising from industrial sources, etc. have been reviewed by Lane (1982). Essentially, atmospheric pressure-chemical ionization spectrometry can provide data on sulfur dioxide and plumes from coal-fired plants; and by single or multiple ion monitoring from a "mobile" laboratory (vehicle or aircraft), it is possible to locate the origin of many pollutant emissions.

Finally, it is appropriate to comment on the use of stable isotopic tracers in studies relating to acid rain. Isotopic labeling can be used for compounds of both nitrogen and sulfur, and Ligon and Dorn (1985) have successfully analyzed nitrate anions in natural precipitation using this technique. Specifically, by using a long-chain alkyltrimethylammonium surfactant in a glycerol solution, they have obtained nanogram-level sensitivities for nitrates

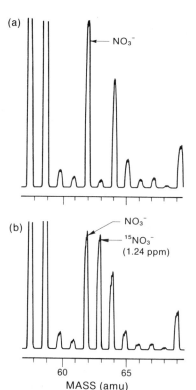

Figure 15.14 Isotope dilution analysis for NO₃⁻ applied to a 1 ml sample of rainwater, using a ¹⁵N tracer, fast-atom bombardment, and a surfactant in a glycerol solution. (*a*) Nitrate anion signal from 1 ml of natural rainwater. (*b*) Nitrate anion signal from an identical 1 ml aliquot of rainwater to which has been added 1.24 ppm ¹⁵NO₃. (Ligon and Dorn, 1985; *Int. J. Mass Spectrom.*, *63*, 315 (1985), by permission of Elsevier Science Publishers.)

via negative secondary ion mass spectrometry. The surfactant serves to bind the anion to the surface of the glycerol droplet, thereby greatly enhancing the secondary ion current produced by primary ion or neutral atom bombardment. Figure 15.14 shows the result of their isotope dilution analysis for NO_3^- of the residue from a 1 ml sample of rainwater. A glycerol solution containing both the surfactant and labeled nitrate was added to the sample before evaporation. The 1.24 ppm nitrate concentration was also observed to be consistent with other data. According to Ligon (1985), this technique should make possible a broad range of studies of NO_x emissions and ultimately provide a better understanding of natural vs. other sources of NO_x.

Hopefully, isotopic tracers can also do more than simply monitor the global environment. As suggested in Chapter 13, mass spectrometry and isotopic labeling can provide much needed detailed data relating to combustion kinetics. Only by such basic studies can combustion parameters be optimized and some undesirable emissions be reduced.

REFERENCES

Baes, C. F., III, and S. B. McLaughlin, *Science, 224,* 494 (1984).

Barber, C., C. P. Young, N. C. Blakey, C. A. M. Ross, and G. M. Williams, *Int. Symp. Groundwater Quality,* Noordwijkerhout, Netherlands (March 23–27, 1981).

Bowen, H. J. M., *Environmental Chemistry of the Elements,* Academic Press, New York (1979), p. 9.

Brown, J. F., Jr., "Human Health Effects of Electrical Grade PCB's," in *Doble. Eng. Conf.,* Watertown, MA (1981), pp 10-901–10-909.

Brown, K. A., *Environ. Pollution, 83,* 47 (1982).

Bumb, R. R., W. . Crummett, S. S. Cutie, J. R. Gledhill, R. H. Hummel, R. O. Kagel, L. L. Lamparski, E. V. Luoma, D. L. Miller, T. J. Nestrick, L. A. Shadoff, R. H. Stehl, and J. S. Woods, *Science, 210,* 385 (1980).

Burke, P., K. R. Jennings, R. P. Morgan, and C. A. Gilchrist, *Anal. Chem., 54,* 1304 (1982).

Buser, H. R., H. P. Bosshardt, and C. Rappe, *Chemosphere, 7,* 165 (1978).

Buser, H. R., and C. Rappe, *Anal. Chem., 52,* 2257 (1980).

Cadle, R. D., *Rev. Geophys. Space Phys., 18,* No. 4, 746 (1980).

Carter, M. H., *J. Chromatogr., 235,* 165 (1982).

Chen, H. N., W. Genuit, A. J. H. Boerboom, and J. Los, *Int. J. Mass Spectrom. Ion Phys., 51,* 207 (1983).

Clarkson, T. W., B. Weise, and C. Cox, *Environ. Health Perspect., 48,* 113 (1983).

Coles, D. G., R. C. Ragaini, and J. M. Ondov, *Environ. Sci. Techol., 12,* No. 4, 442 (1978).

Cothern, C. R., and W. L. Lappenbusch, *Health Physics, 45,* No. 1, 89 (1983).

Cowan, G. A., D. G. Ott, A. Turkevich, L. Machta, G. J. Ferber, and N. R. Daly, *Science, 191,* 1048 (1976).

Dasch, J. M. *Environ. Sci. Technol., 16,* No. 10, 639 (1982).

Davis, W. D., *Environ. Sci. Technol., 11,* 593 (1977).

Egorov, V. V., T. N. Zhigalovskaya, S. G. Malakhov and A. I. Shilina, *Trudy Inst. Eksp. Meteorol., 14,* 73 (1971).

Engelen, G. B., *Sci. Total Environ., 21,* 1 (1981).

EPA Report 520/7-79-006, *Radiological Impact Caused by Emissions of Radionuclides into Air in the United States* (1979).

Fowler, M. M., *Report LA UR-80-1342,* Los Alamos National Laboratory, Los Alamos, NM (1979).

Freudenthal, J., *Int. J. Mass Spectrom. Ion Phys., 45,* 247 (1982).

Gray, A. L., and A. R. Date, *The Analyst, 108,* No. 1290, 1033 (1983).

Gray, A. L., and A. R. Date, private communication (1985).

Gey, D. C., and M. L. Jensen, *Science, 177,* 1099 (1972).

Gross, M. L., T. Sun, P. A. Lyon, S. F. Wojinski, and D. R. Hilker, *Anal. Chem., 53,* 1902 (1981).

Gross, M. L., *Chem. and Eng. News, September 12,* 29 (1983).

Halverson, J. E., *Nucl. Instr. and Methods in Phys. Res., 223,* 349 (1984).

Hanson, W. C., ed., *Transuranic Elements in the Environment,* DOE/TIC-22800, Springfield, VA (1980).

Hartman, B., and D. E. Hammond, *Geochim. et Cosmochim. Acta, 45,* 309 (1981).

Hass, J. R., M. M. Bursey, L. A. Levy, and D. J. Harvan, *Org. Mass. Spectrom., 14,* No. 6, 319 (1979).

Hites, R. A., and J. M. Czuczwa, *Chem. and Eng. News, September 12,* 26 (1983).

Huisman, L., *Sci. Total Environ., 21,* 13 (1981).

Humphrey, H. E. B., "Evaluation of Humans Exposed to Halogenated Biphenyls," in *Am. Chem. Soc., Div. Env. Chem. Preprints, 20,* No. 2, 272 (1980).

Hunt, D. F., J. Shabanowitz, T. M. Harvey, and M. L. Coates, *J. Chromatogr., 271,* 93 (1983).

Jaakkola, T., O. J. Heinonen, M. Keinonen, A. Salmi, and J. K. Miettinen, *Int. J. Mass Spectrom. Ion Phys., 48,* 347 (1983).

Karasek, F. W., and F. I. Onuska, *Anal. Chem., 54,* No. 2, 309A (1982).

Kut, D., and Y. Sarikaya, *J. Radioloanal. Chem., 62,* No. 1–2, 161 (1981).

Lane, D. A., *Environ. Sci. Technol., 16,* No. 1, 38A (1982).

Lewis, E., and W. D. Jamieson, *Int. J. Mass Spectrom. Ion Phys., 48,* 303 (1983).

Ligon, W. V., Jr., private communication (1985).

Ligon, W. V., Jr., and S. B. Dorn, *Int. J. Mass Spectrom. Ion Phys., 63,* 315 (1985).

Linton, R. W., in *Monitoring Toxic Substances,* ed. D. Schuetzle, American Chemical Society, Washington, D. C. (1979), p. 156.

Lovelock, J. E., and G. J. Ferber, *Atmos. Environ., 16,* No. 6, 1467 (1982).

Lunde, G., and A. Bjorseth, *Nature, 268,* No. 5620, 518 (1977).

McEachern, P., W. G. Myers, and F. A. White, *Environ. Sci. Technol., 5,* 700 (1971).

McLafferty, F. W., *Science, 214,* 280 (1981).

Merian, E., *Toxicol. Environ. Chem., 5,* 167 (1982).

Miller, D., O. Braids, and W. Walker, *U. S. EPA Report 68-01-2966,* 1977.

Mitchum, R. K., W. A. Korfmacher, and G. F. Moler, *Anal. Chem., 54,* 719 (1982).

Ng, A., and C. Patterson, *Geochim. Cosmochim. Acta, 45,* 2109 (1981).

Oakes, D. B., C. P. Young, and S. S. D. Foster, *Sci. Total Environ., 21,* 17 (1981).

O'Brien, B. J., "Application to Exposure of Man to Lead Pollution," in *MARC Report No. 13,* Monitoring and Assessment Research Center, London (1979).

Ogata, M., and Y. Miyake, *J. Chromatogr. Sci., 18,* 594 (1980).

Pearson, D. E., *Chem. Eng. News, June 1,* 37 (1981).

Pereira, W. E., C. E. Rostad, H. E. Taylor, and J. M. Klein, *Environ. Sci. Technol., 16,* No. 7, 387 (1982).

Pollack, J. B., *Science, 211,* 815 (1981).

Pye, V. I., and R. Patrick, *Science, 221,* 713 (1983).

Quilliam, M. A., R. J. Gergely, M. S. Lant, W. D. Jamison, E. Lewis, and R. M. Gershey, "HPLC-MS Determination of Polycyclic Aromatic Hydrocarbons in Marine Sediments," presented at the *32nd Ann. Conf. on Mass Spectrom. and Allied Topics,* San Antonio, (May 27 to June 1, 1984).

Rabinowitz, M. B., and G. W. Wetherill, *Environ. Sci. Technol., 6,* 705 (1972).

Rasmussen, R. A., M. A. K. Khalil, A. J. Crawford, and P. J. Fraser, *Geophys. Res. Lett., 9,* No. 6, 704 (1982).

Roessler, D. M., and F. R. Faxvog, *Combust. Sci. Technol., 26,* 225 (1981).

Sakuma, T., N. Gurprasad, S. D. Tanner, A. Ngo, W. R. Davidson, H. A. McLeod, B. P-Y. Lau, and J. J. Ryan, *Symposium on Chlorinated Dioxins in the Total Environment,* Vol. II, Washington, D. C. (August, 1983).

Salmon, L., D. H. F. Atkins, E. M. R. Fisher, C. Healy, and D. V. Law, *U. K. Atomic Energy Authority Report AERE-R 8680,* H. M. Stationery Office, London (1977).

Schaule, B. K., and C. C. Patterson, *Earth and Planet. Sci. Lett., 54,* 97 (1981).

Schmidt, J. A., and A. W. Andren, in *Nickel in the Environment,* ed. J. O. Nriagu, Wiley, New York (1980).

Schuetzle, D., T. L. Riley, T. J. Prater, T. M. Harvey, and D. F. Hunt, *Anal. Chem., 54,* 265 (1982).

Shields, W., personal communication (1982).

Seiler, W., C. Eberling, and F. Slemr, *Pageoph., 118,* 964 (1980).

Shirahata, H., R. W., Elias, C. C. Patterson, and M. Koide, *Geochim. et Cosmochim. Acta, 44,* 149 (1980).

Shushan, B., J. E. Fulford, B. A. Thomson, W. R. Davidson, L. M. Danylewych, A. Ngo, S. Nacson, and S. D. Tanner, *Int. J. Mass. Spectrom. Ion Phys., 46,* 225 (1983).

Sinha, M. P., C. E. Giffin, D. D. Norris, T. J. Estes, V. L. Vilker, and S. K. Friedlander, *J. Colloid and Interface Sci., 87,* No. 1, 140 (1982).

Springer, K. J., and T. M. Baines, *Soc. Automobile Eng., Paper No. 770818,* MECCA, Milwaukee, WI (September 12, 1977).

Stafford, G. C., P. G. Kelley, and D. C. Bradford, *Am. Laboratory, June* (1983).

Stothers, R. B., *Science, 224,* 1191 (1984).

Sugimae, A., *Atmos. Environ., 14,* 1171 (1980).

Tan, Y. L., and M. Heit, *Geochim. Cosmochim. Acta, 45,* 2267 (1981).

Thompson, R. J., in *Ultratrace Metal Analysis in Biological Sciences and Environment,* ed. T. H. Risby, American Chemical Society, Washington, D. C. (1979), p. 60.

Thomson, B. A., T. Sakuma, J. Fulford, D. A. Lane, N. M. Reid, and J. B. French, in *Advances in Mass Spectrometry,* Vol. 8B, ed A. Quayle, Heyden and Son Ltd., London (1980), p. 1422.

Vallyathan, V., M. S. Mentnech, J. H. Tucker, and F. H. Y. Green, *Environ. Res., 30,* 361 (1983).

Vouk, V. B., and W. T. Piver, *Environ. Health Perspect., 47,* 201 (1983).

Yergey, J. A., T. H. Risby, and S. S. Lestz, *Anal. Chem., 54,* 354 (1982).

Young, C. P., *Sci. Total Environ., 21,* 61 (1981).

Zoeteman, B. C. J., E. DeGreef, and P. J. J. Brinkmann, *Sci. Total Environ., 21,* 187 (1981).

Zoller, W. H., J. R. Parrington, and J. M. Phelan Kotra, *Science, 222,* 1118 (1983).

CHAPTER SIXTEEN

AGRICULTURE AND FOOD SCIENCE

The production, processing, preservation, and distribution of food is a so-cietal problem of the highest priority. In a world population of 4.9 billion, some 500 million suffer varying degrees of hunger and malnutrition. This number of undernourished is expected to double by the year 2000, when the projected population is 6 billion. However, during 1975–1977, the actual daily production of food was approximately 2590 calories per person (Zet-tlemeyer, 1981); this quantity was well above the required minimum, but the distribution of this food was wholly inadequate. Furthermore, 50% of stored grains in developing countries is currently lost to pests each year, and worldwide food losses due to drought and disease are large. In the United States alone, crop yields are substantially reduced by 160 species of bacteria, 250 kinds of viruses, 8000 species of pathogenic fungi, 8000 types of insects, and 2000 species of weeds (Batra, 1982). Thus, agronomists, plant pathol-ogists, and biochemists are confronted with the challenge of astute soil man-agement and of developing improved fertilizers, pesticides, and herbicides. Fortunately, significant progress is already being made. For example, much of the present increased food production is due to the introduction of fer-tilizer-responsive cereal varieties, especially wheat and rice, that convert a much higher proportion of their photosynthate to grain and less to stem (Brady, 1982). In many instances, the planting of these new varieties has resulted in a doubling of crop yields.

There are also two major thrusts in agricultural research that are being addressed by mass spectrometry and other analytical techniques. One is the search for a new chemical method to extract the most abundant element in the atmosphere—nitrogen—directly from the air and make it available as a fertilizer via appropriate catalysis. A second relates to the growing area of crop genetics, whereby plant cells can be genetically engineered to adapt to the soil, rather than treating the soil to suit the plants. The goal is not only to produce plants that can provide their own nitrogen fixation, but to develop strains that are resistant to pests, drought, and disease.

SOIL FERTILITY

Soil fertility is not a precisely defined quality, as the availability of nutrients to plants and the rates of nitrification depend upon such factors as a favorable combination of temperature and moisture, the geological substrate from which the soil is derived, effects of chemical additives from many sources, and the action of soil micro-organisms. Soils also acquire organic compounds by way of leaf leachates, root exudates, and other decomposition products; phenolic acids are among many of these compounds that can enter the soil and affect seed germination and plant growth (Dalton et al., 1983). In connection with these inputs, measurements of *nutrient dynamics* are now being quantified by (1) analyzing for soil pH, organic carbon, total nitrogen, exchangeable aluminum, potassium, calcium, and magnesium; (2) measuring available phosphorus, zinc, copper, iron, and manganese; and (3) monitoring harvested crops. Specific gas chromatographic methods have been developed for estimating oxygen, nitrous oxide, and carbon dioxide in soil atmospheres (Hall and Dowdell, 1981); and GC/MS analyses have been used to characterize aquatic humic material. In a typical study, natural products have been found to consist of acidic, hydrophilic, complex products that range in molecular weight from a few hundred to a few thousand (Liao et al., 1982).

Many metals that affect soil fertility are ubiquitous elements that occur in trace quantities throughout the lithosphere. Arsenic is such an element, but its toxicity depends upon the nature of the chemical compound. Hence, chromatography has been employed to detect low concentrations of several arsenic compounds in crops grown in soils to which 100 ppm of arsenic acid was deliberately added (Pyles and Woolson, 1982). Such studies are especially relevant, inasmuch as organic arsenicals have found extensive use as herbicides and pesticides and have largely replaced inorganic arsenicals. Results of other studies in both the United States and Europe are indicating that combinations of acid rain and metals (especially aluminum) affect plant and forest growth. In particular, it has been suggested that acid rain solubilizes the aluminum, thereby causing it to be taken up by tree roots.

In many instances, of course, the roots of plants are assisted by micro-organisms that promote an absorption of the essential nutrient elements of nitrogen, phosphorus, potassium, calcium, magnesium, sulfur, boron, copper, manganese, iron, zinc, and molybdenum (Olsen et al., 1981). During the growing season, most crops will require iron from land having iron concentrations of approximately 5–10 kg/ha (1 ha $= 10,000$ m^2 $= 2.471$ acres), but most soils will contain orders of magnitude greater than this amount. However, the availability of iron for plant growth depends upon solubility. In the case of other metals, the plants may have narrow tolerance limits; boron concentrations in the range of 0.04 ppm are required in the ambient solution of many plants, but 1 ppm levels can be injurious (Eaton, 1944).

Other soil-related studies include comparative measurements of residues in fruits in fumigated and nonfumigated orchards. Dibromochloropropane (DBCP) was found by GC/MS at a level of ~25 ppb in ripened peaches from orchards that had been fumigated 144 days prior to harvest. This relatively high concentration is in contrast to the 0.3 ppb found in peaches from trees in soil that was fumigated 270 days prior to harvest (Carter and Riley, 1982). While this measurement may be atypical, it points out the essential role of chromatography-mass spectrometry in tracing chemical compounds throughout the complete food cycle.

Investigations relating to the availability of nitrogen to crops is of continuing concern, as much of the nitrogen mineralized by soil micro-organisms or added in fertilizers is lost by leaching or volatilization. For wetland rice, as much as two-thirds of the applied nitrogen is volatilized as ammonia gas or through denitrification as N_2O or N_2, or it is immobilized in clays or soil organic matter (Brady, 1982). Direct field measurements have been made of the evolution of N_2 and N_2O in soil profiles. For example, precise isotope ratio mass spectrometry has been employed using labeled ^{15}N to measure the isotopic composition in terms of the per mil notation, that is,

$$^{15}\delta N(\text{‰}) = \left(\frac{R_s - R_{\text{ref}}}{R_{\text{ref}}}\right) 1000$$

where R_s is the isotope ratio (29/28) of the sample and R_{ref} is the isotope ratio of the reference (Limmer et al., 1982). In this work, a double-walled gas lysimeter (~62 mm tubing) was driven into the soil with a hydraulic jack, and the soil core was initially purged of nitrogen by flushing with an inert gas. Subsequently, N_2 enriched in ^{15}N (33.3 atom%) was admitted to the headspace of the lysimeter, and diffusion rates were calculated using Fick's law. A sample probe located beneath the base of the lysimeter was connected to sample collecting flasks containing a "carbsorb" getter that removed > 95% of the CO_2. Gas samples were taken at 2 h intervals and analyzed in an isotope ratio spectrometer. By careful control of the ^{15}N-enriched spike gas and measurement of m/z peaks at 28, 29, and 30, the rates of N_2 evolved from microbial denitrification were estimated. The authors suggest that this in situ technique of flushing with an N-free gas combined with a ^{15}N-enriched spike permits concentrations of N_2 in soil atmospheres to be detected in the range of 2000–5000 ppm. Other $^{15}N/^{14}N$ isotope ratio measurements on soils from 20 states in the United States have indicated that most soils are slightly, but significantly, enriched in ^{15}N compared to atmospheric nitrogen (Shearer and Kohl, 1981). In particular, the abundance of ^{15}N in plants that use biologically fixed nitrogen is lower than in soil nitrogen—and this difference has enabled both studies to be made of N_2-fixing plants and estimations of proportions of nitrogen derived from soils or the atmosphere (Turner et al., 1983). ^{15}N-labeled fertilizer has also been used by Witty (1983) in investigating N_2 fixation during the growing season and in obtaining uptake data using peas, French beans, field beans, and clover.

Other studies have been targeted on diseases, crop rotation, crop genetics, and the eradication of soil-borne fungi. Strawberry yields of 19–22 tons/acre have been reported in fumigated soil compared to only 5–6 tons in untreated acreage, and comparable results have been reported by employing seed-coating procedures and seed inoculations (Schroth and Hancock, 1982). Increased plant growth is also attributed to the elimination of allelopathic chemicals from the soil that are normally released from the standing crop. Removal of vegetation also affects soil fertility through the mineralization of certain chemicals such as Ca^{2+}, Mg^{2+}, K^+, NH_4^+, and nitrate nitrogen (Lohdi, 1981). Recent soil studies have been extended to isolating these allelochemicals, that is, those compounds generated by plants that are transported to the soil and water and that can affect the growth of adjacent plants, even of the same species (Waller et al., 1984).

An additional problem of contemporary concern relates to the estimated 7 million metric tons of sewage sludge that is presently produced annually

TABLE 16.1 Elemental Content of Sludge Sample as a Potential Soil Additive

Element	ppm	Element	ppm
Ag	33	Mg	8881
Al	31,340	Mn	395
As	116	Mo	7.2
Au	0.3	N	8800
B	12	Na	8701
Ba	266	Ni	169
Br	9.3	P	11,200
Ca	62,580	Pb	653
Cd	81	Rb	54
Ce	67	Sb	6.5
Cl	2173	Sc	1.3
Co	5.1	Se	1.1
Cr	111	Sm	12
Cs	1.8	Sn	870
Cu	1112	Sr	1479
Dy	6.3	Ta	0.9
Eu	2.7	Th	13
Fe	10,710	Ti	1376
Hf	5.2	U	2.1
Hg	4.9	V	42
I	35	W	18
In	2	Yb	1
K	1200	Zn	4127
La	19		
Lu	13		

Source: Furr et al. (1981).

in the United States. The possible use of this sludge as a soil conditioner or fertilizer in agriculture has been viewed as an alternative to ocean dumping or burial in sanitary landfills. The evaluation made by Furr et al. (1981) points out both the potential benefits and hazards of this option. Table 16.1 lists the concentrations (parts per million, dry weight) of some 48 elements measured in a sludge sample, which includes both plant nutrients and potentially toxic metals. Specifically, all vegetables grown over 2 years on soils with this sludge as an additive showed higher concentrations of Cd, Cu, Ni, and Zn than the corresponding control soil, although the results for Pb were variable. Studies of residues and the uptake of heavy metals in various crops will continue, as methods for the disposal and/or recycling of municipal wastes are examined further. The use of enriched stable isotopes and mass spectrometry is not needed for the gross chemical assay of wastes, but it can provide detailed data on the uptake of all elements in all crops from sludge-amended soils at even the lowest concentrations. This topic is discussed further in a subsequent section of this chapter.

PHOTOSYNTHESIS AND PLANT GROWTH

"It is generally agreed that the net reaction for green plant photosynthesis can be represented by the equation

$$CO_2 + H_2O + h\nu \xrightarrow{\text{chlorophyll}} O_2 + (1/n)(C \cdot H_2O)n$$

and also that very little is known about the actual mechanism. It would be of considerable interest to know how and from what substance the oxygen is produced. Using ^{18}O as a tracer we have found that the oxygen evolved in photosynthesis comes from water rather than from the carbon dioxide." These introductory sentences of a paper that appeared in the *Journal of the American Chemical Society* in 1941 (Ruben et al., 1941) provided an early stimulus to studies of photosynthesis and plant growth using labeled isotopes. The heavy oxygen water used in this experiment was prepared by fractional distillation, the *Chlorella* cells were suspended in this water, and the $^{18}O/^{16}O$ ratio of the evolved oxygen was measured by a spectrometer and shown to be identical to that of the water.

At present, it is estimated that about 22×10^9 tons/yr of cellulose are generated by photosynthesis worldwide, with approximately 20% of this vast amount being potentially available for conversion to fuels, building materials, and chemicals—as well as foods, so that photosynthesis is the basis for our most abundant and renewable reservoir of raw materials (Detroy and St. Julian, 1983). It is also known now that photosynthesis is dependent on many parameters (i.e., sufficient water, light, available mineral nutrients, and carbon dioxide), and that a reduced rate of photosynthesis can result from pollutants and other pathogens. Furthermore, experiments with controlled atmospheres, using isotopically enriched hydrogen, carbon, nitrogen, and oxygen, have permitted many detailed observations to be made of plant physiology and of the conversion of CO_2 from the atmosphere into organic matter. A method for the direct analysis of ^{13}C abundance in several kinds of plant carbohydrates by GC/MS has been reported by Kouchi (1982). By measuring $^{13}CO_2/^{12}CO_2$ ratios, determinations have been made of the incorporation of ^{13}C assimilated photosynthetically into carbohydrates of soybean plants and their translocation into plant roots. Specific experiments have also been conducted to determine hydrogen isotope fractionations in both photosynthesis and respiration, that is,

$$CO_2 + H_2O + HDO \xrightleftharpoons[dark]{light} \text{organic matter}$$

where D (deuterium) indicates the heavy isotope. By growing microalgae under controlled conditions in continuous light, the forward reaction greatly exceeds the reverse one, and the organic matter must derive its hydrogen from the water. The quantitative measurement of the organic matter is then determined by a mass spectrometric comparison of the hydrogen with a standard reference gas (Estep and Hoering, 1980). Numerous other plant growth studies have monitored the oxygen isotopes. For example, $^{18}O/^{16}O$ ratios in tree ring cellulose follow closely the $^{18}O/^{16}O$ values in atmospheric precipitation—and hence mean annual temperature and humidity (Burk and Stuiver, 1981).

The use of ^{18}O-labeled water has provided the most direct and accurate method for the study of water movement in plants. ^{18}O is preferred to tritiated water (^3H labeling), and water of high deuterium content may induce anomalous effects. Even at low ^{18}O water enrichments, Förstel and Hützen (1981) were able to measure the water exchange between the nutrient solution (soil), plant root, stem, and leaves. Using apple trees 1.5 m in height ("Bitterfield seedling" type), they found that the stem was labeled very rapidly and that it was uniformly labeled within 3 h, indicating that water was moving at a velocity of ~0.5 m/h. The $^{18}O/^{16}O$ ratio in the leaves changed more slowly, which the authors attribute to the transport of "free" intercellular and cellular water, and diffusion effects. In any event, for this particular experiment, they concluded that two-thirds of the water molecules in the leaves were derived from the stem and one-third from the water in the ambient air.

Mass spectrometry has also been widely utilized to quantify the uptake of heavy metals in plant growth. In recent years, certain aquatic plants, such as the water hyacinth, have also been utilized to remove heavy metals from metal-polluted waterways in tropical and subtropical regions (Wells et al., 1982). Factors that affect the heavy metal uptake in these plants include genetic differences, pH, oxygen concentration, amounts of heavy metals in solution, temperature, etc. Metals whose uptake has been studied in various plant species include arsenic, rubidium, zinc, cesium, nickel, chromium, barium, and selenium. It is also apparent that the use of aquatic or terrestrial plants as "nutrient sinks" provides one of the most practical methods for pinpointing sources of pollution at trace levels, and for monitoring long-term changes in the heavy metal concentrations in the ecosystem. The localization of nutrient ions at the subcellular level has been studied by several techniques. One method has been the in vivo fixation and precipitation of ions by use of a reagent containing a heavy metal that provides contrast in electron microscopy. For example, Tl^+ has been suggested as serving as a "physiological isotope" for K^+; Tl^+ is an electron-dense ion (of average mass number 204) that can be easily precipitated with several inorganic and organic reagents (Van Iren and Van der Spiegel, 1975). Ultra-thin sections of samples can also be analyzed by secondary ion mass spectrometry (SIMS) to measure precipitates, even though sample preparation may be difficult. Laser time-of-flight (TOF) instrumentation also appears to be especially applicable in this area, inasmuch as it has a microbeam or spatial resolution of ~1 μm and a very high sensitivity.

In an additional instrumental development relating to photosynthesis, ^{252}Cf fission fragment-induced mass spectra have been reported for chlorophyll (Chl a) produced with a flux of 3000 fission fragments/s (Chait and Field, 1982). In their TOF analysis, they determined that only 1.3% of the positive and 4.7% of the negative Chl a molecular ions remained intact during the 67 μs flight time to the ion detector. However, they were able to obtain highly resolved fragmentation patterns and determine data on fragmentation rate constants.

HERBICIDES, INSECTICIDES, AND FUNGICIDES

In the context of federal regulations, the term "pesticide" includes many types of chemicals used in food production in addition to herbicides, insecticides, and fungicides. Defoliants used in harvesting cotton, plant growth regulators, and soil sterilants are examples. The nomenclature "biocide" is also now an umbrella term to cover all products used to destroy organisms or tissue. In this sense, it includes wood preservatives and a number of industrial chemicals. The largest class of pesticides or biocides, however, consists of herbicides that are now applied to over 250 million acres in the United States alone for weed control. Synthetic organic herbicides were introduced in the 1930s and were utilized on a large scale after World War II. Nevertheless, the U.S. Department of Agriculture estimates about 10% of U.S. agricultural production is still currently lost to some 1500 species of weeds. Herbicides are relatively safe, inasmuch as they suppress weeds by mechanisms that do not exist in humans, that is, by disrupting photosynthesis or destroying a plant's chlorophyll. Their lifetime in the soil is also

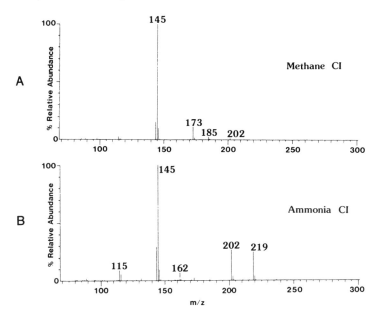

Figure 16.1   Mass spectra of suspected carbaryl as analyzed by (*A*) methane chemical ionization and (*B*) ammonia chemical ionization. (Cairns et al., 1983; *Bull. Environm. Contam. Toxicol.*, *30*, 93 (1983), by permission of Springer-Verlag.)

limited, because they are gradually decomposed by micro-organisms or other means (Sanders, 1981). Herbicides that are used most widely on major crops are: (1) atrazine and alachlor (corn); (2) alachlor and trifluralin (soybeans); (3) 2,4-D (wheat); and (4) trifluralin and fluometuron (cotton).

In searching for pesticide residues, compounds are frequently encountered that require mass spectrometric confirmation and identification in addition to chromatographic data. The Food and Drug Administration (FDA) analysis by Cairns et al. (1983) on the insecticide carbaryl is cited as an example. In this investigation, an extract of marionberries was found to contain an unknown organonitrogen compound, so the sample was subjected to GC/MS using both methane and ammonia as reagent gases (Fig. 16.1). The spectral data shows the increased relative abundance of the MH^+ ion at $m/z = 202$ when ammonia was used as the reagent gas, and the appearance of the peak at $m/z = 219$ represents the adduct ion $M + NH_4^+$. The proposed fragmentation pathway of carbaryl under methane chemical ionization is shown in Figure 16.2. The significant difference of the methane and ammonia chemical ionization spectra point out the desirability of using several reagent gas species to maximize the structural information that can be obtained about the molecule.

The analysis of the phenolic metabolites of carbofuran in peppermint hay and peppermint oil (Inman et al., 1983) is another typical example of the

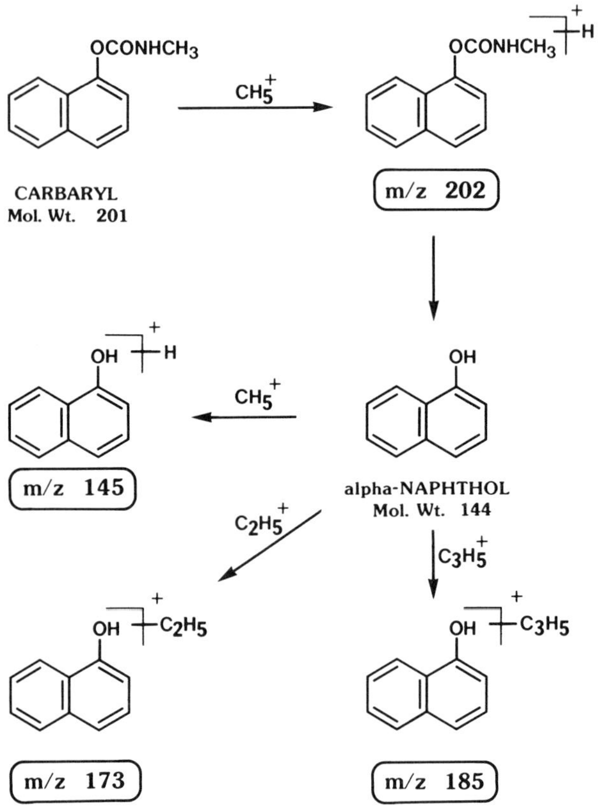

Figure 16.2 Proposed fragmentation pathway of the insecticide carbaryl (1-naphthyl-N-methylcarbamate) under methane chemical ionization. (Cairns et al., 1983; *Bull. Environm. Contam. Toxicol., 30,* 93 (1983), by permission of Springer-Verlag.)

specificity of mass spectrometry for detecting residues in agricultural products. Carbofuran controls strawberry root weevil larvae and mint root borer larvae in peppermint, and phenolic metabolites are produced by plants as oxidation or hydrolysis products. Inman et al. (1983) report that by multiple-ion detection, a sensitivity for several of these metabolites is 0.1 ppm. In a related study by Nelsen et al. (1983), GC/MS assays were made of 12 different crop matrices in which sensitivities in the range of 0.05–0.10 ppm for carbofuran phenols were achieved. Crops included alfalfa, alfalfa hay, grapefruit, oranges, peanuts, peanut hulls, potatoes, soybeans, sunflowers, and sweet corn. A multiresidue method for determining N-methylcarbamate insecticides in fruits and vegetables (Krause, 1980) has also been adopted for assaying coumaphos residues in eggs and milk (Krause et al., 1983). Coumaphos is registered for use as a topical insecticide for livestock and poultry.

Toxaphene, a broad-spectrum pesticide, is another extensively used insecticide in the United States, with 1.8×10^7 kg reportedly being applied to major U.S. crops in 1977. As expected, residues of this insecticide have been found at trace levels in many locations. Quantitation has been made by GC/MS analysis, with residues as low as 0.1 µg/g being observed in fish tissues. Since a portion of the toxaphene applied to crops is lost to the atmosphere through volatilization and dust, measurements have been made of residues at locations remote from its area of application. Concentrations of 0.6 ng/m^3 were observed in air samples over the north Atlantic taken more than 1600 km from its nearest known use; and concentrations of ~3 and ~0.3 ng/m^3 have been reported in air samples over the Canadian northwest and in Maryland rainwater, respectively (Ribick et al., 1982). Kepone $(C_{10}Cl_{10}O)$ and mirex $(C_{10}Cl_{12})$ are closely related insecticides. Interest in the former is continuing because of the presence of residues detected in fish in the James River in Virginia. Levels of 0.1–20 µg/g and a bioconcentration factor of 20,000 has been reported. Mirex has been used for the control of fire ant infestation in nine southeastern states. The residue dynamics of both of these compounds in soils and the aquatic environment have been studied extensively by GC/MS (Huckins et al., 1982).

Other studies have been made of carbaryl and its hydrolysis products, because it has been found in irrigation and drainage water from both aerial spraying and surface runoff; and it has been shown to be toxic to five major fish families (Shea and Berry, 1983). A group of 19 carbamate pesticides has also been analyzed by high-performance liquid chromatography (HPLC)-mass spectrometry, and a comparison of sensitivity has been made to ultraviolet absorption (Wright, 1982). Chemical ionization mass spectra of the widely used pesticide aldicarbe and its toxic metabolites (sulfoxide and sulfone) have also been reported by Muszkat and Aharonson (1982). This ionization method provides abundant $(M + H)^+$ and diagnostic ions for residue and metabolism assay, in contrast to the complex spectra produced by electron-impact ionization.

The utilization of mass spectrometry in the development of fungicides and in the detection of fungicide residues has closely paralleled that of insecticides. Many metals, in the form of metal ions, are essential for the proper functioning of life processes in cells and tissues. Fungi also require metal for growth, but when supplied in more than optimal amounts, compounds of zinc and copper can have potent fungicidal effects. Other metal ions without a known physiological function (i.e., Ag, Hg, Cd, Ni, and Pb) may also promote a fungicidal activity. Modern organic fungicides have largely replaced the previous use of sulfur, copper sulfate, and copper-hydrated lime, and there are now some 40 registered fungicides (Horsfall, 1977). A general overview of the design, synthesis, and utilization of these fungicides has been provided by Hedin (1982). Major groups of fungicides are the dithiocarbamates, metallo-organics, benzimidazoles, oxathins, surfactants, sulfenimides, and natural antibiotics. Some compounds that have been synthesized for herbicidal activity have also proven effective as fungicides, and many metals have been investigated. Strobel and Myers (1980) were also successful in isolating a bacterium normally found on leaves of wheat, barley, and oats that was effective in controlling Dutch elm disease fungus.

Environmental assays of fungicides in soils are made because rain and dew transport the residues from leaves and fruits to the ground, where both the active ingredients and transformation products can accumulate. In one gas chromatographic assay, the fungicide vinclozolin was taken up by benzene and pyrophosphate, with a recovery of 93% at concentrations of 0.01– 10 ppm. In this investigation, a detection limit of 0.001 ppm was also reported (Boccelli et al., 1982).

PHEROMONES AND SYNTHETIC ATTRACTANTS

In 1959, the term *pheromone* was coined from the Greek *pherin* (to carry) and *hormon* (to excite) by a German biochemist and a Swiss entomologist (Karlson and Luscher, 1959). Pheromones are now identified as the chemicals released by one member of a species that generate a behavioral or physiological response in other members of the same species. Pheromones are further subdivided into two major categories: (1) *releaser pheromones*, and (2) *primer pheromones*. The first category causes an instantaneous behavioral response in recipients and includes sex pheromones, alarm or trail pheromones, and trail-marking chemicals. The latter category causes longer term physiological effects in the receiving organism (Wood, 1983).

Insect pheromones are among the most potent, physiologically active substances, and they are produced and detected in very small quantities (Jacobson, 1965). The sensitivity of detection for pheromones by some species is truly remarkable. It has been estimated that the threshold of sensitivity may be as low as two or three molecules per cubic centimeter. One species of male moths is attracted to a female 2.5 miles away, and only 30 molecules (10^{-20} g) of pure sex hormones are sufficient to excite a male American cockroach (Volimer and Gordon, 1974). This amazing chemosensory power of insects also serves as a basis for both their husbandry and control (Lovrien et al., 1981). For this reason, highly sensitive GC/MS techniques offer the greatest potential for studying the mechanisms of pheromone biosynthesis. The use of sex pheromones and food attractants has been applied to both growing crops and stored food products, and pheromones are now an important factor in pest management.

One of the first commercial applications of pheromone traps was initiated in British Columbia, Canada in 1975 to protect both growing timber and sawed lumber from the ambrosia beetle *Gnathotrichus sulcatus*. In 1980, a large-scale population control program was directed against the pink bollworm that threatened the cotton industry in the United States and South America, and similar programs have been initiated throughout the world. A gypsy moth infestation in the San Francisco bay area was also brought under control with an early detection being attributed to pheromone traps (Silverstein, 1981). Traps containing the aggregation pheromone of the spruce bark beetle have been used as part of a pest control program in Norway; about 600,000 such traps were deployed, resulting in the estimated capture of 2.9 billion beetles in 1979 and 4.5 billion in 1980 (Bakke, 1981). Today, parapheromones (chemicals that mimic pheromones) or antipheromones (chemicals that suppress responses) are being widely studied as lures for direct trapping and to disrupt mate-finding or aggregation for many species.

A pheromone may consist of a single compound, or it may be comprised of several compounds—aliphatic alcohols, aldehydes, acetates, expoxides, ketones, and hydrocarbons. In the GC/MS analysis of substances containing hydrocarbon, chemical ionization (CI) is used for molecular weight determination and electron-impact (EI) ionization is used for structure elucidation (Hogge and Olson, 1982). The GC/MS analysis also provides both the mass spectral and retention-time data to characterize specific isomeric components.

An analysis of the volatiles in wheat germ oil by chromatography has revealed several compounds responsible for initiating aggregating activity of *Trogoderma glabrum* larvae (Nara et al., 1981). Synthetic compounds, such as octanoic acid in mineral oil, were prepared based on the analysis of the natural fractions, but the larvae response was not equivalent to that of the intact wheat germ oil. Honeybee queens produce pheromones that influence the behavior and physiology of worker bees (Winston et al., 1982), and an identification has been made of trail-marking pheromones produced by leaf-cutting ants while foraging (Cross et al., 1982).

Recent GC/MS studies have also been made of the volatile compounds from blossoms that are known to be attractants. Flath and Ohinata (1982) have analyzed the volatiles of the orchid *Dendrobium superbum*, because the blossoms are attractive to the male melon fly. Twenty-five components, mostly methyl ketones and 2-alkyl acetates, were identified as listed in Table 16.2. The aromatic hydrocarbons were detected at extremely low concentrations, and the authors questioned whether they evolved from the blossoms or were detected as atmospheric contaminants. GLC/MS studies have also been reported on the aroma components of honeydew melons in connection with the possible formulation of synthetic, attractive mixtures for insect trapping and population control (Buttery et al., 1982).

Another important physical measurement recently made by a gas chromatographic method has been to determine vapor pressures of sex pheromone components at biologically relevant temperatures (Olsson et al., 1983). Clearly, volatility and vapor pressure data are important in preparing and dispensing pheromone components in connection with field-trapping studies and controlled release systems. In a related study, chromatography has been employed to measure the "half-lives" for the evaporation of insect pheromones. Half-lives for pheromone compounds of varying volatility were re-

TABLE 16.2   Volatile Components Identified From Blossoms That Attract a Melon Fly

Alcohols		Ketones	
Ethanol	$+$[a]	Acetone	$+ +$
Linalool	$+ +$	2-pentanone	$+ + +$
Esters		2-heptanone	$+ + +$
Methyl acetate	$+ +$	2-nonanone	$+ +$
Ethyl acetate	$+ + + +$	4-phenylbutan-2-one	$+ +$
2-propyl acetate	$+$	2-undecanone	$+ +$
2-heptyl acetate	$+$	2-tridecanone	$+ + + +$
2-nonyl acetate	$+ +$	2-pentadecanone	$+ +$
2-undecyl acetate	$+$		
2-tridecyl acetate	$+$	Miscellaneous	
Hydrocarbons		Acetaldehyde	$+$
Toluene	$+$	Carbon disulfide	$+$
Ethylbenzene	$+$	Indole	$+ +$
o-xylene	$+$		
p-xylene	$+$		
Limonene	$+$		

Source: Flath and Ohinata (1982).

[a] Plusses indicate the relative amounts of each component in the trapped volatiles mixture.

ported to range from 12 days to greater than 1000 days (McDonough and Butler, 1983).

FLAVORS AND AROMAS

One of the most intriguing applications of mass spectrometry in recent years relates to its use for food flavor analysis. The mass spectrometer is now not only the primary tool for identifying natural, biologically produced flavors; it is indispensable for assaying the qualitative and quantitative changes that take place in food processing and the fermentation of beverages. In a commercial context, flavor research also addresses such problems as (1) natural flavor enhancement; (2) the masking or removal of undesirable off-flavors; and (3) the creation of synthetic flavors (Kolor, 1980). Synthetic flavors may not be as desirable as natural ones, but as one food chemist has aptly remarked, "strawberry flavor is a worldwide favorite, but there just are not enough strawberries to go around." The chocolate flavor business is also important from the standpoint of economics and mass spectrometry. The synthetic chocolate market is $25 million per year in the United States alone, and over 1200 chemical compounds have been identified in natural chocolate (Freedman, 1985).

Developments in both natural and synthetic flavor chemistry have been collinear with progress in instrumentation: (1) gas chromatography in the 1950s; (2) capillary column and GC/MS in the 1960s; (3) head space volatiles and distillation techniques in the 1970s; and (4) flexible silica capillary columns and advanced GC/MS/MS systems in the 1980s. With the aid of computers and reference spectra, much GC/MS flavor and aroma chemistry is now routine. Furthermore, isotopic measurements permit a distinction to be made between many natural food flavors and their synthetic equivalents. And when supplemented by infrared, nuclear magnetic resonance (NMR), and ultraviolet spectrometry, GC/MS analysis provides the link between highly quantitative measurements and the subjective sense of taste. Many chemicals are now known to affect the flavor of a food at/or below the parts per billion range. The classic example is 2-methoxy-3-isobutylpyrazine, which is primarily responsible for the flavor of green pepper, and which can be detected (by the human nose) at a concentration of 0.002 ppb (Buttery et al., 1969).

Hence, quantitative mass spectrometry is essential for controlling, within narrow limits, either natural or synthetic flavor additives. And no other type of analysis approaches the specificity provided by mass spectra in identifying the hundreds of compounds relating to sensory perceptions of freshness and quality that are formed during natural ripening, heating, processing, packaging, or storage of fruits (Van Straten et al., 1981).

Fruits and Juices

In some fruits, a single organic compound may be a dominant contributor to the flavor. For example, methyl anthranilate has been identified in grapes (Holley et al., 1955) as having a strong grape-like aroma, and amyl acetate in bananas was one of the first important flavor components to be identified (Kleber, 1913) by laborious organic chemistry. In other fruits, the flavor reflects the blending of a number of chemicals. Kolor (1979) states that the evaluation of a GC effluent by smelling the emerging components is a valuable subjective method for determining the contribution of the various chemicals to the overall flavor.

Typical of the GC/MS analysis of fruit flavor volatiles is the work reported by Parliment and Kolor (1975) on blueberry essence and the recent analysis by Iwamoto et al. (1983) of watermelon flavor. In the former work, some 900 g of high-bush blueberries (*Vaccinium corymbosum*), a fruit grown commercially in the middle eastern United States, were placed in a blending apparatus with 300 ml of water, and the system was macerated for 30 s. The slurry was then vacuum distilled at 54°C for 1 h, and the volatiles were trapped in a dry ice/acetone bath. After an ether extraction, the solution was dried and concentrated by slow distillation to a volume of about 1 ml. A separation of the volatile components was accomplished by a gas-liquid chromatograph that was programmed from 60–257°C at 6°C/min, and the column effluent was passed into the ion source of a mass spectrometer, where the

TABLE 16.3   Volatile Components Identified in Blueberry Essence

Peak No.	Identity	Relative Value[a]	Concentration (%)
1	Ethyl acetate	0.18	0.29
2	Ethanol	0.24	5.74
3	Ethyl isovalerate	0.56	0.13
4	1-Hexanal	0.59	4.86
5	1-Penten-3-ol	0.77	0.30
8	Limonene	0.96	0.97
9	trans-2-Hexenal	1.00	71.13
10	2-Penten-1-ol	1.28	0.24
11	1-Hexanol	1.40	2.00
12	cis-3-Hexenol	1.50	0.41
13	trans-2-Hexenol	1.56	12.39
14	1-Heptanol	1.72	0.01
15	2-Ethyl-1-hexanol	1.81	0.04
16	Linalool	2.00	0.67
18	1-Nonanol	2.35	0.02
19	α-Terpineol	2.45	0.03
20	Nerol	2.75	0.04
21	Geraniol	2.85	0.61

Source: Parliment and Kolor (1975).

[a] GC Retention time relative to trans-2-hexenal = 1.

source temperature was maintained at 250°C. Identification of sample components was accomplished by a comparison of mass spectral data and GC retention times to that of known standards. Table 16.3 shows the components identified in blueberry essence for 21 GC peaks and their relative concentrations. The authors point out that a large percentage ($> 91\%$) of the volatile components are composed of six carbon compounds (hexanols, hexenols, hexanals, and hexenals), all of which were judged to contribute to the fruity-fresh character of blueberries, and that linalool (a floral woody flavor), which is used in imitation blueberry flavors, also makes an important flavor contribution even though its concentration in the essence is less than 1%. Recent GC/MS analysis of the volatile constituents of other blueberry types have been reported by Horvat et al. (1983).

Among the flavor components of many varieties of apples, Flath et al. (1969) have reported the inclusion of 16 esters, 24 aldehydes and ketones, 9 ethers and acetals, 33 alcohols, and 16 acids. They also suggest that flavor differences can be attributed primarily to different concentrations of these components and not to the unique presence or absence of them. A list of volatile components in a variety of fruits (grapefruit, lemon, lime, orange, passion fruit, black currant, apple, pear, peach, raspberry, pineapple, and banana) has been compiled by Nursten and Williams (1967), together with 175 references. Also, authoritative reviews of flavor chemistry, flavor isolation and concentration, and GC/MS measurements have been published by Kolor (1972, 1979).

The aroma components of both cultivated and wild strawberries have been reported by several food chemists. In a publication by Dirinck et al. (1981), flavor quality has been evaluated by subjective taste and aroma and by GC/MS assay for different strawberry varieties; and comparisons have been made relative to berries that were field-grown, greenhouse-grown, or grown under plastic funnels. In addition to the high concentration of volatile esters, sulfur-containing compounds at much lower concentrations were found to influence the flavor. Also, the authors conclude that important differences in flavor quality occur between different varieties, but that unfortunately the most commercially produced varieties do not rank highest in flavor; that is, a trade-off of flavor is sometimes made for optimal production and external quality.

The volatile constituents of the muscadine grape have been studied by Welch et al. (1982). This large, dark, thick-skinned grape grows abundantly in the southeastern United States and is used in wines, preserves, juices, and jellies. The authors have identified 49 compounds and suggested that the grape's distinctive aroma and flavor is due to the presence of isoamyl alcohol, hexanol, benzaldehyde, and 2-phenylethanol and its derivatives. Furthermore, for verification, these major volatile constituents were blended synthetically to produce the muscadine grape aroma. The recent identification by Kolor (1983) of a new flavor compound in Concord grapes points out the importance of volatiles even at the parts per million level. Figure 16.3 is the gas chromatogram by Kolor of volatiles from a Concord grape essence in which 114 compounds were characterized. However, the trace amount of ethyl 3-mercaptopropionate, whose mass spectrum appears in Figure 16.4, has been found to contribute significantly to the pleasant fruity and fresh grape quality of Concord grape aroma. Kolor (1982) reports that this compound, in large concentrations, actually has an objectionable aroma. However, at levels of 100 ppb to 50 ppm, this newly identified flavoring ingredient (in combination with other compounds) provides a very satisfying Concord grape flavor and aroma to foodstuffs such as beverages, ice cream, candy, baked goods, syrups, jellied products, and gelatin desserts.

The production of lemons in the United States now represents about 25% of the world crop; currently, some 40–50 million, 38 pound cartons of lemons are harvested from the citrus-growing regions of California and Arizona (Staroscik and Wilson, 1982). These authors also reported that climatic differences between the two growing regions are probably responsible for the differences of cold-pressed lemon oils—analyzed by capillary gas chromatography—in which 38 components were identified. The Arizona region produces a desert fruit that differs from the fruit grown in the coastal region, and the difference is reflected in the chemical composition of the extracted lemon oils that are commercially blended and marketed. The composition of lemon leaf oils has also been studied in connection with the fragrance

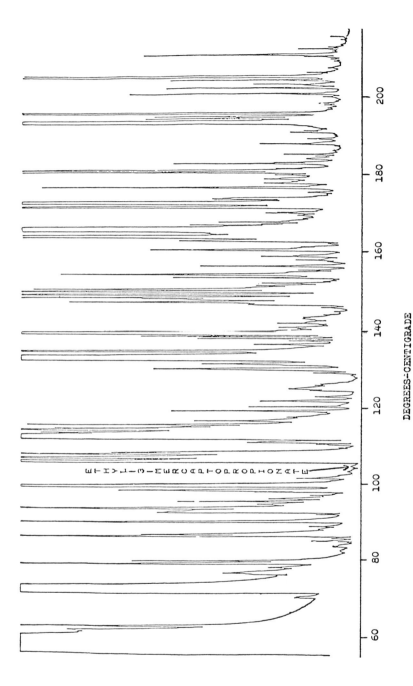

ETHYL-3-MERCAPTOPROPIONATE

DEGREES-CENTIGRADE

Figure 16.3 Gas chromatogram of volatile compounds isolated from Concord grape essence. (Kolor, 1983; reprinted with permission from *J. Agric. Food Chem.*, *31*, 1125 (1983), copyright 1983 American Chemical Society.)

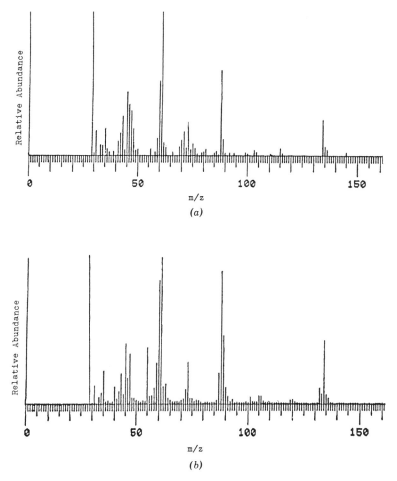

Figure 16.4 Mass spectra of (*a*) an unknown compound isolated from Concord grape essence and (*b*) a reference sample of ethyl 3-mercaptopropionate. (Kolor, 1983; reprinted with permission from *J. Agric. Food Chem., 31,* 1125 (1983), copyright 1983 American Chemical Society.)

components associated with attractants or repellants relating to the citrus blackfly (Lund et al., 1981).

Laboratory testing of both fresh and processed fruit juices is an ongoing activity. Juices from fresh fruit possess a "fresh-squeezed flavor" that changes upon standing or after heat treatment. Kirchner et al. (1950) first identified hydrogen sulfide in fresh orange juice, and they estimated its concentration at 1.6 ppm in fresh orange juice and 0.9 ppm in fresh grapefruit juice. Volatile sulfides have now been identified in head space gases of fresh oranges, grapefruit, lemon, lime, tangelo, and tangor juices; and variations have been related to time after juice extraction and the harvesting season

(Shaw and Wilson, 1982). In the sampling of commercial orange juice, many 10 ml samples were taken from the head space gases above 2500 gal of fresh orange juice in an enclosed 3000 gal tank. The most marked differences in hydrogen sulfide levels were noted in Hamlin and Temple oranges and in Marsh grapefruit after freeze damage. The hydrogen sulfide level was generally lower in freeze-damaged and over-ripe fruit. A change of 1.2 ppm for firm grapefruit to 0.1 ppm for soft grapefruit was measured in a comparison test. In contrast to fresh citrus juices containing hydrogen sulfide at levels as high as 2.9 ppm, this compound was not detected in the head space gases of processed orange, grapefruit, and tangerine juice—although measurable quantities of dimethyl sulfide were noted in canned orange and grapefruit juices. Furthermore, in all cases where dimethyl sulfide was found, the concentration increased when the juice was allowed to stand for 60 min in a closed container (Shaw and Wilson, 1982). Overall, hydrogen sulfide in the head space gases of several citrus fruits was found to be about 1000 times its threshold level in ambient air—and the authors conclude that it probably contributes to the aroma of fresh citrus juices.

The bitter components, naringin (in grapefruit) and limonin (in oranges and grapefruit), have also been quantified together with pulp, pH, acid, oil, and vitamin C (ascorbic acid) in single-strength and reconstituted concentrated juices (Albach et al., 1981). The mean values for a 3 year period were (1) 3 ppm of limonin for oranges; (2) 7 ppm of limonin for grapefruit; (3) 585 ppm of naringin for grapefruit; and (4) in both fruits, the limonin concentration decreased rapidly as the season progressed. Three year mean values of vitamin C in oranges and grapefruit were also reported of 43.8 mg/100 ml and 31.3 mg/100 ml, respectively.

Detailed mass spectrometric analyses have been made of Valencia orange oils. Table 16.4 shows major and minor constituents identified by Vora et al. (1983) as a function of vacuum distillation processing. For example, the concentration of d-limonene decreased from 95% to 57% in 25-fold concentrate. Seasonal variations in these constituents have also been directly related to flavor.

Recently, atmospheric pressure tandem mass spectrometry (API/MS/MS) has been successfully used to analyze volatiles from intact fruits. Labows and Shushan (1983) have produced the profile for bananas shown in Figure 16.5, with all major peaks indicative of esters and a large peak for acetic acid. The ion peak at m/z 131 was shown to be isoamylacetate, which is a characteristic banana aroma component. A parent ion scan for fragment ions m/z 43 and 61 revealed only those masses corresponding to acetate esters.

Meats and Vegetables

As cited in the previous section, the analysis of some food aromas has included the direct introduction of sample vapors into an atmospheric pressure chemical ionization (APCI) inlet of a tandem mass spectrometer. In this arrange-

TABLE 16.4   Quantitative Composition of Cold-Pressed and
Concentrated Valencia Orange Oils

Compound	Concentration (%)		
	Cold-Pressed	10-Fold	25-Fold
α-Pinene	0.42	0.23	0.17
Sabinene	0.24	0.14	0.11
Myrcene/octanal	1.86	1.05	0.77
Phellandrene	0.05	0.03	—
d-Limonene	95.17	79.91	57.09
I-Octanol	0.03	0.03	0.03
Nonanal	0.02	0.03	0.02
Linalool	0.25	0.52	0.51
Citronellal	0.05	0.14	0.16
α-Terpineol	0.03	0.17	0.26
Decanal	0.28	1.48	2.15
Neral	0.07	0.41	0.65
Geranial	0.10	0.68	1.13
Perillaldehyde	0.01	0.06	0.08
Octyl acetate	0.01	0.06	0.07
Undecanal	0.02	0.16	0.29
Dodecanal	0.07	0.92	1.94
β-Caryophyllene	0.02	0.22	0.47
β-Copaene	0.02	0.31	0.65
β-Farnesene	0.02	0.14	0.30
Valencene	0.05	0.72	1.65
β-Sinensal	0.03	0.55	1.20
α-Sinensal	0.02	0.37	0.86
Nootkatone	0.01	0.26	0.59
Total oxygenated compounds	1.00	5.84	9.94

Source: Vora et al. (1983).

ment, the first mass spectrometer is used for separating or profiling the volatiles and the second analyzer serves to identify the molecular structure. This APCI/MS/MS system reduces losses normally associated with chromatographic separations, and it is reported to more accurately characterize aroma composition for some foods. Spectra of meat samples using this multiple analyzer have been reported by Labows and Shushan (1983), and they indicate that (1) Teewursts are high in ethanol (m/z 47$^+$, 65$^+$); (2) sausage shows mass peaks of propanal (m/z 61$^+$, 79$^+$); and (3) salami contains terpenes (m/z 81$^+$, 137$^+$). Knockwurst profiles include acetone and alcohols as major components, and the probable structural assignments are listed in Table 16.5.

The isolation and identification of volatile compounds of roast beef have been reported (Lee et al., 1981a), approximately 600 components have been

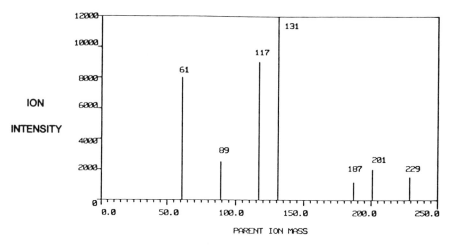

Figure 16.5 Banana volatiles obtained with API/MS/MS, using a parent ion scan for constant daughters m/z 43 and 61. The peak at m/z 131 (isoamylacetate) exhibits a chararacteristic banana aroma. (Labows and Shushan, 1983; reprinted from *American Laboratory, 15*, No. 3, 56 (1983), copyright 1983 by International Communications, Inc.)

reported in meat flavors. A significant complication is introduced by cooking in the case of beef, as important nonvolatile flavor components are produced with heating (Alabran, 1982). In the case of the sukizahi volatiles obtained from beef and heated with vegetables and seasonings (sugar and soy sauce), 44 compounds were identified that included furans, ketones, aldehydes, es-

TABLE 16.5   Probable Structural Assignments of Meat Volatiles

$M + 1$	Structure	$M + 1$	Structure
47	Ethanol	97	Furfural
59	Acetone	99	Furfuryl alcohol 3-hex-en-1-ol
63	—	101	Hexanone
71	—	103	Hexanol
73	Butanone	105	Styrene (?)
75	Butanol	111	Acetyl-furan
77	—	113	—
79	—	115	Ethyl butenoate
81	Terpene fragment	117	Ethyl butyrate
85	Pentanone cyclopentanone	119	—
87	Pentanone	127	Methylheptanone
89	Acetoin	137	Monoterpene
91	Lactic acid	155	Terpene alcohol
93	Toluene	163	Methyl cinnamate

Source: Labows and Shushan (1983).

ters, acids, alcohols, pyrazines, hydrocarbons, and pyrroles (Shibamoto et al., 1981).

A detailed mass spectrometric study has also been reported on the volatile flavor compounds isolated from 120 pounds of fried bacon (Ho et al., 1983). A total of 135 compounds were identified that included (1) hydrocarbons, alcohols, and carbonyl compounds; (2) phenols; (3) pyrazines; (4) furans; (5) thiazoles, oxazoles, and oxazolines; (6) pyrroles; and (7) pyridines. A significant portion of smoke flavor is associated with phenols; alkylpyrazines produce a roasted nut impression, and furans contribute to a sweet, nutty, and caramel-like aroma (Ohloff and Flament, 1979).

Most of the popular vegetables have been subject to extensive flavor analysis, including effects produced in cooking and baking. Coleman et al. (1981) have reported on the volatiles isolated from 540 pounds of Idaho baked potatoes, and they identified 228 compounds from 420 chromatographic fractions. The compounds included hydrocarbons, acids, alcohols, aldehydes, esters, lactones, ethers, furans, halogenated hydrocarbons, ketones, pyrazines, oxazoles, thiazoles, and several heterocycles. The most important components of baked potato flavor were attributed to the pyrazines, oxazoles, thiazoles, and a furanone. Other recent GC/MS analyses on the free amino acids identified in potatoes have been reported by Golan-Goldhirsh et al. (1982).

Tomatoes, their juices, purées, and pastes, have been analyzed resulting in the identification of 154 compounds; that is, 3 sulfides, 10 aromatic hydrocarbons, 4 terpene hydrocarbons, 27 alcohols, 26 aldehydes, 13 ketones, 17 esters, 2 lactones, 7 furans, 4 pyrroles, 2 pyridines, 2 pyrazines, 2 thiazoles, 12 phenols, 16 carboxylic acids, plus 7 others (Chung et al., 1983). Dimethyl sulfide, which is an important component of the heat-induced odor of tomatoes, was observed to increase in tomato juices. The volatiles of fresh mushrooms have also been analyzed by Tressl et al. (1982), who showed that five C_8 and eight C_{10} (C_{11}) components were enzymic breakdown products of linoleic and linolenic acids. A major aroma component from mushrooms (1-octen-3-ol) is known as "mushroom alcohol" and has a threshold in the parts per billion range.

Beverages

Tandem gas chromatography-mass spectrometry makes possible a definitive identification of the many volatile components in beverages. Table 16.6 lists the major components in the head space of a coffee sample, together with GC retention times (Kolor, 1979). Approximately 500 additional chemicals have been identified in coffee flavor. The chemistry of tea has also been reviewed, not only with respect to taste and aroma, but in the context of its pharmacological and physiological properties (Bokochava and Skobeleva, 1980). The soft drink industry, of course, relies heavily upon synthetic flavors, and laboratory analyses have been undertaken of all major beverages.

TABLE 16.6 Coffee Head Space Volatiles

Identity	GC Retention Time (min)
Methane	0.7
Ethylene	3.1
Ethane	3.8
Propene	6.8
Acetaldehyde	8.8
Methyl formate	10.4
Butene	11.0
Acetone + furan	13.4
Dimethyl sulfide	14.3
Isoprene	15.5
Isobutyraldehyde	16.5
Diacetyl + methyl furan	17.5
Isovaleraldehyde	20.4

Source: Kolor (1979).

In one recent report, the sweetness of cocoa drinks was also investigated in terms of other physicochemical properties that affect taste response (Ogunmoyela and Birch, 1982).

Alcoholic beverages have been tested extensively. Some flavor analysts agree that the senses of smell and taste are "analytical" like color vision. In this context, Meilgaard (1982) suggests that a complex flavor is perceived "in a manner in which we hear a symphony." Thus, for beers, he suggests nine "groups of instruments" whose varying contributions can be related to flavor differences throughout the world: (1) bitterness, (2) alcoholic flavor, (3) carbonation, (4) hop character, (5) caramel flavor, (6) fruity/estery flavor, (7) sweetness, (8) acidity, and (9) dimethyl sulfide (DMS) flavor. These principal flavor elements of various beer types have then been related to the 27 official flavor reference standards of the European Brewery Convention and the American Society of Brewing Chemists (Meilgaard, 1982). This report also contains an extensive list of organic compounds that are added to beers and beer mixtures, together with their flavor thresholds. Primary flavor constituents in typical U.S. lagers are ethanol, carbon dioxide, and hop bitter substances; secondary flavor constituents are listed as humolenes, banana esters, fusel alcohols, dialkyl sulfides, and fatty acids. Martin et al. (1981) have also described a rapid GC method for resolving 10 specific compounds that are found as congeners in beer and naturally fermented alcoholic products, and that have a significant effect on quality and flavor: (1) acetaldehyde, (2) methanol, (3) ethanol, (4) ethyl acetate, (5) n-propanal, (6) isobutanol, (7) n-butanol, (8) acetic acid, (9) 2-methyl propanol, and (10) 3-methyl-1-butanol.

Extensive literature exists on the flavor composition of wines (Schreier,

1979). Studies by Shinohara and Watanabe (1981) have compared the volatile esters in commercial wines. White and rosé wines generally contain a higher level of certain esters than red wines; sparkling wines have comparable concentrations to table wines, while lower levels of esters are found in sherries. The ester content of new wines are also generally larger than those of matured wines—confirming that various ester contents from the fermentation and aging process are reflected in the aroma and wine quality. Minor constituents such as the volatile phenols have been evaluated for their contribution to overall aroma (Etievant, 1981) and Shimitzu and Watanabe (1982) have reported on the volatile components of the phenolic fractions of wines and identified 13 volatile phenols, 11 fatty acids, 8 lactones, 4 ketones, and many alcohols. The volatile amines identified in grapes and wines have also been reported by Ough et al. (1981) from experimental vineyards at the University of California, and Almy et al. (1983) have characterized two additional amines: 1-pyrroline and 4,5-dimethyl-1,3-dioxolane-2-propanamine. A detailed study of dry white wines by Yankov et al. (1982) has evaluated the terpene compounds, esters, aromatic alcohols, etc. that are responsible for specific aromas and taste.

Miscellaneous Products

Additional studies have been made of the wide variety of structural features that are found in peptide sweeteners (Kawai et al., 1982), and relationships between molecular structure and taste quality (sweet or bitter potencies) have been reported by Takahashi et al. (1982). The volatiles in eggs and egg products have been assayed by Maga (1982), as well as the influence of storage and feed source on flavor. Eggs serve as a unique flavor model, since all possible reactants in the formation of its flavor are contained in an intact shell. Some 65 volatile components have been identified in hen's eggs, and the synthesis of some of these volatiles has also been undertaken (Gil and MacLeod, 1981). The isolation of the volatile compounds of 70 kg of roasted peanuts by Lee et al. (1981b) revealed eight thiazoles. It is suggested that the mechanism for their formation would be the interaction of sulfur-containing amino acids with carbohydrates or carbonyls. The occurrence of oxazoles, which contribute to a pleasant nutty aroma, have been reported in only five other food systems, that is, coffee, cocoa, barley, potatoes, and meats. The direct characterization of nutmeg constituents by MS/MS in the mass range 100–700 amu has been made by using different reagent gases in the chemical ionization mode and by employing both magnetic sector and triple-quadrupole spectrometers (Davis and Cooks, 1982). This type of analysis will be increasingly attractive in flavor chemistry because of its speed, sensitivity, minimal sample work-up, and its ability to provide detailed structural information. The isobutane CI spectrum of nutmeg, shown in Figure 16.6, corresponds to the intact, protonated molecular spectrum.

It is also interesting to note that some 3000 components have been found

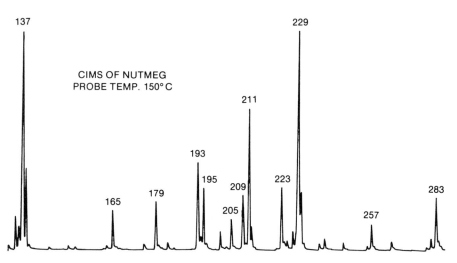

Figure 16.6 Isobutane CI mass spectrum of nutmeg. (Davis and Cooks, 1982; reprinted with permission from *J. Agric. Food Chem.*, *30*, 495 (1982), copyright 1982 American Chemical Society.)

in the flavor oils and smoke of various tobacco types. Furthermore, International Flavors and Fragrances Inc., reports that it is now prepared to provide a new generation of blended components and synthetic flavors (honey, woody, etc.) that can be added to reconstituted tobacco for flavor enhancement (Mookherjee, 1982).

FOOD COLORANTS

The distinctive color of foods ranks close to aroma and taste as a highly desirable attribute, conveying a concept of ripeness, freshness, and quality. Important coloring pigments that occur in nature include carotenoids, anthocyanins, porphyrins, and chlorophylls. Furthermore, it is estimated that some 200 billion pounds of carotenoids are produced in nature (Isler et al., 1961), and this unique coloring substance not only provides many of the red, orange, and yellow colors of fruits, vegetables, and other foods, but it functions as a source of provitamin A. Carotenoid addition to foods for coloring (from natural extracts) predates the commercial production of carotenoids by chemical synthesis in 1954; pure carotenoids are now among the most potent coloring agents presently available—being generally used in a concentration range of 1–25 ppm of the food product (Gordon and Bauernfeind, 1982).

Carotenoids are aliphatic or aliphatic-alicyclic fat-soluble pigments that occur as (1) carotenes, consisting of carbon and hydrogen only; or (2) oxy-

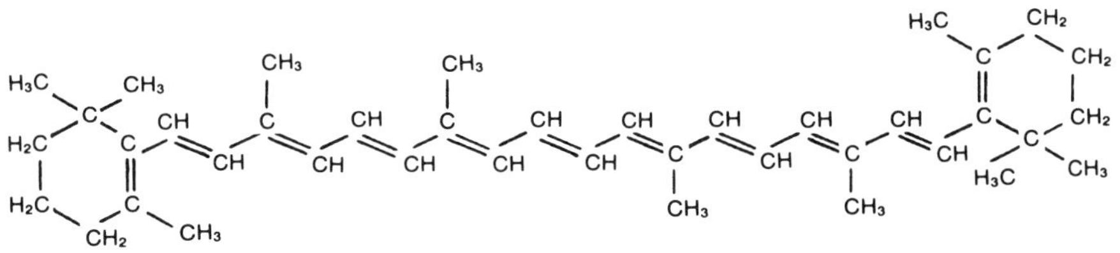

Figure 16.7   Structural formula for beta-carotene ($C_{40}H_{56}$) (Gordon and Bauernfeind, 1982).

carotenoids, containing C, H, and O. The structural formula for one important carotenoid, beta-carotene ($C_{40}H_{56}$), is shown in Fig. 16.7. This substance naturally occurs in butter, cheese, egg yolk, asparagus, lettuce, spinach, tomatoes, pineapples, oranges, figs, peaches, strawberries, and many fruits and vegetables. It is also added as an artificial coloring to cheese, ice cream, sherbet, oils, baked goods, frostings, gelatin mixes, confections, and other food products. Synthetic beta-carotene is now an approved colorant in 60 countries; the natural and synthetic pigments, of course, yield identical mass spectral and optical spectrometric signatures for comparably prepared samples.

In the dairy industry, beta-carotene is added to winter butter to enhance the color, as the natural beta-carotene content of milk fluctuates with the seasonal diet of the cow. Adding beta-carotene to winter butter can also add approximately 3000–5000 IU of vitamin activity per pound, thus providing a nutritive increment as well as color improvement. The carotenoid levels in citrus fruits also vary according to season, as well as with specific variety. Hence, coloration is added to juices and beverages to ensure a uniform product throughout the year. Gordon and Bauernfeind (1982) have cited the uses of carotenoids in many other foods and summarized their specific advantages:

1.   They possess high tinctorial potency as colorants.
2.   They are not foreign substances, but exist widely in nature.
3.   Some produce vitamin A activity and, hence, have a dual function.
4.   They are stable over a wide pH range.
5.   They can be used in conjunction with other coloring agents to extend hue and color spectrum.
6.   They are available in constant supply with specified quality.

VERIFICATION OF NATURAL FOODS AND EXTRACTS

The high economic value of natural fruit juices and extracts has been accompanied by some adulteration of these products with less costly products.

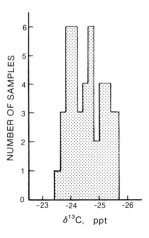

Figure 16.8 Distribution of $\delta^{13}C$ values among 42 orange juice samples from Florida, California, Texas, Arizona, Mexico, Spain, and South Africa. (Doner and Bills, 1981; reprinted with permission from *J. Agric. Food Chem., 29,* 803 (1981), copyright 1981 American Chemical Society.)

When properly labeled, this practice is not fraudulent, nor does it necessarily result in a food of low nutritional content. However, analytical methods are needed to verify the authenticity of certain foods, and a *stable carbon isotope ratio analysis* (SCRA) has recently been proven to be one of the most powerful methods for detecting adulteration. Specifically, it has been applied to a wide variety of fruit juices and concentrates (apple, orange, and berry), maple syrups, honeys, wines, and vanilla extracts. The stable carbon isotope method is applicable to foods, because the $^{13}C/^{12}C$ ratio of a plant-derived material reflects the particular pathway of photosynthesis in the plant. Green plants presumably use the same mechanism for the final fixation of CO_2 into carbohydrates, but alternative pathways exist for the initial extraction of CO_2 from the atmosphere—resulting in isotopic abundances that are different than in atmospheric CO_2 (see also Chapter 8, pp. 253–4).

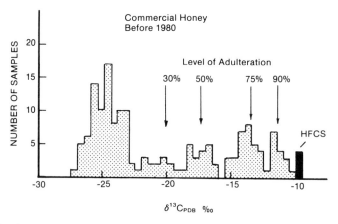

Figure 16.9 Carbon isotopic analyses of commercial honey prior to 1980 (Krueger and Reesman, 1982).

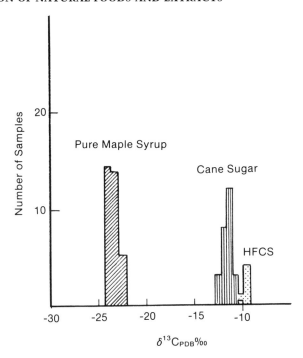

Figure 16.10 Carbon isotopic analyses of pure maple syrup, cane sugars, and high-fructose corn syrup (HFCS) (Krueger and Reesman, 1982).

Plants governed by the Calvin (C_3) photosynthesis cycle have $\delta^{13}C$ values in the range of -22 to -33 ppt (parts per thousand); plants utilizing the Hatch-Slack (C_4) cycle have $\delta^{13}C$ values of -10 to -20 ppt. A majority of plants, including most trees and shrubs, utilize the C_3 pathway, while both sugar cane and corn are C_4-type plants. Hence, inexpensive products such as cane sugar or high-fructose corn syrup (HFCS) can be detected by their isotopic composition (Krueger and Reesman, 1982). Plants such as pineapples and cactus have both types of carbon dioxide fixation, depending upon climatic conditions, and they have intermediate $\delta^{13}C$ values (Doner and Bills, 1981).

Figure 16.8 shows the distribution of $\delta^{13}C$ (ppt) values for a total of 42 samples of pure orange juices from Florida, California, Texas, Arizona, Mexico, Spain, and South Africa. The reference standard is CO_2 produced from Peedee belemnite (PDB) calcium carbonate. The small variation is remarkable and confirms that there was no adulteration for any of the samples. Figure 16.9 is a histogram of commercial honeys evaluated by Krueger and Reesman (1982) showing varying degrees of adulteration with high fructose corn syrup (HFCS). The Association of Official Analytical Chemists (AOAC) now sets a limit of $\delta^{13}C = -21.5$ ppt as a proof of adulteration (White and Robinson, 1983). Other isotopic measurements of hundreds of

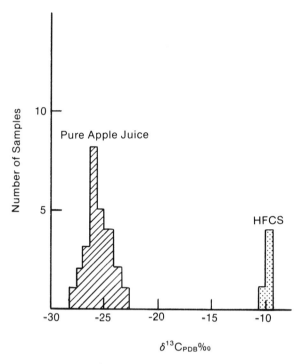

Figure 16.11   Distribution of $\delta^{13}C$ values among 40 pure apple juice samples in contrast to representative analyses of HFCS (Doner et al., 1980).

maple syrups have revealed $\delta^{13}C$ values ranging from -22.37 to -24.81 ppt; cane sugar $\delta^{13}C$ values range from -10.50 to -12.20 ppt, and $\delta^{13}C$ corn syrup values range from -10.40 to -12.40 ppt (Carro et al., 1980). Figure 16.10 shows carbon isotopic analyses by Krueger and Reesman (1982) of pure maple syrup, cane sugars, and HFCS. Maple syrup is produced in New York, Vermont, and Canada, and approximately 40 gal of sap are required to produce 1 gal of 100% pure product. The high price of pure syrup leads to many commercial syrups, with maple syrup contents in the range of only 2%. Clearly, the isotopic method provides an unambiguous marker for identifying these cheaper syrups.

Figure 16.11 shows the data by Doner et al. (1980) obtained for 40 pure apple juice samples representing 18 varieties that gave a mean value for $\delta^{13}C$ of -25.3 ppt, in contrast with representative analyses of HFCS.

The authenticity of pure vanilla extract has been the subject of analysis for many years. Recently, the carbon isotope test has been used in interlaboratory tests of pure Madagascar and Java extracts. No addition of artificial vanillin is permitted in products designated vanilla extracts, and it is suggested that a sample more negative than $\delta^{13}C = -21$ ppt be considered as vanillin from a source other than authentic vanilla beans (Guarino, 1982).

The carbon isotope ratio analysis has even made possible a distinction between the natural vanillas depending upon their source, that is, pure Madagascar and Java vanillas measure $\delta^{13}C = -20.7$ ppt and $\delta^{13}C = -18.7$ ppt, respectively (Martin et al., 1981b). Recently, some concern has been expressed that the carbon isotopic test might be invalidated by the addition of [methyl-^{13}C] vanillin—that is, using enriched ^{13}C in order to simulate the isotopic values of carbon found in natural vanillas (Krueger and Krueger, 1983). Hence, current research is underway to detect "fraudulent" isotopic tagging that attempts to make synthetic vanillins appear as natural products.

METALS IN THE FOOD CHAIN

The U.S. Food and Drug Administration (FDA) maintains a continuing program for monitoring levels of metals in foods, with special attention being given to potentially toxic elements. Table 16.7 lists cadmium concentrations in six broad food categories that represent approximately 90% of total cadmium intake. Additional cadmium in the diet comes from the consumption of various beverages (Travis and Etnier, 1982). Phosphate fertilizers and atmospheric fallout add cadmium to the soil, and cadmium can enter the food chain from industrial sources and via the byproducts of combustion. For example, coal contains cadmium in amounts ranging from 0.02–10 μg/g and heating oil has concentrations of 0.53–10 μg/g. The practice of disposing of municipal wastes and effluent from municipal wastewater treatment plants on agricultural land has both solved and created problems. It is a rather cost-effective and environmentally acceptable method of waste disposal, and the practice provides nutrients to the soil. However, with increasing amounts of sludge being applied to land, the potential for heavy metals entering the food chain has been given serious evaluation (Ryan et al., 1982).

One survey in the United Kingdom (Chumbley and Unwin, 1982) has generated information on the uptake of cadmium and lead by vegetable crops harvested during a 12 month period from land that had been previously

TABLE 16.7 Concentration of Cadmium in Selected
Food Categories

Food Category	Cd Concentration (μg/g)
Grain and cereal products	0.032
Fruits	0.042
Potatoes	0.045
Dairy products	0.005
Vegetables	0.026
Meat, fish, and poultry	0.009

Source: Travis and Etnier (1982).

TABLE 16.8   Cadmium and Lead Concentrations in Vegetables Grown on
Contaminated Soils (mg/kg)

Crop	Soil Cd	Soil Pb	Plant Cd	Plant Pb
Lettuce	7.8	155	4.2	2.3
Cabbage	4.4	117	0.3	0.3
Leeks	3.1	97	0.8	0.8
Salad onions	4.8	107	1.0	0.6
Spring greens	9.3	214	1.1	2.3
Spinach	4.6	124	4.6	3.7
Cauliflower	3.5	137	0.7	2.0
Potatoes	10.8	176	0.6	0.2
Sweet corn	7.8	156	1.5	0.1
Radish	2.7	110	1.7	2.9
Beetroot	6.5	103	2.0	0.4

Source: Chumbley and Unwin (1982).

treated with sewage sludge over a period of years. Table 16.8 shows the mean concentrations of metals in soils and the corresponding concentrations in 11 crops from 172 samples. In this study, there is no significant correlation between soil and plant lead levels, but plant cadmium is efficiently translocated to the edible tissues of vegetable crops—with increased soil concentrations being reflected in generally increased plant levels. General guidelines in England and Wales for the application of sewage sludge to agricultural land now recommend that no more than 5 kg·ha^{-1} of cadmium and 1000 kg·ha^{-1} of lead should be added to previously uncontaminated soils. Selected results from a recent survey by the FDA Bureau of Foods (Jelinek, 1982) of lead levels in foods are also given in Table 16.9. The data show mean value and 90th percentile values in parts per million for a large number of samples. This survey also reports that the mean levels of lead in many canned foods decreased from 0.35–0.40 ppm to 0.20–0.25 ppm within the last decade.

   Selenium is another metal of interest in the food chain; it is a trace metal in fly ash, and this element is trapped by electrostatic precipitators in coal-burning power plants. Inasmuch as fly ash contains many elements and is often alkaline, serious consideration has been given to its use in agriculture. On soils containing 10% fly ash by weight, produce such as beans, cabbage, carrots, onions, potatoes, and tomatoes have shown elevated concentrations of selenium (up to 1 ppm) compared to control plants having 0.02 ppm (Gutenmann et al., 1976).

   Other measurements of trace metals have focused on determining "baseline" concentrations in relatively uncontaminated areas. One such study was made by the U.S. FDA in major growing areas uncontaminated by human activities "other than normal agricultural practices." Mean concentrations

TABLE 16.9 Lead Content of Selected Foods
(ppm)

Product	Mean	90th Percentile
Milk, whole	0.02	0.04
Butter	0.07	0.16
Evaporated milk	0.09	0.15
Eggs, fresh	0.17	0.47
Hamburger	0.25	0.59
Chicken, muscle	0.13	0.49
Bacon	0.10	0.17
Frankfurters	0.20	0.75
Salmon, fresh/ frozen	0.39	0.49
Salmon, canned	0.72	1.50
Tuna, canned	0.45	1.00
Clams, soft shell	0.21	0.96
Flour, white	0.05	0.17
Bread, white	0.08	0.18
Cereals, breakfast	0.11	0.26
Sugar, refined	0.03	0.10

Source: Jelinek (1982).

of Pb and Cd (μg/g wet weight), respectively, for six crops were found to be: (1) for lettuce, 0.013 and 0.025; (2) for peanuts, 0.10 and 0.078; (3) for potatoes, 0.009 and 0.031; (4) for soybeans, 0.042 and 0.059; (5) for sweet corn, 0.0033 and 0.0031; and (6) for wheat, 0.037 and 0.043 (Wolnik et al., 1983a). As part of the same study, mean values (in μg/g wet weight) for other metals were determined in these same crops. For one of these, lettuce, the mean values were measured to be: Ca, 210; Cu, 0.26; Fe, 3.1; K, 1700; Mg, 83; Mn, 1.8; Mo, 0.013; Na, 74; P, 220; Se, 0.0016; and Zn, 1.9 (Wolnik et al., 1983b). Another "baseline" study has reported on Minnesota soils for naturally occurring concentrations (Pierce et al., 1982). Concentrations in these soils for total Cd, Cr, Cu, Ni, Pb, and Zn measured 0.31 ($\pm$0.21), 43 ($\pm$28), 26 ($\pm$9), 21 ($\pm$14), < 25, and 54 ($\pm$18), respectively.

Studies of other trace metals in soils, plants, vegetables, and foods are international in scope, and measurements have been made using many analytical techniques. Metals in Egyptian crops (Sherif et al., 1980) that have been analyzed by neutron activation include Al, As, Au, Br, Ca, Cd, Cr, Cu, Fe, La, Mn, Mo, Sb, Se, W, and Zn—in agricultural products that include African tea, ginger, canella bark, black pepper, and a variety of vegetable seeds. Ca, Mn, Fe, Cu, Zn, Co, and Mo are essential to all plants, and Cr, Mn, Fe, Co, Cu, Zn, and Se are regarded as important to plant growth and as components of enzyme systems that are transferred to humans and animals from the soil. Twenty-four elements have also been measured in the six most commonly used spices in an east Indian food research program that purports to relate concentrations with dietary aspects of food spices and to correlate trace elements with taste (Ila and Jagam, 1980).

Forty-six elements have been monitored in an extended study (1968–1980) of the chemical composition of bananas imported from Ecuador, Colombia, Panama, Honduras, Guatemala, Jamaica, Haiti, the Canary Islands, and several countries in central Africa (Cowgill, 1981). Some of these elements are added to the fruit either through the use of pesticides or chemicals employed to adjust the length of the ripening period. Among the many findings, this specific study indicates a rising K content, relatively unchanged Pb and Hg concentrations during the last 3 decades, greater B than Al concentrations, and particularly that bananas cultivated near an ocean will contain more boron in their tissues than those grown inland. Toxic elements such as Be and Te, when infrequently detected, were present only in trace amounts. Silver is also now detectable in some fruit, as silver compounds are used to regulate the rate of ripening.

A limited number of studies have addressed the atmospheric deposition of metals to forest vegetation in terms of seasonal and climatic changes. Measurements by Lindberg et al. (1982) show that the rates of deposition on deciduous forests during wet and dry cycles vary widely, with rates for Pb, Cd, Zn, and Mn during rainfall ranging as high as 2500, 310, 3700, and 4600 $pg/cm^2/h$, respectively. Inasmuch as forest vegetation is the initial point of contact between the atmosphere and soils in many areas, a sizeable percentage of trace metals will enter the ground via this route.

The accumulation of heavy metals (Fe, Mn, Zn, Cu, Pb, Ni, Cd, and Hg) in marine organisms has been studied extensively because of metal pollution in the marine environment. Mercury, in particular, has been monitored because it is known to be concentrated in predators in the marine food chain, such as seals, dolphins, tuna, and swordfish (Honda et al., 1983a). The concentration of metals in the organs and tissues has also been shown to change with the growth of fish, and small fish appear to have faster uptake and excretion rates for metals than large ones (Honda et al., 1983b).

FOOD TOXICANTS

Many food toxicants are naturally occurring. For example, a substance called ergot (the sclerotia of the fungus *Claviceps purpurea*) has been observed on crops of rye that are grown on moist, shaded land. Food poisoning of this type in the United States is tentatively traced back to 1692 in Massachusetts (Matossian, 1982). Parsnip roots have been reported to contain appreciable levels (~40 ppm) of three phototoxic, mutagenic, and photocarcinogenic compounds, and these chemicals are not destroyed in normal cooking (Ivie et al., 1981). Other toxicants have been observed in dairy products, poultry, peanut products, corn, damaged fruits, and other foods (Stoloff, 1981). Accordingly, the FDA maintains an ongoing program for identifying a wide range of compounds that can find their way into the food chain. Sphon et al. (1980) have reviewed the following potentially toxic substances and cited the appropriate mass spectrometric technique used for their analysis:

1. *Sulfa Drugs*: These chemicals are used for the prevention or treatment of disease in animals that are destined for human consumption. Increased specificity has been achieved by using pulsed positive-negative ion CI and recording both spectra simultaneously.

2. *Mycotoxins*: Naturally occurring mycotoxin residues can be produced as secondary metabolites of molds. Electron-impact (EI) and positive-negative CI sources have provided complementary information relative to molecular weight, structure, and fragmentation.

3. *Dioxins*: The extreme toxicity of chlorinated dibenzo-p-dioxins has stimulated mass spectrometric procedures for their assay down to the parts per trillion (ppt) level. By using an OV-101 coated glass capillary column for sample introduction with a 25 eV EI source, fish extracts were monitored by multiple-ion monitoring (m/z of 257, 259, 261, 305, 307, 320, 321, 322, 324, 326, 332, and 334). And in a feeding study of dairy cattle, higher chlorinated dioxins in milk were observed using negative-ion CI with methane as the reagent gas. An even higher sensitivity and specificity for dioxin analysis has been demonstrated with GC/MS/MS techniques. Sakuma et al. (1983) have detected 2,3,7,8-TCDD in a salmon extract at a 20 ppt level as shown in Figure 16.12. The dioxin was clearly identified by monitoring the loss of COCl, mass 63, from the parent ion.

4. *Cyanogenic Glycosides*: These compounds, which appear in some edible and nonedible plants, are sufficiently polar so that CI sources are not practical. However, molecular weight and structural information have been obtained with a specialized field desorption-CI probe.

Chaytor and Saxby (1982) have developed a method for the analysis of T2 toxin in maize by GC/MS. The O-Trimethylsilyl T2 toxin is detected by monitoring ions at $m/z = 350$ and $m/z = 436$, with a quantitative detection limit of 5 ppb. Intoxications of both human and animal populations have been observed primarily in temperate zones, where crops are subject to cool wet periods. All illnesses appear to result from the consumption of moldy food, due to the prolonged exposure to wet weather in the field.

The general threat of mycotoxin contamination of food for humans and feed for animal consumption has also generated considerable research on field and storage fungi since the discovery of the aflatoxin problem in 1960.

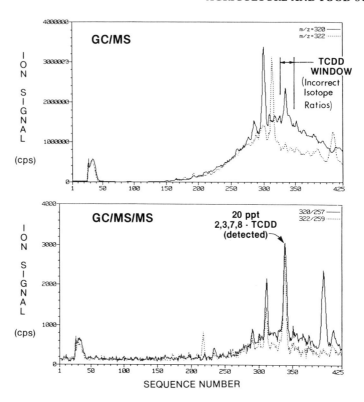

Figure 16.12   High-sensitivity detection and quantitation of TCDD in salmon. (Sakuma et al., 1983; courtesy of Sciex, Inc.)

A summary of important mycotoxins and an overview of multimycotoxin analysis have been documented by Steyn (1981). Furthermore, tandem mass spectrometry (MS/MS) is being increasingly employed when toxic compounds of interest are not amenable to separation by gas chromatography or when an extensive purification procedure is required prior to a mass spectral analysis. For example, Plattner et al. (1983) have applied quadrupole MS/MS techniques and CI to differentiate 12 ergot peptide alkaloids. Feeds contaminated with such alkaloids are known to adversely impact both grain utilization and livestock production.

NITROSAMINES

"Mass spectrometry has been considered the most reliable procedure for confirming the presence and identity of N-nitrosamines—and the International Agency for Research on Cancer has recommended that findings which are not confirmed by mass spectrometry be labeled *apparent*" (Walker et al., 1978; Hotchkiss, 1981). Many foods are now known to contain trace quantities of one or more N-nitrosamines (NAs) in addition to nitrite- and nitrate-cured products. Furthermore, human exposure to trace levels of NAs has been attributed to industrial environments, cigarette smoke, agricultural chemicals, alcoholic beverages, cosmetics, drugs, and even new automobile interiors (Rounbehler, 1980).

In contrast to many other pollutants that come from a limited number of sources, nitrosamines are generated from precursor chemicals that exist almost everywhere throughout the environment. They are produced when amines react with nitrous acid, and this reaction can take place under many circumstances—in cooking, combustion, the atmosphere, and the human digestive system. The existence of nitrosamine contaminants in certain pesticide formulations has also been firmly established. Their occurrence arises from either chemical synthesis or nitrosation in the product due to nitrate (Kearney, 1980). Their uptake in plants has been noted, but the level of concentration in plants decreases rapidly when nitrosamine-containing water is replaced by clean water, so nitrosamines do not appear to accumulate in green plants. In foods, nitrite is a widely used food additive that is effective in preventing botulism, and nitrate is used in meats and poultry to add color to these products. Bacon and nitrite-cured meats such as hot dogs, corned beef, and sausages are also known to form nitrosamines upon cooking. However, nitrite in cured meats contributes only a small proportion of the total exposure to nitrosamines, and vitamin C (ascorbic acid) and vitamin E (α-tocopherol) in the diet provide protection by inhibiting the conversion of nitrite to nitrosamines. In fact, aside from occupational exposures, cigarette smoke has been cited as the largest single source of nitrosamines for many individuals.

Nitrosamines have been found in cured meats, fish, cheese, fried bacon, instant skim milk powder, malt, beer, and whiskies. In these products, nitrosamine formation is attributed to the interaction of amines with either an added nitrite or the gaseous nitrogen oxides in hot flue gases that are used for commercial processing (Sen and Seaman, 1981). Recent data published by the U.S. Agricultural Research Service (Kimoto and Fiddler, 1982), shown in Table 16.10, indicate nitrosamine concentrations found in some of the products obtained by gas chromatography-low-resolution quadrupole mass spectrometry (QMS). Ions are measured at m/z 30, 42, and 74 for NDMA and m/z 30, 42, and 100 for NPYR (see Table 16.10). The method also takes into account the fact that nitrosamines are photolabile, so that the characteristic ions are monitored by GC/MS before and after photolysis by ultraviolet light at 365 nm. Figure 16.13 also shows both a mass spectrum of NTHZ N-nitrosothiazolidine isolated from a fried bacon sample and a comparison spectrum of the NTHZ standard (Kimoto et al., 1982). High-resolution mass spectrometry is also employed in nitrosamine identification and confirmation, in contrast to low-resolution analysis. Andrzejewski et al. (1981) have obtained data (at 12,000 resolution) over a mass range m/z 25–

TABLE 16.10 Nitrosamine Concentrations in Products Determined by GC/MS Multiple Ion Analysis

Sample	n	Nitrosamine Analyzed[a]	(ppb)
Bacon	18	NPYR	3.4–11.8
Bacon	3	NPYR	4.4–8.8
Malt	4	NDMA	2.8–7.4
Beer	1	NDMA	1.5

Source: Kimoto and Fiddler (1982).

[a] NPYR, N-nitrosopyrrolidine; NDMA, N-nitrosodimethylamine.

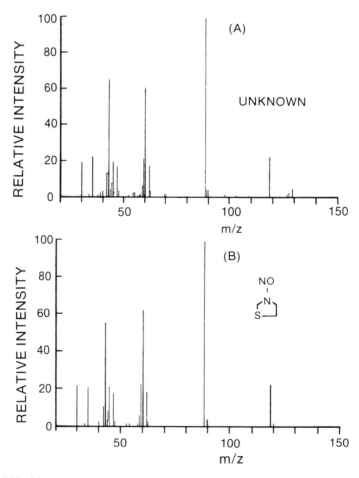

Figure 16.13 Mass spectrum of (A) N-nitrosothiazolidine (NTHZ) isolated from fried bacon compared to (B) the NTHZ standard (Kimoto et al., 1982).

150 and at a scan rate of 200 amu/s in a survey of 22 imported and 42 domestic beers. N-nitrosodimethylamine (NDMA) was found by chromatography in 60 samples at levels up to 7.7 ppb and confirmed by GC/MS in five samples. The GC/MS method in this work had a limit of confirmation of 5 ppb. A comparison of the relative merits of high-resolution vs. low-resolution spectrometry for nitrosamines has been reviewed by Hotchkiss (1981).

CANNING, PACKAGING, AND STORAGE

Food-packaging materials of all types (e.g., meat wraps, paper products, soft drink bottles and cans, etc.) are now regulated by the FDA to warrant their safe use and to insure that they conform to established standards. Packaging materials for meats are also tested by manufacturers for their high oxygen barrier properties in order to minimize spoilage. Canned foods acquire some trace metals from their particular modes of packaging, with lead being an undesired additive. Most cans are constructed of carbon steel sheet with a thin coating of tin, but the source of lead is the solder (98% lead, 2% tin) used to seal the seams. The U.S. Department of Health and Human Services (1979) has estimated that ~14% of the total lead ingested by humans comes from the solder used in can fabrication. Using thermal ionization isotope dilution techniques, Settle and Patterson (1980) have compared lead concentrations in fresh tuna to levels in lead-soldered cans. Their measurements indicate levels of 1400 ng of lead per gram of canned tuna vs. 0.3 ng/g of lead in fresh tuna muscle. The lead contents of commercially canned orange juice is also attributed primarily to the solder; the analysis of 168 samples showed levels within a narrow range of 0.02–0.32 mg/kg of juice (Nagy and Rouseff, 1981).

Paper-based packages are often fabricated with an inner layer that serves to block the transfer of oxygen and water vapor, while the outer layer provides mechanical strength and a printable surface. Polymeric materials used for boil-in-bag use also require testing, as these materials have components that can migrate out of the polymer matrix at high temperatures. An outside reinforcing layer of polyester or nylon film is often used for these bags that is stable to 200°C, with an inner lining being fabricated from a lower temperature polymer. Adhesives used for laminating and sealing these food pouches are also regulated, and a specific study of 2,4-toluenediamine (TDA) migration has been reported by the FDA (Snyder and Breder, 1982). Data on the transport of aromatic amines at a low parts per billion level has also been provided by manufacturers in their requests for approvals of adhesives in food use. Waxes are also an important component in food packaging because of their high stability and resistance to both chemical and biological degradation. Natural waxes are primarily composed of straight-chain carboxylic acids and alcohols in the region of C_{12}-C_{36}; these and other waxes and wax constituents have been characterized by high-temperature gas chromatography (Lawrence et al., 1982). GC/MS has also been used to isolate and identify oligomers from vinylchloride (VC) polymers in food-grade PVC resins (Gilbert et al., 1982).

Furthermore, changes in the quality of both unpackaged and packaged foods are now being analyzed in some detail. Mycotoxin production in whole tomatoes, apples, oranges, and lemons has been reported (Stinson et al., 1981), as commercial fruits and vegetables often become infected after prolonged storage or refrigeration. The dehydrochlorination of some organochlorine pesticides in freeze-dried egg and egg fat during storage has also been reported by the Ministry of Agriculture, Fisheries, and Food in England (Hill and Smart, 1981), and the storage effects on carbohydrates of peanuts has been documented (Pattee et al., 1981). Research that relates to preserving the taste, flavor, and general quality of foods, juices, and beverages during storage at various temperatures is substantial. Not only are extensive assays made to identify flavor degradation, but enzymes have been isolated that can function to counteract bitterness. For example, bacteria have been identified for removing limonin bitterness after its formation in orange juice (Vaks and Lifshitz, 1981), and volatile aldehydes formed by the oxidative degradation of unsaturated fatty acids in meats during storage have been related to rancidity (Reindl and Stan, 1982).

The flavor quality of Golden Delicious apples as a function of controlled atmosphere storage time has been studied by Willaert et al. (1983) using GC/MS aroma analysis as an objective measurement. Important flavor components have been identified as volatile esters, alcohols, aldehydes, and estragole (4-methoxyallylbenzene). In the case of apricots, Snapyan et al. (1982) measured the content of soluble dry substances, sugars, titratable acids, ethanol, and acetaldehyde during development, plant ripening, and post-harvest storage. A marked decrease in respiration intensity (μl/h of CO_2) was also noted during storage. Some attention has also been focused on the biochemical characteristics of seeds during prolonged storage in sealed packages (Kazakov et al., 1983). Germinating capacity and other factors can be affected by moisture content, prefreezing, convective or sublimation drying, etc.

REFERENCES

Alabran, D. M., *J. Agric. Food Chem., 30,* 486 (1982).

Albach, R. F., G. H. Redman, R. R. Cruse, and H. D. Petersen, *J. Agric. Food Chem., 29,* 805 (1981).

Almy, J., C. S. Ough, and E. A. Crowell, *J. Agric. Food Chem., 31,* 911 (1983).

Andrzejewski, D., D. C. Havery, and T. Fazio, *J. Assoc. Off. Anal. Chem., 64,* No. 6, 1457 (1981).

Bakke, A., *Chem. Eng. News, September 7,* 68 (1981).

Batra, S. W. T., *Science, 215,* 134 (1982).

Boccelli, R., G. P. Molinari, and A. Del Re, *J. Agric. Food Chem., 30,* 1233 (1982).

Bokochava, M. A., and H. I. Skobeleva, *CRC Crit. Rev. Food Sci. Nutr., 12,* No. 4, 303 (1980).

Brady, N. C., *Science, 218,* 847 (1982).

Burk, R. L., and M. Stuiver, *Science, 211,* 1419 (1981).

Buttery, R. G., R. M. Seifert, D. G. Guadagni, and L. C. Ling, *J. Agric. Food Chem., 17,* 1322 (1969).

Buttery, R. G., R. M. Seifert, L. C. Ling, E. L. Soderstrom, J. M. Ogawa, and J. C. Turnbaugh, *J. Agric. Food Chem., 30,* 1208 (1982).

Cairns, T., E. G. Siegmund, and G. M. Doose, *Bull. Environ. Cantam. Toxicol., 30,* 93 (1983).

Carro, O., C. Hillaire-Marcel, and M. Gagnon, *J. Assoc. Off. Anal. Chem., 64,* 85 (1980).

Carter, G. E., Jr., and M. B. Riley, *J. Agric. Food Chem., 30,* 647 (1982).

Chait, B. T., and F. H. Field, *J. Am. Chem. Soc., 104,* 5519 (1982).

Chaytor, J. P., and M. J. Saxby, *J. Chromatogr., 237,* 107 (1982).

Chumbley, C. G., and R. J. Unwin, *Environ. Poll., 4,* Series B, 231 (1982).

Chung, T. Y., F. Hayase, and H. Kato, *Agric. Biol. Chem., 47,* No. 2, 343 (1983).

Coleman, E. C., C. T. Ho, and S. S. Chang, *J. Agric. Food Chem., 29,* 42 (1981).

Cowgill, U. M., *Biol. Trace Element Res., 3,* 33 (1981).

Cross, J. H., J. R. West, R. M. Silverstein, A. R. Jutsum, and J. M. Cherrett, *J. Chem. Ecol., 8,* No. 8, 1119 (1982).

Dalton, B. R., U. Blum, and S. B. Weed, *J. Chem. Ecol., 9,* No. 8, 1185 (1983).

Davis, D. V., and R. G. Cooks, *J. Agric. Food Chem., 30,* No. 3, 495 (1982).

Detroy, R. W., and G. St. Julian, *Crit. Rev. Microbiol., 10,* No. 3, 203 (1983).

Dirinck, P. J., H. L. DePooter, G. A. Willaert, and N. M. Schamp, *J. Agric. Food Chem., 29,* 316 (1981).

Doner, L. W., H. W. Krueger, and R. H. Reesman, *J. Agric. Food Chem., 28,* 362 (1980).

Doner, L. W., and D. D. Bills, *J. Agric. Food Chem., 29,* 803 (1981).

Eaton, F. M., *J. Agric. Res., 69,* 237 (1944).

Estep, M. F., and T. C. Hoering, *Geochim. Cosmochim. Acta, 44,* 1197 (1980).

Etievant, P. X., *J. Agric. Food Chem., 29,* 65 (1981).

Flath, R. A., D. R. Black, R. R. Forrey, G. M. McDonald, T. R. Mon, and R. Teranishi, in *Advances in Chromatography,* ed. A. Zlatkis, Preston, Evanston, IL (1969), p. 211.

Flath, R. A., and K. Ohinata, *J. Agric. Food Chem., 30,* 841 (1982).

Förstel, H., and H. Hützen, in *Recent Developments in Mass Spectrometry in Biochemistry, Medicine, and Environmental Research,* ed. A. Frigerio, Elsevier, Amsterdam (1981), p. 205.

Freedman, A. M., *Wall Street J., February 5,* 1 (1985).

Furr, A. K., T. F. Parkinson, D. C. Elfving, C. A. Bache, W. H. Gutenmann, G. J. Doss, and D. J. Lisk, *J. Agric. Food Chem., 29,* 156 (1981).

Gil, V., and A. J. MacLeod, *J. Agric. Food Chem., 29,* 484 (1981).

Gilbert, J., M. J. Shepherd, J. R. Startin, and M. A. Wallwork, *J. Chromatogr., 237,* 249 (1982).

Golan-Goldhirsh, A., A. M. Hogg, and F. H. Wolfe, *J. Agric. Food Chem., 30,* 320 (1982).

Gordon, H. T., and J. C. Bauernfeind, *CRC Crit. Rev. Food Sci. Nutri., 18,* No. 1, 59 (1982).

Guarino, P. A., *J. Assoc. Off. Anal. Chem., 65,* No. 4, 835 (1982).

Gutenmann, W. H., C. A. Bache, W. D. Youngs, and D. J. Lisk, *Science, 191,* 966 (1976).

Hall, K. C., and R. J. Dowdell, *J. Chromatogr. Sci., 19,* 107 (1981).

Hedin, P. A., *J. Agric. Food Chem., 30,* No. 2, 201 (1982).

Hill, A. R., and N. A. Smart, *J. Agric. Food. Chem., 29,* 677 (1981).

Ho, C. T., K. N. Lee, and Q. Z. Jin, *J. Agric. Food Chem., 31,* 336 (1983).

Hogge, L. R., and D. J. H. Olson, *Industr. Res., May,* 144 (1982).

Holley, R. W., B Stoyla, and A. D. Holley, *Food Res., 20,* 326 (1955).

Honda, K., R. Tatsukawa, K. Itano, N. Miyazaki, and T. Fujiyama, *Agric. Biol. Chem., 47,* No. 6, 1219 (1983a).

Honda, K., M. Sahrul, H. Hidaha, and R. Tatsukawa, *Agric. Biol. Chem., 47,* No. 11, 2521 (1983b).

Horsfall, J. G., in *Pesticide Chemistry in the 20th Century,* ed. J. R. Plimmer, ACS Sym. Series No. 37, American Chemical Society, Washington, D.C. (1977), p. 115.

Horvat, R. J., S. D. Senter, and E. D. Dekazos, *J. Food Sci., 48,* 278 (1983).

Hotchkiss, J. H., *J. Off. Anal. Chem., 64,* No. 5, 1055 (1981).

Huckins, J. N., D. L. Stalling, J. D. Petty, D. R. Buckler, and B. T. Johnson, *J. Agric. Food Chem., 30,* 1020 (1982).

Ila, P., and P. Jagam, *J. Radioanal. Chem., 57,* No. 1, 205 (1980).

Inman, R. D., U. Kiizemagi, D. A. Griffin, and M. L. Deinzer, *J. Agric. Food Chem., 31,* No. 4, 722 (1983).

Isler, O., R. Rüegg, and P. Schudel, *Chimica, 15,* 208 (1961).

Ivie, G. W., D. L. Holt, and M. C. Ivey, *Science, 213,* 909 (1981).

Iwamoto, M., Y. Takagi, K. Kogami, and K. Hayashi, *Agric. Biol. Chem., 47,* No. 1, 115 (1983).

Jacobson, M., *Insect Sex Attractants,* Wiley, New York (1965) p. 154.

Jelinek, C. F., *J. Assoc. Off. Anal. Chem., 65,* No. 4, 942 (1982).

Karlson, P., and M. Luscher, *Nature, 183,* 55 (1959).

Kawai, M., R. Nyfeler, J. M. Berman, and M. Goodman, *J. Med. Chem., 25,* 397 (1982).

Kazakov, E. D., N. V. Popova, and N. G. Khoroshailov, *Appl. Biochem. Microbiol., 19,* No. 3, 315 (1983).

Kearney, P. C., *Pure Appl. Chem., 52,* 499 (1980).

Kimoto, W. I., and W. Fiddler, *J. Assoc. Anal. Chem., 65,* No. 5, 1162 (1982).

Kimoto, W. I., J. W. Pensabene, and W. Fiddler, *J. Agric. Food Chem., 30,* 757 (1982).

Kirchner, J. G., R. G. Rice, J. M. Miller, G. J. Keller, *Arch. Biochem. Biophys., 25,* 231 (1950).

Kleber, C., *Am. Perfum., 7,* 235 (1913).

Kolor, M. G., "Flavor Components," in *Biochemical Applications of Mass Spectrometry,* ed. G. R. Waller, Wiley, New York (1972) pp 701–722.

Kolor, M. G., in *Mass Spectrometry,* Part A, ed. C. Merritt, Jr., and C. N. McEwen, Marcel Dekker, New York (1979), chapter 3, p. 67.

Kolor, M. G., in *Biochemical Applications of Mass Spectrometry,* Vol. 1, First Suppl., eds. G. R. Waller, and O. C. Dermer, Wiley, New York (1980), chapter 25.

Kolor, M. G., U.S. Patent 4,329,372 (May 11, 1982).

Kolor, M. G., *J. Agric. Food Chem., 31,* 1125 (1983).

Kouchi, H., *J. Chromatogr., 241,* 305 (1982).

Krause, R. T., *J. Assoc. Off. Anal. Chem., 63,* 1114 (1980).

Krause, R. T., Z. Min, and S. H. Shotkin, *J. Off. Anal. Chem., 66,* No. 6, 1353 (1983).

Krueger, D. A., and H. W. Krueger, *J. Agric. Food Chem. 31,* No. 6, 1265 (1983).

Krueger, H. W., and R. H. Reesman, *Mass Spectrom. Rev., 1,* 205 (1982).

Labows, J. N., and B. Shushan, *Am. Laboratory, March,* 56 (1983).

Lawrence, J. F., J. R. Iyengar, B. D. Page, and H. B. S. Conacher, *J. Chromatogr., 236,* 403 (1982).

Lee, K. N., C. T. Ho, C. S. Giorlando, R. J. Peterson, and S. S. Chang, *J. Agric. Food Chem.*, *29*, 834 (1981a).

Lee, M. H, C. T. Ho, and S. S. Chang, *J. Agric. Food Chem.*, *29*, 684 (1981b).

Liao, A., R. F. Christman, J. D. Johnson, D. S. Millington, and J. R. Hass, *Environ. Sci. Technol.*, *16*, 403 (1982).

Limmer, A. W., K. W. Steele, and A. T. Wilson, *J. Soil Sci.*, *33*, 499 (1982).

Lindberg, S. E., R. C. Harriss, and R. R. Turner, *Science, 215*, 1609 (1982).

Lodhi, M. A. K., *J. Chem. Ecol.*, *7*, No. 4, 685 (1981).

Lovrien, R. E., T. J. Kurtti, R. Tsang, and M. Brooks-Wallace, *J. Biochem. Biophys. Methods*, *5*, 307 (1981).

Lund, E. D., P. E. Shaw, and C. L. Kirkland, *J. Agric. Food Chem.*, *29*, 490 (1981).

Maga, J. A., *J. Agric. Food Chem.*, *30*, No. 1, 9 (1982).

Martin, G. E., J. M. Burggraff, R. H. Dyer, and P. C. Buscemi, *J. Assoc. Off. Anal. Chem.*, *64*, No. 1, 186 (1981a).

Martin, G. E., F. C. Alfonso, D. M. Figert, and J. M. Burggraff, *J. Assoc. Off. Anal. Chem.*, *64*, No. 5, 1149 (1981b).

Matossian, M. K., *Am. Scientist, 70*, 355 (1982).

McDonough, L. M., and L. I. Butler, *J. Chem. Ecol.*, *9*, No. 11, 1491 (1983).

Meilgaard, M. C., *J. Agric. Food Chem.*, *30*, 1009 (1982).

Mookherjee, B. D., *Chem. and Eng. News, October 4*, 27 (1982).

Muszkat, L., and N. Aharonson, *J. Agric. Food Chem.*, *30*, 615 (1982).

Nagy, S., and R. L. Rouseff, *J. Agric. Food Chem.*, *29*, 890 (1981).

Nara, J. M., R. C. Lindsay, and W. E. Bukholder, *J. Agric. Food Chem.*, *29*, 68 (1981).

Nelsen, T. R., R. F. Cook, M. H. Gruenauer, M. D. Gilbert, and S. Witkonton, *J. Agric. Food Chem.*, *31*, No. 6, 1147 (1983).

Nursten, H. E., and A. A. Williams, *Chem. Ind.* (*London*), 486 (1967).

Ogunmoyela, O. A., and G. G. Birch, *J. Agric. Food Chem.*, *30*, 77 (1982).

Ohloff, G., and I. Flament, *Fortschr. Chem. Org. Naturst.*, *36*, 231 (1979).

Olsen, R. A., R. B. Clark, and J. H. Bennett, *Am. Scientist, 69*, 378 (1981).

Olsson, A., J. A. Jonsson, B. Thelin, and T. Liljefors, *J. Chem. Ecol.*, *9*, No. 3, 375 (1983).

Ough, C. S., C. E. Daudt, and E. A. Crowell, *J. Agric. Food Chem.*, *29*, 938 (1981).

Parliment, T. H., and M. G. Kolor, *J. Food Sci.*, *40*, 762 (1975).

Pattee, H. E., C. T. Young, and F. G. Giesbrecht, *J. Agric. Food Chem.*, *29*, 800 (1981).

Pierce, F. J., R. H. Dowdy, and D. F. Grigal, *J. Environ. Qual.*, *11*, No. 3, 416 (1982).

Plattner, R. D., S. G. Yates, and J. K. Porter, *J. Agric. Food Chem.*, *31*, No. 4, 785 (1983).

Pyles, R. H., and E. A. Woolson, *J. Agric. Food Chem.*, *30*, 866 (1982).

Reindl, B., and H. Stan, *J. Agric. Food Chem.*, *30*, 849 (1982).

Ribick, M. A., G. R. Dubay, J. D. Petty, D. L. Stalling, and C. J. Schmitt, *Environ. Sci. Technol.*, *16*, 310 (1982).

Rounbehler, D. P., J. Reisch, and D. H. Fine, *Food Cosmet. Toxicol.*, *18*, 147 (1980).

Ruben, S., M. Randall, M. Kamen, and J. L. Hyde, *J. Am. Chem. Soc.*, *63*, 777 (1941).

Ryan, J. A., H. R. Pahren, and J. B. Lucas, *Environ. Res.*, *28*, 251 (1982).

Sakuma, T., N. Gurprasad, S. D. Tanner, A. Ngo, W. R. Davidson, H. A. McLeod, B. P-Y. Lau, and J. J. Ryan, *ACS Symposium on Chlorinated Dioxins in the Total Environment II*, Washington, D.C. (August 1983).

Sanders, H. J., *Chem. and Eng. News, August 3*, 20 (1981).

Schreier, P., *CRC Crit. Rev. Food Sci. Nutr.*, *12*, No. 1, 59 (1979).

Schroth, M. N., and J. G. Hancock, *Science, 216*, 1376 (1982).

Sen, N. P., and S. Seaman, *J. Agric. Food Chem., 29*, 787 (1981).

Settle, D. M., and C. C. Patterson, *Science, 207*, 1167 (1980).

Shaw, P. E., and C. W. Wilson, III, *J. Agric. Food Chem., 30*, 685 (1982).

Shea, T. B., and E. S. Berry, *Bull. Environ. Contam. Toxicol., 30*, 99 (1983).

Shearer, G., and D. H. Kohl, in *Recent Developments in Mass Spectrometry in Biochemistry in Medicine,* ed. A. Frigerio, Plenum Press, New York (1981), chapter 17.

Sherif, M. K., R. M. Awadallah, and A. H. Amrallah, *J. Radioanal. Chem., 57*, No. 1, 53 (1980).

Shibamoto, T., Y. Kamiya, and S. Mihara, *J. Agric. Food Chem., 29*, 57 (1981).

Shimitzu, J., and M. Watanabe, *Agric. Biol. Chem., 56*, No. 6, 1447 (1982).

Shinohara, T., and M. Watanabe, *Agri. Biol. Chem. 45*, No. 12, 2903 (1981).

Silverstein, R. M., *Science, 213*, 1326 (1981).

Snapyan, G. C., R. M. Geravetova, and T. A. Mkrtchyan, *Appl. Biochem. Microbiol., 18*, No. 5, 492 (1982).

Snyder, R. C., and C. V. Breder, *J. Chromatogr., 236*, 429 (1982).

Sphon, J., D. Andrzejewski, W. Brumley, P. Dreifuss, and J. Roach, in *Advances in Mass Spectrometry,* Vol. 8B, ed. A. Quayle, Heyden and Son Ltd., London (1980), p. 1490.

Staroscik, J. A., and A. A. Wilson, *J. Agric. Food Chem., 30*, 835 (1982).

Steyn, P. S., *Pure and Appl. Chem., 53*, 891 (1981).

Stinson, E. E., S. F. Osman, E. G. Heisler, J. Siciliano, and D. D. Bills, *J. Agric. Food Chem., 29*, 790 (1981).

Stoloff, L., *J. Assoc. Off. Anal. Chem., 64*, No. 2, 373 (1981).

Strobel, G., and D. F. Myers, *Sci. News, 117*, 362 (1980).

Takahashi, Y., Y. Migashita, Y. Tanaka, H. Abe, and S. Sasaki, *J. Med. Chem., 25*, 1245 (1982).

Travis, C. C., and E. L. Etnier, *Environ. Res., 27*, 1 (1982).

Tressl, R., D. Bahri, and K. Engel, *J. Agric. Food Chem., 30*, 89 (1982).

Turner, G. L., F. J. Bergersen, and H. Tantala, *Soil Biol. Biochem., 15*, No. 4, 495 (1983).

U. S. Department of Health and Human Services, *Fed. Register, 44*, 51233 (1979).

Vaks, B., and A. Lifshitz, *J. Agric. Food Chem., 29*, 1258 (1981).

Van Iren, F., and A. Van der Spiegel, *Science, 187*, 1210 (1975).

Van Straten, S., F. L. deVrijer, and J. C. deBeauveser, *Volatile Compounds in Food,* 4th ed., Central Institute for Nutrition and Food Research, Zeist, The Netherlands, 1981.

Volimer, J. J., and S. A. Gordon, *Chemistry, 47*, No. 10, 6 (1974).

Vora, J. D., R. F. Matthews, P. G. Crandell, and R. Cook, *J. Food Sci., 48*, No. 4, 1197 (1983).

Walker, E. A., M. Castegnato, L. Griciute, and R. E. Lyle, eds. *Environmental Aspects of N-Nitroso Compounds,* Sci. Publ. No. 19, International Agency for Research on Cancer, Lyon, France (1978), p. 555.

Waller, G. R., C. R. Ritchey, E. E. Krenzer, G. Smith, and M. Hamming, "Natural Products from Soil," in 32nd Ann. Conf. on Mass Spectrom. and Allied Topics, San Antonio (May 27 to June 1, 1984).

Welch, R. C., J. C. Johnston, and G. L. K. Hunter, *J. Agric. Food Chem., 30*, 681 (1982).

Wells, J. R., P. B. Kaufman, J. D. Jones, G. F. Estabrook, and N. S. Ghosheh, *J. Radioanal. Chem., 71*, Nos. 1–2, 97 (1982).

White, J. W., Jr., and F. A. Robinson, *J. Assoc. Off. Anal. Chem., 66*, No. 1, 1 (1983).

Willaert, G. A., P. J. Dirinick, H. L. DePooter, and N. W. Schamp, *J. Agric. Food Chem., 31*, No. 4, 809 (1983).

Winston, M. L., K. H. Slessor, M. J. Smirle, and A. A. Kandil, *J. Chem. Ecol.*, *8*, No. 10, 1283 (1982).

Witty, J. F., *Soil Biol. Biochem.*, *15*, No. 6, 631 (1983).

Wolnick, K. A., F. L. Fricke, S. G. Capar, G. L. Braude, M. W. Meyer, R. D. Statzger, and E. Bonnin, *J. Agric. Food Chem.*, *37*, No. 6, 1240 (1983a).

Wolnick K. A., F. L. Fricke, S. G. Capar, G. L. Braude, M. W. Meyer, R. D. Statzger, and R. W. Kuennen, *J. Agric. Food Chem.*, *31*, No. 6, 1244 (1983b).

Wood, W. F., *J. Chem. Educ.*, *60*, No. 7, 531 (1983).

Wright, L. H., *J. Chromatogr. Sci.*, *20*, 1 (1982).

Yankov, L. K., M. A. Krysteva, and M. N. Kamburov, *Appl. Biochem. Microbiol.*, *18*, No. 2, 221 (1982).

Zettlemeyer, A. C., *Chem. Eng. News, January 5*, 3 (1981).

CHAPTER SEVENTEEN

BIOMEDICAL APPLICATIONS

MASS SPECTROMETRY AS A BIOMEDICAL IMAGING SYSTEM

By definition, an imaging system provides spatially resolved "contrast" in an object or even in a small aggregation of molecules. In some instances, time-dependent information is also provided. Most life science imaging systems have been derived from research in the physical sciences, that is, x-rays, radionuclides, electron microscopy, laser optics, solid-state photodiode arrays, and ultrasonics. Nuclear magnetic resonance (NMR) (a discovery that merited the Nobel Prize in physics) provides contrast from the absorption of radiofrequency (RF) energy by nuclear spins that interact with a high magnetic field. This imaging system is especially useful because nuclei in different chemical environments give rise to resonances at different frequencies; the availability of enriched stable isotopes has further increased the scope of this form of spectroscopy. Mass spectrometry should also be considered as an imaging system, because it can map concentrations of all the chemical elements in blood, organs, and even single cells. It is true that a sample must be transferred to an instrument for assay, but the required sample size is now so small that many parts of the human anatomy can be "nondestructively" sampled. Furthermore, hundreds of compounds can be detected by the real-time monitoring of the most readily available biological material, that is, human breath.

Today, high-resolution and tandem mass spectrometers are essential in metabolic profiling and in identifying molecular abnormalities in fluids and tissues that accompany the pathological state. In addition, there are approximately 60 *trace metals* in the human anatomy that can be monitored at sensitivities of 10^{-12} to 10^{-16} g or less. The potential for their use in diagnostics, in part, lies in the *multiplicity of the abundance measurements that can be made of these elements*, and changes in the abundance of these elements, for providing early clues to the incidence of disease. In some instances, the *elemental profiling* of metals in biological matrices, such as bone or teeth enamel, can be performed by borrowing techniques that are routinely used in semiconductor device diagnostics. Indeed, the quantitative ion probe microanalysis of biological mineralized tissues has already been demonstrated (Lodding, 1983), and some 300 naturally occurring stable or very long-lived isotopes are available for monitoring transport phenomena and growth. Both elemental (isotopic) distributions and ion transport in cultured cells have also been imaged by secondary ion mass spectrometry, with a spatial resolution approaching 0.5 μm being achieved for sodium, potassium, calcium, and magnesium (Chandra and Morrison, 1985). Thus, mass spectrometry must be viewed as a unique and powerful imaging system for sensing the elements, the compounds, and molecular structures that comprise all biological matter.

STABLE VS. RADIOISOTOPES

The advent of radioisotopic labeling in biology and medicine began on August 2, 1946, when the Oak Ridge National Laboratory shipped a few millicuries of ^{14}C to the Barnard Free Skin and Cancer Hospital in St. Louis. Today, tagging with radiotracers is a standard technique; the inventory of radioisotopes is large, and a broad range of instrumentation is available to detect these radionuclides. However, isotopic labeling by stable tracers is now a viable alternative, and several factors point to their increased use in both clinical and research studies:

1. With stable isotopes, the biological system will not be damaged. Use of a radiotracer may cause damage from the primary ionizing radiation, the recoil kinetic energy of the parent nuclei, or both. Nuclear radiations are known to disrupt chromosomes and to create harmful free radicals.

2. Stable tracers introduce no new chemical species such as those introduced by radionuclides in their decay. Some daughter decay products are potentially toxic, and the lethal effect via transmutations occurring within DNA molecules has been documented for ^{14}C, ^{32}P, ^{33}P, and ^{125}I (Apelgot, 1983).

3. All radiopharmaceuticals and compounds are subject to radiolytic decomposition. Therefore, such compounds have a definite "shelf life." Stable nuclei have an infinite lifetime. Thus, they can be used in studies extending over many years, since there is no loss in analytical sensitivity as a function of time.

4. In continuous infusion experiments, a multiplicity of isotopes (if available in a single chemical element) can provide time-dependent data on turnover rates.

5. Stable isotopes appear to be useful in trace nutrient studies and in

investigations of metal toxicity, where the availability of radionuclides with appropriate half-lives is somewhat restricted.

6. It is unlikely that radioisotopes will be used extensively for statistical studies on human subjects, even if radiation effects are deemed negligible.

7. In favorable cases, the sensitivity of stable isotopes and mass spectrometry already exceeds standard radiotracer detection.

8. With ion accelerator-mass spectrometry, certain trace analyses can be determined at the subparts per billion level.

Stable isotopes can be measured in two ways. One method is to administer a tracer element, and after sampling, subject the specimen to neutron activation. Thus, in order to follow the clearance of an element from the blood, a stable isotope can be injected into a vein and its subsequent concentration in timed blood samples can be determined by activation analysis. (Burkinshaw, 1982). The second method is to use enriched isotopes, high-sensitivity mass spectrometry, and the isotopic dilution method. Elemental and isotopic abundance sensitivity with multistage instruments is now so high that many trace metals can be assayed in the 10^{-12} to 10^{-16} g range. With both neutron activation and high-sensitivity isotopic assays, the mass spectrometer is the ultimate instrument for calibrating standards, determining the degree of enrichment of separated isotopes, and distinguishing elements from interfering isobars or radionuclides.

Stable tracer methodology has been applied to (1) determine pool size; (2) provide estimates of the rate of transport of amino acids across the cell membrane; and (3) characterize aberrations in amino acid metabolism that reflect various pathological conditions (Lapidot and Nissam, 1980). A recent study has even demonstrated that stable isotope tracers are practical for analyzing collagen synthesis and proline recycling in scar tissue (Wolfe et al., 1984). Furthermore, if "the ultimate aim of the biochemist and biologist is to know the number of tracer atoms in any particular cell" (Bowen, 1965), then mass spectrometry will shortly be able to provide such data.

TRACE ELEMENTS IN NUTRITION

Trace elements have been the object of intense study by solid-state physicists largely because semiconducting materials are "impurity sensitive" systems (Chapter 11). In a single-crystal semiconductor, an impurity whose concentration may be only 1 ppm can drastically change its electric conductivity. In an analogous manner, biochemists are now exploring the role of trace elements, their biological availability, and their interaction with other dietary constituents, and are finding that even at very low concentrations, the biological response is significant. Also, as with semiconducting materials, the discovery of essential functions for some trace elements was made possible only with the advent of "ultraclean" reagents and of analytical methods of very high sensitivity. Among the naturally occurring elements, some 82 comprise a composite concentration of less than 0.02% of the total number of atoms in the human body. If the most abundant of these "trace" elements is excluded (Mg = 0.01 atom%), the atom% abundance of the remaining 81 elements is less than 0.01%.

Among these trace elements considered to have either known or potential nutritional significance, most can be analyzed by mass spectrometry via the stable isotope dilution technique; namely, Mg, Si, Ti, V, Cr, Fe, Ni, Cu, Zn, Se, Sr, Mo, Ag, Cd, Sn, Sb, Ba, W, Hg, and Pb. Micronutrients (or toxins) that are elements of ongoing research include Cd, Bi, Os, Tl, and Pb (Lambrecht, 1981). Some trace elements are required in the daily diet in amounts that range from 50 μg to 18 mg. In several instances, their specific function is unclear, but they act as catalysts and a number have been identified as structural components of large molecules (Mertz, 1981). Recent research also indicates that specific nutrients are quantitatively important in processes linked to cell division in addition to their role in cell maintenance. Also, three trace elements previously considered only as toxic— selenium, chromium, and arsenic—are now listed as essential. The biological availability of these elements depends upon their chemical state, their interactions with other dietary constituents, and other parameters. Recommended daily allowances for iron, iodine, zinc, and copper, etc. have been established, but other trace metals of nutritional interest continue to be analyzed, that is, silicon, vanadium, chromium, manganese, nickel, arsenic, selenium, tin, and cadmium. In studies with animals, it is now known that some of these elements can be beneficial or toxic depending upon their interactions with other elements. For example, selenium protects animals against cadmium poisoning, and an equal intake of copper can result in a deficiency, a normal state, or toxicity, depending upon the occurrence of molybdenum in the diet (Mertz, 1979).

Recommended safe and adequate dietary intakes of important trace nutrients for adults are given in Table 17.1. In a 150 pound adult, the total body content of these elements is approximately: Fe, 4.5 g; F, 2.5 g; Zn, 1.85 g; Cu, 125 mg; Mn, 16 mg; I, 15 mg; Se, 13 mg; Mo, 9 mg; and Cr, 3.7 mg. Cobalt, at 1.5 mg, is also an essential nutrient. The concentrations of trace elements in foods can vary over a wide range, depending upon geographical location, the trace elements in regional water supplies, commercial or domestic preparation of foods, etc.

Iron is an essential element in both humans and plants. Although this element is the fourth most abundant within the earth's crust, most forms of life have difficulty in assimilating an adequate amount (Emery, 1982). As part of the hemoglobin component of red blood cells, it transports oxygen and carbon dioxide, and it functions as a catalyst in so many enzymatic reactions that biologists consider it to have played an essential role from the

TABLE 17.1   Recommended Safe and Adequate
Adult Dietary Intakes of Trace Elements

Element	Intake (mg/day)
Iron (males)	10
Iron (females)	18
Zinc	15
Manganese	2.5–5
Fluorine	1.5–4
Copper	2–3
Molybdenum	0.15–0.5
Chromium	0.05–0.2
Selenium	0.05–0.2
Iodine	0.15

Source: National Academy of Sciences (1980).

earliest evolution of life on earth (Schapira, 1964). Bioavailability studies of iron have been extensive (Schricker et al., 1982), but only recently have stable isotopes of iron been utilized as tracers in human subjects. Iron has four naturally occurring isotopes (^{54}Fe, ^{56}Fe, ^{57}Fe, and ^{58}Fe), with ^{56}Fe being slightly over 91% in abundance. A specific study of iron absorption with ^{58}Fe has also confirmed that a stable isotope is a valid extrinsic label that can be used instead of the ^{59}Fe radioisotope (Fairweather-Tait et al., 1983). In humans, stable isotopes of iron have been used as tracers of mineral metabolism in combination with other trace elements (Johnson, 1982). In this study, samples were prepared with known amounts of ^{54}Fe, ^{57}Fe, ^{65}Cu, ^{67}Zn, and ^{70}Zn dissolved in chloroform and analyzed with an electron-impact ion source. Separation of the iron from zinc and copper was unnecessary, since the molecular ion clusters in the mass spectrum did not overlap. The quantitative absorption of the trace metals that was administered orally to the volunteer subjects was calculated by determining the difference in the amount of stable isotope ingested compared to that excreted in the feces in excess of the natural abundance. Enriched stable isotopes of ^{58}Fe and ^{70}Zn have also been used by Turnlund et al. (1982) to determine iron and zinc absorption in a group of elderly men. Daily diets of the subjects contained 10 mg of iron and 15 mg of zinc; mean absorption measurements based on thermal ionization mass spectrometry showed 0.8 mg of iron and 2.6 mg of zinc. These results approximate other absorption data and confirm that 10 mg of iron and 15 mg of zinc are sufficient to replace endogenous Fe and Zn losses in most elderly men.

In humans, it is estimated that the body contains about 10 mg of nickel, of which 18% is contained in the skin, with tissue concentrations ranging from 0.04–2.8 µg/g on a dry basis (Schroeder and Nason, 1971). Nickel compounds at high concentrations are, of course, known to be toxic, and there is a marked difference in the carcinogenic activity between amorphous NiS and crystalline Ni_3S_2 (Costa and Mollenhaver, 1980). Continued interest in the bioavailability of nickel is related to the differential responses of inorganic nickel to specific foods, beverages, and biologically important interactions between nickel and other trace metals.

Levels of nickel in common foods vary over a large range: (1) beans, nuts, and cocoa (>2.0 $\mu g/g$); (2) wheat products, processed meats, and vegetables (0.2–2.0 $\mu g/g$); and (3) milk, fresh fruits, and eggs (<0.2 $\mu g/g$). The radioisotope ^{63}Ni ($\sim$90 year half-life) has been used as a tracer in animal studies, but its use is precluded in humans. ^{64}Ni (0.9% abundance), however, can be used as an enriched isotope with the isotope dilution method and a thermal ionization-type ion source. Recent research has been reported on human subjects in whom the change in plasma nickel concentration was measured at hourly intervals following ingestions of 22.4 mg of nickel sulfate hexahydrate, $NiSO_4 \cdot 6H_2O$ (5 mg of elemental nickel), in 250 ml of five beverages: milk, orange juice, tea, coffee, and Coca Cola®. The relative increase in plasma nickel concentration was greatest for Coca Cola at each hourly measurement, and other tests were made to identify parameters relating to nickel uptake (Solomons et al., 1982).

Zinc, like some other elements, is toxic in large concentrations, but it is essential in trace amounts for a number of physiological processes, including the conformation of enzymes, the formation of RNA templates for protein synthesis, and the stabilization of membranes of circulating cells. Zinc has five stable isotopes (64, 66, 67, 68, and 70), so mass spectrometry has been used extensively to determine zinc absorption from isotope-labeled food sources. The abundance of this element in the earth's crust is estimated to be in the range of 5–200 mg/kg, so that zinc inputs to the food chain come from both natural and anthropogenic sources. Among the richest suppliers of zinc are oysters, which can contain up to 1000 mg Zn/kg, followed by wheat germ (40–120 mg/kg). Meat has a high zinc content, with concentrations ranging from 11 mg/kg in chicken breast to 64 mg/kg in roast beef (Taylor et al., 1982). The bioavailability of zinc in a variety of soy-based foods has also been measured by using the stable isotope ^{70}Zn as a tracer, plus neutron activation after excretion. One study has focused on the concentrations of Zn and Cu relative to vegetarian and nonvegetarian diets (Abraham, 1982). The vegetarian diet, high in both phytates and fiber which binds available zinc, can result in lower Zn levels than normal. Furthermore, the general availability of trace elements in the diet of vegetarians has been found to be significantly less than in the diets of nonvegetarians.

Only a few mass spectrometric studies have been made of copper metabolism by measuring $^{65}Cu/^{63}Cu$ ratios. In one investigation, copper was separated from organic matrices by dry-ashing blood plasma, liver, and feces samples. The isotope ratios were then determined by electron-impact ionization mass spectrometry after chelation of copper with tetraphenylporphine (Buckley et al., 1982). The stable ^{65}Cu tracer permits long-duration nutritional and metabolic studies to be made without imposing restrictions on its use with either animals or humans, as with radioisotopes.

The physiological importance of chromium was established in 1959, when a chromium deficiency was characterized by impaired growth, corneal lesions, and a defect in sugar metabolism. A chromium deficiency is now associated as a risk factor for cardiovascular disease, and a marginal chromium deficiency may exist in the United States among malnourished children, pregnant women, and the elderly. The chromium intake for a typical U.S. diet is estimated to be adequate at 50–100 μg/day (Dubick and Rucker, 1983). Among the different chemical states of chromium, trivalent chromium complexed with an organic molecule of low molecular weight appears to be the metabolically active form in normal glucose metabolism. Chromium has four stable isotopes (50, 52, 53, and 54), and the element has a relatively low ionization potential, so that thermal ionization mass spectrometry is applicable in some nutrient monitoring studies. GC/MS measurements of chromium in biological materials by stable isotope dilution have been reviewed by Veillon et al. (1979).

Selenium is now considered an essential nutrient. In early studies with rats, protection against death from liver necrosis was observed in diets containing 0.1 ppm selenium. There is now evidence of a correlation between selenium deficiency and cardiovascular disease in the United States. Specifically, Colgan (1981) has reported that in selenium-deficient areas (where 80% of the vegetables contain ≤ 0.05 ppm of this element), a disease risk is increased by a factor of 3 compared to the risk in areas where selenium is "adequate" (80% of vegetables contain ≥ 0.1 ppm). "Deficient" areas have been identified as most of Connecticut, Delaware, Illinois, Indiana, Massachusetts, New York, Ohio, Oregon, Pennsylvania, Rhode Island, and the District of Columbia.

Stable selenium isotopes have recently been used in metabolic tracer studies in a double-isotope dilution method that includes rapid sample digestion, chelation, and measurement by combined gas chromatography-mass spectrometry. The method utilizes ^{76}Se as the tracer and ^{82}Se as the internal standard, compared to ^{80}Se naturally occurring in the sample (Reamer and Veillon, 1983). Selenium has also been the object of dietary studies of the elderly, whose reduced caloric intake compared to younger populations could lead to less than adequate selenium levels. However, at least one study at the University of Texas (Lane et al., 1983) showed an average intake of 94 μg/day, which is within the limits cited in Table 17.1. This same report confirms, however, that there is considerable variability in selenium concentrations in many foods, depending on the geographical origin of the food and the type of food processing. Typical of the recent application of stable selenium tracers for studying metabolism during pregnancy is the work of Swanson et al. (1983), in which measurements were made of selenium retention for women in early and late periods. A controlled diet provided about 150 μg/day of Se for 20 days, and the selenium balance was measured during

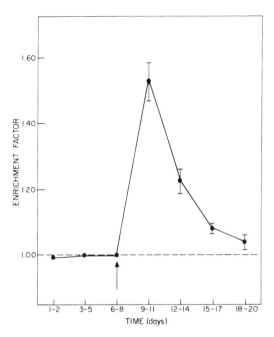

Figure 17.1   Appearance and disappearance of a ^{76}Se tracer in the feces of 16 subjects; the arrow indicates the day in which the stable tracer was administered. (Swanson et al., 1983; *Am. J. Clin. Nutrition, 38,* No. 2, 169 (1983), copyright American Society for Clinical Nutrition.)

the last 12 days. Precautions were taken to insure that no selenium entered the diet from environmental contamination, and the selenium was administered to 16 persons comprising three groups: (1) a nonpregnant control group; (2) women in early pregnancy; and (3) women in late pregnancy. On the eighth day of this metabolic study, an enriched source of ^{76}Se was fed to all groups in egg yolk and egg white, and the appearance and disappearance of ^{76}Se in urine and feces were monitored for 12 days. Quantitation was achieved by comparison with reference samples from the National Bureau of Standards (NBS) and by use of the double-isotope dilution method (Reamer and Veillon, 1983).

Figure 17.1 shows the appearance and disappearance of the enriched stable isotope ^{76}Se in the feces of 16 subjects. The arrow indicates the day in which each woman received approximately 40 μg of ^{76}Se—about 36 μg from labeled egg yolk and 4 μg from labeled egg white. However, the total selenium, protein, and energy content of the diet were maintained at a constant level. More important were the differences between the nonpregnant (NP), early pregnant (EP), and late pregnant (LP) subjects. Mean apparent selenium retention of the NP, EP, and LP women were 11, 21, and 34 μg/day, respectively.

Vanadium is essential in mammals; it is now known to promote the mineralization of teeth, and it is effective as a catalyst in the oxidation of several biological substances. Recent interest in this element also results from a finding that vanadium may inhibit the synthesis of cholesterol. Vanadium has two stable isotopes, 50 and 51; the lighter isotope has the same mass number as one of the chromium isotopes (^{50}Cr), but a mass spectrometric differentiation can be made by standard techniques.

The role of magnesium in nutrition and health has been studied since the 1920s (Day, 1982). Magnesium is an essential prothetic ion in many enzymatic reactions. It is important in protein synthesis and neuromuscular transmission, and it has important inter-relationships with potassium and calcium in blood as well as in bone. The element has three stable isotopes (24, 25, and 26), which have permitted extensive mass spectrometric studies to be made in geochemistry as well as in the life sciences.

Quantitative studies of some toxic metals (Cd, Hg, and Pb) in human milk have been reported by several investigators. In general, the lead content has been found to be lower than in commercial milk formulas. A higher blood lead level has also been observed in formula-fed infants, compared to breast-fed infants (Jensen, 1983).

DNA ANALYSIS

Stable isotope dilution gas chromatography-mass spectrometry for DNA analyses has been utilized in many investigations. Crain and McCloskey (1983) have applied the GC/MS method with selected ion monitoring for determining "methylated bases in picogram-low nanogram amounts, at concentrations at or below 0.02% in DNA." In this work, t-butyl-dimethylsilyl derivatives of thymine (m/z 354) and 5-methylcytosine (m/z 353) were assayed. The molecular ions were not observed, and the most abundant ion in each spectrum is due to the loss of a t-butyl radical (57 amu), as shown in Figure 17.2. This work is significant, as it provides a sensitive and direct measurement of DNA-methylated bases—and their possible correlation with DNA biological function.

Research in DNA molecular structure has also required the determination of minor and trace elements present in the DNA molecules. Neutron activation has been used to measure metal concentrations in commercially obtained DNA preparations and other samples, as the presence of metals in nucleic acids is not completely understood. Wilczok et al. (1982) have measured concentration ranges (expressed in µg/g) for: Fe, 40.2–281.1; Zn, 3.0–86.8; Cr, 4.2–29.6; Co, 0.023–0.225; Ag, 0.045–1.014; Hg, 0.167–4.3; Se, 0.128–0.317; and Rb, 1.55–8.3. These investigators suggest that variations in metal concentration probably relate to biogeochemical and environmental conditions in the life cycle of the animals, which affect the metal accumulations in the analyzed tissue. Wilczok et al. (1982) further suggest that a limitation in trace metal DNA analysis is the short half-life of some

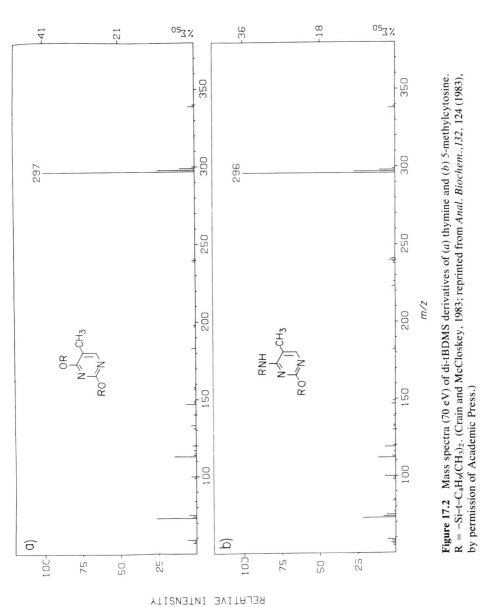

Figure 17.2 Mass spectra (70 eV) of di-tBDMS derivatives of (*a*) thymine and (*b*) 5-methylcytosine. R = -Si-t-C$_4$H$_9$(CH$_3$)$_2$. (Crain and McCloskey, 1983; reprinted from *Anal. Biochem.*, *132*, 124 (1983), by permission of Academic Press.)

of the nuclides; thus, mass spectrometry may be a preferred method of analysis in future investigations. Isotope dilution mass spectrometry can also minimize the effect of trace contaminants that may be introduced into the DNA samples during the isolation procedure.

The role of metals in the development of certain tumors has been discussed in some detail by Andronikashvili and Yesipova (1982), who conclude that "the presence of a metal ion leads to the appearance of a dipole moment in the DNA molecule which, in turn, cannot fail to affect processes of replication and transcription." Increased concentrations of certain metal ions have also been observed during tumor growth, and both irradiations and chemotherapeutics have been known to lower these concentrations in the DNA of tumor tissues.

BODY FLUIDS

The quantification of metals in body fluids (plasma, cerebrospinal fluid, saliva, urine, etc.) is an increasingly important application for mass spectrometry in its relation to physiological/toxicological investigations and to the general detection of metals in complex biological materials. Almost all metals in body fluids are detectable by mass spectrometry at concentrations below those observable by other analytical techniques. In addition, through the use of enriched isotopes and the isotopic dilution method, the quantification is usually more accurate. The selection of the ion source for metal ion production will depend on sample size, chemical preparation, and other parameters; but assays can be made using fast-atom bombardment, surface ionization, field desorption, etc. Some 63 metallic elements have been investigated with this latter technique; and for some analyses, the surface ionization-diffusion method (Chapter 2) offers the specific advantage of producing pure atomic spectra without interference from molecular ion peaks.

Recently, it has also become desirable to undertake the multielemental assay of specimens on a time-dependent basis. With the advent of very high-sensitivity spectrometry requiring only small amounts of sample, this type of analysis will eventually have widespread clinical as well as research application. In both human and animal plasma, lithium, calcium, strontium, rubidium, barium, copper, cobalt, nickel, manganese, and zinc are among the many metals that have been quantified in various investigations. In one interesting measurement of lithium concentrations in the physiological fluids of rabbits, lithium levels were observed to change in the serum of the mother during periods of gestation and lactation (Palavinskas et al., 1982). Reported values were between 2.61–5.02 μg/l, and the higher level in the third week of gestation was correlated with bone growth in the fetus; and at a later stage, to the start of milk production.

In humans, correlations are also being made of patients according to various categories and types of disease. Table 17.2 shows the concentration of

TABLE 17.2 Concentrations of Li, Rb, Mg, and Sr in Urine of a Multiple Sclerosis Patient Over a 24 H Period

Sampling Interval	Lithium (μg/l)	Rubidium (mg/l)	Magnesium (mg/l)	Strontium (μg/l)	Amount of Urine (10^{-3} l)
8–10	15.77	0.619	222.5	262.3	67
10–12	3.87	0.503	59.9	279.8	167
12–14	4.32	0.669	32.3	132.2	270
14–16	4.10	1.109	18.4	145.3	188
16–18	8.91	0.672	25.1	91.5	300
18–20	3.70	0.548	36.2	128.3	213
20–22	11.83	1.822	20.5	95.4	120
22–24	15.70	1.966	48.9	141.0	51
24– 2	11.39	0.482	75.1	180.9	100
2–4	17.37	0.762	92.0	132.1	70
4–6	14.87	1.386	108.5	220.0	50
6–8	15.31	2.410	110.1	222.5	29

Source: Schulten et al. (1983).

lithium, rubidium, magnesium, and strontium in urine of a multiple sclerosis patient who was monitored by field desorption mass spectrometry during a 24 h period (Schulten et al., 1983). The investigators reported that for rubidium, the average concentration was below that of patients in the same hospital but not suffering from multiple sclerosis. Also, the highest levels of rubidium and strontium were excreted during the daytime, while magnesium levels were lower during the night hours. Many other studies, of course, continue to be focused on the distribution and transport of lithium into various cells and tissues, since lithium is presumed to exert a psychotherapeutic action within the brain (Gorkin and Richelson, 1981).

Mass spectrometry and the isotope dilution method have been used for a broad spectrum of other investigations. For example, Dan et al. (1983) have measured glucose production and utilization in children, using dideuteroglucose as a stable tracer. This analysis utilized isothermal injection into a gas chromatograph with a quadrupole spectrometer programmed to monitor selected ions.

The metabolic profiling of urinary carboxylic acids by triple-quadrupole spectrometry has also been reported by Hunt et al. (1982). Inasmuch as extraction, derivatization, and gas chromatographic separations are replaced by a direct analysis of lyophilized urine samples, the authors report a sample analysis time of only 15 min and the detection of more than 100 different organic acids in a typical sample.

Other mass spectral measurements have been made of cerebrospinal and amniotic fluids. For example, chromatography-mass spectrometry has been used to quantify 1-methylimidazole-4-acetic acid in human cerebrospinal fluid (CSF). In this work, Swahn and Sedvall (1983) have achieved a sensitivity of 0.2 ng/ml and have reported on the elevated levels of this histamine in patients with various neurological diseases, compared to concentrations less than 1 ng/ml in healthy volunteers. Gas chromatography-mass spectrometry analysis of 3-methoxy-4-hydroxyphenylglycol in human CSF has also been reported by Holdiness (1983) at picogram sensitivity.

In connection with amniotic fluids and fetal diagnostics, a widely accepted test for the prediction of fetal lung maturation has been the determination of the lecithin/sphingomyelin (L/S) ratio (Gluck et al. 1971). The validity of this ratio measurement in the past, however, has been limited by (1) false predictions—even though the neonatal respiratory function is normal; (2)

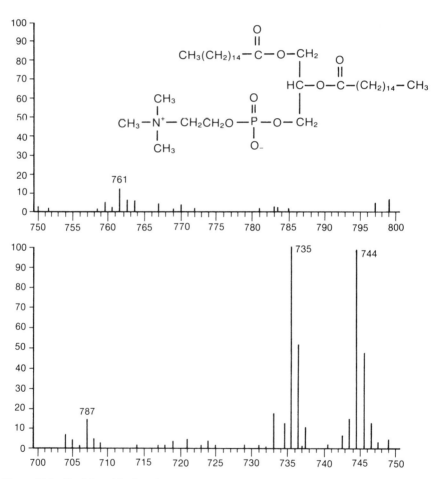

Figure 17.3 Partial positive ion, fast-atom bombardment mass spectrum of a chloroform extract of amniotic fluid to which d9-dipalmitoylphosphatidylcholine was added as an internal standard. (Ho et al., 1983; reprinted from *Clin. Chem., 29,* 1349 (1983), by permission of American Assoc. for Clinical Chemistry.)

anomalous L/S ratio changes in complicated pregnancies; and (3) available sample size. A significant advance in quantifying dipalmitoylphosphatidyl-choline (DPC) in amniotic fluids, and thus in evaluating fetal lung maturity, has been reported by Ho et al. (1983). These investigators have (1) used stable isotope-labeled d_9-DPC as an internal standard; and (2) applied the fast-atom bombardment ionization technique with a double focusing mass spectrometer. Specifically, seven mixtures of labeled d_9-DPC and unlabeled d_0-DPC were analyzed to determine their relative abundances, with the molecular ion peaks at m/z 735 and 744 serving to quantify the results. A partial positive ion, fast-atom bombardment spectrum of a chloroform extract of amniotic fluid to which d_9-DPC was added as an internal standard is shown in Figure 17.3. This approach can clearly separate different lecithin species by molecular masses, and the authors report that their mass spectral analysis indicated a d_9 incorporation into > 99% of the molecules. In addition, increased sample sensitivity and accuracy were obtained in comparison to conventional field desorption, based on sample mixtures of 1, 0.5, 0.25, and 0.1 μg.

Finally, it is appropriate to point out the essential role of mass spectrometry in metabolic profiling in connection with other analytical methods,

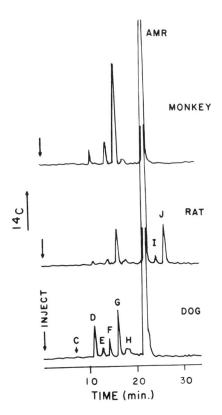

Figure 17.4 Reversed-phase HPLC of urine from dogs, rats, and monkeys that received ^{14}C-amrinone. (Baker et al., 1982; reprinted from *Drug Metabolism and Disposition, 10,* 168 (1982), by permission of American Society for Pharmacology and Experimental Therapeutics.)

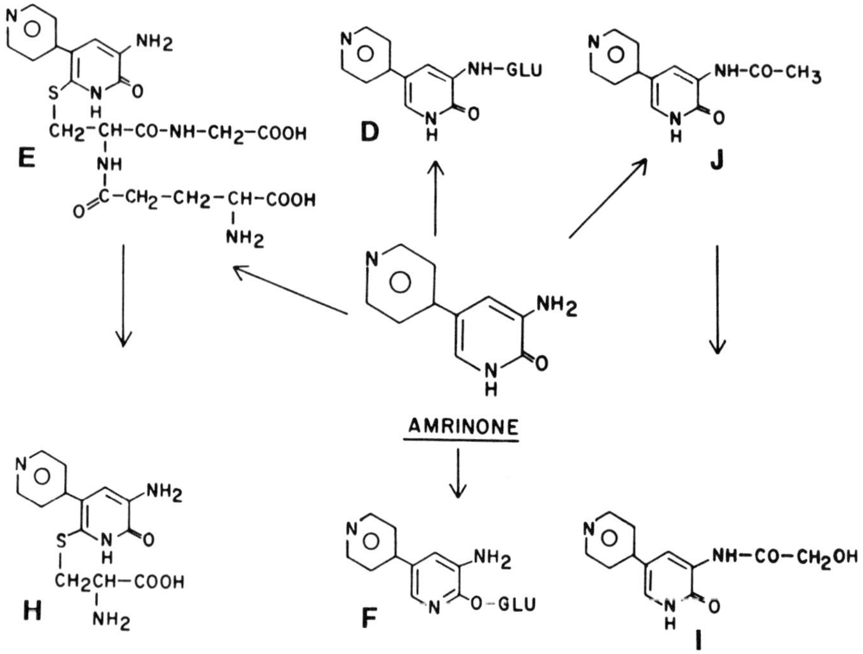

Figure 17.5   Biotransformation of the cardiotomic agent amrinone. (Baker et al., 1982; reprinted from *Drug Metabolism and Disposition, 10,* 168 (1982), by permission of American Society for Pharmacology and Experimental Therapeutics.)

that is, NMR, ^{14}C counting, and high-performance liquid chromatography (HPLC). The metabolism of the cardiotonic agent amrinone 5-amino-3,4'-bipyridine-6(1H)-1 is an example (Baker et al., 1982). Figure 17.4 shows reverse phase gradient HPLC elution profiles of urine from dogs, rats, and monkeys that received ^{14}C amrinone. The metabolic pathways elucidated for amrinone are shown in Figure 17.5.

SCREENING OF DISEASES

The importance of gas-liquid chromatography (GLC) for the rapid diagnosis of infectious diseases has been reviewed by Edman and Brooks (1983). Gas-liquid chromatography provides a powerful method for classifying micro-organisms on the basis of carboxylic acids, hydroxy acids, amines, etc., and the term *GLC-chemotaxonomy* has even been coined to denote the application of GLC for the study of microbial cells and metabolic products in culture media. Edman and Brooks (1983) further predict that the GLC methodology holds the greatest promise for understanding the etiology of infectious diseases caused by bacteria, viruses, fungi, and parasites through the analysis of a variety of body fluids or excretion products of the infected host. When GLC is also complemented by high-resolution mass spectrometry, an even more definitive correlation to diseased states can be made of changes that occur in body fluids. Clearly, GLC-mass spectrometry will never be a substitute for other highly specific and sophisticated laboratory analyses of blood and body fluids, but GLC profiles have already been successful for rapidly differentiating clinically similar diseases during the early stages (1–5 days) of infection. Specifically, Brooks et al. (1981) have used GLC with a frequency-pulsed (FP) modulated electron-capture detector (ECD) and high-resolution columns to analyze serum specimens from human controls and patients with the following diseases: (1) Rocky Mountain spotted fever—a rickettsial disease; (2) chicken pox caused by herpes virus; (3) measles; (4) rubcola and rubella; (5) enterovirus infections; and (6) *Neisseria meningitidis* infection. All of these diseases generate a rash and may be confused in a perfunctory diagnosis. However, chromatograms of derivatized extracts of sera taken from various patients have shown remarkable differences—sufficient to demonstrate the validity of this technique as a rapid screening aid in grouping viruses.

Figure 17.6 shows GLC-FP-ECD profiles of 2′,2,2,-trichloroethanol (TCE) derivatized acidic chloroform extracts of cerebrospinal fluid (CSF) from patients and controls as reported by Brooks et al. (1980). Other infectious diseases that have been investigated with chromatographic profiles include septic arthritis and encephalitis. In one study, the CSF specimens from acutely ill Egyptian patients revealed large amounts of a new amine compound that disappeared after successful therapy. Mass spectrometry was used to identify this amine, which had neither been reported in biological materials nor synthesized chemically (Brooks et al., 1977). Thus, GLC/MS becomes an additional tool for both screening disease or confirming recovery.

GC/MS studies have also been employed in the diagnostics of diabetes. A comparative study of 270 patients with diabetes mellitus, 28 healthy individuals, and 143 nondiabetic patients has indicated that the urinary excretion of total 4-heptanone (the sum of a β-oxocarboxylic acid, 2-ethyl-3-oxohexanoic acid, and its decarboxylation product, 4-heptanone) increases markedly with diabetes. Mean values of 1073 μg/24 h have been reported for diabetics, compared to 207 μg/24 h and 246 μg/24 h for healthy persons and nondiabetic patients, respectively (Liebich, 1983). It has thus been suggested that an increased urinary excretion of total 4-heptanone can provide an additional marker for the disease, using gas chromatography-mass spectrometry for analysis.

In this investigation, the GC 25 m glass capillary column was coated with OV-17 (Bodenseewerk Perkin-Elmer Corp.), helium was used as the carrier gas, and the column temperature was programmed at 2°C/min after an initial heating to 40°C for 10 min. The gas chromatograph and the mass spectrometer were interfaced by a 30 cm $\times$ 0.1 mm ID platinum capillary with both

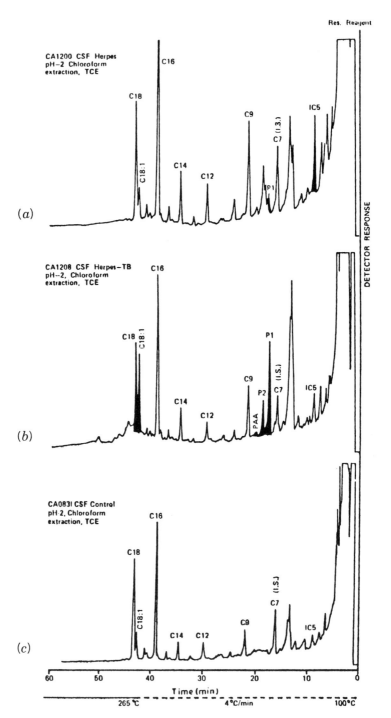

Figure 17.6 GLC-FP-ECD profiles of TCE-derivatized, acidic chloroform extracts of cerebrospinal fluid from patients and a control specimen. (Edman and Brooks, 1983; reprinted from *J. Clin. Microbiol.*, *12*, 108 (1980), by permission of American Society for Microbiology.)

the interface and ion source temperature being held at 220°C. The mass spectrometer parameters included: (1) electron-impact ionization at 70 eV; (2) accelerating voltage of 3 kV; (3) resolution 700; (4) recording mode-automatic repetitive scanning; and (5) the mass spectra were recorded over the mass range 15–380 and stored on magnetic tape. The quantitative determination of total 4-heptanone in urine was based on the ratios of the peak heights of 4-heptanone and an internal standard, and also on calibration graphs from eight aqueous standard solutions with 4-heptanone concentrations between 15–3200 μg/l. In particular, Liebich (1983) reports that the β-oxocarboxylic acid in urine and serum (especially in the form of its methyl ester) was detected with good sensitivity and specificity based on an intense ion peak at m/z 71, plus the molecular ion m/z 172 and fragment ions m/z 129 and m/z 141. This profiling of metabolites in urine and serum now appears to be a valid indicator of the incidence of diabetes. The author further states that other ketonic substances can be monitored by GC/MS—which reflects an abnormality in the metabolism of diabetics even when the glucose levels of such patients are well controlled.

In rare instances, trace metals may provide the only clue to infection or a neurological disorder. For example, in the case of Newcastle disease virus in mice, many metabolic and neurochemical parameters have been observed to be normal, and even the central nervous system concentrations of Na, K, Fe, Cu, Zn, Mg, and Se were normal (Murray et al., 1983). However, an analysis of Rb revealed a significantly elevated concentration of this element compared to the Rb in a control group.

BLOOD GASES

A major problem in blood gas analysis is sampling. Gas samples are acquired via an in vivo catheter or a transcutaneous probe. In either case, the quantities of gas are sufficiently small (10^{-5} to 10^{-7} ml/s), so that the sample can be transferred directly to the mass spectrometer ion source. However, all probes and transfer lines must be made from materials having very low outgassing rates, be vacuum tight, and be kept clean so that the total background gas level is less than ~1% of the sample gas throughout. The reaction of oxygen and molecules containing oxygen with inlet and ion source surfaces will also limit the accuracy of blood gas measurements. The first reported membrane-tipped catheter for monitoring continuous in vivo partial pressures of blood gases was reported by Woldring et al. (1966). Commercial catheters are now available with fittings for connecting to mass spectrometers. Their response time to blood gases is a function of membrane thickness, etc., but times of ~0.1–1 min are typical.

With the transcutaneous probe, gases are sampled through the skin as shown in the schematic of Fig. 17.7 (Weaver, 1982). Prerequisites for significant measurements include (1) a sufficient permeability of the skin to the

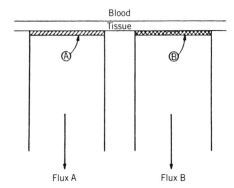

Figure 17.7 Noninvasive probes of tissue metabolism (Weaver, 1982).

specific gas; (2) a blood concentration high enough to provide a measurable sample to the mass spectrometer; and (3) freedom from interference by background gases. Measurement errors may also arise due to the depletion of gases in the vicinity of the diffusion membrane. Blood oxygen, carbon dioxide, nitrogen, nitrous oxide, and several volatile anesthetics have been measured with this noninvasive technique. Under normal conditions, oxygen delivery to adult skin has been reported as $\sim 1.5 \times 10^{5}$ ml/cm^2/s, although this figure makes no allowance for the consumption of oxygen by the tissues. It is thus suggested that when a probe is placed on the skin, it must consume substantially less gas than this rate (perhaps 10^{-7} ml/cm^2/s) both to avoid depleting local tissue and to minimize signal variations caused by local blood flow fluctuations (Delpy and Parker, 1979). It has also been suggested that with slight modifications, transcutaneous measurements could be extended to the assay of volatile compounds in organs by placing an interface to the mass spectrometer on the surface of the organ, or even to an exposed blood vessel (Weaver, 1982). Such a measurement would be less invasive than the insertion of a probe into the vessel, and it would avoid clotting problems.

Blood gas and respiration have also been studied by having a patient inhale small amounts of two gases—one soluble in the blood (freon) and the other insoluble (argon)—using a new technique that requires the use of both a fiber bronchoscope and a mass spectrometer (Catena et al., 1983). The continuous analysis of gases in a particular lung section can be carried out through the suction channel of the fiber bronchoscope that is connected to a mass spectrometer. Because freon is soluble in the blood and its concentration in alveolar air decreases in direct relationship to alveolar blood flow, a single breath assay of freon to argon provides useful information about the functional efficiency of a particular lung section.

RESPIRATORY GASES

Complete mass spectrometer computer systems have now been developed for clinical uses and measurements of physical fitness, in addition to their

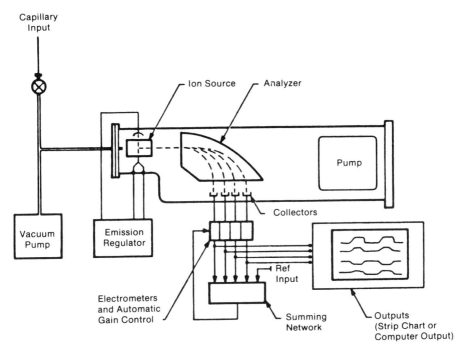

Figure 17.8 Schematic of a medical mass spectrometer for monitoring respiratory gases and cardiovascular parameters (Perkin-Elmer Corp.).

use in research and metabolism. Specific respiratory and cardiocirculatory measurements require the measurement of ventilation, oxygen uptake, O_2 and CO_2 output, etc.; and a recently reported system includes a respiratory mass spectrometer, a pneumotachometer and thermometer, a heart rate meter, a breath checker, and a computer system (Nishi et al., 1983). A schematic of a pulmonary mass spectrometer is shown in Figure 17.8, and a few examples of gas chromatography-mass spectrometry will be cited.

Interest in ethane has increased since the discovery that low-level monitoring of this gas could be correlated with lipid peroxidation in the livers of CCl_4-treated mice. A current long-range goal is to apply this technique to human subjects to monitor various pathological conditions that may be indexed to in vivo lipid peroxidation. Lawrence and Cohen (1982) have developed a simple breath collection chamber for concentrating ethane from large volumes of air, where the sensitivity limits for detecting ethane are increased 100-fold. Using activated charcoal traps that fully absorb at 10°C and fully desorb at 220°C, they were able to obtain gas chromatographic analyses from samples with ~100% ethane recovery. The collecting apparatus is presently being applied to laboratory animals for concentrating ethane, ethylene, and longer chain hydrocarbons, but the authors suggest

TABLE 17.3   CO in Breath of Smokers and Nonsmokers

Smoking Habits	CO in Breath (ppm)
Nonsmokers	4.5
Mild smokers (10–20 cigarettes/day)	18.0
Heavy smokers (> 20 cigarettes/day)	42.4

Source: Sannolo et al. (1983).

that the system is applicable to human subjects—and possibly to the analysis of ethylene that is evolved from plant tissues.

The action of dietary fibers in promoting the excretion of intestinal fermentation gases is also being evaluated, inasmuch as epidemiological evidence has related a low-fiber diet to a greater incidence of diseases that include cancer of the colon, diabetes, appendicitis, etc. Thus, both breath and flatus gases have been analyzed in determining hydrogen, methane, and CO_2 levels in low-fiber and high-fiber diets (Marthinsen and Fleming, 1982). In a related development, a new chromatographic instrument for measuring trace concentrations of breath hydrogen has been reported that has been used to evaluate the intestinal absorption of sugars (Christman and Hamilton, 1982). The production of hydrogen results from fermentation of nonabsorbed sugars by intestinal bacteria, and a fraction of this hydrogen is absorbed and excreted from the lungs. The key element in the instrument is the solid-state sensor, which consists of an *n-type* semiconducting material (sintered SnO_2) that exhibits a decrease in electric resistance when combustible or reducing gases are adsorbed on its surface. Since the detector is insensitive to the major gases in expired air (oxygen, nitrogen, argon, and carbon dioxide), the apparatus can utilize room air as the carrier gas; and the molecular sieve column is only required to separate hydrogen from other gases that might interfere, such as carbon monoxide. Essentially, this solid-state sensor has allowed the sieve column to be reduced by a factor of 10, and has shortened the intersample time interval to approximately 2 min.

Breath analysis has also included the administration of ^{13}C-labeled precursors followed by $^{13}C/^{12}C$ measurements of the expired CO_2. This type of measurement has been used (1) as an indirect measure of a metabolic disorder; (2) to detect fat malabsorption; and (3) for investigations relating to diabetic patients (Lockhart, 1979). Furthermore, H_2 and carbon monoxide measurements provide data of carbohydrate malabsorption and CO production or exposure. For example, Sannolo et al. (1983) have reported on the carbon monoxide levels of smokers and nonsmokers, as shown in Table 17.3.

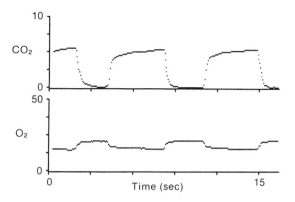

Figure 17.9 Waveform of expired CO_2 and O_2. The respiratory gas signals from the mass spectrometer are received by the microprocessor every 25 ms. The end of expiration is shown by the sharp fall of CO_2 and the rise of expired O_2 (Perkin-Elmer Corp.).

The automatic monitoring of the respiratory function is also becoming important in both surgical and respiratory intensive care, as pulmonary insufficiency is one of the leading causes of death in critically ill patients. Specifically, the simultaneous partial pressures of expired gases can be detected as the patient breathes room air through a mouthpiece in series with a mass spectrometer and a pneumotachograph, both of which are connected to a microprocessor that converts analog to digital signals. The microprocessor acquires and stores partial pressure and gas flow data from expired gases during each successive 2 min assay (Gui et al., 1981). Figure 17.9 shows CO_2 and O_2 signals from expired breath; the end of the respiration is marked by a sharp drop in CO_2 and a rise in O_2 levels. Signals are received by the microprocessor every 25 ms so that a measurement is made 40 times in every second of analysis. The strong correlation between expiratory CO_2 and arterial CO_2 makes it possible to obtain fundamental respiratory function indices in persons who are critically ill.

Another instrumental development that is somewhat unique is a mass spectrometer designed for the simultaneous measurement of both respiratory and blood gases (Nishi et al., 1980). The instrument is a voltage-scanning, 180° magnetic sector type, with separate respiratory gas and blood gas inlets to two ion sources. A dual-channel detector system is also provided, so that selected ions (N_2, NO, O_2, Ar, CO_2, and N_2O) can be simultaneously monitored from these two sources with full resolution. The respiratory gas inlet uses conventional differential pumping, and the blood gas is introduced through a diffusion membrane. The response to the gas introduction is 60 ms for a 90% signal amplitude in standard respiratory samplings, and the introduction of blood gas is on the order of 1×10^{-5} torr·l/s. The schematic diagram of Figure 17.10 shows an integrated system, designed by Perkin-Elmer Corporation, providing pulmonary, anesthetic, and cardiovascular

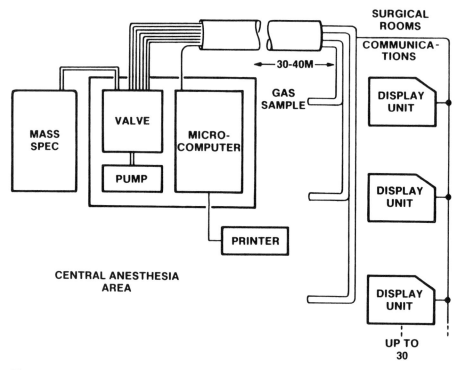

Figure 17.10 Schematic diagram of a computer-controlled system for providing pulmonary anesthesia and cardiovascular data (Perkin-Elmer Corp.).

information. It is typically configured to monitor eight gases with the capability of being connected to 16 or 30 patients. By means of computer control, CRT displays, and a printer, it can furnish data on both inspired and expired gases in percentages or mm Hg. CO_2 waveforms can be printed and arterial blood gases can be compared to respiratory CO_2 and O_2.

Future breath analysis instrumentation is proceeding in several directions. In research applications, high-resolution double focusing magnetic spectrometers will be used where a real-time differentiation must be made for gases having the same nominal mass number, for example, N_2 and CO at mass 28 and CO_2 and N_2O at mass 44. However, quadrupole instruments can provide adequate resolution in many applications; they can be programmed to monitor many gases with switching times that are fast compared to physiological changes, so that the measurement is essentially real time. Finally, new clinical and cardiovascular instrumentation of simplified design and portability is becoming available, so that breath analysis may eventually become as routine as a blood pressure measurement.

BREATH METABOLITES

Mass spectrometry not only surpasses all other techniques for quantifying respiratory gases; it also provides the means for identifying breath metabolites that can be correlated with disease, toxicants, drugs, and absorbed pollutants. Breath is also the most available biological material; and because breath analysis is noninvasive and real time, increasing use will be made of this diagnostic method. Over 200 compounds have been detected in human breath, and a further utilization of atmospheric ionization mass spectrometry and tandem mass spectrometry should identify additional compounds by their fragmentation pattern. Manolis (1983) has classified the metabolites excreted in breath into the following five groups: (1) lipid degradation products; (2) aromatic compounds; (3) thio compounds; (4) ammonia and amines; and (5) halogenated compounds.

Breath acetone has been used to monitor diabetics and patients on diets, and breath analysis in the field of toxicology is already extensive. In the case of alcohol abuse, it has also been suggested that acetaldehyde can serve as a marker because it persists in breath considerably longer than ethanol (Needham, 1980). Other specific areas where breath analysis has been correlated with physiological and biological variables include physical exercise, cardiac output, body fat, ovulation, renal and hepatic disease, and sex-related metabolic studies (Manolis, 1983). This author has also cited the several advantages of breath diagnostics in comparison with serum or urine analyses:

1. Samples are readily obtained without invasion.
2. Breath samples reflect the arterial concentrations of many biological substances; hence, breath assays may provide an option when many arterial samples are required (e.g., in patient monitoring).
3. In contrast to serum or urine samples, no work-up of a breath sample is required.
4. With tandem (MS/MS) spectrometry, real-time monitoring of metabolism yields information that is not available by other techniques.

ENERGY EXPENDITURE MEASUREMENTS

In a number of metabolic studies, a need exists for monitoring daily energy expenditure. One method is based on the assumption that a linear relationship exists between O_2 consumption ($\dot{V}O_2$) and heart rate (HR), and thus between HR and energy expenditure (Christensen et al., 1983). However, a doubly labeled water method, first reported by Lifson et al. (1955), is now an attractive alternative because of (1) the increased precision of isotopic measurements; and (2) the greatly reduced cost of ^2H and ^{18}O tracers. It is

TABLE 17.4   Characteristics of Hypothetical Human Subjects vs. Energy Expenditure

Subject	Wt (kg)	TBW kg	TBW mol	H_2O Output (mol/day)	CO_2 Output (mol/day)	Energy Expenditure (kcal/day)
Neonate	3.5	2.8	155	31	1.6	210
8 yr	28	18	1010	150	19	2410
25 yr (M)	70	43	2400	170	25	3170
65 yr (F)	50	23	1300	90	10	1310

Source: Schoeller (1983).
M, male; F, female.

based on the reported observation that oxygen atoms in exhaled carbon dioxide and body water are in isotopic equilibrium (Lifson et al., 1949), and that the kinetics of respiration and water elimination are interdependent. "After a loading dose of stable isotope labeled water ($^2H^1H^{18}O$), the $H_2{}^{18}O$ is eliminated from the body as both CO_2 and H_2O, while the $^2H^1HO$ is eliminated only as H_2O. Thus, the $^2H^1HO$ elimination rate is a measure of H_2O flux, the $H_2{}^{18}O$ elimination is a measure of H_2O plus CO_2 flux, and the difference between the elimination rates is a measure of CO_2 flux. The latter can be related to energy expenditure by standard calorimetric calculations" (Schoeller, 1983). Thus, this particular ^2H and ^{18}O tracer method measures CO_2 production and not O_2 consumption.

Mass spectrometric measurements using this methodology have been reported for 17 subjects between the ages of 8–34 years. Briefly, urine samples were collected at suitable intervals and 1.5 μl aliquots were equilibrated with 1 ml STP of CO_2 at 25°C for 48 h, and the $^{18}O/^{16}O$ isotope ratio was measured on a differential isotope ratio spectrometer (Schoeller et al., 1980). For the $^2H/^1H$ analysis, 2 μl aliquots were vacuum distilled and reduced to H_2 over uranium at 700°C prior to spectrometric assay. An overall assessment of this method and its possible errors has been made by Schoeller (1983), who has concluded that human energy measurements can be made with relatively small quantities of ^2H and ^{18}O tracers. Table 17.4 shows his calculated parameters of "hypothetical human subjects" relating weight, total body water (TBW), H_2O output, CO_2 output, and energy expenditure. For the limited experimental measurements using human subjects, the author estimates an accuracy of 2%. Also, although some additional error will accompany changes in the TBW pool, he estimates this error will be less than 2% unless a pool size changes by more than 4%.

MASS SPECTROMETRY OF BACTERIA PARTICLES

The assay of atmospheric aerosols by surface-ionization mass spectrometry has been cited in Chapter 15. In this pioneering work of Davis (1977), both

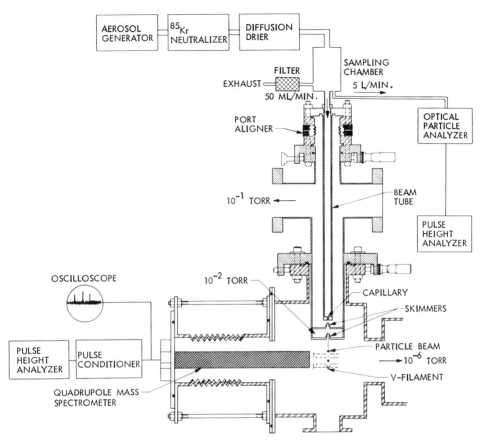

Figure 17.11 Schematic of system for analyzing biological particles. (Sinha et al., 1984; *Int. J. Mass Spectrom.*, *57*, 125 (1984), by permission of Elsevier Science Publishers.)

inorganic and organic species were identified by the direct impact and pyrolysis of airborne particles on a heated rhenium filament. This general technique has now been extended to include the analysis of individual biological particles. Sinha et al. (1984) have reported a system in which the biological particles are introduced in aerosol form into a quadrupole mass spectrometer (QMS); the schematic is shown in Figure 17.11. The aerosol is generated by nebulizing an ethanol suspension of bacteria. The beam of particles is produced by the expansion of the aerosol through a capillary nozzle into the vacuum, with N_2 being the carrier gas and ^{85}Kr functioning as a charge neutralizer. The particles are then volatilized upon impact on a rhenium filament and ionized by electron bombardment.

Figure 17.12 shows the mass spectra obtained by Sinha et al. (1984) for several bacteria particles: *Pseudomonas putida*, *Bacillus cereus*, and *Bacillus subtilis*, where the mass peaks extend to almost 300 amu. The authors

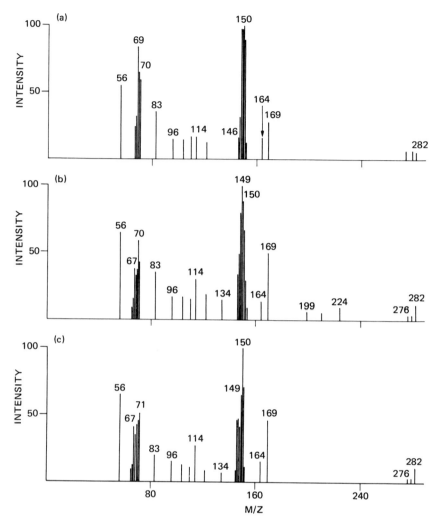

Figure 17.12   Mass spectra of bacteria particles. (*a*) *Pseudomonas putida,* (*b*) *Bacillus cereus,* and (*c*) *Bacillus subtilis.* (Sinha et al., 1984; *Int. J. Mass Spectrom., 57,* 125 (1984), by permission of Elsevier Science Publishers.)

suggest that the similarity of the three spectra "is not surprising in view of the fact that the micro-organisms have essentially the same major chemical building blocks." Also, they report that for the small particles analyzed ($\sim 10^{-13}$ g; 0.5 μm in diameter and 1.5 μm in length), vaporization is complete and pyrolysis is accomplished in ~ 30 μs.

This type of pyrolysis mass spectrometry of bacteria is of considerable significance. First, it permits the detection and identification of biological particles on a continuous, real-time basis in air, and without the need for conventional sample collection. Second, the pyrolysis spectra seem to be reproducible, and they convey substantial structural information for distinguishing between different micro-organisms. A further development of this method should permit the assay of even smaller samples ($\sim10^{-15}$ g), and it should lead to applications relevant to both clinical medicine and biological research.

APPLICATIONS IN INDUSTRIAL MEDICINE

Industrial medicine embraces a broad field, and one of its many professional concerns relates to the environments of quasihazardous occupations. Routinely or on special demand, potentially toxic metals must be machined and chemical compounds processed. As an analytical tool, mass spectrometry is unique because its versatility and sensitivity permit it to function (1) as a continuous or laboratory monitor; (2) in studies relating to the deposition and physiological fate of toxic materials acquired in the workplace; and (3) in post-accident investigations. It also has special relevance in measuring radioactive nuclides, because it can unambiguously identify specific isotopes, thereby distinguishing between naturally occurring radionuclides and those from other sources. In addition, its sensitivity for trace metals and gases is so high that it can detect airborne contaminants at concentration levels that are orders of magnitude below permissible working concentrations. Hence, it can materially assist the industrial hygienist in observing the build-up of toxicants before they reach unacceptable levels.

Overall, occupational hygiene programs include air and surface monitoring of the workplace, the analysis of body deposits (skin wipes) and excreta (urine, feces), and other samples such as hair or blood. The range of metal concentrations in "normal" or "control" subjects varies widely, as metals are continuously assimilated in the human body through food, drink, and the ambient air, as well as from industrial environments. Nevertheless, metals constitute a primary airborne hazard in several industries, and a reduction in exposure to certain metals has been mandated in both consumer products and worker exposure by various government agencies. The maximum concentration of lead permitted in consumer paints has been lowered from 0.5% to 0.06%, and the auto industry has until 1987 to eliminate solder from its products. The Occupational Safety and Health Administration (OSHA) has adopted a reduced standard for lead by limiting worker exposure to 50 μg/ m^3 averaged over 8 h in most industries. In a related ruling, it has lowered the blood-lead level from 70 μg of Pb per gram of blood to 60 μg/g; at this concentration, a worker must be moved to another job (OSHA, 1981). An employee exposure maximum of 2 μg/m^3 of arsenic in air was also recommended to replace an earlier 500 μg/m^3 ceiling due to the known carcinogenic effect of arsenic, and a limit of 1 μg/m^3 has been recommended for carcinogenic hexavalent chromium. An exposure maximum of 15 μg/m^3 has also been reported for nickel because of evidence of lung, nasal, and kidney cancer among exposed workers, and a limit of 0.5 $\mu g/m^3$ has been proposed for beryllium. Cadmium is another element of industrial concern due to its toxicity over long periods and its relatively high concentrations in some manufacturing environments. As a result, OSHA has proposed an occupational maximum of 40 $\mu g/m^3$.

With the exception of beryllium (^9Be), which is monoisotopic, the above-cited metals have multiple isotopes that can be used via the isotopic dilution technique and mass spectrometry for measuring airborne concentrations. In some future studies, it may be desirable to employ the analytical technique of particle-induced x-ray emission (Chapter 20). This technique has been applied (during autopsy) to determine both overall and regional concentrations of metals within the human lung. Because the lung is in contact with more than 15,000 l of air daily, it is sensitive to environmental changes and it keeps a "record" of its exposure. In the autopsies performed by Bartsch et al. (1982), homogeneous distributions were consistently observed for potassium, calcium, iron, copper, zinc, and rubidium, but a very different distribution was noted for titanium, chromium, nickel, and strontium.

Recent research instrumentation that also has relevance to industrial medicine is the GC/GC/MS system first reported by Ligon and May (1980). For the detection and quantification of organic compounds, this "two-dimensional" chromatographic-mass spectrometric system has two salient advantages. First, the system provides an effective dynamic range at the secondary chromatograph from 100% to 0.01% (all peaks on scale) of any compound amenable to GC analysis. A second advantage is that particular regions of interest in the primary chromatograph can be rechromatographed to produce much improved separations and higher resolved spectra. As a consequence, the instrument can quickly identify a wide range of compounds for samples in the 10–100 ng range. Furthermore, by applying standard sampling techniques, it can provide quantitative data on industrial chemicals at sub–parts per million concentrations.

Other mass spectrometric instrumentation that relates to industrial medicine includes (1) the laser microprobe; and (2) the atmospheric pressure-type spectrometer. The first permits particulates and fibers to be analyzed at very high sensitivity, that is, a single laser shot focused on a particle can produce a complete mass spectrum. The second development makes possible the real-time monitoring of the ambient air and industrial environments that are suspected of having high concentrations of hazardous gases or organic compounds. This atmospheric pressure-chemical ionization-type instrument has also been used as a breath analyzer in connection with human exposure to organic volatiles (Davidson, 1981). Specifically, breath analyses by direct exhalation have been made for methanol, toluene, and tetrachloroethylene in an investigation of the uptake and elimination of volatiles in the respiratory system. Average times for the disappearance of the organic compounds from the breath were 24 min for methanol, 27 min for toluene, and 79 min for tetrachlorethylene, indicating a fairly rapid clearance following exposure. While this test was not routine, it indicates that organic vapors in industrial environments can be monitored via direct breath analysis, thus eliminating the conventional work-up of samples that is associated with gas chromatography-mass spectrometry. Essentially, the solvent concentration in the capillary blood vessels of the lung is a fraction of the total amount of the solvent absorbed by the whole body. In rare instances, solvent concentrations have been detected in breath even after a period of 100 h has elapsed from the time of exposure.

In the future, the analysis of selected organic species may be required in the part per billion and even part per trillion (10^{12}) range. Also, it is likely that some environmental and biological extracts will require a multistep purification prior to a mass spectral assay. For such measurements, the two-stage chromatograph-mass spectrometer described by Ligon and May (1984) provides desirable features, and the apparatus is less expensive than MS/MS systems. Essentially, the apparatus consists of tandem chromatographs in which specific components eluting from the first chromatograph can be switched into a cold trap and subsequently transferred to the second chromatograph. The second chromatograph is then directly coupled to the electron-impact ionization source of a high-resolution mass spectrometer. The elimination of solvents and high-level contaminants before they reach the spectrometer are advantages of this GC/GC/MS instrument.

The chromatogram and mass spectrum of Figure 17.13 illustrate the selectivity and sensitivity of the system. This example relates to an industrial product, but the method applies to many compounds in biological and other complex matrices. Briefly, Ligon and May (1986) analyzed a dielectric fluid (PCB, Aroclor 1242) for 2,3,7,8-tetrachlorodibenzofuran (2,3,7,8-TCDF). A 1 mg sample was spiked before injection into the system with 99 atom % ^{13}C 2,3,7,8-TCDF, and the mass spectrometer was operated in the selected ion monitoring mode. A level of 40 ppb was determined in the analysis that was completed in less than 45 min. For the 1 mg PCB sample, this represents the quantification of a minor contaminant at 40 pg. Crude extracts of soil, animal tissue, and other samples have also been examined by these investigators, who aptly remark that while mass spectrometry can now assay almost any compound at high sensitivity, the physiological significance of such compounds in biosystems must be ascertained by toxicologists.

Finally, it is appropriate to note that highly automated mass spectrometer systems can now provide essential data for the corporate concerns of the medical director, industrial hygienist, health physicist, and environmental engineer. With increasing emphasis being given to maintaining employee health in the workplace, instrumentation is desired that can monitor both the environment and the employee in that environment. One medical director has also suggested that the specificity and sensitivity of mass spectrometry

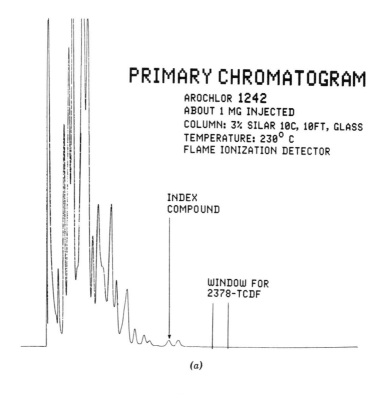

PRIMARY CHROMATOGRAM

AROCHLOR 1242
ABOUT 1 MG INJECTED
COLUMN: 3% SILAR 10C, 10FT, GLASS
TEMPERATURE: 230° C
FLAME IONIZATION DETECTOR

INDEX
COMPOUND

WINDOW FOR
2378-TCDF

(a)

AROCHLOR 1242 FROM A CAPACITOR
100 ppb ^{13}C 2,3,7,8-TCDF ADDED
ONE mg PCB FLUID INJECTED

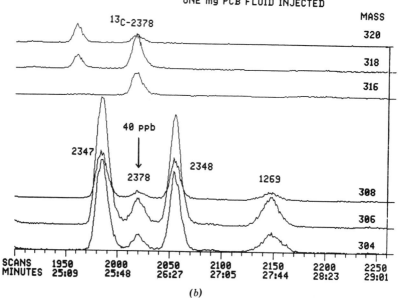

^{13}C-2378

MASS

320

318

316

40 ppb

2347

2378

2348

1269

308

306

304

SCANS   1950   2000   2050   2100   2150   2200   2250
MINUTES 25:09 25:48 26:27 27:05 27:44 28:23 29:01

(b)

may be useful in providing evidence in legal proceedings. In both of these instances, a mass spectrometer may be costly, but also cost-effective.

REFERENCES

Abraham, R., *Proc. Nutr. Soc., 41,* 261 (1982).

Andronikashvili, E. L., and N. G. Yesipova, *Biophysics, 27,* No. 6, 1070 (1982).

Apelgot, S., *Int. J. Rad. Biol., 43,* No. 1, 95 (1983).

Baker, J. F., B. W. Chalecki, D. P. Benziger, P. E. O'Melia, S. D. Clemans, and J. Edelson, *Drug Metab. Dispos., 10,* No. 2, 168 (1982).

Bartsch, P., A. Collignon, G. Weber, G. Robaye, J. M. Delbrouck, I. Roelandts, J. Yujie, *Arch. Environ. Health, 37,* No. 2, 111 (1982).

Bowen, H. J. M., in *Activation Analysis,* ed. J. M. A. Lenihan, and S. J. Thomson, Academic Press, London (1965), p. 146.

Brooks, J. B., G. Choudhary, R. B. Craven, C. C. Alley, J. A. Liddle, D. C. Edman, and J. D. Converse, *J. Clin. Microbiol., 5,* 625 (1977).

Brooks, J. B., D. C. Edman, C. C. Alley, R. B., Craven, and N. I. Girgis, *J. Clin, Microbiol., 12,* 208 (1980).

Brooks, J. B., J. D. McDade, and C. C. Alley, *J. Clin. Microbiol., 14,* 165 (1981).

Buckley, W. T., S. N. Huckin, J. J. Budac, and G. K. Eigendorf, *Anal. Chem., 54,* 504 (1982).

Burkinshaw, L., *J. Radioanal. Chem., 69,* No. 1–2, 27 (1982).

Catena, E., S. Marcatili, and R. Maselli, *Int. J. Mass Spectrom. Ion Phys., 48,* 109 (1983).

Chandra, S., and G. H. Morrison, *Science, 228,* 1543 (1985).

Christensen, C. C., H. M. M. Frey, E. Foenstelein, E. Aadland, and H. E. Refsum, *Am. J. Clin. Nutr., 37,* No. 3, 468 (1983).

Christman, N. T., and L. H. Hamilton, *J. Chromatogr., 229,* 259 (1982).

Colgan, M., *Science, 214,* 744 (1981).

Costa, H., and H. H. Mollenhaver, *Science, 209,* 515 (1980).

Crain, P. F., and J. F. McCloskey, *Anal. Chem., 132,* 124 (1983).

Dan, P., P. M. Clemons, M. I. Sperling, M. J. Gelfand, I. W. Chen, M. A. Sperling, and E. J. Norman, *Anal. Lett., 16,* B9, 655 (1983).

Davidson, W., A. Lovett, S. Nacson, and A. Ngo, in *29th Ann. Conf. on Mass Spectrom. and Allied Topics,* Minneapolis (May 1981).

Davis, W. D., *Environ. Sci. Technol., 11,* 593 (1977).

Day, H. G., *Trends Biochem. Sci., 7,* No. 3, 112 (1982).

Delpy, D. T., and D. Parker, in *The Medical and Biological Application of Mass Spectrometry,* ed. J. P. Payne, J. A. Bushman, and D. W. Hill, Academic Press, New York (1979), p. 179.

Dubick, M. A., and R. B. Rucker, *J. Nutr., 15,* No. 2, 47 (1983).

←――

Figure 17.13 GC/MS assay of a capacitor dielectric fluid (Aroclor, 1242) for 2,3,7,8-TCDF: (*a*) primary chromatogram and (*b*) selected ion mass; spectrum. The analyte was monitored at m/z 304, 306, 308, and the internal standard was monitored at 316, 318, and 320. The spectrometer resolution was 3000, and a level of 40 ppb was determined (Ligon and May, 1986; courtesy of W. V. Ligon).

Edman, D. C., and J. B. Brooks, *J. Chromatogr., 274*, 1 (1983).

Emery, T., *Am. Scientist, 70*, 626 (1982).

Fairweather-Tait, S. J., M. J. Minski, and D. P. Richardson, *Brit. J. Nutr., 50*, 51 (1983).

Gluck, L., M. V. Kulovich, and R. D. Borer, *Am. J. Obstet. Gynecol., 109*, 440 (1971).

Gorkin, R. A., and E. Richelson, *Neuropharmacology, 20*, 791 (1981).

Gui, D., G. Boldrini, G. Sganga, G. Tramutola, and M. Castagneto, in *Recent Developments in Mass Spectrometry in Biochemistry, Medicine and Environmental Research*, Vol. 7, ed. A. Frigerio, Elsevier, Amsterdam (1981), p. 189.

Ho, B. C., C. Fenselau, G. Hansen, J. Larsen, and A. Daniel, *Clin. Chem., 29*, No. 7, 1349 (1983).

Holdiness, M. R., *Anal. Lett., 16*, B14, 1067 (1983).

Hunt, D. F., A. B. Giordani, G. Rhodes, and D. A. Herold, *Clin. Chem., 28*, No. 12, 2387 (1982).

Jensen, A. A., *Residue Rev., 87*, 99 (1983).

Johnson, P. E., *J. Nutr., 112*, 1414 (1982).

Lambrecht, R. M., *Biomedical Applications of Stable Isotopes*, Brookhaven National Laboratory Report, Upton, NY (December 1981) p. 26.

Lane, H. W., D. C. Warren, B. J. Taylor, and E. Stool, *Proc. Soc. Exp. Biol. Med., 173*, 87 (1983).

Lapidot, A., and I. Nissam, in *Advances in Mass Spectrometry*, Vol. 8B, ed. A. Quayle, Heyden and Son Ltd., London (1980), p. 1142.

Lawrence, G. D., and G. Cohen, *Anal. Biochem., 122*, 283 (1982).

Liebich, H. M., *J. Chromatogr., 273*, 67 (1983).

Lifson, N., G. B. Gordon, M. B. Visscher, and A. O. Nier, *J. Biol. Chem., 180*, 803 (1949).

Lifson, N., G. B. Gordon, and R. McClintock, *J. Appl. Physiol., 7*, 704 (1955).

Ligon, W. V., and R. J. May, *Anal. Chem., 52*, 90 (1980).

Ligon, W. V., and R. J. May, *J. Chromatogr., 294*, 77 (1984).

Ligon, W. V., and R. J. May, *Anal. Chem., 58*, 558 (1986).

Lockhart, J. M., in *The Medical and Biological Application of Mass Spectrometry*, ed. J. P. Payne, J. A. Bushman, and D. W. Hill, Academic Press, New York (1979), p. 78.

Lodding, A., *Scan. Electr. Microsc., 3*, 1229 (1983).

Manolis, A., *Clin. Chem., 27*, No. 1, 5 (1983).

Marthinsen, D., and S. E. Fleming, *J. Nutr., 112*, No. 6, 1133 (1982).

Mertz, W., in *Ultratrace Metals Analysis in Biological Sciences and Environment*, ed. T. H. Risby, American Chemical Society, Washington, D. C. (1979), p. 5.

Mertz, W., *Science, 213*, 1332 (1981).

Murray, R. S., J. S. Burks, W. R. Smythe, N. Miller, A. C. Alfrey, and J. C. Gerdes, *J. Neurochem., 41*, 1011 (1983).

National Academy of Sciences, *Recommended Dietary Allowances*, Washington, D. C. (1980).

Needham, D., *Lancet, i*, 825 (1980).

Nishi, I., G. Tomizawa, H. Shibuya, H. Osawa, and A. Yamazawa, in *Advances in Mass Spectrometry*, Vol. 8B, ed. A. Quayle, Heyden and Sons Ltd., London (1980), p. 1926.

Nishi, I., G. Tomizawa, H. Ishii, Y. Suzuki, H. Shibuya, and L. Tatsuta, *Int. J. Mass Spectrom. Ion Phys., 46*, 167 (1983).

OSHA, *Chem. Eng. News, May 18*, 30 (1981).

Palavinskas, R., U. Bahr, K. Kriesten, and H. R. Schulten, *Comp. Biochem. Physiol., 73A*, 223 (1982).

Reamer, D. C., and C. Veillon, *J. Nutr., 113,* 786 (1983).

Sannolo, N., P. Vajro, G. Dioguardi, R. Mensitieri, and D. Longo, *J. Chromatogr., 276,* 257 (1983).

Schapira, G., in *Iron Metabolism,* ed. P. Gross, Springer-Verlag, New York (1964), pp. 1–9.

Schoeller, D. A., E. Van Santen, D. W. Peterson, W. Dietz, J. Jaspen, and P. D. Klein, *Am. J. Clin. Nutr., 33,* 2686 (1980).

Schoeller, D. A., *Am. J. Clin. Nutr., 38,* No. 6, 999 (1983).

Schricker, B. R., H. D. Gilbert, D. D. Miller, and D. Van Campen, *J. Nutr., 112,* 151 (1982).

Schroeder, H. A., and A. P. Nason, *Clin. Chem., 17,* 461 (1971).

Schulten, H. R., U. Bahr, and P. B. Monkhouse, *J. Biochem. Biophys. Methods, 8,* 239 (1983).

Sinha, M. P., R. M. Platz, V. L. Vilker, and S. K. Friedlander, *Int. J. Mass Spectrom. Ion Processes, 57,* 125 (1984).

Solomons, N. W., F. Viteri, T. R. Shuler, and F. H. Nielsen, *J. Nutr., 112,* 39 (1982).

Swahn, C., and G. Sedvall, *J. Neurochem., 40,* No. 3, 688 (1983).

Swanson, C. A., D. C. Reamer, C. Veillon, J. C. King, and O. A. Levander, *Am. J. Clin. Nutr., 38,* No. 2, 169 (1983).

Taylor, M. C., A. Demayo, and K. W. Taylor, *CRC Crit. Rev. Environ. Control, April,* 113 (1982).

Turnlund, J. R., M. C. Michel, W. R. Keyes, J. C. King, and S. Margen, *Am. J. Clin. Nutr., 35,* 1033 (1982).

Veillon, C., W. R. Wolf, and B. E. Guthrie, *Anal. Chem., 51,* 1022 (1979).

Weaver, J. C., in *Noninvasive Probes of Tissue Metabolism,* ed. J. S. Cohen, Wiley, New York (1982), p. 28.

Wilczok, T., G. Siotwinska-Palugnick, A. Kochanoka-Dziurowicz, L. Mosulishvili, and N. Kharabadze, *Radiochem. Radioanal. Lett., 56,* No. 3, 131 (1982).

Woldring, S., G. Owens, and D. Woodford, *Science, 153,* 885 (1966).

Wolfe, M. H., J. A. Molnar, and R. R. Wolfe, "Stable Isotope Analysis of Collagen and Proline Recycling in Scar Tissue," presented at the *32nd Ann. Conf. on Mass Spectrom. and Allied Topics,* San Antonio (May 27 to June 1, 1984).

PHARMACOLOGY

More than any other single analytical method, mass spectrometry provides the data base for developing the products of the pharmaceutical industry and for testing their efficacy in biological systems. Spectrometry is a prerequisite for measurements relating to: (1) the identification of drugs from natural products; (2) molecular structure; (3) drug synthesis; (4) the assay of complex mixtures; (5) isotopic tracer studies; (6) pharmacological activity and drug metabolism; (7) drug certification; (8) quality control of products; and (9) patents. The relevance of spectrometry to patents and patent infringements arises from the fact that many newly developed drugs are protected by patents detailing synthetic routes and potency, rather than by patents covering the compound, per se, or its usage. In general, high-resolution instrumentation is needed to examine the biological response of drugs that are presumed to result from the interaction of the compound with a specific receptor. Thus, multistage systems are utilized, that is, coupled gas chromatography-mass spectrometry, metastable ion techniques, and unique combinations of quadrupole, magnetic, and electrostatic analyzers.

The scope of pharmacological products continues to grow, and along with product expansion is the collinear need for an even more detailed analysis by mass spectrometry. By 1980, the worldwide production of antibiotics alone had reached 25,000 tons—including 17,000 tons of penicillins, 5000 tons of tetracyclines, 1200 tons of cephalosporins, and 800 tons of erythromycins (Demain, 1981). These and a broad spectrum of other pharmaceutical products have been developed and brought into production by a modification of natural antibiotic processes (semisynthesis), so-called directed biosynthesis, and synthetic manufacture. An important phase of GC/MS effort continues to be focused on the analysis of marine natural products that serve as sources of antimicrobial, antiviral, or antineoplastic activity (Rinehart, 1981). Thermospray LC/MS and MS/MS are also among the salient techniques in contemporary pharmaceutical research (Unger and Warrack, 1986).

ISOTOPICALLY LABELED SPECIES

Isotopically labeled compounds are essential in pharmacology for providing both quantitative and structural information. Lockhart (1979) has reviewed the three major areas where stable isotopes are of interest to the pharmacologist:

1. As tracers for the identification of metabolites.
2. As internal standards for quantifying drugs in biological fluids.
3. In studies relating to drug bioavailability.

A partial list of stable isotopes useful in pharmacological studies appears in Table 18.1. These isotopes are needed not only to follow the pathway of a drug, but to distinguish between a true drug metabolite and a naturally oc-curring molecular compound. Indeed, only with labeled species can the de-tection of many drugs be assayed at subpicogram levels; with multiple-la-beled compounds, it is now possible to determine picogram amounts in 1 μg samples. Furthermore, using ^2H, ^{13}C, ^{15}N, or ^{18}O tracers, some metab-olites can be detected in a biological isolate without the need for extensive purification and isolation procedures. Drugs in urine, plasma, breast milk, and amniotic fluid have been assayed in the picogram range by using gas chromatography-mass spectrometry computer systems and by employing stable isotope-labeled drugs as internal standards (Halliday and Lockhart, 1978).

TABLE 18.1   Stable Isotopes Used in
Pharmacological Studies

Element	Atomic Mass	Relative Abundance
^1H	1.00783	99.985
^2H	2.01410	0.015
^{12}C	12.00000	98.89
^{13}C	13.00335	1.11
^{14}N	14.00307	99.634
^{15}N	15.00011	0.366
^{16}O	15.99491	99.763
^{17}O	16.99914	0.037
^{18}O	17.99916	0.200
^{32}S	31.97207	95.02
^{34}S	33.96786	4.21
^{35}Cl	34.96885	75.77
^{37}Cl	36.96590	24.23
^{79}Br	78.91834	50.69
^{81}Br	80.91634	49.31

Some concern has been expressed with possible adverse effects of labeled species due to the mass difference. Any effect would be greatest for deuterium, where a factor of 2 exists. The administration of 200 g of D_2O to an adult male human subject will increase the plasma deuterium concentration to about 0.5%; but even at this level, there are few reported ill effects (Halliday and Lockhart, 1978). ^{13}C isotope effects have been investigated in algae (with labeling up to 90%) with no observed harmful effect, and mice have been fed for 6 months with yeast (with 90 atom% ^{13}C), resulting in a body carbon content of 60% ^{13}C. However, no anatomical abnormalities were observed in a post-mortem examination of the mice (Gregg et al., 1973). ^{13}C labeling has been applied by Rubio et al. (1982a) for measuring − *threo*-chlorocitric acid in human plasma—the ^{13}C analog being added as an internal standard. The simultaneous labeling of a drug with radioactive ^{14}C and stable ^{13}C at the same position has also been employed in animals.

In the case of nitrogen and oxygen, there are no radioisotopes with a sufficiently long half-life to be of practical value as labels. Also, "pulse labeling" of endogenous metabolites through inhalation of $^{18}O_2$-enriched atmospheres has been reported as a noninvasive approach for studying both in vivo turnover of compounds of pharmacological interest and the effects of external pathways of intermediary metabolism (Baillie, 1981). Sulfur isotopes can also be used as a label, and labeled sulfur compounds are available from commercial sources. The kinetics of stable ^{46}Ca and ^{48}Ca have also been compared in parallel studies with radioactive ^{47}Ca, and in vivo comparisons have been made between stable ^{58}Fe and radioactive ^{59}Fe. Overall, it has been shown that stable tracers provide equally accurate data as radiotracers without the limitations imposed by short half-lives or radiation hazards. In some instances, it may be desirable to use *post-irradiation* neutron activation of stable tracers in addition to mass spectrometry.

Figure 18.1 shows experimental results of an investigation in which plasma levels of both labeled and unlabeled barbital were monitored by gas-liquid chromatography (GLC)-mass spectrometry. A dose of 1.5 mg of $[^{15}N_{1,3}, {}^{13}C_2]$ barbital-sodium/kg was injected intravenously (within 20 s) into a human volunteer (65 kg); at the same time, the subject swallowed a capsule containing 4.5 mg of unlabeled barbital-sodium/kg. Blood (7 ml) was drawn at intervals over 6 days, and aliquots of plasma (1 ml) were assayed for barbital. The similar curves support the thesis that isotopes behave similarly in an organism, and that mass spectrometric techniques permit studies over extended periods.

Deuterium, because it is relatively inexpensive and easy to incorporate into molecules, has been the most widely used isotope, and deuterium-labeled analogs result in compounds that can be mass spectrometrically identified in blood or urine samples via the appearance of appropriately spaced doublets. The use of computer programs then substantially assists in relating isotope patterns and intensities to the metabolic fate and pharmacokinetics of drugs (Anderegg, 1983).

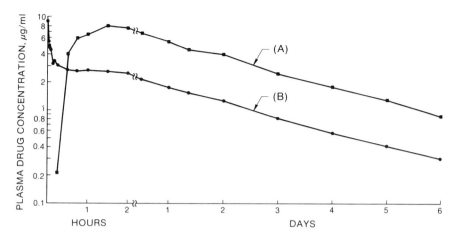

Figure 18.1 Barbital (*A*) and labeled [$^{13}N_{1,3}$, $^{13}C_2$] barbital (*B*) concentrations in the plasma of a volunteer after receiving simultaneous oral and intravenous doses. Measurements were made during the first few hours and at intervals over a 6 day period. (Varin et al., 1980; *J. Pharm. Science, 69*, 640 (1980), by permission of American Pharmaceutical Association.)

For example, the analysis of methadone and its metabolites in biological fluids has been reported by Kang and Abbott (1982), using deuterated methadone and metabolites as internal standards. Their method permitted the determination of 20 ng of methadone in 0.5 ml of plasma or saliva. For tracers other than deuterium, the labeled analog should have a difference of 2 or 3 mass units in order for there to be minimum interference from unlabeled molecules of the naturally occurring heavier isotopes, that is, ^{13}C, ^{37}Cl, or ^{81}Br. In addition, the isotopic labels should be incorporated at a chemically "stable" position in the molecule, and the internal standard should be labeled to a high degree of isotopic purity. A comprehensive overview of stable tracers in pharmacological research has been provided by Baillie (1981).

STRUCTURAL ELUCIDATION OF DRUG METABOLITES

Structural elucidation of metabolites by MS/MS is now possible for many drugs based on the assay of a single plasma extract. The triple-quadrupole mass spectrometer (triple QMS) or tandem reversed geometry, double focusing instruments make such an analysis feasible; and inasmuch as metabolites usually contain a substantial portion of the parent drug in their structure, one or more daughter ions should appear in the metabolite spectrum. Perchalski et al. (1982) have reviewed the common operational modes for triple-stage quadrupoles and provided the following detailed procedure for metabolite identification:

1. "Run a normal CI spectrum for the pure parent drug. Chemical ionization is used in preference to electron impact because of increased sensitivity and spectral simplicity."

2. "Obtain daughter spectra for the molecular ion and most representative fragment ions of the pure parent drug. Include the primary CI addition product (e.g., $(M + C_2H_5)^+$ for methane reagent gas)."

3. "Evaluate the summed (all parents) and split (single parent) daughter spectra and choose the most abundant daughter ions for the parent ion experiment."

4. "Run the parent ion experiment on the plasma or urine extract, selecting the daughter ions chosen in step 3."

5. "Evaluate the split parent spectra and choose, for the daughter ion experiment, ions that have masses greater than the mass of the selected daughter but are not contained in the normal mass spectrum of the pure drug."

6. "Run the daughter ion experiment on the plasma or urine extract, selecting the parent ions chosen in step 5."

7. "Determine the structure of each parent from the split daughter spec-

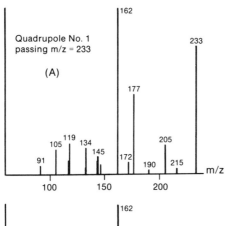

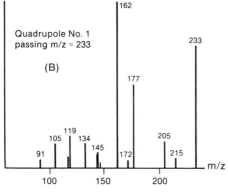

Figure 18.2 Daughter spectra of (A) pure phenobarbital and (B) phenobarbital from plasma extract. (Perchalski et al., 1982; reprinted with permission from *Anal. Chem.*, 54, 1466 (1982), copyright 1982 American Chemical Society.)

trum either by interpretation or by comparison with the daughter spectrum of the pure reference standard.''

Figure 18.2 shows the daughter spectrum of the pure drug—phenobarbital—and phenobarbital from a plasma extract. The daughter spectrum shows the appropriate molecular ion plus the fragmentation ion peaks. DiCorcia et al. (1982) have also reported on a GC method for determining therapeutic levels of phenobarbital. Volumes of 100 μl were sufficient for analysis, and greater than 98% recovery of phenobarbital was realized. The fragmentographic analyses of xanthine and hypoxanthine, and of allopurinol and its major metabolite oxipurinol, have been reported by Lartigue-Mattei et al. (1982). In this work, the serum and urine were monitored of patients who were treated daily with 200 mg of allopurinol for 1 week.

Yost et al. (1983) have pointed out that fragmentographic analysis is a general method whereby daughter identification permits elucidation of the complete metabolite structure. Also, inasmuch as "metabolites that are not present in plasma are not generally bioavailable, this determination gives a good indication of drug byproducts that may have pharmacological activity.''

DRUG RECEPTOR INTERACTION

At present, a salient area of drug research requiring high-resolution mass spectrometry relates to drug-receptor interaction. In order to trigger a biological response, a drug must fulfill the specific requirements of the receptor. One form of drug may be highly effective, whereas a stereoisomer of it may completely lack the potential for interacting. Bentley (1981) has cited the importance of (1) the drug route to the receptor; (2) complementary drug-receptor three-dimensional shapes; and (3) the positions and extent of various forces that are prerequisite to a successful molecular interaction, for example, dipole-dipole, hydrogen bonds, and hydrophobic forces. Direct measurements can now be made on receptors—their number, molecular weight, and relative preferences for different compounds—using both stable isotope and radioactive-labeled drugs. For example, the insulin receptor is a glycoprotein with a molecular weight of 300,000, and the response to insulin appears to be optimum when only about 5% of the receptors are occupied. Bentley (1981) has also commented on the role of Na^+, Ca^{2+}, and Mg^{2+} ions for drug-receptor interactions—relative to "the possible importance of electrical changes on the receptor and its natural rigidity and ability to change its shape." Furthermore, not only has the architecture of receptors been elucidated in some detail, but the continual process of degradation and renewal ("turnover") of receptors has been observed—in response to disease and effects of other drugs and hormones. Properties affecting the action of drugs are now attributed to lipophilicity, electron distribution, and shape (Albert, 1982).

The ideal cytotoxic drug should also be effective against target cells, but produce no undesirable side effects. Although this goal of high potency and selectivity is difficult to achieve, many forms of instrumental analysis, including mass spectrometry, are contributing to a more detailed understanding of the mode of action of various drugs. In the case of certain bacterial and plant toxins, Edwards and Thorpe (1981) have concluded that their toxicity has three distinct stages: (1) binding to the cell surface via a recognition site; (2) penetration into the cytosol; and (3) inhibition and termination of protein synthesis. Thus, they suggest that effective chemotherapy strategy involves the discrete elements of recognition, interaction, and destruction.

PHARMACOKINETICS BY SELECTED ION MONITORING

The quantitation of drugs and/or drug metabolites in a biological matrix by selected ion monitoring (SIM) is a technique in which selected groups of ions and their relative intensities uniquely reflect the identity of the drug from which they are generated. A comprehensive review of this method and its low limit of quantitation (LOQ) for most drugs has been published by Garland and Powell (1981). A known volume or weight of blood, plasma, tissue, etc. containing an unknown amount of the drug is spiked with a known amount of an internal standard that often includes a stable isotope analog (SIA) of the drug. In some instances, a carrier will be added to enhance extraction efficiency, and the SIA can serve as both the internal standard and carrier. After extraction, derivatization, or chromatography, the drug sample is admitted to the ion source of a mass spectrometer, and the instrument is programmed to monitor only ion peaks that can clearly characterize the drug and its internal standard. Drug/standard ion ratio measurements then provide data that can be normalized by appropriate calibrations to yield a quantitative drug assay.

Most drug assays use electron-impact ionization (EI) or chemical ionization (CI), but field desorption and fast-atom bombardment provide alternative modes of ionization for thermally sensitive or highly polar substances. Ideally, the ionization efficiency shoud be relatively insensitive to other chemical compounds that enter the ion source, and either positive or negative ions can be monitored by analog current monitoring or pulse-counting detectors and circuitry. Then, by having either magnetic or quadrupole analyzers focus on only a few selected ion peaks that uniquely characterize the drug, quantitation can be realized at concentrations of 1 ng/ml. In contrast to the "full mass scan," a sensitivity enhancement of 1000 may be realized. However, if the mass spectrometer must quantitate many compounds in the same sample, this sensitivity advantage will be decreased.

Garland and Powell (1981) have reported on both the sensitivity and precision limits that have been realized by selective ion monitoring for the major

TABLE 18.2   Major Drugs Analyzed by Selected Ion Monitoring

Analgesic/antipyretic agents	Anti-inflammatory agents
Anesthetic/antiarrhythmic agents	Antineoplastic agents
Antianxiety agents	Antispsychotic agents
Antibacterial/antimicrobial agents	Hormones
Anticoagulant agents	Prostaglandins
Antidepressants	Psychotomimetic/hallucinogenic agents
Antiepileptic/anticonvulsant agents	Sympathomimetic agents
Sedative hypnotics	Cholinergic/anticholinergic agents
Cardiovascular agents	Stimulants

Source: Garland and Powell (1981).

drug categories listed in Table 18.2, together with the specific mode of sample ionization.

In a further overview of the selected ion methodology, Vanden Heuvel et al. (1983) have discussed its application for (1) determining the pharmacokinetics of beta blockers in humans; (2) measuring residue levels of forced molting agents in the yolk and albumen of chicken eggs; (3) monitoring drugs in animal plasma and tissues; (4) detecting ^{13}C- and ^{15}N-labeled biosynthesis products in human urine; and (5) measuring estradiol levels in biological samples. In a specific administration of the beta-blocking drug timolol to volunteers, SIM was used to determine timolol and [$^{13}C_3$] timolol concentrations in plasma and urine. Figure 18.3 shows the structure of the drug that has been used to control hypertension and angina and to prevent second heart attacks. Figure 18.4 shows the plasma concentrations after a simultaneous oral and intravenous (IV) administration of this drug and its $^{13}C_3$-labeled analog.

The plasma levels were determined by SIM at m/z 86, 89, and 95; and at the 1 ng drug/ml plasma concentration, relative standard deviations of 3% and 2% were reported for the timolol and [$^{13}C_3$] timolol, respectively. More important, the determination of the bioavailability of the drug by simultaneous oral and IV administration and SIM avoids some of the difficulties inherent in the sequential administration of drugs. The authors also point out that while radioactive labeling is an alternative method of assay, it is

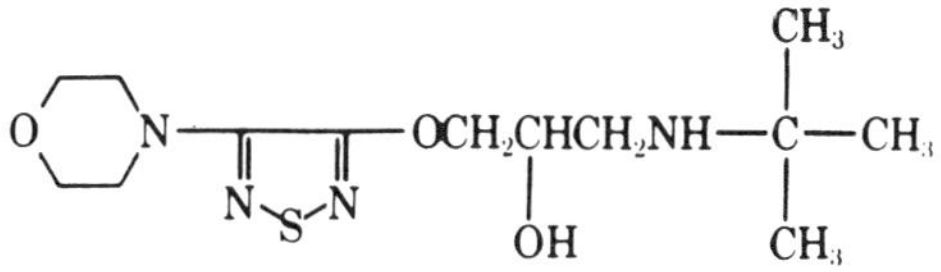

Structure 1. Timolol

Figure 18.3   Structure of the beta-blocking agent timolol (Vanden Heuvel, 1983).

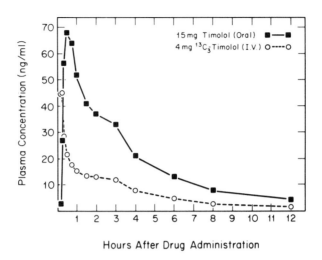

Plasma Concentration–Time Curves After Simultaneous Oral and I.V. Doses

15mg Timolol (Oral) ■——■
4 mg $^{13}C_3$ Timolol (I.V.) ○----○

Hours After Drug Administration

Figure 18.4 Human plasma levels of timolol and [$^{13}C_3$] timolol after simultaneous oral and intravenous doses. (Vanden Heuvel, 1983; reprinted from *J. Chromatogr. Sci.*, *21*, 119 (1983), by permission of Preston Publications, Inc.)

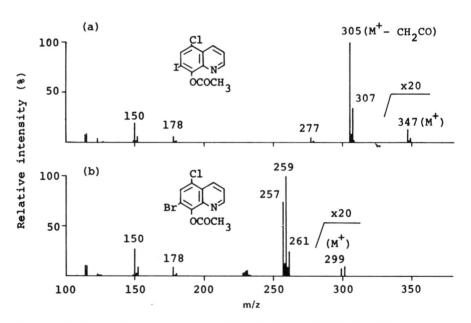

Figure 18.5 Electron-impact mass spectra of (*a*) chinoform and (*b*) 5-chloro-7-bromo-8-quinolinol used as the internal standard. (Kotaki et al., 1983; reprinted from *Chem. Pharm. Bull.*, *31*, 299 (1983), by permission of the Pharmaceutical Society of Japan.)

somewhat less specific in that an additional analytical step is required to insure that the radioactivity was related to the drug, per se, and not to a metabolite. Other workers have also reported the use of capillary column GLC/SIM for assaying timolol, using propranolol as an internal standard (Fourtillon et al., 1981).

A GC/MS procedure using selected ion monitoring has been employed by Kotaki et al. (1983) for the determination of chinoform (CF) and its two metabolites (glucuronide and sulfate) in plasma. Their EI mass spectra of the acetyl derivatives of authentic CF and 5-chloro-7-bromo-8-quinolinol are shown in Figure 18.5. The acetyl derivative of CF yielded a molecular ion (M^+) peak at m/z 347 and a base peak at m/z 305 (M^+-CH$_2$CO). The ion peak at m/z 259 is an isotope peak related to the parent fragment ion at m/z 257 (M^+-CH$_2$CO).

ANALGESICS AND ANESTHETICS

The pharmacokinetic behavior of analgesic and anesthetic drugs has been investigated via chromatography-mass spectrometry by several investigators. Some of these drugs undergo extensive biotransformations after IV or oral administration, so that mass spectrometry has been essential for elucidating the structures of metabolites extracted from the plasma and urine samples of patients. For codeine, a narcotic analgesic and antitussive drug, GLC and mass spectrometric methods have also been used to determine disposition and to quantify therapeutic plasma levels (10–200 ng/ml) after a single dose of codeine phosphate (60 mg). For example, Visser et al. (1983) have reported a rapid and accurate determination of codeine by using straight phase, high-performance liquid chromatography (HPLC) with a detection limit of 5 ng/ml and by using methadone hydrochloride as an internal standard. This method also showed a mean recovery of extraction from plasma greater than 99% and a linear response over a concentration range of 10–160 ng/ml.

The quantitation in serum of dibucaine has been performed at the nanogram level by Shinka et al. (1978), using SIM and deuterium-label dibucaine as an internal standard. The mass spectra of the unlabeled and labeled drug are shown in Figure 18.6. Dibucaine, 2-butoxy-N-(2-diethylaminoethyl) quineoline carboxyamide, is a potent local anesthetic having a long-term effect in comparison with other local anesthetics. Because of its potency and toxicity, it is important to determine dibucaine concentrations in body fluids and in human organs. In the investigation by Shinka et al. (1978), the deuterated compound served as a carrier as well as an internal standard. Also, a plot of m/z ratios of 326/335 and 326/336 for various amounts of drug and a fixed amount (100 ng) of the deuterated specie showed a linear response from 1–100 ng. In measurements of three patients taken 30 min after dibucaine administration, the serum concentrations determined from the 326/

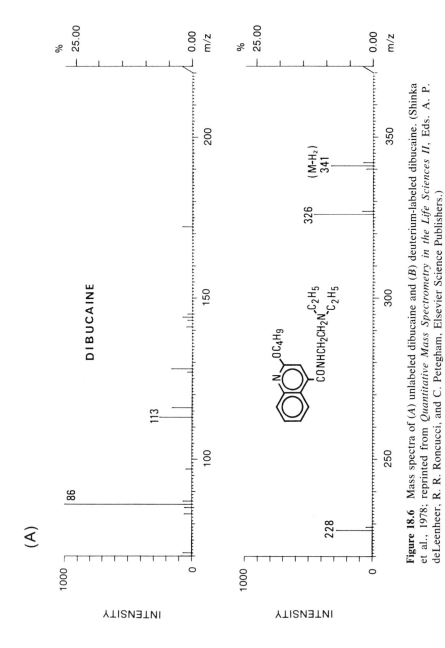

Figure 18.6 Mass spectra of (*A*) unlabeled dibucaine and (*B*) deuterium-labeled dibucaine. (Shinka et al., 1978; reprinted from *Quantitative Mass Spectrometry in the Life Sciences II*, Eds. A. P. deLeenheer, R. R. Roncucci, and C. Petegham, Elsevier Science Publishers.)

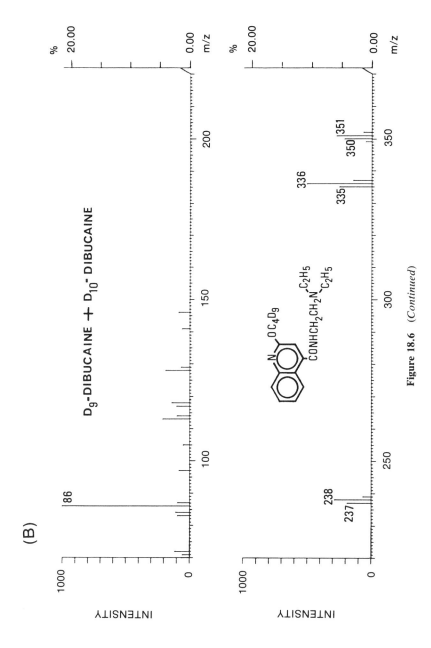

Figure 18.6 (*Continued*)

335 ratio were 0.19, 0.15, and 0.21, µg/ml and from the 326/336 ratio, 0.18, 0.15, and 0.22 µg/ml, respectively.

A GC/MS method for measuring midazolam (MDZ), an anesthesia induction agent and hypnotic, has been described by Rubio et al. (1982b). Methane electron-capture negative ionization was used, and a limit of quantitation of 1 ng/ml was obtained for all three compounds, namely, drug, metabolites, and the internal standard.

CHARACTERIZATION OF ANTIBIOTICS

The biosynthesis of antibiotics has been studied by many techniques in order to determine how micro-organisms can effect the conversion of primary metabolites into antibiotic molecules. A general review of this field has been provided by Rinehart (1983), who has also discussed the applicability of both mass spectrometry and NMR to the biosynthesis of neomycin, ribostamycin, hybomycin, erythromycin, and molecules containing the so-called "m-C$_7$N" units. In this series of studies, a ^{13}C label was used to follow the incorporation of the carbons of glucose and glucosamine into neomycin B$_7$, and NMR was used to observe the chemical shifts. As noted in Table 18.3, ^{13}C has a naturally occurring abundance of ~1%. This nuclide also has a nuclear spin of $\frac{1}{2}$ and a reasonably high relative sensitivity. Deuterium was employed in the study of the biosynthesis of ribostamycin, which is an antibiotic related to neomycin. ^{15}N and ^{18}O were also used as labels; and although there was only a 0.2 ppm frequency difference (chemical shift) for carbon atoms with ^{18}O attached relative to carbons coupled to ^{16}O, this difference was observable. Isotopic labeling with ^{13}C and deuterium was also employed in this investigation, using mass spectrometry with fast-atom bombardment to ion-

TABLE 18.3   Selected Isotopes and Nuclear Properties

Nuclide	Natural Abundance (%)	NMR Frequency in MHz for 50 kG Magnetic Field	Relative Sensitivity[a]	Spin (I)
^{1}H	99.985	212.885	1.000	1/2
^{2}H	1.5×10^{-2}	32.680	0.409	1
^{3}H[a]	—	227.070	1.07	1/2
^{13}C	1.11	53.525	0.251	1/2
^{14}N	99.634	15.380	0.193	1
^{15}N	0.366	21.575	0.101	1/2
^{17}O	3.7×10^{-2}	28.860	1.58	5/2
^{19}F	100	200.275	0.941	1/2
^{31}P	100	86.175	0.405	1/2

Source: Adapted from Rinehart (1981).

[a] For equal number of nuclei at constant frequency.

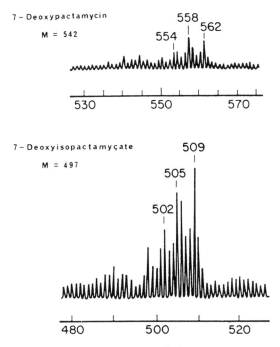

Figure 18.7 Fast-atom bombardment spectrum of $^{13}C^2H_3$-labeled 7-deoxyisopactamycate. (Rinehart, 1983; reprinted with permission from *Bio/Technology*, copyright 1984.)

ize the molecules. Figure 18.7 shows a spectrum of 7-deoxyisopactamycate having a normal molecular weight of 497, but having a molecular weight of 508 for the fully labeled compound (i.e., a total of eight deuterium and three ^{13}C atoms). With fast-atom bombardment, the compound appears as an $M + H$ ion at $m/z = 509$.

The characterization of other antibiotics has also utilized fission fragment ionization as well as fast-atom bombardment. For example, Chait et al. (1982) generated positive and negative ^{252}Cf fission fragment ionization spectra for a natural and synthetic sample of alamethicin I. The positive spectra are identical, and the negative spectra are almost identical except for a trace impurity that was evident in the natural sample. In fact, the primary analytical goal of this work was to demonstrate that fission fragment ionization spectrometry could be used to compare separated and purified alamethicin I from natural sources (produced by the fungus *Trichoderma viride*) to the synthesized peptide. Specifically, by comparing ion peaks at m/z values of 1963 and 1987, these investigators showed the common amino acid identity of the natural and synthetic compound. In other studies, mass spectra of penicillins have been obtained by using "in-beam" electron ionization (Ohashi et al., 1983) and by MS/MS techniques.

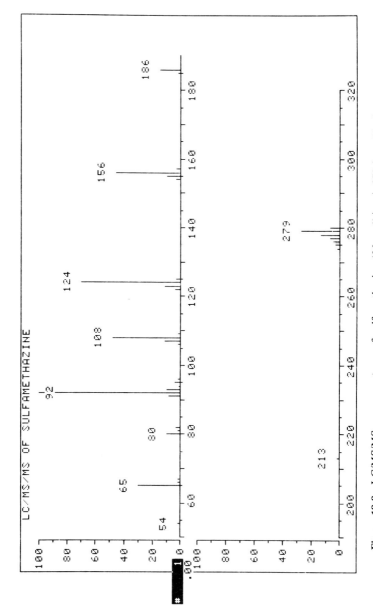

Figure 18.8 LC/MS/MS mass spectrum of sulfamethazine $(M + 1)^+$ m/z 279 ion. (Henion et al., 1982; reprinted with permission from *Anal. Chem., 54*, 451 (1982), copyright 1982 American Chemical Society.)

Sulfa Drugs by LC/MS/MS

Although valid claims are made for MS/MS systems that can identify drugs from "raw" samples, a fast clean-up can often (1) protect the mass spectrometric instrumentation; (2) eliminate the majority of the matrix material; and (3) provide improved mass spectral data. Indeed, some type of chromatographic separation will usually enhance the probability of successful drug identification when exceedingly low concentrations are involved. The LC/MS/MS technique thus provides collision-induced dissociation (CID) spectra for all components of interest. The analysis of sulfa drugs in biological fluids by LC/MS/MS is an example (Henion et al., 1982). In this work, an atmospheric pressure CI, triple QMS was used in both (1) a full mass scan; and (2) SIM modes.

Specifically, a series of sulfa drugs were analyzed that included sulfamethazine (SM), sulfadiazine (SD), and sulfadimethoxine (SDM). In the triple-quadrupole system used, nitrogen was the collision gas with an effective "target" thickness of $\sim 2 \times 10^{14}$ molecules/cm^2. Also, cryogenic pumping was employed to insure that few collisions occurred in quadrupoles 1 and 3, so that the fragmentation region was well defined. The CID mass spectrum of SM is shown in Figure 18.8, and it was produced by focusing the $(M + 1)^+$ ion (m/z 279) in the first quadrupole Q_1, performing CID in Q_2, and separating the resulting daughter ions with Q_3. These ions appear at m/z 186, 156, 124, 108, 92, and 65 and provide a high degree of structural information. Henion et al. (1982) state that the m/z 186 ion appears to result from the loss of aniline from the $(M + 1)^+$ ion, and that the m/z 124 is diagnostic for the 4,6-dimethyl-2-pyrimidyl portion of SM. This LC/MS/MS technique has yielded abundant fragment ions for other sulfa drugs, and the authors claim a "reduction of chemical noise" for the lower portion of the mass spectrum.

ANTIARRHYTHMIC, ANTIHYPERTENSIVE, AND ANTIDEPRESSANT DRUGS

Among the antiarrhythmic drugs, both aprindine hydrochloride [N-phenyl-N-(3-diethylaminopropyl)-2-indanylamine hydrochloride] and propafenone [2'-(3-propylamino-2-hydroxy-propoxy)-3-phenyl-propiophenone]   have been assayed by GC/MS. The first drug has been used for several years in Europe and is being further evaluated in Japan; and because it has a narrow therapeutic index (Zipes and Troup, 1978), plasma level monitoring is important. The second is a new drug that has been reported to be effective in recurrent supraventricular and ventricular tachycardias, and that undergoes extensive biotransformation (Marchesini et al., 1983). It is reported that less than 1% of propafenone is excreted in urine after its oral administration, suggesting that one or more metabolites may relate to the antiarrhythmic

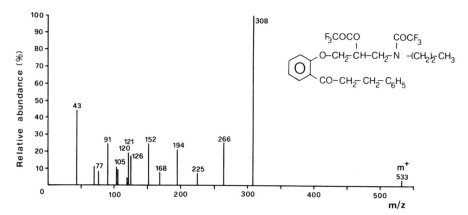

Figure 18.9 Mass spectrum of propafenone-TFA and its structural formula. (Marchesini et al., 1983; reprinted from *J. Chromatogr. Biomed. Appl., 278*, 173 (1983), by permission of Elsevier Science Publishers.)

action of the drug. Hence, Marchesini et al. (1983) have utilized gas chromatography and SIM by electron-impact mass spectrometry to elucidate the nature of metabolites of propafenone, whose mass spectrum is shown in Figure 18.9. The low relative abundance of the propafenone-trifluoroacetyl (TFA) molecular ion at m/z 533 is consistent with the formation of a di-(trifluoroacetyl) derivative, according to the authors; and the spectrum shows fragment ions characteristic of the spectra of TFA beta blockers, that is, ion peaks of high abundance at m/z 308, 266, and 43 and lower peaks at m/z 194, 168, 152, and 126. Other detailed mass spectral studies are consistent with other experimental findings that propafenone undergoes extensive metabolic transformations, and that a hydroxylate derivative is a primary metabolite.

A chromatograph-mass fragmentographic method has also been reported by Clemans et al. (1977) for the simultaneous assay of 17-monochloroacetyl-ajmaline (MCAA) and its hydrolysis product, ajmaline. The mass spectral analysis of derivatives showed that MCAA had formed the mono-trimethylsilyl derivative (m/z 474.2), while ajmaline and estradiol had both converted to the *bis*-trimethylsilyl derivatives m/z 470.30 and 416.2, respectively. The structural formula and a typical plot of specific ion intensity vs. time is shown in Figure 18.10.

[252]Cf plasma desorption has been used by Jungclas et al. (1982) to investigate the bioavailability and kinetics of other antiarrhythmic drugs, such as quinidine and verapamil. One advantage of this specialized form of spectrometry is that an analysis can be made nondestructively by evaporating less than 10^{-8} of the sample molecules.

A combined gas chromatographic-mass spectrometric method has also

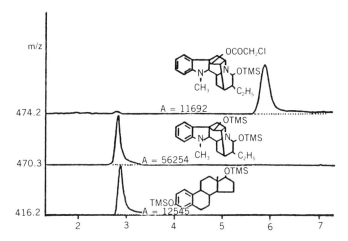

Figure 18.10   Typical GC/MS computer output showing retention times, areas under specific ion curves, and structural formulae of trimethylsilyl (TMS) derivatives (Clemans et al., 1977).

been reported for the simultaneous measurement of the antihypertensive drug captopril and several metabolites: captopril disulfide dimer and S-methyl captopril sulfone (Drummer et al., 1984). The method uses conventional gas chromatography for separating the derivatives in one temperature-programmed run, with quantitation by selected ion monitoring. The drug 1-[(2D)-3-mercapto-2-methyl-1-oxopropyl]-L-proline is a thiol-containing dipeptide based on proline (Ondetti et al., 1977). This GC/MS method utilized a 2 m × 2 mm ID-packed glass column with helium as the carrier gas and both EI and CI mass spectrometry. Sample preparation and derivatization included the use of hexafluorisopropanol (HFI) esters. An EI mass spectrum of the S-methyl captopril HFI ester is shown in Figure 18.11a, with significant ion peaks at m/z 381 (M^+), m/z 366 $(M\text{-}CH_3)^+$, m/z 334 $(M\text{-}SCH_3)^+$, and m/z 264 (proline moiety)$^+$. The CI mass spectrum yielded the dominant $(MH)^+$ ion peak at m/z 382. A spectrum of S-methyl captopril sulfone HFI ester is shown in Figure 18.11b. This GC/MS method demonstrated both high sensitivity and reproducibility for this drug in urine or plasma. Specifically, captopril, S-methyl captopril, and the disulfide dimer of captopril were detected at concentrations of 1, 10, and 25 ng/ml, respectively. Measurements were also made of the excretion of free captopril in rats and humans that were reasonably consistent with another GC/MS study showing that ~21% of a 10 mg/kg dose of captopril is excreted as unchanged captopril in rats over a 24 h period (Matsuki et al., 1982).

A sensitive and highly quantitative method has been developed by Idzu et al. (1982) for monitoring glyceryl trinitrate (GTN) in human plasma. This drug is used to decrease peripheral vascular resistance after acute myocardial infarction, and it has recently been used as a hypotensive drug during

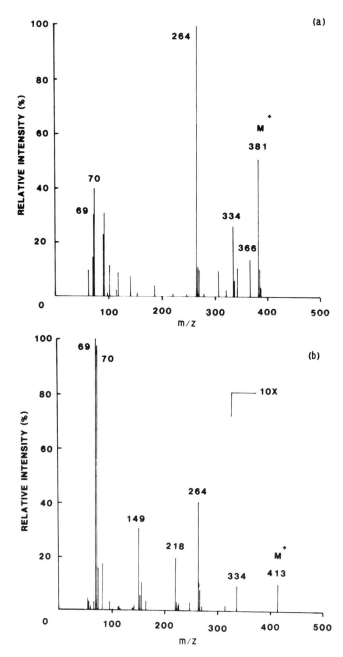

Figure 18.11 Electron-impact mass spectrum of (*a*) S-methyl captopril HFI ester and (*b*) S-methyl captopril sulfone HFI ester. (Drummer et al., 1984; reprinted from *J. Chromatogr., 305*, 83 (1984), by permission of Elsevier Science Publishers.)

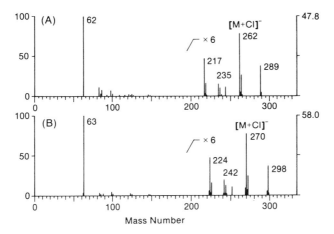

Figure 18.12   NICI mass spectra of glyceryl trinitrate (A) and its [$^2H_5$, $^{15}N_3$] variant (B) using dichloromethane as a reagent gas. (Idzu et al., 1980; reprinted from *J. Chromatogr. Biomed. Appl., 229*, 327 (1982), by permission of Elsevier Science Publishers.)

general anesthesia. Using SIM and negative ion chemical ionization (NICI), a quantitation limit was attained of 0.1 ng/ml for human plasma. Figure 18.12 shows the NICI spectrum of glyceryl trinitrate (Fig. 18.12*a*) and its [$^2H_5$, $^{15}N_3$] variant (Fig. 18.12*b*) using dichloromethane as a reagent gas.

Indalpine, or 4-[2-(3-indolyl)ethyl] piperidine, has been shown to be an effective agent in the treatment of chronically ill depressed patients. The EI mass spectrum of this drug is shown in Figure 18.13, and Jozefczak et al. (1982) have developed an HPLC method for determining indalpine and its major metabolite in human plasma.

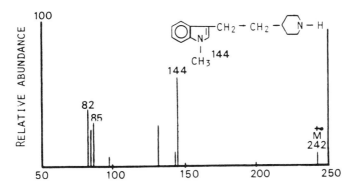

Figure 18.13   Electron-impact mass spectrum of 4-[2-(1-methyl-3-indolyl)ethyl] piperidine. (Jozefczak et al., 1982; reprinted from *J. Chromatogr. Biomed. Appl., 230*, 87 (1982), by permission of Elsevier Science Publishers.)

CHELATING AND ANTITUMOR AGENTS

Mass spectrometry and the isotopic dilution technique can also be used in studying the efficacy of chelation therapy in cases of toxic metals. Cantilena and Klaasen (1982) have studied the effect of several chelators in the distribution and excretion of cadmium, using the ^{109}Cd radioisotope in mice to observe decreased Cd concentrations in the kidney, spleen, and blood. These decreases were accompanied by an increased urinary and fecal elimination and by a redistribution of Cd to the testes, muscle, and brain. However, stable isotopes provide a viable option for detecting the effectiveness of chelating agents in human subjects, as very small changes in toxic metal concentrations can be observed by means of precision isotope ratio measurements. In the case of both Cd and Pb, surface-ionization mass spectrometry is applicable. Also, because of the very low ionization potential of lithium (5.39 eV), surface-ionization mass spectrometry is a very sensitive technique for monitoring this element. At toxic levels in rats (2 × 2.5 mmol/ kg/day for 7 days), Li has produced overt signs of lithium intoxication and reduced both orthodromic and antidromic nerve conduction velocities. This reduction appears to be inversely related to lithium levels in serum and red blood cells (Ebara et al., 1981).

In an investigation of antitumor agents, Marunaka and Umeno (1982) have used GC/mass fragmentography to determine active substances in biological materials. Their detection limit for 5-fluorouracil was reported to be 1 ng/ml of plasma and 1–5 ng/g wet weight of visceral tissues.

Shiba and Weinkam (1982) have described a direct-sample insertion mass spectrometric method for analyzing both the anticancer drug procarbazine and eight known metabolites—including those known to have cytotoxic activity. These investigators followed the procarbazine appearance and metabolite appearance in rat plasma, and they also monitored active metabolites in the plasma of a patient who received orally administered procarbazine during a 14 day course of therapy.

COMPARISON OF DRUG BLOOD LEVELS (EI/CI)

In bioavailability studies, comparisons have been made of blood levels monitored by different mass spectrometric techniques. In one such investigation, Cohen et al. (1982) determined captopril in both human blood and urine by (1) EI, GLC-selected ion monitoring; and (2) direct CI. The GLC/MS system was programmed for EI from 69–502 amu, and the ion source was controlled at a constant temperature. The positive CI GLC/MS used methane as the GLC carrier gas. Otherwise, the chromatographic method was identical to the EI method. The subjects were administered captopril orally either as a compression tablet or a 100 mg solution. Table 18.4 shows the captopril blood level of one subject after administration of the drug.

TABLE 18.4 Comparison of Captopril Blood Levels Determined by Electron-Impact (EI) and Chemical Ionization (CI) GLC/MS

Time (h)	Blood Level (ng/ml)	
	EI	CI
0.0	2.3	7.4
0.25	21.1	26.1
0.50	272.0	265.0
0.75	432.0	437.0
1.0	375.0	378.0
1.5	186.0	176.0
2.0	104.0	98.3
3.0	31.4	28.4
4.0	21.1	14.8
6.0	7.8	16.6
8.0	1.8	1.6
12.0	0.8	2.2

Source: Cohen et al. (1982).

PERCUTANEOUS ABSORPTION OF DRUGS

Investigations of the percutaneous absorption of drugs have been conducted by dermatologists and pharmaceutical scientists to assess the efficacy and safety of externally applied medications. Because the amount of penetrating compounds is small, most studies have used radioactive tracers on experimental animals. However, an extrapolation of the absorbability of drugs from animals to humans is not justified, because the structure and physiology of the skin in humans differs from that of animals (Higuchi, 1960). Hence, stable tracers are now being increasingly used with high-sensitivity mass spectrometry to evaluate absorption, metabolism, and excretion of hydrophilic ointments and other drugs. The recent study at the Tokyo College of Pharmacy employing the drug paeonal (I), [2-hydroxy-4-methoxyacetophenone] is representative of research employing stable isotope-labeled compounds (Mimura and Baba, 1983).

An ointment containing 5% (I) was topically applied on the forearms of healthy male volunteers for 6, 12, and 24 h. The amounts of metabolites (2,5-dihydroxy-4-methoxyacetophenone, II, resacetophenone III, and unchanged substrate I) were determined by selected ion monitoring using I [methoxy-d_3], II [methoxy-d_3], and III [acetyl-$^{13}C_2$] as internal standards. After 6 h applications, the amounts of I, II, and III in 0–24 h urine were determined to be 3.35%, 16.28%, and 5.17% of the applied dose, respectively. Overall, the data correlated well with the Higuchi (1960) equation relating the amount of drug absorbed to the urinary excretion parameters, that is,

$$A_a = (K_{el})^{-1} \times (dA_e/dt) + A_e$$

where A_a is the amount of drug absorbed, K_{el} is the elimination rate constant, dA_e/dt is the urinary excretion rate, and A_e is the amount of drug excreted in the urine.

This work also confirmed the general rule that a drug applied topically is easily reserved in skin tissue, and that the absorption of a drug continues for a longer period than after oral administration. Urinary excretion of I metabolites continued for 6 days from the time of application, although the total amount of metabolites excreted in the first 0–24 h period was 89% of that excreted in 0–144 h.

PHARMACEUTICAL PACKAGING MATERIALS

Because of the nature and end-use of the product, pharmaceuticals must be subjected to special analyses—including their packaging materials. GC/MS is now a preferred analytical method for assaying elastomeric and other materials relative to:

1. Insuring product uniformity.
2. Identifying trace contaminants.
3. Characterizing surfaces affected by product curing, oxidation, decomposition, or drug reaction products.
4. Investigations relating to drug/packaging compatibility.
5. Quality control of seals.

An overview of various analytical methods for packaging materials has been reported by Kiang (1981) using (1) direct solid-probe insertion mass spectrometry; (2) GC/MS; and (3) pyrolysis GC/MS. The direct solid-probe insertion is useful for microscopic-size solid samples; the solids are inserted into the ion source, which is maintained at a high temperature and low pressure. Relatively nonvolatile compounds are amenable to assay by this technique. For example, elemental sulfur has been identified on the closures used to package thiamin or vitamin B_1. Conventional GC/MS is applicable to many packaging materials, the monitoring of manufacturing and curing processes, and the identification of minor contaminants that might affect the stability, compatibility, or purity of a drug. In the pyrolysis GC/MS of rubbers and other high polymers, a multistep assay has been used (Hu, 1977).

The sample is first heated to ~270°C, where volatile components and accelerators are analyzed. The temperature is then raised to 900–1000°C to pyrolyze the polymer. Finally, the remaining nonvolatile organics are analyzed by x-ray and atomic absorption.

Some drugs also warrant strict specifications of packaging materials relative to vapor pressure, adsorption, and permeability, and all of these parameters can be quantified by mass spectral measurements.

REFERENCES

Albert, A., *J. Med. Chem., 25*, No. 1, 2 (1982).

Anderegg, R. J., *J. Chromatogr., 275*, 154 (1983).

Baillie, T. A., *Pharmacol. Rev., 33*, No. 2, 81 (1981).

Bentley, P. J., *Elements of Pharmacology*, Cambridge University Press, London (1981), pp. 44–71.

Cantilena, L. R., Jr., and C. D. Klaasen, *Toxicol. Appl. Pharmacol., 66*, 361 (1982).

Chait, B. T., B. F. Gisin, and F. H. Field, *J. Am. Chem. Soc., 104*, 5157 (1982).

Clemans, S. D., C. Davison, R. F. Koss, B. Dorrbecker, P. E. O'Melia, R. W. Ross, Jr., and J. Edelson, *Arzneim-Forsch/Drug Res., 27*, No. 6, 1128 (1977).

Cohen, A. I., R. G. Devlin, E. Ivashkiv, P. T. Funke, and T. McCormick, *J. Pharm. Sci., 71*, No. 11, 1251 (1982).

Demain, A. L., *Science, 214*, 987 (1981).

DiCorcia, A., L. Ripani, and R. Samperi, *J. Chromatogr. Biomed. Appl., 229*, 365 (1982).

Drummer, O. H., B. Jarrott, and W. J. Louis, *J. Chromatogr., 305*, 83 (1984).

Ebara, T., K. Nakayama, S. Otsuki, and S. Watanabe, *Intr. Pharmacopsychiat., 16*, 129 (1981).

Edwards, D. C., and P. T. Thorpe, *Trends in Biochem. Sci., 6*, 313 (1981).

Fourtillon, J. B., M. A. Lefebvre, J. Girault, and P. H. Courtois, *J. Pharm. Sci., 70*, 573 (1981).

Garland, W. A., and M. L. Powell, *J. Chromatogr. Sci., 19*, 392 (1981).

Gregg, C. T., J. Y. Hutson, J. R. Prine, D. G. Ott, and J. E. Furchner, *Life Sci., 13*, 775 (1973).

Halliday, D., and I. M. Lockhart, *Progr. Med. Chem., 15*, 2 (1978).

Henion, J. D., B. A. Thomson, and P. H. Dawson, *Anal. Chem., 54*, 451 (1982).

Higuchi, T., *J. Soc. Cosmet. Chem., 11*, 85 (1960).

Hu, J. C., *Anal. Chem., 49*, 537 (1977).

Idzu, G., M. Ishibashi, H. Miyazaki, and K. Yamamoto, *J. Chromatogr. Biomed. Appl., 229*, 327 (1982).

Jozefczak, C., N. Ktorza, and A. Uzan, *J. Chromatogr. Biomed. Appl., 230*, 87 (1982).

Jungclas, H., H. Danigel, and L. Schmidt, *Org. Mass Spectrom., 17*, No. 2, 86 (1982).

Kang, G., II, and F. S. Abbott, *J. Chromatogr. Biomed. Appl., 231*, 311 (1982).

Kiang, P. H., *J. Parent. Sci. Technol., 35*, 152 (1981).

Kotaki, H., Y. Yamamura, Y. Tanimura, Y. Saitoh, F. Nakagawa, and Z. Tamura, *Chem. Pharm. Bull., 31*, No. 1, 299 (1983).

Lartigue-Mattei, C., J. L. Chabard, H. Bargnoux, J. Petit, and J. A. Berger, *J. Chromatogr. Biomed. Appl., 229*, 211 (1982).

Lockhart, I. M., in *Isotopes: Essential Chemistry and Applications*, Special Publ. No. 35, eds. J. A. Elvidge, and J. R. Jones, The Chemical Society, London (1979), p. 25.

Marchesini, B., S. Boschi, and C. Berti, *J. Chromatogr. Biomed. Appl., 278*, 173 (1983).

Marunaka, T., and Y. Umeno, *Chem. Pharm. Bull., 30*, No. 5, 1868 (1982).

Matsuki, Y., T. Ito, K. Fukuhara, T. Nakamura, M. Kimura, and H. Ono, *J. Chromatogr., 239*, 585 (1982).

Mimura, K., and S. Baba, *Chem. Pharm. Bull., 31*, No. 10, 3698 (1983).

Ohashi, M., R. P. Barron, and W. R. Benson, *J. Pharm. Sci., 72*, No. 5 (1983).

Ondetti, M. A., B. Rubin, and D. W. Cushman, *Science, 196*, 441 (1977).

Perchalski, R. J., R. A. Yost, and B. J. Wilder, *Anal. Chem., 54*, 1466 (1982).

Rinehart, K. L., Jr., *Pure and Appl. Chem., 53*, 795 (1981).

Rinehart, K. L., Jr., *Bio/Technology, 1*, No. 6, 495 (1983).

Rubio, F., F. DeGrazia, B. J. Miwa, and W. A. Garland, *J. Chromatogr. Biomed. Appl., 233*, 149 (1982a).

Rubio, F., B. J. Miwa, and W. A. Garland, *J. Chromatogr. Biomed. Appl., 233*, 157 (1982b).

Shiba, D. A., and R. J. Weinkam, *J. Chromatogr. Biomed. Appl., 229*, 397 (1982).

Shinka, T., T. Kuhara, and I. Matsumoto, in *Quantitative Mass Spectrometry in the Life Sciences II*, eds. A. Leenheer, R. Roncucci, and C. Van Peteghem, Elsevier, Amsterdam (1978), pp. 315–321.

Unger, S. E., and B. M. Warrack, *Spectroscopy, 1*, No. 3, 33 (1986).

Vanden Heuvel, W. J. A., J. R. Carlin, and R. W. Walker, *J. Chromatogr. Sci., 21*, 119 (1983).

Varin, F., C. Marchand, P. Larochelle, and K. K. Midha, *J. Pharm. Sci., 69*, No. 6, 640 (1980).

Visser, J., G. Grasmeijer, and F. Moolenaar, *J. Chromatogr. Biomed. Appl., 274*, 372 (1983).

Yost, R. A., H. O. Brotherton, and R. J. Perchalski, *Int. J. Mass Spectrom. Ion Phys., 48*, 77 (1983).

Zipes, D. P., and F. P. J. Troup, *Am. J. Cardiol., 41*, 1005 (1978).

CHAPTER NINETEEN

TOXICOLOGY AND FORENSIC SCIENCE

Forensic science encompasses several technical disciplines that are applied to matters of law. Forensic toxicology is principally concerned with the examination of body fluids, tissues, and other biological material in order to detect poisons and to provide quantitative analyses that may relate to the cause and manner of death. Contemporary forensic toxicology also includes blood alcohol determinations, urine drug screening, etc. relative to the enforcement of certain specific laws (Saferstein, 1982). Forensic chemistry is principally the application of analytical chemistry within a legal context. Subspecialties include arson chemistry, explosives, trace evidence (e.g., glass, paint, etc.), and drugs. "The largest single chromatograph-mass spectrometer user group in forensic science is forensic drug chemists. They analyze all sorts of tablets, capsules, powders, liquids, vegetable matter, and drug paraphernalia for the presence of controlled substances and the amount present. They also characterize diluents and excipients in the illicit drug trade" (Campbell, 1984).

In an even broader context, toxicology has been defined as the study of the adverse effects of chemical substances in living systems. In this sense, toxicology is neither a new discipline nor is it limited to humans. Some poisons were well known in ancient Greece, and many drugs have been associated with various religions and cultures. However, chemical methods for quantifying poisons were not in evidence until the nineteenth century, when arsenic was first extracted from human organs and techniques began to be developed for the post-mortem assay of alkaloids and metals. The first practical book, *The Detection of Poisons and Powerful Drugs*, did not appear until 1905 (Finkle, 1982). At present, advances in instrumentation are greatly expanding the scope and number of analyses that can be performed by forensic laboratories, and greater weight is currently being given to purely technical evidence in the courts. Furthermore, among all types of instruments, mass spectrometry may prove to be the most indispensable. It pro-

693

vides isotopic data as well as detailed information for "fingerprinting" molecules, and at a sensitivity that usually exceeds that of other techniques.

QUANTITATION OF DRUGS IN BLOOD

For quantifying drugs and their metabolites in the blood of persons suspected of intoxication or poisoning, an initial screening may be performed on plasma extracts via chromatography alone, ultraviolet (UV) spectrophotometry, or other methods. However, when these analyses provide only limited information, chromatography-mass spectrometry with a chemical ionization source is used for positive identification and quantification. In a special publication issued by the National Institute on Drug Abuse, Foltz et al. (1980) have provided detailed procedures for using GC/MS in analyzing several important drugs, namely,

1. Phencyclidine (PCP).
2. Methaqualone.
3. Methadone.
4. Δ^9-Tetrahydrocannabinol (THC) and two of its metabolites.
5. Cocaine and its major metabolite, Benzoylecgonine.
6. Morphine.
7. Diazepam and its major metabolite, N-Desmethyl-diazepam.
8. Amphetamine.
9. Methamphetamine.
10. 2,5-Dimethoxy-4-Methylamphetamine (DOM).
11. Mescaline.

The electron-impact (EI) and chemical ionization (CI) mass spectra of one of these, methaqualone, is shown in Figure 19.1. For maximum sensitivity, CI with methane and ammonia as reagent gases have been recommended. At high ion source temperatures (> 160°C) under methane CI conditions, fragment ions at m/z 118 become more abundant; and with deuterium labeling, an ion peak will appear at m/z 122. It has thus been suggested that ions at m/z 118 and 122 can be monitored in addition to the protonated molecular ions at m/z 251 and 255. Confirmational data for this drug is thereby enhanced, and the same technique applies to other substances. Methaqualone was introduced in Europe in 1956 and in the United States in 1965 "as a nonbarbiturate, nonaddictive sedative hypnotic," but it has been widely abused and it is particularly toxic when consumed with alcohol.

Approximately 100 drugs have been analyzed by Cailleux et al. (1981) using EI and CI. In this series of investigations, the different drugs were added to drug-free plasma in the range corresponding to their toxic con-

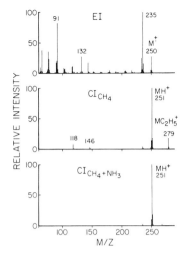

Figure 19.1 Electron-impact and chemical ionization spectra of methaqualone (Foltz et al., 1980).

centrations: (1) antidepressants, 0.1–1 mg/l; (2) benzodiazepine drugs, 0.5–5 mg/l; (3) neuroleptic and cardiovascular drugs, 1–10 mg/l; and (4) analgesic and antipyretic drugs, 5–10 mg/l. Other blood samples were taken from patients suspected of drug intoxication and admitted to the Anti-Poison Center at Angers, France. Most of the intoxications were attributed to benzodiazepine absorption ($n = 210$) followed by barbiturates ($n = 110$) and antidepressants ($n = 50$). These investigators emphasized that blood quantification is essential for the treatment of acute drug intoxication, and that urine or gastric juice levels are not adequate for estimating the severity of intoxication.

In some instances, it is also important for the forensic drug chemist to measure both active components and diluents in order to ascertain whether the drugs originated from a common source, for example, the opiate and sugar content of illicit heroin preparations (White et al., 1983).

ILLEGAL DRUGS IN SPORTS

Currently, mass spectrometry is being widely employed to monitor drug use in athletic events, either for humans or animals. The statement has even been made that "while GLC and radioimmuno methods are used for screening purposes, the unequivocal identification of the ingested drug requires GLC-MS analysis, and no international meeting (sporting event) can be properly organized which does not have at its disposal appropriate GLC-mass spectrometry facilities" (Frigerio, 1980). At the 1984 Olympic Games, three mass spectrometers and several gas chromatographs were used by the UCLA/Olympic Analytical Laboratory for checking on some 1500 athletes. Lists of prohibited drugs now include anabolic steroids, narcotic analgesics,

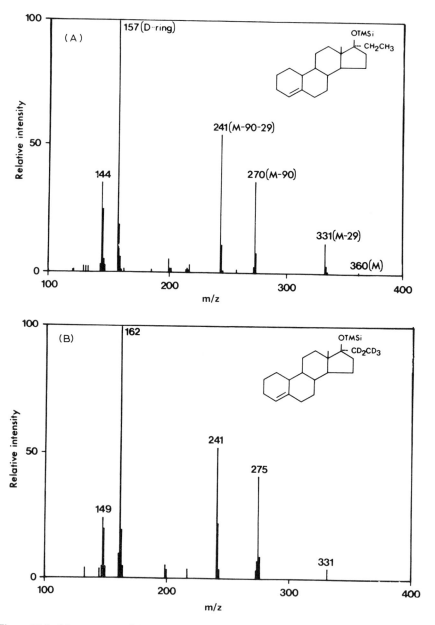

Figure 19.2 Mass spectra of the TMS ethers of (A) unlabeled and (B) deuterium-labeled ethylestrenol. (Björkhem et al., 1982; reprinted from *J. Chromatogr. Biomed. Appl., 232*, 154 (1982), by permission of Elsevier Science Publishers.)

696

psychomotor stimulators, beta blockers, etc. However, in view of the serious legal and other consequences involved, all analytical methods of detection must be accurate and highly specific.

Typical of the GC/MS approach is the assay of ethylestrenol in urine by isotope dilution mass spectrometry, as reported by Björkhem et al. (1982), in which unlabeled and deuterium-labeled samples were compared. In this investigation, urine samples were obtained from healthy male and female subjects. In the control group, concentrations of ethylestrenol never exceeded 1 ng/ml of urine. In the group that was provided with a 6 mg oral administration of the drug, the concentration of ethylestrenol in the first 24 h portion of the urine was 16 ng/ml. The mass spectra of the trimethylsilyl (TMS) derivatives of unlabeled and deuterium-labeled ethylestrenol are shown in Figures 19.2a and 19.2b, respectively. In the unlabeled spectrum, the peak at m/z 157 is a base peak derived from the D ring of the molecule. Peaks at m/z 241 and 270 are attributed, respectively, to (1) the loss of the TMS group and the ethyl group from the molecular ion, and (2) the loss of the TMS group from the molecular ion. In the mass spectrum of the deuterium-labeled compound, the corresponding peaks are seen to be m/z 162, 241, and 275. For the quantitative urine assay, the ions at m/z 270 and 275 were chosen, and 100 ng of deuterium-labeled ethylestrenol was added to the urine sample as an internal standard. However, the authors conclude that the ion peak at m/z 331 is also a "diagnostic" peak for the presence of ethylestrenol; and due to the extensive metabolism of this steroid, the detection of its illegal use "should preferably be directed towards the metabolites rather than towards unmetabolized ethylestrenol."

TANDEM MASS SPECTROMETRY OF DRUGS

A growing number of toxicology and forensic investigations are now using tandem mass spectrometry for drug detection and confirmation. Liquid chromatography-tandem mass spectrometry (LC/MS/MS) has been applied to the detection of administered drugs in equine plasma and urine extracts (Henion et al., 1982). Thin-layer chromatography-tandem spectrometry (TLC/MS/MS) has also been used for on-line analysis in connection with collision-induced dissociation (CID). Advantages and limitations of various separation and tandem spectrometric methods have been reported by (Henion et al., 1983) for caffeine and nicotine in human urine and for butorphanol, betamethasone, and clenbuterol in equine urine. Included in this review are (1) spectra produced by direct insertion probes and atmospheric pressure spectrometry, and (2) CID spectra that are essential for "unequivocal" identification. According to the authors, the choice of investigated drugs was dictated by "a combination of their relative importance in horse racing and the degree of difficulty in their detection and confirmation by conventional methods."

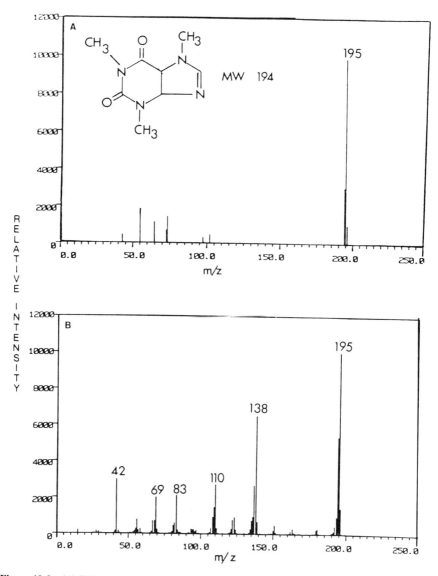

Figure 19.3 (A) Full-scan, direct insertion probe API spectrum of standard caffeine. (B) API-CID spectrum of standard $(M + 1)^+$ ion, $m/z = 195$. (Henion et al., 1983; reprinted from J. Chromatogr., 271, 107 (1983), by permission of Elsevier Science Publishers.)

Caffeine was extracted from the urine of a human volunteer who was a moderate coffee drinker and cigarette smoker, and other drugs were isolated from equine urine after administration of the drug. Authentic standards of each drug were obtained from commercial sources. All samples were introduced into the atmospheric pressure ionization (API) source on glass sur-

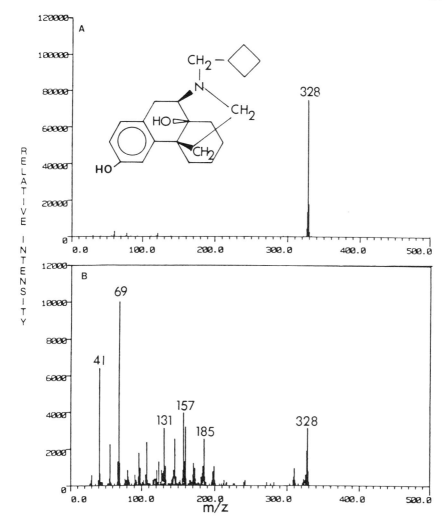

Figure 19.4 (*A*) Full-scan, direct insertion probe API spectrum of standard butorphanol. (*B*) API-CID spectrum of standard butorphanol $(M + 1)^+$ ion, $m/z = 328$. (Henion et al., 1983; reprinted from *J. Chromatogr.*, *271*, 107 (1983), by permission of Elsevier Science Publishers.)

faces in a direct insertion probe; for liquid extracts, 2 µl aliquots were deposited onto the glass within which a small heating coil was embedded. Both full-scan mass spectra and CID spectra of the parent MH$^+$ ions were obtained.

Figure 19.3*a* shows the API MS/MS spectrum of authentic caffeine with the protonated molecule ion at m/z 195, with only a few minor fragment ions. Figure 19.3*b* shows the CID spectrum of the same sample using argon collision gas with a laboratory collision energy of 44 V; the daughter ion spectrum shows the abundant fragment ion peaks at m/z 138, 110, 83, 69, 56, and 42 that characterize the structure of caffeine. A comparison spectrum of raw human urine showed a quite different full-scan spectrum, but the CID spectrum was remarkably similar and definitely verified the identity of the caffeine in the human extract.

A full-scan spectrum of standard butorphanol is shown in Figure 19.4a and an API-CID mass spectrum of the $(M + 1)^+$ ion (m/z 328) is shown in Figure 19.4b. The administered drug appearing in the equine urine showed the 328 ion peak, but only CID spectra provided verification via the fragment m/z ions at 41, 69, 131, 157, and 185. Thus, there is ample evidence that MS/MS and CID techniques provide powerful means for screening drugs with a specificity that is unmatched by most other analyses. Tandem mass spectrometry is also effective in characterizing the new products of illicit drug synthesis (Cooper, 1985).

TOXICITY OF METAL COMPOUNDS

The potential hazard of metal compounds as human carcinogens, and their demonstrated carcinogenicity in certain animal experiments, has caused an increased priority to be assigned to their identification and role in cell transformations. Organic chemical carcinogens have, to date, been the object of more intensive study; but there are now a number of investigations focused upon metal-induced cancers, including exposures to metal-containing particulates of respirable dimensions (< 5 μm). Compounds of arsenic, chromium, and nickel are known to be carcinogenic, but their association with increased tumor incidence is known to be highly dependent upon the specific compounds. In a recent in vitro assessment of the toxicity of inorganic compounds, Heck and Costa (1982) have listed 25 metals and associated compounds (see Table 19.1) together with a qualitative indication of their "transforming activity." In the Table 19.1 classification, $+ +$ indicates potent transforming activity; $+$ = significant activity; $+/-$ = weak or inconclusive activity; $-$ = inactive in inducing a transformation.

The importance of assessing metal compounds on an individual basis in relationship to their biological activity is especially evident for the element nickel. Crystalline nickel sulfides are potent inorganic carcinogens, but amorphous nickel sulfide and some soluble nickel salts exhibit little or no carcinogenic activity (Sunderman, 1979). Hence, while mass spectrometry is an important tool in exploring the toxicity of inorganic compounds, other forms of spectroscopy and bioassays are also essential.

POST-MORTEM ASSAYS OF BODY FLUIDS AND TISSUES

In the Los Angeles County Coroner's Report for 1980, lidocaine was identified in the blood of 84 persons, including 17 drug overdose victims and 16

TABLE 19.1 Mammalian Cell Transformation by Metal Compounds

Element	Compound	Transformation Activity
Aluminum	$AlCl_3$	−
	$Al_2(SO_4)_3$	−
Antimony	$Sb(C_2H_3O_2)_3$	+ +
	$WSbO_4$	+
Arsenic	$AsCl_3$	+ +
	$NaAsO_2$	+ +
	Na_2HAsO_4	+
Barium	$BaCl_2$	−
Beryllium	$BeSO_4$	+
Cadmium	$CdCl_2$	+ +
	$Cd(C_2H_3O_2)_3$	+ +
Calcium	$CaCl_2$	−
Chromium	$CrCl_2$	+
	CrO_3	+ +
	$CaCrO_4$	+ +
	$K_2Cr_2O_7$	+ +
	K_2CrO_4	+ +
	Na_2CrO_4	+ +
	$PbCrO_4$	+ +
	$ZnCrO_4$	+ +
Cobalt	$Co(C_2H_3O_2)_2$	+
	$CoMoO_4$	+
	CoS_2, crystalline	+
	CoS, amorphous	+ / −
Copper	$CuSO_4$	+
	Cu_2S	+
	CuS	+
Iron	$FeCl_2$	+ / −
	$FeSO_4$	+ / −
	Fe_2O_3	−
Lead	PbO	+
	$PbCrO_4$	+ +
	$Pb(C_2H_3O_2)_2$	+
Lithium	$LiCl$	−
Magnesium	$Mg(C_2H_3O_2)_2$	−
Manganese	Mn, metallic	−
	$MnCl_2$	+
Mercury	H_2Cl_2	+
Nickel	Ni, metallic	+ / −
	$NiCl_2$	+ / −
	$NiSO_4$	+
	Ni_2O_3	+ / −
	NiO	−
	NiS, amorphous	−
	NiS, crystalline	+ +
	Ni_3S_2	+ +
	Ni_3Se_2	+ +

TABLE 19.1 *(Continued)*

Element	Compound	Transformation Activity
Platinum	$PtCl_4$	$++$
Silver	$AgNO_3$	$+$
Sodium	Na_2SO_4	$-$
	NaOH	$-$
Strontium	$SrCl_2$	$-$
Thallium	TlCl	$+$
	$Tl(C_2H_3O_2)$	$+$
Titanium	TiO_2	$-$
	$(C_3H_5)_2TiCl$	$+$
Tungsten	WCl_6	$-$
	Na_2WO_4	$-$
	$WSbO_4$	$+$
Zinc	$ZnCl_2$	$+/-$
	$ZnSO_4$	$+/-$
	$ZnCrO_4$	$++$

Source: Heck and Costa (1982).

other drug-related deaths. In most of these cases, the presence of lidocaine was attributed to the probable administration of the drug to revive the victim, and not as a cause of death. This drug has long been used as a local anesthetic, and it has also been employed in coronary care units and emergency room situations (Bernstein et al., 1972). However, in 1981, some suspicions arose as to the possible administration of lidocaine by hospital personnel to end the lives of terminally ill patients. This concern stimulated increased analytical efforts to determine accurately the post-mortem levels of lidocaine in order to assess whether a therapeutic dose or overdose was given (Liu

TABLE 19.2   Lidocaine Blood Levels in 74 Coroner's Cases, 1979–1981

Level (μg/ml)[a]	No. Cases	%
0.01–2.0	45	60.8
2.1–5.9	21	28.4
6.0–10.0	5	6.8
$\geq$ 11.0	3	4.1

Source: Liu et al. (1983).

[a] Therapeutic level, 1.5–2.5 μg/ml; toxic level, 8–10 μg/ml; fatal level, 12 μg/ml.

TABLE 19.3   Post-mortem Tissue and Body Fluid Comparative Concentrations of Lidocaine for Six Exhumed Cases

| | Case No. | | | | | |
Sample	1	2	3	4	5	6
Blood	—	7.40	1.40	3.30	—	—
Brain	2.80	9.00	2.90	1.90	0.90	—
Liver	4.80	6.08	5.60	10.30	0.70	1.40
Heart muscle	5.80	10.44	0.88	4.00	—	—
Kidney	3.84	4.88	ND	3.30	—	—
Lung	9.80	—	2.86	—	0.70	17.70
Spleen	4.25	6.02	0.94	3.10	—	0.30
Urine	—	—	0.30	—	—	—
Adipose tissue	2.30	2.80	ND	—	—	—

Source: Liu et al. (1983).
ND, not detected; — = no data.

et al., 1983). Because of its sensitivity, accuracy, and specificity, computerized GC/MS was selected to determine the lidocaine levels that are tabulated in Table 19.2.

In these analyses, lidocaine was separated from other drugs by an initial extraction that removed opiates and acidic and neutral drugs, and the lidocaine mass spectrum was compared to a standard. Also reported by the Los Angeles Medical Examiner-Coroner's office were lidocaine levels in various tissues in six exhumed post-mortem cases where "the victims had received lidocaine prior to death" (Liu et al., 1983). Table 19.3 shows comparative levels found in the heart, blood, brain, spleen, adipose tissue, kidney, lung, and liver.

Post-mortem estimates of the time of death are also now being made, based on biochemical indicators as well as on physical factors such as body temperature, rigor mortis, etc. Specific chemical analyses in forensic medicine now include chemistry of the blood, cerebrospinal fluid (CSF), and nonprotein nitrogen (NPN) measurements in brain, lung, liver, and kidney (Sasaki et al., 1983).

DIOXIN ANALYSES

There now exists a substantial overlap between environmental measurements (Chapter 15) and forensic investigations. In the tragic industrial incident at Seveso, north of Milan, Italy, in 1976, there was an accidental escape of vapor from a factory producing 2,4,5-trichlorophenol. The vapor was released for about 1 h, and the subsequent fallout occurred over a nearby agricultural area. Soil sampling revealed the presence of highly toxic 2,3,7,8-

TABLE 19.4 Post-mortem
Concentrations of 2, 3, 7, 8-
Tetrachlorodibenzo-p-dioxin
(TCDD)

Sample	TCDD/wet tissue (pg/g)
Fat	1840
Pancreas	1040
Liver	150
Thyroid	85
Brain	60
Lung	60
Kidney	40
Blood	6

Source: Facchetti et al. (1980).

tetrachlorodibenzo-p-dioxin (TCDD) at various concentrations ranging from 0.75–5000 $\mu g/m^?$. One woman who had been exposed by inhalation, and by the consumption of contaminated vegetables from her garden, etc., died 7 months after the accident and a post-mortem was carried out. Analyses were made by gas chromatography-mass spectrometry using both capillary columns and a mass spectrometer of medium-to-high resolution, operating in the multiple-ion detection mode (Facchetti et al., 1980). Tissue concentrations, in picograms per gram of wet tissue, are shown in Table 19.4.

The authors reported that, as a whole, the observed distribution of TCDD was consistent with that described in the literature for nonhuman primates. This analysis is also of special interest because it documented, for the first time, a detailed distribution of TCDD in cadaveric human tissue subsequent to a lethal exposure to this dioxin. In addition, the investigation demonstrated that mass spectrometers of reasonably high resolution are essential for the unambiguous assay of specific isomers in biological samples.

MARINE TOXINS

Marine life species have produced a variety of organic compounds that have been studied for their unique pharmacological and toxicological activities (Kaul, 1982). Some of these show significant effects on neurotransmitters and cardiovascular systems; others have been investigated for their presumed action as tumor promoters, or as possible sources of new antineoplastic agents (Moore, 1982). One of the most deadly marine poisons associated with tropical sea animals, palytoxin, has been studied extensively by mass spectrometry, proton nuclear magnetic resonance (NMR), and other forms of spectroscopy. This substance, whose chemical formula is $C_{129}H_{223}N_3O_{54}$, has a molecular weight of approximately 2700 (Fox, 1982). Historically, this poison is said to have been applied to spears and other weapons by natives on Maui in the Hawaiian Islands. The molecular structure of this compound is very complex—containing 64 asymmetric carbon centers—and it is apparently made by organisms that have grown in different habitats. In any event, it is lethal at very low doses in animal tests; the LD_{50} in rabbits is 0.25 µg/kg. Structure elucidation of the molecule has been very difficult because of size, number of asymmetric centers, and the absence of

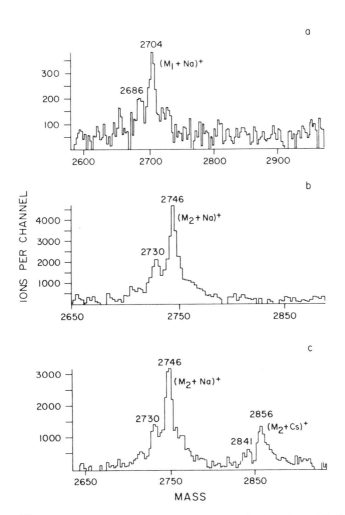

Figure 19.5 ^{252}Cf plasma desorption, positive ion spectra of palytoxin and derivatives. (*a*) Palytoxin, $T = 10,000$ s. (*b*) N-acetylpalytoxin, $T = 112,000$ s. (*c*) N-acetylpalytoxin deposited over a film of CsI, $T = 50,000$ s. (Macfarlane et al., 1980; reprinted with permission from *J. Am. Chem. Soc.*, *102*, 875 (1980), copyright 1980 American Chemical Society.)

any repeating units. Furthermore, "because of its high mass, low volatility, and thermal instability," it has been categorized as "one of the most difficult problems ever encountered for a mass spectrometric measurement of a natural product" (Macfarlane et al., 1980).

These investigators, however, have successfully applied ^{252}Cf plasma desorption spectroscopy (see Chapter 2) to obtain the spectra shown in Figure 19.5. The specimens were irradiated with a ^{252}Cf fission fragment source of 2500 fissions/cm^2/s for periods up to 31 h. Figure 19.5a shows a significant peak detected at $M = 2704$ and Figure 19.5b shows the spectrum for N-acetylpalytoxin with prominent peaks occurring at $M = 2746$ and $M = 2730$, presumably being adducts of $(M + Na)^+$. Figure 19.5c was obtained after a thin film of the sample was electrosprayed over a layer of CsI—with the additional peak at $M = 2856$ being observed, whose mass is 110 higher than the primary peak and corresponds to the (Cs-Na) mass difference. The structure of other marine micro-organisms and their potential use for antibiotics and enzyme production has been reviewed by Okami (1982).

GEOGRAPHICAL SOURCE OF DRUGS BY ISOTOPIC "FINGERPRINTING"

As discussed in Chapter 16, variations in isotopic ratios have made it possible to detect adulterated foods and, in limited instances, to determine the geographical regions where the food is grown. Using the same techniques, efforts are now being made to apply isotopic fingerprinting to locate the origin of illicit drugs. By measuring ^{13}C/^{12}C ratios, Liu et al. (1979) attempted to determine the source of *Cannabis sativa* by analyzing both leaf and flower samples from Mississippi and Illinois. Their objective was to find a possible correlation between the ambient CO_2, temperature, humidity, photoperiod, etc. and the ^{13}C/^{12}C ratio of the plant. The limited number of samples assayed, however, permitted no significant conclusions to be drawn. The recent investigation by Dunbar and Wilson (1982) is a further attempt to establish a source location based upon three stable isotopes (^{13}C, ^{2}H, and ^{18}O), rather than on isotopic ratio measurements for carbon only.

This approach at least affords a much greater degree of discrimination. For a feasibility study, the authors chose to use caffeine rather than morphine or cocaine for two reasons: "(1) it contains the same elements as does morphine, i.e., carbon, oxygen, hydrogen, (2) it can be extracted from tea and coffee, and from known origins (i.e., from specific countries)."

Their mass spectrometric results are shown in Table 19.5, where the carbon isotope ratios were measured relative to the Peedee belemnite (PDB) standard (see Chapter 16), that is,

$$\delta^{13}C_{PDB} \, \%o = \left[\frac{(^{13}C/^{12}C)_{sample} - (^{13}C/^{12}C)_{standard}}{(^{13}C/^{12}C)_{standard}} \right] \times 1000$$

TABLE 19.5 Geographical Origin and Isotopic Assays of Caffeine Samples

Origin	Source	δ $^{13}C_{PDB}$ ‰	δ D_{SMOW} ‰	δ $^{18}O_{SMOW}$ ‰
Jamaica	Coffee	−28.8 ± 0.6	−132.5 ± 3.8	+9.6 ± 1.8
Kenya		−29.8 ± 0.6	−136.5 ± 3.5	+3.6 ± 0.6
Brazil		−28.2 ± 0.2	−157.3 ± 3.9	+4.9 ± 0.7
Sri Lanka	Tea	−31.7 ± 0.8	−223.6 ± 2.8	+1.8 ± 0.2
Darjeeling		−29.6 ± 0.2	−195.9 ± 2.5	−4.3 ± 0.8
China		−32.4 ± 0.6	−226.8 ± 4.1	+1.2 ± 0.3
Lab grade[a]		−35.8 ± 0.2	−237.1 ± 1.7	+13.0 ± 0.3

Source: Dunbar and Wilson (1982).

[a] Caffeine prepared commercially.

and D/H and $^{18}O/^{16}O$ ratios were measured relative to the SMOW (standard mean ocean water) standard.

It will be noted that the $^{13}C/^{12}C$ ratios of samples from Jamaica, Kenya, and Brazil are within limits of error, but that the D/H and $^{18}O/^{16}O$ ratios are different. Furthermore, the authors point out that because Darjeeling is a mountainous country and has a higher altitude than most countries, its water transmitted to plants should be depleted in deuterium and ^{18}O, in contrast to Jamaica and Kenya—and this difference is experimentally observed. Other differences in Table 19.5, are tentatively attributed to other phenomena. These results are not conclusive, but they indicate that a *multiplicity of isotopic ratios* may be a potentially useful forensic technique for tracing drugs from either their natural or laboratory (synthetic) point of origin.

FORENSIC GEOLOGY

In 1904, the German forensic scientist Dr. Georg Popp presented evidence of matching minerals and soils that was sufficient to bring an admission of guilt in a murder trial. This is believed to be the first legal acceptance of geological evidence in the courts (Murray and Tedrow, 1975). In the United States, the Federal Bureau of Investigation (FBI) began assaying soils and minerals in the 1930s, so that by World War II, mineral identification became a standard procedure in its laboratory. The uniqueness of soil analysis is derived from its basic complexity; soil composition reflects glacial movements and climatic conditions in addition to the inclusion of minerals, fossils, trace metals, and dissolved rocks. Hence, in principle, it is possible to determine the point of origin of sand in a soil sample by matching not a few, but many parameters. In this sense, geochemical specimens provide more clues than manufactured products such as glasses, fibers, and paints.

In the forensic studies of soil, the size of available samples will usually be small—and it may contain only a few of the 2000 individual minerals that have been identified. Some minerals are magnetic, and some can be identified on the basis of their crystal structure and density as well as chemistry (a distinction is usually made between "light" and "heavy" minerals at 2.89 g/cm^3). The mineral/organic ratio of soils will vary widely. The organic content in desert sands may be less than 1%, but be 90% in some bogs; a value of 3% organic content is often quoted for "average" mineral soils. Overall, variations in soils are so extensive that a list of 10,466 soils in the United States had been classified by 1965, and these classifications have been utilized in both agriculture and forensic contexts. In one celebrated kidnap and murder case, some 360 soil samples collected near Denver, CO, where the victim disappeared, were compared to soil samples under the fender of an automobile that was abandoned on a dump in Atlantic City, NJ. (Murray and Tedrow, 1975). This soil-matching evidence contributed to the eventual conviction of the suspect.

The trace element composition in soils, of course, provides an even greater potential for identification, when coupled with more general data on minerals and major chemical constituents. These trace elements not only provide data on virgin soils, their abundances reflect inputs such as fertilizers and local environmental anomalies. For example, the lead levels of soils along major highways will be somewhat affected by automobile emissions, and the soils in the immediate vicinity of a residential dwelling can reflect the high lead composition of various paints. The report of Chaperlin (1981) is of special interest in a forensic context, and Table 19.6 indicates the large variation of soil concentrations for this single element. Lead concentrations around houses built about 1900 range up to 1400 ppm.

High-performance liquid chromatographic (HPLC) profiles also reveal

TABLE 19.6 Lead Content
of Garden Soils Adjacent to
Eight Houses of Different
Ages in Icklesham, Sussex,
England

Date of House	Lead Content of Soil (ppm)
1928	426
1928	191
1928	160
1960	67
1960	64
1965	47
1965	42
1965	40

Source: Chaperlin (1981).

TABLE 19.7   Trace Element Concentration of Soil and Soil Clay (ppm)

	Cu	Mn	Zn	Ni	Sn	Pb	Cr	V	Ga
Total soil	4	20	16	3	5	8	17	11	2
Soil clay	135	80	98	15	20	44	101	133	27

Source: Connor et al. (1957).

that, in general, soil taken from the same area produces similar chromatograms (Reuland and Trinler, 1981). However, mass spectrometry is mandated if a definitive analysis is required of many elements. Table 19.7 includes spectrographic data of Connor et al. (1957) for selected trace metals that they found in total soil and soil clay. With modern mass spectrometry, the relative abundance of at least 20 metals could be measured for even small samples.

EXPLOSIVES

The analysis of explosive residues and volatile constituents of explosives requires sophisticated mass spectrometric techniques for several reasons. First, the amount of material available may be limited to nanogram or subnanogram amounts. Second, post-explosion residues in debris will usually be contaminated with interfering compounds. And third, the detection of concealed explosives in baggage or mail by sensing the volatile vapors of explosives, per se, represents another instrumental problem. Recently, attention has also been focused on the possible environmental impact of explosives due to both the disposal of obsolete munitions in the oceans and the potential contamination of surface and ground waters in the environs of munitions plants.

Explosive materials can be grouped into two broad categories: (1) high explosives and related compounds and (2) low explosives and propellants, such as gunpowders and rocket fuels. High explosives are triggered by a primary explosive, and their detonation results in an almost instantaneous molecular rearrangement accompanied by the generation of a large volume of hot gases, such as H_2, N_2, CO, CO_2, CH_4, etc. (Yinon and Zitrin, 1981). These high explosives may include aliphatic and aromatic nitro compounds, nitrate esters, nitramines, salts of inorganic acids, plus nonexplosive stabilizers and plasticizers. Commercial and military explosives (e.g., dynamites) include mixtures of these and other materials. Low explosives and propellants contain ingredients such as potassium and sodium nitrates, sulfates, mixtures of perchlorates, and combustible binders.

The general procedure for detecting traces of explosives at nanogram levels in contaminated samples, such as handswab extracts, usually involves some preliminary clean-up. One method described by Douse (1982) utilizes a selective charge transfer extraction of the explosives from solutions of extracts from pentane onto Amberlite XAD-7 porous polymer beads. The explosive chemicals are then removed from the surface of the beads by ethyl acetate to yield suitable samples for direct analysis by capillary column gas chromatography. The technique removes interfering lipid material and is reported to be applicable to important commercial and military explosives.

Chemical ionization (CI) is now a commonly used technique for analysis of explosives; reagent gases that have been used include methane, isobutane, and ammonia. Electron-impact (EI) ionization is also used. This latter mode of ionization is especially relevant, inasmuch as research has revealed some interesting parallels between the early stages of decomposition by explosive shock and the decomposition occurring by EI in the mass spectrometer source. For example, the main fragment ions in the EI mass spectrum of tetranitromethane—$C(NO_2)_4$—and the gaseous products of its detonation include NO, CO_2, and CO (Yinon, 1982). Thus, for an unambiguous identification of an unknown explosive or explosive mixture, a mass spectrometer with a dual CI/EI source is preferred. Both positive and negative CI spectra are useful; the negative ion spectra are simple and include M^- or $(M - H)^-$ ions, or both, as well as typical fragmentations (Yinon, 1980; 1985).

The trinitrotoluene (TNT-$C_7H_5N_3O_6$) analyses of Bulusu and Axenrod (1979) merit special mention because these TNT samples were synthesized using highly enriched ^{15}N and ^2H, and precursors of the major ions in the EI spectrum were identified by observing metastable peaks and isotope shifts with a double focusing spectrometer. The isotopic labeling made possible a clearer interpretation of EI-induced fragmentations of nitroaromatics and the elucidation of molecular structures. Table 19.8 shows the calculated and observed mass measurements for selected ions of two TNT isomers.

The detection of hidden explosives poses special instrumentation problems, as explosive vapors must be detected at atmospheric pressures. A portable quadrupole mass spectrometer (QMS) specifically designed for this application has been reported (Evans and Arnold, 1975), having an inlet system with a multistage dimethylsilicone membrane separator. This membrane has a higher permeability to organic molecules than to air, thereby increasing the organic vapor-to-air ratio at each stage of the separator. A sample enrichment of 10^6 is reported with a sensitivity to mononitrotoluene of 2 ppb. The detection of TNT has also been reported at an even higher sensitivity (~1 ppt) using an API source, with ambient air being drawn directly into the source at flow rates of 10 l/s (Reid et al., 1978). Cryogenic pumping was used for the vacuum system, with a rated speed of 20,000 l/s. This same system was employed to detect nitroglycerin (NG) in an aircraft cabin in real time in a concentration range of a few parts per trillion (Buckley et al., 1978). The mass spectral detection was based on observing the m/z peaks 212 $(M_{NG} - H + O_2 - NO_2)^-$ and 213 $(M_{NG} - O_2 - NO_2)^-$.

TABLE 19.8   Exact Masses of Selected Ions in the Spectra of TNT Isomers

Nominal m/z	Most Probable Formula	Exact Mass	
		Calculated	Observed
2,4,6-TNT			
179	$C_7H_3N_2O_4$	179.00928	179.00951
166	$C_6H_2N_2O_4$	166.00145	166.00133
149	$C_7H_3NO_3$	149.01129	149.01136
134	$C_7H_4NO_2$	134.02420	134.02211
120	$C_6H_2NO_2$	120.00855	120.00866
105	C_6H_3NO	105.02148	105.02266
105	C_7H_5O	105.03404	105.03461
89	C_7H_5	89.03912	89.04035
88	C_7H_4	88.03130	88.03236
77	C_6H_5	77.03910	77.03634
77	C_5H_3N	77.02654	77.02903
76	C_6H_4	76.03130	76.03040
63	C_5H_3	63.02347	63.02259
62	C_5H_2	62.01560	62.01643
3,4,5-TNT			
211	$C_7H_5N_3O_5$	211.02292	211.02234
197	$C_7H_5N_2O_5$	197.01984	197.04140
135	$C_7H_5NO_2$	135.03203	135.03126
134	$C_7H_4NO_2$	134.02420	134.03298
107	C_6H_5NO	107.03711	107.03687

Source: Bulusu and Axenrod (1979).

Mass spectrometric analyses also provide the most detailed data relative to gunshot residues, and the FBI laboratory has compiled a list of 23 organic compounds that occur in smokeless gunpowder (Table 19.9). Simple "dermal nitrate" tests that were introduced in 1933 for firearms residues have been superceded because nitrates are ubiquitous in the environment, that is, nitrates are common to tobacco ash, pharmaceuticals, fertilizers, leguminous plants, urine, and certain fingernail polishes (Maloney and Thorton, 1982). Recent research by Yinon and Hwang (1984) has also focused on the metabolism of explosives in the human body. In forensic applications, an ultimate objective is to be able to detect trace amounts of explosives that may be absorbed through the pores of a suspect's hand and detected as metabolites in blood and urine.

The technical monograph by Yinon and Zitrin (1981) reviews all the major analytical techniques for the analysis of explosives, and an additional report (Yinon, 1982) focuses specifically on the mass spectrometry of explosives according to the mode of ionization. High-performance liquid chromatography (HPLC) together with mass spectrometry have been effectively ap-

TABLE 19.9   FBI List of Organic Compounds That May be Found in Smokeless Gunpowder

Cresol	RDX (cyclonite)
Resorcinol	Diethyl phthalate
Carbazole	Nitroglycerin
Diphenylamine	Trinitrotoluene
Dimethyl phthalate	Dimethylsebacate
N-Nitrosodiphenylamine	N,N-Dimethylcarbanilide (methylcentralite)
Dinitrocresol	2,4-Dinitrodiphenylamine
Carbanilide	N,N-Diethylcarbanilide (ethylcentralite)
Nitroaiphenylamine	Dibutyl phthalate
Triacetin	PETN (Pentaerythritol tetranitrate)
Nitrocellulose	N,N-Dibutylcarbanilide (butylcentralite)
Nitrotoluene	

Source: Maloney and Thorton (1982).

plied to the assay of pure compounds of NG, diethyleneglycol dinitrate (DEGN), 2,4,6-trinitrotoluene (TNT), and 2,4,6-N-tetranitro-N-methylaniline (tetryl), and to technical mixtures including TNT, 2,4-dinitrotoluene (2,4-DNT), 1,3,5-trinitro-1,3,5-triazacyclohexane (RDX), pentaerythritol tetranitrate (PETN), NG, and ammonium nitrate (Yinon and Hwang, 1983). The mass spectrum of TNT shown in Figure 19.6 is characterized by the MH^+ ion at m/z 228, the adduct ion $(M + CH_3OH + H)^+$ at m/z 260, the molecular ion M^+ at m/z 227, plus fragment ions $(M - OH)^+$ at m/z 210 and $(MH - 30)^+$ m/z 198.

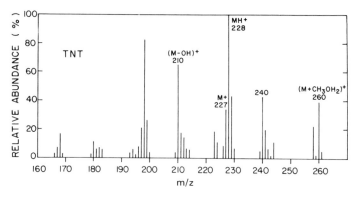

Figure 19.6   LC-MS spectrum of TNT with methanol/water (1:1) as reagent. (Yinon and Hwang, 1983; reprinted from *J. Chromatogr.*, *268*, 45 (1983), by permission of Elsevier Science Publishers.)

ARSON ACCELERANTS

Forensic chemists associated with the New Jersey Forensic Science Bureau report that 9772 fires were due to arson in that state within a single year (Saferstein and Park, 1982). This high arson rate has, in part, stimulated both the forensic application of computerized GC/MS instruments equipped with high-speed data recording systems and the use of computer searches of the 60,000 spectrum library of the National Bureau of Standards (NBS). The recent work of Smith (1983) lists the ions selected to represent major hydrocarbon families normally present in common accelerants, including m/z values of 57 + 71 + 85 + 99 (aliphatics), 55 + 69 + 83 + 97 (alicyclics and olefinics), 91 + 105 + 119 + 33 (alkylbenzenes), 104 + 118 + 132 + 146 (alkylstyrenes/alkyldihydroindenes), 128 + 142 + 156 + 170 (alkylnaphthalenes), 178 + 192 + 206 (alkylanthracenes), and 93 + 136 (monoterpenes). Additional molecular ion-counting measurements often identify single components.

The experimental procedures for this investigation included (1) use of standard accelerants procured from commercial outlets; (2) sample injection by the wet-needle technique with injection sizes of ~0.1 μl; (3) combined glass capillary GC/MS with temperature programming to 250°; (4) EI mode (75 eV) mass spectrometry; and (5) spectral scan rates of 128 amu/s over a range of m/z 35–250. The work clearly demonstrated that a distinction could be made between assortments of standard accelerants and their evaporated residues, although some difficulties will always remain in the identification of samples recovered from debris and other objects taken from fire sites.

Recent studies at the FBI laboratory (Kelly and Martz, 1984) have also shown that advanced chromatography-mass spectrometry techniques provide a clear distinction between common fuels that are used as accelerants, that is, leaded and unleaded gasolines, lighter fluid, kerosene, and fuel oil. Furthermore, interfering pyrolysis products that often hinder the identification of accelerants have been minimized by employing reconstructed chromatograms of specific ions that are unique to different accelerants.

DYES AND PROTECTIVE SPRAYS

The detection and identification of submicrogram quantities of various dyes and sprays (tear gases) usually calls for some type of chromatography-mass spectrometry. Specifically, exploding money packets that are designed to eject aerosols of dyes and tear gas can be remotely activated after a robber leaves a bank or store. In theory, the dye can link the money and the suspect to the robbery and the tear gas could lead to an abandonment of the money or getaway vehicle, as well as linking chemical residues to clothing, automobile upholstry, or other objects. Electron-impact (EI) ionization, positive ion chemical ionization (PICI), and negative ion chemical ionization (NICI)

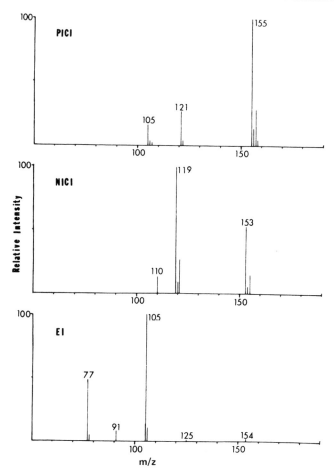

Figure 19.7 Mass spectra of chloroacetophenone (CN tear gas) using PICI, NICI, and EI ionization. (Martz et al., 1983; reprinted with permission from *J. Forensic Sciences, 28*, 200 (1983), copyright ASTM.)

have all been applied to determine the lower limits of detection for trace quantities of 1-methyl-aminoanthraquinone (MAAQ), orthochlorobenzal-malononitrile (CS tear gas), and chloroacetophenone (CN tear gas), which are the commonly used red dye and lachrymators, respectively (Martz et al., 1983). The gas chromatograph-mass spectrometer used in this work was a quadrupole that was equipped with a detector for both positive and negative ions. Figure 19.7 shows mass spectra for CN tear gas when PICI, NICI, and EI are used. The authors conclude that the choice of ionization will depend upon the dye or tear gas of interest and the chemical nature of the interfering substances. The advantage of multiple modes of ionization for high sensitivity and selectivity, however, has been amply demonstrated.

This type of "residue analysis" is becoming of increasing importance despite the fact that a number of states have prohibited the carrying of protective spray devices and characterized them as concealed weapons. Ultraviolet (UV) and infrared (IR) spectroscopy are also used for dye and spray diagnostics, and gas chromatography, per se, is an adequate presumptive test. For positive identification, however, mass spectrometry must be coupled to chromatography in most practical situations (Nowicki, 1982). The characterization of organic dyes by secondary ion mass spectrometry (SIMS) and fast-atom bombardment (FAB) has already been demonstrated at the 10 ng level of sensitivity (Scheifers et al., 1983). Both techniques furnish abundant molecular and structurally diagnostic ions for dye identification with little or no sample preparation, and the sputtering mode of analysis automatically yields a depth profile of multilayered dyed specimens.

METALS, GLASSES, WAXES, PAINTS, AND OILS

As indicated in Chapter 10, mass spectrometry can be used to characterize metals and alloys in great detail. In forensic applications, it has been used to analyze bulk and surface compositions, and SIMS is especially useful for depth profiling either large or fragmented specimens. Mass spectrometry has even been applied to the assay of faked gold ingots, and the laser-produced spectrum obtained by Dennemont and Landry (1983) is shown in Figure 19.8. The spectrum of Figure 19.8a was obtained from surface scrapings removed with a scalpel, and the presence of gold is indicated at m/z 197. Sodium, potassium, and calcium are common surface contaminants when objects are exposed to the atmosphere, and the authors suggest that chromium, iron, and cobalt may have been deposited during the electrolytic-plating process. Figure 19.8b is the laser-produced spectrum (0.2 mm depth) clearly revealing the "gold ingot" as brass. The spectra also indicates how, in general, a multiplicity of elements can be used to identify a specimen and how "isotopic tagging" or "coding" (with enriched isotopes) can be employed to provide an unambiguous "fingerprint" to a valuable object, that is, by ion implantation.

The identification of glass has always received considerable attention for obvious reasons. Microscopic examination can reveal matching edges of broken headlamps, windows, containers, etc., and the bulk chemistry of glasses is not likely to be altered by exposure to heat, contamination, or other external exposures. In the case of stolen objects, analytical data can identify types of glasses and a possible source, that is, glass plant, fabricator, and retail merchant. The refractive index of glasses together with their dispersion and density have been used by many forensic laboratories (Slater and Fong, 1982), and a distinction can even be made between bulk- and surface-refractive indices that assists markedly in identification. For example, window glasses made by a "float" process were consistently found to have a higher index at the surface, which had been in contact with molten

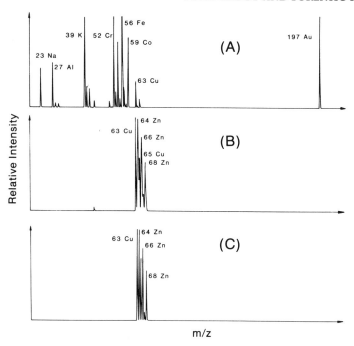

Figure 19.8 Laser-produced spectra of a faked gold ingot revealing brass: (A) Surface analysis, (B) 0.2 mm below the surface, and (C) 1 mm below the surface. (Dennemont and Landry, 1983; courtesy of J. Dennemont, J.-Cl. Landry, and the Leybold-Heraeus Co.)

tin during manufacture, than in the bulk. A tin concentration as high as 2% has been reported in the tin contact surface, and it is this change in the chemistry of a 10 μm layer that yields a measurable change in the index of refraction (Davies et al., 1980).

The chemical assay of glasses (bulk or powdered specimens) is routinely determined, and trace element analysis plus refractive index has been reported for 349 glass samples (Hickman, 1981). In this work undertaken by the Metropolitan Police Forensic Science Laboratory (London), blind trials produced definite classifications of glass types at the 91% level, and no wrong classifications were made. Selected data from this large number of glass samples are given in Table 19.10. Concentrations of Mn and Ba are given in parts per million, and those for Fe, Mg, and Al are reported in percentages.

Other investigators have suggested that Cs, Rb, Sr, and Sb might be useful discriminating elements for glass identification. In a recent study, various forms of spectroscopy were employed to measure the absolute concentrations of 22 elements in a group of sheet, container, and tableware glasses of refractive index 1.5177–1.5183 (Hickman et al., 1983). Elements determined were Al, As, Ba, Ca, Ce, Co, Cr, Cs, Eu, Fe, K, La, Li, Mg, Mn, Na, Rb, Sc, Sr, Th, U, and Zn. Thus, it is apparent that trace element assays

TABLE 19.10   Trace Element Assay and Refractive Indices of Glass Specimens

Specimen	Mn (ppm)	Fe (%)	Mg (%)	Al (%)	Ba (ppm)	Refractive Index
Vehicle window	91.0	0.069	2.04	0.53	101.0	1.5161
Beer mug	18.0	0.015	1.18	0.60	217.0	1.5185
Shop window	85.0	0.056	1.88	0.09	32.0	1.5174
Container	35.0	0.036	0.28	0.58	148.0	1.5174
Coffee jar	58.0	0.193	0.44	0.86	156.0	1.5219
Tableware	17.0	0.005	0.01	0.08	152.0	1.5082
Wine glass	9.0	0.018	1.11	0.51	14.0	1.5127
Milk bottle	112.0	0.026	0.06	0.74	244.0	1.5218
Shop door	88.0	0.062	1.77	0.50	101.0	1.5178
Church windows	3420.0	0.620	0.45	0.84	4300.0	1.5377
Headlamp	5.0	0.022	0.04	0.91	10.0	1.4778
Ashtray	5.0	0.010	0.03	0.19	2500.0	1.5086

Source: Hickman (1981).

can establish unambiguously the type or class to which an unknown glass sample belongs. Furthermore, if a surface assay by an ion probe method can also be made on a specimen, this combined surface-bulk analysis may provide a more positive identification than any other analytical method.

Mass spectrometry has also been used for distinguishing between pencil markings, in which a discrimination was made on the basis of the relative intensities of various ions. In a specific study by Zoro and Totty (1980), 17 pencils of various brand names and hardness designation were analyzed by EI ionization. The *m/z* species identified were essentially the mixed spectra of the waxes used in manufacturing the pencil cores. For each pencil mark on clear paper, an unmarked area of paper was also analyzed in order to eliminate the contribution from traces of volatile organics residing in the paper. The pencil marks remained easily distinguishable for several months, but some difficulty was encountered when the pencil markings were made on soiled papers. The identification of specific sealing waxes has also been made for various seals and official documents.

Paints and petroleum products are other common materials that are subject to scrutiny in criminal investigations. In many cases, automobile paints are made using similar polymers that differ only slightly from one another, but particular paints may be different with respect to pigments, modifiers, and additives (Zieba and Pomianowski, 1981). Automobile lubricants are also frequently encountered at the scene of a crime, and therefore are submitted along with standards from the defendant's vehicle in order to establish a common source (Kubic et al., 1983). Oil spills from tankers are also being investigated more frequently because of extensive environmental damage, and in most instances oils from different sources can be distinguished by mass spectral assays.

DETECTION OF FRAUDULENT DOCUMENTS

"No other instrument of crime is so prevalent in our society as the document. Other crimes of violence draw attention from the media—however, crimes committed with documents involve billions of dollars each year and actually have a greater impact on our nation's economy" (Brunelle, 1982). Thus, sophisticated chemical analyses are now complementing the many physical methods that are employed for detecting forged checks, wills, altered legal documents, bogus stock certificates, counterfeit currency, credit cards, etc. Thin-layer chromatography has been used to compare inks, and other chromatographic methods have been applied to ink components, dyes, and typewriter ribbon inks. Also, in the 1970s, several ink manufacturers began "tagging" their inks in a continuing program that allows inks to be dated to the exact year of their manufacture. However, verification of a document via a paper analysis may prove to be an even more effective and unambiguous test. Large amounts of material are usually available, so that samples can be obtained for many tests. And as in other forensic investigations, mass spectrometry offers the advantage of multielement analysis at a very high sensitivity.

Table 19.11 shows the range of trace metal concentrations that have been reported by Brunelle et al. (1971), via neutron activation, in the characterization of various papers. Clearly, all of these concentrations are well within the detection limits of modern mass spectrometry, and comparative measurements between papers would be of special significance. Furthermore,

TABLE 19.11   Trace Metal Concentrations in Papers[a]

Element	Concentration Range (ppm)	Element	Concentration Range (ppm)
Tantalum	0.1–200	Rubidium	9.0–60
Copper	0.1–110	Iron	100.0–620
Arsenic	1.0–9	Barium	37.0–10,200
Manganese	0.2–510	Zinc	22.0–90.0
Samarium	0.01–150	Scandium	0.1–18.0
Sodium	50.0–1840	Molybdenum	8.0–330
Lanthanum	0.3–570	Antimony	0.2–245
Gold	0.01–90	Cobalt	1.0–4.1
Chromium	1.2–245	Mercury	0.1–90

Source: Brunelle et al. (1971).

[a] Partial list for various white and colored papers.

all but five of these elements (As, Na, Au, Sc, and Co) have multiple isotopes, so that an accurate quantification is possible with the isotopic dilution method.

Many types of spectroscopy have been employed to detect spurious works of art. The most common error of forgers in classic paintings has been to use a pigment that was simply not available when the original artist was active (Fleming, 1980). The lead "age" in a pigment is also reflected by its isotopic ratio, and galena ores from different mining regions throughout the world contain different isotopic abundances. Hence, mass spectrometry has been useful in analyzing this element in a work of art that is suspect.

REFERENCES

Bernstein, V., M. Bernstein, J. Griffiths, and D. Peretz, *J.A.M.A., 219*, 1927 (1972).

Björkhem, I., H. Ek, and O. Lantto, *J. Chromatogr. Biomed. Appl., 232*, 154 (1982).

Brunelle, R. L., W. Washington, C. Hoffman, and M. Pro, *J. Assoc. Off. Anal. Chem., 54*, 920 (1971).

Brunelle, R. L., in *Forensic Science Handbook*, ed. R. Saferstein, Prentice-Hall, Englewood Cliffs, NJ (1982), p. 673.

Buckley, J. A., J. B. French, and N. M. Reid, *Proc. New Concepts Symp. Workshop on the Detection and Identification of Explosives*, Reston, VA (1978), pp. 109–117.

Bulusu, S., and T. Axenrod, *Organic Mass Spectrom., 14*, No. 11, 585 (1979).

Cailleux, A., A. Turcant, A. Premel-Cabic, and P. Allein, *J. Chromatogr. Sci., 19*, 163 (1981).

Campbell, R. J., personal communication (1984).

Chaperlin, K., *Forensic Sci. Int., 18*, 79 (1981).

Connor, J., N. F. Shrimp, and J. F. C. Tedrow, *Soil Sci., 83*, 65 (1957).

Cooper, D., *Spectra, 10*, No. 3, 14 (1985).

Davies, M. M., R. J. Dudley, and K. W. Smalldon, *Forensic Sci. Int., 16*, 125 (1980).

Dennemont, J., and J. C. Landry, "Applications of the Laser Microprobe Mass Analyzer (LAMMA) to Problems of Forensic Science," in *Lamma Workshop*, Forschungsintitut Borstel, F.R.G. (September 1983), p. 149.

Douse, J. M. F., *J. Chromatogr., 234*, 415 (1982).

Dunbar, J., and A. T. Wilson, *Anal. Chem., 54*, 590 (1982).

Evans, J. E., and J. R. Arnold, *Environ. Sci. Technol., 9*, 1134 (1975).

Facchetti, S., A. Fornari, and M. Montagna, in *Advances in Mass Spectrometry*, Vol. 8B, ed. A. Quayle, Heyden and Son Ltd., London (1980), p. 1405.

Finkle, B. S., *Anal. Chem., 54*, No. 3, 433A (1982).

Fleming, S., *Physics Today, April*, 34 (1980).

Foltz, R. L., A. F. Fentiman, Jr., and R. B. Foltz, *GC/MS Assays for Abused Drugs in Body Fluids*, NIDA Monograph No. 32, U. S. Government Printing Office, Washington, D. C. (1980).

Fox, J. L., *Chem. Eng. News, January 4*, 19 (1982).

Frigerio, A., in *Advances in Mass Spectrometry*, Vol. 8B, ed. A. Quayle, Heyden & Son Ltd., London (1980), p. 1076.

Heck, J. D., and M. Costa, *Biol. Trace Element Res., 4*, 71 (1982).

Henion, J. D., B. A. Thomson, and P. H. Dawson, *Anal. Chem., 54*, 451 (1982).

Henion, J. D., G. A. Maylin, and B. A. Thomson, *J. Chromatogr., 271*, 107 (1983).

Hickman, D. A., *Forensic Sci. Int., 17*, 265 (1981).

Hickman, D. A., G. Harbottle, and E. V. Sayre, *Forensic Sci. Int., 23*, 189 (1983).

Kaul, P. M., *Pure Appl. Chem., 54*, No. 10, 1963 (1982).

Kelly, R. L., and R. M. Martz, *J. Forensic Sci., 29*, No. 3, 714 (1984).

Kubic, T. A., G. M. Lasher, and J. Dwyer, *J. Forensic Sci., 28*, No. 1, 186 (1983).

Liu, J. H., W. F. Lin, M. P. Fitzgerald, S. C. Saxena, and Y. N. Shied, *Forensic Sci., 24*, 814 (1979).

Liu, Y., E. C. Griesemer, R. D. Budd, and T. T. Noguchi, *J. Chromatogr., 268*, 329 (1983).

Macfarlane, R. D., D. Uemura, K. Ueda, and Y. Hirata, *J. Am. Chem. Soc., 102*, No. 2, 875 (1980).

Maloney, R. S., and J. I. Thorton, *J. Forensic Sci., 27*, No. 2, 318 (1982).

Martz, R. M., R. J. Reutter, and L. D. Lasswell, III, *J. Forensic Sci., 28*, No. 1, 200 (1983).

Moore, R. E., *Pure Appl. Chem., 54*, No. 10, 1919 (1982).

Murray, R. C., and J. C. F. Tedrow, *Forensic Geology*, Rutgers University Press, New Brunswick, NJ (1975), pp. 7, 73.

Nowicki, J., *J. Forensic Sci., 27*, No. 3, 704 (1982).

Okami, Y., *Pure and Appl. Chem., 54*, No. 10, 1951 (1982).

Reid, N. M., J. A. Buckley, D. A. Lane, A. M. Lovett, J. B. French, and C. Poon, conference papers presented at the *26th Ann. Conf. Am. Soc. Mass Spectrom. and Allied Topics*, St. Louis, (1978), pp. 662–664.

Reuland, D. J., and W. A. Trinler, *Forensic Sci. Int., 18*, 201 (1981).

Saferstein, R., ed., *Forensic Science Handbook*, Prentice-Hall, Englewood Cliffs, NJ (1982), p. 125.

Saferstein, R., and S. A. Park, *J. Forensic Sci., 27*, No, 3, 484 (1982).

Sasaki, S., S. Tsunenari, and M. Kanda, *Forensic Sci. Int., 22*, 11 (1983).

Scheifers, S. M., S. Verma, and R. G. Cooks, *Anal. Chem., 55*, 2260 (1983).

Slater, D. P., and W. Fong, *J. Forensic Sci., 27*, No. 1, 474 (1982).

Smith, R. M., *J. Forensic Sci., 28*, No. 2, 318 (1983).

Sunderman, F. W., Jr., in *Environmental Carcinogenesis*, ed. P. Emmelot, and E. Krick, Elsevier, Amsterdam, (1979), p. 165.

White, P. C., I. Jane, A. Scott, and B. E. Connett, *J. Chromatogr., 265*, 293 (1983).

Yinon, J., *J. Forensic Sci., 25*, No. 2, 401 (1980).

Yinon, J., *Mass Spectrom. Rev., 1*, 257 (1982).

Yinon, J., *Spectra, 10*, No. 3, 5 (1985).

Yinon, J., and S. Zitrin, *The Analysis of Explosives*, Pergamon, New York (1981), p. 4.

Yinon, J., and D. G. Hwang, *J. Chromatogr., 268*, 45 (1983).

Yinon, J., and D. G. Hwang, "Metabolic Studies of Explosives," in *32nd Ann. Conf. on Mass Spectrom. and Allied Topics*, San Antonio, (May 27 to June 1, 1984).

Zieba, J. and A. Pomianowski, *Forensic Sci. Int., 17*, 101 (1981).

Zoro, J. A., and R. N. Totty, *J. Forensic Sci., 25*, No. 3, 675 (1980).

CHAPTER TWENTY

NEW FRONTIERS IN MASS SPECTROMETRY AND ION BEAM TECHNOLOGY

In this concluding chapter, it is appropriate to comment briefly on both new analytical techniques and engineering applications for ion beam technology.

At present, spectrometers of high sensitivity, resolution, and mass range can now characterize single atoms, molecular aggregates, and reaction products. In the future, the combined use of primary ion beam probes and tuned lasers promises to enhance the sensitivity of surface impurity analysis by a factor of 100. A new frontier in mass spectrometry also includes the study of many transient phenomena—the dynamics of highly excited plasmas, time-resolved mechanisms of energy transfer, and atomic/molecular processes produced by picosecond excitation. Such studies make use of the fact that pulsed electromagnetic radiation and particle beams often provide more detailed information than continuous excitation sources. New insights into molecular structure will also be forthcoming from experiments relating to collision-induced dissociation (CID). One new development relates to collisions at solid surfaces. According to Cooks (1984)—"it is possible to effectively dissociate ions in this way, and most importantly—the energy transferred to the dissociating ion greatly exceeds that available from collisions with gaseous targets. The reason this is significant is that larger (bio) molecules require considerable internal energy if they are to fragment, and getting them to fragment is essential to structural characterization. The problem in the future will not be achieving ionization; it will be in structural characterization."

Several analytical methods that have been borrowed from nuclear physics are undergoing a continuing development: (1) Fourier transform ion cyclotron resonance; (2) accelerator-mass spectrometry; (3) Rutherford backscattering; and (4) particle-induced x-ray emission. As indicated in Chapter 3, Fourier transform spectrometry is rapidly expanding the scope of research in the physical sciences, and tandem instruments permit significant new data to be acquired. The ion accelerator-mass spectrometer has already made feasible many unique measurements; $^{12}C/^{14}C$ ratio determinations in the range of $10^{15}:1$ have been made, and platinum has been detected in ores at concentrations of 1 part in 10^{12}. Rutherford backscattering (RBS) is another specialized extension of mass spectrometry, as it can not only reveal the mass of the scattering atom at an interface, it can also indicate the depth at which the atom is located. Particle-induced x-ray emission (PIXE) combines energetic particle bombardment with energy-dispersive x-ray detection to provide a nondestructive, multielement type of assay.

Closely paralleling these advances in instrumentation are many other new uses for ion beams. Ion beam technology now encompasses such diverse fields as spacecraft propulsion, plasma heating, ion beam modification of materials, ion beam lithography, and ion beam radiotherapy. Hence, a brief discussion of these applications is included in the final pages of this book.

ARCHAEOLOGY

Subtle differences in the elemental composition of archaeological artifacts are now providing important new information as to their source and chronology, and isotopic measurements are revealing an even more detailed scenario of their origin and history. The trace element compositions as well as major alloying elements show distinct patterns that can be related to ancient civilizations, technologies, and even trade routes. For example, ratios of bismuth to gold in coins have revealed a distinction between Anglo-Saxon and Oriental coinage (Widemann, 1980). The copper alloy containing zinc used by the ancient Romans for coins and other metallic objects was also not of fixed composition. With the passage of time, the proportion of zinc gradually decreased, and concentrations of tin and lead increased from being minor impurities to being major components (Glascock and Cornman, 1983). Chemical assays of glasses and ceramics have also revealed substantial data on origin and age.

The basis of this trace element "fingerprinting" is that "if the concentrations of fifteen or more trace elements, in the parts-per-billion range vary in essentially the same way in two samples of lithic material, the samples logically can be considered to have come from the same geographic area" (Rapp and Gifford, 1982). In some potteries, the B/Li ratio has even indicated whether the clay used was a freshwater or marine sediment (Fink, 1981).

Variations in lead isotopic abundances have been correlated with the sources of lead and silver artifacts. The isotopic composition of lead, an element present in many copper ores and bronze objects, is unchanged during metallurgical processing, and so is an important tracer for determining the source of Bronze Age artifacts (Gale and Stos-Gale, 1982). In addition, by analyzing the isotopic differences of carbon and oxygen in marble quarries, Craig and Craig (1972) were able to assign marble objects to the quarry of their origin. And in one instance involving a question of authenticity, a sculpture in the Fogg Art Museum was shown to have been assembled from five unrelated pieces of other statues (Rapp and Gifford, 1982). A determination of Pb, Fe, Co, Mn, and Al has also been made in lead-glazed ceramic pottery shards of archaeological significance (Dolnikowski et al., 1984). In these recent analyses, fast-atom bombardment was employed, and lead ratios were compared to a glass (National Bureau of Standards, Reference Material No. 981) containing "common" lead.

Isotopic measurements of the uranium series are also now being used to date human bones, as well as carbonate materials of geochemical interest. Specifically, analyses of skeletal remains from the Del Mar and Sunnyvale, California sites indicate ages of 11,000 and 8300 years, respectively. These new findings support conjectures that such remains derive from a civilization that came across the Bering Strait during the last lowering of sea level about 13,000 years ago (Bischoff and Rosenbauer, 1981).

Various pathways of photosynthesis also result in differences in carbon isotopic ratios that make it possible to study prehistoric human diets. The $^{15}N/^{14}N$ ratio in plants varies, depending on whether the plants use atmospheric nitrogen directly or whether the uptake is primarily from the soil. Trace metal abundances also vary in foods such as nuts, berries, wild foods, and animal tissues, so that anthropologists are attempting to correlate diets with organic and inorganic samples from human fossils (DeNiro and Epstein, 1981). The reconstruction of past environments, however, is very complex, as many parameters are involved. Recent studies, for example, show that $\delta^{13}C$ values of leaves in a forest can vary with the height at which the sample is taken. Essentially, the CO_2 assimilated by leaves at the top of a forest canopy undergoes some mixing with the CO_2 of the atmosphere, in contrast to the environment of vegetation and residue on the forest floor (Van der Merwe, 1982).

Finally, one of the most significant advances in archaeological chemistry is ^{14}C dating by accelerator-mass spectrometry. By producing negative ions via bombarding the sample with a cesium ion beam coupled with electron stripping, it is now possible to discriminate against ^{14}N atoms and molecular species such as $^{12}CH_2$ and ^{13}CH. At the National Science Foundation facility at the University of Arizona, a 1 mg sample can be dated with an accuracy of ± 100 years in about 1 h (for ages $< 50,000$ years). In the future, a ^{14}C dating of samples as old as 100,000 years may be achieved by this technique (Zurer, 1983).

ASTROPHYSICS AND COSMOLOGY

The 1983 Nobel Prize in physics was shared by two astrophysicists: S. Chandrasekhar (1984) of the University of Chicago and W. A. Fowler (1984) of the California Institute of Technology, for their contributions to the under-standing of stellar evolution and the nuclear reactions involved in the for-mation of chemical elements in the universe. In particular, much of the work of Fowler and collaborators focused on nuclear reactions, isotopic abun-dance ratios, and the sequence of events responsible for the build-up of helium from hydrogen, the carbon-nitrogen cycle, and the energy production rates in the sun and distant stars. In the sun, the dominant energy production is now attributed to the proton-proton reaction.

$$H + H = D + e^{+} + \nu$$

and associated reactions that include

$$D + H = {}^{3}He + \gamma$$
$${}^{3}He + {}^{3}He = {}^{4}He + H + H$$

On the basis of these and other reactions (Bethe, 1983), the temperature at the center of the sun is now estimated to be about 14 million K. Fur-thermore, temperatures and nuclear reaction rates that have been measured in the laboratory are now used to compute the build-up of the elements heavier than helium; elements of atomic number greater than iron are gen-erated by successive neutron capture. Theoretical models for the formation of the chemical elements have also been compared to experimentally mea-sured isotopic abundances of the rare earth elements, whose existence should be inversely proportional to their neutron capture cross-sections. Mass spectrometry has been the primary tool for measuring the abundance of all stable and long-lived isotopes, and it has been employed extensively to measure neutron cross-sections (Chapter 13). Measured isotope ratios of uranium and thorium for lunar samples are comparable to terrestrial mea-surements, but spectrometric measurements of some meteorites reveal dif-ferent isotopic abundances for some elements and, hence, a different nuclear evolution (Prombo and Clayton, 1985). As more samples are collected that have entered our solar system, there is little doubt that additional isotopic measurements will contribute to an expanding data base and to refined as-trophysical models.

THE SUPERHEAVY ELEMENTS AND RARE PARTICLES

In 1966, based on theoretical calculations, an island of nuclear stability was predicted for nuclides in the region of atomic number 114 and having an atomic mass of approximately 300. Presumably, such nuclei may have been produced in the stars by the same processes that produce Th, U, and Pu. Also, the discovery of elements 107 and 109 suggested that new neutron-rich isotopes might exist, and that stable superheavy elements having a closed shell with N near 184 and Z in the range 110–120 might eventually be detected (Keller et al., 1984). Thus, the present search for superheavy elements, includes attempts to synthesize them by heavy ion reactions, as well as to detect such nuclei in terrestrial and extraterrestrial samples. The superheavy elements may indeed exist in the earth's crust or meteorites, but their average concentration has already been established as being less than 10^{-12} g/g (Flerov and Ter-Akopian, 1983). Experimental techniques employed to obtain direct or indirect evidence of these superheavy elements have included the ion microprobe, x-ray energy analysis, selective solvent extractions based on predicted ionic radii, high energy-heavy ion accelerators, and specialized spectrometers for measuring heavy ion reaction products. In the latter category are instruments whereby the combined action of magnetic and electric fields plus time-of-flight (TOF) measurements have permitted narrow limits to be placed upon ion energy, momentum, and charge state.

In the initial testing of one of these facilities (Maidikov et al., 1982), a 120 MeV ^{132}Xe beam from a cyclotron was used to adjust the operating parameters of the spectrometer, timing circuits, and detectors. Subsequently, the selectivity was tested by measuring the ^{197}Au recoil nuclei elastically scattered from a gold target at $0 \pm 5°$ by a 175 MeV ^{40}Ar ion beam. With selectivity defined as the ratio of the number of particles detected (N_{Au}) to the total number of incident particles (N_{Ar}), a spectrometer selectivity of $\sim 10^{-13}$ was measured. This very high selectivity is a prerequisite for detecting the products of heavy ion reactions, including the identification of nuclides that have been categorized as the superheavy elements.

This effort to develop new high-energy mass spectrometers has been collinear with advances in heavy ion accelerators. For example, in the "Bevelac" accelerator at the Lawrence Berkeley Laboratory, the total accelerating ion flight path of $\sim 10^8$ m required a cryogenically pumped liner in the vacuum tank to achieve pressures of $\sim 10^{-10}$ torr to minimize collisions with gas atoms. The technique of charge stripping has been successfully applied to very heavy nuclei, so that even uranium nuclei can be accelerated at very high charge states. As a result, the production of energetic ions is now making possible a whole new series of experiments relating to nuclear science and astrophysics. Table 20.1 shows selected species and intensities reported by the Berkeley Laboratory staff (Alonso et al., 1982; Alonso, 1985) for the upgraded heavy ion accelerator. The "mass spectrometry" of the reaction products that can be produced by these ions clearly presents a future challenge—for determining isotopic species and energies, and at signal-to-noise ratios (SNR) quite beyond that of conventional instrumentation. Contemporary experiments at the Heavy Ion Research Establishment in Darmstadt, West Germany typify this new thrust in mass spectrometry, where the reported identification of a single atom of element 109 has been compared to "finding a single grain of sand in a whole trainload."

TABLE 20.1   Energetic Ions Produced by the Bevalac Accelerator

Ion	Intensity[a]	Ion	Intensity[a]
Hydrogen-1	2×10^9	Manganese-55	1×10^6
Helium-4	3×10^9	Iron-56	2×10^8
Carbon-12	5×10^9	Krypton-84	1×10^7
Oxygen-16	6×10^9	Niobium-93	2×10^6
Neon-20	1×10^{10}	Xenon-129	3×10^6
Aluminum-27	5×10^8	Lanthanum-139	4×10^7
Silicon-28	6×10^9	Gold-197	1×10^7
Argon-40	1×10^9	Uranium-238	1×10^6
Calcium-40	4×10^7		
Calcium-48	1×10^7		

Source: Alonso et al. (1982); Alonso (1985).

[a] Particles per pulse in the external beam channel; particles are accelerated in a variety of high charge states.

One attempt to identify a nuclide of $Z = 110$, $A = 294$ was made by Stephens et al. (1980) with a tandem accelerator. A platinum nugget from Alaska was used as the source material and a TOF search was made for identifying mass 294. A limit on the possible presence of $^{294}110$ in platinum was established of 1 part in 10^{11}. A search for anomalously heavy isotopes has also been made to determine whether nuclides of known elements exist having masses beyond what can be formed by simply adding neutrons, for example, by the addition of a heavy particle to the nucleus (Kutschera, 1984). In an investigation at the Rutherford Laboratory (Smith et al., 1982), an accelerator-mass spectrometer permitted abundance limits to be set that are well below theoretical predictions of some big bang cosmological models.

A correlated problem to finding rare particles of anomalous mass or charge is to detect particles produced in rare processes. For example, the inverse beta decay induced by solar neutrinos on earth produces about 1 atom/ton/yr of target material (Kutschera, 1984), and the production rates of some so-called cosmogenic isotopes also depend upon sample depth.

ION BEAM LITHOGRAPHY

Conventional microcircuit lithography involves exposing thin films of polymers or "resists" to light in order that subsequent processing can produce high-resolution patterns. In a "negative resist," the radiation generates cross-linking in the polymer, so that exposed regions do not dissolve in the developer and the exposed pattern remains. In a "positive resist," polymer links are broken by the radiation, so that the resist dissolves more readily in the developer and the light exposed areas are removed. Below 1 μm, the photolithographic process becomes limited by diffraction, and practical problems associated with masking and etching often limit line resolution to 2 μm. Diffraction effects are negligible for energetic electrons, as their deBroglie wavelength is less than 1 Å (10^{-4} μm), and electron beams can be finely focused and modulated by electrostatic and magnetic fields. Hence, electron beam lithography can be employed to create a pattern on a substrate by beam scanning or by flood beam exposure of a resist-coated substrate through a mask. Furthermore, electron beam lithography provides an option for making one-step, high-quality photolithographic master masks having very high resolution and low defect densities. Although electron beams can be focused to spot sizes of 1000 Å or less, actual resolution is still limited by electron scattering. When an electron is incident upon a positive resist, such as polymethylmethacrylate (PMMA), it loses energy by both elastic and inelastic collisions; and the resultant electron scattering imposes a fundamental limit on electron lithography (Bowden, 1981) in addition to factors associated with the processing of the resist.

Hence, ion beams are being seriously considered for fabricating integrated circuits containing all the transistors, capacitors, resistors, and their metallic interconnects. In principle, ion beams could produce circuits of higher packing density and complexity, and certain physical constraints imposed by either photolithography or electron beam lithography could be circumvented (Brown et al., 1981). For future very large-scale integrated circuitry (VLSIC), ion beam diameters of 0.25 μm must be raster scanned within exceedingly narrow limits, and interferometer-controlled stages will have to move target wafers over an area of many square centimeters. Hence, the technical problems are indeed formidable. Nevertheless, the potential advantages of ion beam lithography are substantial:

1. Focused ion beams provide a maskless means for etching and ion implantation.
2. There is a lower limit to feature size, and the slope on etched features can be controlled.
3. The process involves the control of only a few instrumental parameters, and it can be highly automated.
4. The high-energy deposition of ion beams make them applicable to conventional resists and many other materials.
5. Ions are subject to less scattering than electrons, produce only secondary electrons of low energies, and thus high spatial resolution is maintained.
6. Microstructures can also be fabricated by projecting an ion beam *through* a mask on a resist, with very low damage to the mask.

The ion beam-assisted maskless etching of GaAs by 50 keV-focused Au ions (Gamo et al., 1982) provides a recently reported example of this ad-

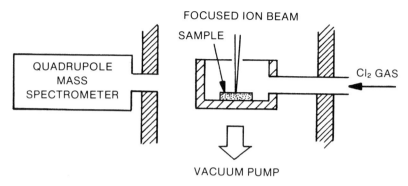

Figure 20.1 Target arrangement for maskless ion beam-assisted etching of GaAs. (Gamo et al., 1982; adapted with permission from *Jpn. J. Appl. Phys., 21,* L782, 1982.)

vanced ion technology. In this instance, the finely focused ion beam was employed to enhance the chemical etching of a sample in a chlorine atmosphere; this schematic diagram is shown in Figure 20.1. Scan rates of the beam were in the range of 0.5 μm/s and the ion current was 105 pA. Most important, the etching rate observed with a chlorine gas flow was more than 100 times that of a physical sputtering etching rate. Furthermore, mass spectra were observed both before and after the introduction of the chlorine gas (Fig. 20.2). The authors report that 75% of the chlorine gas reacted with the residual H_2O to produce HCl, so that both chlorine and HCl contributed to the etching process. Based on other data, they also conclude that this ion etching is enhanced by a kinetic collision process and not by a chemical interaction with the ion, and that the general methodology could also be applied to the fabrication of integrated optical elements such as gratings and waveguides.

Resolution in ion beam lithography is dependent upon the atomic number Z of the ion, its energy, and the scattered secondary particle distribution. For ions of high Z and high energy, where the range of the particle greatly exceeds the resist thickness, resolution will be primarily affected by the range of the secondary electrons. For the same energy, heavy ions will generate secondary electrons of lower velocities and range than light ions. For example, the distribution of energy from secondary electrons resulting from O^+ ion impact at 1.5 MeV is more localized than from He^+ or H^+, as schematically shown in Figure 20.3. However, for ions of both low Z and low kinetic energy (e.g., 50 keV H^+), the secondary electron range is quite small (≤ 50 Å), and resolution will be primarily determined by ion scattering (Bowden, 1981).

The energy deposited per unit path length (dE/dx) through a resist is, of course, much greater for an ion than an electron, and the trajectory of an ion is less susceptible to scattering. Polymer resists, mainly PMMA, have been investigated for ion beam sensitivity—for argon, gallium, helium, and

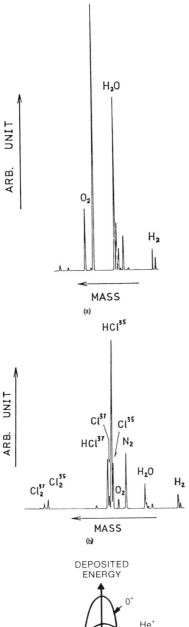

Figure 20.2 Mass spectra obtained of gases in target chamber (*a*) prior to and (*b*) after the introduction of chlorine gas. (Gamo et al., 1982; reprinted with permission from *Jpn. J. Appl. Phys.*, *21*, L782, 1982.)

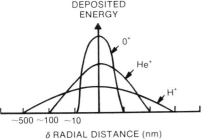

Figure 20.3 Radial distribution of secondary electron energy resulting from O^+, He^+, and H^+ impact at 1.5 MeV. (Hall et al., 1979; Bowden, 1981; reprinted by permission of The Electrochemical Society.)

hydrogen ions (Ryssel et al., 1983). At energies in the range of 40–120 keV, resists have been found to be ~100 times more sensitive to hydrogen ions and ~300 times more sensitive to argon than 10–20 keV electrons.

Adesida et al. (1983) have also compared the radiation yields or G values of PMMA for electrons, protons, helium, lithium, beryllium, and boron ions. Patterns with linewidths of <0.1 μm have also been replicated with proton beams (Economou et al., 1981). An inherent potential advantage over all other microlithographic techniques is that ion beams can be used for many functions, that is, producing a latent image, etching, implanting impurities, fabricating metallic interconnects, and generating multilayer films. Focused ion beams from Au-Si, Au-Be, and B-Pt liquid metal alloy ion sources have been used to implant GaAs and Si; and a 2000 Å diameter Au-Si focused ion beam has been used to implant the doped regions of GaAs metal semiconductor gate, field effect transistors (Kubena et al., 1981).

From a practical point of view, several types of "high brightness" ion beams are needed: (1) relatively light ions are appropriate for exposing a thick resist layer; (2) heavy ions are suitable for high etching rates; and (3) a variety of dopant ions are required for maskless implantation (Gamo et al., 1983). Be-Si-Au ternary alloys are among the useful alloy sources, because Be and Si are p-type and n-type dopants, respectively for GaAs. Beryllium is also a light mass ion that can be generated in a stable and finely focused beam from a liquid metal ion source, and Gamo et al. (1983) have reported high-intensity beams operating for 100 h with an energy spread limited to 10 eV full width at half maximum (FWHM). Somewhat lower energy spreads have been reported by other investigators. Wilkens and Venkatesan (1983) suggest that liquid metal ion sources are characterized by angular intensities ($dI/d\Omega$) of 20–30 μA/steradian, with a typical 5 eV energy spread compared to E beams of up to 1 mA/steradian with a ΔE of 1 eV.

Other Be-Si-Au liquid metal ion sources have yielded average ion emission currents of ~2 μA for 25 h, with measured ion energy spreads (FWHM) of 6.5, 7.0, and 18 eV for Be^{2+}, Si^{2+}, and Au^+, respectively (Miyauchi et al., 1983). The authors conclude, based on these and other measurements, that 0.1 μm diameter focused ion beams for Be^{2+} and Si^{2+} can be obtained from liquid metal sources at practical accelerating potentials (tens of kV). Such finely focused beams thus have dimensions comparable to the smallest features, yet replicated by contact photolithography, wherein narrow resist lines (~150 nm) have been produced by a molecular (F_2 dimer) laser operating at 157 nm as a vacuum ultraviolet source (White et al., 1984).

A number of facilities have now been successfully built for direct write-focused ion beam fabrication of semiconductors, with many investigators reporting on metal ion sources and finely focused beams. A eutectic alloy, liquid metal ion source for producing boron and arsenic beams for direct implantation of silicon devices has also been developed by Wang et al. (1981). A field emission, liquid Al ion source has also been reported with maximum currents of 380 μA/steradian at 900 μA (Torii and Yamada, 1982). Kumuro

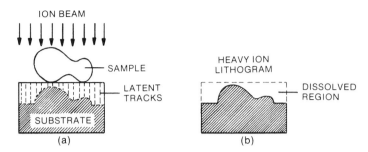

Figure 20.4 Ion beam microlithography, wherein the mass distribution of an ion-imaged object is registered in a nuclear track-sensitive material. The latent image is then dissolved to produce a relief-like replica that can be observed by a scanning transmission microscope. (Adapted from Fischer and Spohr, *Nucl. Instr. and Methods, 168*, 241 (1980), by permission of North-Holland Physics Publishing.)

et al. (1983) have reported finely focused beams of 40–50 nm for gallium and approximately 70 nm for indium, and tin sources at emission currents of ~4–12 μA. Such beams have been used for ion-enhanced etching of Si and GaAs substrates, as well as a masking etching process. Resists are very sensitive to heavy ion beams, for example, about 1×10^{-6} C/cm² of a 55 kV Ga beam is sufficient to expose PMMA resist. Kumuro et al. (1983) have also evaluated other parameters relating to ion irradiations, linewidths, and writing speeds. Mass spectrometry has also been used extensively to investigate many basic phenomena relative to liquid metal ion sources, per se, that is, clusters, charge states, and angle and energy distributions that ultimately place constraints on such sources for lithography. For example, the maximum number of atoms/cluster has been observed to range from 6 for Ga to 12 for In.

At the present state-of-the-art, finely focused ion beams from liquid metal ion sources can now provide current densities that are 10^3 to 10^6 times greater than from unfocused beams in commercial ion implantation facilities. Ion lithographic patterns have now been reported (for Si in GaAs) with center-to-center beam lines of 0.4 μm, and with beam diameters estimated to be ~0.2 μm (Bamba et al., 1983); implant time to raster scan a 400×400 μm pattern with this beam is reported to range from 38 s to 21 min, corresponding to doses of 2×10^{12} to 1×10^{14}/cm². For larger area implant, beam deflection scanning is supplemented by a motor-driven translation of the sample stage.

In addition to microcircuit applications, ion beams can also be used to replicate submicroscopic objects—including high-resolution micrographs of intact living cells. Sometimes termed "heavy ion microlithography," the technique generates relief-like replicas using irradiation by accelerated heavy ions followed by a nuclear track-etching step (Fig. 20.4). As cited by Fischer and Spohr (1980), advantages over x-ray techniques include:

1. Small accelerators/ion implanters can be used.
2. The technique offers high contrast due to the well-defined ion range.
3. A wide range of materials can be directly imaged by the technique.

The ultimate lateral resolution possible with heavy ions is about 100 Å. An additional advantage of heavy ions for microlithography results from the fact that all insulating solids and some semiconductors can, in principle, be used as "ion resists." The problem of radiation damage in a mask is much more severe for heavy ions than for electrons or x-rays, but Fischer and Spohr (1980) report that a mask can survive at least 10^{16} particles/cm^2, which is sufficient for about 10^4 exposures of a PMMA layer.

ION BEAM MODIFICATION OF MATERIALS

In addition to the specialized use of ion implantation for ion beam lithography, this technology continues to contribute to many fields. Clearly, applications in semiconductor materials will be dominant, because such materials are inherently impurity-sensitive, and their electric properties can be drastically altered by the addition of a relatively small number of atoms. For example, the scheduled NASA-sponsored Galileo mission to Jupiter will use a 640,000 element charge-coupled device (CCD) fabricated, in part, by ion implantation. After its arrival at Jupiter, this sensor must function in a high radiation field and provide a dynamic range of 3500 (faintest to brightest). Ion implant technology makes this feasible by providing permanent potential wells to trap photon-generated electrons, and yielding a much higher signal-to-noise ratio (ranging from 70–300) compared to conventional three-phase CCD sensors (Lerner, 1984). The production of buried oxide layers in silicon has also been made feasible through the development of ion sources that can furnish high currents of atomic oxygen to silicon wafers (Shubaly, 1985).

Furthermore, ion implantation is being used in conjunction with other techniques. For example, ion implantation has recently been employed with the metal-organic vapor deposition of InP (Favennec et al., 1983). These authors suggest that ion implantation may be useful in selected area epitaxy—by a judicious selection of ions, their energy, and their dose. Ion implantation can also produce advanced magnetic bubble memory devices of high density, reliability, and short access times (Kryder and Bortz, 1984).

Ion implantation is also rapidly becoming important in metallurgy, as "metallurgists are now able to take any element in the periodic table and implant it in the surface region of virtually any metal" (Picraux, 1984). In order to achieve thicker surface layers and higher concentrations, investigators are now exploring processes involving both ion implantation and film deposition. This combination of techniques permits mixing on an atomic scale to produce truly unique surface claddings. Hence, the prediction has been made by Gibson (1984) that the tailoring of the chemistry, structure, and topology of materials other than semiconductors by ion implantation will eventually have a comparable technological and economic impact. As mentioned in previous chapters, ion implantation of metals can result in dramatic improvements in the wear life of engineering components, such as cutting tools, discs, bearings, etc. And although considerable emphasis has been given to nitrogen implantation, many other species are being used effectively. Titanium-implanted steels, using implants at various energies, and dual implants of carbon and titanium show considerable promise. For example, improvements in sliding wear have been observed by Sioshansi and Au (1985) for bearing-grade steel implanted with C and Ti.

In aerodynamic research, either energetic electron or ion beams can also be used to modify wing surfaces in such a manner as to control laminar flow and reduce airfoil drag (Pearce, 1982). Specifically, Nagamatsu et al. (1984) have shown that at a Mach number of 0.81, the profile drag coefficient can be dramatically reduced if a solid surface airfoil is replaced with a porous surface. Essentially, an array of very small holes ($\sim$2% porosity) in a titanium airfoil, suppresses shock waves that form over airfoils at high subsonic speeds. Reductions of fuel consumption as high as 15–20% have been projected by means of this surface texturing. Perforated wing sections can, of course, be fabricated by several techniques. However, ion beam technology can provide a greater range of options for modifying both the physical and metallurgical characteristics of airfoils and other critical structures. [The natural texturing of the outer skins of marine life, e.g., fast sharks with longitudinal ridges and microgrooves of 40–100 μm, also results in a drag reduction of $\sim$10% (Bechert and Reif, 1984).]

Ion implants are also being examined in some detail in connection with the life sciences, where passive metals are being sought that do not release undesirable metal ions into the body. In the case of artificial hip joints, implanted specimens are relatively inert to corrosive body fluids, and wear rates lower by two orders of magnitude have been indicated. Ion beam sputtering is also an important technique for modifying the surface morphology (topography) and chemistry of biological implant materials (Kowalski, 1983). The sputtering process produces a microroughening of a material due either (1) to variations in the sputtering yield of the elements comprising the surface, or (2) to variations in sputtering due to the use of a screen mesh mask. This latter technique produces "pattern texturing," resulting in an ordered array of pores of constant dimension. Investigators are currently evaluating this "texturing" technique with respect to possible improvements of firm attachment and response of the surrounding tissue to the implant material.

The synthesis of some novel inorganic films by ion beam irradiation of polymer films has also been reported (Venkatesan et al., 1983). The investigators suggest that the effect of the ion beam irradiation "is to produce more Si-C bonds at the expense of the C-H and Si-Si bonds, with $\sim$10% of the original hydrogen being present in the film at high (Ar^+) doses."

Ion implantation has also been considered as a unique means for "tagging" or branding a valuable component or product. By implanting several elements with distinctive isotopic ratios, an unambiguous "fingerprint" of the component can be embedded in a surface layer. At General Electric (1983), researchers have even exposed diamonds to an ion beam (with a suitable mask) and demonstrated that a conductivity change could be retained within the diamond. By means of a subsequent dusting with an electrostatic powder that adhered to the ion-implanted regions, the ion-implanted pattern was made visible and photographed. Thus, it has been suggested that a photograph of a diamond might be retained by both customer and retailer for future identification.

Finally, the potential of ion beams in analysis cannot be overemphasized. By a combination of ion doping, diffusion, the isotopic dilution method, and ancillary techniques, trace metals in metals and other materials can be quantified at exceedingly low levels. By employing simultaneous multielemental implants, this approach will become a very powerful and general method for trace element assay. Quantitative implantation, to produce a homogeneously distributed internal standard element in heterogeneous biological soft tissue, has already been demonstrated (Harris et al., 1983).

NEUTRAL BEAMS FOR FUSION AND SPACE PROPULSION

In marked contrast to the use of highly focused ion beams in ion beam lithography, high-powered ion beam systems continue to be developed for (1) auxiliary plasma heating in fusion reactors, and (2) ion engines for space propulsion. In both applications, the essential elements are (1) the ion source, (2) an array of accelerating electrodes, and (3) a beam neutralizer. For fusion, ion beams must be neutralized in order to traverse the reactor's magnetic field and to deposit their energy by collision with the confined plasma's ions and electrons. In the propulsion application, if the ions emerged from the ion thruster into space without neutralization, the immediate build-up of electric charge on the space vehicle would cause some ions to return, resulting in an electrostatic "drag."

Engineering progress in both fields has been substantial. In the early 1970s, programs at the Oak Ridge National Laboratory and the Lawrence Berkeley Laboratory produced neutral beams of approximately 100 kW of power, with an energy of ~10 kV, but limited to a pulse length of less than 0.05 s. At present, beam systems can generate 10,000 kW of beam power at energies of 100 keV for intervals of about 30 s. Furthermore, a single negative ion beam line has been designed that is capable of quasi-DC operation for delivering a 1 MW, 200 keV beam of neutral deuterium atoms to a confined plasma. Charge neutralization is accomplished by gas stripping with a neutralizer "thickness" of $\sim 4 \times 10^{15}$ molecules/cm^2. Consideration is also being given to a laser photodetachment neutralization scheme,

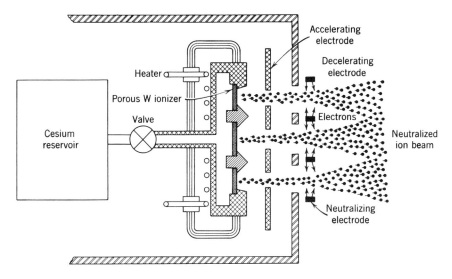

Figure 20.5 Schematic diagram of an ion engine for spacecraft propulsion, having a porous tungsten ionizer and utilizing cesium as a propellant.

whereby an optical power density of roughly 30 kW/cm² along a 1.5 m long neutralizer would have a neutralization efficiency of 99% (Goldberg et al., 1983).

An additional area of ion beam technology relates to the potential use of heavy ions to directly implode deuterium-tritium capsules for producing fusion power. In this heavy ion fusion (HIF) concept, an accelerator is needed to provide high power through the use of multiple beams and a sophisticated "bunching" of ion beams (Keefe, 1985). Experiments with intense cesium ions have demonstrated that very high currents can be transported in a long quadrupole system, and an advanced ion induction accelerator is presently being fabricated at the Lawrence Berkeley Laboratory in a proof-of-principle investigation.

The use of ions for space propulsion systems is intriguing to the mass spectrometrist in several respects. The basic ion engine can be considered to be a scaled-up version of a mass spectrometer or an isotope separator ion source, and the accelerating voltages are on the same order of magnitude. In addition, the mass spectrometer is one of the more basic analytical tools required to investigate surface ionization, charge exchange, and other phenomena that are specifically related to advanced ion thruster design. A typical ion engine is shown in Figure 20.5. Ions are produced from the "propellant fuel" in the ion source region by surface ionization, electron bombardment, or arc discharge (Boden, 1968). They are subsequently accelerated by a high-voltage electrode that provides them with an energy of 10–30 keV. After emerging from an exit aperture or nozzle, these ions are

TABLE 20.2   Comparison of Propulsion Systems

Propulsion System	Specific Impulse (s)
Chemical rockets	400
Nuclear rockets (with H_2)	800
Ion engines	15,000

Source: Michelson (1960).

neutralized by recombining with electrons, so that a high-velocity neutralized beam is discharged from the engine to produce thrust.

The thrust developed by the continuous expulsion of a neutralized atomic beam from an ion rocket operating in a vacuum can be computed directly from Newton's second law:

$$F = \frac{d(m\bar{v}_e)}{dt},$$

where $m\bar{v}_e$ is the momentum of the ion exhaust in the frame of reference of the space vehicle. Carrying out the indicated differentiation results in

$$F = \bar{v}_e \frac{dm}{dt} + m \frac{d\bar{v}_e}{dt}$$

$$= \dot{m}\bar{v}_e \text{ (for a constant exhaust velocity)}.$$

Thus the net thrust of an ion engine is directly proportional to $\dot{m}$, the rate at which propellant mass is discharged from the nozzle, and the terminal velocity of the electrically accelerated ions.

It is the exceedingly large ion velocity that makes the ion engine an attractive candidate for certain applications. In particular the ion engine has a very high *specific impulse*, a parameter whose defining equation is

$$I_{sp} = \frac{F}{\dot{m}g_0} = \frac{\dot{m}\bar{v}_e}{\dot{m}g_0} = \frac{\bar{v}_e}{g_0},$$

where g_0 is the gravitational constant. Specific impulse is thus the thrust generated per rate of propellant discharged (i.e., the number of pounds thrust per pound of fuel per second). Specific impulse then has the dimensions of seconds, and its magnitude is a direct measure of propellant economy. In this respect the ion rocket is superior to all other practical space propulsion systems. A rough comparison of the ion engine with other systems, using practical temperatures, is suggested in Table 20.2.

Nuclear-powered ion propulsion is envisioned for reducing substantially the transit times for unmanned vehicles to reach distant planets. And in the next century, nuclear-electric propulsion could make possible a manned expedition to Mars. A 6 megawatt electric power plant would enable a five man crew to travel to Mars in about 600 days, stay for 30 days, and return in less than 300 days (Truscello and Davis, 1984).

HYPERSONIC BOUNDARY LAYER ANALYSIS

Future applications of mass spectrometry to space vehicles relate to the interaction of the vehicles with the rarefied atmosphere during reentry. Data is specifically needed in the transition region between lower altitudes where balloon measurements have been made and very high altitudes where data has been acquired with sounding rockets and satellites. This transition region extends roughly from 125 km ($\sim 10^{-5}$ torr) to 55 km ($\sim 10^{-1}$ torr). At hypersonic reentry velocities, technical concerns include aerodynamics, surface heating and ablation, and topics that include:

1. Identification of gaseous species.
2. Measurements of dissociated molecules.
3. Determination of wall catalysis and nonequilibrium chemistry.
4. Measurements of rapid fluctuations of other near-surface parameters.

All of these measurements must be made within a fraction of a millimeter of the vehicle surface (Wood et al., 1983).

A major problem is in sampling the boundary layer without disturbing the flow or altering the gas composition. Hence, a flush-mounted inlet system has been proposed, consisting of an array of microchannel capillaries through which the gases can infuse, similar to the geometrical configuration used in microchannel plates. In addition, the simultaneous monitoring of several gaseous species is desired, because of the short time for vehicle reentry. Among species of interest are atomic and molecular oxygen, N_2, NO, NO_x, CO, CO_2, and recombination products. For this type of analysis, atmospheric argon ($\sim 0.9\%$) will be useful for normalizing the data. Comparable instrumentation is also being developed in hypersonic high-enthalpy wind tunnels and shock tube facilities. These differ from the flight experiments in that run times are very short (~ 1 ms to ~ 40 s) and wall effects must be considered.

PLASMA ION BEAM PROBES

Techniques that have been widely utilized for plasma diagnostics include microwaves, interferometers, and Langmuir probes. Ion beams represent

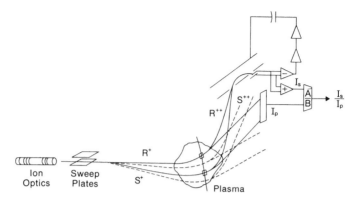

Figure 20.6 An ion beam probe method for measuring plasma electron density and electron temperature (adapted from Reinovsky et al., 1974; courtesy of R. L. Hickok; © 1974 IEEE.)

an attractive alternative because they can probe a plasma with high spatial resolution and leave the plasma undisturbed. Basically, a beam of singly charged ions (Fig. 20.6) can be swept through a plasma volume, and some of these primary ions will undergo nonmomentum transfer collisions to form doubly charged ions. The number of doubly charged ions formed is related to the number density and temperature of the electrons in the plasma, so that a detailed analysis of primary and secondary ion trajectories and currents can yield information on many plasma parameters, that is, plasma density, electron temperature, space potential, and plasma current density (Jobes and Hickok, 1970; Hickok and Jobes, 1971; Reinovsky et al., 1974). The latter report describes an ion probe apparatus for determining electron temperature, T_e, in which Na^+ and K^+ ions are simultaneously emitted from a surface-ionization source. The sample volume is probed along a "detector line" by sweeping the primary ion beam by means of electrostatic deflection plates. The secondary ion current (I_s) that arrives at the detector is

$$I_s = \gamma I_p \, n_e \, f(T_e)$$

where γ is a known geometrical factor, I_p is the primary ion beam current, n_e is the electron density at the point where the alkali ions become doubly charged, and $f(T_e)$ represents the effective cross-section for electron-impact (EI) ionization from the 1^+ to 2^+ charge state.

For the case where two beams of different species are programmed to probe the same point in the plasma, the ratio of the normalized secondary current is given by

$$\frac{(I_s/I_p)_1}{(I_s/I_p)_2} = \frac{\gamma n_e \, f_1 \, (T_e)}{\gamma n_e \, f_2 \, (T_e)} = g \, (T_e)$$

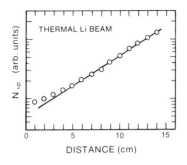

THERMAL Li BEAM

N_{vp} (arb. units)

DISTANCE (cm)

Figure 20.7 Semilogarithmic plot of the photon flux (N_{vp}) emitted from Li atoms thermally injected into a plasma. The electron density estimated from the slope of the straight line is $4.4 \times 10^{11}/\text{cm}^3$. (Kadota et al., 1982; reprinted with permission from *Jpn. J. Appl. Phys., 21,* L260, 1982.)

Therefore, if the effective cross-sections are known, T_e may be evaluated. Also, once T_e is known, this equation can be used for either ion species to calculate the electron density n_e. In the schematic diagram, the ion gun voltage is selected so the momentum of the R^+ species brings the R^{2+} secondaries onto a detector through a simple energy filter, while the second species (S^+) and its doubly charged ions (S^{2+}) do not reach the detector, so that only the I_s/I_p ratio is measured for the R beam. Rapid ion gun voltage switching then brings the S beam onto the detector and the cycle is repeated. This experimental apparatus was designed for alkali metal ions in the range of 10–100 keV. A salient feature of this work is that it provides a spatial resolution of approximately 0.1 cm^3 and a time resolution of 10^{-5} s. A complete radial temperature profile is obtained in 10 ms.

Spatial profiles of electron densities in plasmas have also been determined with both fast (several keV) and thermal ($\sim 1.5 \times 10^5$ cm/s) neutral Li beams. This method is especially useful in low-density plasmas, $n_e < 10^{12}/\text{cm}^3$ (Kadota et al., 1982). The lithium atoms are excited by electron impact, and the photon emission from the resonance line of the Li atom (6708 Å) is monitored via a monochromater and photomultiplier detector. Figure 20.7 shows a semilog plot of photon flux N_{vp} vs. distance (x) for thermally injected neutral lithium atoms into a plasma. If the cross-section for electron excitation and other parameters are known, the electron density can be determined from the slope of the line.

The injection of intense beams of neutral Xe and Hg atoms into plasmas, and a collimated x-ray detector system that can analyze the energy distribution of the emitted photons, represent a similar diagnostic method (Moses et al., 1984).

ACCELERATOR-MASS SPECTROMETRY

The advent of accelerator analyzers specifically designed for ultrasensitive mass spectrometry is making possible a greatly expanded scenario of stable and radioisotopic measurements. As indicated in Chapters 4 and 8, accelerator-mass spectrometry is now an important tool for ^{10}Be, ^{14}C, ^{36}Cl detection and dating, and it is indispensable for the detection of some radio-nuclides having very long half-lives.

One recent accelerator-mass spectrometric study has indicated that this type of analysis is even applicable to short-lived tritium (^{3}H) with a half-life of only 12.3 years. Jiang et al. (1984) have reported the detection of tritium at a concentration of 1 part in 10^{14} of deuterium and a detection efficiency of two orders of magnitude greater than by beta counting. This unique type of spectrometry affords salient advantages and utilizes a multiplicity of tech-niques that can be summarized as follows:

1. The production of a pure atomic spectrum with no overlapping mo-lecular spectrum.
2. The ability to resolve isobars such as ^{14}C and ^{14}N with a discrimination factor greater than $10^{15}:1$.
3. The measurement of very small isotopic ratios (e.g., 10^{-15} for ^{14}C/^{12}C).
4. The combined use of energy, mass, and charge analyzers, with ion range, dE/dx, and TOF detectors for identifying and counting single ions.
5. The assay of small samples with reasonably high ionization efficien-cies and transmission.

A fundamental advantage of accelerator-mass spectrometry is that it makes use of not one but several atomic parameters in order to resolve nuclides of nearly identical mass. The analytical power of this spectrometry is also related to its value in assaying the entire Periodic table. As aptly pointed out by Litherland (1980, 1984) in his excellent overview of this field, the only element that has no stable isotope with a unique mass number is indium. Essentially, because of the very long half-lives (over 10^{14} years) of ^{113}Cd and ^{115}In, only indium cannot be uniquely identified by observing one of the stable isotopes. In the case of carbon, a sputter ion source (such as Cs$^+$) has been used to generate a high yield (~5%) of C$^-$ ions. This source also makes possible the discrimination against N$^-$ ions, which have a very short lifetime compared to the flight time through the instrument. In the contemporary tandem accelerator-mass spectrometer described in Chapter 4 (Fig. 4.22), the negative carbon ions are first analyzed by a 90° magnet and then accelerated to 2 MeV, where four electrons are removed from the negative ion in the high-voltage terminal gas stripper to produce C^{3+} ions. These positive ions are then further accelerated, focused, and analyzed by a combination of electrostatic and magnetic deflectors. Further particle iden-tification (E/q) is provided by having the ions pass through an electrostatic analyzer. With this type of ion source discrimination and multistage filtering, ratios of 10^{-15} for ^{14}C/^{12}C can be determined. For very heavy ions, a TOF detection system provides an additional means of discrimination.

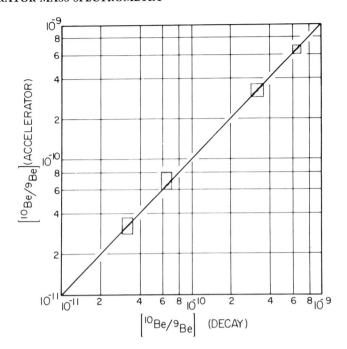

Figure 20.8 Comparison of the $^{10}Be/^9Be$ ratios determined by beta ray counting (DECAY) and atom counting (ACCELERATOR). (Turekian et al., 1979; reprinted with permission from *Geophys. Res. Lett.*, *6*, 417 (1979), copyright American Geophysical Union.)

The use of this type of spectrometry for ^{10}Be is equally impressive (Kilius et al., 1980), with ratios of 7×10^{-15} being obtained for $^{10}Be/^9Be$ and ^{10}Be being detected in a sample containing as few as 10^7 atoms of ^{10}Be. Because ^{10}Be is produced by cosmic rays in the atmosphere, geological investigations of this nuclide have included measurements of ice from Antarctica, the search for ^{10}Be variations in marine sediments, and its abundance in ocean surface layers. Furthermore, in a series of laboratory experiments where ^{10}Be was generated by neutrons in a $^9Be(n, \gamma)$ ^{10}Be reaction, a comparison was made of the irradiated samples by both beta decay measurements and accelerator-mass spectrometry (Turekian et al., 1979). As shown in Figure 20.8, a linearity of better than 1% was noted over three orders of magnitude in concentration. Other nuclides that have been detected with the accelerator technique include ^{26}Al, ^{36}Cl, and ^{129}I. ^{129}I has also been studied by conventional negative ion mass spectrometry; it is a radioactive nuclide (half-life, 1.6×10^7 years) that is created in meteorites by cosmic rays, and it is also produced by the spontaneous fission of ^{238}U and the neutron-induced fission of ^{235}U. Table 20.3 shows the half-lives of these nuclides together with the sensitivity reached by single ion counting (Litherland and Rucklidge, 1981).

The accelerator-mass spectrometer at the University of Toronto provides

TABLE 20.3   Mass Spectrometry of Rare Isotopes

Nuclide	Half-Life (yr)	Sensitivity[a]
^{10}Be	1.6×10^6	7 ppq
^{14}C	5730.	0.3 ppq
^{26}Al	7.2×10^5	10 ppq
^{36}Cl	3.1×10^5	0.2 ppq
^{129}I	16.0×10^6	300 ppq
Pt	Stable	10 ppt

Source: Litherland and Rucklidge (1981).

[a] ppt, parts per trillion (10^{12}); ppq, parts per quadrillion (10^{15}).

some unique options (Kilius et al., 1984). A simplified schematic of this facility is shown in Figure 20.9. Heavy element analysis is performed with ion source 1 and a 90° magnet (M1) that has a mass resolution of $M/\Delta M = 400$. Ions that emerge from the accelerator first pass through a 15° electrostatic analyzer (ESA3) and a pair of 45° strong focusing magnets. The ion detection system includes a surface barrier detector for ion energy determination and an optional thin-foil TOF detector (TF1) that can be inserted between the two magnets. Ion source 2 (for the production of positive ions) is followed by a pair of 45° spherical electrostatic analyzers (ESA1 and ESA2) and a metal vapor charge changing canal. Mass identification after molecular disintegration is also obtained by means of TOF measurements through an appropriate set of toroidal electrodes (ESA5 and ESA6) that compensate for time degradation introduced by the start detector. This *isochronator* provides a mass resolution of >1000.

This high-energy mass spectrometer at Toronto has been employed for trace heavy element analysis, with detection sensitivities of >1 count/s/ppb for platinum and silver isotopes in a variety of matrices. Sputtered negative light ions that have been measured include carbon, aluminum, and beryllium.

Several cyclotrons have been converted to operate as mass spectrometers, with appropriate modifications of ion sources and the addition of ion beam tubes and ion detectors. Figure 20.10 is a schematic diagram of the sector-focused cyclotron at the Naval Research Laboratory after its conversion for single-particle counting (Glagola et al., 1984). Another recent accelerator-mass spectrometer makes use of the cyclotron principle, but the instrument is very small (the final orbit radius is only 10.5 cm) and the final ion energy is only 40 keV (Welch et al., 1984). This latter instrument is being developed at the Lawrence Berkeley Laboratory, where negative ions are used for radiocarbon measurements in order to eliminate interference from ^{14}N, and a mass resolution of 30,000 is anticipated. The small size of this apparatus is expected to reduce the cost to that of conventional mass spec-

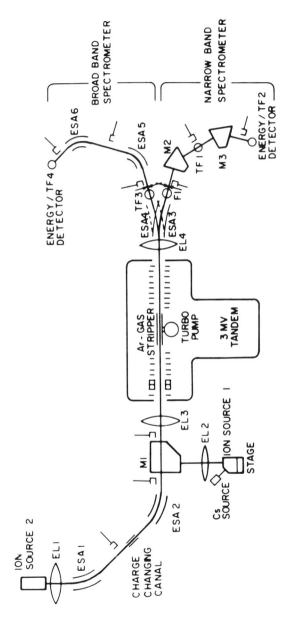

Figure 20.9 Dual ion beam transport system for the accelerator-mass spectrometer at the IsoTrace Laboratory of the University of Toronto. (Kilius et al., 1984; courtesy of the Isotrace Laboratory, University of Toronto, and permission of North-Holland Physics Publishing.)

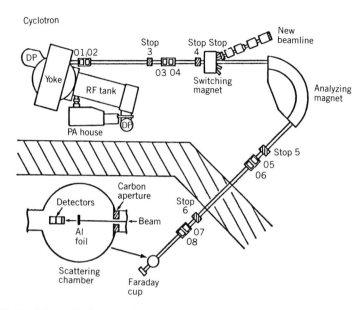

Figure 20.10 Schematic diagram of the cyclotron-mass spectrometer used for the direct atom counting of tritium. (Glagola et al., 1984; reprinted from *Nucl. Instr. and Methods*, *B5*, 221 (1984), by permission of North-Holland Physics Publishing.)

trometers, and the cyclotron analyzer is especially effective in suppressing spurious ions that arise from various scattering processes.

Worldwide, there are now some 30 accelerators being used as ultrasensitive mass spectrometers, and this number is expected to increase. Many of these are being used for ^{14}C dating, especially for small samples, because the specific ^{14}C activity of natural carbon is only 13.56 disintegrations/min/ g of carbon. A limited number of applications are cited in Table 20.4. Wilson et al. (1984) have also reported the development of an ion microprobe for accelerator-mass spectrometers, and this microprobe technique is expected to be useful for the assay of many geochemical specimens and particles. The feasibility of detecting ^{41}Ca ($T_{1/2} = 103,000$ yr) has been demonstrated by Elmore et al. (1984); this research group has also demonstrated the separation of ^{98}Tc and ^{98}Mo for use as a solar neutrino detector by accelerating ions in a high-charge state (18) to 215 MeV.

Among other specialized accelerator-mass spectrometer systems is the on-line recoil mass spectrometer (RMS) at the tandem Van de Graaff Laboratory at the University of Rochester (Figure 20.11). This is a true recoil spectrometer, in which the usual ion source is replaced by a nuclear reaction target itself, in contrast to on-line isotope separators where reaction products are transported via helium jet or diffusion (Cormier et al., 1983). This system reduces the flight times by a factor of 1000 compared to the usual times involved for on-line ionization, extraction, and flight time to the focal plane

TABLE 20.4   Applications of Archaeological and Geological Interest for Accelerator-Mass Spectrometry

Meteorites	Historical artifacts
Earthquake samples	Amino acids
Ground water supplies	Pottery
Marine reservoirs	Bones
Deep sea sediments	Teeth
Manganese nodules	Ivory
Lunar samples	Shells
Lavas	Textiles
Coral	Lunar samples
Tektites	Atmospheric particles
Oil	Polar ice
Coal	Atmospheric CO_2
Tree rings	Soils

Source: Mook (1984).

(5–500 ms). The RMS system times are only 1–4.8 μs, which allows a 1:1 correlation to be established in coincidence experiments relating cascade radiation detected at the target and ions detected at the focal plane. This instrument has a mass resolution of approximately 700 and is currently being used to investigate subcoulomb fusion, particle-γ coincidence measurements, and other nuclear structure research.

Finally, two accelerators working in tandem at the Lawrence Berkeley Laboratory have also produced the first completely ionized uranium atoms, that is, bare ^{238}U nuclei with no orbital electrons. This combined accelerator system (Bevalac) is able to generate heavy ions having up to 87% of the speed of light. By subsequently passing these ions through thin films, ions can be produced that are either totally ionized or stripped of any desired number of electrons (Gould et al., 1984). Magnetic analyzers are then able to separate ions of various charge states. This new technique makes beams

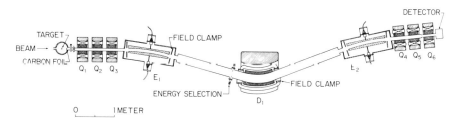

Figure 20.11   On-line recoil mass spectrometer for nuclear research at the University of Rochester. (Cormier et al., 1983; reprinted from *Nucl. Instr. Methods, 212,* 185 (1983), by permission of North-Holland Physics Publishing.)

of highly ionized nuclides available for basic studies in atomic and nuclear physics. The newly constructed ATLAS heavy ion accelerator at the Argonne National Laboratory also merits special mention. This radiofrequency accelerator contains 42 modular niobium split ring resonators cooled by liquid helium, so that only a few watts of power are required to accelerate ion beams of ^{7}Li to ^{127}I to energies as high as 25 MeV/nucleon. Ultimately, this system will be augmented to accelerate ions up to ^{238}U.

ANALYSIS BY RUTHERFORD BACKSCATTERING

Rutherford backscattering (RBS) of MeV He$^+$ ions is now one of the most important spectrometric techniques for the quantitative elemental analysis of the near-surface regions of solids. It is also a complementary technique to secondary ion mass spectrometry (SIMS) for studying shallow implantation profiles, the stoichiometry of oxides and silicides, impurity distributions, and diffusion. RBS can be employed with other high-energy ions such as protons or ^{16}O$^+$, but ^{4}He$^+$ beams are the most widely used probe. The specialized use of ion beams to explore the phenomenon of channeling has been discussed in Chapter 12.

In support of microelectronics research, RBS is being employed to determine the uniformity and stoichiometry of solid-phase epitaxial films. For example, silicides such as $CoSi_2$, NiS_2, Pd_2Si, and PtSi grow epitaxially in silicon substrates, and RBS is a unique nondestructive method for determining the compositional profile of such structures (Ishibashi and Furukawa, 1983). RBS is also important in revealing implantation defects in semiconductors, very shallow depth distributions, and changes induced by annealing (Wilson et al., 1983; Evans and Strathman, 1983).

The theoretical basis for RBS is coulomb scattering between two bare (unscreened) nuclear point charges, so that for an incident or projectile ion of atomic number Z_1 interacting with a target nucleus of atomic number Z_2, the normal differential Rutherford cross-section $(d\sigma/d\Omega)_R^{lab}$ is characterized by the expression:

$$\left(\frac{d\sigma}{d\Omega}\right)_R^{lab} = \left(\frac{Z_1 Z_2 e^2}{4\,E\,\sin^2\theta/2}\right)^2 [1 + A\gamma^2 + B\gamma^4 + C\gamma^6 \ldots]$$

where e is the electronic charge, γ is the mass ratio (M_1/M_2) of the incident to target atom, E is the incident particle energy, and θ is the scattering angle in laboratory coordinates. For the commonly used backscattering angle (150°), the constants have values of $A = 1.741$, $B = 0.544$, $C = 0.150$, and the E^{-2} energy dependence has been shown to be valid for ^{4}He backscattered from O, Al, and Si atoms in the energy range 0.6–2.3 MeV to within 2% (MacDonald et al., 1983). These investigators have also produced the energy

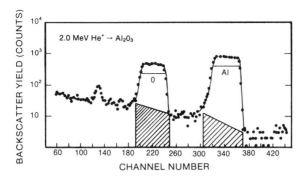

Figure 20.12 Energy spectrum of 2 MeV ^{4}He backscattered from a 58 μg/cm^2 self-supporting Al$_2$O$_3$ film. The shaded area under each peak indicates the magnitude of the background correction. (MacDonald et al., 1983; reprinted from *J. Appl. Phys.*, *54*, 1800 (1983), by permission of the American Institute of Physics.)

spectrum (Fig. 20.12) of 2.0 MeV ^{4}He$^+$ ions backscattered at 150° from a self-supporting Al$_2$O$_3$ film. The O and Al peaks are clearly resolved. In an analogous series of backscattering-channeling experiments, Tatsuta et al. (1982) have used 330 keV He$^+$ ions to study plasma-enhanced, thermally grown nitride films on (100) Si.

For ions backscattered by surface atoms through an angle of 180°, a simple relationship holds for the initial energy E_0 and mass m of the incident particle, the mass M of the target atom, and the energy E of the backscattered particle

$$E = \left[\frac{M - m}{M + m} \right]^2 E_0$$

Thus, it is easy to correlate an energy spectrum to a mass spectrum. For target atoms below the surface in a solid, an additional energy loss of the backscattered ion provides depth information.

The 180° geometry has been utilized by Jackman et al. (1981) in an experimental arrangement whereby a steering magnet permits ^{4}He$^+$ ions to strike the target at normal incidence, and reflected ions (at 180°) can be captured by a surface barrier detector. This scheme permits the detection of *all* backscattered ions within a small (~0.04°) angular cone centered at 180°, and it results in a large enhancement of the backscattered ion yield. Also, in a recent development (Gibson, 1984), ion beams from an accelerator have been focused to a diameter of only 10^{-3} mm to provide highly detailed spatial as well as depth information. Thus, RBS is being used as a "nuclear microprobe" that can scan samples in a manner analogous to the ion microprobe that uses low-energy ions.

One other salient advantage of RBS in contrast to the SIMS technique is that ion counting can be correlated to impurity concentrations in an absolute

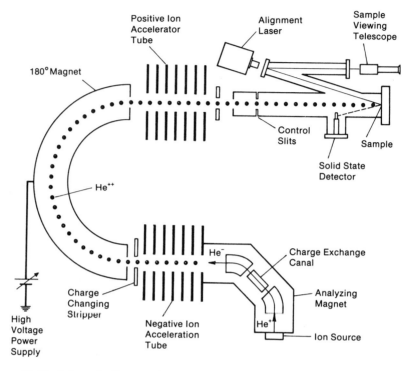

Figure 20.13   Schematic diagram of a dedicated Rutherford backscattering surface analyzer for nondestructive depth and mass analysis (General Ionex Corp.).

sense. A single reference target containing a precisely known amount of a well-chosen high Z nucleus/cm² suffices to calibrate all RBS spectra (Amsel and Davies, 1983). In another recent application, RBS has been used to observe the structure and composition of a Pt (100) surface, and to dynamically monitor kinetic oscillations in the CO oxidation rate (Norton et al., 1984). A schematic diagram of a commercial instrument specifically designed for RBS analyses is shown in Figure 20.13 (Purser, 1984).

PARTICLE INDUCED X-RAY EMISSION

The use of ion beams for producing x-rays that can be correlated with an elemental analysis was first reported in 1970 (Johansson et al., 1970). At present, the use of energetic (MeV) protons or heavier ions combined with energy-dispersive x-ray detection permits a nondestructive multielement analysis to be made at a sensitivity approaching 10^{-18} g. Also, in contrast to electron beam probes having a relative elemental sensitivity in the 1000 ppm range, particle-induced x-ray emission (PIXE) analysis permits some trace analyses at the 1 ppm level. Unlike RBS analysis, however, PIXE provides poor depth information. A general review of PIXE methodology and its many applications has been reported by Cahill (1980), who aptly characterizes this ion/x-ray technique as being at the interface between basic atomic physics and the discipline of analytical chemistry.

There are several valid reasons for the increasing attractiveness of this ion-induced x-ray technique. First, the probability of x-ray production by MeV ions is large for all elements, that is, the cross-sections for K or L shell x-rays usually exceeds 100 barns. Second, these cross-sections vary gradually from element to element as a function of ion energy, unlike thermal neutron cross-sections that apply to neutron activation analysis. Third, ion beam energies are well within the range of commercially available accelerators, that is, 2–10 MeV protons can be effectively used for the analysis of aluminum to uranium. Fourth, increasing use is being made of "external" beams on targets or samples by passing the ion beam through Be or plastic foils. This makes possible the ion beam excitation of samples of biological interest, liquid samples, and large specimens such as archaeological artifacts or meteorites. Fifth, the use of microbeams (< 100 μm in diameter), either by collimation or employment of ion focusing schemes, has made PIXE an attractive candidate for microprobe analysis. Minimum spot sizes approaching 1 μm have been reported at some laboratories in the United States and Europe, using multiple magnetic or electrostatic lenses (e.g., Cookson, 1979). Finally, advances in energy-dispersive x-ray detectors and associated electronics have resulted in improved resolution, better signal-to-noise ratios, and higher counting rates. Standard reference samples, of course, are required for quantitative work.

As with classical mass spectrometry, applications include many disciplines, for example,

1. Atmospheric chemistry (particles and aerosols).
2. Environmental science (water samples).
3. Biomedical assays (trace levels in tissue).
4. Materials.
5. Forensic science (residues).
6. Geochemistry (rocks and soils).

Successful microprobe assays have been reported of geological specimens, and one report has suggested the dating of minerals by the measurement of U/Th/Pb ratios. In the future, it is also likely that ion implantation will be used in conjunction with the PIXE methodology for specialized "nondestructive" assays.

Figure 20.14 is a schematic diagram of the proton-induced x-ray apparatus employed by Jolly et al. (1975) to analyze aerosols emitted from aircraft engines. In this investigation, the aerosols were separated into eight size

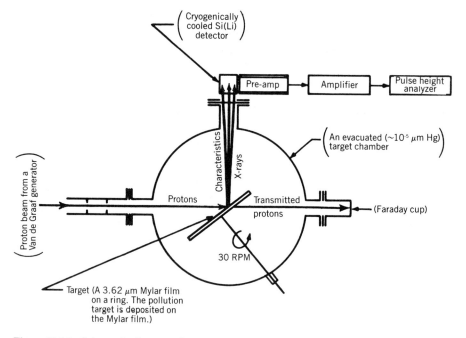

Figure 20.14 Schematic diagram of a proton-induced x-ray apparatus for measuring aerosols emitted from aircraft engines. (courtesy of J. J. Singh and the NASA-Langley Research Center.)

groups and deposited on a mylar film. A proton beam was then directed onto this film, which was rotated, and characteristic x-rays were detected by a liquid nitrogen-cooled Si (Li) detector. Si, P, S, Cl, K, Ca, Ti, V, Fe, Ni, Cu, Zn, Br, Sr, Nb, and Pb were identified and quantified, and six elements (S, Ca, Fe, Zn, Sn, and Pb) showed a fractional concentration increase with decreasing aerosol size.

In the medical context, PIXE has been applied to determine metallic elements in patients having neurological diseases, and in forensic studies, this technology has been applied to assaying gunshot residues, hair, shreds of fabrics, glass fragments, etc.

Other recent applications of PIXE have been made feasible by having samples bombarded by an *external* particle beam, with the beam penetrating a thin window so that a sample is excited in air or in a helium environment. In the field of marine biology, shells of living oysters have been analyzed by this technique (Swann, 1983). The analysis of aerosols in the workplace has also been successfully demonstrated by this external beam technique (Williams and Punyasena, 1983). PIXE assays with low-current external beams have also been used in connection with many manuscripts, art works, and archaeological artifacts where vacuum environments cannot be tolerated (Eldred, 1983). Thin specimens such as inks and paints are especially amenable to elemental analysis.

In-air PIXE has also been applied to such practical concerns as the corrosion or erosion of metal surfaces and corrosive deposits that are found in steam turbine systems (Pronko et al., 1983).

SPECTROMETRY OF VERY LARGE MOLECULES

"The most important frontier of this decade will be the extension of mass spectrometry to the analysis of increasingly heavier molecules, first of middle mass molecules (mass range 1,000–10,000) and then beyond 10,000 amu to heavy molecules" (Fenselau, 1984). Professor Fenselau also envisions this new frontier as an interdisciplinary challenge, demanding significant mass spectrometric advances "in order to provide structural information with enough accuracy and sensitivity to be truly useful to polymer, industrial, biological, and pharmaceutical chemists." Progress within the last decade has already been impressive. For example, the analysis of vitamin B_{12} (1355 amu) is now considered routine, and many biochemicals in the range of 2000–4000 have been analyzed at the Middle Atlantic Mass Spectrometry Laboratory at Johns Hopkins University and elsewhere. The detection of insulin has also been considered something of a milestone in mass spectrometry, because this peptide is nearly large enough to be called a protein. Thus, a future goal of 100,000 amu does not seem to be unreasonable.

First, however, there must be a greater recognition that advanced instrumentation, per se, is a prerequisite to further advances in mass spectrometry. Other disciplines, such as high-energy physics and astronomy, have always operated on the premise that the discovery of new elementary particles and the detection of electromagnetic radiations from distant stars required advanced facilities—some demanding a decade of design and development. Mass spectrometry, as a discipline, merits a similar type of commitment to advanced instrumentation quite beyond some of the excellent spectrometers that are now available from commercial sources.

It would appear that several facets of instrumentation could be borrowed directly from nuclear physics laboratories. For the production of heavy ions, an extension could be made of the existing techniques of fission fragment (^{252}Cf) and fast-atom bombardment. Fluxes of energetic heavy ions are available from various high-voltage accelerators, so that *ion-produced ions* (with or without charge neutralization) are possible beyond the constraints of a ^{252}Cf source. Field desorption and laser field desorption will probably be developed further, and selected combinations of sample ablation and ionization techniques (laser, chemical ionization, electrospray, multiphoton, thermal transients, pulsed-focused radiations, etc.) can be explored. Also, some research can be profitably pursued with very low ion counting rates.

The mass filtering or mass analysis of singly charged ions of mass 100,000 is already feasible, providing that conventional mass resolution requirements are substantially relaxed. Thus, while good resolution is desirable, it is not always essential for many exploratory investigations. "The quadrupole mass filter provides one option. One such instrument has been built, with a higher than normal rf voltage level operating at a lower than normal frequency, driving a mass filter with $\frac{3}{8}''$ rods to a mass of 80,000 amu and with $\frac{1}{4}''$ diameter rods to 150,000 amu" (Fite, 1984). Large magnetic analyzers are also attractive, and they afford several advantages. And in comparison with some magnetic systems used in nuclear experiments, they can be of uncomplicated design and modestly priced. For example, a 100,000 amu singly charged ion could be decelerated to 1 keV after extraction from an ion source and then mass analyzed. At 1 keV, a magnet of only 144 cm radius of curvature operating at 10^4 gauss would provide a rough mass-range identification for the heavy ion. At a more realistic ion kinetic energy of 10 keV, the magnetic radius of curvature would have to be increased to approximately 4.5 meters. This dimension is still "small" compared to the dimensions of contemporary accelerators and analyzers in high-energy physics. Another option for heavy ion analysis exists by utilizing cyclotron resonance with superconducting magnets, even though its ultimate analytical potential has yet to be demonstrated (Fenselau, 1984).

In addition, a large magnetic analyzer combined with a TOF technique would appear to merit a serious appraisal. A system comprising a pulsed ion source, magnetic m/z identification, plus submicrosecond measurements in a long flight path would yield data that could not be acquired with a single mode of analysis. Inhomogeneous and multistage systems afford further options.

The detection of a single molecule or ion cluster of 100,000 amu is difficult, but not impossible. It is feasible because only the arrival of a single ion must be registered at the detector, that is an "event" detector, but not an energy-sensitive detector, is required. Nevertheless, some type of post-acceleration of a mass-analyzed ion appears to be essential. Even with post-acceleration, however, the velocity of the particle would probably be too low to produce kinetic secondary electron emission from a surface, and the range of the particle would be much too short to generate electron-hole pairs in the depletion region of a reverse-biased p-n junction. Thus, the detection of a single heavy ion—after it has passed through the final defining aperture—must be registered (1) via the collection of electrons that may be emitted by the charged particle itself on impact with a detector surface; (2) via charged particle counting where a high local electric field (e.g., surrounding a wire in a gas ionization chamber) breaks down producing ions and free electrons; or (3) via an actual momentum (mv) sensor in contrast to charge amplification and collection. The first scheme has not been tested for very heavy ions, but Ligon (1983) has suggested that this method should be explored, because very large ions might be expected to have relatively low energy requirements for multiple ionization (< 30 eV) and because such a process is entirely compatible with existing mass spectrometers that utilize detection by electron multipliers. The second scheme assumes differential pumping at the detector, because a spark-type counter (such as those formerly used for registering single alpha particles) requires some partial pressure of gas to provide charge amplification. By ingenious means, however, the gas could probably diffuse through the walls of the wire counter, so as to impose a very small differential pumping load. A third "momentum"-sensing scheme has been mentioned in Chapter 5 in connection with detecting beams (not single molecules) of neutrals by means of a very thin (~100 Å) membrane microphonic sensor.

Furthermore, the somewhat specialized heavy ion detector described by Fite and Myers (1978) may also be useful in certain applications. Provided that the macromolecule contains alkali atoms (as molecular constituents or impurities), detection may be accomplished by having the particle impact a heated surface of appropriate work function, with subsequent pyrolysis, surface ionization, and the detection of positively charged ions. Fragments that can form negative ions by this technique include halogen atoms, F, Cl, Br, I, and certain radicals.

ION BEAM RADIOTHERAPY

The use of energetic particle beams for the treatment of diseases dates back to the use of neutron beams in the late 1930s at the Lawrence Berkeley Laboratory, and to the subsequent use of protons and deuterons in the 1950s. At present, energetic ion beams are being utilized in experimental cancer therapy; and over 400 patients at the Berkeley laboratory have been treated with beams of helium, carbon, neon, argon, or silicon. The rationale for using charged particle beams in radiotherapy is somewhat analogous to that of ion-implanted species in electronic materials. Not only is there greater control with ion-implanted profiles, but the choice of an ion (in selected instances) may be an important biochemical parameter in the same sense that local "doping" drastically alters the electric properties of semiconductors; that is, both biological materials and semiconductors are impurity- and radiation-sensitive systems. Specific potential advantages of energetic ion beams in radiotherapy have been reviewed by Chen et al. (1983):

1. Beams can be focused to limit the high-dose region to the tumor target volume.
2. Dose localization or penetration can be achieved in three dimensions by programming ion beam energy.
3. Charged particles with $Z \geq 10$ have enhanced biological properties for eradicating hypoxic tumor cells.
4. With the Lawrence Berkeley Laboratory *Bevalac*, heavy ions can be accelerated to an energy of 1 GeV/amu, with depths of penetration up to 25 cm at reasonable dose rates.

5. Broad, uniform beams can be generated, as well as highly localized beams that furnish doses to irregularly-shaped target volumes.

6. Ion beam characterization and modeling can be aided by the use of computerized axial tomographic scans in pretreatment diagnostics.

7. Accelerators can provide radioactive beams (such as ^{11}C and ^{19}Ne), to confirm dose localization.

Some excellent results have been obtained in the treatment of small tumors in the eye (ocular melanomas) and in precision dose-localization to tumors wrapped around the spinal cord, where minimal exposure of the cord itself is crucial (Alonso, 1985). In any event, it is clear that ion accelerators primarily designed for basic physics research will be increasingly utilized in radiobiology and biomedical studies. Furthermore, a design study is currently underway for a heavy ion accelerator that could function as a dedicated radiotherapy facility (Elioff, 1982; Gough, 1985). Features of this facility include fast switching between ion species and ion energies and a beam transport system for bringing the ion beam to multiple treatment rooms.

REFERENCES

Adesida, I., C. Anderson, and E. D. Wolf, *J. Vac. Sci. Tech., B1,* No. 4, 1182 (1983).

Alonso, J. R., private communication (1985).

Alonso, J. R., R. T. Avery, T. Elioff, R. J. Force, H. A. Grunder, H. D. Lancaster, E. J. Lofgren, J. R. Meneghetti, F. B. Selph, R. R. Stevenson, and R. B. Yourd, *Science, 217,* 1135 (1982).

Amsel, G., and J. A. Davies, *Nucl. Instr. Methods Phys. Res., 218,* 177 (1983).

Bamba, Y., E. Miyauchi, H. Arimoto, K. Kuramoto, A. Takamori, and H. Hashimoto, *Jpn. J. Appl. Phys., 22,* No. 10, Part 2, L650 (1983).

Bechert, D. W., and W. E. Reif, *Bull. Am. Phys. Soc., 29,* No. 9, 1515 (1984).

Bethe, H. A., *Science, 222,* 881 (1983).

Bischoff, J. L., and R. J. Rosenbauer, *Science, 213,* 1003 (1981).

Boden, R. H., in *Jet, Rocket, Nuclear, Ion and Electric Propulsion,* ed. W. H. T. Loh, Springer-Verlag, New York (1968), pp. 463–483.

Bowden, M. J., *J. Electrochem. Soc., 128,* 195C (1981).

Brown, W. L., T. Venkatesan, and A. Wagner, *Nucl. Instr. and Methods, 191,* 157 (1981).

Cahill, T. A., *Ann. Rev. Nucl. Part. Sci., 30,* 211 (1980).

Chandrasekhar, S., *Science, 226,* 497 (1984).

Chen, C. T. Y., J. R. Castro, J. M. Collier, and W. M. Saunders, *IEEE Trans. Nucl. Sci., NS-30,* No. 2 1813 (1983).

Cooks, R. G., private communication (1984).

Cookson, J. A., *Nucl. Instr. and Methods, 165,* 477 (1979).

Cormier, T. M., M. G. Herman, B. S. Lin, and P. M. Stwertka, *Nucl. Instr. and Methods, 212,* 185 (1983).

Craig, H., and V. Craig, *Science, 176,* 401 (1972).

DeNiro, M. J., and S. Epstein, *Geochim. et Cosmochim. Acta, 45,* 341 (1981).

Dolnikowski, G. G., J. T. Watson, and J. Allison, *Anal. Chem., 56,* 197 (1984).

Economou, H. P., D. C. Flanders, and J. P. Donnelly, *J. Vac. Sci. Tech., 19,* No. 4, 1172 (1981).

Eldred, R. A., *IEEE Trans. Nucl. Sci., NS-30,* No. 2, 1276 (1983).

Elioff, T., *A Dedicated Heavy Ion Accelerator for Radiotherapy,* LBL-15043, Lawrence Berkeley Laboratory, Berkeley, CA (1982).

Elmore, D., P. W. Kubik, L. E. Tubbs, H. E. Gove, R. Teng., T. Hemmick, B. Chrunyk, and N. Conard, *Nucl. Instr. and Methods in Phys. Res., B5,* 109 (1984).

Evans, C. A., Jr., and M. D. Strathman, *Industr. Res. and Develop., December,* 99 (1983).

Favennec, P. M., M. Salvi, M. A. DiForte Poisson, and J. P. Duchemin, *Appl. Phys. Lett., 43,* No. 8, 771 (1983).

Fenselau, C., Mass Spectrometric Analysis of Heavy Molecules: An Interdisciplinary Frontier, The Johns Hopkins University School of Medicine, Department of Pharmacology and Experimental Therapeutics, Baltimore, MD (private communication, 1984).

Fink, D., *Nucl. Instr. and Methods, 191,* 408 (1981).

Fischer, B. E., and R. Spohr, *Nucl. Instr. and Methods, 16B,* 241 (1980).

Fite, W. L., personal communication (1984).

Fite, W. L., and R. L. Myers, U. S. Patent 4,093,855 (1978).

Flerov, G. N., and G. M. Ter-Akopian, *Rep. Prog. Phys., 46,* No. 7, 819 (1983).

Fowler, W. A., *Science, 226,* 922 (1984).

Gale, N. H., and Z. A. Stos-Gale, *Science, 216,* 11 (1982).

Gamo, K., Y. Ochiai, and S. Namba, *Jpn. J. Appl. Phys., 21,* No. 12, L792 (1982).

Gamo, K., T. Matsui, and S. Namba, *Jpn. J. Appl. Phys., 22,* No. 11, Part 2, L692 (1983).

General Electric Co., *IEEE Spectrum, September,* 22 (1983).

Gibson, W. M., private communication (1984).

Glagola, B. G., G. W. Phillips, K. W. Marlow, L. T. Myers, and R. J. Omohundro, *Nucl. Instr. and Methods in Phys. Res., B5,* 221 (1984).

Glascock, M. D., and M. F. Cornman, *Radiochem. Radioanal. Lett., 57,* No. 2, 73 (1983).

Goldberg, D. A., O. A. Anderson, W. S. Cooper, and J. T. Tanabe, *J. Fusion Energy, 3,* No. 1, 67 (1983).

Gould, H., D. Greiner, P. Lindstron, T. J. M. Symons, and H. Crawford, *Phys. Rev. Lett., 52,* 180, 1654 (1984).

Gough, R. A., *Bull. Am. Phys. Soc., 30,* No. 5, 953 (1985).

Hall, T. M., A. Wagner, and L. F. Thompson, *J. Vac. Sci. Tech., 16,* 1889 (1979).

Harris, W. C., S. Chandra, and G. H. Morrison, *Anal. Chem., 55,* 1959 (1983).

Hickok, R. L., and F. C. Jobes, *Bull. Am. Phys. Soc., 16,* 1231 (1971).

Ishibashi, K., and S. Furukawa, *App. Phys. Lett., 43,* No. 7, 660 (1983).

Jackman, T. E., J. A. Davies, W. Eckstein, and J. A. Moore, *Nucl. Instr. and Methods, 191,* 527 (1981).

Jiang, S., A. Yu, Y. Cui, Z. Zhou, D. Luo, and Q. Li, *Nucl. Instr. Methods Phys. Res., B5,* 226 (1984).

Jobes, F. C., and R. L. Hickok, *Nucl. Fusion, 10,* 195 (1970).

Johansson, T. B., R. Akselsson, and S. A. E. Johansson, *Nucl. Instr. and Methods, 84,* 141 (1970).

Jolly, R. K., S. K. Gupta, G. Randers-Pehrson, D. C. Buckle, W. B. Thornton, H. Aceto, Jr. Singh, J. J., and D. C. Woods, *J. Appl. Phys., 46,* No. 10, 4590 (1975).

Kadota, K., K. Matsunaga, H. Iguchi, M. Fujiwara, K. Tsuchida, and J. Fujita, *Jpn. J. Appl. Phys., 21*, No. 5, Part 2, L260 (1982).

Keefe, D., *Bull. Am. Phys. Soc., 30*, No. 5, 953 (1985).

Keller, O. L., Jr., D. C. Hoffman, R. A. Penneman, and G. R. Choppin, *Physics Today, March,* 35 (1984).

Kilius, L. R., R. P. Beukens, K. H. Chang, H. W. Lee, A. E. Litherland, D. Elmore, R. Ferraro, H. E. Gove, and K. H. Purser, *Nucl. Instr. and Methods, 171*, 355 (1980).

Kilius, L. R., J. C. Rucklidge, G. C. Wilson, H. W. Lee, K. H. Chang, A. E. Litherland, W. E. Kieser, R. P. Beukens, and M. P. Gorton, *Nucl. Instr. Methods Phys. Res., B5*, 185 (1984).

Kowalski, Z. W., *J. Materials Sci., 18*, 2531 (1983).

Kryder, M. H., and A. B. Bortz, *Physics Today, 37*, No. 12, 20 (1984).

Kubena, R. L., C. L. Anderson, R. L. Selizer, R. A. Jollens, and E. H. Stevens, *J. Vac. Sci. Tech., 19*, No. 4, 916 (1981).

Kumuro, M., H. Hiroshima, H. Tanoue, and T. Kanayama, *J. Vac. Sci. Tech., B1*, No. 4, 984 (1983).

Kutschera, W., *Nucl. Instr. Methods Phys. Res., B5*, 420 (1984).

Lerner, E. J., *Aerospace Am., September,* 29 (1984).

Ligon, W. V., personal communication (1983).

Litherland, A. E., *Ann. Rev. Nucl. Part. Sci., 30*, 437 (1980).

Litherland, A. E., and J. C. Rucklidge, *EOS, 62*, No. 11, 105 (1981).

Litherland, A. E., *Nucl. Instr. and Methods in Phys. Res., 233*, B5, 100 (1984).

MacDonald, J. R., J. A. Davies, T. E. Jackman, and L. C. Feldman, *J. Appl. Phys., 54*, No. 4, 1800 (1983).

Maidikov, V. Z., N. T. Surovitskaya, N. K. Skobelev, and W. Neubert, *Nucl. Instr. Methods, 192*, 295 (1982).

Michelsen, W. R., *Aerospace Eng., 19*, 7 (1960).

Miyauchi, E., H. Hashimoto, and T. Utsumi, *Jpn. J. Appl. Phys., 22*, No. 4, Part 2, L225 (1983).

Mook, W. G., *Nucl. Instr. Methods Phys. Res.* B5, 297 (1984).

Moses, K. G., B. H. Quon, and W. F. DiVergilio, *Bull. Am. Phys. Soc., 29*, No. 8, 1189 (1984).

Nagamatsu, H. T., R. D. Oro, and D. C. Ling, "Porosity Effect on Supercritical Airfoil Drag Reduction by Shock Wave/Boundary Layer Control," in *AIAA 17th Fluid Dynamics Conf.,* Snowmass, CO (June 25–27, 1984); *Aerospace Am., September,* 26 (1984).

Norton, P. R., P. E. Bindner, K. Giffiths, T. E. Jackman, J. A. Davies, and J. Rustig, *J. Chem. Phys., 80*, No. 8, 3859 (1984).

Pearce, W. E., "Progress at Douglas on Laminar Flow Control Applied to Commercial Transport Aircraft," in *Aircraft Systems and Technol. Conf., Thirteenth Congr. of the ICAS/ AIAA,* Seattle, WA (August 22–27, 1982).

Picraux, S. T., *Physics Today, 37*, No. 11, 38 (1984).

Prombo, C. A., and R. N. Clayton, *Science, 230*, 935 (1985).

Pronko, J. G., T. T. Bardin, I. V. Chapman, and D. Kohler, *IEEE Trans. Nucl. Sci., NS-30,* No. 2, 1305 (1983).

Purser, K. H., personal communication (1984).

Rapp, G., Jr., and J. A. Gifford, *Am. Scientist, 70*, 45 (1982).

Reinovsky, R. E., J. C. Glowienka, A. E. Seaver, W. C. Jennings, and R. L. Hickok, *IEEE Trans. Plasma Sci., PS-2*, 250 (1974).

Ryssel, H., K. Haberger, and H. Kranz, *J. Vac. Sci. Tech., 19*, No. 4, 1358 (1983).

Shubaly, M. R., *Bull. Am. Phys. Soc., 30,* No. 5, 944 (1985).

Sioshansi, P., and J. J. Au, *Materials Sci. Eng., 69,* 161 (1985).

Smith, P. F., J. R. J. Bennet, G. J. Homer, J. D. Lewin, H. E. Walford, and W. A. Smith, *Nucl. Phys., B206,* 333 (1982).

Stephens, W., J. Klein, and R. Zurmühle, *Phys. Rev., C2,* 1664 (1980).

Swann, C. P., *IEEE Trans. Nucl. Sci., NS-30,* No. 2, 1298 (1983).

Tatsuta, S., H. Nishi, and T. Ito, *Jpn. J. Appl. Phys., 21,* No. 2, Part 2, L113 (1982).

Torii, Y., and H. Yamada, *Jpn. J. Appl. Phys., 21,* No. 3, Part 2, L132 (1982).

Truscello, V. C., and H. S. Davis, *IEEE Spectrum, 21,* No. 12, 58 (1984).

Turekian, K. K., J. K. Cochran, S. Krishnaswami, W. A. Lanford, P. D. Parker, and K. A. Bauer, *Geophys. Res. Lett., 6,* 417 (1979).

Van der Merwe, N. J., *Am. Scientist, 70,* 596 (1982).

Venkatesan, F., T. Wolf, D. Allara, B. J. Wilkens, and G. N. Taylor, *Appl. Phys. Lett., 43,* No. 10, 934 (1983).

Wang, V., J. W. Ward, and R. L. Seliger, *J. Vac. Sci. Tech., 19,* No. 4, 1158 (1981).

Welch, J. J., K. J. Bertsche, P. G. Friedman, D. E. Morris, R. A. Muller, and P. P. Tans, *Nucl. Instr. Methods Phys. Res., B5,* 230 (1984).

White, J. C., R. E. Craighead, L. D. Howard, R. E. Behringer, R. W. Epworth, D. Henderson, and J. E. Sweney, *Appl. Phys. Lett., 44,* No. 1, 22 (1984).

Widemann, F., *J. Radioanal. Chem., 55,* No. 2, 271 (1980).

Wilkens, B., and T. Venkatesan, *J. Vac. Sci. Tech., B1,* No. 4, 1132 (1983).

Williams, E. T., and L. W. Punyasena, *IEEE Trans. Nucl. Sci., NS-30,* No. 2, 1294 (1983).

Wilson, G. C., J. C. Rucklidge, W. E. Kieser, and R. P. Beukens, *Nucl. Instr. Methods Phys. Res., B5,* 200 (1984).

Wilson, R. G., D. K. Sadana, T. W. Sigimon, and C. A. Evans, Jr., *Appl. Phys. Lett., 43,* No. 6, 549 (1983).

Wood, G. M., B. W. Lewis, B. T. Upchurch, R. J. Nowak, D. G. Eide, and P. A. Paulin, *IEEE Int. Congr. on Instr. in Aerospace Simulation,* ICISF Record, St. Louis, France (September 1983), p. 259.

Zurer, P. S., *Chem Eng. News, February 21,* 26 (1983).

Appendix 1   Isotopic Abundances of the Elements

Element	Mass No.	Relative Abundance	Element	Mass No.	Relative Abundance
H	1	99.985	Ar	36	0.34
	2	0.015		38	0.06
He	3	0.00014		40	99.60
	4	100.0	K	39	93.26
Li	6	7.50		40	0.01
	7	92.50		41	6.73
Be	9	100.0	Ca	40	96.94
B	10	19.82		42	0.65
	11	80.18		43	0.14
C	12	98.889		44	2.08
	13	1.11		46	0.004
N	14	99.634		48	0.19
	15	0.366	Sc	45	100.0
O	16	99.763	Ti	46	8.01
	17	0.037		47	7.33
	18	0.200		48	73.81
F	19	100.0		49	5.50
Ne	20	90.51		50	5.35
	21	0.27	V	50	0.25
	22	9.22		51	99.75
Na	23	100.0	Cr	50	4.35
Mg	24	78.99		52	83.79
	25	10.00		53	9.50
	26	11.01		54	2.36
Al	27	100.0	Mn	55	100.0
Si	28	92.23	Fe	54	5.81
	29	4.67		56	91.75
	30	3.10		57	2.15
P	31	100.0		58	0.29
S	32	95.02	Co	59	100.0
	33	0.75	Ni	58	68.27
	34	4.21		60	26.10
	36	0.02		61	1.13
Cl	35	75.77		62	3.59
	37	24.23		64	0.91
			Cu	63	69.17
				65	30.83

Appendix 1 (*Continued*)

Element	Mass No.	Relative Abundance	Element	Mass No.	Relative Abundance
Zn	64	48.63	Mo	92	14.84
	66	27.90		94	9.25
	67	4.10		95	15.92
	68	18.75		96	16.68
	70	0.62		97	9.55
Ga	69	60.08		98	24.13
	71	39.92		100	9.63
Ge	70	20.52	Ru	96	5.52
	72	27.43		98	1.86
	73	7.76		99	12.74
	74	36.53		100	12.60
	76	7.76		101	17.05
As	75	100.0		102	31.57
				104	18.66
Se	74	0.88	Rh	103	100.0
	76	8.95	Pd	102	1.02
	77	7.65		104	11.14
	78	23.51		105	22.33
	80	49.62		106	27.33
	82	9.39		108	26.46
Br	79	50.69		110	11.72
	81	49.31	Ag	107	51.84
Kr	78	0.36		109	48.16
	80	2.28	Cd	106	1.25
	82	11.58		108	0.89
	83	11.52		110	12.49
	84	56.96		111	12.80
	86	17.30		112	24.13
Rb	85	72.17		113	12.22
	87	27.83		114	28.73
Sr	84	0.56		116	7.49
	86	9.86	In	113	4.33
	87	7.00		115	95.67
	88	82.58	Sn	112	1.01
Y	89	100.0		114	0.67
Zr	90	51.45		115	0.38
	91	11.32		116	14.76
	92	17.19		117	7.75
	94	17.28		118	24.30
	96	2.76		119	8.55
Nb	93	100.0		120	32.38

Appendix 1 (*Continued*)

Element	Mass No.	Relative Abundance	Element	Mass No.	Relative Abundance
Sn	122	4.56	Nd	145	8.30
	124	5.64		146	17.17
Sb	121	57.25		148	5.74
	123	42.75		150	5.62
Te	120	0.10	Sm	144	3.07
	122	2.60		147	15.00
	123	0.91		148	11.24
	124	4.81		149	13.82
	125	7.14		150	7.38
	126	18.95		152	26.74
	128	31.69		154	22.75
	130	33.80	Eu	151	47.77
I	127	100.0		153	52.23
Xe	124	0.09	Gd	152	0.20
	126	0.09		154	2.18
	128	1.92		155	14.80
	129	26.44		156	20.47
	130	4.08		157	15.65
	131	21.18		158	24.84
	132	26.89		160	21.86
	134	10.44	Tb	159	100.0
	136	8.87	Dy	156	0.06
Cs	133	100.0		158	0.10
Ba	130	0.11		160	2.34
	132	0.10		161	18.91
	134	2.42		162	25.50
	135	6.59		163	24.90
	136	7.85		164	28.19
	137	11.23	Ho	165	100.0
	138	71.70	Er	162	0.14
La	138	0.09		164	1.61
	139	99.91		166	33.60
Ce	136	0.19		167	22.93
	138	0.25		168	26.79
	140	88.48		170	14.93
	142	11.08	Tm	169	100.0
Pr	141	100.0	Yb	168	0.13
Nd	142	27.16		170	3.04
	143	12.18		171	14.28
	144	23.83		172	21.83
				173	16.13

Element	Mass No.	Relative Abundance	Element	Mass No.	Relative Abundance
Yb	174	31.83	Pt	190	0.01
	176	12.76		192	0.79
Lu	175	97.40		194	32.9
	176	2.60		195	33.8
Hf	174	0.16		196	25.3
	176	5.21		198	7.2
	177	18.56	Au	197	100.0
	178	27.10	Hg	196	0.16
	179	13.75		198	10.12
	180	35.22		199	16.99
Ta	180	0.012		200	23.07
	181	99.988		201	13.23
W	180	0.13		202	29.64
	182	26.31		204	6.79
	183	14.28	Tl	203	29.52
	184	30.64		205	70.48
	186	28.64	Pb	204	1.4
Re	185	37.40		206	24.1
	187	62.60		207	22.1
Os	184	0.02		208	52.4
	186	1.59	Bi	209	100.0
	187	1.64	Th	232	100.0
	188	13.27	U	234	0.0055
	189	16.14		235	0.72
	190	26.38		238	99.27
	192	40.96			
Ir	191	37.3			
	193	62.7			

Source: *Pure Appl. Chem.*, *55*, No. 7, 1119–1136 (1983).

Appendix 2 Ionization Potentials, Electron Affinities, Work Functions, and Melting Points

Atomic Number	Element	IP (eV)	EA (eV)	Work Function (eV)	Melting Point (°C)
1	H	13.60	0.80		−259.14
2	He	24.59			−272[a]
3	Li	5.39	0.6	2.9	180.54
4	Be	9.32		4.98	1278
5	B	8.30	0.4	4.45	2079
6	C	11.27	1.12	5.0	3652
7	N	14.55			−209.86
8	O	13.61	1.466		−218.4
9	F	17.43	3.448		−219.62[d]
10	Ne	21.56			−248.67
11	Na	5.14	0.3	2.75	97.81
12	Mg	7.64		3.66	648.8
13	Al	5.99	1.0	4.28	660.37
14	Si	8.15	1.9	4.91	1410
15	P	10.98	1.1		44.1[b]
16	S	10.36	2.07		112.8[c]
17	Cl	13.02	3.613		−100.98[d]
18	Ar	15.60			−189.2[d]
19	K	4.34	0.9	2.3	63.25
20	Ca	6.11		2.87	839
21	Sc	6.56		3.5	1539
22	Ti	6.82	0.4	3.6–4.3	1660
23	V	6.74	0.9	4.1	1890
24	Cr	6.76	1.0	4.5	1857
25	Mn	7.43		4.1	1244
26	Fe	7.90	0.6	4.5	1535
27	Co	7.87	0.9	5.0	1495
28	Ni	7.64	1.3	5.15	1453
29	Cu	7.73	1.8	4.65	1083.4
30	Zn	9.39		4.4	419.58
31	Ga	6.00		4.2	29.78
32	Ge	7.89		5.0	937.4
33	As	9.81		3.75	817[e]
34	Se	9.75		5.9	217[e]
35	Br	11.85	3.363		−7.2
36	Kr	14.00			−156.6
37	Rb	4.18	0.6	2.16	38.89
38	Sr	5.69		2.59	769
39	Y	6.40			1523
40	Zr	6.84		4.0	1852
41	Nb	6.88		4.3	2468
42	Mo	7.10		4.6	2617
43	Tc	7.28			2172

Atomic Number	Element	IP (eV)	EA (eV)	Work Function (eV)	Melting Point (°C)
44	Ru	7.37		4.71	2310
45	Rh	7.46		4.48	1965
46	Pd	8.33		5.12	1554
47	Ag	7.57	(2.0)	4.26	961.93
48	Cd	8.99		4.22	320.9
49	In	5.79		4.12	156.61
50	Sn	7.34		4.42	231.968
51	Sb	8.64		4.7	630.74
52	Te	9.01		4.8	449.5
53	I	10.45	3.063		113.5
54	Xe	12.11			−111.9
55	Cs	3.89	0.6	2.14	28.4
56	Ba	5.21		2.7	725
57	La	5.61		3.5	920
58	Ce	5.65		2.9	798
59	Pr	5.42		2.7	931
60	Nd	5.49		3.2	1010
61	Pm	5.55			1080
62	Sm	5.70		2.7	1072
63	Eu	5.68		2.5	822
64	Gd	6.16		3.1	1311
65	Tb	5.85		3.0	1360
66	Dy	5.80			1409
67	Ho	6.19			1470
68	Er	6.38			1522
69	Tm	6.22			1545
70	Yb	6.25			824
71	Lu	6.41		2.9	1656
72	Hf	6.8		3.9	2227
73	Ta	7.88		4.15	2996
74	W	7.98		4.5	3410
75	Re	7.87		4.96	3180
76	Os	8.70		4.83	3045
77	Ir	9.1		5.27	2410
78	Pt	9.0		5.65	1772
79	Au	9.22	(2.8)	5.1	1064.43
80	Hg	10.44	1.8	4.49	−38.87
81	Tl	6.11		3.7	303.5
82	Pb	7.42		4.25	327.50
83	Bi	7.29		4.22	271.3
84	Po	8.43			254
85	At	9.5			302
86	Rn	10.75			−71
87	Fr	4.0			27

Atomic Number	Element	IP (eV)	EA (eV)	Work Function (eV)	Melting Point (°C)
88	Ra	5.28			700
89	Ac	6.9			1050
90	Th	7.0		3.73	1750
91	Pa				1600
92	U	5.8		3.08	1132
93	Np				640
94	Pu	5.1			641

Source: These data for Appendix 2 were kindly supplied by R. E. Honig and E. L. Wildner.

[a] 26 atm.
[b] White.
[c] Rhombic.
[d] Freezing point.
[e] Gray.

INDEX